Lecture Notes in Computer Science 13676

More information about this series at https://link.springer.com/bookseries/558

Shai Avidan · Gabriel Brostow ·
Moustapha Cissé · Giovanni Maria Farinella ·
Tal Hassner (Eds.)

Computer Vision – ECCV 2022

17th European Conference
Tel Aviv, Israel, October 23–27, 2022
Proceedings, Part XVI

 Springer

Editors
Shai Avidan
Tel Aviv University
Tel Aviv, Israel

Gabriel Brostow (ID)
University College London
London, UK

Moustapha Cissé
Google AI
Accra, Ghana

Giovanni Maria Farinella (ID)
University of Catania
Catania, Italy

Tal Hassner (ID)
Facebook (United States)
Menlo Park, CA, USA

ISSN 0302-9743   ISSN 1611-3349 (electronic)
Lecture Notes in Computer Science
ISBN 978-3-031-19786-4   ISBN 978-3-031-19787-1 (eBook)
https://doi.org/10.1007/978-3-031-19787-1

This Springer imprint is published by the registered company Springer Nature Switzerland AG
The registered company address is: Gewerbestrasse 11, 6330 Cham, Switzerland

Foreword

Organizing the European Conference on Computer Vision (ECCV 2022) in Tel-Aviv during a global pandemic was no easy feat. The uncertainty level was extremely high, and decisions had to be postponed to the last minute. Still, we managed to plan things just in time for ECCV 2022 to be held in person. Participation in physical events is crucial to stimulating collaborations and nurturing the culture of the Computer Vision community.

There were many people who worked hard to ensure attendees enjoyed the best science at the 16th edition of ECCV. We are grateful to the Program Chairs Gabriel Brostow and Tal Hassner, who went above and beyond to ensure the ECCV reviewing process ran smoothly. The scientific program includes dozens of workshops and tutorials in addition to the main conference and we would like to thank Leonid Karlinsky and Tomer Michaeli for their hard work. Finally, special thanks to the web chairs Lorenzo Baraldi and Kosta Derpanis, who put in extra hours to transfer information fast and efficiently to the ECCV community.

We would like to express gratitude to our generous sponsors and the Industry Chairs, Dimosthenis Karatzas and Chen Sagiv, who oversaw industry relations and proposed new ways for academia-industry collaboration and technology transfer. It's great to see so much industrial interest in what we're doing!

Authors' draft versions of the papers appeared online with open access on both the Computer Vision Foundation (CVF) and the European Computer Vision Association (ECVA) websites as with previous ECCVs. Springer, the publisher of the proceedings, has arranged for archival publication. The final version of the papers is hosted by SpringerLink, with active references and supplementary materials. It benefits all potential readers that we offer both a free and citeable version for all researchers, as well as an authoritative, citeable version for SpringerLink readers. Our thanks go to Ronan Nugent from Springer, who helped us negotiate this agreement. Last but not least, we wish to thank Eric Mortensen, our publication chair, whose expertise made the process smooth.

October 2022

Rita Cucchiara
Jiří Matas
Amnon Shashua
Lihi Zelnik-Manor

Preface

Welcome to the proceedings of the European Conference on Computer Vision (ECCV 2022). This was a hybrid edition of ECCV as we made our way out of the COVID-19 pandemic. The conference received 5804 valid paper submissions, compared to 5150 submissions to ECCV 2020 (a 12.7% increase) and 2439 in ECCV 2018. 1645 submissions were accepted for publication (28%) and, of those, 157 (2.7% overall) as orals.

846 of the submissions were desk-rejected for various reasons. Many of them because they revealed author identity, thus violating the double-blind policy. This violation came in many forms: some had author names with the title, others added acknowledgments to specific grants, yet others had links to their github account where their name was visible. Tampering with the LaTeX template was another reason for automatic desk rejection.

ECCV 2022 used the traditional CMT system to manage the entire double-blind reviewing process. Authors did not know the names of the reviewers and vice versa. Each paper received at least 3 reviews (except 6 papers that received only 2 reviews), totalling more than 15,000 reviews.

Handling the review process at this scale was a significant challenge. To ensure that each submission received as fair and high-quality reviews as possible, we recruited more than 4719 reviewers (in the end, 4719 reviewers did at least one review). Similarly we recruited more than 276 area chairs (eventually, only 276 area chairs handled a batch of papers). The area chairs were selected based on their technical expertise and reputation, largely among people who served as area chairs in previous top computer vision and machine learning conferences (ECCV, ICCV, CVPR, NeurIPS, etc.).

Reviewers were similarly invited from previous conferences, and also from the pool of authors. We also encouraged experienced area chairs to suggest additional chairs and reviewers in the initial phase of recruiting. The median reviewer load was five papers per reviewer, while the average load was about four papers, because of the emergency reviewers. The area chair load was 35 papers, on average.

Conflicts of interest between authors, area chairs, and reviewers were handled largely automatically by the CMT platform, with some manual help from the Program Chairs. Reviewers were allowed to describe themselves as senior reviewer (load of 8 papers to review) or junior reviewers (load of 4 papers). Papers were matched to area chairs based on a subject-area affinity score computed in CMT and an affinity score computed by the Toronto Paper Matching System (TPMS). TPMS is based on the paper's full text. An area chair handling each submission would bid for preferred expert reviewers, and we balanced load and prevented conflicts.

The assignment of submissions to area chairs was relatively smooth, as was the assignment of submissions to reviewers. A small percentage of reviewers were not happy with their assignments in terms of subjects and self-reported expertise. This is an area for improvement, although it's interesting that many of these cases were reviewers hand-picked by AC's. We made a later round of reviewer recruiting, targeted at the list of authors of papers submitted to the conference, and had an excellent response which helped provide enough emergency reviewers. In the end, all but six papers received at least 3 reviews.

The challenges of the reviewing process are in line with past experiences at ECCV 2020. As the community grows, and the number of submissions increases, it becomes ever more challenging to recruit enough reviewers and ensure a high enough quality of reviews. Enlisting authors by default as reviewers might be one step to address this challenge.

Authors were given a week to rebut the initial reviews, and address reviewers' concerns. Each rebuttal was limited to a single pdf page with a fixed template.

The Area Chairs then led discussions with the reviewers on the merits of each submission. The goal was to reach consensus, but, ultimately, it was up to the Area Chair to make a decision. The decision was then discussed with a buddy Area Chair to make sure decisions were fair and informative. The entire process was conducted virtually with no in-person meetings taking place.

The Program Chairs were informed in cases where the Area Chairs overturned a decisive consensus reached by the reviewers, and pushed for the meta-reviews to contain details that explained the reasoning for such decisions. Obviously these were the most contentious cases, where reviewer inexperience was the most common reported factor.

Once the list of accepted papers was finalized and released, we went through the laborious process of plagiarism (including self-plagiarism) detection. A total of 4 accepted papers were rejected because of that.

Finally, we would like to thank our Technical Program Chair, Pavel Lifshits, who did tremendous work behind the scenes, and we thank the tireless CMT team.

October 2022

Gabriel Brostow
Giovanni Maria Farinella
Moustapha Cissé
Shai Avidan
Tal Hassner

Organization

General Chairs

Rita Cucchiara   University of Modena and Reggio Emilia, Italy
Jiří Matas   Czech Technical University in Prague, Czech Republic
Amnon Shashua   Hebrew University of Jerusalem, Israel
Lihi Zelnik-Manor   Technion – Israel Institute of Technology, Israel

Program Chairs

Shai Avidan   Tel-Aviv University, Israel
Gabriel Brostow   University College London, UK
Moustapha Cissé   Google AI, Ghana
Giovanni Maria Farinella   University of Catania, Italy
Tal Hassner   Facebook AI, USA

Program Technical Chair

Pavel Lifshits   Technion – Israel Institute of Technology, Israel

Workshops Chairs

Leonid Karlinsky   IBM Research, Israel
Tomer Michaeli   Technion – Israel Institute of Technology, Israel
Ko Nishino   Kyoto University, Japan

Tutorial Chairs

Thomas Pock   Graz University of Technology, Austria
Natalia Neverova   Facebook AI Research, UK

Demo Chair

Bohyung Han   Seoul National University, Korea

Social and Student Activities Chairs

Tatiana Tommasi   Italian Institute of Technology, Italy
Sagie Benaim   University of Copenhagen, Denmark

Diversity and Inclusion Chairs

Xi Yin   Facebook AI Research, USA
Bryan Russell   Adobe, USA

Communications Chairs

Lorenzo Baraldi   University of Modena and Reggio Emilia, Italy
Kosta Derpanis   York University & Samsung AI Centre Toronto,
                            Canada

Industrial Liaison Chairs

Dimosthenis Karatzas   Universitat Autònoma de Barcelona, Spain
Chen Sagiv   SagivTech, Israel

Finance Chair

Gerard Medioni   University of Southern California & Amazon,
                            USA

Publication Chair

Eric Mortensen   MiCROTEC, USA

Area Chairs

Lourdes Agapito   University College London, UK
Zeynep Akata   University of Tübingen, Germany
Naveed Akhtar   University of Western Australia, Australia
Karteek Alahari   Inria Grenoble Rhône-Alpes, France
Alexandre Alahi   École polytechnique fédérale de Lausanne,
                            Switzerland
Pablo Arbelaez   Universidad de Los Andes, Columbia
Antonis A. Argyros   University of Crete & Foundation for Research
                            and Technology-Hellas, Crete
Yuki M. Asano   University of Amsterdam, The Netherlands
Kalle Åström   Lund University, Sweden
Hadar Averbuch-Elor   Cornell University, USA

Matthijs Douze	Facebook AI Research, USA
Mohamed Elhoseiny	King Abdullah University of Science and Technology, Saudi Arabia
Sergio Escalera	University of Barcelona, Spain
Yi Fang	New York University, USA
Ryan Farrell	Brigham Young University, USA
Alireza Fathi	Google, USA
Christoph Feichtenhofer	Facebook AI Research, USA
Basura Fernando	Agency for Science, Technology and Research (A*STAR), Singapore
Vittorio Ferrari	Google Research, Switzerland
Andrew W. Fitzgibbon	Graphcore, UK
David J. Fleet	University of Toronto, Canada
David Forsyth	University of Illinois at Urbana-Champaign, USA
David Fouhey	University of Michigan, USA
Katerina Fragkiadaki	Carnegie Mellon University, USA
Friedrich Fraundorfer	Graz University of Technology, Austria
Oren Freifeld	Ben-Gurion University, Israel
Thomas Funkhouser	Google Research & Princeton University, USA
Yasutaka Furukawa	Simon Fraser University, Canada
Fabio Galasso	Sapienza University of Rome, Italy
Jürgen Gall	University of Bonn, Germany
Chuang Gan	Massachusetts Institute of Technology, USA
Zhe Gan	Microsoft, USA
Animesh Garg	University of Toronto, Vector Institute, Nvidia, Canada
Efstratios Gavves	University of Amsterdam, The Netherlands
Peter Gehler	Amazon, Germany
Theo Gevers	University of Amsterdam, The Netherlands
Bernard Ghanem	King Abdullah University of Science and Technology, Saudi Arabia
Ross B. Girshick	Facebook AI Research, USA
Georgia Gkioxari	Facebook AI Research, USA
Albert Gordo	Facebook, USA
Stephen Gould	Australian National University, Australia
Venu Madhav Govindu	Indian Institute of Science, India
Kristen Grauman	Facebook AI Research & UT Austin, USA
Abhinav Gupta	Carnegie Mellon University & Facebook AI Research, USA
Mohit Gupta	University of Wisconsin-Madison, USA
Hu Han	Institute of Computing Technology, Chinese Academy of Sciences, China

Bohyung Han	Seoul National University, Korea
Tian Han	Stevens Institute of Technology, USA
Emily Hand	University of Nevada, Reno, USA
Bharath Hariharan	Cornell University, USA
Ran He	Institute of Automation, Chinese Academy of Sciences, China
Otmar Hilliges	ETH Zurich, Switzerland
Adrian Hilton	University of Surrey, UK
Minh Hoai	Stony Brook University, USA
Yedid Hoshen	Hebrew University of Jerusalem, Israel
Timothy Hospedales	University of Edinburgh, UK
Gang Hua	Wormpex AI Research, USA
Di Huang	Beihang University, China
Jing Huang	Facebook, USA
Jia-Bin Huang	Facebook, USA
Nathan Jacobs	Washington University in St. Louis, USA
C.V. Jawahar	International Institute of Information Technology, Hyderabad, India
Herve Jegou	Facebook AI Research, France
Neel Joshi	Microsoft Research, USA
Armand Joulin	Facebook AI Research, France
Frederic Jurie	University of Caen Normandie, France
Fredrik Kahl	Chalmers University of Technology, Sweden
Yannis Kalantidis	NAVER LABS Europe, France
Evangelos Kalogerakis	University of Massachusetts, Amherst, USA
Sing Bing Kang	Zillow Group, USA
Yosi Keller	Bar Ilan University, Israel
Margret Keuper	University of Mannheim, Germany
Tae-Kyun Kim	Imperial College London, UK
Benjamin Kimia	Brown University, USA
Alexander Kirillov	Facebook AI Research, USA
Kris Kitani	Carnegie Mellon University, USA
Iasonas Kokkinos	Snap Inc. & University College London, UK
Vladlen Koltun	Apple, USA
Nikos Komodakis	University of Crete, Crete
Piotr Koniusz	Australian National University, Australia
Philipp Kraehenbuehl	University of Texas at Austin, USA
Dilip Krishnan	Google, USA
Ajay Kumar	Hong Kong Polytechnic University, Hong Kong, China
Junseok Kwon	Chung-Ang University, Korea
Jean-Francois Lalonde	Université Laval, Canada

Ivan Laptev   Inria Paris, France
Laura Leal-Taixé   Technical University of Munich, Germany
Erik Learned-Miller   University of Massachusetts, Amherst, USA
Gim Hee Lee   National University of Singapore, Singapore
Seungyong Lee   Pohang University of Science and Technology,
                            Korea
Zhen Lei   Institute of Automation, Chinese Academy of
                            Sciences, China
Bastian Leibe   RWTH Aachen University, Germany
Hongdong Li   Australian National University, Australia
Fuxin Li   Oregon State University, USA
Bo Li   University of Illinois at Urbana-Champaign, USA
Yin Li   University of Wisconsin-Madison, USA
Ser-Nam Lim   Meta AI Research, USA
Joseph Lim   University of Southern California, USA
Stephen Lin   Microsoft Research Asia, China
Dahua Lin   The Chinese University of Hong Kong,
                            Hong Kong, China
Si Liu   Beihang University, China
Xiaoming Liu   Michigan State University, USA
Ce Liu   Microsoft, USA
Zicheng Liu   Microsoft, USA
Yanxi Liu   Pennsylvania State University, USA
Feng Liu   Portland State University, USA
Yebin Liu   Tsinghua University, China
Chen Change Loy   Nanyang Technological University, Singapore
Huchuan Lu   Dalian University of Technology, China
Cewu Lu   Shanghai Jiao Tong University, China
Oisin Mac Aodha   University of Edinburgh, UK
Dhruv Mahajan   Facebook, USA
Subhransu Maji   University of Massachusetts, Amherst, USA
Atsuto Maki   KTH Royal Institute of Technology, Sweden
Arun Mallya   NVIDIA, USA
R. Manmatha   Amazon, USA
Iacopo Masi   Sapienza University of Rome, Italy
Dimitris N. Metaxas   Rutgers University, USA
Ajmal Mian   University of Western Australia, Australia
Christian Micheloni   University of Udine, Italy
Krystian Mikolajczyk   Imperial College London, UK
Anurag Mittal   Indian Institute of Technology, Madras, India
Philippos Mordohai   Stevens Institute of Technology, USA
Greg Mori   Simon Fraser University & Borealis AI, Canada

Vittorio Murino — Istituto Italiano di Tecnologia, Italy
P. J. Narayanan — International Institute of Information Technology, Hyderabad, India
Ram Nevatia — University of Southern California, USA
Natalia Neverova — Facebook AI Research, UK
Richard Newcombe — Facebook, USA
Cuong V. Nguyen — Florida International University, USA
Bingbing Ni — Shanghai Jiao Tong University, China
Juan Carlos Niebles — Salesforce & Stanford University, USA
Ko Nishino — Kyoto University, Japan
Jean-Marc Odobez — Idiap Research Institute, École polytechnique fédérale de Lausanne, Switzerland
Francesca Odone — University of Genova, Italy
Takayuki Okatani — Tohoku University & RIKEN Center for Advanced Intelligence Project, Japan
Manohar Paluri — Facebook, USA
Guan Pang — Facebook, USA
Maja Pantic — Imperial College London, UK
Sylvain Paris — Adobe Research, USA
Jaesik Park — Pohang University of Science and Technology, Korea
Hyun Soo Park — The University of Minnesota, USA
Omkar M. Parkhi — Facebook, USA
Deepak Pathak — Carnegie Mellon University, USA
Georgios Pavlakos — University of California, Berkeley, USA
Marcello Pelillo — University of Venice, Italy
Marc Pollefeys — ETH Zurich & Microsoft, Switzerland
Jean Ponce — Inria, France
Gerard Pons-Moll — University of Tübingen, Germany
Fatih Porikli — Qualcomm, USA
Victor Adrian Prisacariu — University of Oxford, UK
Petia Radeva — University of Barcelona, Spain
Ravi Ramamoorthi — University of California, San Diego, USA
Deva Ramanan — Carnegie Mellon University, USA
Vignesh Ramanathan — Facebook, USA
Nalini Ratha — State University of New York at Buffalo, USA
Tammy Riklin Raviv — Ben-Gurion University, Israel
Tobias Ritschel — University College London, UK
Emanuele Rodola — Sapienza University of Rome, Italy
Amit K. Roy-Chowdhury — University of California, Riverside, USA
Michael Rubinstein — Google, USA
Olga Russakovsky — Princeton University, USA

Mathieu Salzmann   École polytechnique fédérale de Lausanne,
                             Switzerland
Dimitris Samaras   Stony Brook University, USA
Aswin Sankaranarayanan   Carnegie Mellon University, USA
Imari Sato   National Institute of Informatics, Japan
Yoichi Sato   University of Tokyo, Japan
Shin'ichi Satoh   National Institute of Informatics, Japan
Walter Scheirer   University of Notre Dame, USA
Bernt Schiele   Max Planck Institute for Informatics, Germany
Konrad Schindler   ETH Zurich, Switzerland
Cordelia Schmid   Inria & Google, France
Alexander Schwing   University of Illinois at Urbana-Champaign, USA
Nicu Sebe   University of Trento, Italy
Greg Shakhnarovich   Toyota Technological Institute at Chicago, USA
Eli Shechtman   Adobe Research, USA
Humphrey Shi   University of Oregon & University of Illinois at
                             Urbana-Champaign & Picsart AI Research,
                             USA
Jianbo Shi   University of Pennsylvania, USA
Roy Shilkrot   Massachusetts Institute of Technology, USA
Mike Zheng Shou   National University of Singapore, Singapore
Kaleem Siddiqi   McGill University, Canada
Richa Singh   Indian Institute of Technology Jodhpur, India
Greg Slabaugh   Queen Mary University of London, UK
Cees Snoek   University of Amsterdam, The Netherlands
Yale Song   Facebook AI Research, USA
Yi-Zhe Song   University of Surrey, UK
Bjorn Stenger   Rakuten Institute of Technology
Abby Stylianou   Saint Louis University, USA
Akihiro Sugimoto   National Institute of Informatics, Japan
Chen Sun   Brown University, USA
Deqing Sun   Google, USA
Kalyan Sunkavalli   Adobe Research, USA
Ying Tai   Tencent YouTu Lab, China
Ayellet Tal   Technion – Israel Institute of Technology, Israel
Ping Tan   Simon Fraser University, Canada
Siyu Tang   ETH Zurich, Switzerland
Chi-Keung Tang   Hong Kong University of Science and
                             Technology, Hong Kong, China
Radu Timofte   University of Würzburg, Germany & ETH Zurich,
                             Switzerland
Federico Tombari   Google, Switzerland & Technical University of
                             Munich, Germany

James Tompkin   Brown University, USA
Lorenzo Torresani   Dartmouth College, USA
Alexander Toshev   Apple, USA
Du Tran   Facebook AI Research, USA
Anh T. Tran   VinAI, Vietnam
Zhuowen Tu   University of California, San Diego, USA
Georgios Tzimiropoulos   Queen Mary University of London, UK
Jasper Uijlings   Google Research, Switzerland
Jan C. van Gemert   Delft University of Technology, The Netherlands
Gul Varol   Ecole des Ponts ParisTech, France
Nuno Vasconcelos   University of California, San Diego, USA
Mayank Vatsa   Indian Institute of Technology Jodhpur, India
Ashok Veeraraghavan   Rice University, USA
Jakob Verbeek   Facebook AI Research, France
Carl Vondrick   Columbia University, USA
Ruiping Wang   Institute of Computing Technology, Chinese
                              Academy of Sciences, China
Xinchao Wang   National University of Singapore, Singapore
Liwei Wang   The Chinese University of Hong Kong,
                              Hong Kong, China
Chaohui Wang   Université Paris-Est, France
Xiaolong Wang   University of California, San Diego, USA
Christian Wolf   NAVER LABS Europe, France
Tao Xiang   University of Surrey, UK
Saining Xie   Facebook AI Research, USA
Cihang Xie   University of California, Santa Cruz, USA
Zeki Yalniz   Facebook, USA
Ming-Hsuan Yang   University of California, Merced, USA
Angela Yao   National University of Singapore, Singapore
Shaodi You   University of Amsterdam, The Netherlands
Stella X. Yu   University of California, Berkeley, USA
Junsong Yuan   State University of New York at Buffalo, USA
Stefanos Zafeiriou   Imperial College London, UK
Amir Zamir   École polytechnique fédérale de Lausanne,
                              Switzerland
Lei Zhang   Alibaba & Hong Kong Polytechnic University,
                              Hong Kong, China
Lei Zhang   International Digital Economy Academy (IDEA),
                              China
Pengchuan Zhang   Meta AI, USA
Bolei Zhou   University of California, Los Angeles, USA
Yuke Zhu   University of Texas at Austin, USA

Todd Zickler   Harvard University, USA
Wangmeng Zuo   Harbin Institute of Technology, China

Technical Program Committee

Davide Abati
Soroush Abbasi
   Koohpayegani
Amos L. Abbott
Rameen Abdal
Rabab Abdelfattah
Sahar Abdelnabi
Hassan Abu Alhaija
Abulikemu Abuduweili
Ron Abutbul
Hanno Ackermann
Aikaterini Adam
Kamil Adamczewski
Ehsan Adeli
Vida Adeli
Donald Adjeroh
Arman Afrasiyabi
Akshay Agarwal
Sameer Agarwal
Abhinav Agarwalla
Vaibhav Aggarwal
Sara Aghajanzadeh
Susmit Agrawal
Antonio Agudo
Touqeer Ahmad
Sk Miraj Ahmed
Chaitanya Ahuja
Nilesh A. Ahuja
Abhishek Aich
Shubhra Aich
Noam Aigerman
Arash Akbarinia
Peri Akiva
Derya Akkaynak
Emre Aksan
Arjun R. Akula
Yuval Alaluf
Stephan Alaniz
Paul Albert
Cenek Albl

Filippo Aleotti
Konstantinos P.
   Alexandridis
Motasem Alfarra
Mohsen Ali
Thiemo Alldieck
Hadi Alzayer
Liang An
Shan An
Yi An
Zhulin An
Dongsheng An
Jie An
Xiang An
Saket Anand
Cosmin Ancuti
Juan Andrade-Cetto
Alexander Andreopoulos
Bjoern Andres
Jerone T. A. Andrews
Shivangi Aneja
Anelia Angelova
Dragomir Anguelov
Rushil Anirudh
Oron Anschel
Rao Muhammad Anwer
Djamila Aouada
Evlampios Apostolidis
Srikar Appalaraju
Nikita Araslanov
Andre Araujo
Eric Arazo
Dawit Mureja Argaw
Anurag Arnab
Aditya Arora
Chetan Arora
Sunpreet S. Arora
Alexey Artemov
Muhammad Asad
Kumar Ashutosh

Sinem Aslan
Vishal Asnani
Mahmoud Assran
Amir Atapour-Abarghouei
Nikos Athanasiou
Ali Athar
ShahRukh Athar
Sara Atito
Souhaib Attaiki
Matan Atzmon
Mathieu Aubry
Nicolas Audebert
Tristan T.
   Aumentado-Armstrong
Melinos Averkiou
Yannis Avrithis
Stephane Ayache
Mehmet Aygün
Seyed Mehdi
   Ayyoubzadeh
Hossein Azizpour
George Azzopardi
Mallikarjun B. R.
Yunhao Ba
Abhishek Badki
Seung-Hwan Bae
Seung-Hwan Baek
Seungryul Baek
Piyush Nitin Bagad
Shai Bagon
Gaetan Bahl
Shikhar Bahl
Sherwin Bahmani
Haoran Bai
Lei Bai
Jiawang Bai
Haoyue Bai
Jinbin Bai
Xiang Bai
Xuyang Bai

Yang Bai
Yuanchao Bai
Ziqian Bai
Sungyong Baik
Kevin Bailly
Max Bain
Federico Baldassarre
Wele Gedara Chaminda
Bandara
Biplab Banerjee
Pratyay Banerjee
Sandipan Banerjee
Jihwan Bang
Antyanta Bangunharcana
Aayush Bansal
Ankan Bansal
Siddhant Bansal
Wentao Bao
Zhipeng Bao
Amir Bar
Manel Baradad Jurjo
Lorenzo Baraldi
Danny Barash
Daniel Barath
Connelly Barnes
Ioan Andrei Bârsan
Steven Basart
Dina Bashkirova
Chaim Baskin
Peyman Bateni
Anil Batra
Sebastiano Battiato
Ardhendu Behera
Harkirat Behl
Jens Behley
Vasileios Belagiannis
Boulbaba Ben Amor
Emanuel Ben Baruch
Abdessamad Ben Hamza
Gil Ben-Artzi
Assia Benbihi
Fabian Benitez-Quiroz
Guy Ben-Yosef
Philipp Benz
Alexander W. Bergman

Urs Bergmann
Jesus Bermudez-Cameo
Stefano Berretti
Gedas Bertasius
Zachary Bessinger
Petra Bevandić
Matthew Beveridge
Lucas Beyer
Yash Bhalgat
Suvaansh Bhambri
Samarth Bharadwaj
Gaurav Bharaj
Aparna Bharati
Bharat Lal Bhatnagar
Uttaran Bhattacharya
Apratim Bhattacharyya
Brojeshwar Bhowmick
Ankan Kumar Bhunia
Ayan Kumar Bhunia
Qi Bi
Sai Bi
Michael Bi Mi
Gui-Bin Bian
Jia-Wang Bian
Shaojun Bian
Pia Bideau
Mario Bijelic
Hakan Bilen
Guillaume-Alexandre
Bilodeau
Alexander Binder
Tolga Birdal
Vighnesh N. Birodkar
Sandika Biswas
Andreas Blattmann
Janusz Bobulski
Giuseppe Boccignone
Vishnu Boddeti
Navaneeth Bodla
Moritz Böhle
Aleksei Bokhovkin
Sam Bond-Taylor
Vivek Boominathan
Shubhankar Borse
Mark Boss

Andrea Bottino
Adnane Boukhayma
Fadi Boutros
Nicolas C. Boutry
Richard S. Bowen
Ivaylo Boyadzhiev
Aidan Boyd
Yuri Boykov
Aljaz Bozic
Behzad Bozorgtabar
Eric Brachmann
Samarth Brahmbhatt
Gustav Bredell
Francois Bremond
Joel Brogan
Andrew Brown
Thomas Brox
Marcus A. Brubaker
Robert-Jan Bruintjes
Yuqi Bu
Anders G. Buch
Himanshu Buckchash
Mateusz Buda
Ignas Budvytis
José M. Buenaposada
Marcel C. Bühler
Tu Bui
Adrian Bulat
Hannah Bull
Evgeny Burnaev
Andrei Bursuc
Benjamin Busam
Sergey N. Buzykanov
Wonmin Byeon
Fabian Caba
Martin Cadik
Guanyu Cai
Minjie Cai
Qing Cai
Zhongang Cai
Qi Cai
Yancheng Cai
Shen Cai
Han Cai
Jiarui Cai

Bowen Cai
Mu Cai
Qin Cai
Ruojin Cai
Weidong Cai
Weiwei Cai
Yi Cai
Yujun Cai
Zhiping Cai
Akin Caliskan
Lilian Calvet
Baris Can Cam
Necati Cihan Camgoz
Tommaso Campari
Dylan Campbell
Ziang Cao
Ang Cao
Xu Cao
Zhiwen Cao
Shengcao Cao
Song Cao
Weipeng Cao
Xiangyong Cao
Xiaochun Cao
Yue Cao
Yunhao Cao
Zhangjie Cao
Jiale Cao
Yang Cao
Jiajiong Cao
Jie Cao
Jinkun Cao
Lele Cao
Yulong Cao
Zhiguo Cao
Chen Cao
Razvan Caramalau
Marlène Careil
Gustavo Carneiro
Joao Carreira
Dan Casas
Paola Cascante-Bonilla
Angela Castillo
Francisco M. Castro
Pedro Castro

Luca Cavalli
George J. Cazenavette
Oya Celiktutan
Hakan Cevikalp
Sri Harsha C. H.
Sungmin Cha
Geonho Cha
Menglei Chai
Lucy Chai
Yuning Chai
Zenghao Chai
Anirban Chakraborty
Deep Chakraborty
Rudrasis Chakraborty
Souradeep Chakraborty
Kelvin C. K. Chan
Chee Seng Chan
Paramanand Chandramouli
Arjun Chandrasekaran
Kenneth Chaney
Dongliang Chang
Huiwen Chang
Peng Chang
Xiaojun Chang
Jia-Ren Chang
Hyung Jin Chang
Hyun Sung Chang
Ju Yong Chang
Li-Jen Chang
Qi Chang
Wei-Yi Chang
Yi Chang
Nadine Chang
Hanqing Chao
Pradyumna Chari
Dibyadip Chatterjee
Chiranjoy Chattopadhyay
Siddhartha Chaudhuri
Zhengping Che
Gal Chechik
Lianggangxu Chen
Qi Alfred Chen
Brian Chen
Bor-Chun Chen
Bo-Hao Chen

Bohong Chen
Bin Chen
Ziliang Chen
Cheng Chen
Chen Chen
Chaofeng Chen
Xi Chen
Haoyu Chen
Xuanhong Chen
Wei Chen
Qiang Chen
Shi Chen
Xianyu Chen
Chang Chen
Changhuai Chen
Hao Chen
Jie Chen
Jianbo Chen
Jingjing Chen
Jun Chen
Kejiang Chen
Mingcai Chen
Nenglun Chen
Qifeng Chen
Ruoyu Chen
Shu-Yu Chen
Weidong Chen
Weijie Chen
Weikai Chen
Xiang Chen
Xiuyi Chen
Xingyu Chen
Yaofo Chen
Yueting Chen
Yu Chen
Yunjin Chen
Yuntao Chen
Yun Chen
Zhenfang Chen
Zhuangzhuang Chen
Chu-Song Chen
Xiangyu Chen
Zhuo Chen
Chaoqi Chen
Shizhe Chen

Xiaotong Chen
Xiaozhi Chen
Dian Chen
Defang Chen
Dingfan Chen
Ding-Jie Chen
Ee Heng Chen
Tao Chen
Yixin Chen
Wei-Ting Chen
Lin Chen
Guang Chen
Guangyi Chen
Guanying Chen
Guangyao Chen
Hwann-Tzong Chen
Junwen Chen
Jiacheng Chen
Jianxu Chen
Hui Chen
Kai Chen
Kan Chen
Kevin Chen
Kuan-Wen Chen
Weihua Chen
Zhang Chen
Liang-Chieh Chen
Lele Chen
Liang Chen
Fanglin Chen
Zehui Chen
Minghui Chen
Minghao Chen
Xiaokang Chen
Qian Chen
Jun-Cheng Chen
Qi Chen
Qingcai Chen
Richard J. Chen
Runnan Chen
Rui Chen
Shuo Chen
Sentao Chen
Shaoyu Chen
Shixing Chen

Shuai Chen
Shuya Chen
Sizhe Chen
Simin Chen
Shaoxiang Chen
Zitian Chen
Tianlong Chen
Tianshui Chen
Min-Hung Chen
Xiangning Chen
Xin Chen
Xinghao Chen
Xuejin Chen
Xu Chen
Xuxi Chen
Yunlu Chen
Yanbei Chen
Yuxiao Chen
Yun-Chun Chen
Yi-Ting Chen
Yi-Wen Chen
Yinbo Chen
Yiran Chen
Yuanhong Chen
Yubei Chen
Yuefeng Chen
Yuhua Chen
Yukang Chen
Zerui Chen
Zhaoyu Chen
Zhen Chen
Zhenyu Chen
Zhi Chen
Zhiwei Chen
Zhixiang Chen
Long Chen
Bowen Cheng
Jun Cheng
Yi Cheng
Jingchun Cheng
Lechao Cheng
Xi Cheng
Yuan Cheng
Ho Kei Cheng
Kevin Ho Man Cheng

Jiacheng Cheng
Kelvin B. Cheng
Li Cheng
Mengjun Cheng
Zhen Cheng
Qingrong Cheng
Tianheng Cheng
Harry Cheng
Yihua Cheng
Yu Cheng
Ziheng Cheng
Soon Yau Cheong
Anoop Cherian
Manuela Chessa
Zhixiang Chi
Naoki Chiba
Julian Chibane
Kashyap Chitta
Tai-Yin Chiu
Hsu-kuang Chiu
Wei-Chen Chiu
Sungmin Cho
Donghyeon Cho
Hyeon Cho
Yooshin Cho
Gyusang Cho
Jang Hyun Cho
Seungju Cho
Nam Ik Cho
Sunghyun Cho
Hanbyel Cho
Jaesung Choe
Jooyoung Choi
Chiho Choi
Changwoon Choi
Jongwon Choi
Myungsub Choi
Dooseop Choi
Jonghyun Choi
Jinwoo Choi
Jun Won Choi
Min-Kook Choi
Hongsuk Choi
Janghoon Choi
Yoon-Ho Choi

Yukyung Choi
Jaegul Choo
Ayush Chopra
Siddharth Choudhary
Subhabrata Choudhury
Vasileios Choutas
Ka-Ho Chow
Pinaki Nath Chowdhury
Sammy Christen
Anders Christensen
Grigorios Chrysos
Hang Chu
Wen-Hsuan Chu
Peng Chu
Qi Chu
Ruihang Chu
Wei-Ta Chu
Yung-Yu Chuang
Sanghyuk Chun
Se Young Chun
Antonio Cinà
Ramazan Gokberk Cinbis
Javier Civera
Albert Clapés
Ronald Clark
Brian S. Clipp
Felipe Codevilla
Daniel Coelho de Castro
Niv Cohen
Forrester Cole
Maxwell D. Collins
Robert T. Collins
Marc Comino Trinidad
Runmin Cong
Wenyan Cong
Maxime Cordy
Marcella Cornia
Enric Corona
Huseyin Coskun
Luca Cosmo
Dragos Costea
Davide Cozzolino
Arun C. S. Kumar
Aiyu Cui
Qiongjie Cui

Quan Cui
Shuhao Cui
Yiming Cui
Ying Cui
Zijun Cui
Jiali Cui
Jiequan Cui
Yawen Cui
Zhen Cui
Zhaopeng Cui
Jack Culpepper
Xiaodong Cun
Ross Cutler
Adam Czajka
Ali Dabouei
Konstantinos M. Dafnis
Manuel Dahnert
Tao Dai
Yuchao Dai
Bo Dai
Mengyu Dai
Hang Dai
Haixing Dai
Peng Dai
Pingyang Dai
Qi Dai
Qiyu Dai
Yutong Dai
Naser Damer
Zhiyuan Dang
Mohamed Daoudi
Ayan Das
Abir Das
Debasmit Das
Deepayan Das
Partha Das
Sagnik Das
Soumi Das
Srijan Das
Swagatam Das
Avijit Dasgupta
Jim Davis
Adrian K. Davison
Homa Davoudi
Laura Daza

Matthias De Lange
Shalini De Mello
Marco De Nadai
Christophe De
    Vleeschouwer
Alp Dener
Boyang Deng
Congyue Deng
Bailin Deng
Yong Deng
Ye Deng
Zhuo Deng
Zhijie Deng
Xiaoming Deng
Jiankang Deng
Jinhong Deng
Jingjing Deng
Liang-Jian Deng
Siqi Deng
Xiang Deng
Xueqing Deng
Zhongying Deng
Karan Desai
Jean-Emmanuel Deschaud
Aniket Anand Deshmukh
Neel Dey
Helisa Dhamo
Prithviraj Dhar
Amaya Dharmasiri
Yan Di
Xing Di
Ousmane A. Dia
Haiwen Diao
Xiaolei Diao
Gonçalo José Dias Pais
Abdallah Dib
Anastasios Dimou
Changxing Ding
Henghui Ding
Guodong Ding
Yaqing Ding
Shuangrui Ding
Yuhang Ding
Yikang Ding
Shouhong Ding

Haisong Ding
Hui Ding
Jiahao Ding
Jian Ding
Jian-Jiun Ding
Shuxiao Ding
Tianyu Ding
Wenhao Ding
Yuqi Ding
Yi Ding
Yuzhen Ding
Zhengming Ding
Tan Minh Dinh
Vu Dinh
Christos Diou
Mandar Dixit
Bao Gia Doan
Khoa D. Doan
Dzung Anh Doan
Debi Prosad Dogra
Nehal Doiphode
Chengdong Dong
Bowen Dong
Zhenxing Dong
Hang Dong
Xiaoyi Dong
Haoye Dong
Jiangxin Dong
Shichao Dong
Xuan Dong
Zhen Dong
Shuting Dong
Jing Dong
Li Dong
Ming Dong
Nanqing Dong
Qiulei Dong
Runpei Dong
Siyan Dong
Tian Dong
Wei Dong
Xiaomeng Dong
Xin Dong
Xingbo Dong
Yuan Dong

Samuel Dooley
Gianfranco Doretto
Michael Dorkenwald
Keval Doshi
Zhaopeng Dou
Xiaotian Dou
Hazel Doughty
Ahmad Droby
Iddo Drori
Jie Du
Yong Du
Dawei Du
Dong Du
Ruoyi Du
Yuntao Du
Xuefeng Du
Yilun Du
Yuming Du
Radhika Dua
Haodong Duan
Jiafei Duan
Kaiwen Duan
Peiqi Duan
Ye Duan
Haoran Duan
Jiali Duan
Amanda Duarte
Abhimanyu Dubey
Shiv Ram Dubey
Florian Dubost
Lukasz Dudziak
Shivam Duggal
Justin M. Dulay
Matteo Dunnhofer
Chi Nhan Duong
Thibaut Durand
Mihai Dusmanu
Ujjal Kr Dutta
Debidatta Dwibedi
Isht Dwivedi
Sai Kumar Dwivedi
Takeharu Eda
Mark Edmonds
Alexei A. Efros
Thibaud Ehret

Max Ehrlich
Mahsa Ehsanpour
Iván Eichhardt
Farshad Einabadi
Marvin Eisenberger
Hazim Kemal Ekenel
Mohamed El Banani
Ismail Elezi
Moshe Eliasof
Alaa El-Nouby
Ian Endres
Francis Engelmann
Deniz Engin
Chanho Eom
Dave Epstein
Maria C. Escobar
Victor A. Escorcia
Carlos Esteves
Sungmin Eum
Bernard J. E. Evans
Ivan Evtimov
Fevziye Irem Eyiokur
    Yaman
Matteo Fabbri
Sébastien Fabbro
Gabriele Facciolo
Masud Fahim
Bin Fan
Hehe Fan
Deng-Ping Fan
Aoxiang Fan
Chen-Chen Fan
Qi Fan
Zhaoxin Fan
Haoqi Fan
Heng Fan
Hongyi Fan
Linxi Fan
Baojie Fan
Jiayuan Fan
Lei Fan
Quanfu Fan
Yonghui Fan
Yingruo Fan
Zhiwen Fan

Zicong Fan
Sean Fanello
Jiansheng Fang
Chaowei Fang
Yuming Fang
Jianwu Fang
Jin Fang
Qi Fang
Shancheng Fang
Tian Fang
Xianyong Fang
Gongfan Fang
Zhen Fang
Hui Fang
Jiemin Fang
Le Fang
Pengfei Fang
Xiaolin Fang
Yuxin Fang
Zhaoyuan Fang
Ammarah Farooq
Azade Farshad
Zhengcong Fei
Michael Felsberg
Wei Feng
Chen Feng
Fan Feng
Andrew Feng
Xin Feng
Zheyun Feng
Ruicheng Feng
Mingtao Feng
Qianyu Feng
Shangbin Feng
Chun-Mei Feng
Zunlei Feng
Zhiyong Feng
Martin Fergie
Mustansar Fiaz
Marco Fiorucci
Michael Firman
Hamed Firooz
Volker Fischer
Corneliu O. Florea
Georgios Floros

Wolfgang Foerstner
Gianni Franchi
Jean-Sebastien Franco
Simone Frintrop
Anna Fruehstueck
Changhong Fu
Chaoyou Fu
Cheng-Yang Fu
Chi-Wing Fu
Deqing Fu
Huan Fu
Jun Fu
Kexue Fu
Ying Fu
Jianlong Fu
Jingjing Fu
Qichen Fu
Tsu-Jui Fu
Xueyang Fu
Yang Fu
Yanwei Fu
Yonggan Fu
Wolfgang Fuhl
Yasuhisa Fujii
Kent Fujiwara
Marco Fumero
Takuya Funatomi
Isabel Funke
Dario Fuoli
Antonino Furnari
Matheus A. Gadelha
Akshay Gadi Patil
Adrian Galdran
Guillermo Gallego
Silvano Galliani
Orazio Gallo
Leonardo Galteri
Matteo Gamba
Yiming Gan
Sujoy Ganguly
Harald Ganster
Boyan Gao
Changxin Gao
Daiheng Gao
Difei Gao

Chen Gao
Fei Gao
Lin Gao
Wei Gao
Yiming Gao
Junyu Gao
Guangyu Ryan Gao
Haichang Gao
Hongchang Gao
Jialin Gao
Jin Gao
Jun Gao
Katelyn Gao
Mingchen Gao
Mingfei Gao
Pan Gao
Shangqian Gao
Shanghua Gao
Xitong Gao
Yunhe Gao
Zhanning Gao
Elena Garces
Nuno Cruz Garcia
Noa Garcia
Guillermo
    Garcia-Hernando
Isha Garg
Rahul Garg
Sourav Garg
Quentin Garrido
Stefano Gasperini
Kent Gauen
Chandan Gautam
Shivam Gautam
Paul Gay
Chunjiang Ge
Shiming Ge
Wenhang Ge
Yanhao Ge
Zheng Ge
Songwei Ge
Weifeng Ge
Yixiao Ge
Yuying Ge
Shijie Geng

Zhengyang Geng
Kyle A. Genova
Georgios Georgakis
Markos Georgopoulos
Marcel Geppert
Shabnam Ghadar
Mina Ghadimi Atigh
Deepti Ghadiyaram
Maani Ghaffari Jadidi
Sedigh Ghamari
Zahra Gharaee
Michaël Gharbi
Golnaz Ghiasi
Reza Ghoddoosian
Soumya Suvra Ghosal
Adhiraj Ghosh
Arthita Ghosh
Pallabi Ghosh
Soumyadeep Ghosh
Andrew Gilbert
Igor Gilitschenski
Jhony H. Giraldo
Andreu Girbau Xalabarder
Rohit Girdhar
Sharath Girish
Xavier Giro-i-Nieto
Raja Giryes
Thomas Gittings
Nikolaos Gkanatsios
Ioannis Gkioulekas
Abhiram
    Gnanasambandam
Aurele T. Gnanha
Clement L. J. C. Godard
Arushi Goel
Vidit Goel
Shubham Goel
Zan Gojcic
Aaron K. Gokaslan
Tejas Gokhale
S. Alireza Golestaneh
Thiago L. Gomes
Nuno Goncalves
Boqing Gong
Chen Gong

Yuanhao Gong
Guoqiang Gong
Jingyu Gong
Rui Gong
Yu Gong
Mingming Gong
Neil Zhenqiang Gong
Xun Gong
Yunye Gong
Yihong Gong
Cristina I. González
Nithin Gopalakrishnan
    Nair
Gaurav Goswami
Jianping Gou
Shreyank N. Gowda
Ankit Goyal
Helmut Grabner
Patrick L. Grady
Ben Graham
Eric Granger
Douglas R. Gray
Matej Grcić
David Griffiths
Jinjin Gu
Yun Gu
Shuyang Gu
Jianyang Gu
Fuqiang Gu
Jiatao Gu
Jindong Gu
Jiaqi Gu
Jinwei Gu
Jiaxin Gu
Geonmo Gu
Xiao Gu
Xinqian Gu
Xiuye Gu
Yuming Gu
Zhangxuan Gu
Dayan Guan
Junfeng Guan
Qingji Guan
Tianrui Guan
Shanyan Guan

Denis A. Gudovskiy
Ricardo Guerrero
Pierre-Louis Guhur
Jie Gui
Liangyan Gui
Liangke Gui
Benoit Guillard
Erhan Gundogdu
Manuel Günther
Jingcai Guo
Yuanfang Guo
Junfeng Guo
Chenqi Guo
Dan Guo
Hongji Guo
Jia Guo
Jie Guo
Minghao Guo
Shi Guo
Yanhui Guo
Yangyang Guo
Yuan-Chen Guo
Yilu Guo
Yiluan Guo
Yong Guo
Guangyu Guo
Haiyun Guo
Jinyang Guo
Jianyuan Guo
Pengsheng Guo
Pengfei Guo
Shuxuan Guo
Song Guo
Tianyu Guo
Qing Guo
Qiushan Guo
Wen Guo
Xiefan Guo
Xiaohu Guo
Xiaoqing Guo
Yufei Guo
Yuhui Guo
Yuliang Guo
Yunhui Guo
Yanwen Guo

Akshita Gupta
Ankush Gupta
Kamal Gupta
Kartik Gupta
Ritwik Gupta
Rohit Gupta
Siddharth Gururani
Fredrik K. Gustafsson
Abner Guzman Rivera
Vladimir Guzov
Matthew A. Gwilliam
Jung-Woo Ha
Marc Habermann
Isma Hadji
Christian Haene
Martin Hahner
Levente Hajder
Alexandros Haliassos
Emanuela Haller
Bumsub Ham
Abdullah J. Hamdi
Shreyas Hampali
Dongyoon Han
Chunrui Han
Dong-Jun Han
Dong-Sig Han
Guangxing Han
Zhizhong Han
Ruize Han
Jiaming Han
Jin Han
Ligong Han
Xian-Hua Han
Xiaoguang Han
Yizeng Han
Zhi Han
Zhenjun Han
Zhongyi Han
Jungong Han
Junlin Han
Kai Han
Kun Han
Sungwon Han
Songfang Han
Wei Han

Xiao Han
Xintong Han
Xinzhe Han
Yahong Han
Yan Han
Zongbo Han
Nicolai Hani
Rana Hanocka
Niklas Hanselmann
Nicklas A. Hansen
Hong Hanyu
Fusheng Hao
Yanbin Hao
Shijie Hao
Udith Haputhanthri
Mehrtash Harandi
Josh Harguess
Adam Harley
David M. Hart
Atsushi Hashimoto
Ali Hassani
Mohammed Hassanin
Yana Hasson
Joakim Bruslund Haurum
Bo He
Kun He
Chen He
Xin He
Fazhi He
Gaoqi He
Hao He
Haoyu He
Jiangpeng He
Hongliang He
Qian He
Xiangteng He
Xuming He
Yannan He
Yuhang He
Yang He
Xiangyu He
Nanjun He
Pan He
Sen He
Shengfeng He

Songtao He
Tao He
Tong He
Wei He
Xuehai He
Xiaoxiao He
Ying He
Yisheng He
Ziwen He
Peter Hedman
Felix Heide
Yacov Hel-Or
Paul Henderson
Philipp Henzler
Byeongho Heo
Jae-Pil Heo
Miran Heo
Sachini A. Herath
Stephane Herbin
Pedro Hermosilla Casajus
Monica Hernandez
Charles Herrmann
Roei Herzig
Mauricio Hess-Flores
Carlos Hinojosa
Tobias Hinz
Tsubasa Hirakawa
Chih-Hui Ho
Lam Si Tung Ho
Jennifer Hobbs
Derek Hoiem
Yannick Hold-Geoffroy
Aleksander Holynski
Cheeun Hong
Fa-Ting Hong
Hanbin Hong
Guan Zhe Hong
Danfeng Hong
Lanqing Hong
Xiaopeng Hong
Xin Hong
Jie Hong
Seungbum Hong
Cheng-Yao Hong
Seunghoon Hong

Yi Hong
Yuan Hong
Yuchen Hong
Anthony Hoogs
Maxwell C. Horton
Kazuhiro Hotta
Qibin Hou
Tingbo Hou
Junhui Hou
Ji Hou
Qiqi Hou
Rui Hou
Ruibing Hou
Zhi Hou
Henry Howard-Jenkins
Lukas Hoyer
Wei-Lin Hsiao
Chiou-Ting Hsu
Anthony Hu
Brian Hu
Yusong Hu
Hexiang Hu
Haoji Hu
Di Hu
Hengtong Hu
Haigen Hu
Lianyu Hu
Hanzhe Hu
Jie Hu
Junlin Hu
Shizhe Hu
Jian Hu
Zhiming Hu
Juhua Hu
Peng Hu
Ping Hu
Ronghang Hu
MengShun Hu
Tao Hu
Vincent Tao Hu
Xiaoling Hu
Xinting Hu
Xiaolin Hu
Xuefeng Hu
Xiaowei Hu

Yang Hu
Yueyu Hu
Zeyu Hu
Zhongyun Hu
Binh-Son Hua
Guoliang Hua
Yi Hua
Linzhi Huang
Qiusheng Huang
Bo Huang
Chen Huang
Hsin-Ping Huang
Ye Huang
Shuangping Huang
Zeng Huang
Buzhen Huang
Cong Huang
Heng Huang
Hao Huang
Qidong Huang
Huaibo Huang
Chaoqin Huang
Feihu Huang
Jiahui Huang
Jingjia Huang
Kun Huang
Lei Huang
Sheng Huang
Shuaiyi Huang
Siyu Huang
Xiaoshui Huang
Xiaoyang Huang
Yan Huang
Yihao Huang
Ying Huang
Ziling Huang
Xiaoke Huang
Yifei Huang
Haiyang Huang
Zhewei Huang
Jin Huang
Haibin Huang
Jiaxing Huang
Junjie Huang
Keli Huang

Lang Huang
Lin Huang
Luojie Huang
Mingzhen Huang
Shijia Huang
Shengyu Huang
Siyuan Huang
He Huang
Xiuyu Huang
Lianghua Huang
Yue Huang
Yaping Huang
Yuge Huang
Zehao Huang
Zeyi Huang
Zhiqi Huang
Zhongzhan Huang
Zilong Huang
Ziyuan Huang
Tianrui Hui
Zhuo Hui
Le Hui
Jing Huo
Junhwa Hur
Shehzeen S. Hussain
Chuong Minh Huynh
Seunghyun Hwang
Jaehui Hwang
Jyh-Jing Hwang
Sukjun Hwang
Soonmin Hwang
Wonjun Hwang
Rakib Hyder
Sangeek Hyun
Sarah Ibrahimi
Tomoki Ichikawa
Yerlan Idelbayev
A. S. M. Iftekhar
Masaaki Iiyama
Satoshi Ikehata
Sunghoon Im
Atul N. Ingle
Eldar Insafutdinov
Yani A. Ioannou
Radu Tudor Ionescu

Umar Iqbal
Go Irie
Muhammad Zubair Irshad
Ahmet Iscen
Berivan Isik
Ashraful Islam
Md Amirul Islam
Syed Islam
Mariko Isogawa
Vamsi Krishna K. Ithapu
Boris Ivanovic
Darshan Iyer
Sarah Jabbour
Ayush Jain
Nishant Jain
Samyak Jain
Vidit Jain
Vineet Jain
Priyank Jaini
Tomas Jakab
Mohammad A. A. K.
    Jalwana
Muhammad Abdullah
    Jamal
Hadi Jamali-Rad
Stuart James
Varun Jampani
Young Kyun Jang
YeongJun Jang
Yunseok Jang
Ronnachai Jaroensri
Bhavan Jasani
Krishna Murthy
    Jatavallabhula
Mojan Javaheripi
Syed A. Javed
Guillaume Jeanneret
Pranav Jeevan
Herve Jegou
Rohit Jena
Tomas Jenicek
Porter Jenkins
Simon Jenni
Hae-Gon Jeon
Sangryul Jeon

Boseung Jeong
Yoonwoo Jeong
Seong-Gyun Jeong
Jisoo Jeong
Allan D. Jepson
Ankit Jha
Sumit K. Jha
I-Hong Jhuo
Ge-Peng Ji
Chaonan Ji
Deyi Ji
Jingwei Ji
Wei Ji
Zhong Ji
Jiayi Ji
Pengliang Ji
Hui Ji
Mingi Ji
Xiaopeng Ji
Yuzhu Ji
Baoxiong Jia
Songhao Jia
Dan Jia
Shan Jia
Xiaojun Jia
Xiuyi Jia
Xu Jia
Menglin Jia
Wenqi Jia
Boyuan Jiang
Wenhao Jiang
Huaizu Jiang
Hanwen Jiang
Haiyong Jiang
Hao Jiang
Huajie Jiang
Huiqin Jiang
Haojun Jiang
Haobo Jiang
Junjun Jiang
Xingyu Jiang
Yangbangyan Jiang
Yu Jiang
Jianmin Jiang
Jiaxi Jiang

Jing Jiang
Kui Jiang
Li Jiang
Liming Jiang
Chiyu Jiang
Meirui Jiang
Chen Jiang
Peng Jiang
Tai-Xiang Jiang
Wen Jiang
Xinyang Jiang
Yifan Jiang
Yuming Jiang
Yingying Jiang
Zeren Jiang
ZhengKai Jiang
Zhenyu Jiang
Shuming Jiao
Jianbo Jiao
Licheng Jiao
Dongkwon Jin
Yeying Jin
Cheng Jin
Linyi Jin
Qing Jin
Taisong Jin
Xiao Jin
Xin Jin
Sheng Jin
Kyong Hwan Jin
Ruibing Jin
SouYoung Jin
Yueming Jin
Chenchen Jing
Longlong Jing
Taotao Jing
Yongcheng Jing
Younghyun Jo
Joakim Johnander
Jeff Johnson
Michael J. Jones
R. Kenny Jones
Rico Jonschkowski
Ameya Joshi
Sunghun Joung

Felix Juefei-Xu
Claudio R. Jung
Steffen Jung
Hari Chandana K.
Rahul Vigneswaran K.
Prajwal K. R.
Abhishek Kadian
Jhony Kaesemodel Pontes
Kumara Kahatapitiya
Anmol Kalia
Sinan Kalkan
Tarun Kalluri
Jaewon Kam
Sandesh Kamath
Meina Kan
Menelaos Kanakis
Takuhiro Kaneko
Di Kang
Guoliang Kang
Hao Kang
Jaeyeon Kang
Kyoungkook Kang
Li-Wei Kang
MinGuk Kang
Suk-Ju Kang
Zhao Kang
Yash Mukund Kant
Yueying Kao
Aupendu Kar
Konstantinos Karantzalos
Sezer Karaoglu
Navid Kardan
Sanjay Kariyappa
Leonid Karlinsky
Animesh Karnewar
Shyamgopal Karthik
Hirak J. Kashyap
Marc A. Kastner
Hirokatsu Kataoka
Angelos Katharopoulos
Hiroharu Kato
Kai Katsumata
Manuel Kaufmann
Chaitanya Kaul
Prakhar Kaushik

Yuki Kawana
Lei Ke
Lipeng Ke
Tsung-Wei Ke
Wei Ke
Petr Kellnhofer
Aniruddha Kembhavi
John Kender
Corentin Kervadec
Leonid Keselman
Daniel Keysers
Nima Khademi Kalantari
Taras Khakhulin
Samir Khaki
Muhammad Haris Khan
Qadeer Khan
Salman Khan
Subash Khanal
Vaishnavi M. Khindkar
Rawal Khirodkar
Saeed Khorram
Pirazh Khorramshahi
Kourosh Khoshelham
Ansh Khurana
Benjamin Kiefer
Jae Myung Kim
Junho Kim
Boah Kim
Hyeonseong Kim
Dong-Jin Kim
Dongwan Kim
Donghyun Kim
Doyeon Kim
Yonghyun Kim
Hyung-Il Kim
Hyunwoo Kim
Hyeongwoo Kim
Hyo Jin Kim
Hyunwoo J. Kim
Taehoon Kim
Jaeha Kim
Jiwon Kim
Jung Uk Kim
Kangyeol Kim
Eunji Kim

Daeha Kim
Dongwon Kim
Kunhee Kim
Kyungmin Kim
Junsik Kim
Min H. Kim
Namil Kim
Kookhoi Kim
Sanghyun Kim
Seongyeop Kim
Seungryong Kim
Saehoon Kim
Euyoung Kim
Guisik Kim
Sungyeon Kim
Sunnie S. Y. Kim
Taehun Kim
Tae Oh Kim
Won Hwa Kim
Seungwook Kim
YoungBin Kim
Youngeun Kim
Akisato Kimura
Furkan Osman Kınlı
Zsolt Kira
Hedvig Kjellström
Florian Kleber
Jan P. Klopp
Florian Kluger
Laurent Kneip
Byungsoo Ko
Muhammed Kocabas
A. Sophia Koepke
Kevin Koeser
Nick Kolkin
Nikos Kolotouros
Wai-Kin Adams Kong
Deying Kong
Caihua Kong
Youyong Kong
Shuyu Kong
Shu Kong
Tao Kong
Yajing Kong
Yu Kong

Zishang Kong
Theodora Kontogianni
Anton S. Konushin
Julian F. P. Kooij
Bruno Korbar
Giorgos Kordopatis-Zilos
Jari Korhonen
Adam Kortylewski
Denis Korzhenkov
Divya Kothandaraman
Suraj Kothawade
Iuliia Kotseruba
Satwik Kottur
Shashank Kotyan
Alexandros Kouris
Petros Koutras
Anna Kreshuk
Ranjay Krishna
Dilip Krishnan
Andrey Kuehlkamp
Hilde Kuehne
Jason Kuen
David Kügler
Arjan Kuijper
Anna Kukleva
Sumith Kulal
Viveka Kulharia
Akshay R. Kulkarni
Nilesh Kulkarni
Dominik Kulon
Abhinav Kumar
Akash Kumar
Suryansh Kumar
B. V. K. Vijaya Kumar
Pulkit Kumar
Ratnesh Kumar
Sateesh Kumar
Satish Kumar
Vijay Kumar B. G.
Nupur Kumari
Sudhakar Kumawat
Jogendra Nath Kundu
Hsien-Kai Kuo
Meng-Yu Jennifer Kuo
Vinod Kumar Kurmi

Yusuke Kurose
Keerthy Kusumam
Alina Kuznetsova
Henry Kvinge
Ho Man Kwan
Hyeokjun Kweon
Heeseung Kwon
Gihyun Kwon
Myung-Joon Kwon
Taesung Kwon
YoungJoong Kwon
Christos Kyrkou
Jorma Laaksonen
Yann Labbe
Zorah Laehner
Florent Lafarge
Hamid Laga
Manuel Lagunas
Shenqi Lai
Jian-Huang Lai
Zihang Lai
Mohamed I. Lakhal
Mohit Lamba
Meng Lan
Loic Landrieu
Zhiqiang Lang
Natalie Lang
Dong Lao
Yizhen Lao
Yingjie Lao
Issam Hadj Laradji
Gustav Larsson
Viktor Larsson
Zakaria Laskar
Stéphane Lathuilière
Chun Pong Lau
Rynson W. H. Lau
Hei Law
Justin Lazarow
Verica Lazova
Eric-Tuan Le
Hieu Le
Trung-Nghia Le
Mathias Lechner
Byeong-Uk Lee

Chen-Yu Lee
Che-Rung Lee
Chul Lee
Hong Joo Lee
Dongsoo Lee
Jiyoung Lee
Eugene Eu Tzuan Lee
Daeun Lee
Saehyung Lee
Jewook Lee
Hyungtae Lee
Hyunmin Lee
Jungbeom Lee
Joon-Young Lee
Jong-Seok Lee
Joonseok Lee
Junha Lee
Kibok Lee
Byung-Kwan Lee
Jangwon Lee
Jinho Lee
Jongmin Lee
Seunghyun Lee
Sohyun Lee
Minsik Lee
Dogyoon Lee
Seungmin Lee
Min Jun Lee
Sangho Lee
Sangmin Lee
Seungeun Lee
Seon-Ho Lee
Sungmin Lee
Sungho Lee
Sangyoun Lee
Vincent C. S. S. Lee
Jaeseong Lee
Yong Jae Lee
Chenyang Lei
Chenyi Lei
Jiahui Lei
Xinyu Lei
Yinjie Lei
Jiaxu Leng
Luziwei Leng

Jan E. Lenssen
Vincent Lepetit
Thomas Leung
María Leyva-Vallina
Xin Li
Yikang Li
Baoxin Li
Bin Li
Bing Li
Bowen Li
Changlin Li
Chao Li
Chongyi Li
Guanyue Li
Shuai Li
Jin Li
Dingquan Li
Dongxu Li
Yiting Li
Gang Li
Dian Li
Guohao Li
Haoang Li
Haoliang Li
Haoran Li
Hengduo Li
Huafeng Li
Xiaoming Li
Hanao Li
Hongwei Li
Ziqiang Li
Jisheng Li
Jiacheng Li
Jia Li
Jiachen Li
Jiahao Li
Jianwei Li
Jiazhi Li
Jie Li
Jing Li
Jingjing Li
Jingtao Li
Jun Li
Junxuan Li
Kai Li

Kailin Li
Kenneth Li
Kun Li
Kunpeng Li
Aoxue Li
Chenglong Li
Chenglin Li
Changsheng Li
Zhichao Li
Qiang Li
Yanyu Li
Zuoyue Li
Xiang Li
Xuelong Li
Fangda Li
Ailin Li
Liang Li
Chun-Guang Li
Daiqing Li
Dong Li
Guanbin Li
Guorong Li
Haifeng Li
Jianan Li
Jianing Li
Jiaxin Li
Ke Li
Lei Li
Lincheng Li
Liulei Li
Lujun Li
Linjie Li
Lin Li
Pengyu Li
Ping Li
Qiufu Li
Qingyong Li
Rui Li
Siyuan Li
Wei Li
Wenbin Li
Xiangyang Li
Xinyu Li
Xiujun Li
Xiu Li

Xu Li
Ya-Li Li
Yao Li
Yongjie Li
Yijun Li
Yiming Li
Yuezun Li
Yu Li
Yunheng Li
Yuqi Li
Zhe Li
Zeming Li
Zhen Li
Zhengqin Li
Zhimin Li
Jiefeng Li
Jinpeng Li
Chengze Li
Jianwu Li
Lerenhan Li
Shan Li
Suichan Li
Xiangtai Li
Yanjie Li
Yandong Li
Zhuoling Li
Zhenqiang Li
Manyi Li
Maosen Li
Ji Li
Minjun Li
Mingrui Li
Mengtian Li
Junyi Li
Nianyi Li
Bo Li
Xiao Li
Peihua Li
Peike Li
Peizhao Li
Peiliang Li
Qi Li
Ren Li
Runze Li
Shile Li

Sheng Li
Shigang Li
Shiyu Li
Shuang Li
Shasha Li
Shichao Li
Tianye Li
Yuexiang Li
Wei-Hong Li
Wanhua Li
Weihao Li
Weiming Li
Weixin Li
Wenbo Li
Wenshuo Li
Weijian Li
Yunan Li
Xirong Li
Xianhang Li
Xiaoyu Li
Xueqian Li
Xuanlin Li
Xianzhi Li
Yunqiang Li
Yanjing Li
Yansheng Li
Yawei Li
Yi Li
Yong Li
Yong-Lu Li
Yuhang Li
Yu-Jhe Li
Yuxi Li
Yunsheng Li
Yanwei Li
Zechao Li
Zejian Li
Zeju Li
Zekun Li
Zhaowen Li
Zheng Li
Zhenyu Li
Zhiheng Li
Zhi Li
Zhong Li

Zhuowei Li
Zhuowan Li
Zhuohang Li
Zizhang Li
Chen Li
Yuan-Fang Li
Dongze Lian
Xiaochen Lian
Zhouhui Lian
Long Lian
Qing Lian
Jin Lianbao
Jinxiu S. Liang
Dingkang Liang
Jiahao Liang
Jianming Liang
Jingyun Liang
Kevin J. Liang
Kaizhao Liang
Chen Liang
Jie Liang
Senwei Liang
Ding Liang
Jiajun Liang
Jian Liang
Kongming Liang
Siyuan Liang
Yuanzhi Liang
Zhengfa Liang
Mingfu Liang
Xiaodan Liang
Xuefeng Liang
Yuxuan Liang
Kang Liao
Liang Liao
Hong-Yuan Mark Liao
Wentong Liao
Haofu Liao
Yue Liao
Minghui Liao
Shengcai Liao
Ting-Hsuan Liao
Xin Liao
Yinghong Liao
Teck Yian Lim

Che-Tsung Lin
Chung-Ching Lin
Chen-Hsuan Lin
Cheng Lin
Chuming Lin
Chunyu Lin
Dahua Lin
Wei Lin
Zheng Lin
Huaijia Lin
Jason Lin
Jierui Lin
Jiaying Lin
Jie Lin
Kai-En Lin
Kevin Lin
Guangfeng Lin
Jiehong Lin
Feng Lin
Hang Lin
Kwan-Yee Lin
Ke Lin
Luojun Lin
Qinghong Lin
Xiangbo Lin
Yi Lin
Zudi Lin
Shijie Lin
Yiqun Lin
Tzu-Heng Lin
Ming Lin
Shaohui Lin
SongNan Lin
Ji Lin
Tsung-Yu Lin
Xudong Lin
Yancong Lin
Yen-Chen Lin
Yiming Lin
Yuewei Lin
Zhiqiu Lin
Zinan Lin
Zhe Lin
David B. Lindell
Zhixin Ling

Zhan Ling
Alexander Liniger
Venice Erin B. Liong
Joey Litalien
Or Litany
Roee Litman
Ron Litman
Jim Little
Dor Litvak
Shaoteng Liu
Shuaicheng Liu
Andrew Liu
Xian Liu
Shaohui Liu
Bei Liu
Bo Liu
Yong Liu
Ming Liu
Yanbin Liu
Chenxi Liu
Daqi Liu
Di Liu
Difan Liu
Dong Liu
Dongfang Liu
Daizong Liu
Xiao Liu
Fangyi Liu
Fengbei Liu
Fenglin Liu
Bin Liu
Yuang Liu
Ao Liu
Hong Liu
Hongfu Liu
Huidong Liu
Ziyi Liu
Feng Liu
Hao Liu
Jie Liu
Jialun Liu
Jiang Liu
Jing Liu
Jingya Liu
Jiaming Liu

Jun Liu
Juncheng Liu
Jiawei Liu
Hongyu Liu
Chuanbin Liu
Haotian Liu
Lingqiao Liu
Chang Liu
Han Liu
Liu Liu
Min Liu
Yingqi Liu
Aishan Liu
Bingyu Liu
Benlin Liu
Boxiao Liu
Chenchen Liu
Chuanjian Liu
Daqing Liu
Huan Liu
Haozhe Liu
Jiaheng Liu
Wei Liu
Jingzhou Liu
Jiyuan Liu
Lingbo Liu
Nian Liu
Peiye Liu
Qiankun Liu
Shenglan Liu
Shilong Liu
Wen Liu
Wenyu Liu
Weifeng Liu
Wu Liu
Xiaolong Liu
Yang Liu
Yanwei Liu
Yingcheng Liu
Yongfei Liu
Yihao Liu
Yu Liu
Yunze Liu
Ze Liu
Zhenhua Liu

Zhenguang Liu
Lin Liu
Lihao Liu
Pengju Liu
Xinhai Liu
Yunfei Liu
Meng Liu
Minghua Liu
Mingyuan Liu
Miao Liu
Peirong Liu
Ping Liu
Qingjie Liu
Ruoshi Liu
Risheng Liu
Songtao Liu
Xing Liu
Shikun Liu
Shuming Liu
Sheng Liu
Songhua Liu
Tongliang Liu
Weibo Liu
Weide Liu
Weizhe Liu
Wenxi Liu
Weiyang Liu
Xin Liu
Xiaobin Liu
Xudong Liu
Xiaoyi Liu
Xihui Liu
Xinchen Liu
Xingtong Liu
Xinpeng Liu
Xinyu Liu
Xianpeng Liu
Xu Liu
Xingyu Liu
Yongtuo Liu
Yahui Liu
Yangxin Liu
Yaoyao Liu
Yaojie Liu
Yuliang Liu

Yongcheng Liu
Yuan Liu
Yufan Liu
Yu-Lun Liu
Yun Liu
Yunfan Liu
Yuanzhong Liu
Zhuoran Liu
Zhen Liu
Zheng Liu
Zhijian Liu
Zhisong Liu
Ziquan Liu
Ziyu Liu
Zhihua Liu
Zechun Liu
Zhaoyang Liu
Zhengzhe Liu
Stephan Liwicki
Shao-Yuan Lo
Sylvain Lobry
Suhas Lohit
Vishnu Suresh Lokhande
Vincenzo Lomonaco
Chengjiang Long
Guodong Long
Fuchen Long
Shangbang Long
Yang Long
Zijun Long
Vasco Lopes
Antonio M. Lopez
Roberto Javier
    Lopez-Sastre
Tobias Lorenz
Javier Lorenzo-Navarro
Yujing Lou
Qian Lou
Xiankai Lu
Changsheng Lu
Huimin Lu
Yongxi Lu
Hao Lu
Hong Lu
Jiasen Lu

Juwei Lu
Fan Lu
Guangming Lu
Jiwen Lu
Shun Lu
Tao Lu
Xiaonan Lu
Yang Lu
Yao Lu
Yongchun Lu
Zhiwu Lu
Cheng Lu
Liying Lu
Guo Lu
Xuequan Lu
Yanye Lu
Yantao Lu
Yuhang Lu
Fujun Luan
Jonathon Luiten
Jovita Lukasik
Alan Lukezic
Jonathan Samuel Lumentut
Mayank Lunayach
Ao Luo
Canjie Luo
Chong Luo
Xu Luo
Grace Luo
Jun Luo
Katie Z. Luo
Tao Luo
Cheng Luo
Fangzhou Luo
Gen Luo
Lei Luo
Sihui Luo
Weixin Luo
Yan Luo
Xiaoyan Luo
Yong Luo
Yadan Luo
Hao Luo
Ruotian Luo
Mi Luo

Tiange Luo
Wenjie Luo
Wenhan Luo
Xiao Luo
Zhiming Luo
Zhipeng Luo
Zhengyi Luo
Diogo C. Luvizon
Zhaoyang Lv
Gengyu Lyu
Lingjuan Lyu
Jun Lyu
Yuanyuan Lyu
Youwei Lyu
Yueming Lyu
Bingpeng Ma
Chao Ma
Chongyang Ma
Congbo Ma
Chih-Yao Ma
Fan Ma
Lin Ma
Haoyu Ma
Hengbo Ma
Jianqi Ma
Jiawei Ma
Jiayi Ma
Kede Ma
Kai Ma
Lingni Ma
Lei Ma
Xu Ma
Ning Ma
Benteng Ma
Cheng Ma
Andy J. Ma
Long Ma
Zhanyu Ma
Zhiheng Ma
Qianli Ma
Shiqiang Ma
Sizhuo Ma
Shiqing Ma
Xiaolong Ma
Xinzhu Ma

Aron Monszpart
Gyeongsik Moon
Suhong Moon
Taesup Moon
Sean Moran
Daniel Moreira
Pietro Morerio
Alexandre Morgand
Lia Morra
Ali Mosleh
Inbar Mosseri
Sayed Mohammad
    Mostafavi Isfahani
Saman Motamed
Ramy A. Mounir
Fangzhou Mu
Jiteng Mu
Norman Mu
Yasuhiro Mukaigawa
Ryan Mukherjee
Tanmoy Mukherjee
Yusuke Mukuta
Ravi Teja Mullapudi
Lea Müller
Matthias Müller
Martin Mundt
Nils Murrugarra-Llerena
Damien Muselet
Armin Mustafa
Muhammad Ferjad Naeem
Sauradip Nag
Hajime Nagahara
Pravin Nagar
Rajendra Nagar
Naveen Shankar Nagaraja
Varun Nagaraja
Tushar Nagarajan
Seungjun Nah
Gaku Nakano
Yuta Nakashima
Giljoo Nam
Seonghyeon Nam
Liangliang Nan
Yuesong Nan
Yeshwanth Napolean

Dinesh Reddy
    Narapureddy
Medhini Narasimhan
Supreeth
    Narasimhaswamy
Sriram Narayanan
Erickson R. Nascimento
Varun Nasery
K. L. Navaneet
Pablo Navarrete Michelini
Shant Navasardyan
Shah Nawaz
Nihal Nayak
Farhood Negin
Lukáš Neumann
Alejandro Newell
Evonne Ng
Kam Woh Ng
Tony Ng
Anh Nguyen
Tuan Anh Nguyen
Cuong Cao Nguyen
Ngoc Cuong Nguyen
Thanh Nguyen
Khoi Nguyen
Phi Le Nguyen
Phong Ha Nguyen
Tam Nguyen
Truong Nguyen
Anh Tuan Nguyen
Rang Nguyen
Thao Thi Phuong Nguyen
Van Nguyen Nguyen
Zhen-Liang Ni
Yao Ni
Shijie Nie
Xuecheng Nie
Yongwei Nie
Weizhi Nie
Ying Nie
Yinyu Nie
Kshitij N. Nikhal
Simon Niklaus
Xuefei Ning
Jifeng Ning

Yotam Nitzan
Di Niu
Shuaicheng Niu
Li Niu
Wei Niu
Yulei Niu
Zhenxing Niu
Albert No
Shohei Nobuhara
Nicoletta Noceti
Junhyug Noh
Sotiris Nousias
Slawomir Nowaczyk
Ewa M. Nowara
Valsamis Ntouskos
Gilberto Ochoa-Ruiz
Ferda Ofli
Jihyong Oh
Sangyun Oh
Youngtaek Oh
Hiroki Ohashi
Takahiro Okabe
Kemal Oksuz
Fumio Okura
Daniel Olmeda Reino
Matthew Olson
Carl Olsson
Roy Or-El
Alessandro Ortis
Guillermo Ortiz-Jimenez
Magnus Oskarsson
Ahmed A. A. Osman
Martin R. Oswald
Mayu Otani
Naima Otberdout
Cheng Ouyang
Jiahong Ouyang
Wanli Ouyang
Andrew Owens
Poojan B. Oza
Mete Ozay
A. Cengiz Oztireli
Gautam Pai
Tomas Pajdla
Umapada Pal

Simone Palazzo
Luca Palmieri
Bowen Pan
Hao Pan
Lili Pan
Tai-Yu Pan
Liang Pan
Chengwei Pan
Yingwei Pan
Xuran Pan
Jinshan Pan
Xinyu Pan
Liyuan Pan
Xingang Pan
Xingjia Pan
Zhihong Pan
Zizheng Pan
Priyadarshini Panda
Rameswar Panda
Rohit Pandey
Kaiyue Pang
Bo Pang
Guansong Pang
Jiangmiao Pang
Meng Pang
Tianyu Pang
Ziqi Pang
Omiros Pantazis
Andreas Panteli
Maja Pantic
Marina Paolanti
Joao P. Papa
Samuele Papa
Mike Papadakis
Dim P. Papadopoulos
George Papandreou
Constantin Pape
Toufiq Parag
Chethan Parameshwara
Shaifali Parashar
Alejandro Pardo
Rishubh Parihar
Sarah Parisot
JaeYoo Park
Gyeong-Moon Park

Hyojin Park
Hyoungseob Park
Jongchan Park
Jae Sung Park
Kiru Park
Chunghyun Park
Kwanyong Park
Sunghyun Park
Sungrae Park
Seongsik Park
Sanghyun Park
Sungjune Park
Taesung Park
Gaurav Parmar
Paritosh Parmar
Alvaro Parra
Despoina Paschalidou
Or Patashnik
Shivansh Patel
Pushpak Pati
Prashant W. Patil
Vaishakh Patil
Suvam Patra
Jay Patravali
Badri Narayana Patro
Angshuman Paul
Sudipta Paul
Rémi Pautrat
Nick E. Pears
Adithya Pediredla
Wenjie Pei
Shmuel Peleg
Latha Pemula
Bo Peng
Houwen Peng
Yue Peng
Liangzu Peng
Baoyun Peng
Jun Peng
Pai Peng
Sida Peng
Xi Peng
Yuxin Peng
Songyou Peng
Wei Peng

Weiqi Peng
Wen-Hsiao Peng
Pramuditha Perera
Juan C. Perez
Eduardo Pérez Pellitero
Juan-Manuel Perez-Rua
Federico Pernici
Marco Pesavento
Stavros Petridis
Ilya A. Petrov
Vladan Petrovic
Mathis Petrovich
Suzanne Petryk
Hieu Pham
Quang Pham
Khoi Pham
Tung Pham
Huy Phan
Stephen Phillips
Cheng Perng Phoo
David Picard
Marco Piccirilli
Georg Pichler
A. J. Piergiovanni
Vipin Pillai
Silvia L. Pintea
Giovanni Pintore
Robinson Piramuthu
Fiora Pirri
Theodoros Pissas
Fabio Pizzati
Benjamin Planche
Bryan Plummer
Matteo Poggi
Ashwini Pokle
Georgy E. Ponimatkin
Adrian Popescu
Stefan Popov
Nikola Popović
Ronald Poppe
Angelo Porrello
Michael Potter
Charalambos Poullis
Hadi Pouransari
Omid Poursaeed

Shraman Pramanick
Mantini Pranav
Dilip K. Prasad
Meghshyam Prasad
B. H. Pawan Prasad
Shitala Prasad
Prateek Prasanna
Ekta Prashnani
Derek S. Prijatelj
Luke Y. Prince
Véronique Prinet
Victor Adrian Prisacariu
James Pritts
Thomas Probst
Sergey Prokudin
Rita Pucci
Chi-Man Pun
Matthew Purri
Haozhi Qi
Lu Qi
Lei Qi
Xianbiao Qi
Yonggang Qi
Yuankai Qi
Siyuan Qi
Guocheng Qian
Hangwei Qian
Qi Qian
Deheng Qian
Shengsheng Qian
Wen Qian
Rui Qian
Yiming Qian
Shengju Qian
Shengyi Qian
Xuelin Qian
Zhenxing Qian
Nan Qiao
Xiaotian Qiao
Jing Qin
Can Qin
Siyang Qin
Hongwei Qin
Jie Qin
Minghai Qin

Yipeng Qin
Yongqiang Qin
Wenda Qin
Xuebin Qin
Yuzhe Qin
Yao Qin
Zhenyue Qin
Zhiwu Qing
Heqian Qiu
Jiayan Qiu
Jielin Qiu
Yue Qiu
Jiaxiong Qiu
Zhongxi Qiu
Shi Qiu
Zhaofan Qiu
Zhongnan Qu
Yanyun Qu
Kha Gia Quach
Yuhui Quan
Ruijie Quan
Mike Rabbat
Rahul Shekhar Rade
Filip Radenovic
Gorjan Radevski
Bogdan Raducanu
Francesco Ragusa
Shafin Rahman
Md Mahfuzur Rahman
  Siddiquee
Hossein Rahmani
Kiran Raja
Sivaramakrishnan
  Rajaraman
Jathushan Rajasegaran
Adnan Siraj Rakin
Michaël Ramamonjisoa
Chirag A. Raman
Shanmuganathan Raman
Vignesh Ramanathan
Vasili Ramanishka
Vikram V. Ramaswamy
Merey Ramazanova
Jason Rambach
Sai Saketh Rambhatla

Clément Rambour
Ashwin Ramesh Babu
Adín Ramírez Rivera
Arianna Rampini
Haoxi Ran
Aakanksha Rana
Aayush Jung Bahadur
  Rana
Kanchana N. Ranasinghe
Aneesh Rangnekar
Samrudhdhi B. Rangrej
Harsh Rangwani
Viresh Ranjan
Anyi Rao
Yongming Rao
Carolina Raposo
Michalis Raptis
Amir Rasouli
Vivek Rathod
Adepu Ravi Sankar
Avinash Ravichandran
Bharadwaj Ravichandran
Dripta S. Raychaudhuri
Adria Recasens
Simon Reiß
Davis Rempe
Daxuan Ren
Jiawei Ren
Jimmy Ren
Sucheng Ren
Dayong Ren
Zhile Ren
Dongwei Ren
Qibing Ren
Pengfei Ren
Zhenwen Ren
Xuqian Ren
Yixuan Ren
Zhongzheng Ren
Ambareesh Revanur
Hamed Rezazadegan
  Tavakoli
Rafael S. Rezende
Wonjong Rhee
Alexander Richard

Christian Richardt
Stephan R. Richter
Benjamin Riggan
Dominik Rivoir
Mamshad Nayeem Rizve
Joshua D. Robinson
Joseph Robinson
Chris Rockwell
Ranga Rodrigo
Andres C. Rodriguez
Carlos Rodriguez-Pardo
Marcus Rohrbach
Gemma Roig
Yu Rong
David A. Ross
Mohammad Rostami
Edward Rosten
Karsten Roth
Anirban Roy
Debaditya Roy
Shuvendu Roy
Ahana Roy Choudhury
Aruni Roy Chowdhury
Denys Rozumnyi
Shulan Ruan
Wenjie Ruan
Patrick Ruhkamp
Danila Rukhovich
Anian Ruoss
Chris Russell
Dan Ruta
Dawid Damian Rymarczyk
DongHun Ryu
Hyeonggon Ryu
Kwonyoung Ryu
Balasubramanian S.
Alexandre Sablayrolles
Mohammad Sabokrou
Arka Sadhu
Aniruddha Saha
Oindrila Saha
Pritish Sahu
Aneeshan Sain
Nirat Saini
Saurabh Saini

Takeshi Saitoh
Christos Sakaridis
Fumihiko Sakaue
Dimitrios Sakkos
Ken Sakurada
Parikshit V. Sakurikar
Rohit Saluja
Nermin Samet
Leo Sampaio Ferraz
    Ribeiro
Jorge Sanchez
Enrique Sanchez
Shengtian Sang
Anush Sankaran
Soubhik Sanyal
Nikolaos Sarafianos
Vishwanath Saragadam
István Sárándi
Saquib Sarfraz
Mert Bulent Sariyildiz
Anindya Sarkar
Pritam Sarkar
Paul-Edouard Sarlin
Hiroshi Sasaki
Takami Sato
Torsten Sattler
Ravi Kumar Satzoda
Axel Sauer
Stefano Savian
Artem Savkin
Manolis Savva
Gerald Schaefer
Simone Schaub-Meyer
Yoni Schirris
Samuel Schulter
Katja Schwarz
Jesse Scott
Sinisa Segvic
Constantin Marc Seibold
Lorenzo Seidenari
Matan Sela
Fadime Sener
Paul Hongsuck Seo
Kwanggyoon Seo
Hongje Seong

Dario Serez
Francesco Setti
Bryan Seybold
Mohamad Shahbazi
Shima Shahfar
Xinxin Shan
Caifeng Shan
Dandan Shan
Shawn Shan
Wei Shang
Jinghuan Shang
Jiaxiang Shang
Lei Shang
Sukrit Shankar
Ken Shao
Rui Shao
Jie Shao
Mingwen Shao
Aashish Sharma
Gaurav Sharma
Vivek Sharma
Abhishek Sharma
Yoli Shavit
Shashank Shekhar
Sumit Shekhar
Zhijie Shen
Fengyi Shen
Furao Shen
Jialie Shen
Jingjing Shen
Ziyi Shen
Linlin Shen
Guangyu Shen
Biluo Shen
Falong Shen
Jiajun Shen
Qiu Shen
Qiuhong Shen
Shuai Shen
Wang Shen
Yiqing Shen
Yunhang Shen
Siqi Shen
Bin Shen
Tianwei Shen

Xi Shen
Yilin Shen
Yuming Shen
Yucong Shen
Zhiqiang Shen
Lu Sheng
Yichen Sheng
Shivanand Venkanna
    Sheshappanavar
Shelly Sheynin
Baifeng Shi
Ruoxi Shi
Botian Shi
Hailin Shi
Jia Shi
Jing Shi
Shaoshuai Shi
Baoguang Shi
Boxin Shi
Hengcan Shi
Tianyang Shi
Xiaodan Shi
Yongjie Shi
Zhensheng Shi
Yinghuan Shi
Weiqi Shi
Wu Shi
Xuepeng Shi
Xiaoshuang Shi
Yujiao Shi
Zenglin Shi
Zhenmei Shi
Takashi Shibata
Meng-Li Shih
Yichang Shih
Hyunjung Shim
Dongseok Shim
Soshi Shimada
Inkyu Shin
Jinwoo Shin
Seungjoo Shin
Seungjae Shin
Koichi Shinoda
Suprosanna Shit

Palaiahnakote
    Shivakumara
Eli Shlizerman
Gaurav Shrivastava
Xiao Shu
Xiangbo Shu
Xiujun Shu
Yang Shu
Tianmin Shu
Jun Shu
Zhixin Shu
Bing Shuai
Maria Shugrina
Ivan Shugurov
Satya Narayan Shukla
Pranjay Shyam
Jianlou Si
Yawar Siddiqui
Alberto Signoroni
Pedro Silva
Jae-Young Sim
Oriane Siméoni
Martin Simon
Andrea Simonelli
Abhishek Singh
Ashish Singh
Dinesh Singh
Gurkirt Singh
Krishna Kumar Singh
Mannat Singh
Pravendra Singh
Rajat Vikram Singh
Utkarsh Singhal
Dipika Singhania
Vasu Singla
Harsh Sinha
Sudipta Sinha
Josef Sivic
Elena Sizikova
Geri Skenderi
Ivan Skorokhodov
Dmitriy Smirnov
Cameron Y. Smith
James S. Smith
Patrick Snape

Mattia Soldan
Hyeongseok Son
Sanghyun Son
Chuanbiao Song
Chen Song
Chunfeng Song
Dan Song
Dongjin Song
Hwanjun Song
Guoxian Song
Jiaming Song
Jie Song
Liangchen Song
Ran Song
Luchuan Song
Xibin Song
Li Song
Fenglong Song
Guoli Song
Guanglu Song
Zhenbo Song
Lin Song
Xinhang Song
Yang Song
Yibing Song
Rajiv Soundararajan
Hossein Souri
Cristovao Sousa
Riccardo Spezialetti
Leonidas Spinoulas
Michael W. Spratling
Deepak Sridhar
Srinath Sridhar
Gaurang Sriramanan
Vinkle Kumar Srivastav
Themos Stafylakis
Serban Stan
Anastasis Stathopoulos
Markus Steinberger
Jan Steinbrener
Sinisa Stekovic
Alexandros Stergiou
Gleb Sterkin
Rainer Stiefelhagen
Pierre Stock

Ombretta Strafforello
Julian Straub
Yannick Strümpler
Joerg Stueckler
Hang Su
Weijie Su
Jong-Chyi Su
Bing Su
Haisheng Su
Jinming Su
Yiyang Su
Yukun Su
Yuxin Su
Zhuo Su
Zhaoqi Su
Xiu Su
Yu-Chuan Su
Zhixun Su
Arulkumar Subramaniam
Akshayvarun Subramanya
A. Subramanyam
Swathikiran Sudhakaran
Yusuke Sugano
Masanori Suganuma
Yumin Suh
Yang Sui
Baochen Sun
Cheng Sun
Long Sun
Guolei Sun
Haoliang Sun
Haomiao Sun
He Sun
Hanqing Sun
Hao Sun
Lichao Sun
Jiachen Sun
Jiaming Sun
Jian Sun
Jin Sun
Jennifer J. Sun
Tiancheng Sun
Libo Sun
Peize Sun
Qianru Sun

Shanlin Sun
Yu Sun
Zhun Sun
Che Sun
Lin Sun
Tao Sun
Yiyou Sun
Chunyi Sun
Chong Sun
Weiwei Sun
Weixuan Sun
Xiuyu Sun
Yanan Sun
Zeren Sun
Zhaodong Sun
Zhiqing Sun
Minhyuk Sung
Jinli Suo
Simon Suo
Abhijit Suprem
Anshuman Suri
Saksham Suri
Joshua M. Susskind
Roman Suvorov
Gurumurthy Swaminathan
Robin Swanson
Paul Swoboda
Tabish A. Syed
Richard Szeliski
Fariborz Taherkhani
Yu-Wing Tai
Keita Takahashi
Walter Talbott
Gary Tam
Masato Tamura
Feitong Tan
Fuwen Tan
Shuhan Tan
Andong Tan
Bin Tan
Cheng Tan
Jianchao Tan
Lei Tan
Mingxing Tan
Xin Tan

Zichang Tan
Zhentao Tan
Kenichiro Tanaka
Masayuki Tanaka
Yushun Tang
Hao Tang
Jingqun Tang
Jinhui Tang
Kaihua Tang
Luming Tang
Lv Tang
Sheyang Tang
Shitao Tang
Siliang Tang
Shixiang Tang
Yansong Tang
Keke Tang
Chang Tang
Chenwei Tang
Jie Tang
Junshu Tang
Ming Tang
Peng Tang
Xu Tang
Yao Tang
Chen Tang
Fan Tang
Haoran Tang
Shengeng Tang
Yehui Tang
Zhipeng Tang
Ugo Tanielian
Chaofan Tao
Jiale Tao
Junli Tao
Renshuai Tao
An Tao
Guanhong Tao
Zhiqiang Tao
Makarand Tapaswi
Jean-Philippe G. Tarel
Juan J. Tarrio
Enzo Tartaglione
Keisuke Tateno
Zachary Teed

Ajinkya B. Tejankar
Bugra Tekin
Purva Tendulkar
Damien Teney
Minggui Teng
Chris Tensmeyer
Andrew Beng Jin Teoh
Philipp Terhörst
Kartik Thakral
Nupur Thakur
Kevin Thandiackal
Spyridon Thermos
Diego Thomas
William Thong
Yuesong Tian
Guanzhong Tian
Lin Tian
Shiqi Tian
Kai Tian
Meng Tian
Tai-Peng Tian
Zhuotao Tian
Shangxuan Tian
Tian Tian
Yapeng Tian
Yu Tian
Yuxin Tian
Leslie Ching Ow Tiong
Praveen Tirupattur
Garvita Tiwari
George Toderici
Antoine Toisoul
Aysim Toker
Tatiana Tommasi
Zhan Tong
Alessio Tonioni
Alessandro Torcinovich
Fabio Tosi
Matteo Toso
Hugo Touvron
Quan Hung Tran
Son Tran
Hung Tran
Ngoc-Trung Tran
Vinh Tran

Phong Tran
Giovanni Trappolini
Edith Tretschk
Subarna Tripathi
Shubhendu Trivedi
Eduard Trulls
Prune Truong
Thanh-Dat Truong
Tomasz Trzcinski
Sam Tsai
Yi-Hsuan Tsai
Ethan Tseng
Yu-Chee Tseng
Shahar Tsiper
Stavros Tsogkas
Shikui Tu
Zhigang Tu
Zhengzhong Tu
Richard Tucker
Sergey Tulyakov
Cigdem Turan
Daniyar Turmukhambetov
Victor G. Turrisi da Costa
Bartlomiej Twardowski
Christopher D. Twigg
Radim Tylecek
Mostofa Rafid Uddin
Md. Zasim Uddin
Kohei Uehara
Nicolas Ugrinovic
Youngjung Uh
Norimichi Ukita
Anwaar Ulhaq
Devesh Upadhyay
Paul Upchurch
Yoshitaka Ushiku
Yuzuko Utsumi
Mikaela Angelina Uy
Mohit Vaishnav
Pratik Vaishnavi
Jeya Maria Jose Valanarasu
Matias A. Valdenegro Toro
Diego Valsesia
Wouter Van Gansbeke
Nanne van Noord

Simon Vandenhende
Farshid Varno
Cristina Vasconcelos
Francisco Vasconcelos
Alex Vasilescu
Subeesh Vasu
Arun Balajee Vasudevan
Kanav Vats
Vaibhav S. Vavilala
Sagar Vaze
Javier Vazquez-Corral
Andrea Vedaldi
Olga Veksler
Andreas Velten
Sai H. Vemprala
Raviteja Vemulapalli
Shashanka
    Venkataramanan
Dor Verbin
Luisa Verdoliva
Manisha Verma
Yashaswi Verma
Constantin Vertan
Eli Verwimp
Deepak Vijaykeerthy
Pablo Villanueva
Ruben Villegas
Markus Vincze
Vibhav Vineet
Minh P. Vo
Huy V. Vo
Duc Minh Vo
Tomas Vojir
Igor Vozniak
Nicholas Vretos
Vibashan VS
Tuan-Anh Vu
Thang Vu
Mårten Wadenbäck
Neal Wadhwa
Aaron T. Walsman
Steven Walton
Jin Wan
Alvin Wan
Jia Wan

Jun Wan
Xiaoyue Wan
Fang Wan
Guowei Wan
Renjie Wan
Zhiqiang Wan
Ziyu Wan
Bastian Wandt
Dongdong Wang
Limin Wang
Haiyang Wang
Xiaobing Wang
Angtian Wang
Angelina Wang
Bing Wang
Bo Wang
Boyu Wang
Binghui Wang
Chen Wang
Chien-Yi Wang
Congli Wang
Qi Wang
Chengrui Wang
Rui Wang
Yiqun Wang
Cong Wang
Wenjing Wang
Dongkai Wang
Di Wang
Xiaogang Wang
Kai Wang
Zhizhong Wang
Fangjinhua Wang
Feng Wang
Hang Wang
Gaoang Wang
Guoqing Wang
Guangcong Wang
Guangzhi Wang
Hanqing Wang
Hao Wang
Haohan Wang
Haoran Wang
Hong Wang
Haotao Wang

Hu Wang
Huan Wang
Hua Wang
Hui-Po Wang
Hengli Wang
Hanyu Wang
Hongxing Wang
Jingwen Wang
Jialiang Wang
Jian Wang
Jianyi Wang
Jiashun Wang
Jiahao Wang
Tsun-Hsuan Wang
Xiaoqian Wang
Jinqiao Wang
Jun Wang
Jianzong Wang
Kaihong Wang
Ke Wang
Lei Wang
Lingjing Wang
Linnan Wang
Lin Wang
Liansheng Wang
Mengjiao Wang
Manning Wang
Nannan Wang
Peihao Wang
Jiayun Wang
Pu Wang
Qiang Wang
Qiufeng Wang
Qilong Wang
Qiangchang Wang
Qin Wang
Qing Wang
Ruocheng Wang
Ruibin Wang
Ruisheng Wang
Ruizhe Wang
Runqi Wang
Runzhong Wang
Wenxuan Wang
Sen Wang

Shangfei Wang
Shaofei Wang
Shijie Wang
Shiqi Wang
Zhibo Wang
Song Wang
Xinjiang Wang
Tai Wang
Tao Wang
Teng Wang
Xiang Wang
Tianren Wang
Tiantian Wang
Tianyi Wang
Fengjiao Wang
Wei Wang
Miaohui Wang
Suchen Wang
Siyue Wang
Yaoming Wang
Xiao Wang
Ze Wang
Biao Wang
Chaofei Wang
Dong Wang
Gu Wang
Guangrun Wang
Guangming Wang
Guo-Hua Wang
Haoqing Wang
Hesheng Wang
Huafeng Wang
Jinghua Wang
Jingdong Wang
Jingjing Wang
Jingya Wang
Jingkang Wang
Jiakai Wang
Junke Wang
Kuo Wang
Lichen Wang
Lizhi Wang
Longguang Wang
Mang Wang
Mei Wang

Min Wang
Peng-Shuai Wang
Run Wang
Shaoru Wang
Shuhui Wang
Tan Wang
Tiancai Wang
Tianqi Wang
Wenhai Wang
Wenzhe Wang
Xiaobo Wang
Xiudong Wang
Xu Wang
Yajie Wang
Yan Wang
Yuan-Gen Wang
Yingqian Wang
Yizhi Wang
Yulin Wang
Yu Wang
Yujie Wang
Yunhe Wang
Yuxi Wang
Yaowei Wang
Yiwei Wang
Zezheng Wang
Hongzhi Wang
Zhiqiang Wang
Ziteng Wang
Ziwei Wang
Zheng Wang
Zhenyu Wang
Binglu Wang
Zhongdao Wang
Ce Wang
Weining Wang
Weiyao Wang
Wenbin Wang
Wenguan Wang
Guangting Wang
Haolin Wang
Haiyan Wang
Huiyu Wang
Naiyan Wang
Jingbo Wang

Jinpeng Wang
Jiaqi Wang
Liyuan Wang
Lizhen Wang
Ning Wang
Wenqian Wang
Sheng-Yu Wang
Weimin Wang
Xiaohan Wang
Yifan Wang
Yi Wang
Yongtao Wang
Yizhou Wang
Zhuo Wang
Zhe Wang
Xudong Wang
Xiaofang Wang
Xinggang Wang
Xiaosen Wang
Xiaosong Wang
Xiaoyang Wang
Lijun Wang
Xinlong Wang
Xuan Wang
Xue Wang
Yangang Wang
Yaohui Wang
Yu-Chiang Frank Wang
Yida Wang
Yilin Wang
Yi Ru Wang
Yali Wang
Yinglong Wang
Yufu Wang
Yujiang Wang
Yuwang Wang
Yuting Wang
Yang Wang
Yu-Xiong Wang
Yixu Wang
Ziqi Wang
Zhicheng Wang
Zeyu Wang
Zhaowen Wang
Zhenyi Wang

Zhenzhi Wang
Zhijie Wang
Zhiyong Wang
Zhongling Wang
Zhuowei Wang
Zian Wang
Zifu Wang
Zihao Wang
Zirui Wang
Ziyan Wang
Wenxiao Wang
Zhen Wang
Zhepeng Wang
Zi Wang
Zihao W. Wang
Steven L. Waslander
Olivia Watkins
Daniel Watson
Silvan Weder
Dongyoon Wee
Dongming Wei
Tianyi Wei
Jia Wei
Dong Wei
Fangyun Wei
Longhui Wei
Mingqiang Wei
Xinyue Wei
Chen Wei
Donglai Wei
Pengxu Wei
Xing Wei
Xiu-Shen Wei
Wenqi Wei
Guoqiang Wei
Wei Wei
XingKui Wei
Xian Wei
Xingxing Wei
Yake Wei
Yuxiang Wei
Yi Wei
Luca Weihs
Michael Weinmann
Martin Weinmann

Congcong Wen
Chuan Wen
Jie Wen
Sijia Wen
Song Wen
Chao Wen
Xiang Wen
Zeyi Wen
Xin Wen
Yilin Wen
Yijia Weng
Shuchen Weng
Junwu Weng
Wenming Weng
Renliang Weng
Zhenyu Weng
Xinshuo Weng
Nicholas J. Westlake
Gordon Wetzstein
Lena M. Widin Klasén
Rick Wildes
Bryan M. Williams
Williem Williem
Ole Winther
Scott Wisdom
Alex Wong
Chau-Wai Wong
Kwan-Yee K. Wong
Yongkang Wong
Scott Workman
Marcel Worring
Michael Wray
Safwan Wshah
Xiang Wu
Aming Wu
Chongruo Wu
Cho-Ying Wu
Chunpeng Wu
Chenyan Wu
Ziyi Wu
Fuxiang Wu
Gang Wu
Haiping Wu
Huisi Wu
Jane Wu

Jialian Wu
Jing Wu
Jinjian Wu
Jianlong Wu
Xian Wu
Lifang Wu
Lifan Wu
Minye Wu
Qianyi Wu
Rongliang Wu
Rui Wu
Shiqian Wu
Shuzhe Wu
Shangzhe Wu
Tsung-Han Wu
Tz-Ying Wu
Ting-Wei Wu
Jiannan Wu
Zhiliang Wu
Yu Wu
Chenyun Wu
Dayan Wu
Dongxian Wu
Fei Wu
Hefeng Wu
Jianxin Wu
Weibin Wu
Wenxuan Wu
Wenhao Wu
Xiao Wu
Yicheng Wu
Yuanwei Wu
Yu-Huan Wu
Zhenxin Wu
Zhenyu Wu
Wei Wu
Peng Wu
Xiaohe Wu
Xindi Wu
Xinxing Wu
Xinyi Wu
Xingjiao Wu
Xiongwei Wu
Yangzheng Wu
Yanzhao Wu

Yawen Wu
Yong Wu
Yi Wu
Ying Nian Wu
Zhenyao Wu
Zhonghua Wu
Zongze Wu
Zuxuan Wu
Stefanie Wuhrer
Teng Xi
Jianing Xi
Fei Xia
Haifeng Xia
Menghan Xia
Yuanqing Xia
Zhihua Xia
Xiaobo Xia
Weihao Xia
Shihong Xia
Yan Xia
Yong Xia
Zhaoyang Xia
Zhihao Xia
Chuhua Xian
Yongqin Xian
Wangmeng Xiang
Fanbo Xiang
Tiange Xiang
Tao Xiang
Liuyu Xiang
Xiaoyu Xiang
Zhiyu Xiang
Aoran Xiao
Chunxia Xiao
Fanyi Xiao
Jimin Xiao
Jun Xiao
Taihong Xiao
Anqi Xiao
Junfei Xiao
Jing Xiao
Liang Xiao
Yang Xiao
Yuting Xiao
Yijun Xiao

Yao Xiao
Zeyu Xiao
Zhisheng Xiao
Zihao Xiao
Binhui Xie
Christopher Xie
Haozhe Xie
Jin Xie
Guo-Sen Xie
Hongtao Xie
Ming-Kun Xie
Tingting Xie
Chaohao Xie
Weicheng Xie
Xudong Xie
Jiyang Xie
Xiaohua Xie
Yuan Xie
Zhenyu Xie
Ning Xie
Xianghui Xie
Xiufeng Xie
You Xie
Yutong Xie
Fuyong Xing
Yifan Xing
Zhen Xing
Yuanjun Xiong
Jinhui Xiong
Weihua Xiong
Hongkai Xiong
Zhitong Xiong
Yuanhao Xiong
Yunyang Xiong
Yuwen Xiong
Zhiwei Xiong
Yuliang Xiu
An Xu
Chang Xu
Chenliang Xu
Chengming Xu
Chenshu Xu
Xiang Xu
Huijuan Xu
Zhe Xu

Jie Xu
Jingyi Xu
Jiarui Xu
Yinghao Xu
Kele Xu
Ke Xu
Li Xu
Linchuan Xu
Linning Xu
Mengde Xu
Mengmeng Frost Xu
Min Xu
Mingye Xu
Jun Xu
Ning Xu
Peng Xu
Runsheng Xu
Sheng Xu
Wenqiang Xu
Xiaogang Xu
Renzhe Xu
Kaidi Xu
Yi Xu
Chi Xu
Qiuling Xu
Baobei Xu
Feng Xu
Haohang Xu
Haofei Xu
Lan Xu
Mingze Xu
Songcen Xu
Weipeng Xu
Wenjia Xu
Wenju Xu
Xiangyu Xu
Xin Xu
Yinshuang Xu
Yixing Xu
Yuting Xu
Yanyu Xu
Zhenbo Xu
Zhiliang Xu
Zhiyuan Xu
Xiaohao Xu

Yanwu Xu
Yan Xu
Yiran Xu
Yifan Xu
Yufei Xu
Yong Xu
Zichuan Xu
Zenglin Xu
Zexiang Xu
Zhan Xu
Zheng Xu
Zhiwei Xu
Ziyue Xu
Shiyu Xuan
Hanyu Xuan
Fei Xue
Jianru Xue
Mingfu Xue
Qinghan Xue
Tianfan Xue
Chao Xue
Chuhui Xue
Nan Xue
Zhou Xue
Xiangyang Xue
Yuan Xue
Abhay Yadav
Ravindra Yadav
Kota Yamaguchi
Toshihiko Yamasaki
Kohei Yamashita
Chaochao Yan
Feng Yan
Kun Yan
Qingsen Yan
Qixin Yan
Rui Yan
Siming Yan
Xinchen Yan
Yaping Yan
Bin Yan
Qingan Yan
Shen Yan
Shipeng Yan
Xu Yan

Yan Yan
Yichao Yan
Zhaoyi Yan
Zike Yan
Zhiqiang Yan
Hongliang Yan
Zizheng Yan
Jiewen Yang
Anqi Joyce Yang
Shan Yang
Anqi Yang
Antoine Yang
Bo Yang
Baoyao Yang
Chenhongyi Yang
Dingkang Yang
De-Nian Yang
Dong Yang
David Yang
Fan Yang
Fengyu Yang
Fengting Yang
Fei Yang
Gengshan Yang
Heng Yang
Han Yang
Huan Yang
Yibo Yang
Jiancheng Yang
Jihan Yang
Jiawei Yang
Jiayu Yang
Jie Yang
Jinfa Yang
Jingkang Yang
Jinyu Yang
Cheng-Fu Yang
Ji Yang
Jianyu Yang
Kailun Yang
Tian Yang
Luyu Yang
Liang Yang
Li Yang
Michael Ying Yang

Yang Yang
Muli Yang
Le Yang
Qiushi Yang
Ren Yang
Ruihan Yang
Shuang Yang
Siyuan Yang
Su Yang
Shiqi Yang
Taojiannan Yang
Tianyu Yang
Lei Yang
Wanzhao Yang
Shuai Yang
William Yang
Wei Yang
Xiaofeng Yang
Xiaoshan Yang
Xin Yang
Xuan Yang
Xu Yang
Xingyi Yang
Xitong Yang
Jing Yang
Yanchao Yang
Wenming Yang
Yujiu Yang
Herb Yang
Jianfei Yang
Jinhui Yang
Chuanguang Yang
Guanglei Yang
Haitao Yang
Kewei Yang
Linlin Yang
Lijin Yang
Longrong Yang
Meng Yang
MingKun Yang
Sibei Yang
Shicai Yang
Tong Yang
Wen Yang
Xi Yang

Xiaolong Yang
Xue Yang
Yubin Yang
Ze Yang
Ziyi Yang
Yi Yang
Linjie Yang
Yuzhe Yang
Yiding Yang
Zhenpei Yang
Zhaohui Yang
Zhengyuan Yang
Zhibo Yang
Zongxin Yang
Hantao Yao
Mingde Yao
Rui Yao
Taiping Yao
Ting Yao
Cong Yao
Qingsong Yao
Quanming Yao
Xu Yao
Yuan Yao
Yao Yao
Yazhou Yao
Jiawen Yao
Shunyu Yao
Pew-Thian Yap
Sudhir Yarram
Rajeev Yasarla
Peng Ye
Botao Ye
Mao Ye
Fei Ye
Hanrong Ye
Jingwen Ye
Jinwei Ye
Jiarong Ye
Mang Ye
Meng Ye
Qi Ye
Qian Ye
Qixiang Ye
Junjie Ye

Sheng Ye
Nanyang Ye
Yufei Ye
Xiaoqing Ye
Ruolin Ye
Yousef Yeganeh
Chun-Hsiao Yeh
Raymond A. Yeh
Yu-Ying Yeh
Kai Yi
Chang Yi
Renjiao Yi
Xinping Yi
Peng Yi
Alper Yilmaz
Junho Yim
Hui Yin
Bangjie Yin
Jia-Li Yin
Miao Yin
Wenzhe Yin
Xuwang Yin
Ming Yin
Yu Yin
Aoxiong Yin
Kangxue Yin
Tianwei Yin
Wei Yin
Xianghua Ying
Rio Yokota
Tatsuya Yokota
Naoto Yokoya
Ryo Yonetani
Ki Yoon Yoo
Jinsu Yoo
Sunjae Yoon
Jae Shin Yoon
Jihun Yoon
Sung-Hoon Yoon
Ryota Yoshihashi
Yusuke Yoshiyasu
Chenyu You
Haoran You
Haoxuan You
Yang You

Quanzeng You
Tackgeun You
Kaichao You
Shan You
Xinge You
Yurong You
Baosheng Yu
Bei Yu
Haichao Yu
Hao Yu
Chaohui Yu
Fisher Yu
Jin-Gang Yu
Jiyang Yu
Jason J. Yu
Jiashuo Yu
Hong-Xing Yu
Lei Yu
Mulin Yu
Ning Yu
Peilin Yu
Qi Yu
Qian Yu
Rui Yu
Shuzhi Yu
Gang Yu
Tan Yu
Weijiang Yu
Xin Yu
Bingyao Yu
Ye Yu
Hanchao Yu
Yingchen Yu
Tao Yu
Xiaotian Yu
Qing Yu
Houjian Yu
Changqian Yu
Jing Yu
Jun Yu
Shujian Yu
Xiang Yu
Zhaofei Yu
Zhenbo Yu
Yinfeng Yu

Zhuoran Yu
Zitong Yu
Bo Yuan
Jiangbo Yuan
Liangzhe Yuan
Weihao Yuan
Jianbo Yuan
Xiaoyun Yuan
Ye Yuan
Li Yuan
Geng Yuan
Jialin Yuan
Maoxun Yuan
Peng Yuan
Xin Yuan
Yuan Yuan
Yuhui Yuan
Yixuan Yuan
Zheng Yuan
Mehmet Kerim Yücel
Kaiyu Yue
Haixiao Yue
Heeseung Yun
Sangdoo Yun
Tian Yun
Mahmut Yurt
Ekim Yurtsever
Ahmet Yüzügüler
Edouard Yvinec
Eloi Zablocki
Christopher Zach
Muhammad Zaigham
    Zaheer
Pierluigi Zama Ramirez
Yuhang Zang
Pietro Zanuttigh
Alexey Zaytsev
Bernhard Zeisl
Haitian Zeng
Pengpeng Zeng
Jiabei Zeng
Runhao Zeng
Wei Zeng
Yawen Zeng
Yi Zeng

Yiming Zeng
Tieyong Zeng
Huanqiang Zeng
Dan Zeng
Yu Zeng
Wei Zhai
Yuanhao Zhai
Fangneng Zhan
Kun Zhan
Xiong Zhang
Jingdong Zhang
Jiangning Zhang
Zhilu Zhang
Gengwei Zhang
Dongsu Zhang
Hui Zhang
Binjie Zhang
Bo Zhang
Tianhao Zhang
Cecilia Zhang
Jing Zhang
Chaoning Zhang
Chenxu Zhang
Chi Zhang
Chris Zhang
Yabin Zhang
Zhao Zhang
Rufeng Zhang
Chaoyi Zhang
Zheng Zhang
Da Zhang
Yi Zhang
Edward Zhang
Xin Zhang
Feifei Zhang
Feilong Zhang
Yuqi Zhang
GuiXuan Zhang
Hanlin Zhang
Hanwang Zhang
Hanzhen Zhang
Haotian Zhang
He Zhang
Haokui Zhang
Hongyuan Zhang

Hengrui Zhang
Hongming Zhang
Mingfang Zhang
Jianpeng Zhang
Jiaming Zhang
Jichao Zhang
Jie Zhang
Jingfeng Zhang
Jingyi Zhang
Jinnian Zhang
David Junhao Zhang
Junjie Zhang
Junzhe Zhang
Jiawan Zhang
Jingyang Zhang
Kai Zhang
Lei Zhang
Lihua Zhang
Lu Zhang
Miao Zhang
Minjia Zhang
Mingjin Zhang
Qi Zhang
Qian Zhang
Qilong Zhang
Qiming Zhang
Qiang Zhang
Richard Zhang
Ruimao Zhang
Ruisi Zhang
Ruixin Zhang
Runze Zhang
Qilin Zhang
Shan Zhang
Shanshan Zhang
Xi Sheryl Zhang
Song-Hai Zhang
Chongyang Zhang
Kaihao Zhang
Songyang Zhang
Shu Zhang
Siwei Zhang
Shujian Zhang
Tianyun Zhang
Tong Zhang

Tao Zhang
Wenwei Zhang
Wenqiang Zhang
Wen Zhang
Xiaolin Zhang
Xingchen Zhang
Xingxuan Zhang
Xiuming Zhang
Xiaoshuai Zhang
Xuanmeng Zhang
Xuanyang Zhang
Xucong Zhang
Xingxing Zhang
Xikun Zhang
Xiaohan Zhang
Yahui Zhang
Yunhua Zhang
Yan Zhang
Yanghao Zhang
Yifei Zhang
Yifan Zhang
Yi-Fan Zhang
Yihao Zhang
Yingliang Zhang
Youshan Zhang
Yulun Zhang
Yushu Zhang
Yixiao Zhang
Yide Zhang
Zhongwen Zhang
Bowen Zhang
Chen-Lin Zhang
Zehua Zhang
Zekun Zhang
Zeyu Zhang
Xiaowei Zhang
Yifeng Zhang
Cheng Zhang
Hongguang Zhang
Yuexi Zhang
Fa Zhang
Guofeng Zhang
Hao Zhang
Haofeng Zhang
Hongwen Zhang

Hua Zhang	Zhizhong Zhang	Bowen Zhao
Jiaxin Zhang	Qilong Zhangli	Pu Zhao
Zhenyu Zhang	Bingyin Zhao	Bingchen Zhao
Jian Zhang	Bin Zhao	Borui Zhao
Jianfeng Zhang	Chenglong Zhao	Fuqiang Zhao
Jiao Zhang	Lei Zhao	Hanbin Zhao
Jiakai Zhang	Feng Zhao	Jian Zhao
Lefei Zhang	Gangming Zhao	Mingyang Zhao
Le Zhang	Haiyan Zhao	Na Zhao
Mi Zhang	Hao Zhao	Rongchang Zhao
Min Zhang	Handong Zhao	Ruiqi Zhao
Ning Zhang	Hengshuang Zhao	Shuai Zhao
Pan Zhang	Yinan Zhao	Wenda Zhao
Pu Zhang	Jiaojiao Zhao	Wenliang Zhao
Qing Zhang	Jiaqi Zhao	Xiangyun Zhao
Renrui Zhang	Jing Zhao	Yifan Zhao
Shifeng Zhang	Kaili Zhao	Yaping Zhao
Shuo Zhang	Haojie Zhao	Zhou Zhao
Shaoxiong Zhang	Yucheng Zhao	He Zhao
Weizhong Zhang	Longjiao Zhao	Jie Zhao
Xi Zhang	Long Zhao	Xibin Zhao
Xiaomei Zhang	Qingsong Zhao	Xiaoqi Zhao
Xinyu Zhang	Qingyu Zhao	Zhengyu Zhao
Yin Zhang	Rui Zhao	Jin Zhe
Zicheng Zhang	Rui-Wei Zhao	Chuanxia Zheng
Zihao Zhang	Sicheng Zhao	Huan Zheng
Ziqi Zhang	Shuang Zhao	Hao Zheng
Zhaoxiang Zhang	Siyan Zhao	Jia Zheng
Zhen Zhang	Zelin Zhao	Jian-Qing Zheng
Zhipeng Zhang	Shiyu Zhao	Shuai Zheng
Zhixing Zhang	Wang Zhao	Meng Zheng
Zhizheng Zhang	Tiesong Zhao	Mingkai Zheng
Jiawei Zhang	Qian Zhao	Qian Zheng
Zhong Zhang	Wangbo Zhao	Qi Zheng
Pingping Zhang	Xi-Le Zhao	Wu Zheng
Yixin Zhang	Xu Zhao	Yinqiang Zheng
Kui Zhang	Yajie Zhao	Yufeng Zheng
Lingzhi Zhang	Yang Zhao	Yutong Zheng
Huaiwen Zhang	Ying Zhao	Yalin Zheng
Quanshi Zhang	Yin Zhao	Yu Zheng
Zhoutong Zhang	Yizhou Zhao	Feng Zheng
Yuhang Zhang	Yunhan Zhao	Zhaoheng Zheng
Yuting Zhang	Yuyang Zhao	Haitian Zheng
Zhang Zhang	Yue Zhao	Kang Zheng
Ziming Zhang	Yuzhi Zhao	Bolun Zheng

Haiyong Zheng
Mingwu Zheng
Sipeng Zheng
Tu Zheng
Wenzhao Zheng
Xiawu Zheng
Yinglin Zheng
Zhuo Zheng
Zilong Zheng
Kecheng Zheng
Zerong Zheng
Shuaifeng Zhi
Tiancheng Zhi
Jia-Xing Zhong
Yiwu Zhong
Fangwei Zhong
Zhihang Zhong
Yaoyao Zhong
Yiran Zhong
Zhun Zhong
Zichun Zhong
Bo Zhou
Boyao Zhou
Brady Zhou
Mo Zhou
Chunluan Zhou
Dingfu Zhou
Fan Zhou
Jingkai Zhou
Honglu Zhou
Jiaming Zhou
Jiahuan Zhou
Jun Zhou
Kaiyang Zhou
Keyang Zhou
Kuangqi Zhou
Lei Zhou
Lihua Zhou
Man Zhou
Mingyi Zhou
Mingyuan Zhou
Ning Zhou
Peng Zhou
Penghao Zhou
Qianyi Zhou

Shuigeng Zhou
Shangchen Zhou
Huayi Zhou
Zhize Zhou
Sanping Zhou
Qin Zhou
Tao Zhou
Wenbo Zhou
Xiangdong Zhou
Xiao-Yun Zhou
Xiao Zhou
Yang Zhou
Yipin Zhou
Zhenyu Zhou
Hao Zhou
Chu Zhou
Daquan Zhou
Da-Wei Zhou
Hang Zhou
Kang Zhou
Qianyu Zhou
Sheng Zhou
Wenhui Zhou
Xingyi Zhou
Yan-Jie Zhou
Yiyi Zhou
Yu Zhou
Yuan Zhou
Yuqian Zhou
Yuxuan Zhou
Zixiang Zhou
Wengang Zhou
Shuchang Zhou
Tianfei Zhou
Yichao Zhou
Alex Zhu
Chenchen Zhu
Deyao Zhu
Xiatian Zhu
Guibo Zhu
Haidong Zhu
Hao Zhu
Hongzi Zhu
Rui Zhu
Jing Zhu

Jianke Zhu
Junchen Zhu
Lei Zhu
Lingyu Zhu
Luyang Zhu
Menglong Zhu
Peihao Zhu
Hui Zhu
Xiaofeng Zhu
Tyler (Lixuan) Zhu
Wentao Zhu
Xiangyu Zhu
Xinqi Zhu
Xinxin Zhu
Xinliang Zhu
Yangguang Zhu
Yichen Zhu
Yixin Zhu
Yanjun Zhu
Yousong Zhu
Yuhao Zhu
Ye Zhu
Feng Zhu
Zhen Zhu
Fangrui Zhu
Jinjing Zhu
Linchao Zhu
Pengfei Zhu
Sijie Zhu
Xiaobin Zhu
Xiaoguang Zhu
Zezhou Zhu
Zhenyao Zhu
Kai Zhu
Pengkai Zhu
Bingbing Zhuang
Chengyuan Zhuang
Liansheng Zhuang
Peiye Zhuang
Yixin Zhuang
Yihong Zhuang
Junbao Zhuo
Andrea Ziani
Bartosz Zieliński
Primo Zingaretti

Nikolaos Zioulis
Andrew Zisserman
Yael Ziv
Liu Ziyin
Xingxing Zou
Danping Zou
Qi Zou

Shihao Zou
Xueyan Zou
Yang Zou
Yuliang Zou
Zihang Zou
Chuhang Zou
Dongqing Zou

Xu Zou
Zhiming Zou
Maria A. Zuluaga
Xinxin Zuo
Zhiwen Zuo
Reyer Zwiggelaar

Contents – Part XVI

StyleGAN-Human: A Data-Centric Odyssey of Human Generation

Jianglin Fu[1], Shikai Li[1], Yuming Jiang[2], Kwan-Yee Lin[1],
Chen Qian[1], Chen Change Loy[2], Wayne Wu[1,3]([✉]), and Ziwei Liu[2]

[1] SenseTime Research, Beijing, China
wuwenyan0503@gmail.com
[2] S-Lab, Nanyang Technological University, Singapore, Singapore
[3] Shanghai AI Laboratory, Shanghai, China

Abstract. Unconditional human image generation is an important task in vision and graphics, enabling various applications in the creative industry. Existing studies in this field mainly focus on "network engineering" such as designing new components and objective functions. This work takes a data-centric perspective and investigates multiple critical aspects in "data engineering", which we believe would complement the current practice. To facilitate a comprehensive study, we collect and annotate a large-scale human image dataset with over $230K$ samples capturing diverse poses and textures. Equipped with this large dataset, we rigorously investigate three essential factors in data engineering for StyleGAN-based human generation, namely data size, data distribution, and data alignment. Extensive experiments reveal several valuable observations $w.r.t.$ these aspects: 1) Large-scale data, more than $40K$ images, are needed to train a high-fidelity unconditional human generation model with a vanilla StyleGAN. 2) A balanced training set helps improve the generation quality with rare face poses compared to the long-tailed counterpart, whereas simply balancing the clothing texture distribution does not effectively bring an improvement. 3) Human GAN models that employ body centers for alignment outperform models trained using face centers or pelvis points as alignment anchors. In addition, a model zoo and human editing applications are demonstrated to facilitate future research in the community. Code and models are publicly available (Project page: https://stylegan-human.github.io/. Code and models: https://github.com/stylegan-human/StyleGAN-Human.)

Keywords: Human image generation · Data-centric · StyleGAN

J. Fu and S. Li—Equal contribution.

Supplementary Information The online version contains supplementary material available at https://doi.org/10.1007/978-3-031-19787-1_1.

Fig. 1. A data-centric odyssey of human generation. With good "data engineering" practices, the StyleGAN-Human model could generate high-resolution photo-realistic human images as presented. Zoom in for the best view.

1   Introduction

Generating photo-realistic images of clothed humans unconditionally can provide great support for downstream tasks such as human motion transfer [8,44], digital human animation [42], fashion recommendation [32,40], and virtual try-on [15,56,85]. Traditional methods create dressed humans with classical graphics modeling and rendering processes [17,31,50,61,63,70,78,93]. Although impressive results have been achieved, these prior works are easy to suffer from the limitation of robustness and generalizability in complex environments. Recent years, Generative Adversarial Networks (GANs) have demonstrated remarkable abilities in real-world scenarios, generating diverse and realistic images by learning from large-quantity and high-quality datasets. [24,33,36,66].

Among the GAN family, StyleGAN2 [37] stands out in generating faces and simple objects with unprecedented image quality. A major driver behind recent advancements [2,34,37,80,90] on such StyleGAN architectures is the prosperous discovery of "network engineering" like designing new components [2,34,90] and loss functions [37,80]. While these approaches show compelling results in generating diverse objects (*e.g.*, faces of humans and animals), applying them to the photo-realistic generation of articulated humans in natural clothing is still a challenging and open problem.

In this work, we focus on the task of Unconditional Human Generation, with a specific aim to train a good StyleGAN-based model for articulated humans from a *data-centric* perspective. First, to support the data-centric investigation, collecting a large-scale, high-quality, and diverse dataset of human bodies in clothing is necessary. We propose the Stylish-Humans-HQ Dataset (SHHQ), which contains $230K$ clean full-body images with a resolution of 1024×512 at least and up to 2240×1920. The SHHQ dataset lays the foundation for extensive experiments on unconditional human generation. Second, based on the proposed SHHQ dataset, we investigate three fundamental and critical questions that were not thoroughly discussed in prior works and attempt to provide useful insights for future research on unconditional human generation.

To extract the questions that are indeed *important* for the community of Unconditional Human Generation, we make an extensive survey on recent literature in the field of general unconditional generation [5,6,20,24,33,36,51]. Based on the survey, three questions that are investigated actively can be concluded as below. **Question-1**: What is the relationship between the *data size* and the generation quality? Several previous works [6,27,34,81,98] pointed out that the quantity of training data is the primary factor to determine the strategy for improving image quality in face and other object generation tasks. In this study, we want to examine the minimum quantity of training data required to generate human images of high quality without any extensive "network engineering" effort. **Question-2**: What is the relationship between the *data distribution* and the generation quality? This question has received extensive attention [14,22,52,71,92] and leads to a research topic dealing with data imbalance [46]. In this study, we aim to exploit data imbalance problem in the human generation task. **Question-3**: What is the relationship between the scheme of *data alignment* and the generation quality? Different alignment schemes applied to uncurated faces [36,38] and non-rigid objects [6,13,72] show success in enhancing training performance. In this study, we seek a better data alignment strategy for human generation.

Based on the proposed SHHQ dataset and observations from our experiments, we establish a Model Zoo with three widely-adopted unconditional generation models, *i.e.*, StyleGAN [36], StyleGAN2 [37], and alias-free StyleGAN [35], in both resolution of 1024×512 and 512×256. Although hundreds of StyleGAN-based studies exist for *face* generation/editing tasks, a high-quality and public model zoo for *human* generation/editing with StyleGAN family is still missing. We believe the provided model zoo has great potentials in many human-centric tasks, *e.g.*, human editing, neural rendering, and virtual try-on.

We further construct a human editing benchmark by adapting previous editing methods based on facial models to human body models (*i.e.*, PTI [68] for image inversion, InterFaceGAN [75], StyleSpace [90], and SeFa [76] for image manipulation). The impressive results in editing human clothes and attributes demonstrate the potential of the given model zoo in downstream tasks. In addition, a concurrent work, InsetGAN [16], is evaluated with our baseline model, further showing the potential usage of our pre-trained generative models.

Here is the summary of the main contributions of this paper: **1)** We collect a large-scale, high-quality, and diverse dataset, Stylish-Humans-HQ (SHHQ), containing $230K$ human full-body images for unconditional human generation task. **2)** We investigate three crucial questions that have aroused broad interest in the community and discuss our observation through comprehensive analysis.

3) We build a model zoo for unconditional human generation to facilitate future research. An editing benchmark is also established to demonstrate the potential of the proposed model zoo.

2   Related Work

2.1   Dataset for Human Generation

Large-scale and high-quality clothed human-centric training datasets are the critical fuel for the training of StyleGAN models. A qualified dataset should conform to the following aspects: **1)** *Image quality*: high-resolution images with rich textures offer more raw detailed semantic information to the model. **2)** *Data volume*: the size of dataset should be sufficient to avoid generative overfitting [4, 97]. **3)** *Data coverage*: the dataset should cover multiple attribute dimensions to guarantee diversity of the model, for instance, gender, clothing type, clothing texture, and human pose. **4)** *Data content*: since this report only focuses on the generation of single full-body human, occlusion caused by other people or objects is not considered here, whereas self-occlusion is taken into account. That is, each image should contain only one complete human body.

Publicly available datasets built particularly for full human-body generation are rare, but there are several practices [30,48,49,77] cooperating with Deep-Fashion [45] and Market1501 [99]. DeepFashion dataset [30,45] with well-labeled attributes and diverse garment categories is satisfactory for image classification and attribute prediction, but not adequate for unconditional human generation since it emphasizes fashion items rather than human bodies. Thus the number of close-up shots of clothing is much higher than that of full-body images. Market1501 dataset [99] fails for human generation tasks due to its low resolution (128×64). There are some human-related datasets in other domains rather than GAN-based applications: datasets related to human parsing [19,43] are limited by scalability and diversity; common datasets for virtual try-on tasks either contain only the upper body [25] or are not public [96]. A detailed comparison of the above datasets in terms of data scale, average resolution, attributes labeling, and proportion of full-body images across the whole dataset is listed in Table 1. In general, there is no high-quality and large-scale full human-body dataset publicly available for the generative purpose.

2.2   StyleGAN

In recent years, the research focus has gradually shifted to generating high-fidelity and high-resolution images through Generative Adversarial Networks [6, 33]. The StyleGAN generator [36] was introduced and became the state-of-the-art network of unconditional image generation. Compared to previous GAN-based architectures [5,24,55], SytleGAN injects a separate attribute factor (i.e., style) into the generator to influence the appearance of generated images. Then StyleGAN2 [37] redesigns the normalization, multi-scale scheme, and regularization method to rectify the artifacts in StyleGAN images. The latest update to

Table 1. Comparison of SHHQ with other publicly available datasets.

Dataset	Total image #	Mean resolution	Labeled attributes	Full-body ratio
ATR [43]	7,700	400×600	✓	76%
Mark1et1501 [99]	32,668	128×64	✓	100%
DeepFashion [45]	146,680	1101×750	✓	6.8%
LIP [19]	50,462	196×345	✓	37%
VITON [25]	16,253	256×192	✗	0%
SHHQ	**231,176**	$\mathbf{1024 \times 512}$	✓	**100%**

StyleGAN [35] reveals the non-ideal case of detailed textures sticking to fixed pixel locations and proposes an alias-free network.

2.3 Image Editing

Benefiting from StyleGAN, one of the significant downstream applications is image editing [1,60,75,90,95]. A standard image editing pipeline usually involves inversion from a real image to the latent space and manipulating the embedded latent code. Existing works for *image inversion* can be categorized into optimization-based [2,79], encoder-based [82,86], and hybrid methods [68], which exploit encoders to embed images into latent space and then refine with optimization. As for *image manipulation*, studies explore the capability of attribute disentanglement in the latent space in supervised [29,75,90] or unsupervised [26,76,83] manners. In specific, Jiang *et al.* [29] proposes to use fine-grained annotations to find non-linear manipulation directions in the latent space, while SeFa [76] searchs for semantic directions without supervision. StyleSpace [90] defines the style space S and proves that it is more disentangled than W and $W+$ space.

3 Stylish-Humans-HQ Dataset

To investigate the key factors in unconditional human generation task from a data-centric perspective, we propose a large-scale, high-quality, and diverse dataset, Stylish-Humans-HQ (SHHQ). In this section, we first present the data collection and preprocessing (Sect. 3.1), in which we construct the SHHQ dataset. Then, we analyze the data statistic (Sect. 3.2) to demonstrate the superiority of SHHQ compared to other datasets from a statistical perspective.

3.1 Data Collection and Preprocessing

Over $500K$ raw data were collected legally in two ways: 1) *From the Internet.* We crawled images, with CC-BY-2.0 licenses available, mainly from Flickr, Unsplash, Pixabay and Pexels, by searching keywords related to humans. 2) *From data*

(a)   (b)   (c)   (d)   (e)

Fig. 2. Data preprocessing. The following types of images will be removed during our data preprocessing pipeline. (a) Low resolution. (b) Not placed in the center. (c) Missing body parts. (d) Extreme posture. (e) Multi-person.

providers. We purchased images from individual photographers, model agencies, and other providers' databases. Images were reviewed by our institute's legal team before the purchase, to ensure the permission of usage in research. We preprocess the data with six factors taken into consideration (*e.g.*, resolution [45], body position [36], body-part occlusion, human pose [36,45], multi-person, and background), which are critical for the quality of a human dataset. After the data preprocessing, we obtain a clean dataset of $231,176$ images with high quality; see Fig. 5(a) for examples. We filter the images according to following aspects. **1) Resolution:** We discard images lower than 1024×512 resolution (Fig. 2(a)). **2) Body Position:** The position of the body varies widely in different images, *i.e.*, Fig. 2(b). We design a procedure in which each person is appropriately cropped based on human segmentation [12], padded and resized to the same scale, and then placed in the image such that the body center is aligned. The body center is defined as the average coordinate of the entire body using segmentation. **3) Body-Part Occlusion:** This work aims at generating full-body human images, images with any missing body parts are removed (*e.g.*, the half-body portrait shown in Fig. 2(c)). We remove images with extreme poses (*e.g.*, lying postures, handstand in Fig. 2(d)) to ensure learnability of the data distribution. We exploit human pose estimation [7] to detect those extreme poses. **4) Multi-Person Images:** Some images contain multiple persons, such as Fig. 2(e). The goal of this work is to generate single full-body person, so we keep unoccluded single-person full-body images, and remove those with occluded people. **5) Background:** Some images contain complicated backgrounds, requiring additional representation ability. To focus on the generation of the human body itself and eliminate the influence of various backgrounds, we use a segmentation mask [12] to modify the image background to pure white. The edges of the mask are then smoothed by Gaussian blur.

3.2   Data Statistics

Table 1 presents the comparison between SHHQ and other public datasets from the following three aspects: **1) Dataset Scale:** As shown in the table, our

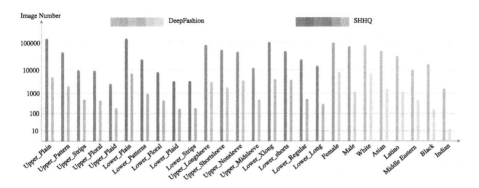

Fig. 3. Attribute distribution. Comparison of different attributes between the pruned DeepFashion and SHHQ dataset: Texture/length of the upper/lower clothing, gender, and ethnicity. (More attributes comparison in supplementary).

proposed SHHQ is currently the largest dataset in scale compared to others. Among them, the data volume of SHHQ is 1.6 times that of DeepFashion [45] dataset and is much larger than that of others. **Resolution.** Images from ATR [43], Market1501 [99], LIP [19], and VITON [25] are lower in resolution, which is insufficient for our generation task, while the proposed SHHQ and DeepFashion provide high-definition images up to 2240 × 1920. **2) Labels:** All datasets beside VITON provide various labeled attributes. Specifically, DeepFashion [30,45] and SHHQ label the clothing types and textures, which is useful for human generation/editing tasks. **3) Full-Body Ratio:** This number denotes the proportion of full-body images in the dataset. Although DeepFashion [45] offers over 146K images with decent resolution, only 6.8% of them are full-body images, while SHHQ achieves a 100% full-body ratio. The visual comparison among these datasets and the proposed SHHQ dataset is shown in supplementary.

In sum, SHHQ covers the largest number of human images with high-resolution, labeled clothing attributes, and 100% full-body ratio. It again confirms that SHHQ is more suitable for full-body human generation than other public datasets.

Of all the datasets compared above, DeepFashion [45] is the most relevant to our human generation task. In Fig. 3, we further present the comparison of different attributes between filtered DeepFashion [45] (full-body only) and SHHQ in a more detailed view. The bar chart depicts the distributions along six dimensions: upper cloth texture, lower cloth texture, upper cloth length, lower cloth length, gender, and ethnicity. In particular, the number of females is approximately 4 times the number of males in filtered DeepFashion [45], while our dataset features a more balanced female-to-male ratio of 1.49. With the help of DeepFace API [74], it is shown that SHHQ is more diverse in terms of ethnicity. Advantages are also shown in the other five attributes. In terms of garment-related attributes, images with specific labels in filtered DeepFashion [45] are too scarce to be used as a training set. The Stylish-Humans-HQ dataset boosts the number of each category by an average of 24.4 times.

4 Systematic Investigation

We conduct extensive experiments to study three factors concerning the quality of generated images: 1) data size (Sect. 4.1), 2) data distribution (Sect. 4.2), and 3) data alignment (Sect. 4.3). Our investigations are all built on the Style-GAN2 architecture and codebase. More implementation details and experimental results can be found in supplementary.

4.1 Data Size

Motivation. Data size is an essential factor that determines the quality of generated images. Previous literature takes different strategies to improve the generation performance according to different dataset sizes: regularization techniques [6] are employed to train a large dataset, while augmentation [34,81,98] and conditional feature transferring [52,88] are proposed to tackle the limited data. Here, we design sets of experiments to examine the relationship between training data size and the image quality of generated humans.

Experimental Settings. To determine the relationship between data size and image quality for the unconditional human GAN, we construct 6 sub-datasets and denoted these subsets as $S0$ ($10K$), $S1$ ($20K$), $S2$ ($40K$), $S3$ ($80K$), $S4$ ($160K$) and $S5$ ($230K$). Here, $S0$ is the pruned DeepFashion dataset. We perform the training on two resolution settings for each set: 1024×512 and 512×256. Considering the case of limited data, we also conduct additional training experiments with adaptive discriminator augmentation (ADA) [34] for small datasets $S0$, $S1$, and $S2$. Fréchet Inception Distance (FID) and Inception Score (IS) are the indicators for evaluating the model performance.

Results. As shown in Fig. 4(a), the FID scores (solid lines) decrease as the size of the training dataset increases for both resolution settings. The declining trend is gradually flattening and tends to converge. $S0$ generates the least satisfactory results, with FID of 7.80 and 7.23 for low- and high-resolution, respectively, while $S1$ achieves corresponding improvements of 42% and 40% on FID with only an additional $10K$ training images. When the training size reaches $40K$ for both resolutions, the FID curves start to converge to a certain extent. The dotted lines indicate the results of ADA experiments with subsets $S0$–$S2$. The employed data augmentation strategy helps to reduce FID when training data is less than $40K$. More quantitative results (FID/IS) are in supplementary.

Discussion. The experiments confirm that ADA can improve the generation quality for datasets smaller than $40K$ images, in terms of FID and IS. However, ADA still cannot fully compensate for the impact of insufficient data. Besides, when the amount of data is less than $40K$, the relationship between image quality and data size is close to linear. As the amount of data increases to $40K$ and more, the improvement in the image quality slows down and is less significant.

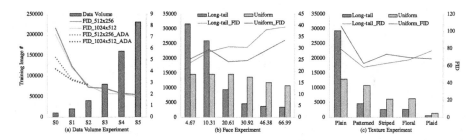

Fig. 4. Experiment results. (a) FIDs for experiments $S0$–$S5$ in 1024×512 and 512×256 resolutions. Dotted lines shows the FIDs of the models trained with ADA. (b) Bin-wise FIDs of long-tailed and uniform distribution in terms of facial yaw angle along with the number of training images. (c) Bin-wise texture FIDs of long-tailed/uniform distribution along with the number of training images. (Color figure online)

4.2 Data Distribution

Motivation. The nature of GAN makes the model inherits the distribution of the training dataset and introduces generation bias due to dataset imbalance [46]. This bias severely affects the performance of GAN models. To address this issue, studies for unfairness mitigation [14,22,52,71,92] have attracted substantial research interest. In this work, we explore the question of data distribution in human generation and conduct experiments to verify whether a uniform data distribution can improve the performance of a human generation model.

Experimental Settings. This study decomposes the distribution of the human body into Face Orientation and Clothing Texture, since face fidelity has a significant impact on visual perception and clothing occupies a large portion of the full-body image. The general features of human faces are relatively symmetrical; thus, we fold yaw distribution vertically along $0°$ and get the long-tailed distribution. For the face and clothing experiments, we collect an equal number of long-tailed and uniformly distributed datasets from SHHQ for face rotation angle and upper-body clothing texture, respectively.

Results. To evaluate the image quality in terms of different distributions, the cropped faces and clothing regions are used to calculate FID, and FID is calculated separately for each bin. Result can be found in Fig. 4(b) and (c).

1) Face Orientation: As for the long-tailed experiment (blue curve in Fig. 4(b)), the FID progressively grows as the face yaw angle increases and remains high when the facial rotation angle is too large. By contrast, the upward trend for the face FID in the uniform experiment (red) is more gradual. In addition, the amount of the training data of the first two bins in the uniform set is greatly reduced compared to the long-tail experiment, but the damage to FID is slight.

2) Clothing Texture: From Fig. 4(c), except for the first bin ("plain" pattern), the FID curve climbs steadily as the amount of training data for the long-tailed experiment decreases, and the FID curve for the uniform experiment

Fig. 5. Example of preprocessed data and different alignment schemes. Part (a) shows processed training data with the consideration of resolution, body position, body-part occlusion, human pose, multi-person and background. (b)–(d) display random samples with three different alignment strategies.

also shows a near-uniform pattern. In particular, FID of the last bin for the uniform experiment is lower than that in the long-tailed setting. We infer that the training samples for "plaid" clothing texture in the long-tailed experiment are too few to be learned by the model. As for the "plain" bin results, the long-tailed distribution has a lower FID score in this bin. The reason may lie in that the number of plain textures in the long-tailed distribution is considerably higher than that in the uniform distribution. Also, it can be observed that the training patches in this bin are mainly textureless color blocks where such patterns may be easier to capture by models.

Discussion. Based on the above analysis, we conclude that the uniform distribution of face rotation angles can effectively reduce the FID of rare training faces while maintaining acceptable image quality for the dominant faces. However, simply balancing the distribution of texture patterns does not always reduce the corresponding FID effectively. This phenomenon raises an interesting question that can be further explored: is the relation between image quality and data distribution also entangled with other factors, *e.g.*, image pattern and data size? Additionally, due to the nature of GAN-based structures, a GAN model memorizes the entire dataset, and usually, the discriminator tends to overfit those poorly sampled images at the tail of the distribution. Consequently, the long-tailed situation accumulated as "tail" images is barely generated. From this perspective, it also can be seen that the uniform distribution preserves the diversity of faces/textures and partially alleviates this problem.

4.3   Data Alignment

Motivation. Recently, researchers have drawn attention to spatial bias in generation tasks. Several works [36,38] align face images with keypoints for face generation, and other studies propose different alignment schemes to preprocess non-rigid objects [6,13,28,47,72]. In this paper, we study the relationship between the spatial deviation of the entire human and the generated image quality.

Experimental Settings. We randomly sample a set of $50K$ images from the SHHQ dataset and align every image separately using three different alignment strategies: aligning the image based on the face center, pelvis, and the midpoint of the whole body, as shown in Fig. 5.

Following are the reasons for selecting these three positions as alignment centers. 1) For the face center, we hypothesize that faces contain rich semantic information that is valuable for learning and may account for a heavy proportion in human generation. 2) For the pelvis, studies related to human pose estimation [53,57,62,84] conventionally predict the body joint coordinates relative to the pelvis. Thus we employ the pelvis as the alignment anchor. 3) For the body's midpoint, the leg-to-body ratio (the proportion of upper and lower body length) may vary among different people; therefore, we try to find the mean coordinates of the full body with the help of the segmentation mask.

Results. Human images are complex and easily affected by various extrinsic factors such as body poses and camera viewpoints. The FID scores for the face-aligned, pelvis-aligned, and mid-body-aligned experiments are 3.5, 2.8, and 2.4, respectively. Figure 5 further interprets this perspective as the human bodies in (b) and (c) are tilted, and the overall image quality is degraded. The example shown in Fig. 5(c) also presents the inconsistent human positions caused by different leg-to-body ratios.

Discussion. Both FID scores and visualizations suggest that the human generative models gain more stable spatial semantic information through the mid-body alignment method than face- and pelvis-centered methods. We believe this observation could benefit later studies on human generation.

4.4   Experimental Insights

Now the questions can be answered based on the above investigations:

For **Question-1** (Data Size): A large dataset with more than $40K$ images helps to train a high-fidelity unconditional human generation model, for both 512×256 and 1024×512 resolution.

For **Question-2** (Data Distribution): The uniform distribution of face rotation angles helps reduce the FID of rare faces while maintaining a reasonable quality of dominant faces. But simply balancing the clothing texture distribution does not effectively improve the generation quality.

For **Question-3** (Data Alignment): Aligning the human by the center of the full body presents a quality improvement over aligning the human by face or pelvis centers.

5   Model Zoo and Editing Benchmark

5.1   Model Zoo

In the field of face generation, a pre-trained StyleGAN [36] model has shown remarkable potential and success in various downstream tasks, including editing [2,90], neural rendering [23], and super-resolution [11,54]. Nevertheless, a

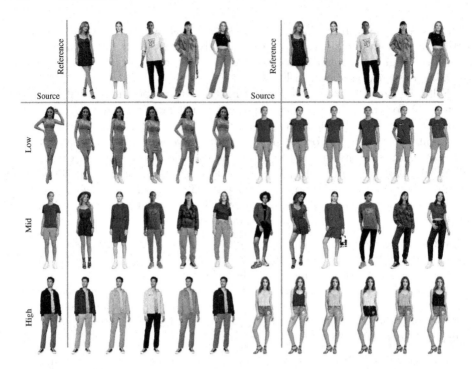

Fig. 6. Style-mixing results. The reference and source images are randomly sampled from the provided baseline model. The rest of the images are generated by style-mixing: borrowing low/mid/high layers in the reference images' latent codes and combining them with the rest layers of latent code in source images.

publicly available pre-trained model is still lacking for the human generation task. To fill this gap, we train our baseline model on SHHQ using the Style-GAN2 [37] model, which provides the best FID of 1.57. As seen in Figs. 1, our model has the ability to generate full-body images with diverse poses and clothing textures under satisfactory image quality. To adapt various application scenarios, we build a model zoo consisting of trained models from different StyleGAN architectures [35–37] in both resolution (1024 × 512 and 512 × 256).

Furthermore, the style mixing results of the baseline model show the interpretability of the corresponding latent space. As seen in the Fig. 6, source and reference images are sampled from the baseline model, and the rest images are the style-mixing results. We see that copying low layers from reference images to source images brings changes in geometry features (pose) while other features such as skin, garments, and identities in source images are preserved. When replicating middle styles, the source person's clothing type and identical appearance are replaced by reference. Finally, we observe that fine styles from high-resolution layers control the clothing color. These results suggest that the provided model's geometry and appearance information are well disentangled.

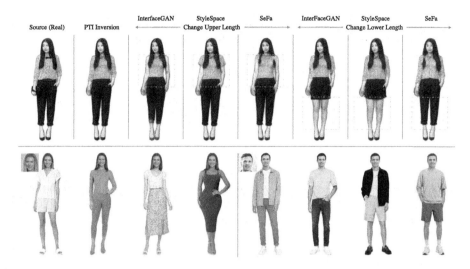

Fig. 7. Image editing and InsetGAN results. *Top row* presents editing results on an real image (left) after PTI inversion. The length of sleeve and pants are edited using different techniques. *Bottom Row* shows the InsetGAN results of different human bodies generated from the given baseline model and two faces generated from the FFHQ [37] model.

5.2 Editing Benchmark

StyleGAN has presented remarkable editing capabilities over faces. We extend it to the full-scale human by using off-the-shelf inversion and editing methods, in which we validate the potential of our proposed model zoo. We also re-implement the concurrent human generation method, InsetGAN [16], to further demonstrate another practical usage.

First, we leverage several SOTA StyleGAN-based facial editing techniques, such as InterFaceGAN [75], StyleSpace [90], and SeFa [76], with multiple editing directions: garment length for tops and bottoms, and global pose orientation. To examine the ability of editing real images with the provided model, we trained the e4e encoder [82] on SHHQ to obtain the inverted latent code as initial pivot. PTI [68] is then used to fine-tune the generator for each specific image.

As illustrated in Fig. 7, PTI presents the ability to invert real full-body human images. For attributes manipulation, StyleSpace [90] expresses better disentanglement compared to InterFaceGAN [75] and SeFa [76], as only the attribute-targeted region has been changed. However, as for the regions to be edited, the results of InterFaceGAN [75] are more natural and photo-realistic. It turns out that the latent space of the human body is more complicated than other domains such as faces, objects, and scenes, and more attention should be paid to disentangle human attributes. More editing results are shown in supplementary.

We re-implement InsetGAN [16] by iteratively optimizing the latent codes for random faces and bodies generated by the FFHQ [37] and our model, respectively. In the bottom row of Fig. 7, we show the fused full-body images with different male and female faces. The optimization procedure blends diverse faces and bodies in a graceful manner. Both the adopted editing methods and the multi-GAN optimization method demonstrate the effectiveness and convenience of our provided model zoo and verify its potential in human-centric tasks.

6 Future Work

In this study, we take a preliminary step towards the exploration of the human generation/editing tasks. We believe many future works can be further explored based on the SHHQ dataset and the provided model zoo. In the following, we discuss three interesting directions, *i.e.*, Human Generation/Editing, Neural Rendering, and Multi-modal Generation.

Human Generation/Editing. Studies in human generation [16,96], human editing [3,21,69], virtual try-on [15,41,56,85], and motion transfer [8,44] heavily rely on large datasets to train or use existing pre-trained models as the first step of transfer learning. Furthermore, editing benchmarks show that disentangled editing of the human body remains challenging for existing methods [75,90]. In this context, the released model zoo could expedite such research progress. Additionally, we further analyze failure cases generated by the provided model and discuss corresponding potential efforts that could be made to human generation tasks in supplementary.

Neural Rendering. Another future research direction is to improve 3D consistency and mitigate artifacts in full-body human generation through neural rendering [9,10,23,58,59,73]. Similar to work such as EG3D [9], StyleNeRF [23], and StyleSDF [59], we encourage researchers to use our human models to facilitate human generation with multi-view consistency.

Multi-modal Generation. Cross-modal representation is an emerging research trend, such as CLIP [65] and ImageBERT [64]. Hundreds of studies are made on text-driven image generation and manipulation [29,30,39,60,64,67,87,91,94], *e.g.*, DALLE [67] and AttnGAN [94]. In the meantime, several studies show interest in probing the transfer learning benefits of large-scale pre-trained models [18,65,89]. Most of these works focus on faces and objects, whereas research fields related to full-scale humans could be explored more, *e.g.*, text-driven human attributes manipulation, with the help of the provided full-body human models.

7 Conclusion

This work mainly probes how to train unconditional human-based GAN models to generate photo-realistic images from a data-centric perspective. By leveraging the $230K$ SHHQ dataset, we analyze three fundamental yet critical issues that the community cares most about: data size, data distribution, and data alignment. While experimenting with StyleGAN and large-scale data, we obtain several empirical insights. Apart from these, we create a model zoo, consisting of six human-GAN models, and the effectiveness of the model zoo is demonstrated by employing several state-of-the-art face editing methods.

Acknowledgements. This work is supported by NTU NAP, MOE AcRF Tier 1 (2021-T1-001-088), and under the RIE2020 Industry Alignment Fund – Industry Collaboration Projects (IAF-ICP) Funding Initiative, as well as cash and in-kind contribution from the industry partner(s).

References

1. Abdal, R., Qin, Y., Wonka, P.: Image2StyleGAN++: how to edit the embedded images? In: CVPR (2020)
2. Abdal, R., Zhu, P., Mitra, N.J., Wonka, P.: StyleFlow: attribute-conditioned exploration of StyleGAN-generated images using conditional continuous normalizing flows. ACM TOG **40**(3), 1–21 (2021)
3. Albahar, B., Lu, J., Yang, J., Shu, Z., Shechtman, E., Huang, J.B.: Pose with style: detail-preserving pose-guided image synthesis with conditional StyleGAN. ACM TOG **40**(6), 1–11 (2021)
4. Arjovsky, M., Bottou, L.: Towards principled methods for training generative adversarial networks. In: ICLR (2017)
5. Arjovsky, M., Chintala, S., Bottou, L.: Wasserstein generative adversarial networks. In: ICML (2017)
6. Brock, A., Donahue, J., Simonyan, K.: Large scale GAN training for high fidelity natural image synthesis. In: ICLR (2019)
7. Cao, Z., Simon, T., Wei, S.E., Sheikh, Y.: Realtime multi-person 2D pose estimation using part affinity fields. In: CVPR (2017)
8. Chan, C., Ginosar, S., Zhou, T., Efros, A.A.: Everybody dance now. In: ICCV (2019)
9. Chan, E.R., et al.: Efficient geometry-aware 3D generative adversarial networks. In: CVPR (2022)
10. Chan, E.R., Monteiro, M., Kellnhofer, P., Wu, J., Wetzstein, G.: Pi-GAN: periodic implicit generative adversarial networks for 3D-aware image synthesis. In: CVPR (2021)
11. Chan, K.C., Wang, X., Xu, X., Gu, J., Loy, C.C.: GLEAN: generative latent bank for large-factor image super-resolution. In: CVPR (2021)
12. MMSegmentation Contributors: MMSegmentation: OpenMMLab semantic segmentation toolbox and benchmark (2020). https://github.com/open-mmlab/mmsegmentation
13. Dhariwal, P., Nichol, A.: Diffusion models beat GANs on image synthesis. In: NeurIPS (2021)
14. Dionelis, N., Yaghoobi, M., Tsaftaris, S.A.: Tail of distribution GAN (TailGAN): GenerativeAdversarial-network-based boundary formation. In: SSPD (2020)
15. Dong, H., et al.: Towards multi-pose guided virtual try-on network. In: ICCV (2019)
16. Frühstück, A., Singh, K.K., Shechtman, E., Mitra, N.J., Wonka, P., Lu, J.: Inset-GAN for full-body image generation. In: CVPR (2022)

17. Gahan, A.: 3ds Max Modeling for Games: Insider's Guide to Game Character, Vehicle, and Environment Modeling (2012)
18. Ghadiyaram, D., Tran, D., Mahajan, D.: Large-scale weakly-supervised pre-training for video action recognition. In: CVPR (2019)
19. Gong, K., Liang, X., Zhang, D., Shen, X., Lin, L.: Look into person: self-supervised structure-sensitive learning and a new benchmark for human parsing. In: CVPR (2017)
20. Goodfellow, I., et al.: Generative adversarial nets. In: NeurIPS (2014)
21. Grigorev, A., et al.: StylePeople: a generative model of fullbody human avatars. In: CVPR (2021)
22. Grover, A., et al.: Bias correction of learned generative models using likelihood-free importance weighting. In: NeurIPS (2019)
23. Gu, J., Liu, L., Wang, P., Theobalt, C.: StyleNeRF: a style-based 3d aware generator for high-resolution image synthesis. In: ICLR (2022)
24. Gulrajani, I., Ahmed, F., Arjovsky, M., Dumoulin, V., Courville, A.C.: Improved training of wasserstein GANs. In: NeurIPS (2017)
25. Han, X., Wu, Z., Wu, Z., Yu, R., Davis, L.S.: VITON: an image-based virtual try-on network. In: CVPR (2018)
26. Härkönen, E., Hertzmann, A., Lehtinen, J., Paris, S.: GanSpace: discovering interpretable GAN controls. In: NeurIPS (2020)
27. Jiang, L., Dai, B., Wu, W., Loy, C.C.: Deceive D: adaptive pseudo augmentation for GAN training with limited data. In: NeurIPS (2021)
28. Jiang, Y., Chan, K.C., Wang, X., Loy, C.C., Liu, Z.: Robust reference-based super-resolution via C2-matching. In: CVPR (2021)
29. Jiang, Y., Huang, Z., Pan, X., Loy, C.C., Liu, Z.: Talk-to-edit: fine-grained facial editing via dialog. In: ICCV (2021)
30. Jiang, Y., Yang, S., Qiu, H., Wu, W., Loy, C.C., Liu, Z.: Text2Human: text-driven controllable human image generation. ACM TOG **41**(4), 1–11 (2022)
31. Jojic, N., Gu, J., Shen, T., Huang, T.S.: Computer modeling, analysis, and synthesis of dressed humans. TCSVT **9**(2), 378–388 (1999)
32. Kang, W.C., Fang, C., Wang, Z., McAuley, J.: Visually-aware fashion recommendation and design with generative image models. In: ICDM (2017)
33. Karras, T., Aila, T., Laine, S., Lehtinen, J.: Progressive growing of GANs for improved quality, stability, and variation. In: ICLR (2017)
34. Karras, T., Aittala, M., Hellsten, J., Laine, S., Lehtinen, J., Aila, T.: Training generative adversarial networks with limited data. In: NeurIPS (2020)
35. Karras, T., et al.: Alias-free generative adversarial networks. In: NeurIPS (2021)
36. Karras, T., Laine, S., Aila, T.: A style-based generator architecture for generative adversarial networks. In: CVPR (2019)
37. Karras, T., Laine, S., Aittala, M., Hellsten, J., Lehtinen, J., Aila, T.: Analyzing and improving the image quality of StyleGAN. In: CVPR (2020)
38. Kazemi, V., Sullivan, J.: One millisecond face alignment with an ensemble of regression trees. In: CVPR (2014)
39. Kocasari, U., Dirik, A., Tiftikci, M., Yanardag, P.: StyleMC: multi-channel based fast text-guided image generation and manipulation. In: WACV (2022)
40. Lei, C., Liu, D., Li, W., Zha, Z.J., Li, H.: Comparative deep learning of hybrid representations for image recommendations. In: CVPR (2016)
41. Lewis, K.M., Varadharajan, S., Kemelmacher-Shlizerman, I.: TryOnGAN: body-aware try-on via layered interpolation. ACM TOG **40**(4), 1–10 (2021)
42. Li, Z., et al.: Animated 3D human avatars from a single image with GAN-based texture inference. CNG **95**, 81–91 (2021)

43. Liang, X., et al.: Deep human parsing with active template regression. PAMI **37**(12), 2402–2414 (2015)

44. Liu, W., Piao, Z., Min, J., Luo, W., Ma, L., Gao, S.: Liquid warping GAN: a unified framework for human motion imitation, appearance transfer and novel view synthesis. In: ICCV (2019)

45. Liu, Z., Luo, P., Qiu, S., Wang, X., Tang, X.: DeepFashion: powering robust clothes recognition and retrieval with rich annotations. In: CVPR (2016)

46. Liu, Z., Miao, Z., Zhan, X., Wang, J., Gong, B., Yu, S.X.: Large-scale long-tailed recognition in an open world. In: CVPR (2019)

47. Liu, Z., Yan, S., Luo, P., Wang, X., Tang, X.: Fashion landmark detection in the wild. In: Leibe, B., Matas, J., Sebe, N., Welling, M. (eds.) ECCV 2016. LNCS, vol. 9906, pp. 229–245. Springer, Cham (2016). https://doi.org/10.1007/978-3-319-46475-6_15

48. Ma, L., Jia, X., Sun, Q., Schiele, B., Tuytelaars, T., Van Gool, L.: Pose guided person image generation. In: NeurIPS (2017)

49. Ma, L., Sun, Q., Georgoulis, S., Van Gool, L., Schiele, B., Fritz, M.: Disentangled person image generation. In: CVPR (2018)

50. Ma, Q., et al.: Learning to dress 3D people in generative clothing. In: CVPR (2020)

51. Mao, X., Li, Q., Xie, H., Lau, R.Y., Wang, Z., Paul Smolley, S.: Least squares generative adversarial networks. In: ICCV (2017)

52. Mariani, G., Scheidegger, F., Istrate, R., Bekas, C., Malossi, C.: BaGAN: data augmentation with balancing GAN. arXiv preprint arXiv:1803.09655 (2018)

53. Martinez, J., Hossain, R., Romero, J., Little, J.J.: A simple yet effective baseline for 3D human pose estimation. In: ICCV (2017)

54. Menon, S., Damian, A., Hu, S., Ravi, N., Rudin, C.: PULSE: self-supervised photo upsampling via latent space exploration of generative models. In: CVPR (2020)

55. Miyato, T., Kataoka, T., Koyama, M., Yoshida, Y.: Spectral normalization for generative adversarial networks. In: ICLR (2018)

56. Neuberger, A., Borenstein, E., Hilleli, B., Oks, E., Alpert, S.: Image based virtual try-on network from unpaired data. In: CVPR (2020)

57. Nie, X., Feng, J., Zhang, J., Yan, S.: Single-stage multi-person pose machines. In: ICCV (2019)

58. Niemeyer, M., Geiger, A.: GIRAFFE: representing scenes as compositional generative neural feature fields. In: CVPR (2021)

59. Or-El, R., Luo, X., Shan, M., Shechtman, E., Park, J.J., Kemelmacher-Shlizerman, I.: StyleSDF: high-resolution 3D-consistent image and geometry generation. In: CVPR (2022)

60. Patashnik, O., Wu, Z., Shechtman, E., Cohen-Or, D., Lischinski, D.: StyleCLIP: text-driven manipulation of StyleGAN imagery. In: ICCV (2021)

61. Patel, C., Liao, Z., Pons-Moll, G.: TailorNet: predicting clothing in 3D as a function of human pose, shape and garment style. In: CVPR (2020)

62. Pavlakos, G., Zhou, X., Derpanis, K.G., Daniilidis, K.: Coarse-to-fine volumetric prediction for single-image 3D human pose. In: CVPR (2017)

63. Pumarola, A., Sanchez-Riera, J., Choi, G., Sanfeliu, A., Moreno-Noguer, F.: 3DPeople: modeling the geometry of dressed humans. In: CVPR (2019)

64. Qi, D., Su, L., Song, J., Cui, E., Bharti, T., Sacheti, A.: ImageBERT: cross-modal pre-training with large-scale weak-supervised image-text data. arXiv preprint arXiv:2001.07966 (2020)

65. Radford, A., et al.: Learning transferable visual models from natural language supervision. In: ICML (2021)

66. Radford, A., Metz, L., Chintala, S.: Unsupervised representation learning with deep convolutional generative adversarial networks (2016)
67. Ramesh, A., et al.: Zero-shot text-to-image generation. In: ICML (2021)
68. Roich, D., Mokady, R., Bermano, A.H., Cohen-Or, D.: Pivotal tuning for latent-based editing of real images. ACM TOG **42**(1), 1–13 (2022)
69. Sarkar, K., Golyanik, V., Liu, L., Theobalt, C.: Style and pose control for image synthesis of humans from a single monocular view. arXiv preprint arXiv:2102.11263 (2021)
70. Sarkar, K., Mehta, D., Xu, W., Golyanik, V., Theobalt, C.: Neural re-rendering of humans from a single image. In: Vedaldi, A., Bischof, H., Brox, T., Frahm, J.-M. (eds.) ECCV 2020. LNCS, vol. 12356, pp. 596–613. Springer, Cham (2020). https://doi.org/10.1007/978-3-030-58621-8_35
71. Sattigeri, P., Hoffman, S.C., Chenthamarakshan, V., Varshney, K.R.: Fairness GAN: generating datasets with fairness properties using a generative adversarial network. IBM JRD **63**(4/5), 3-1 (2019)
72. Sauer, A., Schwarz, K., Geiger, A.: StyleGAN-XL: scaling StyleGAN to large diverse datasets. ACM TOG (2022)
73. Schwarz, K., Liao, Y., Niemeyer, M., Geiger, A.: GRAF: generative radiance fields for 3D-aware image synthesis. In: NeurIPS (2020)
74. Serengil, S.I., Ozpinar, A.: HyperExtended LightFace: a facial attribute analysis framework. In: ICEET (2021)
75. Shen, Y., Yang, C., Tang, X., Zhou, B.: InterFaceGAN: interpreting the disentangled face representation learned by GANs. PAMI **44**(4), 2004–2018 (2020)
76. Shen, Y., Zhou, B.: Closed-form factorization of latent semantics in GANs. In: CVPR (2021)
77. Siarohin, A., Sangineto, E., Lathuilière, S., Sebe, N.: Deformable GANs for pose-based human image generation. In: CVPR (2018)
78. Song, D., Tong, R., Chang, J., Yang, X., Tang, M., Zhang, J.J.: 3D body shapes estimation from dressed-human silhouettes. In: CGF (2016)
79. Tewari, A., et al.: PIE: portrait image embedding for semantic control. ACM TOG **39**(6), 1–14 (2020)
80. Tewari, A., et al.: StyleRig: rigging StyleGAN for 3D control over portrait images. In: CVPR (2020)
81. Toutouh, J., Hemberg, E., O'Reilly, U.-M.: Data dieting in GAN training. In: Iba, H., Noman, N. (eds.) Deep Neural Evolution. NCS, pp. 379–400. Springer, Singapore (2020). https://doi.org/10.1007/978-981-15-3685-4_14
82. Tov, O., Alaluf, Y., Nitzan, Y., Patashnik, O., Cohen-Or, D.: Designing an encoder for StyleGAN image manipulation. ACM TOG **40**(4), 1–14 (2021)
83. Tzelepis, C., Tzimiropoulos, G., Patras, I.: WarpedGANSpace: finding non-linear RBF paths in GAN latent space. In: ICCV (2021)
84. Véges, M., Lőrincz, A.: Absolute human pose estimation with depth prediction network. In: IJCNN (2019)
85. Wang, B., Zheng, H., Liang, X., Chen, Y., Lin, L., Yang, M.: Toward characteristic-preserving image-based virtual try-on network. In: Ferrari, V., Hebert, M., Sminchisescu, C., Weiss, Y. (eds.) ECCV 2018. LNCS, vol. 11217, pp. 607–623. Springer, Cham (2018). https://doi.org/10.1007/978-3-030-01261-8_36
86. Wang, T., Zhang, Y., Fan, Y., Wang, J., Chen, Q.: High-fidelity GAN inversion for image attribute editing. In: CVPR (2022)
87. Wang, T., Zhang, T., Lovell, B.: Faces a la carte: text-to-face generation via attribute disentanglement. In: WACV (2021)

88. Wu, C., Li, H.: Conditional transferring features: scaling GANs to thousands of classes with 30% less high-quality data for training. In: IJCNN (2020)
89. Wu, Q., Li, L., Yu, Z.: TextGAIL: generative adversarial imitation learning for text generation. In: AAAI (2021)
90. Wu, Z., Lischinski, D., Shechtman, E.: StyleSpace analysis: disentangled controls for StyleGAN image generation. In: CVPR (2021)
91. Xia, W., Yang, Y., Xue, J.H., Wu, B.: TediGAN: text-guided diverse face image generation and manipulation. In: CVPR (2021)
92. Xu, D., Yuan, S., Zhang, L., Wu, X.: FairGAN: fairness-aware generative adversarial networks. In: IEEE BigData (2018)
93. Xu, H., Bazavan, E.G., Zanfir, A., Freeman, W.T., Sukthankar, R., Sminchisescu, C.: GHUM & GHUML: generative 3D human shape and articulated pose models. In: CVPR (2020)
94. Xu, T., et al.: AttnGAN: fine-grained text to image generation with attentional generative adversarial networks. In: CVPR (2018)
95. Xu, Y., et al.: TransEditor: transformer-based dual-space GAN for highly controllable facial editing. In: CVPR (2022)
96. Yildirim, G., Jetchev, N., Vollgraf, R., Bergmann, U.: Generating high-resolution fashion model images wearing custom outfits. In: ICCVW (2019)
97. Zhang, D., Khoreva, A.: Progressive augmentation of GANs. In: NeurIPS (2019)
98. Zhao, S., Liu, Z., Lin, J., Zhu, J.Y., Han, S.: Differentiable augmentation for data-efficient GAN training. In: NeurIPS (2020)
99. Zheng, L., Shen, L., Tian, L., Wang, S., Wang, J., Tian, Q.: Scalable person re-identification: a benchmark. In: ICCV (2015)

ColorFormer: Image Colorization via Color Memory Assisted Hybrid-Attention Transformer

Xiaozhong Ji[1], Boyuan Jiang[1], Donghao Luo[1], Guangpin Tao[1], Wenqing Chu[1], Zhifeng Xie[2], Chengjie Wang[1(✉)], and Ying Tai[1(✉)]

[1] Youtu Lab, Tencent, Shanghai, China
{xiaozhongji,byronjiang,michaelluo,guangpintao,wenqingchu,jasoncjwang, yingtai}@tencent.com
[2] Shanghai University, Shanghai, China
zhifeng_xie@shu.edu.cn

Abstract. Automatic image colorization is a challenging task that attracts a lot of research interest. Previous methods employing deep neural networks have produced impressive results. However, these colorization images are still unsatisfactory and far from practical applications. The reason is that semantic consistency and color richness are two key elements ignored by existing methods. In this work, we propose an automatic image colorization method via color memory assisted hybrid-attention transformer, namely ColorFormer. Our network consists of a transformer-based encoder and a color memory decoder. The core module of the encoder is our proposed global-local hybrid attention operation, which improves the ability to capture global receptive field dependencies. With the strong power to model contextual semantic information of grayscale image in different scenes, our network can produce semantic-consistent colorization results. In decoder part, we design a color memory module which stores various semantic-color mapping for image-adaptive queries. The queried color priors are used as reference to help the decoder produce more vivid and diverse results. Experimental results show that our method can generate more realistic and semantically matched color images compared with state-of-the-art methods. Moreover, owing to the proposed end-to-end architecture, the inference speed reaches 40 FPS on a V100 GPU, which meets the real-time requirement.

Keywords: Automatic colorization · Vision transformer · Global and local attention · Memory network

---

X. Ji and B. Jiang—Equal contribution.

Supplementary Information The online version contains supplementary material available at https://doi.org/10.1007/978-3-031-19787-1_2.

1 Introduction

Colorization is a challenging task since there are many possible color images for a grayscale image. Recent colorization approaches can be divided into two categories: *reference-based* and *automatic*. Reference-based colorization methods require user assistance or colorful reference images to reduce uncertainty. In this paper, we focus on automatic colorization task, which requires no additional reference and is therefore more widely applicable. With the development of deep learning, automatic colorization is simply modeled as a learning task. However, it is very challenging to achieve reasonable and natural colorization in a fully automatic setting.

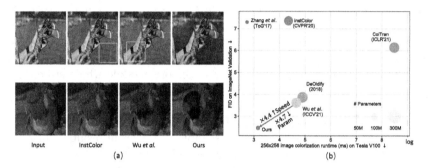

Fig. 1. (a) Visual comparison. The first row indicates InstColor [26] and Wu *et al.* [30] produce unreasonable colors in the water surface with inconsistent color and color-bleeding in the yellow rectangle areas. The second row shows our result is more vivid and colorful. **(b) Speed, parameters and FID comparison.** Our method can colorize images at 40 FPS with the best FID. (Color figure online)

Some classic methods [22,34] based on convolutional neural networks suffer from color confusion because of lacking effective semantic understanding. To locate and learn meaningful semantics, recent methods [26,28,36] combine other tasks (classification, detection, and segmentation) to enhance global or object-level semantic representation, but they still fail to build long-range visual dependencies, resulting in unreasonable colorization results such as green water surface with a lobster held by a person, as shown in the first row of Fig. 1(a).

Besides the limitations in semantic plausibility, these methods also fail to produce photo-realistic results. To improve the color richness, Wu *et al.* [30] combines image synthesis models and ref-based methods for automatic colorization, which is affected by the results of Generative Adversarial Network (GAN) inversion leading to a lack of vividness as shown in the second row of Fig. 1(a). Furthermore, these methods split colorization into multiple stages or branches, which affects the inference efficiency.

In summary, the automatic colorization methods mainly face two difficulties: 1) **Semantic consistency**: The color of an object should be semantically consistent with itself as well as its environment. 2) **Color richness**: The color of objects with different semantics should be diverse. To better address these challenges, we propose a novel colorization approach via hybrid attention and color memory, termed ColorFormer. ColorFormer is divided into two parts: transformer-based encoder to extract contextual semantic, and memory decoder for diverse color acquisition.

For the encoder part, we argue that capturing local and global visual dependencies through self-attention is crucial to reduce the uncertainty of semantic and produce natural results in different scenes. However, the global self-attention operation will bring challenges due to quadratic computation complexity. Therefore, we propose the Global-Local hybrid Multi-head Self-Attention (GL-MSA) operation to build a transformer-based encoder, which enjoys both efficient computation and global attention receptive field.

For the decoder part, we design a Color Memory (CM) module for semantic-level diverse color acquisition. Reference-based methods often produce better results than automatic due to the extra reference images or user guidance. However, searching for suitable reference images is time-consuming and difficult. Inspired by this, we propose a color memory module which stored multiple groups of semantic-color mapping. The decoder can adaptively query the semantic-related color priors from CM, used as reference information to help the decoder produce vivid results. As in Fig. 1, compared to state-of-the-art competitors, our network achieves more reasonable and colorful results, along with faster speed due to the more practical one-stage architecture design.

In general, our contributions can be summarized as follows:

- An effective transformer-based architecture for context-aware semantic extraction in automatic image colorization.
- A novel memory network for diverse color prior acquisition at semantic level.
- Comprehensive experiments demonstrate the effectiveness and efficiency of our method compared to state-of-the-art methods.

2   Related Work

Research on image colorization has developed rapidly in recent years, mainly due to the proposal of high-performance visual backbone network and the inspiration of semantic information for upstream vision tasks. In the following, we focus on the related work in reference-based and automatic colorization methods and briefly introduce existing progress in Vision Transformers.

Reference-Based Colorization. Reference-based colorization integrates the grayscale input with color knowledge transferred from a given reference. The earliest work [29] learns to transfer color by matching brightness and texture within the pixel's neighborhood, but the results are unsatisfactory due to the lack of spatial semantic consistency. To overcome this problem, recent works [12,31] employ deep neural networks to improve spatial correspondence and colorization results. These methods achieve remarkable results but are time-consuming and challenging for automatic retrieval systems [30]. Instead, we employ the color memory to automatically query color priors without reference images.

Automatic Colorization. Automatic colorization is inherently a highly ill-posed problem. The emergence of large-scale datasets makes it possible with deep learning in a data-driven manner. Cheng *et al.* [5] propose the first deep learning based image colorization method. Zhang *et al.* [34] learn the color distribution of every pixel. The network is trained with a multinomial cross-entropy loss with rebalanced rare classes allowing unusual colors to appear. Yoo *et al.* [32] present a memory-augmented colorization model along with threshold triplet loss that can produce high-quality colorization with limited data. Su *et al.* [26] propose instance-aware colorization, which leverages object detectors to crop and extract object-level features. Instance and full-image colorization share the same network, but are trained separately, and a fusion module is applied to predict the final colors. Wu *et al.* [30] propose to recover vivid colors by exploiting the rich and diverse color priors contained in pre-trained Generative Adversarial Networks (GANs). Specifically, matching features are first retrieved through a GAN encoder, and then incorporated into the colorization process through feature modulation. Different from these methods, we focus on integrating accurate semantic understanding and diverse color acquisition into a single network, which can produce reasonable and colorful results with a faster inference speed, and 72% improvement on FID as well, shown in the right part in Fig. 1.

Vision Transformers. Vision Transformers have achieved rapid development and are widely used in various high-and-low level tasks since Dosovitskiy *et al.* [8] successfully introduce Transformer from natural language processing to image recognition. Recently, Liu *et al.* [24] propose a hierarchical Transformer, namely Swin-Transformer, that capably serves as a general-purpose backbone for computer vision and achieves encouraging performance in many computer vision tasks. The representation is computed with shifted windows, which brings greater efficiency by limiting self-attention computation to non-overlapping local windows while also allowing for cross-window connection. Kumar *et al.* [21] propose ColTran, which uses axial self-attention [15] to capture global receptive field attention to conditional produce a low-resolution coarse coloring of the grayscale image and then upsample the coarse colored low-resolution image into a finely colored high-resolution image. However, the inference speed of ColTran is quite slow. It takes about 4.5 s to colorize one image on a V100 GPU, which is unaccepted in real-time applications. Our work revisits the local window attention module by utilizing global information outside the local window, achieving ×180 faster than ColTran.

3 Method

3.1 Overview

Grayscale image colorization is to restore the missing ab channel $x^{ab} \in \mathbb{R}^{H \times W \times 2}$ with only l channel $x^l \in \mathbb{R}^{H \times W \times 1}$, where the l, ab channels represent the lumi-

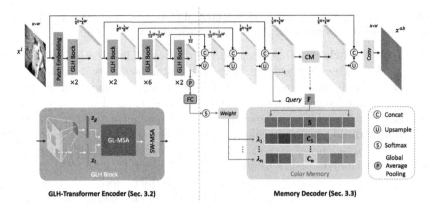

Fig. 2. The framework of our approach. Our network is divided into two parts: *GLH-Transformer Encoder* consisting of multiple GL-MSA modules to model contextual information from grayscale input, and *Memory Decoder* incorporated with CM module to generate colorful results.

nance and chrominance in CIELAB color space, respectively [2,16,23]. We construct an encoder-decoder network for automatic colorization. The encoder is to extract four hierarchical semantic information from the input gray image using stacked Global-Local Hybrid (GLH) Transformer blocks. Each block has the ability to capture global receptive field dependencies with the proposed Global-Local hybrid Multi-head Self-Attention (GL-MSA) module. The decoder consists of four upsampling stages with shortcuts from the corresponding stage of the encoder. Between the third and the last stage, we design a Color Memory (CM) module to introduce adaptive color priors providing relevant color reference for the decoder. As described in Fig. 2. The proposed model mainly consists of the GLH-Transformer encoder and the CM-incorporated decoder.

3.2   GLH-Transformer Encoder

Given a grayscale input image, the encoder first splits it into non-overlapping patches (tokens) and then a linear embedding layer is applied to project patch features to an arbitrary dimension (denoted as C). Here we use a patch size of 4×4 therefore the number of tokens is $\frac{H}{4} \times \frac{W}{4}$. After patch embedding, several GLH-Transformer blocks with global-local hybrid attention computation are applied on these patch tokens. To produce a hierarchical representation, we build four transformer stages. The number of tokens is reduced by a multiple of 2×2 ($2\times$ downsampling of spatial resolution) and the feature dimension is double by a patch merging layer between two stages. In the following we will detailly introduce how to design the GLH-Transformer block.

GL-MSA. The GL-MSA is designed for efficiently building long-range dependencies. As illustrated in Fig. 3(a), supposing the input feature map $\mathbf{z} \in \mathbb{R}^{h \times w \times c}$,

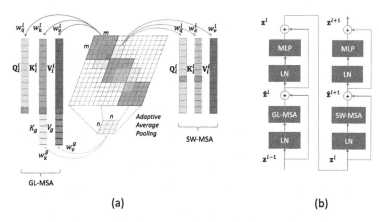

(a)   (b)

Fig. 3. (a) Attention receptive field of GL-MSA and SW-MSA. GL-MSA has global and local hybrid attention receptive field by fusing global and local information to key and value. **(b) Detailed architecture of the GLH-Transformer block.**

GL-MSA first divides $\mathbf{z}$ into non-overlapped $m \times m$ patches, producing the number of $L = \lceil \frac{h}{m} \rceil \times \lceil \frac{w}{m} \rceil$ patches. For patch $\mathbf{z}_l^i \in \mathbb{R}^{mm \times c}, i = 1 \ldots L$, we can obtain local query $\mathbf{Q}_l^i \in \mathbb{R}^{mm \times d}$, key $\mathbf{K}_l^i \in \mathbb{R}^{mm \times d}$ and value $\mathbf{V}_l^i \in \mathbb{R}^{mm \times d}$ with three projection layers. We then apply a $n \times n$ adaptive average pooling layer to the input feature map $\mathbf{z}$, resulting $\mathbf{z}_g \in \mathbb{R}^{nn \times c}$. $n \times n \ll h \times w$ and we set $n = 8, 4, 2, 1$ for four stages respectively. With $\mathbf{z}_g$ we can compute global key $\mathbf{K}_g \in \mathbb{R}^{nn \times d}$ and global value $\mathbf{V}_g \in \mathbb{R}^{nn \times d}$ with two projection layers. To compute global and local hybrid self-attention, we fuse local and global key and value by concatenation, which are formed as:

$$\begin{aligned} \mathbf{K}^i &= [\mathbf{K}_l^i, \mathbf{K}_g], \\ \mathbf{V}^i &= [\mathbf{V}_l^i, \mathbf{V}_g], \end{aligned} \tag{1}$$

where $\mathbf{K}^i \in \mathbb{R}^{(mm+nn) \times d}$ and $\mathbf{V}^i \in \mathbb{R}^{(mm+nn) \times d}$ are global and local hybrid key and value. We then calculate GL-MSA by:

$$\text{GL-MSA}(\mathbf{Q}_l^i, \mathbf{K}^i, \mathbf{V}^i) = \text{softmax}(\frac{\mathbf{Q}_l^i \mathbf{K}^{iT}}{\sqrt{d}}) \mathbf{V}^i. \tag{2}$$

GLH-Transformer Block. Although GL-MSA is able to capture global and local hybrid dependencies, we experimentally find that equipped with Shift Window based Multi-head Self-Attention (SW-MSA) [24] for cross-window connection, our model can produce more semantic reasonable and consistent colorful images. As illustrated in Fig. 3(b), GLH-Transformer blocks are computed as:

$$\begin{aligned} \hat{\mathbf{z}}^l &= \text{GL-MSA}(\text{LN}(\mathbf{z}^{l-1})) + \mathbf{z}^{l-1}, \\ \mathbf{z}^l &= \text{MLP}(\text{LN}(\hat{\mathbf{z}}^l)) + \hat{\mathbf{z}}^l, \\ \hat{\mathbf{z}}^{l+1} &= \text{SW-MSA}(\text{LN}(\mathbf{z}^l)) + \mathbf{z}^l, \\ \mathbf{z}^{l+1} &= \text{MLP}(\text{LN}(\hat{\mathbf{z}}^{l+1})) + \hat{\mathbf{z}}^{l+1}, \end{aligned} \tag{3}$$

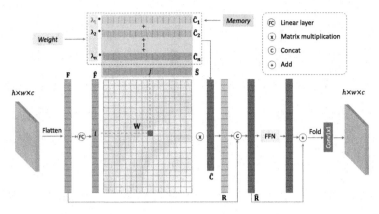

Fig. 4. Detailed architecture of the CM module. The input features $\hat{\mathbf{F}}$ query weighted *values* $\hat{\mathbf{C}}$ based on similarity to *keys* $\hat{\mathbf{S}}$.

where LN is layer normalization [3] and MLP is multi-layer perceptrons.

3.3   Memory Decoder

The memory decoder consists of four cascaded stages progressively enlarging the spatial resolution, where each stage is made up of an upsampling layer and a concatenation layer. In detail, the upsampling layer is implemented by convolution and pixel-shuffle, and the concatenation layer also combines a convolution to integrate the feature from the corresponding stage of the encoder by short-cut connections. Between the third and the last stage, the Color Memory (CM) module is applied to provide diverse color acquisition. We choose to calculate on feature maps of size $\frac{H}{4} \times \frac{W}{4}$ for the balance of computation complexity and representational capacity. At the end of the network are a residual block and a convolutional layer to get the final ab value. In the following part, we describe the detailed architectural design of the CM module.

Architecture of CM Module. For automatic image coloring, introducing various colors is the key to the diversity of the results. The CM module is used to provide the decoder with semantically matched color priors. We construct color memory to store two types of information: *keys* is the patch-level semantic representation $\mathbf{S} \in \mathbb{R}^{m \times k}$ of color images, and *values* store the corresponding color priors $\mathbf{C}_1, \mathbf{C}_2, \ldots, \mathbf{C}_n \in \mathbb{R}^{m \times 2}$, where k is the dimension of semantic representation, m is the number of semantic-color prior correspondence pairs, each color prior with two values (*i.e.* ab in CIELAB color space). We adopt multiple groups of color priors to enlarge the capacity of color memory and n is the number of groups. We first describe the network structure here, and then describe how to obtain the *keys* and *values* in the next Subsect. 3.4. Due to the fact that one object may have different colors in different scenes, we set multiple color values for one semantic, the proportion of which is determined by global semantics. In our network, different color priors are fused together via the output weights from encoder. Note that Yoo *et al.* [32] proposes a memory-augmented colorization model, which stores the one-to-one mapping of the whole image-level spatial feature and color histogram for few-shot or zero-shot colorization. Different from this, our CM stores multiple groups of color prior and provide fine-grained guidance at feature-level. Next, we introduce the specific operation of the CM module illustrated in Fig. 4.

Given the input feature $\mathbf{F} \in \mathbb{R}^{h \times w \times c}$, we first flatten its spatial dimensions and use a linear function to transform it into $\hat{\mathbf{F}} \in \mathbb{R}^{hw \times d_1}$ as the *query*. Then, the semantic representation $\mathbf{S}$ is mapped to $\hat{\mathbf{S}} \in \mathbb{R}^{m \times d_1}$ as the *keys* to match the dimension of the *query*. The color priors $\mathbf{C}_1, \mathbf{C}_2, \ldots, \mathbf{C}_n$ are mapped to $\hat{\mathbf{C}}_1, \hat{\mathbf{C}}_2, \ldots, \hat{\mathbf{C}}_n \in \mathbb{R}^{m \times d_2}$ as the *values*, where d_2 is the dimension of color prior embedding. To integrate multiple color prior vectors, we propose an image-adaptive color priors fusion mechanism, which is based on the global semantic information of the input image. Specifically, we first apply a global average pooling layer, a linear layer, and a softmax layer to generate the fusion weights λ at the end of the encoder, and then fuse color priors via the weights:

$$\hat{\mathbf{C}} = \sum_{l=1}^{n} \lambda_l \hat{\mathbf{C}}_l. \tag{4}$$

Then, we compute the attention weight between each query $\hat{\mathbf{F}}_i \in \mathbb{R}^{d_1}$ and each key $\hat{\mathbf{S}}_j \in \mathbb{R}^{d_1}$ and normalize them through a softmax layer along dimension j, which can be formulated as matrix multiplication:

$$\mathbf{W} = \text{softmax}(\hat{\mathbf{F}}\hat{\mathbf{S}}^T). \tag{5}$$

The weight $\mathbf{W} \in \mathbb{R}^{hw \times m}$ indicates the correspondence between query locations and stored semantic embeddings. The stored color prior is then transferred to query location according to the correspondence. Then we get the cross attention result $\mathbf{R} \in \mathbb{R}^{hw \times d_2}$, calculated as:

$$\mathbf{R} = \mathbf{W}\hat{\mathbf{C}}. \tag{6}$$

We then concatenate $\mathbf{R}$ with the input feature $\mathbf{F}$ at the channel dimension. The concatenated feature $\hat{\mathbf{R}} \in \mathbb{R}^{hw \times (c+d_2)}$ is regarded as the enhanced feature by color prior, which can help produce more diverse and vividly colorful images. To further enhance the generation ability of CM module, the concatenated result is then fed into an FFN [27] layer which consists of two linear transformations with a GELU [13] activation in between. Finally the output of CM is the addition of the output of the FFN layer and the original input together with a 1×1 convolution to the recover feature dimension to c:

$$\text{CM}(\mathbf{F}) = \text{Conv}_{1 \times 1}(\text{FFN}(\hat{\mathbf{R}}) + \hat{\mathbf{R}}). \tag{7}$$

3.4   Memory Build

In this subsection, we describe the detailed build process of the *keys* and *values* in the CM. To better provide semantic-aware colorization guidance, we propose to establish the correspondence between semantic representations (*i.e. keys*) and color priors (*i.e. values*) before network training. The construction process can be divided into two steps: semantic clustering and color clustering.

We first represent local regions of images as specific semantic embeddings and divide them into specific clusters. We randomly select $N = 10,000$ colorful images from ImageNet training set [7] to balance the semantic richness and computational complexity of clustering, and then use a pre-trained classification network (*e.g.*, GLH-Transformer) to extract semantic features. All input images are resized to a fixed size of 256, then we obtain feature maps of size 8×8 by the pre-trained network. In total, we collect $64N$ semantic features, each representing a local patch. In order to reduce the computational complexity of clustering while ensuring that the projected features are as dispersed as possible, we use the Principal Component Analysis (PCA) algorithm to reduce the feature space dimension to k. To represent these features sparsely, we use K-means clustering algorithm to divide them into m categories, and regard the geometric center of each cluster as a semantic representation of image local information.

For regions with similar semantics, we also divide their multiple corresponding colors into n categories. First, we scale the image to the same size as the feature map, and transform it from RGB space to CIELAB space. These divided local areas of the same semantic cluster have various ab values which can be aggregated to form n clusters. Similarly, the geometric centers are regarded as possible color candidates for the current semantic. Furthermore, the n centers are ordered in a counter-clockwise order within the ab plane to ensure all clusters are in the same order. Through the above two clustering steps, we establish a mapping relationship between the semantic embedding and the corresponding color priors, loaded by the CM as the *keys* and the *values*.

3.5 Objectives

During the training, we adopt three losses: *Content loss* to provide pixel-level supervision, *Perceptual loss* to align semantic feature, and *Adversarial loss* to improve authenticity.

Content Loss. The content loss is L_1 distance between the colorized image $\hat{y}$ and the ground-truth colorful image y:

$$\mathcal{L}_c = \|y - \hat{y}\|_1. \tag{8}$$

This loss encourages the generator to output similar color as the given images.

Perceptual Loss. To make generated images with better visual quality, we use pre-trained VGG16 network [25] to extract deep features of $\hat{y}$ and y [18] and calculate the distance between them:

$$\mathcal{L}_p = \sum_{l=1}^{5} w_l \|\Phi_l(\hat{y}) - \Phi_l(y)\|_1, \tag{9}$$

where $\Phi_l(\cdot)$ denotes the layer $conv\_l\_1$ of the VGG16 network, w_l is the weight for the corresponding layer and set to $\frac{1}{16}$, $\frac{1}{8}$, $\frac{1}{4}$, $\frac{1}{2}$, and 1.0, respectively.

Adversarial Loss. Our model is a GAN-based [9] network, where the generator G and the discriminator D are trained alternately. We use the popular PatchGAN discriminator [17], consisting of 4 convolutions with a stride of 2. The loss can be formulated as:

$$\mathcal{L}_d = \|1 - D(y)\|_1 + \|D(\hat{y})\|_1,$$
$$\mathcal{L}_g = \|1 - D(\hat{y})\|_1. \tag{10}$$

Table 1. Quantitative results with SOTA methods on benchmark datasets. ΔCF is the absolute difference of CF between generated colorization images and ground-truth images. $\uparrow$ and $\downarrow$ mean higher or lower is desired.

Method	ImageNet				COCO-Stuff				CelebA-HQ			
	FID↓	CF↑	ΔCF↓	PSNR↑	FID↓	CF↑	ΔCF↓	PSNR↑	FID↓	CF↑	ΔCF↓	PSNR↑
CIC [34]	19.17	**43.92**	4.83	20.86	27.88	33.84	3.01	22.73	14.97	38.21	4.54	24.54
ChromaGAN [28]	5.16	27.49	11.6	23.12	25.65	27.86	8.99	23.56	14.43	**45.93**	3.18	24.54
ColTran [21]	6.14	35.50	3.59	22.30	14.94	36.27	0.58	21.72	10.05	43.62	0.87	22.98
Zhang et al. [35]	7.30	27.23	11.86	**24.13**	17.43	25.95	10.9	**24.66**	11.81	36.98	5.77	**26.25**
DeOldify [1]	3.87	22.83	16.26	22.97	13.86	24.99	11.86	24.19	9.48	43.93	1.18	25.20
InstColor [26]	7.36	27.05	12.04	22.91	13.09	27.45	9.4	23.38	13.28	37.08	5.67	24.77
Wu et al. [30]	3.62	35.13	3.96	21.81	–	–	–	–	–	–	–	–
Ours	**1.71**	39.76	**0.67**	23.00	**8.68**	**36.34**	**0.51**	23.91	**7.54**	42.43	**0.32**	25.62

Full Objectives. Therefore the full objective for the generator is formed as:

$$\mathcal{L}(y, \hat{y}) = \lambda_c \mathcal{L}_c + \lambda_p \mathcal{L}_p + \lambda_g \mathcal{L}_g, \tag{11}$$

where λ_c, λ_p, and λ_g represent weights for different terms, respectively.

4 Experiments

4.1 Datasets and Implementation Details

Dataset. We conduct experiments on datasets: ImageNet [7], COCO-Stuff [4] and CelebA-HQ [19]. We use the training part of ImageNet to train our method and evaluate it on the validation part. To show the generalization of our method, we also test on COCO-Stuff and CelebA-HQ validation sets without fine-tuning.

Evaluation Metrics. We mainly use Fréchet inception distance (FID) [14] and Colorfulness Score (CF) [10] to measure the performance of our method, where FID measures the distribution similarity between generated colorization images and ground truth color images, and CF reflects the vividness of generated colorization images. We also provide PSNR for reference. Although such pixel-wise measurements may not well reflect the actual performance [30].

Implementation Details. We train our network with Adam optimizer [20] and set $\beta_1 = 0.9$, $\beta_2 = 0.99$, weight-decay $= 0$ and initial learning rate $= 1e^{-4}$. For three loss terms, we set $\lambda_c = 0.1$, $\lambda_p = 5.0$ and $\lambda_g = 1.0$. For the GLH-Transformer encoder, we set the window size of GL-MSA and SW-MSA to 7. The feature dimension after four transformer stages is $96, 192, 384$, and 768, respectively. For color memory module, we set $m = 512$, $k = 64$, $n = 4$, $d_1 = 512$ and $d_2 = 256$. The network is trained for 200,000 iterations with batch size of 16 and the learning rate is decayed by 0.5 at 80,000, 120,000 and 160,000 iterations. The training and evaluation images are resized to 256×256. We conduct all experiments with 4 T V100 GPUs.

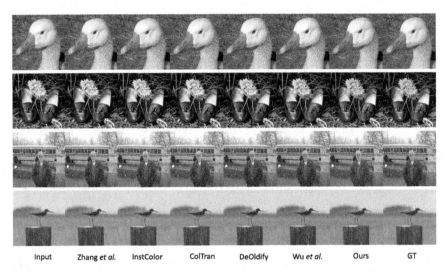

| Input | Zhang *et al.* | InstColor | ColTran | DeOldify | Wu *et al.* | Ours | GT |

Fig. 5. Visual comparisons with previous automatic colorization methods. Our method is able to generate semantic consistent and color vivid images.

4.2   Comparison with State-of-the-Art Methods

Quantitative Comparison. We compare our method with previous methods on three datasets and list the quantitative results in Table 1. All competing methods use codes and model parameters provided by authors. Our method achieves the lowest FID on the ImageNet, indicating that our method could generate realistic and natural color images. On COCO-Stuff and CelebA-HQ datasets, our method also gains the lowest FID, demonstrating the generalization of our method. For the colorfulness score, some methods are higher than ours. However, as mentioned in [30,33], the higher CF may be because they encourage rare colors, leading to unreal colorization results, which are also reflected in their high FID. Therefore, we provide the absolute CF difference between the colorization images and the ground truth images. We exclude all grayscale images

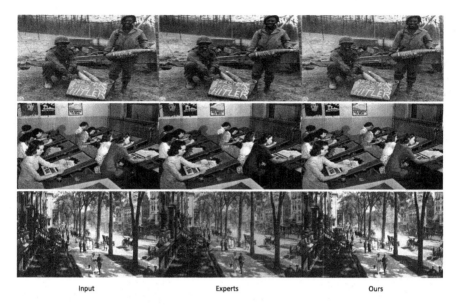

| Input | Experts | Ours |

Fig. 6. Visual results on real-world black-and-white photos.

in the ground truth images to calculate ground truth CF, which is different from [30]. The lower ΔCF indicates more precise colorization results, and we achieve the lowest ΔCF on all three datasets.

Qualitative Comparison. We then visualize the grayscale image colorization results in Fig. 5. Here we display comparisons of images in different scenes from ImageNet validation dataset. Note that the GT images are provided for reference only but the evaluation criterion should not be color similarity. Overall, our results are more reasonable and vivid compared to other competitors. We can see that InstColor produces wrong colors due to miscalculating the semantics, as shown by the duckbill in the first row and the dog ear in the penultimate row. Zhang *et al.* and DeOldify tend to obtain results with uneven and unstable colors in surface like bus affected by luminance change in the grayscale input, while our results look more natural. ColTran and Wu *et al.* produce colorful results, but with unpleasant chromaticity appearing in blue or yellow. Instead, our method can generate semantic-consistent and vivid colorization in complex scenes such as shoes and flowers displayed together, and the bokeh grass.

User Study. We conduct a subjective user study to evaluate which colorization approach is preferred by human observers. We choose InstColor [26], ColTran [21], DeOldify [1] and Wu *et al.* [30] as competing methods for their low FID. We randomly select 50 images from the ImageNet validation set. For each participant, we show him/her five shuffled colorization images at one time and ask for the participant to choose the preferred one. We totally invite 20 volunteers to participate in the user study. The result is shown in Fig. 7 through boxplots. Our method is preferred by 35.16% of users, outperforming all other

Table 2. Quantitative comparisons for ablation studies on the ImageNet dataset. CM_1 means one group of color prior and CM_4 means four. $*$ indicates that the CM is initialized randomly without Memory Build.

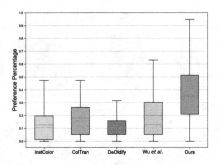

Encoder	Decoder	FID↓	CF↑	ΔCF↓	#Param.
ResNet50	w/o CM	4.10	34.27	4.82	44.5M
Swin-Trans.	w/o CM	2.68	36.03	3.06	41.1M
Twins	w/o CM	2.48	36.46	2.63	36.1M
GLH-Trans.	w/o CM	1.85	36.70	2.39	43.4M
GLH-Trans.	w/CM$_1$	**1.71**	36.96	2.13	44.4M
GLH-Trans.	w/CM$_4^*$	1.80	38.25	0.84	44.8M
GLH-Trans.	w/CM$_4$	**1.71**	**39.76**	**0.67**	44.8M

Fig. 7. Boxplots of user study for five methods. Green dash lines represent the mean preference percentage by users. Our method outperforms all other competing methods by a large margin. (Color figure online)

competing methods (InstColor 12.74%, ColTran 18.11%, DeOldify 14%, Wu *et al.* 20%), which is consistent with the FID score.

Runtime and Model Parameters. We illustrate speed, parameters and FID comparison among SOTA colorization methods in Fig. 1(b). Our method colorizes 256×256 gray images at 40 FPS with 44.8M model parameters, which is ×4.4 speed faster and ×4.7 parameters fewer than the previous SOTA method [30].

Visual Results on Real-World Black-and-White Photos. We collect some historical black-and-white pictures from a website[1] and compare our results with manually colorized ones by human experts, as shown in Fig. 6. The results demonstrate the practicality of our method.

4.3 Ablation Studies

To inspect the effectiveness of the proposed GLH-Transformer encoder and Color Memory decoder in the image colorization task, we conduct a series of ablation studies and list results in Table 2. We select ResNet50 [11] encoder as baseline model for its comparable parameters to our full model. We also adopt Twins [6] as our backbone, which also combines global and local attention.

GLH-Transformer Encoder. Semantic consistency of color is one key point to image colorization. Traditional CNNs are weak in building long-range dependencies. As shown in the left part of Fig. 8, with ResNet50 as encoder, some areas are not reasonable and semantic consistent, which is also reflected by its high FID. Transformer is notable for its use of attention to model long-range dependencies

---

[1] https://www.boredpanda.com/colorized-history-black-and-white-pictures-restored-in-color/.

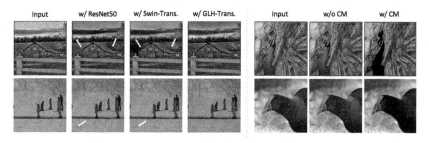

Fig. 8. Illustrations of ablation studies. GLH-Transformer helps produce semantic reasonable and consistent results, and CM leads to vivid results.

in the data. When replacing ResNet50 with Swin-Transformer in the baseline model, FID is reduced to 2.68 from 4.1. However, to reduce the computation complexity, Swin-Transformer calculates self-attention within each non-overlap local window, leading to the receptive field being relatively small in low-level features. Therefore the colorization results are still unreasonable in some areas. By introducing GLH-Transformer, with the help of building global and local hybrid dependencies for each window, FID is further reduced to 1.85, resulting in reasonable and natural visual results. Compared to the baseline model, our GLH-Transformer reduces FID by 55% with fewer parameters. We also compare our GLH with other backbone, Twins [6], and GLH achieves lower FID score. The keys and values of Global sub-sampled attention (GSA) in Twins only come from global features, while the keys and values of GL-MSA have both global and local patch features, which provide more effective feature fusion.

Memory Decoder. The color richness of colorization images is another key point. As in Table 2 and right part of Fig. 8, with CM, our method tends to generate color diversity and vivid images, and the colorfulness score improves from 36.7 to 39.76. We also explore the effectiveness of multiple groups of *values*. Compared to one group, multi groups of *values* can improve colorfulness by 7% with only 0.4M extra parameters. In addition, the CM performs better with pre-stored memory, although it can be trained from random scratch.

Hyperparameters. We conduct more ablation studies on m, k, d_1, d_2, and list quantitative results in Table 3 and Table 4. Increasing these hyperparameters can slightly improve performance, therefore we set them according to the principle of complexity balance.

Table 3. Quantitative results of CM under different m **and** k.

m	512	512	512	256	1024
k	64	32	128	64	64
FID↓	1.71	1.87	1.78	1.93	**1.68**
CF↑	39.76	39.12	39.57	38.56	**39.89**

Table 4. Quantitative results of CM under different d_1 **and** d_2.

d_1	512	256	768	512	512
d_2	256	256	256	128	512
FID↓	1.71	1.85	1.74	1.91	**1.65**
CF↑	39.76	38.94	39.81	38.88	**39.86**

5   Conclusion

In this work, we design a colorization network based on hybrid attention and color memory to improve semantic consistency and color richness. On one hand, we propose the GL-MSA operation, suitable to capture long-range dependencies along with efficient computation. On the other hand, proposed color memory module introduces image-adaptive color priors for feature queries. The experimental results show that the model's accurate understanding of semantics and the introduction of more color priors help obtain more vivid results. What's more, instead of splitting the colorization into multiple steps, we verify that the end-to-end architecture can achieve better results while ensuring efficiency.

Limitations. When dealing with extreme low-quality old images with difficult scenes, our method may produce unreasonable artifacts or incoherent colors, which are also hard cases for recent works. Fortunately, this might be alleviated to some extent by performing image restoration beforehand.

References

1. Antic, J.: A deep learning based project for colorizing and restoring old images (2018)
2. Anwar, S., Tahir, M., Li, C., Mian, A., Khan, F.S., Muzaffar, A.W.: Image colorization: a survey and dataset. arXiv preprint arXiv:2008.10774 (2020)
3. Ba, J.L., Kiros, J.R., Hinton, G.E.: Layer normalization. arXiv preprint arXiv:1607.06450 (2016)
4. Caesar, H., Uijlings, J., Ferrari, V.: COCO-Stuff: thing and stuff classes in context. In: Proceedings of the IEEE Conference on Computer Vision and Pattern Recognition, pp. 1209–1218 (2018)
5. Cheng, Z., Yang, Q., Sheng, B.: Deep colorization. In: Proceedings of the IEEE International Conference on Computer Vision, pp. 415–423 (2015)
6. Chu, X., et al.: Twins: revisiting the design of spatial attention in vision transformers. In: Advances in Neural Information Processing Systems, vol. 34 (2021)
7. Deng, J.: A large-scale hierarchical image database. In: Proceedings of IEEE Computer Vision and Pattern Recognition 2009 (2009)
8. Dosovitskiy, A., et al.: An image is worth 16×16 words: transformers for image recognition at scale. arXiv preprint arXiv:2010.11929 (2020)
9. Goodfellow, I., et al.: Generative adversarial nets. In: Advances in Neural Information Processing Systems (NeurIPS), pp. 2672–2680 (2014)

10. Hasler, D., Suesstrunk, S.E.: Measuring colorfulness in natural images. In: Human Vision and Electronic Imaging VIII, vol. 5007, pp. 87–95. International Society for Optics and Photonics (2003)

11. He, K., Zhang, X., Ren, S., Sun, J.: Deep residual learning for image recognition. In: Proceedings of the IEEE Conference on Computer Vision and Pattern Recognition (CVPR), June 2016

12. He, M., Chen, D., Liao, J., Sander, P.V., Yuan, L.: Deep exemplar-based colorization. ACM Trans. Graph. (TOG) **37**(4), 1–16 (2018)

13. Hendrycks, D., Gimpel, K.: Gaussian error linear units (GELUs). arXiv preprint arXiv:1606.08415 (2016)

14. Heusel, M., Ramsauer, H., Unterthiner, T., Nessler, B., Hochreiter, S.: GANs trained by a two time-scale update rule converge to a local Nash equilibrium. In: Advances in Neural Information Processing Systems, vol. 30 (2017)

15. Ho, J., Kalchbrenner, N., Weissenborn, D., Salimans, T.: Axial attention in multi-dimensional transformers. arXiv preprint arXiv:1912.12180 (2019)

16. Huang, Y.C., Tung, Y.S., Chen, J.C., Wang, S.W., Wu, J.L.: An adaptive edge detection based colorization algorithm and its applications. In: Proceedings of the 13th Annual ACM International Conference on Multimedia, pp. 351–354 (2005)

17. Isola, P., Zhu, J.Y., Zhou, T., Efros, A.A.: Image-to-image translation with conditional adversarial networks. In: Proceedings of the IEEE Conference on Computer Vision and Pattern Recognition, pp. 1125–1134 (2017)

18. Johnson, J., Alahi, A., Fei-Fei, L.: Perceptual losses for real-time style transfer and super-resolution. In: Leibe, B., Matas, J., Sebe, N., Welling, M. (eds.) ECCV 2016. LNCS, vol. 9906, pp. 694–711. Springer, Cham (2016). https://doi.org/10.1007/978-3-319-46475-6_43

19. Karras, T., Aila, T., Laine, S., Lehtinen, J.: Progressive growing of GANs for improved quality, stability, and variation. In: International Conference on Learning Representations (2018)

20. Kingma, D.P., Ba, J.: Adam: a method for stochastic optimization. In: ICLR (Poster) (2015)

21. Kumar, M., Weissenborn, D., Kalchbrenner, N.: Colorization transformer. In: International Conference on Learning Representations (2021)

22. Larsson, G., Maire, M., Shakhnarovich, G.: Learning representations for automatic colorization. In: Leibe, B., Matas, J., Sebe, N., Welling, M. (eds.) ECCV 2016. LNCS, vol. 9908, pp. 577–593. Springer, Cham (2016). https://doi.org/10.1007/978-3-319-46493-0_35

23. Levin, A., Lischinski, D., Weiss, Y.: Colorization using optimization. In: ACM SIGGRAPH 2004 Papers, pp. 689–694 (2004)

24. Liu, Z., et al.: Swin transformer: hierarchical vision transformer using shifted windows. In: Proceedings of the IEEE/CVF International Conference on Computer Vision, pp. 10012–10022 (2021)

25. Simonyan, K., Zisserman, A.: Very deep convolutional networks for large-scale image recognition. arXiv preprint arXiv:1409.1556 (2014)

26. Su, J.W., Chu, H.K., Huang, J.B.: Instance-aware image colorization. In: Proceedings of the IEEE/CVF Conference on Computer Vision and Pattern Recognition, pp. 7968–7977 (2020)

27. Vaswani, A., et al.: Attention is all you need. In: Advances in Neural Information Processing Systems, vol. 30 (2017)

28. Vitoria, P., Raad, L., Ballester, C.: ChromaGAN: adversarial picture colorization with semantic class distribution. In: Proceedings of the IEEE/CVF Winter Conference on Applications of Computer Vision, pp. 2445–2454 (2020)

29. Welsh, T., Ashikhmin, M., Mueller, K.: Transferring color to greyscale images. In: Proceedings of the 29th Annual Conference on Computer Graphics and Interactive Techniques, pp. 277–280 (2002)

30. Wu, Y., Wang, X., Li, Y., Zhang, H., Zhao, X., Shan, Y.: Towards vivid and diverse image colorization with generative color prior. In: Proceedings of the IEEE/CVF International Conference on Computer Vision, pp. 14377–14386 (2021)

31. Xu, Z., Wang, T., Fang, F., Sheng, Y., Zhang, G.: Stylization-based architecture for fast deep exemplar colorization. In: Proceedings of the IEEE/CVF Conference on Computer Vision and Pattern Recognition, pp. 9363–9372 (2020)

32. Yoo, S., Bahng, H., Chung, S., Lee, J., Chang, J., Choo, J.: Coloring with limited data: few-shot colorization via memory-augmented networks. IEEE (2019)

33. Zhang, B., et al.: Deep exemplar-based video colorization. In: Proceedings of the IEEE/CVF Conference on Computer Vision and Pattern Recognition, pp. 8052–8061 (2019)

34. Zhang, R., Isola, P., Efros, A.A.: Colorful image colorization. In: Leibe, B., Matas, J., Sebe, N., Welling, M. (eds.) ECCV 2016. LNCS, vol. 9907, pp. 649–666. Springer, Cham (2016). https://doi.org/10.1007/978-3-319-46487-9_40

35. Zhang, R., et al.: Real-time user-guided image colorization with learned deep priors. ACM Trans. Graph. (TOG) 36(4), 1–11 (2017)

36. Zhao, J., Liu, L., Snoek, C.G., Han, J., Shao, L.: Pixel-level semantics guided image colorization. arXiv preprint arXiv:1808.01597 (2018)

EAGAN: Efficient Two-Stage Evolutionary Architecture Search for GANs

Guohao Ying[2] , Xin He[1] , Bin Gao[3] , Bo Han[1] ,
and Xiaowen Chu[1,4,5(✉)]

[1] Hong Kong Baptist University, Hong Kong SAR, China
[2] University of Southern California, Los Angeles, USA
[3] National University of Singapore, Singapore, Singapore
[4] The Hong Kong University of Science and Technology (Guangzhou),
Guangzhou, China
xwchu@ust.hk
[5] The Hong Kong University of Science and Technology, Hong Kong SAR, China

Abstract. Generative adversarial networks (GANs) have proven successful in image generation tasks. However, GAN training is inherently unstable. Although many works try to stabilize it by manually modifying GAN architecture, it requires much expertise. Neural architecture search (NAS) has become an attractive solution to search GANs automatically. The early NAS-GANs search only generators to reduce search complexity but lead to a sub-optimal GAN. Some recent works try to search both generator (G) and discriminator (D), but they suffer from the instability of GAN training. To alleviate the instability, we propose an efficient two-stage evolutionary algorithm-based NAS framework to search GANs, namely **EAGAN**. We decouple the search of G and D into two stages, where stage-1 searches G with a fixed D and adopts the many-to-one training strategy, and stage-2 searches D with the optimal G found in stage-1 and adopts the one-to-one training and weight-resetting strategies to enhance the stability of GAN training. Both stages use the non-dominated sorting method to produce Pareto-front architectures under multiple objectives (e.g., model size, Inception Score (IS), and Fréchet Inception Distance (FID)). EAGAN is applied to the unconditional image generation task and can efficiently finish the search on the CIFAR-10 dataset in 1.2 GPU days. Our searched GANs achieve competitive results (IS = 8.81 ± 0.10, FID = 9.91) on the CIFAR-10 dataset and surpass prior NAS-GANs on the STL-10 dataset (IS = 10.44 ± 0.087, FID = 22.18). Source code: https://github.com/marsggbo/EAGAN.

G. Ying and X. He—Equal contributions.

Supplementary Information The online version contains supplementary material available at https://doi.org/10.1007/978-3-031-19787-1_3.

1   Introduction

Generative adversarial networks (GANs) [11] have obtained remarkable achievements on image generation tasks. A GAN consists of two networks (i.e., generator (G) and discriminator (D)) that contest with each other in a zero-sum game. G learns to generate semantic images from real data distributions, while D distinguishes real data from generated data. Since G and D have conflicting optimization objectives, GAN training is unstable and prone to collapse. Therefore, many efforts have been made to manually enhance architectures of GANs [3,29], but this requires much professional knowledge. Recently, neural architecture search (NAS) has proven to be effective in automatically finding superior models in various tasks [8,14], including GANs. The early NAS-GAN works [10,35] search only generator with a fixed discriminator to reduce search difficulty, but this may lead to a sub-optimal GAN. Although some recent works have searched both G and D, they suffer from the instability of GAN training. For example, AdversarialNAS [9], which is the first gradient-based NAS-GAN, proposes an adversarial loss function to search G and D simultaneously, but the architectures of G and D are deeply coupled, which increases search complexity and the instability of GAN training. A subsequent gradient-based NAS-GAN work [32] also demonstrates that simultaneously searching both G and D hampers the search of optimal GANs. DGGAN [25] alleviates instability by progressively growing G and D but takes 580 GPU days to search on the CIFAR-10 dataset [20].

In this paper, we propose an efficient two-stage **E**volutionary **A**rchitecture search framework for **G**enerative **A**dversarial **N**etworks (**EAGAN**) on the unconditional image generation task. First, to alleviate the instability of GAN training during the search, we decouple the search of G and D into two stages. In stage-1, we fix the architecture of discriminator and search only generators. All generators are paired with the same discriminator, i.e., the candidate generators and the fixed discriminator are in a *many-to-one* relationship. In stage-2, the best generator of stage-1 is used to provide supervision signals for searching discriminators. Specifically, in stage-2, we create multiple copies of the best generator architecture of stage-1, and each generator copy is paired with a different discriminator and trained independently. Thus, the generators and candidate discriminators of stage-2 are in a *one-to-one* relationship. Because we indirectly evaluate the discriminators of stage-2 via IS (Inception Score [31]) and FID (Fréchet Inception Distance [15]) based on generators, the one-to-one strategy has a potential problem, i.e., if some generators have mode collapse at some time, then subsequently searched discriminators paired with these generators will be evaluated unfairly. To solve this problem, we propose the *weight-resetting* strategy, where all generators inherit the weights of the best generator of the previous search round before a new search round starts. The results in Sect. 5.3 show that our simple yet effective weight-resetting strategy can stabilize GAN searching. We summarize our contributions as follows.

1. We greatly reduce the instability of GAN training by decoupling the search of generator and discriminator into two stages, where stage-1 and stage-2 adopt the *many-to-one* and *one-to-one* training strategy, respectively.

2. We propose the *weight-resetting* strategy, which is simple yet effective to avoid mode collapse when searching discriminators in stage-2 and ensure fair evaluations of different discriminators.
3. EAGAN is efficient and takes 1.2 GPU days on the CIFAR-10 dataset to finish searching GANs. EAGAN achieves competitive results on the CIFAR-10 dataset and outperforms the prior NAS-GANs on the STL-10 dataset [4].

2 Related Work

2.1 Generative Adversarial Network (GAN)

Generative Adversarial Networks (GANs) are first proposed in [11] and have been widely used in the various generation and synthesis tasks. A GAN comprises a generator (G) that generates plausible new data and a discriminator (D) that distinguishes the generator's fake data from real data. Suppose D and G are parameterized by θ and ϕ, respectively, their loss functions are defined as

$$L^D(\phi, \theta) = -E_{x \sim p_{data}(x)}[\log D_\theta(x)] - E_{z \sim p(z)}[\log(1 - D_\theta(G_\phi(z)))] \quad (1)$$
$$L^G(\phi, \theta) = E_{z \sim p(z)}[\log(1 - D_\theta(G_\phi(z)))] \quad (2)$$

where p_{data} is the real data distribution and p_z is a prior distribution. In other words, G and D play a min-max game with value function V, formulated below

$$\min_G \max_D V(G, D) = E_{x \sim p_{data}}[\log D(x)] + E_{z \sim p_z}[\log(1 - D(G(z)))] \quad (3)$$

The mix-max optimization incurs that GAN training suffers from multiple instability issues, such as mode collapse and gradient vanishing. To alleviate these problems, many efforts have been made [2] from the perspective of loss functions [1, 16, 36], normalization and constraint [12, 26], conditional techniques [18, 27], and validation methods [15, 31]. Besides, architecture enhancements have been proven effective to improve GANs performance in many works [3, 17, 29].

2.2 Neural Architecture Search (NAS)

NAS aims at automatic architecture design and has achieved remarkable results in various fields [8, 14]. It can be formulated as a bilevel optimization problem as below

$$\alpha^* = \arg \min_\alpha L_{\text{val}} (\alpha | w^*)$$
$$\text{s.t.} \ \ w^* = \arg \min_w L_{\text{train}} (w | \alpha) \tag{4}$$

where L_{train} and L_{val} indicate the training and validation loss; w and α indicate the weight and architecture of neural network. This process aims to select the architecture α^* performing best on the validation set, conditioned on the optimal network weights w on the training set. There are mainly four approaches in NAS: 1) Reinforcement learning (RL) [28,39] based methods train an RNN controller to generate neural networks; 2) Gradient-based methods [24] apply softmax function to relax the discrete search space, allowing differential optimization of architectures; 3) Surrogate model-based optimization (SMBO) [23] builds a surrogate model of the objective function to predict the searched model's performance, which can substantially improve search efficiency; 4) Evolutionary algorithm (EA) based methods [30,38] maintain and evolve a large population of neural architectures to produce the Pareto-front architectures.

Table 1. Comparison of our EAGAN and the existing NAS-GAN methods. The third column indicates whether the method supports searching discriminators. † indicates a linear combination of metrics. ‡ indicates the Pareto-front of multiple metrics.

Method	Type	search D?	Multi-objective?	Evaluation Metric(s)
AGAN [35]	RL	✗	✗	IS
AutoGAN [10]		✗	✗	IS
E2GAN [33]		✗	✓	IS+FID†
DEGAS [7]	Gradient	✗	✗	Loss
AdversarialNAS [9]		✓	✗	Loss
AlphaGAN [32]		✓	✗	Loss
EGAN [34]	EA	✗	✓	Loss
EAS-GAN [22]		✗	✗	Loss
COEGAN [5]		✓	✗	FID (G); Loss (D)
EAGAN		✓	✓	Pareto-front(IS,FID,#size)‡

2.3 NAS for GANs

Due to the great success of NAS in searching neural networks, many works have also applied NAS to search GANs, summarized in Table 1. AGAN [35] and Auto-GAN [10] are among the first RL-based NAS methods to search GANs, but they only use IS as the reward to guide the search. E2GAN [33] is rewarded by a linear combination of IS and FID. However, to avoid the notorious instability of GAN training, these early NAS-GAN methods only search generator (G) with a fixed discriminator (D) architecture, resulting in a sub-optimal GAN. AdversarialNAS [9] proposes to search G and D simultaneously in a differentiable way. However, it results in highly coupled architectures of G and D. The ablation study in [32] has demonstrated that simultaneously searching G and D would potentially increase the negative impact of inferior discriminators and hinder finding the optimal GANs. Liu et al. [25] propose to progressively grow the architectures of G and D in an alternating fashion, but this is only a remedy to alleviate the issue of architecture coupling and causes huge computational costs (580 GPU days on the CIFAR-10 [20] dataset). COEGAN [5] is very relevant to our work, which also uses an evolutionary algorithm to search G and D in two separate groups of architectures (called populations), but the two populations' architectures are coupled during the search. To reduce the search difficulty, COEGAN only explores a simple search space and experiments on a small dataset (MNIST [21]). The final results show that COEGAN fails to outperform the previous human-designed GANs. In summary, since coupling G and D is not conducive to searching for the optimal GAN, we decouple them into two stages.

3   Preliminary

3.1   Weight-Sharing Based Neural Architecture Search

The early NAS methods first retrain the searched models from scratch and then evaluate their performance [30,39], which obtains accurate evaluation but consumes huge resources, e.g., [30] took 3,150 GPU days to search. To improve search efficiency, the weight-sharing strategy [28] was proposed to allow all subnets to share weights within a super network, so they can be evaluated without retraining by inheriting the weights from SuperNet. In our work, we also adopt the weight-sharing method to search generators and discriminators from SuperNet-G $\mathcal{N}_G$ and SuperNet-D $\mathcal{N}_D$, respectively. To simplify the notations, we use $\mathcal{N}$ to refer to both $\mathcal{N}_G$ and $\mathcal{N}_D$. Denote the loss of the i-th subnet $\mathcal{N}_i$ as L_i, and the weights of $\mathcal{N}$ as W. The gradients of SuperNet loss L with respect to W is

$$\nabla_W L = \frac{1}{N} \sum_{i=1}^{N} \nabla_{W_i} L_i = \frac{1}{N} \sum_{i=1}^{N} \frac{\partial L_i}{\partial W_i} \tag{5}$$

where W_i is the weights of $\mathcal{N}_i$, and N is the total number of subnets. However, it is not practical to accumulate all subnets' gradients in each batch. An alternative way is to use mini-batch subnets to update weights W. In our experiments, we find that randomly sampling one subnet (i.e., $N = 1$) per batch can also work.

3.2   Search Space

To ensure a fair comparison, we use the same search space as in [9] since it also searches both generators and discriminators. The search space is given in Fig. 1.

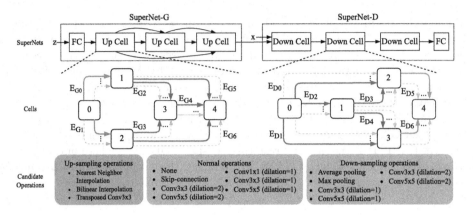

Fig. 1. Overview of search space. E_{G0} and E_{G1} are up-sampling operations, E_{D5} and E_{D6} are down-sampling operations, and the other edges are normal operations.

SuperNet-G. $\mathcal{N}_G$ comprises a fully-connected (FC) layer and three Up-Cells. Each cell contains five ordered nodes (0–4), where node 0 is the output of the previous cell. There are multiple candidate operations between two nodes, each represented by an edge, and only one operation will be activated (solid edge). The edges E_{G0} and E_{G1} indicate up-sampling operations. The rest edges (E_{G2} to E_{G6}) are normal operations, where "None" indicates no connection between two nodes. We encode each edge by a one-hot sequence. For example, [0,1,0] for edge E_{G0} indicates that the bilinear interpolation operation is activated. **SuperNet-D** $\mathcal{N}_D$ comprises three Down-Cells and an FC layer. The Down-Cell is the inverted structure of the Up-Cell. The edges E_{D0} to E_{D4} are normal operations, and E_{D5} and E_{D6} are down-sampling operations. Thus, searching the architecture of G and D is transformed into searching a set of one-hot sequences.

4   Methods

EAGAN comprises two stages, each having two steps: *weights training* and *architecture evolution*. The *many-to-one* and *one-to-one* training strategies tailored for two stages are detailed in Sect. 4.1 and Sect. 4.2, respectively. Sect. 4.3 describes the steps for evolving architectures, which is the same in both stages.

4.1   Stage-1: Searching Generator

Many-to-One GAN Training. As shown in Fig. 2 (left), in stage-1, we search generators (G) with a fixed discriminator (D) that has 0.91M parameters and the same architecture as that of [9]. We adopt the *many(G)-to-one(D)* training strategy. Specifically, the fixed discriminator $\bar{D}$ is denoted by architecture and weights variables, i.e., $\bar{D} \sim (\bar{\beta}, w_{\bar{D}})$. During each round, we produce P candidate generators to form the *population-G* $\mathcal{A}_G$, where all candidate generators share the weights W_G of SuperNet-G, and each candidate G_i is parameterized with architecture and weights variables, i.e., $G_i \sim (\alpha_i, w_{G_i})$, where $w_{G_i} = W_G(\alpha_i)$. We then pair each candidate generator with the fixed discriminator $\bar{D}$ to form P GANs, i.e., $\{(G_1, \bar{D}), ..., (G_P, \bar{D})\}$. Stage-1 can be formalized as below

$$\alpha^* = \arg\min_{\alpha_i}\{V_{val}\left(\alpha_i \mid w_{G_i}^*, w_{\bar{D}}^*, \bar{\beta}\right), i \in \{1, ..., P\}\} \tag{6}$$

$$\text{s.t.} \quad w_{G_i}^* = \arg\min_{w_{G_i}} E_{z \sim p(z)}\left[\log\left(1 - \bar{D}\left(G_i(z)\right)\right)\right] \tag{7}$$

$$w_{\bar{D}}^* = \arg\max_{w_{\bar{D}}} \sum_{i=1}^{P} E_{x \sim p_{\text{data}}(x)}[\log \bar{D}(x)] + E_{z \sim p(z)}[\log(1 - D(G_i(z)))] \tag{8}$$

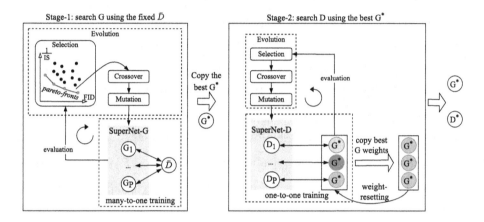

Fig. 2. Two-stage pipeline of EAGAN.

where the inner (Eq. (7)–(8)) is to optimize weights of P GANs on the training set via the many-to-one strategy, and the outer (Eq. (6)) is to obtain the optimal architecture of G according to the value function on the validation set (i.e., V_{val}). The inner and outer optimizations are solved by iterative procedures, outlined in Algorithm 1. These P GANs share the same discriminator and are trained for multiple epochs for each round. To get a fair comparison between generators, for each training batch, we uniformly draw a generator from P candidate generators and train it with the fixed discriminator (lines 4 to 10 in Algorithm 1). The many-to-one training mechanism can bring two benefits. First, the fixed discriminator $\bar{D}$ is trained with various generators, which can be viewed as an ensemble method to some extent, avoiding that $\bar{D}$ is over-fitted and much stronger than generators. Second, different generators are trained with the same discriminator, so we can fairly compare the performance of these generators to find the optimal one. Besides, a generator with mode collapse will not interfere with other generators because the selection step will eliminate it from the population (see Sect. 4.3).

4.2   Stage-2: Searching Discriminator

After stage-1, we obtain an optimal generator G^* with architecture α^*. In stage-2, we use it to guide searching discriminators (D). There are two major challenges in searching D: the lack of evaluation metrics for discriminators and the instability of GAN training. Next, we describe our approaches to these two challenges.

One-to-One GAN Training. Unlike generators, discriminators are difficult to be assessed directly. For example, the accuracy of discriminators does not reflect the overall performance of GANs, as high accuracy may indicate that generators are too weak to fool discriminators, and low accuracy may indicate that generator has mode collapse, with no way to analyze the real cause. Some works [5,9, 32] use the reconstructed loss (e.g., Eq. (1)) to monitor discriminator, but the loss is not a reliable monitor metric as GAN training is a dynamic equilibrium process. An alternative solution is to *indirectly* assess the discriminator via IS and FID metrics calculated based on a generator, so we cannot simply imitate the training strategy of stage-1 (e.g., many(D)-to-one(G)) in stage-2; otherwise, all discriminators are paired with the same generator and not comparable. To this end, we propose the *one-to-one* training strategy. Specifically, we create P copies of G^*, each paired with a candidate discriminator from *population-D* $\mathcal{A}_D$. Thus, we obtain P GANs, i.e., $\{(G_i, D_i), i \in \{1, ..., P\}\}$, where $G_i \sim (\alpha^*, w_{G_i})$ and $D_i \sim (\beta_i, w_{D_i})$. Each GAN is independently trained as a regular GAN via Eqs. (1)–(3). Therefore, stage-2 can be formalized as follows

$$\beta^* = \arg\min_{\beta_i}\{V_{val}\left(\beta_i \mid w^*_{G_i}, w^*_{D_i}, \alpha^*\right), i \in \{1, ..., P\}\} \tag{9}$$

$$\text{s.t. } w^*_{G_i}, w^*_{D_i} = \min_{G_i}\max_{D_i} E_{x \sim p_{\text{data}}(x)}[\log D_i(x)] + E_{z \sim p(z)}[\log(1 - D_i(G_i(z)))] \tag{10}$$

Weight-Resetting. The second challenge of stage-2 is that the one-to-one training strategy does not fully guarantee a fair comparison between different discriminators. Since P generators are trained independently, each generator will have different weights after a round of one-to-one training, presented with different colors (see Fig. 2(right)). If some generators have mode collapse due to combination with unsuitable discriminators, then subsequent discriminators paired with these generators will obtain unfair and biased estimation. To alleviate this problem, we propose the *weight-resetting* strategy, which is to first copy the weights of best generator in the current round, and then initialize all generators in the next round with the copied weights. In the first round, all generators are initialized with the weights of G^* found in stage-1. In summary, the one-to-one training strategy allows each discriminator to be paired with an independent generator, and the weight-resetting strategy ensures a fair comparison between different discriminators and alleviates the instability of GAN training.

Algorithm 1 EAGAN.

Input: SuperNet-G $\mathcal{N}_G$, SuperNet-D $\mathcal{N}_D$, population-G $\mathcal{A}_G$, population-D $\mathcal{A}_D$, population size $P = |\mathcal{A}_G| = |\mathcal{A}_D|$, multi-objective set $\mathcal{F}$, total search rounds R, each round contains E epochs of training.

Output: G^* and D^*

1   $\bar{D} \sim (\bar{\beta}, w_{\bar{D}}) \leftarrow$ Initialize a discriminator with weights $w_{\bar{D}}$ and fixed architecture $\bar{\beta}$;

2   $\mathcal{A}_G^{(0)} = \{G_1^{(0)}, ..., G_P^{(0)}\} \leftarrow$ Warm-up($\mathcal{N}_G, \bar{D}$);

3   $\{(G_i^{(0)}, \bar{D}), i \in \{1, ..., P\}\} \leftarrow$ Initialize P GANs that share the same discriminator;

4   **for** $r=0:R-1$ **do**

5   **for** $e=0:E-1$ **do**

6   **for** *batch* $x = \{x_1, ..., x_m\}$ *in training set* **do**

7   Sample noise data $z = \{z_1, ..., z_m\}$;

8   Uniformly sample $G_i^{(r)}$ from $\mathcal{A}_G^{(r)}, i \in \{1, ..., P\}$;

9   Update weights of $\bar{D}$ via Eq. (8);

10   Update weights of $G_i^{(r)}$ via Eq. (7);

11   **end**

12   **end**

13      $\mathcal{A}_G^{(r)} \leftarrow$ Select Pareto-front generators under $\mathcal{F}$ based on validation set;

14      $\mathcal{A}_G^{(r)} \leftarrow$ Crossover&Mutation($\mathcal{A}_G^{(r)}$);

15 **end**

16   $G^* \sim (\alpha^*, w_{G^*}) \leftarrow$ the best generator with architecture α^* and weights w_{G^*};

17   $\mathcal{A}_D^{(0)} = \{D_1^{(0)}, ..., D_P^{(0)}\} \leftarrow$ Warm-up($G^*, \mathcal{N}_D$);

18   $\{(G_i, D_i^{(0)}), i \in \{1, ..., P\}\} \leftarrow$ Initialize P GANs, where G_i is a copy of G^*;

19   **for** $r=0:R-1$ **do**

20   **for** $e=0:E-1$ **do**

21   **for** *batch* $x = \{x_1, ..., x_m\}$ *in training set* **do**

22   Sample noise data $z = \{z_1, ..., z_m\}$;

23   Uniformly sample a GAN $(G_i, D_i^{(r)})$ from P GANs;

24   Update weights of G_i and $D_i^{(r)}$ via Eq. (10);

25   **end**

26   **end**

27      $\mathcal{A}_D^{(r)} \leftarrow$ Select Pareto-front discriminators under $\mathcal{F}$ based on validation set;

28      $\mathcal{A}_D^{(r)} \leftarrow$ Crossover&Mutation($\mathcal{A}_D^{(r)}$);

29      $w_{G^*} \leftarrow$ the generator weights of the best GAN;

30      $w_{G_1} = ... = w_{G_P} = w_{G^*} \leftarrow$ Weight-resetting;

31 **end**

32   $D^* \sim (\beta^*, w_{D^*}) \leftarrow$ the best discriminator with architecture β^* and weights w_{D^*};

4.3 Architecture Evolution

As shown in Fig. 2, after weights training, stage-1 and stage-2 perform the same steps to evolve generators and discriminators, respectively. To simplify notations, we use $\mathcal{N}$, $\mathcal{N}_i$, and $\mathcal{A}$ to denote the SuperNet, the i-th subnet, and population, of candidate generators (stage-1) and discriminators (stage-2), respectively.

Selection. This step is equivalent to Eq. (6) of stage-1 and Eq. (9) of stage-2. In our work, we use IS [31] and FID [15] metrics to evaluate the performance of individual (i.e., subnet). FID is inversely correlated with IS, so we adopt the *non-dominated sorting strategy* [6] as the value function to produce the Pareto-front individuals during each round. An individual $\mathcal{N}_i$ is said to be dominated by another individual $\mathcal{N}_j$ when Eq. (11) satisfies.

$$
\begin{aligned}
\mathcal{F}_k\left(\mathcal{N}_i\right) \geq \mathcal{F}_k\left(\mathcal{N}_j\right) \ \forall k \in \{1, \ldots, K\} \\
\mathcal{F}_k\left(\mathcal{N}_i\right) > \mathcal{F}_k\left(\mathcal{N}_j\right) \ \exists k \in \{1, \ldots, K\}
\end{aligned}
\tag{11}
$$

where $\mathcal{F}_k$ indicates the objective (e.g., FID, and $\frac{1}{IS}$[1]). We split the population with P individuals into a number of disjoint subsets (or ranks) $\Omega = \{\Omega_0, \Omega_1, \ldots\}$ by comparing the number of times each individual being dominated by other individuals, where the length of Ω and each subset may be different for each search round. After non-dominated sorting, individuals in the same subset are regarded as equally important and better than those in a larger rank. For example, the individuals in the subset Ω_0 outperform all other subsets of individuals. Finally, we sequentially select $\frac{P}{2}$ individuals from lower to higher ranks.

Crossover&Mutation. As detailed in Sect. 3.2, the architecture of each subnet is encoded by a set of one-hot sequences, where the one-hot sequence indicates an edge and the position of 1 indicates the candidate operation activated on that edge. Thus, the basic unit of crossover and mutation is the one-hot sequence. We set $\frac{P}{2}$ Pareto-front individuals obtained from the selection step as parents. Then, we repeatedly perform crossover and mutation on these parents with probabilities of 0.3 and 0.5, respectively, until we generate $\frac{P}{2}$ new individuals. For crossover, we randomly choose two parents and exchange a single one-hot sequence (i.e., an edge). For mutation, we also randomly choose the one-hot sequence of an edge and change the position of 1 on it.

5   Experiments

5.1   Implementation Settings

Datasets. Following the previous NAS-GANs [9,10,34], we search on the CIFAR-10 [20] and evaluate on both CIFAR-10 and STL-10 [4] datasets. CIFAR-10 has 50,000 training images and 10,000 test images with 32×32 resolutions. STL-10 has 100,500 images with 96×96 resolutions, but we resize them to 48×48.

Warm-up Stage. We set up a warm-up stage before the start of stage-1 and stage-2 to ensure a fair competition for all candidate subnets. Specifically, all candidate operations in search space are activated uniformly and trained equally. The warm-up stage has 50 epochs. After that, we randomly sample P subnets to form the first round of population.

---

[1] The higher the IS value, the better the GAN performance.

Two-Stage Search. For both stage-1 and stage-2, we use the hinge loss [26] and Adam optimizer [19] with an initial learning rate of 0.0002. The total number of search rounds is 18, each containing 10 epochs. The noise data is sampled from the Gaussian distribution. A population of $P = 32$ individuals is trained and evolved during each round. The batch sizes for generator and discriminator are 40 and 80, respectively. Besides, we adopt a low-fidelity evaluation strategy, i.e., the number of images used to calculate FID and IS is reduced to 5,000, which greatly reduces the evaluation time and keeps the performance of the searched architectures. Stage-1 and stage-2 take 0.8 and 0.4 GPU days, respectively.

Fully-Train Stage. After the two-stage search, we fully train the best-performing GAN (G^*, D^*) from scratch. For the CIFAR-10 dataset, the batch size and learning rate are the same as the search stage, but the total number of training epochs is 600. For the STL-10 dataset, the batch size and the learning rate are 128 and 0.0003 for the generator, and 64 and 0.0002 for the discriminator, respectively. Following the previous NAS-GAN works [9,10], we generate 50,000 images to calculate IS and FID metrics.

5.2   Results and Analysis

Search only Generator (EAGAN-G). Our searched generator G^* is shown in Fig. 3. Note that the generators for the CIFAR-10 (G_C with 7.14M parameters) and STL-10 (G_S with 11.55M parameters) datasets have the same architecture but different input channels, so their sizes are different. We can see that 1) bi-linear operation is preferred for up-sampling, which is also observed in previous NAS-GANs [9,33]; 2) there are 6 "None" operations and 3 "skip-connect" operations among 15 total normal operations, and the normal convolution with kernel size 3×3 is preferred, which is probably because the low-resolution images do not need complicated convolutions to generate. The results in Table 2 show that, compared with AdversarialNAS [9], our EAGAN can find a better generator with similar time overhead, given the same search space and fixed discriminator. Specifically, our discovered generator achieves a highly competitive FID (10.14) and IS (8.76 ± 0.09) on the CIFAR-10 dataset. In terms of IS, there is a certain gap between NAS-GANs and BigGAN [3] because BigGAN additionally introduces category information as input into the generator's architecture, while NAS-GANs only receive noise data as input. Besides, our generator G_S achieves remarkable results (IS 10.02 ± 0.11, FID = 23.34) on the STL-10 dataset, showing an excellent transferability.

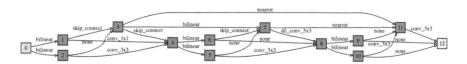

Fig. 3. The architecture of the searched generator ($G_C = G_S = G^*$).

Table 2. Results on the CIFAR-10 and STL-10 datasets. † indicates searching both generators (G) and discriminators (D).

Method	Search method	GPU days	CIFAR-10		STL-10	
			IS↑	FID↓	IS↑	FID↓
DCGANs [29]	Manual	–	6.64 ± 0.14	37.7	–	–
WGAN-GP [12]			7.86 ± 0.07	29.3	–	–
Progressive GAN [17]			8.80 ± 0.05	18.33	–	–
SN-GAN [26]			8.22 ± 0.05	21.7	9.16 ± 0.12	40.1
ProbGAN [13]			7.75	24.60	8.87 ± 0.09	46.74
Improv MMD GAN [36]			8.29	16.21	9.34	37.63
BigGAN [3]			**9.22**	14.73	–	–
AGAN [35]	RL	1200	8.29 ± 0.09	30.5	9.23 ± 0.08	52.7
AutoGAN [10]		2	8.55 ± 0.10	12.42	9.16 ± 0.12	31.01
E2GAN [33]		0.3	8.51 ± 0.13	11.26	9.51 ± 0.09	25.35
DEGAS [7]	Gradient	1.167	8.37 ± 0.08	12.01	9.71 ± 0.11	28.76
AlphaGAN [32]		0.13	8.98 ± 0.09	10.35	10.12 ± 0.13	22.43
AlphaGAN [32]†		–	8.70 ± 0.11	15.56	–	–
AdversarialNAS [9]		1	7.86 ± 0.08	24.04	8.52 ± 0.05	38.85
AdversarialNAS [9]†		1	8.74 ± 0.07	10.87	9.63 ± 0.19	26.98
DGGAN [25]	Heuristic	580	8.64 ± 0.06	12.10	–	–
EGAN [34]	EA	1.25	6.9 ± 0.09	–	–	–
EAS-GAN [22]		1	7.45 ± 0.08	33.2	–	38.84
EAGAN-G	EA	0.8	8.76 ± 0.09	10.14	10.02 ± 0.11	23.34
EAGAN-GD1†		0.8+0.4	8.81 ± 0.10	**9.91**	**10.44 ± 0.08**	**22.18**
EAGAN-GD2†		0.75+0.37	8.63 ± 0.09	12.84	9.76 ± 0.06	26.52
EAGAN-GD3†		1.55+0.73	8.69 ± 0.10	10.53	10.14 ± 0.11	24.22

Search both Generator and Discriminator (EAGAN-GD1). In stage-2, we use the best generator G^* found in stage-1 to help search a set of Pareto-front discriminators, from which we select the optimal discriminators for the CIFAR-10 (D_C with 0.91M parameters) and STL-10 (D_S with 1.58M parameters) datasets, respectively, shown in Fig. 4. We can see a subtle difference (marked in red) between them, i.e., D_S prefers convolutions with a larger kernel size (5 × 5), while D_C selects skip-connection and a smaller convolution. A possible reason is that the resolution of STL-10 (48 × 48) is larger than CIFAR-10 (32 × 32), so it needs a larger kernel size to obtain larger receptive fields.

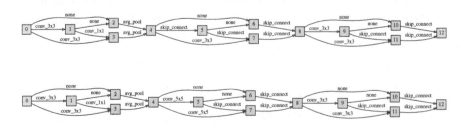

Fig. 4. The searched discriminators on CIFAR-10 (top) and STL-10 (bottom).

After two-stage search, we retrain two GANs (i.e., (G_C, D_C) and (G_C, D_S)) on the CIFAR-10 and STL-10 datasets, respectively, and report their results in Table 2. We can see that none of existing NAS-GANs can guarantee to find excellent GANs in both search scenarios: (a) searching only generators; and (b) searching both generators and discriminators. For example, AdverearialNAS [9] performs poorly (IS = 7.86 ± 0.08, FID = 24.04) in scenario (a), and Alpha-GAN [32] suffers from instability in scenario (b), as its performance drops significantly from (IS = 8.89 ± 0.09, FID = 10.35) in scenario (a) to (IS = 8.70 $\pm$ 0.11, FID = 15.56) in scenario (b). However, our EAGAN performs well in both search scenarios, and the discriminators searched in stage-2 can further improve the performance of the optimal generator discovered in stage-1. Specifically, we achieve a competitive IS value (8.81 ± 0.10) and the best FID (9.91) on the CIFAR-10 dataset. Besides, our EAGAN achieves remarkable performance (IS = 10.44 ± 0.08, FID = 22.18) on the STL-10 dataset, which outperforms the existing NAS-searched GANs. In Fig. 5, we present 50 images randomly generated by generators trained on the CIFAR-10 and the STL-10 datasets without cherry-picking, respectively. The generated images are of rich diversity and high quality.

(a) CIFAR-10   (b) STL-10

Fig. 5. The generated images by EAGAN in random without cherry-picking.

5.3 Ablation Study

Search G or D First? EAGAN searches G first and then searches D. *What about search D first?* Our experiments show that searching D first in stage-1 will make the searched D much stronger than candidate G in stage-2, which in turn causes the gradients of G to vanish. Thus, we should search G first.

Initialize Different D in Stage-1. Our above experiment (i.e., EAGAN-GD1) uses the discriminator of [9] in stage-1. We further implement two experiments to explore the effect of initializing different D in stage-1. EAGAN-GD2 uses a simple network with 0.92M parameters, comprising five normal convolutions and a linear layer, as the initial D in stage-1. EAGAN-GD3 is to repeat the two-stage search several times, i.e., the optimal D of the previous stage-2 is set as the initial D of the next stage-1. From Table 2, we can see that both EAGAN-GD2 and EAGAN-GD3 achieve competitive results on the CIFAR-10 and STL-10 datasets, indicating that EAGAN does not require strong prior knowledge to design the initial state of D and that searching once is sufficient to find good models, balancing search overhead and model performance.

Decoupled vs. Coupled. To validate the effectiveness of our decoupled search method, we perform a coupled search experiment as the baseline, i.e., the architectures of G and D are evolved simultaneously for each search round. Figure 6 presents the learning curves of the baseline and our EAGAN, which shows that coupled search is unstable as it fluctuates throughout the search. In contrast, the overall performance of our decoupled search is better and significantly improved, especially in stage-2 of searching discriminators. Besides, the decoupled search also fluctuates in stage-1 due to the competition among candidate generators incurred by the weight-sharing strategy, and how to address the negative impact of weight-sharing is still an open problem [37].

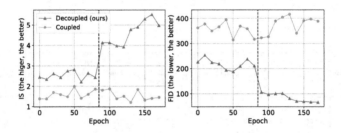

Fig. 6. Learning curves when generators and discriminators are coupled/decoupled. The dashed line indicates the boundary between the two decoupled stages of EAGAN.

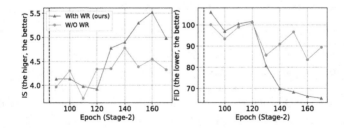

Fig. 7. Learning curves with and without (W/O) the weight-resetting (WR) strategy in stage-2.

Weight-Resetting Strategy. We conduct another experiment on the CIFAR-10 dataset, which differs from our EAGAN only in that the weights of P generators in stage-2 are continuously and independently trained without weight-resetting (WR) strategy. Figure 7 presents the learning curves with and without the WR strategy in stage-2, which shows that our proposed WR strategy can effectively enhance the stability of GAN training and obtain better IS and FID scores in stage-2 of searching discriminators.

6   Conclusion and Future Work

This paper proposes an efficient two-stage evolutionary algorithm-based NAS framework to search GANs, namely EAGAN. We demonstrate that decoupling the search of the generator and discriminator into two stages can significantly improve the stability of searching GANs via the GAN training strategies (many-to-one and one-to-one) tailored for both stages and the weight-resetting strategy. EAGAN is very efficient and takes 1.2 GPU days to finish the search on CIFAR-10. Our searched GANs achieve competitive performance (IS and FID) on the CIFAR-10 dataset and outperform previous NAS-GANs on the STL-10 dataset.

We believe our work deserves more in-depth study and may benefit other potential fields. For example, our decoupled paradigm and tailored training strategies are well suited for large-scale parallel search when architectures require adversarial training. Further, we shall investigate reducing the interference of weight-sharing in search and explore high-resolution generative tasks.

Acknowledgements. Thanks to the NVIDIA AI Technology Center (NVAITC) for providing the GPU cluster to support our work. BH was supported by the NSFC Young Scientists Fund No. 62006202, Guangdong Basic and Applied Basic Research Foundation No. 2022A1515011652, RGC Early Career Scheme No. 22200720, RGC Research Matching Grant Scheme No. RMGS2022_11_02 and HKBU CSD Departmental Incentive Grant.

References

1. Arjovsky, M., Chintala, S., Bottou, L.: Wasserstein generative adversarial networks. In: International Conference on Machine Learning, pp. 214–223. PMLR (2017)
2. Bissoto, A., Valle, E., Avila, S.: The six fronts of the generative adversarial networks. arXiv preprint arXiv:1910.13076 (2019)
3. Brock, A., Donahue, J., Simonyan, K.: Large scale GAN training for high fidelity natural image synthesis. In: ICLR (2019)
4. Coates, A., Ng, A., Lee, H.: An analysis of single-layer networks in unsupervised feature learning. In: Proceedings of the Fourteenth International Conference on Artificial Intelligence and Statistics (2011)
5. Costa, V., Lourenço, N., Machado, P.: Coevolution of generative adversarial networks. In: Kaufmann, P., Castillo, P.A. (eds.) EvoApplications 2019. LNCS, vol. 11454, pp. 473–487. Springer, Cham (2019). https://doi.org/10.1007/978-3-030-16692-2_32

6. Deb, K., Agrawal, S., Pratap, A., Meyarivan, T.: A fast elitist non-dominated sorting genetic algorithm for multi-objective optimization: NSGA-II. In: Schoenauer, M., Deb, K., Rudolph, G., Yao, X., Lutton, E., Merelo, J.J., Schwefel, H.-P. (eds.) PPSN 2000. LNCS, vol. 1917, pp. 849–858. Springer, Heidelberg (2000). https://doi.org/10.1007/3-540-45356-3_83

7. Doveh, S., Giryes, R.: Degas: Differentiable efficient generator search. arXiv preprint arXiv:1912.00606 (2019)

8. Elsken, T., Metzen, J.H., Hutter, F.: Neural architecture search: a survey. arXiv preprint arXiv:1808.05377 (2018)

9. Gao, C., Chen, Y., Liu, S., Tan, Z., Yan, S.: Adversarialnas: adversarial neural architecture search for GANs. In: Proceedings of the CVPR (2020)

10. Gong, X., Chang, S., Jiang, Y., Wang, Z.: Autogan: neural architecture search for generative adversarial networks. In: Proceedings of the ICCV (2019)

11. Goodfellow, I., et al.: Generative adversarial nets. Advances in neural information processing systems 27 (2014)

12. Gulrajani, I., Ahmed, F., Arjovsky, M., Dumoulin, V., Courville, A.C.: Improved training of wasserstein gans. Advances in neural information processing systems 30 (2017)

13. He, H., Wang, H., Lee, G.H., Tian, Y.: Probgan: towards probabilistic gan with theoretical guarantees. In: ICLR (2018)

14. He, X., Zhao, K., Chu, X.: Automl: a survey of the state-of-the-art. Knowl.-Based Syst. **212**, 106622 (2021)

15. Heusel, M., Ramsauer, H., Unterthiner, T., Nessler, B., Hochreiter, S.: Gans trained by a two time-scale update rule converge to a local nash equilibrium. In: Proceedings of the NeurIPS (2017)

16. Hjelm, R.D., Jacob, A.P., Che, T., Trischler, A., Cho, K., Bengio, Y.: Boundary-seeking generative adversarial networks. arXiv preprint arXiv:1702.08431 (2017)

17. Karras, T., Aila, T., Laine, S., Lehtinen, J.: Progressive growing of gans for improved quality, stability, and variation. arXiv preprint arXiv:1710.10196 (2017)

18. Karras, T., Laine, S., Aila, T.: A style-based generator architecture for generative adversarial networks. In: Proceedings of the IEEE/CVF Conference on Computer Vision and Pattern Recognition, pp. 4401–4410 (2019)

19. Kingma, D.P., Ba, J.: Adam: A method for stochastic optimization. arXiv preprint arXiv:1412.6980 (2014)

20. Krizhevsky, A., Hinton, G., et al.: Learning multiple layers of features from tiny images. Technical report (2009)

21. LeCun, Y., Bottou, L., Bengio, Y., Haffner, P.: Gradient-based learning applied to document recognition. Proc. IEEE **86**(11), 2278–2324 (1998)

22. Lin, Q., Fang, Z., Chen, Y., Tan, K.C., Li, Y.: Evolutionary architectural search for generative adversarial networks. IEEE Trans. Emerging Top. Comput. Intell. (2022)

23. Liu, C., Zoph, B., Neumann, M., Shlens, J., Hua, W., Li, L.-J., Fei-Fei, L., Yuille, A., Huang, J., Murphy, K.: Progressive neural architecture search. In: Ferrari, V., Hebert, M., Sminchisescu, C., Weiss, Y. (eds.) ECCV 2018. LNCS, vol. 11205, pp. 19–35. Springer, Cham (2018). https://doi.org/10.1007/978-3-030-01246-5_2

24. Liu, H., Simonyan, K., Yang, Y.: Darts: differentiable architecture search. arXiv preprint arXiv:1806.09055 (2018)

25. Liu, L., Zhang, Y., Deng, J., Soatto, S.: Dynamically grown generative adversarial networks. In: Proceedings of the AAAI Conference on Artificial Intelligence, vol. 35(10), pp. 8680–8687 (2021)

26. Miyato, T., Kataoka, T., Koyama, M., Yoshida, Y.: Spectral normalization for generative adversarial networks. arXiv preprint arXiv:1802.05957 (2018)
27. Odena, A., Olah, C., Shlens, J.: Conditional image synthesis with auxiliary classifier gans. In: International Conference on Machine Learning, pp. 2642–2651. PMLR (2017)
28. Pham, H., Guan, M.Y., Zoph, B., Le, Q.V., Dean, J.: Efficient neural architecture search via parameter sharing. arXiv preprint arXiv:1802.03268 (2018)
29. Radford, A., Metz, L., Chintala, S.: Unsupervised representation learning with deep convolutional generative adversarial networks. arXiv preprint arXiv:1511.06434 (2015)
30. Real, E., Aggarwal, A., Huang, Y., Le, Q.V.: Regularized evolution for image classifier architecture search. In: Proceedings of the AAAI, vol. 33 (2019)
31. Salimans, T., Goodfellow, I., Zaremba, W., Cheung, V., Radford, A., Chen, X.: Improved techniques for training gans. In: Proceedings of the NeurIPS (2016)
32. Tian, Y., Shen, L., Su, G., Li, Z., Liu, W.: Alphagan: Fully differentiable architecture search for generative adversarial networks. arXiv preprint arXiv:2006.09134 (2020)
33. Tian, Y., Wang, Q., Huang, Z., Li, W., Dai, D., Yang, M., Wang, J., Fink, O.: Off-policy reinforcement learning for efficient and effective GAN architecture search. In: Vedaldi, A., Bischof, H., Brox, T., Frahm, J.-M. (eds.) ECCV 2020. LNCS, vol. 12352, pp. 175–192. Springer, Cham (2020). https://doi.org/10.1007/978-3-030-58571-6_11
34. Wang, C., Xu, C., Yao, X., Tao, D.: Evolutionary generative adversarial networks. IEEE Trans. Evol. Comput. **23**(6), 921–934 (2019)
35. Wang, H., Huan, J.: Agan: Towards automated design of generative adversarial networks. arXiv preprint arXiv:1906.11080 (2019)
36. Wang, W., Sun, Y., Halgamuge, S.: Improving MMD-GAN training with repulsive loss function. In: ICLR (2019)
37. Xie, L., Chen, X., Bi, K., Wei, L., Xu, Y., Wang, L., Chen, Z., Xiao, A., Chang, J., Zhang, X., et al.: Weight-sharing neural architecture search: a battle to shrink the optimization gap. ACM Comput. Surv. (CSUR) **54**(9), 1–37 (2021)
38. Yang, Z., et al.: Cars: continuous evolution for efficient neural architecture search. In: Proceedings of the CVPR (2020). https://doi.org/10.1109/CVPR42600.2020.00190
39. Zoph, B., Le, Q.V.: Neural architecture search with reinforcement learning. arXiv preprint arXiv:1611.01578 (2016)

Weakly-Supervised Stitching Network for Real-World Panoramic Image Generation

Dae-Young Song[1], Geonsoo Lee[1], HeeKyung Lee[2], Gi-Mun Um[2], and Donghyeon Cho[1(✉)]

[1] Chungnam National University, Daejeon, South Korea
{201501747.o,geonsoo.o,cdh12242}@cnu.ac.kr
[2] Electronics and Telecommunication Research Institute, Daejeon, South Korea
{lhk95,gmum}@etri.re.kr

Abstract. Recently, there has been growing attention on an end-to-end deep learning-based stitching model. However, the most challenging point in deep learning-based stitching is to obtain pairs of input images with a narrow field of view and ground truth images with a wide field of view captured from real-world scenes. To overcome this difficulty, we develop a weakly-supervised learning mechanism to train the stitching model without requiring genuine ground truth images. In addition, we propose a stitching model that takes multiple real-world fisheye images as inputs and creates a 360° output image in an equirectangular projection format. In particular, our model consists of color consistency corrections, warping, and blending, and is trained by perceptual and SSIM losses. The effectiveness of the proposed algorithm is verified on two real-world stitching datasets.

Keywords: Image stitching · 360° panoramic image

1 Introduction

Image stitching is a task that combines multiple images obtained from different viewpoints to generate a single panoramic image with a larger field of view (FOV). By exploiting this advantage, image stitching technique can be used in various applications such as street view service, virtual reality [27], video surveillance [16], and Mars exploration [6]. Traditional stitching proceeds in the order of feature point extraction, feature matching, homography estimation, warping, and blending. For instance, Brown and Lowe [3] proposed an automatic stitching method that finds correspondence of feature points using SIFT [35], estimates

Project page is at https://eadcat.github.io/WSSN.

Supplementary Information The online version contains supplementary material available at https://doi.org/10.1007/978-3-031-19787-1_4.

© The Author(s), under exclusive license to Springer Nature Switzerland AG 2022
S. Avidan et al. (Eds.): ECCV 2022, LNCS 13676, pp. 54–71, 2022.
https://doi.org/10.1007/978-3-031-19787-1_4

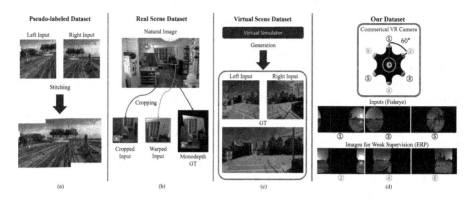

Fig. 1. Comparison of existing stitching dataset for training. (a) The pseudo-labeled dataset is constructed by existing stitching methods. (b) The real scene dataset is generated by cropping with global homography. (c) Virtual scene dataset with a simulator. (d) Our dataset for weak supervisions.

global homography by RANSAC [11], aligns two images using estimated homography, and combines them by multi-band blending. Since then, a lot of following methods have been developed for creating high-quality panoramic images, and the main research issue of them is to deal with parallax distortion caused by depth differences. To overcome parallax distortion, spatially varying multi-homography estimation methods [13,14,19,26,33,52] and non-uniform mesh-based warping methods [17,54,55] have been introduced. Additionally, another challenge to consider in real-world stitching is to overcome visually unpleasant artifacts such as structural distortion and the difference in overall tones between input images. To seamlessly combine input images without suffering these distortions, constraints on line and structure can be explicitly included in the stitching algorithm [21,30,51], and color consistency correction can be applied to input images based on a parametric color model [10,50]. However, the aforementioned approaches depend on the performance of the algorithm that estimates the correspondence of feature points between the input images. Therefore, when the overlapping area between input images is too small or there are many repetitive patterns, feature matching becomes challenging, resulting in parallax distortion and visually unpleasant artifacts, or stitching itself may fail. In other words, the success rate of stitching depends on the performance of the matching algorithm.

Recently, the limitations of these traditional approaches have been solved by the CNN-based feature matching technique [44] and deep homography estimation methods [8,24,38,48,56]. Furthermore, researches on modeling the entire stitching process as a single pipeline based on neural network are being introduced [7,23,28,39,40,45,47]. Unlike feature matching and homography estimation, it is difficult to construct the inputs-GT pairs for training the end-to-end deep stitching model. To get inputs-GT pairs, Shen *et al.* [45] built a unique hardware system that can capture the real-world scene with fixed viewpoints, but it cannot contain dynamic objects due to its systemic limitation. In addition, there were several efforts [7,27,28] to make pseudo GT labels by applying existing stitching methods to real-world images. However, pseudo GT labels may be sensitive to the methods used to create them. In [39], inputs-GT pairs were constructed by cropping sub-images from natural images with random geometric transformations. However, it cannot cover various depths because it is a crop-based method. To handle multiple depth layers and moving objects, there have been studies [23,47] using a virtual simulator such as CARLA [9] to generate inputs-GT pairs. However, it is difficult to use stitching models trained with synthetic datasets on real-world images without the help of domain adaptation. In summary, constructing real inputs-GT pairs that take the depth of the scene into account for training an end-to-end stitching model is a very challenging problem. Therefore, in this paper, we present a weakly-supervised learning method for training a deep stitching model. To this end, we use a commercial camera to capture six fisheye images uniformly rotated at 60° intervals. We use half of the captured images (0°, 120°, 240°) as inputs and the other remaining images (60°, 180°, 300°) as weak supervisions. Note that all images are captured simultaneously, thus dynamic scenes and objects can be covered in our dataset. Then, we introduce a novel mechanism to train an end-to-end stitching model using our dataset. Meanwhile, we develop a deep stitching model that performs color consistency corrections, warping, and blending. In addition, our model and training mechanism can be applied to the existing pseudo GT-based dataset [27]. Comparisons of training datasets for the stitching model are shown in Fig. 1. Our contributions can be summarized as follows:

- We introduce a novel weak-supervised method for training a stitching network to create real-world 360° panoramic images.
- Our stitching model can effectively deal with parallax distortion due to depth differences as well as inconsistent colors between input images.
- We provide a variety of ablation studies, including the results of training the proposed model using the existing CROSS dataset [27].

2   Related Works

In this section, we review both traditional stitching methods and recent deep learning-based stitching methods.

2.1   Traditional Stitching Methods

After Brown and Lowe [3] introduced an automatic stitching method using SIFT feature [35], RANSAC [11], and multi-band blending, lots of follow-up studies addressing various issues have been introduced.

Parallax Distortion. To handle multiple depth layers in the scene, Gao *et al.* [13] proposed a method that estimates dual homography for two separate regions: ground plane and distant plane. Lin *et al.* [33] introduced a spatially varying affine field to adaptively align pixels. Zaragoza *et al.* [54] proposed as-projective-as-possible (APAP) image stitching based on moving direct linear transformation (Moving DLT) for allowing local non-projective deviations. In [55], input images are aligned by estimated homography, then content-preserving warping is applied to solve local parallax distortion. However, the existing methods have problems such as perspective distortion when stitching multiple images. Perazzi et al. [43] proposed a video stitching technique using multiple scenes from unstructured camera arrays. It deals well with parallax and perspective distortions, but takes a long time due to its large computational complexity. In addition, seam-driven image stitching that finds the best homography based on the quality of seam-cut was introduced in [14], while seam-guided local alignment methods were proposed in [31]. Herrmann et al. [19] proposed a robust stitching method that generates multiple registrations and combines them using Markov random field (MRF) with energy terms discouraging duplication and tearing effects. Recently, Lee and Sim [26] introduced a novel concept of warping residual to deal with large parallax using locally optimal warping. For a similar goal, our network includes warping operations that take global and local information into account.

Visually Unpleasant Distortion. In human visual perception, distortion tends to be particularly noticeable on thin objects such as lines or curves. Xiang et al. [51] proposed a line-guided local warping method with a global similarity constraint to overcome projective distortions. Liao et al. [30] presented two single-perspective warpings consisting of parametric warping and mesh-based warping for enhancing the naturalness of stitched images. Jia et al. [21] presented a structure-preserving method based on line-guided warping and line-point constraint. Also, methods exploiting semantic information about pedestrians [12], faces [41], human perception [29] and objects [18] were introduced for natural stitching. In addition, color and tone differences between input images are noticeable distortions. Especially in the near of the seam lines, the distortion becomes more prominent. Doutre and Nasiopoulos [10] proposed a method that corrects differences in color between images using simple linear regression. A more advanced color consistency correction method using convex quadratic programming for the stitching problem is proposed in [50]. To satisfy human visual perception, we utilize perceptual loss [25] in the training step and include color correction operation in our stitching model.

2.2   Deep Learning-Based Stitching Methods

To train a deep stitching model, it is necessary to construct pairs of input images with a narrow FOV and a GT image with a wide FOV. Shen et al. [45] built the hardware system with a flat mirror to create the dataset and trained a stitching model using the constructed dataset. Since it is not practical to use a specialized camera, there have been several studies that make inputs-GT pair as follows.

Dataset with Pseudo GT. Li et al. [27] captured 4 fisheye images taken by lenses rotated at 90° intervals. Then, two images facing opposite directions are used as inputs, and the stitched image using the other two images is used as a pseudo GT image. To create the stitched images, a method with the highest mean opinion score (MOS) among existing stitching methods is used for each image. Using this dataset, Li *et al.* [28] introduced an attentive deep stitching approach consisting of two modules for deformation and resolution. Similarly, Dai *et al.* [7] generated pseudo GT images using existing stitching methods and used them to train an edge-guided composition network. An example of pseudo GT is illustrated in Fig. 1(a). However, pseudo GT labels are sensitive to the methods used to generate them.

Dataset with Only Global Homography. Nie *et al.* [39] presented a deep learning-based view-free stitching model consisting of global homography estimation, structure stitching, and content revision. For the training, they constructed inputs-GT pairs using natural images such as the COCO dataset [32] as shown in Fig. 1(b). Specifically, given an image, two sub-images having overlapping regions are extracted, then geometric transformation is applied to one of them. Thus, these two sub-images have different perspectives, and their geometric relationship can be modeled by a global homography. These two sub-images are used as inputs to the stitching model while an image containing both sub-images is used as GT. However, the problem with their dataset is that depth is not considered when generating two input images. It means that a single depth layer is assumed, which is unrealistic in the real-world scenario. As a result, there is a limitation to stitching images containing scenes with multiple depth layers. Furthermore, parallax distortion caused by depth differences that may occur in the real-world environment cannot be dealt with. Recently, Nie *et al.* [40] proposed an unsupervised learning method for a view-free stitching model composed of coarse alignment and image reconstruction. However, the unsupervised coarse alignment module is performed by a global homography. Thus, parallax distortion induced by depth difference still causes visual artifacts, even though the image reconstruction module enhances the quality of the output image.

Dataset Using Virtual Simulator. There have been several studies that train a stitching model by using inputs-GT pairs generated from a virtual simulator such as CARLA [9]. Since it is possible to control camera configuration and the scene in the virtual space, depth information can be included in the relationship between inputs as shown in Fig. 1(c). Thus, with these virtual datasets, parallax distortion due to depth differences can be covered. Using the virtual dataset, Lai *et al.* [23] proposed a pushbroom stitching network that estimates flow maps in fixed view, and Song *et al.* [47] developed an end-to-end virtual image stitching network via multi-homography estimation. However, these stitching models trained with virtual dataset has limitations in applying them to real-world images, and additional techniques such as domain adaptation may be required. In summary, it is hard to obtain real-world datasets that take the depth information of the scenes into account. In this paper, we use real-world images captured at different viewpoints themselves as inputs to the stitching model. In this case, there are no GT images with a wide FOV, thus we propose a new mechanism for training the stitching model.

3   Approach

In this section, we first describe the procedure of generating training data using real-world fisheye images. Then, we define the problem setup for creating a 360° panoramic image and introduce an architecture of the proposed stitching model to solve the defined problems. Finally, loss functions for training the proposed stitching model are explained.

3.1   Dataset Preparation

To construct the training data for learning the real-world stitching model, we use a commercial VR camera called Kandao Obsidian R [1] to acquire fisheye images. It can capture six fisheye images simultaneously using six lenses rotated at 60° intervals. We use three fisheye images rotated by 0°, 120°, 240° as inputs to our stitching model while the remaining three images rotated by 60°, 180°, 300° are utilized as weak supervisions. As shown in Fig. 1(d), overlapping areas between the input images correspond to the central regions of the images for weak supervisions. Therefore, when training a stitching model using three input images, the remaining three images can be used as weak supervisions. Input images are used as themselves whereas pre-processing is applied on images for weak supervisions. Two types of pre-processing are performed on images for weak supervisions as follows.

Geometric Calibration. We represent the GT 360° panoramic image in equirectangular projection (ERP) format. Since there are no genuine GT images in our setup, it is required to register images for weak supervisions as much as possible in the GT format in advance. Therefore, we transform images for weak supervisions into ERP coordinates. To this end, we perform geometric calibration for fisheye cameras to compute intrinsic and extrinsic parameters by utilizing NVIDIA VRWorks 360 Video SDK [42]. As shown in Fig. 1(d), three fisheye images for weak supervisions are well projected on ERP coordinates.

Color Consistency Correction. Multi-view images captured in a real-world environment may have different color tones. To utilize three images for weak supervisions as GT, the color tones of them should be matched consistently. Therefore, we correct the color consistency of three images for weak supervisions in advance. We use the polynomial curve mapping function to convert the color values of two images (called query image) to the those of remaining one image (called reference image) as

$$\bar{x} = ax^2 + bx + c, \tag{1}$$

where x is the original pixel value in query images, $\bar{x}$ is the corrected pixel value, and a, b, and c are learnable parameters of the polynomial model, respectively. We estimate a, b, and c as follows. First, we find correspondences between query and reference images using SuperGlue [44], then extract patches centered at matched points from both images. Then, we minimize mean squared error (MSE) between corrected patches from query images by (1) and extracted patches from the reference images. Then, we obtain three images with consistent color tones in ERP format by using (1) with the learned a, b, and c. These three images are used for weak supervisions to train our stitching model.

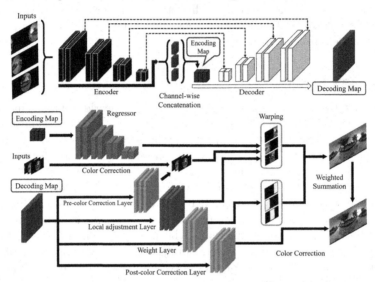

Fig. 2. The entire pipeline of our stitching model. Our model takes N inputs and produces global warping maps, pre- and post-color correction maps, local adjustment maps, and weight maps. After extracting an encoding map (red) and a decoding map (blue), a final warping map U_n is created by adding a global warping map and a local adjustment map, and then the input is warped with U_n. All warped images are weighted by weight maps and merged into a panorama. Color correction is applied once before warping and once after weighted summation using color correction maps. (Color figure online)

3.2   Problem Definition

In this paper, we aim to create a 360° panoramic image by stitching N adjacent images taken with fisheye lenses rotated at different angles. Our stitching model $\mathbf{S}(\cdot)$ takes N fisheye images I_n as inputs and generates a pre-color correction map C_n^{pre}, a global warping map G_n, a local warping adjustment map L_n, and a weight map W_n for each input image. It also produces a post-color correction map C^{post} for an input pair of N images. It is defined as

$$\mathbf{S}(I_1, ..., I_N; \theta) \rightarrow (C_n^{pre}, G_n, L_n, W_n, C^{post}), \qquad (2)$$

where θ and n are learnable parameters of $\mathbf{S}(\cdot)$ and the index of input images, respectively. In our experiment, N is 3, the vertical FOV of each fisheye image is 185°, and the lens for each input is rotated 60° from each other. The panoramic image is created by applying all estimates from the stitching model in (2) to the input fisheye images.

3.3 Architecture

Our stitching model $\mathbf{S}(\cdot)$ is composed of an encoder $\mathbf{E}(\cdot)$, a regressor $\mathbf{R}(\cdot)$, and a decoder $\mathbf{D}(\cdot)$ as illustrated in Fig. 2. The role and details of each component are described as follows.

Encoder. In our stitching model, there are N encoders to extract visual features f_n of each input fisheye image as

$$f_n = \mathbf{E}(I_n; \theta_e),\qquad(3)$$

where I_n is one of the input fisheye images and θ_e is the learnable parameters of the encoder. Our encoder consists of a series of convolutional layers, batch normalization layers, and ELU activations [5]. Learnable parameters of each encoder are shared. Visual features extracted from each input image are concatenated along the channel axis and used as input for a regressor and a decoder.

Regressor. The purpose of the regressor is to find affine transformation matrices that can warp the pixel values of each input image to the pixel coordinates of the output image globally. The regressor takes the visual features f_n as input and generates affine matrices A_n as follows.

$$A_n = \mathbf{R}(f_n; \theta_r),\qquad(4)$$

where θ_r is the learnable parameters of the regressor. Using the estimated affine matrices, a global warping map G_n for each input image is created. The global warping map contains x- and y-direction information on where the pixels of the input image are moved to the coordinates of the output pixels. Global warping can be viewed as global registration by a single homography.

Decoder. Except for the global warping map G_n, the remaining components in (2) needed to make the final output are generated by the decoder as

$$\mathbf{D}(f_n; \theta_d) \to (C_n^{pre}, L_n, W_n, C^{post}),\qquad(5)$$

where θ_d is the learnable parameters of the decoder. Specifically, the decoder consists of a shared decoder and four private decoders for each output component. The shared decoder takes the visual features obtained from the encoder as inputs and generates shared features. Shared features are passed as input to each private decoder to create each output component. A shared decoder consists of a series of convolutional blocks and upsampling layers, and each private decoder consists of several convolutional blocks.

Output Generation Processes. First, the color of the N input images with different color tones is corrected by using the estimated pre-color correction map C_n^{pre}. Inspired by Zero-DCE [15], we convert the color intensity values of input images by a monotonic quadratic curve as follows:

$$\hat{I}_n = I_n + C_n^{pre} I_n (1 - I_n).\qquad(6)$$

Weak Supervision Masking

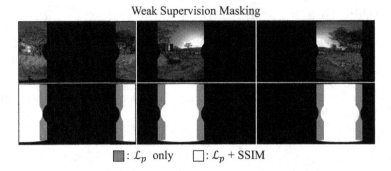

■: $\mathcal{L}_p$ only   □: $\mathcal{L}_p$ + SSIM

Fig. 3. The specific application area of loss functions. Note that SSIM loss is valid only in the white area.

The color-corrected images $\hat{I}_n$ will have color tones harmonized with each other. Then, each color-corrected input fisheye image is warped to the output pixel grid. Our output image is in 360° ERP format. Warping is performed by a global warping map G_n and a local warping adjustment map L_n for each input image. Since the entire depth layer of the scene cannot be covered with only a single global warping map, the local warping adjustment map is used to supplement it as follows.

$$U_n = G_n + \alpha L_n, \tag{7}$$

where α is a balancing factor and set to 0.3 in our experiment. Using the final warping map U_n, color-corrected fisheye images $\hat{I}_n$ are warped as

$$\bar{I}_n = \mathbf{warp}(\hat{I}_n, U_n), \tag{8}$$

where $\mathbf{warp}(\cdot)$ is a pixel mapping function. After that, all warped images are weighted and merged to create a panoramic image P as follow:

$$P = \sum_{n=1}^{N} \bar{I}_n W_n, \tag{9}$$

where W_n is a per-pixel weight map for fusing warped images. Finally, a post-color correction map C^{post} is applied to generate the final panoramic image O as

$$O = P + C^{post}P(1 - P). \tag{10}$$

Detailed formulations of the architecture are in the supplementary material.

3.4   Training

Learnable parameters of our stitching model $\mathbf{S}(\cdot)$ are trained using the images for weak supervisions generated by the method described in Sect. 3.1. Since genuine

GT images do not exist in our settings, we use perceptual loss [25] instead of pixel-wise loss as follows:

$$\mathcal{L}_p(\theta) = \sum_{n=1}^{N} \sum_{i=3}^{5} \mathcal{L}_1(\phi_i(\bar{O}_n), \phi_i(M_n O)), \tag{11}$$

where $\mathcal{L}_1(\cdot)$ and $\phi_i(\cdot)$ are functions of L_1 distance and feature extractor at i-th maxpooling layer of VGG16 [46], respectively. $\bar{O}_n$ represents the image for weak supervisions and M_n is the mask representing the valid pixels of $\bar{O}_n$. Note that M_n is the union of the red and white areas in the bottom row of Fig. 3. Also, for the consistency in color tone and contrast of the input images, we use SSIM loss as follows:

$$\mathcal{L}_{SSIM}(\theta) = \sum_{n=1}^{N} [(1 - SSIM(\hat{M}\bar{O}_n, \hat{M}O))], \tag{12}$$

where $SSIM(\cdot)$ is a function of the structural similarity [49], and $\hat{M}$ is a mask representing non-overlapping regions between the images for weak supervisions. By using this loss function, our model can harmonize the color tone in the overlapping regions between inputs. Note that $\hat{M}$ is only white areas in the bottom row of Fig. 3. Overall loss for training our stitching model is defined as

$$\mathcal{L}(\theta) = (1 - \lambda)\mathcal{L}_p(\theta) + \lambda\mathcal{L}_{SSIM}(\theta), \tag{13}$$

where λ represents the balancing factor between two losses. We set λ to 0.4 in our experiments.

4 Experiments

4.1 Implementation Details

For the experiments, we use our dataset as well as the CROSS dataset [27]. For our dataset, we use 47,063 sets of images for the training and 1,400 for the test. Each training set includes three input fisheye images, three ERP images for weak supervisions, and three masks. For the CROSS, we divide the dataset into 1,146 for the training and 128 for the test. Each set of the CROSS includes two fisheye inputs, and a GT that is pseudo-labeled by SamsungGear. SamsungGear's MOS obtain the highest in most data, thus we choose it as our pseudo-labeling method. For both datasets, all images have a resolution of 1024×512, and data augmentations such as brightness and tone adjustments are randomly applied during the training. Our model was trained by Adam Optimizer [22] with a learning rate of 0.0004. The number of epochs for our dataset and the CROSS is set to 20 and 1200, respectively. Our method is implemented using Pytorch 1.8.1 with CUDA 11.1 on Ubuntu 18.04.

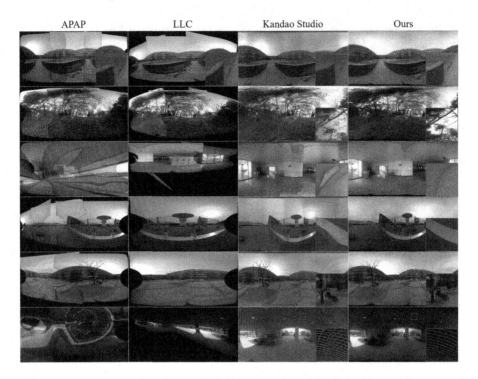

Fig. 4. Qualitative comparisons on our dataset. Please refer to the supplementary material for the rest of the test examples.

4.2 Comparisons

Results on Our Dataset. Since there are no genuine GT images in our datasets, we utilize a perceptual distance P_d using VGG16 as an evaluation metric. The perceptual distance is computed by using making in the same way as in the training step, but there is a difference that unlike in training, the distance is calculated using all five feature maps from five max pooling layers as follows:

$$P_d = \sum_{n=1}^{N} \sum_{i=1}^{5} \mathcal{L}_1(\phi_i(\bar{O}_n), \phi_i(M_n O)), \tag{14}$$

As a result, P_d can evaluate low-level features such as edges. Since our model is trained in the same way, this evaluation can be unfair. Therefore, to compensate for this, we also utilize SIQE [36], LPIPS [57], and FID [20] as quantitative evaluation metrics. As for the competition methods, APAP [54], LLC [21], and Kandao Studio [1] are selected for which the softwares are publicly available. We use ERP format input images for the APAP and the LLC because they were not developed for fisheye inputs. Qualitative comparisons are shown in Fig. 4. Our method produces the most natural, high-quality 360° panoramic images without structural distortions and color inconsistency. In Table 1, there are quantitative

Table 1. Quantitative result of our 1,400 test dataset. **bold**: best. Note that Ours† is our model without the post-color correction map.

Metric	APAP [54]	LLC [21]	Kandao [1]	Ours† (CPU/GPU)	Ours (CPU/GPU)
Time spent (s)	8.5887	16.5126	0.8275	1.3010/**0.0347**	1.3870/0.0363
P_d (↓)	6.498	6.625	5.308	2.773	**2.731**
LPIPS (Alex) (↓)	0.647	0.722	0.266	0.122	**0.118**
LPIPS (VGG16) (↓)	0.652	0.690	0.408	0.178	**0.175**
SIQE [36] (↑)	22.644	20.602	29.399	**39.528**	37.714
FID [20] (↓)	585.6	608.1	224.0	**132.4**	140.8

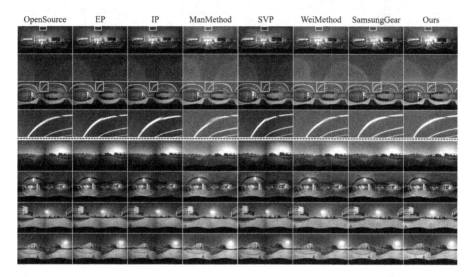

Fig. 5. Qualitative results on CROSS dataset. Top: our method well preserves structural patterns compared to existing stitching models. Bottom: our method produces more color-consistent results than other existing methods.

comparisons with the existing methods. As ablation studies, we also compare our model without a post-color correction map. In addition, we measure the average running time per image for all methods. Note that the running time of the kandao studio includes time for saving a 1920 × 960 image because there are no open-source codes. As reported in Table 1, our method performs better and much faster than the existing methods. However, the results were not significantly different according to a post-color correction map. Also, as expected, the proposed method using GPU acceleration is much faster than other algorithms, including the commercial kandao studio.

Results on the CROSS. To validate the versatility of the proposed method, we evaluate the proposed method on the CROSS dataset, which contains pseudo GT 360° panoramic images as supervisions. As shown in Fig. 5, our method produces more visually pleasing results compared to the existing stitching methods.

Table 2. Quantitative comparisons on CROSS dataset [27]. **bold**: best.

Metric	OpenSource [2]	EP [34]	IP [4]	ManMethod	SVP [37]	WeiMethod [53]	Ours
PSNR (↑)	16.417	15.908	15.177	18.943	16.110	18.730	**22.440**
SSIM (↑)	0.589	0.565	0.546	0.611	0.562	0.595	**0.736**
P_d (↓)	3.31	3.55	3.79	3.12	3.69	3.23	**2.53**

W/O Color Correction   Pre-color Correction   Post-color Correction   Dual-color Correction

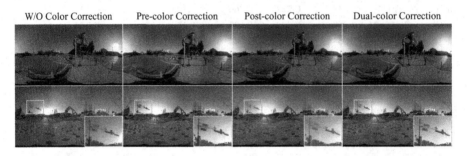

Fig. 6. Ablation studies of our model. The w/o color correction and the pre-color correction model have the unpleasant boundaries. The post-color correction model is suffered from fading.

In particular, our results demonstrate robustness to structural distortion and vignetting artifacts. To measure PSNR, SSIM, and P_d, we use the pseudo-labeled GT images, because the SamsungGear method obtains the highest MOS in [27]. Note that $M_n = 1$ in all pixels because masking is not required. As reported in Table 2, the proposed model outperforms the existing methods.

4.3   Ablation Studies

Effects of Color Correction. We conduct experiments depending on whether the pre-color correction map C_n^{pre} and the post-color correction map C^{post} are used. As shown in Fig. 6, the color tone around boundary lines is inconsistent when only pre-color correction is applied. In addition, results of only using post-color correction suffer from fading effects as shown in the second row of Fig. 6. Overall, the dual-color correction model using both pre-color correction and post-color correction produces the most comfortable results.

Table 3. Self-comparisons according to the number of warpings K on our full test dataset (14 sets). The number of epochs is set to 10. **bold**: best.

Metrics	$K = 1$	$K = 2$	$K = 3$	$K = 4$	$K = 5$
P_d (↓)	4.001	3.936	3.944	**3.904**	3.962
SIQE (↑)	**22.89**	21.16	15.28	17.89	18.29

Fig. 7. Effects of loss. Utilizing L_1 loss instead of $\mathcal{L}_p$ (left). Ours (right).

Effects of Loss. Since the images for weak supervisions have parallax between themselves, it may not be appropriate to use a common pixel-wise regression loss. Considering this point, we adopt the perceptual loss as in (11). Therefore, as ablation studies, we train our model with a pixel-wise L_1 loss instead of (11). As shown in Fig. 7, models trained using L_1 loss are vulnerable to parallax distortion, which causes noticeable distortion.

Effects of the Number of Warpings. Inspired by [47], we modified our model to perform multiple K warpings. However, as shown in Table 3, multiple warpings do not have a significant effect on the quantitative results. We guess that it is because our model uses different input and output coordinates from the stitching model in [47]. Note that cylindrical coordinates are used in [47] while our model is operated on fisheye input and ERP output. Based on the above results, we use the simplest model with $K = 1$ for all experiments.

5   Limitations and Future Works

Even though our model can be trained without genuine GTs, our research does not take view-free inputs into account. We believe that subsequent studies based on this paper can be extended to studies on view-free stitching. Another promising future work is video stitching to cover dynamic scenes. Although the proposed method is developed for a static scene, it can be extended to video, and we believe that temporal artifacts such as waving effects can be solved by temporal consistency loss as in [23].

6   Conclusion

In this paper, we present a weakly supervised method for training the real-world stitching model. Our model takes multiple fisheye images as inputs and generates a 360° panorama image. For training, we generate images of weak supervisions and utilize them for perceptual and SSIM losses. We verify the proposed method on our stitching dataset as well as the CROSS dataset. Through the various experiments, we demonstrate superior stitching performance over existing methods. In particular, it is more robust to structural artifacts and color inconsistency problems compared to existing methods.

Acknowledgement. This work was supported by Institute of Information & Communications Technology Planning & Evaluation (IITP) grant funded by the Korea government(MSIT) (No. 2018-0-00207, Immersive Media Research Laboratory) and the National Research Foundation of Korea(NRF) grant funded by the Korea government(MSIT) (No.2021R1A4A1032580, No.2022R1C1C1009334).

References

1. Kandao. https://www.kandaovr.com/. Accessed 05 Mar 2022
2. Dualfisheye (2016). https://github.com/ooterness/DualFisheye
3. Brown, M., Lowe, D.G.: Automatic panoramic image stitching using invariant features. Intl. J. Comput. Vis. (IJCV) **74**(1), 59–73 (2007). https://doi.org/10.1007/s11263-006-0002-3
4. Cai, D., He, X., Han, J.: Isometric projection. In: Association for the Advancement of Artificial Intelligence (AAAI), pp. 528–533. AAAI Press (2007)
5. Clevert, D.A., Unterthiner, T., Hochreiter, S.: Fast and accurate deep network learning by exponential linear units (ELUS). In: International Conference on Learning Representation (ICLR) (2016)
6. Coates, A., et al.: The PanCam instrument for the ExoMars rover. Astrobiology **17**(6–7), 511–541 (2017)
7. Dai, Q., Fang, F., Li, J., Zhang, G., Zhou, A.: Edge-guided composition network for image stitching. Pattern Recogn. (PR) **118**, 108019 (2021)
8. DeTone, D., Malisiewicz, T., Rabinovich, A.: Deep image homography estimation. CoRR abs/1606.03798 (2016)
9. Dosovitskiy, A., Ros, G., Codevilla, F., Lopez, A., Koltun, V.: CARLA: an open urban driving simulator. In: Proceedings of Conference on Robot Learning (CoRL), pp. 1–16 (2017)
10. Doutre, C., Nasiopoulos, P.: Fast vignetting correction and color matching for panoramic image stitching. In: IEEE International Conference on Image Processing (ICIP), pp. 709–712 (2009)
11. Fischler, M.A., Bolles, R.C.: Random sample consensus: a paradigm for model fitting with applications to image analysis and automated cartography. Commun. ACM **24**(6), 381–395 (1981)
12. Flores, A., Belongie, S.: Removing pedestrians from Google street view images. In: Proceedings of Computer Vision and Pattern Recognition Workshops (CVPRW), pp. 53–58. IEEE (2010)
13. Gao, J., Kim, S.J., Brown, M.S.: Constructing image panoramas using dual-homography warping. In: Proceedings of Computer Vision and Pattern Recognition (CVPR), pp. 49–56. IEEE (2011)
14. Gao, J., Li, Y., Chin, T.J., Brown, M.S.: Seam-driven image stitching. In: Eurographics, pp. 45–48 (2013)
15. Guo, C., et al.: Zero-reference deep curve estimation for low-light image enhancement. In: Proceedings of Computer Vision and Pattern Recognition (CVPR), pp. 1780–1789 (2020)
16. He, B., Yu, S.: Parallax-robust surveillance video stitching. Sensors **16**(1), 7 (2016)
17. He, K., Chang, H., Sun, J.: Rectangling panoramic images via warping. ACM Trans. Graph. (ToG) **32**(4), 1–10 (2013)

18. Herrmann, C., Wang, C., Bowen, R.S., Keyder, E., Zabih, R.: Object-centered image stitching. In: Ferrari, V., Hebert, M., Sminchisescu, C., Weiss, Y. (eds.) ECCV 2018. LNCS, vol. 11207, pp. 846–861. Springer, Cham (2018). https://doi.org/10.1007/978-3-030-01219-9_50

19. Herrmann, C., et al.: Robust image stitching with multiple registrations. In: Ferrari, V., Hebert, M., Sminchisescu, C., Weiss, Y. (eds.) ECCV 2018. LNCS, vol. 11206, pp. 53–69. Springer, Cham (2018). https://doi.org/10.1007/978-3-030-01216-8_4

20. Heusel, M., Ramsauer, H., Unterthiner, T., Nessler, B., Hochreiter, S.: GANs trained by a two time-scale update rule converge to a local Nash equilibrium. In: Proceedings of Neural Information Processing Systems (NeurIPS), pp. 6626–6637 (2017)

21. Jia, Q., et al.: Leveraging line-point consistence to preserve structures for wide parallax image stitching. In: Proceedings of Computer Vision and Pattern Recognition (CVPR), pp. 12186–12195 (2021)

22. Kingma, D.P., Ba, J.: Adam: a method for stochastic optimization. In: International Conference on Learning Representation (ICLR) (2014)

23. Lai, W.S., Gallo, O., Gu, J., Sun, D., Yang, M.H., Kautz, J.: Video stitching for linear camera arrays. In: British Machine Vision Conference (BMVC), pp. 1–12 (2019)

24. Le, H., Liu, F., Zhang, S., Agarwala, A.: Deep homography estimation for dynamic scenes. In: Proceedings of Computer Vision and Pattern Recognition (CVPR), pp. 7652–7661 (2020)

25. Ledig, C., et al.: Photo-realistic single image super-resolution using a generative adversarial network. In: Proceedings of Computer Vision and Pattern Recognition (CVPR), pp. 4681–4690 (2017)

26. Lee, K.Y., Sim, J.Y.: Warping residual based image stitching for large parallax. In: Proceedings of Computer Vision and Pattern Recognition (CVPR), pp. 8198–8206 (2020)

27. Li, J., Yu, K., Zhao, Y., Zhang, Y., Xu, L.: Cross-reference stitching quality assessment for 360 omnidirectional images. In: Proceedings of the 27th ACM International Conference on Multimedia, pp. 2360–2368 (2019)

28. Li, J., Zhao, Y., Ye, W., Yu, K., Ge, S.: Attentive deep stitching and quality assessment for 360° omnidirectional images. IEEE J. Sel. Top. Sig. Process. **14**(1), 209–221 (2019)

29. Li, N., Liao, T., Wang, C.: Perception-based seam cutting for image stitching. Sig. Image Video Process. **12**, 967–974 (2018). https://doi.org/10.1007/s11760-018-1241-9

30. Liao, T., Li, N.: Single-perspective warps in natural image stitching. IEEE Trans. Image Process. (TIP) **29**, 724–735 (2019)

31. Lin, K., Jiang, N., Cheong, L.-F., Do, M., Lu, J.: SEAGULL: seam-guided local alignment for parallax-tolerant image stitching. In: Leibe, B., Matas, J., Sebe, N., Welling, M. (eds.) ECCV 2016. LNCS, vol. 9907, pp. 370–385. Springer, Cham (2016). https://doi.org/10.1007/978-3-319-46487-9_23

32. Lin, T.-Y., et al.: Microsoft COCO: common objects in context. In: Fleet, D., Pajdla, T., Schiele, B., Tuytelaars, T. (eds.) ECCV 2014. LNCS, vol. 8693, pp. 740–755. Springer, Cham (2014). https://doi.org/10.1007/978-3-319-10602-1_48

33. Lin, W.Y., Liu, S., Matsushita, Y., Ng, T.T., Cheong, L.F.: Smoothly varying affine stitching. In: Proceedings of Computer Vision and Pattern Recognition (CVPR), pp. 345–352. IEEE (2011)

34. Ling, S., Cheung, G., Le Callet, P.: No-reference quality assessment for stitched panoramic images using convolutional sparse coding and compound feature selection. In: IEEE International Conference on Multimedia and Expo (ICME), pp. 1–6 (2018). https://doi.org/10.1109/ICME.2018.8486545

35. Lowe, D.G.: Distinctive image features from scale-invariant keypoints. Intl. J. Comput. Vis. (IJCV) **60**(2), 91–110 (2004). https://doi.org/10.1023/B:VISI.0000029664.99615.94

36. Madhusudana, P.C., Soundararajan, R.: Subjective and objective quality assessment of stitched images for virtual reality. IEEE Trans. Image Process. (TIP) **28**(11), 5620–5635 (2019)

37. Maneshgar, B., Sujir, L., Mudur, S., Poullis, C.: A long-range vision system for projection mapping of stereoscopic content in outdoor areas. In: VISIGRAPP (1: GRAPP), pp. 290–297, January 2017. https://doi.org/10.5220/0006258902900297

38. Nguyen, T., Chen, S.W., Shivakumar, S.S., Taylor, C.J., Kumar, V.: Unsupervised deep homography: a fast and robust homography estimation model. IEEE Robot. Autom. Lett. (RAL) **3**(3), 2346–2353 (2018)

39. Nie, L., Lin, C., Liao, K., Liu, M., Zhao, Y.: A view-free image stitching network based on global homography. J. Vis. Commun. Image Represent. **73**, 102950 (2020)

40. Nie, L., Lin, C., Liao, K., Liu, S., Zhao, Y.: Unsupervised deep image stitching: reconstructing stitched features to images. IEEE Trans. Image Process. (TIP) **30**, 6184–6197 (2021)

41. Ozawa, T., Kitani, K.M., Koike, H.: Human-centric panoramic imaging stitching. In: Proceedings of the 3rd Augmented Human International Conference, pp. 1–6 (2012)

42. Patil, T., Turkowski, K.: Calibrating stitched videos with VRWorks 360 video SDK (2018). https://developer.nvidia.com/blog/calibrating-videos-vrworks-360-video/

43. Perazzi, F., et al.: Panoramic video from unstructured camera arrays. In: Computer Graphics Forum, vol. 34, pp. 57–68. Wiley Online Library (2015)

44. Sarlin, P.E., DeTone, D., Malisiewicz, T., Rabinovich, A.: SuperGlue: learning feature matching with graph neural networks. In: Proceedings of Computer Vision and Pattern Recognition (CVPR), pp. 4938–4947 (2020)

45. Shen, C., Ji, X., Miao, C.: Real-time image stitching with convolutional neural networks. In: 2019 IEEE International Conference on Real-Time Computing and Robotics (RCAR), pp. 192–197. IEEE (2019)

46. Simonyan, K., Zisserman, A.: Very deep convolutional networks for large-scale image recognition. In: International Conference on Learning Representation (ICLR) (2015)

47. Song, D.Y., Um, G.M., Lee, H.K., Cho, D.: End-to-end image stitching network via multi-homography estimation. IEEE Sig. Process. Lett. (SPL) **28**, 763–767 (2021)

48. Wang, C., Wang, X., Bai, X., Liu, Y., Zhou, J.: Self-supervised deep homography estimation with invertibility constraints. Pattern Recogn. Lett. (PRL) **128**, 355–360 (2019)

49. Wang, Z., Bovik, A.C., Sheikh, H.R., Simoncelli, E.P.: Image quality assessment: from error visibility to structural similarity. IEEE Trans. Image Process. (TIP) **13**(4), 600–612 (2004)

50. Xia, M., Yao, J., Xie, R., Zhang, M., Xiao, J.: Color consistency correction based on remapping optimization for image stitching. In: Proceedings of the International Conference on Computer Vision Workshops (ICCVW), pp. 2977–2984 (2017)

51. Xiang, T.Z., Xia, G.S., Bai, X., Zhang, L.: Image stitching by line-guided local warping with global similarity constraint. Pattern Recogn. (PR) **83**, 481–497 (2018)

52. Xu, B., Jia, Y.: Wide-angle image stitching using multi-homography warping. In: IEEE International Conference on Image Processing (ICIP), pp. 1467–1471 (2017)
53. Ye, W., Yu, K., Yu, Y., Li, J.: Logical stitching: a panoramic image stitching method based on color calibration box. In: 2018 14th IEEE International Conference on Signal Processing (ICSP), pp. 1139–1143 (2018). https://doi.org/10.1109/ICSP.2018.8652363
54. Zaragoza, J., Chin, T.J., Brown, M.S., Suter, D.: As-projective-as-possible image stitching with moving DLT. In: Proceedings of the Computer Vision and Pattern Recognition (CVPR), pp. 2339–2346 (2013)
55. Zhang, F., Liu, F.: Parallax-tolerant image stitching. In: Proceedings of Computer Vision and Pattern Recognition (CVPR), pp. 3262–3269 (2014)
56. Zhang, J., et al.: Content-aware unsupervised deep homography estimation. In: Vedaldi, A., Bischof, H., Brox, T., Frahm, J.-M. (eds.) ECCV 2020. LNCS, vol. 12346, pp. 653–669. Springer, Cham (2020). https://doi.org/10.1007/978-3-030-58452-8_38
57. Zhang, R., Isola, P., Efros, A.A., Shechtman, E., Wang, O.: The unreasonable effectiveness of deep features as a perceptual metric. In: Proceedings of Computer Vision and Pattern Recognition (CVPR), pp. 586–595 (2018)

DynaST: Dynamic Sparse Transformer for Exemplar-Guided Image Generation

Songhua Liu, Jingwen Ye, Sucheng Ren, and Xinchao Wang[✉]

National University of Singapore, Singapore, Singapore
{songhua.liu,suchengren}@u.nus.edu, {jingweny,xinchao}@nus.edu.sg

Abstract. One key challenge of exemplar-guided image generation lies in establishing fine-grained correspondences between input and guided images. Prior approaches, despite the promising results, have relied on either estimating *dense attention to compute per-point matching*, which is limited to only coarse scales due to the quadratic memory cost, or *fixing the number of correspondences* to achieve linear complexity, which lacks flexibility. In this paper, we propose a dynamic sparse attention based Transformer model, termed *Dynamic Sparse Transformer (DynaST)*, to achieve fine-level matching with favorable efficiency. The heart of our approach is a novel dynamic-attention unit, dedicated to covering the variation on the optimal number of tokens one position should focus on. Specifically, DynaST leverages the multi-layer nature of Transformer structure, and performs the dynamic attention scheme in a cascaded manner to refine matching results and synthesize visually-pleasing outputs. In addition, we introduce a unified training objective for DynaST, making it a versatile reference-based image translation framework for both supervised and unsupervised scenarios. Extensive experiments on three applications, pose-guided person image generation, edge-based face synthesis, and undistorted image style transfer, demonstrate that DynaST achieves superior performance in local details, outperforming the state of the art while reducing the computational cost significantly. Our code is available here.

Keywords: Dynamic sparse attention · Transformer · Exemplar-guided image generation

1 Introduction

Semantic-based conditional image generation refers to synthesizing a photo-realistic image with aligned semantic information, and finds its application across a wide spectrum of scenarios including label-to-scene [10, 16, 37, 44, 48, 54, 55, 70], sketch-to-photo [11, 24, 25], and landmark-to-face [42, 47, 61]. Exemplar-guided

Supplementary Information The online version contains supplementary material available at https://doi.org/10.1007/978-3-031-19787-1_5.

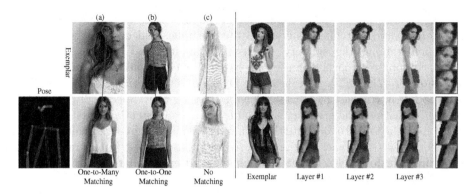

Fig. 1. Left: Different numbers of matching are required for different query locations in the exemplar-guided image generation task. This fact has been largely overlooked by prior methods that impose only static number of matching. The proposed DynaST, by contrast, is dedicated to handling such variations. Right: Details in faces and edges can be refined with the propagation in the multi-layer structure of Transformer. Warping results of the exemplar images using attention maps in each layer are shown here.

image generation, as a mainstream approach along this route, provides users with the flexibility to specify an image as reference to control the appearance, style, or identity for the output image, and has recently received wide attentions across academic and industrial communities.

The core problem of exemplar-guided image generation lies in guiding input semantics to focus on appropriate context in exemplars. Early methods have largely relied on holistic convolution, normalization, and non-linear transformation [1,36,37,44,46,52,67,71]. Despite the promising results on global-style migration and the favorable efficiency, these methods overlook the fine-grained local details and thus lead to coarse results. To account for local context in the exemplar-guided image generation, recent works [45,62,64] employ per-point attention mechanism to model spacial correspondence between input and reference images. Nevertheless, hindered by the quadratic time and memory complexity, such dense matching operations, unfortunately, again limit themselves to coarse scales, making them difficult to capture fine-grained details in reference images.

To alleviate this issue, the works of [63,69] propose to fix the number of feature points that each query position focuses on, thereby achieving a linear-complexity model. The reason behind such a design lies in that, each query position in target images is, in reality, independent to and dissimilar to most points in the reference. This rationale, in turn, implies that the matching between the target and reference image is intrinsically sparse.

Unfortunately, such the static number of correspondences, in many cases, fail to capture the dynamic nature of matching. As shown in Fig. 1, different queries may end up having different numbers of necessary matching: in Fig. 1 Left (a) and (b), due to the scale variations, the highlighted query location in the target image corresponds to different numbers of locations in the reference; in Fig. 1 Left (c), however, the query has no correspondence at all, meaning that all points in the exemplar would be negative samples.

These facts motivate us to explore a more sophisticated technique to account for the dynamic matching inherent to the exemplar-guided image generation, while maintaining the sparsity to ensure computational efficiency. To this end, we propose a novel Transformer-based model, termed *Dynamic Sparse Transformer (DynaST)*. The heart in DynaST is the dynamic sparse attention module in contrast to previous dense and static ones. Specifically, since the adopted attention strategy is sparse, a large number of potential matching candidates are dismissed. To alleviate this issue, we are inspired by the architecture of Transformer [51] and back up the proposed attention strategy with a multi-layer structure, which enables DynaST to explore and evolve matching results in a cascade manner. As shown in Fig. 1 Right, we visualize the matching results by warping exemplar images using attention maps in each layer and observe that with the feature propagation in the multi-layer structure of Transformer, the model produces finer matching results, especially in local details such as faces and edges. To refine matching results progressively, each current matching candidate would pass through a differentiable and learnable attention link pruning unit in each layer, to predict whether it is an irrelevant correspondence.

Consequently, (1) such dynamic pruning manner encourages more precise and cleaner matching results; (2) due to the effective higher-order dependency modeling capability of Transformer, DynaST is competent in aggregating relevant features and synthesizing high-quality outputs; and (3) sparse attention in DynaST guarantees the efficiency of even full-resolution matching construction. Moreover, we introduce a unified training objective, so that DynaST is readily applicable for universal exemplar-guided image generation under both supervised and unsupervised settings.

We conduct extensive evaluations on three challenging tasks: pose-guided person image generation, edge-based face synthesis, and undistorted image style transfer. In all experiments, DynaST outperforms state-of-the-art exemplar-guided image generation models significantly, by up to 36.7% in quantitative metrics, and achieves near-real-time inference efficiency, with more than 2× speedup compared with previous state-of-the-art full-resolution matching solutions.

2   Related Works

2.1   Exemplar-Guided Image Generation

Exemplar-guided image generation has recently emerged as a popular task in the computer vision community. Park *et al.* [37] propose spatially-adaptive normalization (SPADE) module that generates normalization parameters based on given semantic information. In their work, exemplar images are fed to a VAE to encode the overall style and appearance, which guide the following generation procedure. Nevertheless, it is difficult to migrate local details in exemplar images due to such global transformation. Similar drawback also exists in works like [1,20,36,46,52,67,71]. To enhance the generation of local textures, Ren *et al.* [40] introduce a neural texture extraction and distribution module. Recent attempts have also been made to introduce point-wise attention to exemplar-guided image generation and achieved superior results. For instance, Zhang *et al.* [64] propose CoCosNet to learn the correspondence between input semantics and reference images. Zhan *et al.* [62] use unbalanced optimal transport to achieve the same goal. However, without any sparse mechanism, the quadratically increasing memory cost prevents these methods from learning fine-grained correspondence, which is important for synthesizing high-quality images. Although Zhou *et al.* [69] propose CoCosNet-v2 that is capable to learn full-resolution correspondence, the iterative global searching process by GRU makes a negative influence on the efficiency. DynaST in this paper brings the best of two worlds: it establishes matching at fine scales based on the dynamic sparse mechanism, which can generate images with high-quality local details while maintaining high efficiency simultaneously.

2.2   Image-Wise Matching

Given a pair of images, image matching such as [17,19,27,31,41,43,49] aims to find pixel-wise correspondence leveraging local features, which is a fundamental problem in computer vision and is one related field to exemplar-guided image generation in this paper. The key difference is that cross-domain matching establishment, semantic map to the exemplar image, is required in the image synthesis problem unlike matching between two highly correlated images. This is also one major difference between the general cross-domain exemplar-guided image generation and reference-based image super-resolution [18,34,56,66].

2.3   Efficient Transformer

The full token-wise attention operation in standard Transformer [9,51] poses high requirements on memory and significantly increases computational cost. Thus, a lot of works are devoted to designing efficient attention mechanisms for Transformer or by extension, graph-based methods [57,58]. On the one hand, some works rely on heuristic strategies to lead a current token only focus on those in a certain local context [3,4,6,26]. Recently, more strategies based on properties of images to boost the efficiency in vision Transformer are explored [39,60]. On the other hand, random sampling based Informer [68], locality sensitive hashing based Reformer [22], and approximated Softmax based Performer [5] achieve lower complexity with a fine theoretical guarantee. Wang *et al.* [53] only involve tokens with top K attention scores for feature aggregation. Similar strategy is also adopted in [44,63,69]. Although effective, it is not flexible enough to fix the number of attentive tokens, which fails to model the complex and changeable matching patterns in practice. Different from all these methods, the sparse mechanism in the attention module of this paper is based on prior knowledge in image matching, targeting at exemplar-guided image generation.

3   Dynamic Sparse Transformer

In this section, we illustrate details of the proposed DynaST model for exemplar-guided image generation. The overview of DynaST is shown in Fig. 2. DynaST

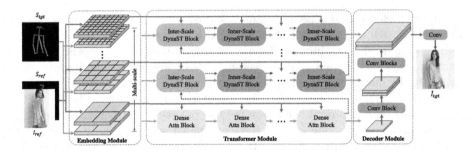

Fig. 2. DynaST overview. Solid arrows denote flows of features and dotted ones denote inheritance of attention maps. Yellow, green, and blue blocks in the middle adopt dense attention, inter-scale sparse attention, and inner-scale sparse attention respectively. (Color figure online)

takes three images as input: a reference image I_{ref}, a corresponding semantic map S_{ref} of I_{ref} (*e.g.*, a pose image or an edge map), and a target semantic map S_{tgt}. It aims at synthesizing the image $\hat{I}_{tgt}$ with the target semantic information specified in S_{tgt} and the appearance as well as the style in I_{ref}.

The proposed DynaST consists of three parts. The first is an *embedding module* (Sect. 3.1), which is established by a set of multi-scale layers to extract and aggregate features at different levels. The second is a *Transformer module* (Sect. 3.2), which restores features of target images with features of the semantic map as target and features of reference information as memory. The last one is a lightweight *decoder module* to synthesize final images, where the multi-scale features generated by the Transformer module are the input. The training objectives and supervised signals for the pipeline are described in Sect. 3.3.

3.1 Embedding Module

Given an input semantic image S_{tgt} and a reference image I_{ref} along with its corresponding semantics S_{ref}, the embedding module produces a set of feature embedding, F_{tgt} and F_{ref}. DynaST adopts a hierarchical patch embedding module as a multi-scale generative model, to enable the scale-wise cascaded matching process. The proposed embedding module is utilized to obtain rich features and contextual representation. Besides, the position embedding is also included to make the network aware of the positional information in the following matching process. Specifically, we use two independent sets of linear transformations: E_{tgt}^i and E_{ref}^i, to obtain multi-scale patch embedding for target semantic map S_{tgt} as well as reference information I_{ref} and S_{ref}, where i denotes the embedding of the i-th scale with patch size $2^i \times 2^i$. The features at the i-th scale, F_{tgt}^i and F_{ref}^i, can then be written as:

$$
\begin{aligned}
F_{tgt}^i &= \mathrm{X}([E_{tgt}^j(S_{tgt})' | 0 < j < M]), \\
F_{ref}^i &= \mathrm{Y}([E_{ref}^j([S_{ref}, I_{ref}])' | 0 < j < M]),
\end{aligned}
\tag{1}
$$

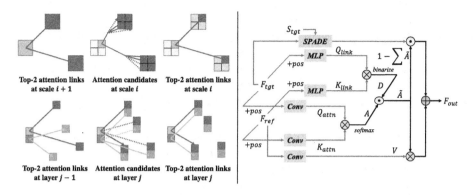

Fig. 3. Left: Intuition of inter-scale (up) and inner scale (bottom) sparse attention mechanism. Here $k = 2$ is used for illustration. The yellow tokens represent queries in the target feature map and the blue ones represent tokens in the reference map. The dotted lines denote links masked by dynamic attention pruning. Right: Illustration of dynamic pruning and feature aggregation in DynaST blocks. (Color figure online)

where $0 \leq i < M$, M is the number of scales, notation $'$ denotes the bilinear interpolation to unify the spacial dimension for scale i, $[\cdot]$ stands for the channel-wise concatenation, and X and Y are two MLPs consisting of two convolutional blocks for non-linear transformation. In other words, features of one specific scale aggregates multi-level patch embedding information, which provides rich features and contextual representations for the following matching and transformation steps. We also concatenate a learnable positional embedding to F_{tgt} and F_{ref} before computing their attention scores, denoted as $F_{tgt}^{i,pos}$ and $F_{ref}^{i,pos}$ at the i-th scale. The details can be found in the supplement.

3.2   Transformer Module

The Transformer module is built for exemplar-guided image generation, by a set of dynamic sparse Transformer blocks (DynaST Block) and dense attention blocks (Dense Attn Block). The construction of these blocks consists of four steps, which are attention map computation, dynamic attention pruning, feature aggregation, and non-linear transformation. In the attention map computation step, Dense Attn Blocks are used at the coarsest scale to construct matching at a low resolution, while DynaST Blocks are used at higher levels to infer high-resolution matching based on previous attention results. At each higher scale, the first DynaST block adopts inter-scale sparse attention, and is therefore named as Inter-Scale DynaST Block; the remaining blocks adopt inner-scale sparse attention and are thus named as Inner-Scale DynaST Blocks. We will give details in the following sections.

Attention Map Computation. Attention map computation is crucial to guide each target semantic feature point to focus on correct positions on reference feature maps and further restore features of target images. In DynaST, at scale

i, the j-th DynaST block takes feature map produced by the $j-1$-th block $F^i_{tgt,j-1}$, the reference feature map F^i_{ref}, and attention map of previous DynaST block as input to compute attention scores. Note that $F^i_{tgt,0}$ is defined as F^i_{tgt} from the multi-scale patch embedding layer. At the coarsest level $i = M - 1$, Dense Attn Blocks with vanilla attention are used to derive attention scores:

$$A^{M-1}_j = \text{softmax}(\tau\alpha(\overline{F^{M-1,pos}_{tgt,j-1}})\beta(\overline{F^{M-1,pos}_{ref}})^\top), \qquad (2)$$

where α and β are implemented as two 1×1 convolutional kernels, τ is a hyper-parameter controlling the smoothness of attention distribution, and $\overline{x}$ denotes channel-independent instance normalization [50].

Then, at finer level $i < M - 1$, for the first DynaST block ($j = 1$), inter-scale sparse attention is proposed to compute the attention map at this layer:

$$A^i_1 = \text{softmax}(\tau\alpha(\overline{F^{i,pos}_{tgt,0}})\beta(\text{TopK}_{A^{i+1}_{-1}}(\overline{F^{i,pos}_{ref}}))^\top), \qquad (3)$$

where $\text{TopK}_{A^{i+1}_{-1}}$ denotes that matching candidates for a point in $F^{i,pos}_{tgt,0}$ come from those with top k large scores in the attention map of the last block in the previous scale. Note that a point in the previous scale would be divided into four in the current one. Therefore, there are $k \times 4$ matching candidates for attention map computation at this layer.

For the following blocks ($j > 1$), inner-scale sparse attention is performed to refine the attention matching at the current scale, based on the prior knowledge that the matching offset in a local area is likely to be the same [2]:

$$A^i_j = \text{softmax}(\tau\alpha(\overline{F^{i,pos}_{tgt,j-1}})\beta(\mathcal{N}(\text{TopK}_{A^i_{j-1}}(\overline{F^{i,pos}_{ref}})))^\top), \qquad (4)$$

where notation $\mathcal{N}$ is the operation to derive the points with the same matching offsets as the neighbors for one target point. For example, for one target point at the current layer, it firstly finds matching results of its right neighbor and then takes the left neighbors of these points as candidates. We define the up, bottom, left, and right points of one position plus the current point itself as the neighboring points in this paper. In this way, the number of matching candidates for each target point in the inner-scale sparse attention layer is $k \times 5$. The intuitions of inter/inner-scale sparse attention layer are illustrated in Fig. 3 Left.

Dynamic Attention Pruning. Considering that not all the matching candidates in attention modules are necessary for feature aggregation, we propose dynamic attention pruning to decide whether an attention link between a point in the target map and that in the reference map is useful. To this end, we use two MLPs Ω and Φ to transform F_{tgt} and F_{ref} into a common feature space. Then, a sign function is applied on the inner product of the transformed results to obtain the decision D for each attention link:

$$P^i_j = \Omega(F^i_{tgt,j-1})\Phi(F^i_{ref})^\top$$
$$D^i_j = \begin{cases} 1, & P^i_j > 0, \\ 0, & otherwise \end{cases} \qquad (5)$$

Note that in the above function, the sign function introduces obstacles for gradient-based optimization in training. To tackle this issue, we alternatively take gradients from *sigmoid* function during the backward propagation:

$$\frac{\mathrm{d}D_j^i}{\mathrm{d}P_j^i} = \frac{\exp(-P_j^i)}{(1 + \exp(-P_j^i))^2}. \tag{6}$$

Thus, attention maps after the dynamic pruning operation are derived by:

$$\tilde{A}_j^i = D_j^i \odot A_j^i, \tag{7}$$

where $\odot$ represents the element-wise multiplication.

Feature Aggregation. One straightforward way to conduct feature aggregation is to use pruned attention matrix $\tilde{A}_j^i$ directly to perform weighted summation over reference features. However, due to the pruning operation, the sum of attention weights for one target point to all reference feature points is no longer guaranteed to be 1, which would result in unbalanced magnitudes in feature aggregation. For example, in the most extreme case, a target point would be untraceable in the reference image and attention decisions would be all 0 for this point. Then the aggregated features for this point would also be all 0, which impedes the synthesis of a plausible image. To alleviate this problem, we use features restored by a SPADE block SP [37] to compensate for the masked part by dynamic attention pruning:

$$F_{out} = (1 - \sum \tilde{A}_j^i) \odot SP(F_{tgt,j-1}^i, S_{tgt}) + \tilde{A}_j^i \eta(F_{ref}^i),$$
$${F_{tgt,j}^i}' = \mathrm{Norm}(F_{out} + F_{tgt,j-1}^i), \tag{8}$$

where the summation is along the dimension of reference feature points, η is another 1×1 convolutional kernel, and $Norm$ denotes the layer normalization. Key steps for feature aggregation in DynaST blocks are shown in Fig. 3 Right.

Non-linear Transformation. Finally, following the standard Transformer architecture, a residual block is added at the end of the DynaST block for non-linear transformation:

$$F_{tgt,j}^i = \mathrm{Norm}({F_{tgt,j}^i}' + \mathrm{Conv}(\mathrm{ReLU}(\mathrm{Conv}({F_{tgt,j}^i}')))). \tag{9}$$

3.3   Training Objectives

DynaST is a universal framework for exemplar-guided image generation, which is compatible for objectives of both supervised and unsupervised tasks. The overall training objective consists of two parts by default: task-specific loss $\mathcal{L}_t$ and matching loss $\mathcal{L}_m$.

Task-Specific Loss. Firstly, task-specific loss $\mathcal{L}_t$ targets at the task itself and is flexible for different forms of loss functions in our model. Typically, for supervised tasks like pose-guided person image generation, the objective is defined as the MSE between generated images $\hat{I}_{tgt}$ and ground truths I_{tgt} in original image space and perceptual feature space plus an adversarial loss term:

$$\mathcal{L}_t = \|\hat{I}_{tgt} - I_{tgt}\|_2^2 + \sum_i \lambda_i \|\phi_i(\hat{I}_{tgt}) - \phi_i(I_{tgt})\|_2^2 + \tag{10}$$

$$\lambda_{adv} \max\{\log \text{Dis}(S_{tgt}, I_{tgt}) + \log(1 - \text{Dis}(S_{tgt}, \hat{I}_{tgt}))\},$$

where ϕ denotes a pretrained feature extractor (e.g., VGG-19), subscript i specifies which layer features come from, λ_i controls the weight of each layer, Dis represents a discriminator to be trained alternatively with the generator, and λ_{adv} is the weight of the adversarial term [1]. For another example, style transfer is an unsupervised task, whose loss function can be written as:

$$\mathcal{L}_t = l_c + \lambda_s l_s, \quad l_c = \|\phi_{4\_1}(I_{cs}) - \phi_{4\_1}(I_c)\|_2^2$$

$$l_s = \sum_{i=1}^{4} (\|\mu(\phi_{i\_1}(I_{cs})) - \mu(\phi_{i\_1}(I_s)))\|_2^2 + \|\sigma(\phi_{i\_1}(I_{cs})) - \sigma(\phi_{i\_1}(I_s)))\|_2^2, \tag{11}$$

where ϕ_{x_1} aims to extract features of ReLU_x-1 layer of a VGG-19 network pretrained on ImageNet [7], and μ and σ denote mean and standard deviation of each feature channel respectively [13,21]. In DynaST, content images I_c and style images I_s are input to the embedding modules E_{tgt} and E_{ref} respectively, and style transfer images I_{cs} are the framework output $\hat{I}_{tgt}$.

Matching Loss. To provide a more direct supervision signal for matching modules and dynamic pruning modules to produce proper attention maps, we introduce a matching loss $\mathcal{L}_m$ that uses the output attention maps to warp the reference images and measures the task-specific loss produced by the warped images. To be specific, we denote the correlation maps, results of Eq. 2, 3, and 4 before $softmax$, by C_j^i. The warp matrices W_j^i are derived taking attention decision D_j^i into consideration:

$$W_j^i = \frac{D_j^i \odot \exp(C_j^i)}{\sum\{D_j^i \odot \exp(C_j^i)\} + \epsilon}, \tag{12}$$

where the summation is over reference feature points and ϵ is a small constant for numerical stability. Then, the warped reference images are derived by:

$$\hat{I}_{warp,j}^i = W_j^i I_{ref}', \tag{13}$$

where I_{ref}' denotes the resized version of the reference images to keep dimension scales of W_j^i and I_{ref} the same. Matching loss is defined by the MSE:

$$\mathcal{L}_m = \sum_i \sum_j \|\hat{I}_{warp,j}^i - I_{tgt}'\|_2^2. \tag{14}$$

Finally, the overall objective is given by a weighted sum of $\mathcal{L}_t$ and $\mathcal{L}_m$:

$$\mathcal{L} = \mathcal{L}_t + \lambda_m \mathcal{L}_m, \tag{15}$$

with λ_m controlling the weight of term $\mathcal{L}_m$.

Fig. 4. Comparisons with state-of-the-art exemplar-guided image generation methods on *DeepFashion* and *CelebA-HQ* datasets.

4 Experiments

Implementation Details. For all the experiments, DynaST are trained under 256×256 resolution. 4 different scales are set to be 32, 64, 128 and 256, where dimensions of the corresponding feature channel are 512, 256, 128 and 64 respectively. Each level of Transformation module is built by 2 blocks, where on the coarsest scale there are 2 *Dense Attn Blocks* and on each upper level there is 1 *Inter-Scale DynaST Block* and 1 *Inner-Scale DynaST Block*. For supervised tasks, hyper-parameters λ_m and λ_{adv} are set as 100 and 10, respectively. For style transfer, λ_m and λ_s are set as 1 and 3, respectively. Matching loss defined in Eq. 14 is not adopted here. The smoothness parameter τ is set as 100 and $k = 4$ is used by default when selecting attention candidates. Training is done on 8 T V100 GPUs with batch size 32.

Datasets. For the pose-guided person image generation and the edge-guided face generation tasks, *DeepFashion* [32] and *CelebA-HQ* [33] datasets are used. Splits of training and validation sets and policies on retrieval of input-exemplar image pairs are consistent with those in [64]. For style transfer, following the common settings, *MS-COCO* [29] and *WikiArt* [38] are adopted as content and style image sets for training respectively. During training, all the images are resized to 512×512 and then randomly cropped to 256×256. Inference results under 512×512 resolution are reported in this paper.

4.1 Supervised Tasks

Comparison with Other Methods. On the pose-guided person image generation and the edge-guided face generation problems, we mainly compare our method with three state-of-the-art attention-based exemplar-guided image generation methods, including UNITE [62], CoCosNet [64], and CoCosNet-v2 [69]. The attention matching in UNITE and CoCosNet is limited on a relatively coarse scale (64×64) by the quadratic memory cost of the dense attention operation. Thus, as shown in Fig. 4, some detailed textures are not good enough, *e.g.*, cloth and face details in the 1st row of comparisons on *DeepFashion* and the

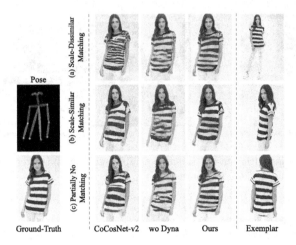

Fig. 5. Challenging scenarios involving (a) scale-dissimilar matching, (b) scale-similar matching, and (c) partially no matching, where the static matching scheme inevitably fails. The proposed DynaST, thanks to its dedicated scheme for handling dynamic numbers of matching, may well handle all such cases and yield visually plausible results.

beard in the 2nd row of comparisons on *CelebA-HQ*. CoCosNet-v2 leverages the Conv-GRU module to predict correspondence at fine scales, where noisy correspondences would be added easily due to the large search space under high resolutions. And there is no pruning to mask irrelevant matching in CoCosNet-v2, which may lead to some artifacts, *e.g.*, straps in the 2nd row of comparisons on *DeepFashion*. As shown in the last column of each example, our method addresses these problems successfully with the proposed dynamic sparse attention based Transformer model, which generates more high-quality results.

Notably, one major difference between our DynaST and CoCosNet-v2 is that DynaST uses a dynamic number of matching points for feature aggregation, while CoCosNet-v2 only considers a fixed number of candidates, which fails to account for the dynamic property of matching in different cases. One illustrative example is shown in Fig. 5, where results by CoCosNet-v2 are less robust to cases when scales of input and exemplar are different, since one location must select a fixed number of matching points. When there are less informative correspondences, noises are inevitably introduced. By comparison, dynamic pruning involved in DynaST is more competent to handle such scale variation robustly.

To further illustrate the advantage of our method on matching construction, we use the attention matrix derived by each method to warp the exemplar image and report the warped results. As shown in Fig. 6, results by UNITE and CoCosNet are blurry due to low-resolution matching, and results by CoCosNet-v2 contain too much noise. Compared with the above methods, our method can generate matching under the full resolution with the best quality.

Quantitatively, leveraging the paired samples in *DeepFashion* dataset, we compare warped results with ground-truth images and show the average L1 loss,

Table 1. Left: Quantitative metrics on the quality and efficiency of matching establishment. Results are measured by comparing warped results with the ground truth on DeepFashion dataset. Running time for one sample ($\times 10^{-2}$ s) is shown here. Right: Quantitative metrics of image quality on two datasets.

Method	DeepFashion				DeepFashion		CelebA-HQ	
	L1↓	PSNR↑	SSIM↑	Time↓	FID↓	SWD↓	FID↓	SWD↓
Pix2PixHD [54]	–	–	–	–	25.2	16.4	62.7	43.3
SPADE [37]	–	–	–	–	36.2	27.8	31.5	26.9
MUNIT [14]	–	–	–	–	74.0	46.2	56.8	40.8
EGSC-IT [35]	–	–	–	–	29.0	39.1	29.5	23.8
UNITE [62]	13.1	<u>16.7</u>	<u>13.2</u>	14.9	13.1	<u>16.7</u>	<u>13.2</u>	14.9
CoCosNet [64]	0.067	18.48	0.80	<u>11.5</u>	14.4	17.2	14.3	15.2
CoCosNet-v2 [69]	<u>0.064</u>	18.24	0.80	21.7	<u>13.0</u>	<u>16.7</u>	<u>13.2</u>	<u>14.0</u>
Ours-32	0.077	18.12	0.73	5.53	8.55	15.4	16.0	17.8
Ours-64	0.063	18.22	0.78	7.24	8.50	12.8	13.1	13.1
Ours-128	0.061	19.13	0.82	8.43	8.41	12.9	12.3	12.7
Ours wo Inner	0.064	18.30	0.82	9.45	8.88	12.0	14.7	17.2
Ours wo Dyna	0.063	19.04	0.81	9.26	9.32	21.8	15.3	19.0
Ours	**0.054**	**19.25**	**0.83**	9.63	**8.36**	**11.8**	**12.0**	**12.4**

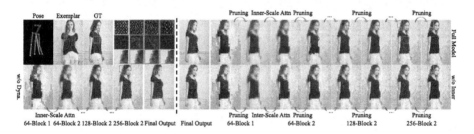

Fig. 6. Ablation studies on dynamic pruning and inner-scale sparse attention. Intermediate warping results using attention maps are visualized to demonstrate the evolving of input-exemplar matching. Zoom-in for better visualization.

PSNR, and SSIM scores in Table 1 Left, where our method performs the best. The time needed for generating matching and synthesizing results for one sample is also included, measured on a single Nvidia 3090 GPU by averaging 1000 samples. Dense attention mechanism and iterative solving of optimal transport problem in CoCosNet and UNITE respectively leave a high computational burden. Recurrent prediction under full resolution in CoCosNet-v2 increases the latency further. Different from previous approaches, the efficient dynamic sparse attention operation in DynaST makes it achieve the most satisfactory computational speed while generating the best matching results impressively.

We report the widely-used FID [12] and SWD [23] metrics to reflect the distance of feature distributions between generated samples and real images, following Zhang *et al.* [64] in Table 1 Right. Our method outperforms previous

Table 2. Quantitative metrics of semantic (Sem.), color (Col.), and texture (Tex.) consistency on two datasets compared with state-of-the-art image synthesis methods.

Method	DeepFashion			CelebA-HQ		
	Sem.↑	Col.↑	Tex.↑	Sem.↑	Col.↑	Tex.↑
Pix2PixHD	0.943	NA	NA	0.914	NA	NA
SPADE	0.936	0.943	0.904	0.922	0.955	0.927
MUNIT	0.910	0.893	0.861	0.848	0.939	0.884
EGSC-IT	0.942	0.945	0.916	0.915	0.965	0.942
UNITE	0.957	0.973	0.930	**0.952**	0.966	0.950
CoCosNet	<u>0.968</u>	**0.982**	**0.958**	<u>0.949</u>	<u>0.977</u>	<u>0.958</u>
CoCosNet-v2	0.959	<u>0.974</u>	0.925	0.948	0.975	0.954
Ours	**0.975**	<u>0.974</u>	<u>0.937</u>	**0.952**	**0.980**	**0.959**

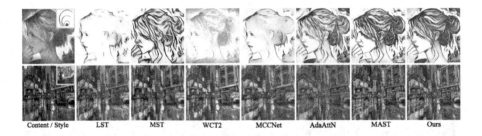

Fig. 7. Comparisons with state-of-the-art undistorted style transfer methods.

ones significantly under both datasets, suggesting the best quality of results. On the other hand, we show the semantic, color, and texture consistency in Table 2, also under the same setting as [64]. Results in Table 2 demonstrate that our method achieves competent semantic restoration and style migration performance.

Ablation Study. We conduct ablation studies on the two core ideas in this paper: dynamic pruning and sparse attention. Based on the full model, we (1) remove the dynamic pruning mechanism, with the corresponding result denoted as *Ours wo Dyna*; (2) replace all inner-scale DynaST blocks with inter-scale ones, denoted as *Ours wo Inner*; and (3) remove inter-scale sparse attention layers at different scales and only use attention up to a coarse scale instead of full resolution, denoted as *Ours-x*, where $x \in \{32, 64, 128\}$ representing the highest resolution used for matching. Evaluations on warped results and final results mentioned above are repeated using the resulting models. The quantitative results in Table 1 indicate that the incomplete models lead to inferior results.

Qualitatively, we visualize the intermediate warping results using attention maps in Fig. 6, to demonstrate how the matching results are evolving through multi-layer dynamic pruning and inter/inner-scale attention. For one thing, dynamic pruning is capable of suppressing noisy matching and helps generate a clearer view. The example in Fig. 5 also demonstrates the importance of dynamic pruning for the robustness to handle scale variation. For another, replacing inner-scale DynaST blocks may lead to some artifacts such as checkerboard, as shown in Fig. 6, due to the missing of local refinement. Removing them would have negative impacts on local details like areas of hair and cloth.

4.2 Undistorted Image Style Transfer

The full resolution matching mechanism makes DynaST well suitable to generate undistorted style transfer results. In order to demonstrate such advantage, we compare our DynaST with 6 state-of-the-art style transfer methods with the same or similar goal, including LST [28], MST [65], WCT2 [59], MCCNet [8], AdaAttN [30], and MAST [15]. As shown in Fig. 7, compared with the photorealistic style transfer method WCT2, our results migrate global and

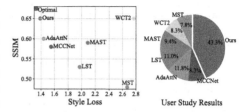

Fig. 8. Left: visualization of content SSIM and style loss of different undistorted style transfer methods; Right: distribution of user preference.

local style patterns better, while compared with other methods, our results preserve texture details of content images best, *e.g.*, hair in the 1st row. In particular, our method is capable of dealing with complicated scenes like the 2nd row without distortion, where other methods fail. We also visualize *SSIM scores* with content images and *style loss* against style images in Fig. 8 Left, using the test dataset in [59]. It is demonstrated that DynaST, as the first full-resolution matching based method in style transfer, can even achieve comparable content preservation ability with photorealistic style transfer methods, while significantly improving the stylization effects. Furthermore, we conduct a *user study* with the same test dataset to reflect user preference. There are 155 users involved and 12 content-style pairs are shown to each randomly. They are invited to select their favorite one for each pair among results by the 7 methods. We receive 1860 votes in total and the preference distribution is shown in Fig. 8 Right, where our preference score outperforms others significantly. In this way, both qualitative and quantitative comparisons demonstrate the superiority of DynaST.

5 Conclusion

In this paper, we introduce a novel multi-scale Transformer model, DynaST, to account for dynamic sparse attention and construct fine-level matching in exemplar-guided image generation tasks. DynaST comes with a unified training objective, making it a versatile model for various exemplar-guided image generation tasks under both supervised and unsupervised settings. Extensive evaluations on multiple benchmarks demonstrate that the proposed DynaST outperforms previous state-of-the-art methods, on both the matching quality and the running efficiency.

Acknowledgement. This project is supported by AI Singapore (Award No.: AISG2-RP-2021-023) and NUS Faculty Research Committee Grant (WBS: A-0009440-00-00).

References

1. AlBahar, B., Huang, J.B.: Guided image-to-image translation with bi-directional feature transformation. In: Proceedings of the IEEE/CVF International Conference on Computer Vision, pp. 9016–9025 (2019)
2. Barnes, C., Shechtman, E., Finkelstein, A., Goldman, D.B.: PatchMatch: a randomized correspondence algorithm for structural image editing. ACM Trans. Graph. **28**(3), 24 (2009)
3. Beltagy, I., Peters, M.E., Cohan, A.: Longformer: the long-document transformer. arXiv preprint arXiv:2004.05150 (2020)
4. Child, R., Gray, S., Radford, A., Sutskever, I.: Generating long sequences with sparse transformers. arXiv preprint arXiv:1904.10509 (2019)
5. Choromanski, K., et al.: Rethinking attention with performers (2021)
6. Dai, Z., Yang, Z., Yang, Y., Carbonell, J.G., Le, Q., Salakhutdinov, R.: Transformer-XL: attentive language models beyond a fixed-length context. In: Proceedings of the 57th Annual Meeting of the Association for Computational Linguistics, pp. 2978–2988 (2019)
7. Deng, J., Dong, W., Socher, R., Li, L.J., Li, K., Fei-Fei, L.: ImageNet: a large-scale hierarchical image database. In: 2009 IEEE Conference on Computer Vision and Pattern Recognition, pp. 248–255. IEEE (2009)
8. Deng, Y., Tang, F., Dong, W., Huang, H., Ma, C., Xu, C.: Arbitrary video style transfer via multi-channel correlation. arXiv preprint arXiv:2009.08003 (2020)
9. Dosovitskiy, A., et al.: An image is worth 16×16 words: transformers for image recognition at scale. arXiv preprint arXiv:2010.11929 (2020)
10. Esser, P., Rombach, R., Ommer, B.: Taming transformers for high-resolution image synthesis. In: Proceedings of the IEEE/CVF Conference on Computer Vision and Pattern Recognition, pp. 12873–12883 (2021)
11. Gao, C., Liu, Q., Xu, Q., Wang, L., Liu, J., Zou, C.: SketchyCOCO: image generation from freehand scene sketches. In: Proceedings of the IEEE/CVF Conference on Computer Vision and Pattern Recognition, pp. 5174–5183 (2020)
12. Heusel, M., Ramsauer, H., Unterthiner, T., Nessler, B., Hochreiter, S.: GANs trained by a two time-scale update rule converge to a local Nash equilibrium. In: Advances in Neural Information Processing Systems, vol. 30 (2017)
13. Huang, X., Belongie, S.: Arbitrary style transfer in real-time with adaptive instance normalization. In: Proceedings of the IEEE International Conference on Computer Vision, pp. 1501–1510 (2017)
14. Huang, X., Liu, M.-Y., Belongie, S., Kautz, J.: Multimodal unsupervised image-to-image translation. In: Ferrari, V., Hebert, M., Sminchisescu, C., Weiss, Y. (eds.) ECCV 2018. LNCS, vol. 11207, pp. 179–196. Springer, Cham (2018). https://doi.org/10.1007/978-3-030-01219-9_11
15. Huo, J., et al.: Manifold alignment for semantically aligned style transfer. In: Proceedings of the IEEE/CVF International Conference on Computer Vision, pp. 14861–14869 (2021)

16. Isola, P., Zhu, J.Y., Zhou, T., Efros, A.A.: Image-to-image translation with conditional adversarial networks. In: Proceedings of the IEEE Conference on Computer Vision and Pattern Recognition, pp. 1125–1134 (2017)
17. Jiang, W., Trulls, E., Hosang, J., Tagliasacchi, A., Yi, K.M.: COTR: correspondence transformer for matching across images. In: Proceedings of the IEEE/CVF International Conference on Computer Vision, pp. 6207–6217 (2021)
18. Jiang, Y., Chan, K.C., Wang, X., Loy, C.C., Liu, Z.: Robust reference-based super-resolution via C2-matching. In: Proceedings of the IEEE/CVF Conference on Computer Vision and Pattern Recognition, pp. 2103–2112 (2021)
19. Jin, Y., et al.: Image matching across wide baselines: from paper to practice. Int. J. Comput. Vis. **129**(2), 517–547 (2021). https://doi.org/10.1007/s11263-020-01385-0
20. Jing, Y., et al.: Dynamic instance normalization for arbitrary style transfer. In: Proceedings of the AAAI Conference on Artificial Intelligence, vol. 34, pp. 4369–4376 (2020)
21. Johnson, J., Alahi, A., Fei-Fei, L.: Perceptual losses for real-time style transfer and super-resolution. In: Leibe, B., Matas, J., Sebe, N., Welling, M. (eds.) ECCV 2016. LNCS, vol. 9906, pp. 694–711. Springer, Cham (2016). https://doi.org/10.1007/978-3-319-46475-6_43
22. Kitaev, N., Kaiser, Ł., Levskaya, A.: Reformer: the efficient transformer. arXiv preprint arXiv:2001.04451 (2020)
23. Lee, C.Y., Batra, T., Baig, M.H., Ulbricht, D.: Sliced wasserstein discrepancy for unsupervised domain adaptation. In: Proceedings of the IEEE/CVF Conference on Computer Vision and Pattern Recognition, pp. 10285–10295 (2019)
24. Lee, J., Kim, E., Lee, Y., Kim, D., Chang, J., Choo, J.: Reference-based sketch image colorization using augmented-self reference and dense semantic correspondence. In: Proceedings of the IEEE/CVF Conference on Computer Vision and Pattern Recognition, pp. 5801–5810 (2020)
25. Li, B., Zhao, F., Su, Z., Liang, X., Lai, Y.K., Rosin, P.L.: Example-based image colorization using locality consistent sparse representation. IEEE Trans. Image Process. **26**(11), 5188–5202 (2017)
26. Li, S., et al.: Enhancing the locality and breaking the memory bottleneck of transformer on time series forecasting. In: Advances in Neural Information Processing Systems, vol. 32, pp. 5243–5253 (2019)
27. Li, X., Han, K., Li, S., Prisacariu, V.: Dual-resolution correspondence networks. In: Advances in Neural Information Processing Systems, vol. 33, pp. 17346–17357 (2020)
28. Li, X., Liu, S., Kautz, J., Yang, M.H.: Learning linear transformations for fast image and video style transfer. In: Proceedings of the IEEE/CVF Conference on Computer Vision and Pattern Recognition, pp. 3809–3817 (2019)
29. Lin, T.-Y., et al.: Microsoft COCO: common objects in context. In: Fleet, D., Pajdla, T., Schiele, B., Tuytelaars, T. (eds.) ECCV 2014. LNCS, vol. 8693, pp. 740–755. Springer, Cham (2014). https://doi.org/10.1007/978-3-319-10602-1_48
30. Liu, S., et al.: AdaAttN: revisit attention mechanism in arbitrary neural style transfer. In: Proceedings of the IEEE/CVF International Conference on Computer Vision, pp. 6649–6658 (2021)
31. Liu, X., et al.: Extremely dense point correspondences using a learned feature descriptor. In: Proceedings of the IEEE/CVF Conference on Computer Vision and Pattern Recognition, pp. 4847–4856 (2020)

32. Liu, Z., Luo, P., Qiu, S., Wang, X., Tang, X.: DeepFashion: powering robust clothes recognition and retrieval with rich annotations. In: Proceedings of the IEEE Conference on Computer Vision and Pattern Recognition, pp. 1096–1104 (2016)
33. Liu, Z., Luo, P., Wang, X., Tang, X.: Deep learning face attributes in the wild. In: Proceedings of the IEEE International Conference on Computer Vision, pp. 3730–3738 (2015)
34. Lu, L., Li, W., Tao, X., Lu, J., Jia, J.: MASA-SR: matching acceleration and spatial adaptation for reference-based image super-resolution. In: Proceedings of the IEEE/CVF Conference on Computer Vision and Pattern Recognition, pp. 6368–6377 (2021)
35. Ma, L., Jia, X., Georgoulis, S., Tuytelaars, T., Van Gool, L.: Exemplar guided unsupervised image-to-image translation with semantic consistency. arXiv preprint arXiv:1805.11145 (2018)
36. Ma, L., Jia, X., Sun, Q., Schiele, B., Tuytelaars, T., Van Gool, L.: Pose guided person image generation. In: Proceedings of the 31st International Conference on Neural Information Processing Systems, pp. 405–415 (2017)
37. Park, T., Liu, M.Y., Wang, T.C., Zhu, J.Y.: Semantic image synthesis with spatially-adaptive normalization. In: Proceedings of the IEEE/CVF Conference on Computer Vision and Pattern Recognition, pp. 2337–2346 (2019)
38. Phillips, F., Mackintosh, B.: Wiki Art Gallery Inc.: a case for critical thinking. Issues Account. Educ. **26**(3), 593–608 (2011)
39. Ren, S., Zhou, D., He, S., Feng, J., Wang, X.: Shunted self-attention via multi-scale token aggregation. In: Proceedings of the IEEE/CVF Conference on Computer Vision and Pattern Recognition (2022)
40. Ren, Y., Fan, X., Li, G., Liu, S., Li, T.H.: Neural texture extraction and distribution for controllable person image synthesis. In: Proceedings of the IEEE/CVF Conference on Computer Vision and Pattern Recognition, pp. 13535–13544 (2022)
41. Sarlin, P.E., DeTone, D., Malisiewicz, T., Rabinovich, A.: SuperGlue: learning feature matching with graph neural networks. In: Proceedings of the IEEE/CVF Conference on Computer Vision and Pattern Recognition, pp. 4938–4947 (2020)
42. Song, L., Lu, Z., He, R., Sun, Z., Tan, T.: Geometry guided adversarial facial expression synthesis. In: Proceedings of the 26th ACM International Conference on Multimedia, pp. 627–635 (2018)
43. Sun, J., Shen, Z., Wang, Y., Bao, H., Zhou, X.: LoFTR: detector-free local feature matching with transformers. In: Proceedings of the IEEE/CVF Conference on Computer Vision and Pattern Recognition, pp. 8922–8931 (2021)
44. Tan, Z., et al.: Efficient semantic image synthesis via class-adaptive normalization. IEEE Trans. Pattern Anal. Mach. Intell. **44**(9), 4852–4866 (2022)
45. Tang, H., Bai, S., Torr, P., Sebe, N.: Bipartite graph reasoning GANs for person image generation (2020)
46. Tang, H., Bai, S., Zhang, L., Torr, P.H.S., Sebe, N.: XingGAN for person image generation. In: Vedaldi, A., Bischof, H., Brox, T., Frahm, J.-M. (eds.) ECCV 2020. LNCS, vol. 12370, pp. 717–734. Springer, Cham (2020). https://doi.org/10.1007/978-3-030-58595-2_43
47. Tang, H., Xu, D., Liu, G., Wang, W., Sebe, N., Yan, Y.: Cycle in cycle generative adversarial networks for keypoint-guided image generation. In: Proceedings of the 27th ACM International Conference on Multimedia, pp. 2052–2060 (2019)
48. Tang, H., Xu, D., Yan, Y., Torr, P.H., Sebe, N.: Local class-specific and global image-level generative adversarial networks for semantic-guided scene generation. In: Proceedings of the IEEE/CVF Conference on Computer Vision and Pattern Recognition, pp. 7870–7879 (2020)

49. Truong, P., Danelljan, M., Timofte, R.: GLU-Net: global-local universal network for dense flow and correspondences. In: Proceedings of the IEEE/CVF Conference on Computer Vision and Pattern Recognition, pp. 6258–6268 (2020)
50. Ulyanov, D., Vedaldi, A., Lempitsky, V.: Instance normalization: the missing ingredient for fast stylization. arXiv preprint arXiv:1607.08022 (2016)
51. Vaswani, A., et al.: Attention is all you need. In: Advances in Neural Information Processing Systems, pp. 5998–6008 (2017)
52. Wang, M., et al.: Example-guided style-consistent image synthesis from semantic labeling. In: Proceedings of the IEEE/CVF Conference on Computer Vision and Pattern Recognition, pp. 1495–1504 (2019)
53. Wang, P., et al.: KVT: k-NN attention for boosting vision transformers. arXiv preprint arXiv:2106.00515 (2021)
54. Wang, T.C., Liu, M.Y., Zhu, J.Y., Tao, A., Kautz, J., Catanzaro, B.: High-resolution image synthesis and semantic manipulation with conditional GANs. In: Proceedings of the IEEE Conference on Computer Vision and Pattern Recognition, pp. 8798–8807 (2018)
55. Wang, Y., Qi, L., Chen, Y.C., Zhang, X., Jia, J.: Image synthesis via semantic composition. In: Proceedings of the IEEE/CVF International Conference on Computer Vision, pp. 13749–13758 (2021)
56. Yang, F., Yang, H., Fu, J., Lu, H., Guo, B.: Learning texture transformer network for image super-resolution. In: Proceedings of the IEEE/CVF Conference on Computer Vision and Pattern Recognition, pp. 5791–5800 (2020)
57. Yang, Y., Feng, Z., Song, M., Wang, X.: Factorizable graph convolutional networks. In: Conference on Neural Information Processing Systems (2020)
58. Yang, Y., Qiu, J., Song, M., Tao, D., Wang, X.: Distilling knowledge from graph convolutional networks. In: Proceedings of the IEEE/CVF Conference on Computer Vision and Pattern Recognition (2020)
59. Yoo, J., Uh, Y., Chun, S., Kang, B., Ha, J.W.: Photorealistic style transfer via wavelet transforms. In: Proceedings of the IEEE/CVF International Conference on Computer Vision, pp. 9036–9045 (2019)
60. Yu, W., et al.: MetaFormer is actually what you need for vision. In: Proceedings of the IEEE/CVF Conference on Computer Vision and Pattern Recognition (2022)
61. Zakharov, E., Shysheya, A., Burkov, E., Lempitsky, V.: Few-shot adversarial learning of realistic neural talking head models. In: Proceedings of the IEEE/CVF International Conference on Computer Vision, pp. 9459–9468 (2019)
62. Zhan, F., et al.: Unbalanced feature transport for exemplar-based image translation. In: Proceedings of the IEEE/CVF Conference on Computer Vision and Pattern Recognition, pp. 15028–15038 (2021)
63. Zhan, F., et al.: Bi-level feature alignment for versatile image translation and manipulation. arXiv preprint arXiv:2107.03021 (2021)
64. Zhang, P., Zhang, B., Chen, D., Yuan, L., Wen, F.: Cross-domain correspondence learning for exemplar-based image translation. In: Proceedings of the IEEE/CVF Conference on Computer Vision and Pattern Recognition, pp. 5143–5153 (2020)
65. Zhang, Y., et al.: Multimodal style transfer via graph cuts. In: Proceedings of the IEEE/CVF International Conference on Computer Vision, pp. 5943–5951 (2019)
66. Zhang, Z., Wang, Z., Lin, Z., Qi, H.: Image super-resolution by neural texture transfer. In: Proceedings of the IEEE/CVF Conference on Computer Vision and Pattern Recognition, pp. 7982–7991 (2019)

67. Zheng, H., Liao, H., Chen, L., Xiong, W., Chen, T., Luo, J.: Example-guided image synthesis using masked spatial-channel attention and self-supervision. In: Vedaldi, A., Bischof, H., Brox, T., Frahm, J.-M. (eds.) ECCV 2020. LNCS, vol. 12359, pp. 422–439. Springer, Cham (2020). https://doi.org/10.1007/978-3-030-58568-6_25
68. Zhou, H., et al.: Informer: beyond efficient transformer for long sequence time-series forecasting. In: Proceedings of AAAI (2021)
69. Zhou, X., et al.: CoCosNet v2: full-resolution correspondence learning for image translation. In: Proceedings of the IEEE/CVF Conference on Computer Vision and Pattern Recognition, pp. 11465–11475 (2021)
70. Zhu, P., Abdal, R., Qin, Y., Wonka, P.: SEAN: image synthesis with semantic region-adaptive normalization. In: Proceedings of the IEEE/CVF Conference on Computer Vision and Pattern Recognition, pp. 5104–5113 (2020)
71. Zhu, Z., Huang, T., Shi, B., Yu, M., Wang, B., Bai, X.: Progressive pose attention transfer for person image generation. In: Proceedings of the IEEE/CVF Conference on Computer Vision and Pattern Recognition, pp. 2347–2356 (2019)

Multimodal Conditional Image Synthesis with Product-of-Experts GANs

Xun Huang$^{(\boxtimes)}$, Arun Mallya, Ting-Chun Wang, and Ming-Yu Liu

NVIDIA, California, USA
Xunh@nvidia.com

Abstract. Existing conditional image synthesis frameworks generate images based on user inputs in a single modality, such as text, segmentation, or sketch. They do not allow users to simultaneously use inputs in multiple modalities to control the image synthesis output. This reduces their practicality as multimodal inputs are more expressive and complement each other. To address this limitation, we propose the Product-of-Experts Generative Adversarial Networks (PoE-GAN) framework, which can synthesize images conditioned on multiple input modalities or any subset of them, even the empty set. We achieve this capability with a single trained model. PoE-GAN consists of a product-of-experts generator and a multimodal multiscale projection discriminator. Through our carefully designed training scheme, PoE-GAN learns to synthesize images with high quality and diversity. Besides advancing the state of the art in multimodal conditional image synthesis, PoE-GAN also outperforms the best existing unimodal conditional image synthesis approaches when tested in the unimodal setting. The project website is available at this link.

Keywords: Image synthesis · Multimodal learning · GAN

1 Introduction

Conditional image synthesis allows users to use their creative inputs to control the output of image synthesis methods. It has found applications in many content creation tools. Over the years, a variety of input modalities have been studied, mostly based on conditional GANs [11,19,27,34]. To this end, we have various single modality-to-image models. When the input modality is text, we have the *text-to-image* model [40,41,51,58,60,62,67]. When the input modality is a segmentation mask, we have the *segmentation-to-image* model [8,19,29,36,45, 54]. When the input modality is a sketch, we have the *sketch-to-image* model [4, 6,10,44].

Supplementary Information The online version contains supplementary material available at https://doi.org/10.1007/978-3-031-19787-1_6.

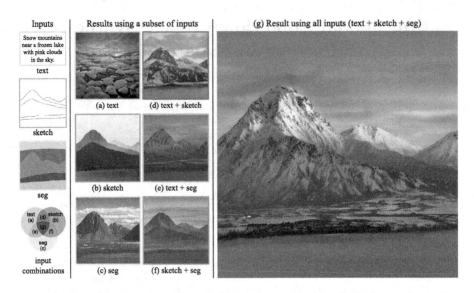

Fig. 1. Given conditional inputs in multiple modalities (the left column), our approach can synthesize images that satisfy all input conditions (the right column (g)) or an arbitrary subset of input conditions (the middle column (a)–(f)) with *a single model*.

However, different input modalities are best suited for conveying different types of conditioning information. For example, as seen in the first column of Fig. 1, segmentation makes it easy to define the coarse layout of semantic classes in an image—the relative locations of sky, cloud, mountain, and water regions. Sketch allows us to specify the structure and details within the same semantic region, such as individual mountain ridges. On the other hand, text is well-suited for modifying and describing objects or regions in the image, which cannot be achieved via segmentation or sketch, *e.g.*, '*frozen* lake' and '*pink* clouds' in Fig. 1. Despite this synergy among modalities, prior work has considered image generation conditioned on each modality as a distinct task and studied it in isolation. Existing models thus fail to utilize complementary information available in different modalities. Clearly, a conditional generative model that can combine input information from all available modalities would be of immense value.

Even though the benefits are enormous, the task of conditional image synthesis with multiple input modalities poses several challenges. First, it is unclear how to combine multiple modalities with different dimensions and structures in a single framework. Second, from a practical standpoint, the generator needs to handle missing modalities since it is cumbersome to ask users to provide every single modality all the time. This means that the generator should work well even when only a subset of modalities are provided. Lastly, conditional GANs are known to be susceptible to mode collapse [19,34], wherein the generator produces identical images when conditioned on the same inputs. This makes it difficult for the generator to produce diverse output images that capture the full conditional distribution when conditioned on an arbitrary set of modalities.

We present Product-of-Experts Generative Adversarial Networks (PoE-GAN), a framework that can generate images conditioned on any subset of the input modalities presented during training, as illustrated in Fig. 1 (a)-(g). This framework provides users unprecedented control, allowing them to specify exactly what they want using multiple complementary input modalities. When users provide no inputs, it falls back to an unconditional GAN model [2,11,20–23,32,39]. One key ingredient of our framework is a novel product-of-experts generator that can effectively fuse multimodal user inputs and handle missing modalities (Sect. 3.1). A novel hierarchical and multiscale latent representation leads to better usage of the structure in spatial modalities, such as segmentation and sketch (Sect. 3.2). Our model is trained with a multimodal projection discriminator (Sect. 3.4) together with contrastive losses for better input-output alignment. In addition, we adopt modality dropout for additional robustness to missing inputs (Sect. 3.5). Extensive experiment results show that PoE-GAN outperforms prior work in both multimodal and unimodal settings (Sect. 4), including state-of-the-art approaches specifically designed for a single modality. We also show that PoE-GAN can generate diverse images when conditioned on the same inputs.

2   Related Work

Image Synthesis. Our network architecture design is inspired by previous work in unconditional image synthesis. Our decoder employs some techniques proposed in StyleGAN [22] such as global modulation. Our latent space is constructed in a way similar to hierarchical variational auto-encoders (VAEs) [7,30,48,52]. While hierarchical VAEs encode the image itself to the latent space, our network encodes conditional information from different modalities into a unified latent space. Our discriminator design is inspired by the projection discriminator [33] and multiscale discriminators [29,54], which we extend to our multimodal setting.

Multimodal Image Synthesis. Prior work [25,46,49,50,53,56] has explored learning the joint distribution of multiple modalities using VAEs [24,42]. Some of them [25,53,56] use a product-of-experts inference network to approximate the posterior distribution. This is conceptually similar to how our generator combines information from multiple modalities. While their goal is to estimate the complete joint distribution, we focus on learning the image distribution conditioned on other modalities. Besides, our framework is based on GANs rather than VAEs and we perform experiments on high-resolution and large-scale datasets, unlike the above work. Recently, Xia et al.. [57] propose a GAN-based multimodal image synthesis method named TediGAN. Their method relies on a pretrained unconditional generator. However, such a generator is difficult to train on a complex dataset such as MS-COCO [26]. Concurrently, Zhang et al.. [65] propose a multimodal image synthesis method based on VQGAN. The way they combine different modalities is similar to our baseline using concatenation and modality dropout (Sect. 4.2). We will show that our product-of-experts generator design significantly improves upon this baseline. Another parallel work by Gafni*et al.* [9] propose a VQGAN model that is conditioned on both text and segmentation. They assume both modalities are always present at inference time and cannot deal with missing modalities.

3   Product-of-Experts GANs

Given a dataset of images x paired with M different input modalities $(y_1, y_2, ..., y_M)$, our goal is to train a single generative model that learns to capture the image distribution conditioned on an arbitrary subset of possible modalities $p(x|\mathcal{Y}), \forall \mathcal{Y} \subseteq \{y_1, y_2, ..., y_M\}$. In this paper, we consider four different modalities including text, semantic segmentation, sketch, and style reference. Note that our framework is general and can easily incorporate additional modalities.

Learning image distributions conditioned on *any subset* of M modalities is challenging because it requires a single generator to simultaneously model 2^M distributions. Of particular note, the generator needs to capture the unconditional image distribution $p(x)$ when $\mathcal{Y}$ is an empty set, and the unimodal conditional distributions $p(x|y_i), \forall i \in \{1, 2, ..., M\}$, such as the image distribution conditioned on text alone. These settings have been popular and widely studied in isolation, and we aim to bring them all under a unified framework.

3.1   Product-of-Experts Modeling

Our generator consists of a decoder G that deterministically maps a latent code z to an output image x, and a set of encoders that estimate the latent distribution $p(z|\mathcal{Y})$ conditioned on a set of modalities $\mathcal{Y}$. The conditional image distribution $p(x|\mathcal{Y})$ is implicitly defined as $x = G(z), z \sim p(z|\mathcal{Y})$. A naive approach would require us to train 2^M different encoder networks, one for each possible combination of modalities. This is highly parameter-inefficient and does not scale to a large number of modalities. Fortunately, if we assume all modalities $(y_1, ..., y_M)$ are conditionally independent given the image (x or equivalently z), i.e., $p(y_1, ..., y_M|z) = \prod_{i=1}^{M} p(y_i|z)$[1], we can prove that the distribution $p(z|\mathcal{Y})$ is proportional to a product of distributions:

$$p(z|\mathcal{Y}) = \frac{p(\mathcal{Y}|z)p(z)}{p(\mathcal{Y})} = \frac{p(z)}{p(\mathcal{Y})} \prod_{y_i \in \mathcal{Y}} p(y_i|z) = \frac{p(z)}{p(\mathcal{Y})} \prod_{y_i \in \mathcal{Y}} \frac{p(z|y_i)p(y_i)}{p(z)}$$

$$= \frac{\prod_{y_i \in \mathcal{Y}} p(z|y_i)}{(p(z))^{|\mathcal{Y}|-1}} \cdot \frac{\prod_{y_i \in \mathcal{Y}} p(y_i)}{p(\mathcal{Y})} \propto \frac{\prod_{y_i \in \mathcal{Y}} p(z|y_i)}{(p(z))^{|\mathcal{Y}|-1}} = p(z) \prod_{y_i \in \mathcal{Y}} \tilde{q}(z|y_i), \quad (1)$$

---

[1] The conditional independence assumption is sound in our setting since an image alone contains sufficient information to infer a modality independent of other modalities. For example, given an image, we do not need its caption to infer its segmentation.

where $\tilde{q}(z|y_i) \equiv \frac{p(z|y_i)}{p(z)}$. Dividing it by the normalization constant, we have

$$p(z|\mathcal{Y}) \propto p(z) \prod_{y_i \in \mathcal{Y}} q(z|y_i), q(z|y_i) = \frac{\tilde{q}(z|y_i)}{\int \tilde{q}(z|y_i)dz}, \quad (2)$$

where $q(z|y_i)$ is a latent distribution only dependent on a single modality y_i and $p(z)$ is the unconditional prior distribution. As a result, we can reduce the number of encoders from 2^M to M, with each encoder estimating the distribution $q(z|y_i)$ from a single modality[2]. This idea of combining several distributions ("*experts*") by multiplying them has been previously referred to as product-of-experts [16].

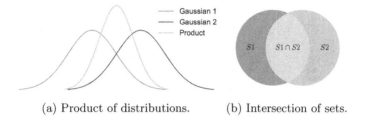

(a) Product of distributions.   (b) Intersection of sets.

Fig. 2. The product of distributions (a) is analogous to the intersection of sets (b).

Figures. 2a and 2b show that the product of distributions is intuitively analogous to the intersection of sets. The product distribution only has a high density in regions where all distributions have a relatively high density. Also, the product distribution is always narrower (of lower entropy) than the individual distributions, just like the intersection of sets is always smaller than the individual set. While each set poses a *hard* constraint, each individual distribution in a product represents a *soft* constraint, which is more amenable to neural network learning. In the multimodal conditional image synthesis setting, the model samples images from the prior $p(z)$ when no modalities are given. Each additional modality y_i specifies a set of images that satisfy a certain constraint and we model that by multiplying the prior with an additional distribution $q(z|y_i)$.

3.2 Multiscale and Hierarchical Latent Space

Some of the modalities we consider (*e.g.*, sketch, segmentation) are two-dimensional and naturally contain information at multiple scales. Therefore, we devise a hierarchical latent space with latent variables at different resolutions. This allows us to directly pass information from each resolution of the encoder to the corresponding resolution of the latent space, so that the high-resolution control signals can be better preserved. Mathematically, our latent

---

[2] With a slight abuse of notation, we will use $q(z|y_i)$ (and similarly $p(z|\mathcal{Y})$) to denote both the "true" distribution and the estimated distribution produced by our network.

code is partitioned into groups $z = (z^0, z^1, ..., z^N)$ where $z^0 \in \mathbb{R}^{c_0}$ is a feature vector and $z^k \in \mathbb{R}^{c_k \times r_k \times r_k}, 1 \leq k \leq N$ are feature maps of increasing resolutions $(r_{k+1} = 2r_k, r_1 = 4, r_N$ is the image resolution). We can therefore decompose the prior $p(z)$ into $\prod_{k=0}^{N} p(z^k|z^{<k})$ and the experts $q(z|y_i)$ into $\prod_{k=0}^{N} q(z^k|z^{<k}, y_i)$, where $z^{<k}$ denotes $(z^0, z^1, ..., z^{k-1})$. Following Eq. (2), we assume the conditional latent distribution at each resolution is a product-of-experts given by

$$p(z^k|z^{<k}, \mathcal{Y}) \propto p(z^k|z^{<k}) \prod_{y_i \in \mathcal{Y}} q(z^k|z^{<k}, y_i), \tag{3}$$

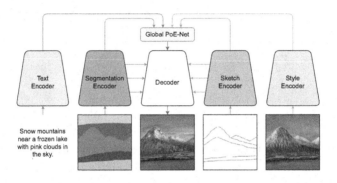

Fig. 3. An overview of our generator. The architecture of Global PoE-Net and decoder are detailed in Fig. 4a and Fig. 4b respectively. The architecture of modality encoders are described in Supplementary Material B.

where $p(z^k|z^{<k}) = \mathcal{N}\left(\mu_0^k, \sigma_0^k\right)$ and $q(z^k|z^{<k}, y_i) = \mathcal{N}\left(\mu_i^k, \sigma_i^k\right)$ are independent Gaussian distributions with mean and standard deviation parameterized by a neural network.[3] It can be shown [55] that the product of Gaussian experts is also a Gaussian $p(z^k|z^{<k}, \mathcal{Y}) = \mathcal{N}(\mu^k, \sigma^k)$, with

$$\mu^k = \frac{\frac{\mu_0^k}{(\sigma_0^k)^2} + \sum_i \frac{\mu_i^k}{(\sigma_i^k)^2}}{\frac{1}{(\sigma_0^k)^2} + \sum_i \frac{1}{(\sigma_i^k)^2}}, \quad \sigma^k = \frac{1}{\frac{1}{(\sigma_0^k)^2} + \sum_i \frac{1}{(\sigma_i^k)^2}}. \tag{4}$$

3.3 Generator Architecture

Fig. 3 shows an overview of our generator architecture. We encode each modality into a feature vector which is then aggregated in *Global PoE-Net*. We use convolutional networks with input skip connections to encode segmentation and sketch maps, a residual network to encode style images, and CLIP [38] to encode

---

[3] Except for $p(z^0)$, which is simply a standard Gaussian distribution.

text. Details of all modality encoders are given in Supplementary Material B. The decoder generates the image using the output of Global PoE-Net and skip connections from the segmentation and sketch encoders.

In Global PoE-Net (Fig. 4a), we predict a Gaussian $q(z^0|y_i) = \mathcal{N}\left(\mu_i^0, \sigma_i^0\right)$ from the feature vector of each modality using an MLP. We then compute the product of Gaussians including the prior $p(z^0) = \mathcal{N}(\mu_0^0, \sigma_0^0) = \mathcal{N}(0, I)$ and sample z^0 from the product distribution. An MLP further convert z^0 to the vector w.

The decoder mainly consists of a stack of residual blocks[4] [14], each of which is shown in Fig. 4b. *Local PoE-Net* samples the latent feature map z^k at the current resolution from the product of $p(z^k|z^{<k}) = \mathcal{N}\left(\mu_0^k, \sigma_0^k\right)$ and $q(z^k|z^{<k}, y_i) = \mathcal{N}\left(\mu_i^k, \sigma_i^k\right), \forall y_i \in \mathcal{Y}$, where $\left(\mu_0^k, \sigma_0^k\right)$ is computed from the output of the last layer and $\left(\mu_i^k, \sigma_i^k\right)$ is computed by concatenating the output of the last layer and the skip connection from the corresponding modality. Note that only modalities that have skip connections (segmentation and sketch, i.e. $i = 1, 4$) contribute to the computation. Other modalities (text and style reference) only provide global information but not local details. The latent feature map z^k produced by *Local PoE-Net* and the feature vector w produced by *Global PoE-Net* are fed to our local-global adaptive instance normalization (LG-AdaIN) layer,

$$\text{LG-AdaIN}(h^k, z^k, w) = \gamma_w \left(\gamma_{z^k} \frac{h^k - \mu(h^k)}{\sigma(h^k)} + \beta_{z^k}\right) + \beta_w, \tag{5}$$

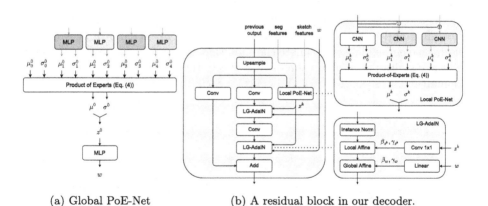

(a) Global PoE-Net   (b) A residual block in our decoder.

Fig. 4. (a) Global PoE-Net. We sample a latent feature vector z^0 using product-of-experts (Eq. (4) in Sect. 3.2), which is then processed by an MLP to output a feature vector w. (b) A residual block in our decoder. Local PoE-Net samples a latent feature map z^k using product-of-experts. Here $\oplus$ denotes concatenation. LG-AdaIN uses w and z^k to modulate the feature activations in the residual branch.

---

[4] Except for the first layer that convolves a constant feature map and the last layer that convolves the previous output to synthesize the output.

where h^k is a feature map in the residual branch after convolution, $\mu(h^k)$ and $\sigma(h^k)$ are channel-wise mean and standard deviation. β_w, γ_w are feature vectors computed from w, while β_{z^k}, γ_{z^k} are feature maps computed from z^k. The LG-AdaIN layer can be viewed as a combination of AdaIN [17] and SPADE [36] that takes both a global feature vector and a spatially-varying feature map to modulate the activations.

3.4 Multiscale Multimodal Projection Discriminator

Our discriminator receives the image x and a set of conditions $\mathcal{Y}$ as inputs and produces a score $D(x, \mathcal{Y}) = \text{sigmoid}(f(x, \mathcal{Y}))$ indicating the realness of x given $\mathcal{Y}$. Under the GAN objective [11], the optimal solution of f is

$$f^*(x, \mathcal{Y}) = \underbrace{\log \frac{q(x)}{p(x)}}_{\text{unconditional term}} + \sum_{y_i \in \mathcal{Y}} \underbrace{\log \frac{q(y_i|x)}{p(y_i|x)}}_{\text{conditional term}} , \qquad (6)$$

if we assume conditional independence of different modalities given x. The projection discriminator (PD) [33] proposes to use the inner product to estimate the conditional term. This implementation restricts the conditional term to be relatively simple, which imposes a good inductive bias that leads to strong empirical results. We propose a multimodal projection discriminator (MPD) that generalizes PD to our multimodal setting. As shown in Fig. 5a, the original PD first encodes both the image and the conditional input into a shared latent space. It then uses a linear layer to estimate the unconditional term from the image embedding and uses the inner product between the image embedding and the conditional embedding to estimate the conditional term. The unconditional term and the conditional term are summed to obtain the final discriminator logits. In our multimodal scenario, we simply encode each observed modality and add its

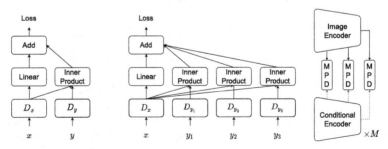

(a) Projection discriminator (PD) (b) Multimodal PD (MPD) (c) Multiscale MPD (MMPD)

Fig. 5. Comparison between the standard projection discriminator and our proposed multiscale multimodal projection discriminator.

inner product with the image embedding to the final loss (Fig. 5b)

$$f(x, \mathcal{Y}) = \text{Linear}(D_x(x)) + \sum_{y_i \in \mathcal{Y}} D_{y_i}^T(y_i) D_x(x) \,. \tag{7}$$

For spatial modalities such as segmentation and sketch, it is more effective to enforce their alignment with the image in multiple scales [29]. As shown in Fig. 5c, we encode the image and spatial modalities into feature maps of different resolutions and compute the MPD loss at each resolution. We compute a loss value at each location and resolution, and obtain the final loss by averaging first across locations then across resolutions. The resulting discriminator is named as the multiscale multimodal projection discriminator (MMPD) and detailed in Supplementary Material B.

3.5 Losses and Training Procedure

Latent regularization. Under the PoE assumption (Eq. (2)), the marginalized conditional latent distribution should match the unconditional prior:

$$\int p(z|y_i)p(y_i)dy_i = p(z|\varnothing) = p(z) \,. \tag{8}$$

| Segmentation | Ground truth | SPADE | OASIS | Ours |

Fig. 6. Visual comparison of segmentation-to-image synthesis on MS-COCO 2017.

| Text | Ground truth | DF-GAN | DM-GAN+CL | Ours |

A man riding a snowboard down a snow covered slope.

A red stop sign sitting on the side of a road.

Fig. 7. Visual comparison of text-to-image synthesis on MS-COCO 2017.

To this end, we minimize the Kullback-Leibler (KL) divergence from the prior distribution $p(z)$ to the conditional latent distribution $p(z|y_i)$ at every resolution

$$\mathcal{L}_{\mathrm{KL}} = \sum_{y_i \in \mathcal{Y}} \omega_i \sum_k \omega^k \mathbb{E}_{p(z^{<k}|y_i)} \left[D_{\mathrm{KL}}(p(z^k|z^{<k}, y_i) \| p(z^k|z^{<k})) \right], \qquad (9)$$

where ω^k is a resolution-dependent rebalancing weight and ω_i is a modality-specific loss weight. We describe both weights in detail in Supplementary Material B.

The KL loss also reduces conditional mode collapse since it encourages the conditional latent distribution to be close to the prior and therefore have high entropy. From the perspective of information bottleneck [1], the KL loss encourages each modality to only provide the minimum information necessary to specify the conditional image distribution.

Contrastive Losses. The contrastive loss has been widely adopted in representation learning [5,13] and more recently in image synthesis [12,28,35,61]. Given a batch of paired vectors $(\mathbf{u}, \mathbf{v}) = \{(u_i, v_i), i = 1, 2, ..., N\}$, the symmetric cross-entropy loss [38,64] maximizes the similarity of the vectors in a pair while keeping non-paired vectors apart

$$\mathcal{L}^{\mathrm{ce}}(\mathbf{u}, \mathbf{v}) = -\frac{1}{2N} \sum_{i=1}^{N} \log \frac{\exp(\cos(u_i, v_i)/\tau)}{\sum_{j=1}^{N} \exp(\cos(u_i, v_j)/\tau)}$$
$$-\frac{1}{2N} \sum_{i=1}^{N} \log \frac{\exp(\cos(u_i, v_i)/\tau)}{\sum_{j=1}^{N} \exp(\cos(u_j, v_i)/\tau)}, \qquad (10)$$

where τ is a temperature hyper-parameter. We use two kinds of pairs to construct two loss terms: the image contrastive loss and the conditional contrastive loss.

The *image contrastive loss* maximizes the similarity between a real image x and a fake image $\tilde{x}$ synthesized given the corresponding conditional inputs:

$$\mathcal{L}_{cx} = \mathcal{L}^{\mathrm{ce}}(E_{\mathrm{vgg}}(\mathbf{x}), E_{\mathrm{vgg}}(\tilde{\mathbf{x}})), \qquad (11)$$

where E_{vgg} is a pretrained VGG [47] encoder. This loss serves a similar purpose to the widely used perceptual loss but has been found to perform better [35,61].

The *conditional contrastive loss* aims to better align images with the corresponding conditions. Specifically, the discriminator is trained to maximize the similarity between its embedding of a real image $\mathbf{x}$ and the conditional input $\mathbf{y}_i$.

$$\mathcal{L}_{cy}^{D} = \sum_{i=1}^{M} \mathcal{L}^{\mathrm{ce}}(D_x(\mathbf{x}), D_{y_i}(\mathbf{y}_i)), \qquad (12)$$

where D_x and D_{y_i} are two modules in the discriminator that extract features from x and y_i, respectively, as shown in Eq. (7) and Fig. 5b. The generator is trained with the same loss, but using the generated image $\tilde{\mathbf{x}}$ instead of the real image to compute the discriminator embedding,

$$\mathcal{L}_{cy}^{G} = \sum_{i=1}^{M} \mathcal{L}^{\mathrm{ce}}(D_x(\tilde{\mathbf{x}}), D_{y_i}(\mathbf{y}_i)) \,. \tag{13}$$

In practice, we only use the conditional contrastive loss for text since it consumes too much GPU memory to use the conditional contrastive loss for the other modalities, especially when the image resolution and batch size are large. A similar image-text contrastive loss is used in XMC-GAN [61], where they use a non-symmetric cross-entropy loss that only includes the first term in Eq. (10).

Full Training Objective. In summary, the generator loss $\mathcal{L}^G$ and the discriminator loss $\mathcal{L}^D$ can be written as

$$\mathcal{L}^G = \mathcal{L}_{\mathrm{GAN}}^{G} + \mathcal{L}_{\mathrm{KL}} + \lambda_1 \mathcal{L}_{cx} + \lambda_2 \mathcal{L}_{cy}^{G}, \; \mathcal{L}^D = \mathcal{L}_{\mathrm{GAN}}^{D} + \lambda_2 \mathcal{L}_{cy}^{D} + \lambda_3 \mathcal{L}_{\mathrm{GP}} \,, \tag{14}$$

where $\mathcal{L}_{\mathrm{GAN}}^{G}$ and $\mathcal{L}_{\mathrm{GAN}}^{D}$ are non-saturated GAN losses [11], $\mathcal{L}_{\mathrm{GP}}$ is the R_1 gradient penalty loss [31], and $\lambda_1, \lambda_2, \lambda_3$ are weights associated with the loss terms.

Table 1. FID Comparison on MM-CelebA-HQ (1024×1024). We evaluate models conditioned on different modalities (from left to right: no conditions, text, segmentation, sketch, and all three modalities). The best scores are highlighted in bold

	Uncond	Text	Seg	Sketch	All
StyleGAN2 [23]	11.7	—	—	—	—
SPADE-Seg [36]	—	—	48.6	—	—
pSp-Seg [43]	—	—	44.1	—	—
SPADE-Sketch [36]	—	—	—	33.0	—
pSp-Sketch [43]	—	—	—	45.8	—
TediGAN [57]	—	38.4	45.1	45.1	45.1
PoE-GAN (Ours)	**10.5**	**10.1**	**9.9**	**9.9**	**8.3**

Table 2. Comparison on MS-COCO 2017 (256×256) using FID. We evaluate models conditioned on different modalities (from left to right: no conditions, text, segmentation, sketch, and all three modalities). The best scores are highlighted in bold

	Uncond	Text	Seg	Sketch	All
StyleGAN2 [23]	43.6	—	—	—	—
DF-GAN [51]	—	45.2	—	—	—
DM-GAN + CL [60]	—	29.9	—	—	—
SPADE-Seg [36]	—	—	22.1	—	—
VQGAN [8]	—	—	21.6	—	—
OASIS [45]	—	—	19.2	—	—
SPADE-Sketch [36]	—	—	—	63.7	—
PoE-GAN (Ours)	**43.4**	**20.5**	**15.8**	**25.5**	**13.6**

Modality Dropout. By design, our generator, discriminator, and loss terms are able to handle missing modalities. We also find that randomly dropping out some input modalities before each training iteration further improves the robustness of the generator towards missing modalities at test time.

4   Experiments

We evaluate the proposed approach on several datasets, including MM-CelebA-HQ [57], MS-COCO 2017 [26] with COCO-Stuff annotations [3], and a proprietary dataset of landscape images. Images are labeled with all input modalities obtained from either manual annotation or pseudo-labeling methods. More details about datasets and the pseudo-labeling procedure are in Supplementary Material A.

4.1   Main Results

We compare PoE-GAN with a recent multimodal image synthesis method named TediGAN [57] and also with state-of-the-art approaches specifically designed for each modality. For text-to-image, we compare with DF-GAN [51] and DM-GAN + CL [60] on MS-COCO. Since the original models are trained on the 2014 split, we retrain their models on the 2017 split using the official code. For segmentation-to-image synthesis, we compare with SPADE [36], VQGAN [8], OASIS [45], and pSp [43]. For sketch-to-image synthesis, we compare with SPADE [36] and pSp [43]. We additionally compare with StyleGAN2 [23] in the unconditional setting. We use Clean-FID [37] for benchmarking due to its reported benefits over previous implementations of FID [15].[5]

Results on MM-CelebA-HQ and MS-COCO are summarized in Table 1 and Table 2, respectively. PoE-GAN obtains a much lower FID than TediGAN in all

Table 3. Ablation study on MM-CelebA-HQ (256×256). The best scores are highlighted in bold and the second best ones are underlined

	Methods	Uncond	Text		Segmentation		Sketch		All	
		FID↓	FID↓	LPIPS↑	FID↓	LPIPS↑	FID↓	LPIPS↑	FID↓	LPIPS↑
(a)	Concatenation + dropout	29.3	26.4	0.33	19.4	0.22	<u>9.2</u>	0.11	**7.7**	0.11
(b)	Ours w/o KL loss	29.1	26.6	0.35	18.1	0.21	**9.1**	0.10	**7.7**	0.12
(c)	Ours w/o modality dropout	30.8	31.6	0.50	21.0	0.39	28.0	0.34	9.5	0.30
(d)	Ours w/o MMPD	21.5	20.8	0.48	18.3	0.40	16.4	0.36	16.2	0.34
(e)	Ours w/o image contrastive	<u>15.4</u>	<u>14.5</u>	0.55	13.5	**0.46**	10.2	**0.44**	9.5	**0.42**
(f)	Ours w/o text contrastive	15.8	15.0	<u>0.56</u>	<u>13.1</u>	0.40	10.0	<u>0.39</u>	8.9	<u>0.38</u>
(g)	Ours	**14.9**	**13.7**	**0.58**	**12.9**	<u>0.43</u>	9.9	0.37	<u>8.5</u>	0.35

---

[5] As a result, the baseline scores differ slightly from those in the original papers.

settings on MM-CelebA-HQ. In Supplementary Material C.3, we compare PoE-GAN with TediGAN in more detail and show that PoE-GAN is faster and more general than TediGAN. When conditioned on a single modality, PoE-GAN surprisingly outperforms the state-of-the-art method designed specifically for that modality on both datasets, although PoE-GAN is trained for a more general purpose. We note that PoE-GAN and TediGAN are trained on multiple modalities while other baselines are trained on an individual modality or unconditionally (StyleGAN2). In Supplementary Material C.4, we further show that PoE-GAN trained on a single modality always outperforms the multimodal-trained PoE-GAN when evaluated on that modality. This shows that the improvement of PoE-GAN over state-of-the-art unimodal image synthesis methods comes from our architecture and training scheme rather than additional annotations. In Figs. 6 and 7, we qualitatively compare PoE-GAN with previous segmentation-to-image and text-to-image methods on MS-COCO. We find that PoE-GAN produces images of much better quality and can synthesize realistic objects with complex structures, such as cats and stop signs. More qualitative comparisons are included in Supplementary Material C.5.

Multimodal Generation Examples. In Fig. 8, we show example images generated by our PoE-GAN using multiple input modalities on the landscape dataset. Our model is able to synthesize a wide range of landscapes in high resolution with photo-realistic details. More results are included in Supplementary Material C.5, where we additionally show that PoE-GAN can generate diverse images when given the same conditional inputs.

4.2   Ablation Studies

In Tables. 3 and 4, we analyze the importance of different components of PoE-GAN. We use LPIPS [63] as an additional metric to evaluate the diversity of images conditioned on the same input. Specifically, we randomly sample two output images conditioned on the same input and report the average LPIPS distance between the two outputs. A higher LPIPS score indicates more diverse outputs.

First, we compare our product-of-experts generator (row (g)) with a baseline that simply concatenates the embedding of all modalities, while performing modality dropout (missing modality embeddings set to zero). As seen in row (a), this baseline only works well when all modalities are available and its FID significantly drops when some modalities are missing. Further, the output images have low diversity as indicated by the LPIPS score. This is not surprising as previous work has shown that conditional GANs are prone to mode collapse [18,59,66].

*A lake in the desert with   A waterfall in sunset.   A mountain with pine trees
mountains at a distance.   in a starry winter night.*

Fig. 8. Examples of multimodal conditional image synthesis results produced by PoE-GAN trained on the 1024×1024 landscape dataset. We show the segmentation/sketch/style inputs on the bottom right of the generated images for the results in the first row. The results in the second row additionally leverage text inputs, which are shown below the corresponding generated images. Please zoom in for details.

Row (b) of Tables. 3 and 4 shows that the KL loss is important for training our model. Without it, our model suffers from low sample diversity and lack of robustness towards missing modalities, similar to the concatenation baseline described above. The variances of individual experts become near zero without the KL loss. The latent code z^k then becomes a deterministic weighted average of the mean of each expert, which is equivalent to concatenating all modality embeddings and projecting it with a linear layer. This explains why our model without the KL loss behaves similarly to the concatenation baseline. Row (c) shows that our modality dropout scheme is important for handling missing modalities. Without it, the model tends to overly rely on the most informative modality, such as segmentation in MS-COCO.

Table 4. Ablation study on MS-COCO 2017 (64×64). The best scores are highlighted in bold and the second best ones are underlined

	Methods	Uncond	Text		Segmentation		Sketch		All	
		FID↓	FID↓	LPIPS↑	FID↓	LPIPS↑	FID↓	LPIPS↑	FID↓	LPIPS↑
(a)	Concatenation + dropout	59.1	30.7	0.40	20.4	0.16	36.1	0.27	**16.6**	0.12
(b)	Ours w/o KL loss	59.3	30.5	0.39	21.5	0.16	33.0	0.27	<u>16.9</u>	0.12
(c)	Ours w/o modality dropout	86.2	87.8	0.58	19.9	0.44	85.1	0.55	18.7	<u>0.47</u>
(d)	Ours w/o MMPD	43.1	40.2	0.64	21.7	0.46	45.5	0.57	21.1	0.42
(e)	Ours w/o image contrastive	**25.7**	**21.3**	0.64	18.0	**0.50**	37.9	**0.61**	18.5	**0.54**
(f)	Ours w/o text contrastive	27.6	26.0	**0.66**	<u>17.4</u>	0.46	<u>33.5</u>	0.55	17.9	0.43
(g)	Ours	<u>26.6</u>	<u>22.2</u>	<u>0.65</u>	**17.1**	<u>0.47</u>	**30.2**	<u>0.58</u>	17.1	0.44

Table 5. User study on text-to-image synthesis. Each column shows the percentage of users that prefer the image generated by our model over the baseline

	DF-GAN [51]	DM-GAN + CL [60]
Ours vs	82.1%	72.9%

Table 6. User study on segmentation-to-image synthesis. Each column shows the percentage of users that prefer the image generated from our model over the baseline

	SPADE [36]	VQGAN [8]	OASIS [45]
Ours vs	69%	66.7%	64.9%

To evaluate the proposed multiscale multimodal discriminator architecture, we replace MMPD with a discriminator that receives concatenated images and all conditional inputs. Row (d) shows that MMPD is much more effective than such a concatenation-based discriminator in all settings.

Finally in rows (e) and (f), we show that contrastive losses are useful but not essential. The image contrastive loss slightly improves FID in most settings, while the text contrastive loss improves FID for text-to-image synthesis.

4.3 User Study

We conduct a user study to compare PoE-GAN with state-of-the-art text-to-image and segmentation-to-image synthesis methods on MS-COCO. We show users two images generated by different algorithms from the same conditional input and ask them which one is more realistic. As shown in Table. 5 and Table. 6, the majority of users prefer PoE-GAN over the baseline methods.

5   Conclusion

We introduce a multimodal conditional image synthesis model based on product-of-experts and show its effectiveness for converting an arbitrary subset of input modalities to an image satisfying all conditions. While empirically superior than the prior multimodal synthesis work, it also outperforms state-of-the-art uni-modal conditional image synthesis approaches when conditioned on a single modality.

Acknowledgements. We thank Jan Kautz, David Luebke, Tero Karras, Timo Aila, and Zinan Lin for their feedback on the manuscript. We thank Daniel Gifford and Andrea Gagliano on their help on data collection.

References

1. Alemi, A.A., Fischer, I., Dillon, J.V., Murphy, K.: Deep variational information bottleneck. In: ICLR (2017)
2. Brock, A., Donahue, J., Simonyan, K.: Large scale GAN training for high fidelity natural image synthesis. In: ICLR (2019)
3. Caesar, H., Uijlings, J., Ferrari, V.: COCO-Stuff: Thing and stuff classes in context. In: CVPR (2018)
4. Chen, S.Y., Su, W., Gao, L., Xia, S., Fu, H.: DeepFaceDrawing: deep generation of face images from sketches. ACM Trans. Graphics (TOG) **72**, 72:1–72:16 (2020)
5. Chen, T., Kornblith, S., Norouzi, M., Hinton, G.: A simple framework for contrastive learning of visual representations. In: ICML (2020)
6. Chen, W., Hays, J.: SketchyGAN: towards diverse and realistic sketch to image synthesis. In: CVPR (2018)
7. Child, R.: Very deep VAEs generalize autoregressive models and can outperform them on images. In: ICLR (2021)
8. Esser, P., Rombach, R., Ommer, B.: Taming transformers for high-resolution image synthesis. In: CVPR (2021)
9. Gafni, O., Polyak, A., Ashual, O., Sheynin, S., Parikh, D., Taigman, Y.: Make-a-scene: Scene-based text-to-image generation with human priors. In: CVPR (2022)
10. Ghosh, A., et al.: Interactive sketch & fill: multiclass sketch-to-image translation. In: ICCV (2019)
11. Goodfellow, I., et al.: Generative adversarial nets. In: NeurIPS (2014)
12. Han, J., Shoeiby, M., Petersson, L., Armin, M.A.: Dual contrastive learning for unsupervised image-to-image translation. In: CVPR (2021)
13. He, K., Fan, H., Wu, Y., Xie, S., Girshick, R.: Momentum contrast for unsupervised visual representation learning. In: CVPR (2020)
14. He, K., Zhang, X., Ren, S., Sun, J.: Deep residual learning for image recognition. In: CVPR (2016)
15. Heusel, M., Ramsauer, H., Unterthiner, T., Nessler, B., Hochreiter, S.: Gans trained by a two time-scale update rule converge to a local nash equilibrium. In: NeurIPS (2017)
16. Hinton, G.E.: Training products of experts by minimizing contrastive divergence. Neural Comput. **14**, 1771–1800 (2002)
17. Huang, X., Belongie, S.: Arbitrary style transfer in real-time with adaptive instance normalization. In: ICCV (2017)

18. Huang, X., Liu, M.-Y., Belongie, S., Kautz, J.: Multimodal unsupervised image-to-image translation. In: Ferrari, V., Hebert, M., Sminchisescu, C., Weiss, Y. (eds.) ECCV 2018. LNCS, vol. 11207, pp. 179–196. Springer, Cham (2018). https://doi.org/10.1007/978-3-030-01219-9_11
19. Isola, P., Zhu, J.Y., Zhou, T., Efros, A.A.: Image-to-image translation with conditional adversarial networks. In: CVPR (2017)
20. Karras, T., Aila, T., Laine, S., Lehtinen, J.: Progressive growing of GANs for improved quality, stability, and variation. In: ICLR (2018)
21. Karras, T., et al.: Alias-free generative adversarial networks. In: NeurIPS (2021)
22. Karras, T., Laine, S., Aila, T.: A style-based generator architecture for generative adversarial networks. In: CVPR (2019)
23. Karras, T., Laine, S., Aittala, M., Hellsten, J., Lehtinen, J., Aila, T.: Analyzing and improving the image quality of StyleGAN. In: CVPR (2020)
24. Kingma, D.P., Welling, M.: Auto-encoding variational bayes. In: ICLR (2014)
25. Kutuzova, S., Krause, O., McCloskey, D., Nielsen, M., Igel, C.: Multimodal variational autoencoders for semi-supervised learning: In defense of product-of-experts. arXiv preprint arXiv:2101.07240 (2021)
26. Lin, T.-Y., et al.: Microsoft COCO: common objects in context. In: Fleet, D., Pajdla, T., Schiele, B., Tuytelaars, T. (eds.) ECCV 2014. LNCS, vol. 8693, pp. 740–755. Springer, Cham (2014). https://doi.org/10.1007/978-3-319-10602-1_48
27. Liu, M.Y., Huang, X., Yu, J., Wang, T.C., Mallya, A.: Generative adversarial networks for image and video synthesis: algorithms and applications. In: Proceedings of the IEEE (2021)
28. Liu, R., Ge, Y., Choi, C.L., Wang, X., Li, H.: DivCo: diverse conditional image synthesis via contrastive generative adversarial network. In: CVPR (2021)
29. Liu, X., Yin, G., Shao, J., Wang, X., Li, H.: Learning to predict layout-to-image conditional convolutions for semantic image synthesis. In: NeurIPS (2019)
30. Maaløe, L., Fraccaro, M., Liévin, V., Winther, O.: BIVA: a very deep hierarchy of latent variables for generative modeling. In: NeurIPS (2019)
31. Mescheder, L., Geiger, A., Nowozin, S.: Which training methods for GANs do actually converge? In: ICML (2018)
32. Miyato, T., Kataoka, T., Koyama, M., Yoshida, Y.: Spectral normalization for generative adversarial networks. In: ICLR (2018)
33. Miyato, T., Koyama, M.: cGANs with projection discriminator. In: ICLR (2018)
34. Odena, A., Olah, C., Shlens, J.: Conditional image synthesis with auxiliary classifier GANs. In: ICML (2017)
35. Park, T., Efros, A.A., Zhang, R., Zhu, J.-Y.: Contrastive learning for unpaired image-to-image translation. In: Vedaldi, A., Bischof, H., Brox, T., Frahm, J.-M. (eds.) ECCV 2020. LNCS, vol. 12354, pp. 319–345. Springer, Cham (2020). https://doi.org/10.1007/978-3-030-58545-7_19
36. Park, T., Liu, M.Y., Wang, T.C., Zhu, J.Y.: Semantic image synthesis with spatially-adaptive normalization. In: CVPR (2019)
37. Parmar, G., Zhang, R., Zhu, J.Y.: On buggy resizing libraries and surprising subtleties in FID calculation. arXiv preprint arXiv:2104.11222 (2021)
38. Radford, A., et al.: Learning transferable visual models from natural language supervision. In: ICML (2021)
39. Radford, A., Metz, L., Chintala, S.: Unsupervised representation learning with deep convolutional generative adversarial networks. In: CVPR (2016)
40. Ramesh, A., et al.: Zero-shot text-to-image generation. In: ICML (2021)
41. Reed, S., Akata, Z., Yan, X., Logeswaran, L., Schiele, B., Lee, H.: Generative adversarial text to image synthesis. In: ICML (2016)

42. Rezende, D.J., Mohamed, S., Wierstra, D.: Stochastic backpropagation and approximate inference in deep generative models. In: ICML (2014)
43. Richardson, E., et al.: Encoding in style: a StyleGAN encoder for image-to-image translation. In: CVPR (2021)
44. Sangkloy, P., Lu, J., Fang, C., Yu, F., Hays, J.: Scribbler: controlling deep image synthesis with sketch and color. In: CVPR (2017)
45. Schönfeld, E., Sushko, V., Zhang, D., Gall, J., Schiele, B., Khoreva, A.: You only need adversarial supervision for semantic image synthesis. In: ICLR (2020)
46. Shi, Y., Siddharth, N., Paige, B., Torr, P.: Variational mixture-of-experts autoencoders for multi-modal deep generative models. In: NeurIPS (2019)
47. Simonyan, K., Zisserman, A.: Very deep convolutional networks for large-scale image recognition. In: ICLR (2015)
48. Sønderby, C.K., Raiko, T., Maaløe, L., Sønderby, S.K., Winther, O.: Ladder variational autoencoders. In: NeurIPS (2016)
49. Sutter, T.M., Daunhawer, I., Vogt, J.E.: Generalized multimodal ELBO. In: ICLR (2020)
50. Suzuki, M., Nakayama, K., Matsuo, Y.: Joint multimodal learning with deep generative models. In: ICLR workshop (2017)
51. Tao, M., et al.: DF-GAN: deep fusion generative adversarial networks for text-to-image synthesis. arXiv preprint arXiv:2008.05865 (2020)
52. Vahdat, A., Kautz, J.: NVAE: A deep hierarchical variational autoencoder. In: NeurIPS (2020)
53. Vedantam, R., Fischer, I., Huang, J., Murphy, K.: Generative models of visually grounded imagination. In: ICLR (2018)
54. Wang, T.C., Liu, M.Y., Zhu, J.Y., Tao, A., Kautz, J., Catanzaro, B.: High-resolution image synthesis and semantic manipulation with conditional GANs. In: CVPR (2018)
55. Williams, C.K., Agakov, F.V., Felderhof, S.N.: Products of gaussians. In: NeurIPS (2001)
56. Wu, M., Goodman, N.: Multimodal generative models for scalable weakly-supervised learning. In: NeurIPS (2018)
57. Xia, W., Yang, Y., Xue, J.H., Wu, B.: TediGAN: text-guided diverse face image generation and manipulation. In: CVPR (2021)
58. Xu, T., et al.: AttnGAN: Fine-grained text to image generation with attentional generative adversarial networks. In: CVPR (2018)
59. Yang, D., Hong, S., Jang, Y., Zhao, T., Lee, H.: Diversity-sensitive conditional generative adversarial networks. In: ICLR (2019)
60. Ye, H., Yang, X., Takac, M., Sunderraman, R., Ji, S.: Improving text-to-image synthesis using contrastive learning. arXiv preprint arXiv:2107.02423 (2021)
61. Zhang, H., Koh, J.Y., Baldridge, J., Lee, H., Yang, Y.: Cross-modal contrastive learning for text-to-image generation. In: CVPR (2021)
62. Zhang, H., et al.: StackGAN: Text to photo-realistic image synthesis with stacked generative adversarial networks. In: ICCV (2017)
63. Zhang, R., Isola, P., Efros, A.A., Shechtman, E., Wang, O.: The unreasonable effectiveness of deep features as a perceptual metric. In: CVPR (2018)
64. Zhang, Y., Jiang, H., Miura, Y., Manning, C.D., Langlotz, C.P.: Contrastive learning of medical visual representations from paired images and text. arXiv preprint arXiv:2010.00747 (2020)
65. Zhang, Z., et al.: M6-UFC: unifying multi-modal controls for conditional image synthesis. In: NeurIPS (2021)

66. Zhu, J.Y., et al.: Toward multimodal image-to-image translation. In: NeurIPS (2017)
67. Zhu, M., Pan, P., Chen, W., Yang, Y.: DM-GAN: dynamic memory generative adversarial networks for text-to-image synthesis. In: CVPR (2019)

Auto-regressive Image Synthesis
with Integrated Quantization

Fangneng Zhan[1,2], Yingchen Yu[1], Rongliang Wu[1], Jiahui Zhang[1],
Kaiwen Cui[1], Changgong Zhang[3], and Shijian Lu[1(✉)]

[1] Nanyang Technological University, Singapore, Singapore
`fzhan@mpi-inf.mpg.de,`
`{yingchen001,ronglian001,jiahui003,kaiwen001}@e.ntu.edu.sg,`
`shijian.lu@ntu.edu.sg`
[2] Max Planck Institute for Informatics, Saarbrücken, Germany
[3] Amazon, Beijing, China
`cgzhang@amazon.com`

Abstract. Deep generative models have achieved conspicuous progress
in realistic image synthesis with multifarious conditional inputs, while
generating diverse yet high-fidelity images remains a grand challenge in
conditional image generation. This paper presents a versatile framework
for conditional image generation which incorporates the inductive bias
of CNNs and powerful sequence modeling of auto-regression that naturally leads to diverse image generation. Instead of independently quantizing the features of multiple domains as in prior research, we design
an integrated quantization scheme with a variational regularizer that
mingles the feature discretization in multiple domains, and markedly
boosts the auto-regressive modeling performance. Notably, the variational regularizer enables to regularize feature distributions in incomparable latent spaces by penalizing the intra-domain variations of distributions. In addition, we design a Gumbel sampling strategy that allows
to incorporate distribution uncertainty into the auto-regressive training
procedure. The Gumbel sampling substantially mitigates the exposure
bias that often incurs misalignment between the training and inference
stages and severely impairs the inference performance. Extensive experiments over multiple conditional image generation tasks show that our
method achieves superior diverse image generation performance qualitatively and quantitatively as compared with the state-of-the-art.

1 Introduction

Conditional image generation aims to generate photorealistic images conditioning on certain guidance which can be semantic segmentation [35], key points
[43], layout [18] as well as heterogeneous guidance such as text [39] and audio
[2]. It has been widely formulated as one-to-one mapping tasks [48], though it is
essentially one-to-many mappings since one conditional input could correspond
to multiple images. Targeting to mimic the true conditional image distribution,

S. Avidan et al. (Eds.): ECCV 2022, LNCS 13676, pp. 110–127, 2022.
https://doi.org/10.1007/978-3-031-19787-1_7

diverse yet high-fidelity image synthesis remains a great challenge in conditional image generation, especially when the conditional inputs come from different visual domains or even heterogeneous domains.

A typical approach to model diverse mapping is to employ extra style exemplars to guide the generation process. For example, [63] build dense correspondences between conditional inputs and style exemplars to transfer textures for diverse generation, while building semantic correspondences essentially requires the exemplars to have similar semantics as the conditional inputs. Without requiring extra exemplars, Variational Autoencoders (VAEs) [5] aim to regularize the latent distribution of encoded features, thus diverse generation can be achieved by directly sampling from the latent distribution. However, VAEs inevitably suffer from *posterior collapse* phenomenon [24] which leads to degraded diverse generation performance. Instead of regularizing the latent feature distribution in VAE, VQ-VAE [33] is designed to auto-regressively model the distributions of image feature sequences. [6] further introduce transformers in VQ-VAE to achieve high-resolution image synthesis. Nevertheless, above auto-regressive generation methods discretize relevant features independently, neglecting the potential association among multi-domain features in latent spaces.

This paper presents an **I**ntegrated **Q**uantization **V**ariational **A**uto-Encoder (**IQ-VAE**) that inherits the merits of CNNs (locality and spatial invariance) for high-fidelity image generation and the powerful sequence modeling of auto-regressive transformer for diverse image generation. Instead of quantizing multi-domain features independently as in [6], we introduce an integrated quantization scheme to quantize the involved features collaboratively in the latent spaces. The integrated quantization scheme provides a sound way to regularize the latent structure of multi-domain distributions, which can facilitate the ensuing auto-regressive modeling of sequence distributions. However, as the conditional inputs and real images often have heterogeneous features with incomparable latent spaces, KL-divergence or Wasserstein distance cannot directly measure their feature discrepancy for regularization. Inspired by the differential circuit which takes the variation between two signals as valid input, we introduce a variational regularizer which penalizes the intra-domain variation between distributions to regularize their structural discrepancy.

In addition, most auto-regressive models are trained with a so-called "teacher forcing" framework where the ground truth of target sequence (i.e., gold sequence) is provided at the training stage. However, such framework is susceptible to exposure bias, i.e., the misalignment between the training stage and the inference stage where the gold target sequence is not available and decisions are conditioned on previous model prediction. We design a Gumbel sampling strategy that greatly mitigates the exposure bias by incorporating the uncertainty of sequence distributions in training stage. Specifically, we adopt a reparameterization trick with Gumbel softmax to samples tokens from the predicted distributions and then mixes them with the gold sequence according to a reliability-based scheduling to make the final prediction. The Gumbel sampling also serves as data augmentation strategy that helps to avoid overfitting and improve the auto-regression performance substantially.

The contributions of this work can be summarized in three aspects. First, we introduce a versatile auto-regression framework with an integrated quantization scheme for conditional image generation. Second, we propose a variational regularizer that exploits intra-domain variations to regularize heterogeneous features in latent spaces. Third, we design a Gumbel sampling strategy with a reliability-based scheduling to mitigate the misalignment between the training and inference stages of auto-regressive models.

2   Related Work

2.1   Conditional Image Generation

Conditional image generation has achieved remarkable progress by learning the mapping among data of different domains. To achieve high-fidelity yet flexible image generation, various conditional inputs have been adopted including semantic segmentation [12,35,48,59,62], scene layouts [18,42,65], key points [26,29,57,61], edge maps [12,55,56], etc. Recently, several studies explored to generate images with cross-modal guidance [53,58]. For example, Qiao et al. [37] propose a novel global-local attentive and semantic-preserving text-to-image-to-text framework based on the idea of redescription. Ramesh et al. [39] handle text-to-image generation by using a transformer that auto-regressively models the text and image tokens. Chen et al. [2] investigated audio-to-visual generation with a conditional GANs. Nevertheless, the aforementioned methods all focus on deterministic image generation with a single generated image.

As an ill-posed problem, conditional image generation is a naturally a one-to-many mapping task as one conditional input could map to multiple diverse and faithful images. Earlier studies [15] manipulate latent feature codes to control the generation outcome, but they struggle to capture complex textures. With the emergence of GANs [7,34,54,60,67], style code injection has been designed to address this issue. For example, Zhu et al. [69] design semantic region-adaptive normalization (SEAN) to control the style of each semantic region individually. Choi et al. [4] employ a style encoder for style consistency between exemplars and the translated images. Huang et al. [11] and Ma et al. [25] transfer style codes from exemplars to source images via adaptive instance normalization (AdaIN) [10]. Recently, Zhang et al. [63] learn dense semantic correspondences between conditional inputs and exemplars, but require the exemplars to have similar semantics with the conditional input.

The aforementioned methods all suffer from low performance in diverse generation or require extra guidance for decent diverse generation. In this work, we propose a versatile auto-regressive framework that introduces a joint quantization scheme to achieve conditional image generation, and it inherently allows to generate diverse yet high-fidelity images as well.

2.2   Auto-regression in Image Generation

Different from VAE or GANs in image generation, auto-regressive models treat image pixels as a sequence and generate pixels one by one conditioning on the previously generated pixels by modeling their conditional distributions. With the recent advance of deep learning, a number of studies explored to use deep auto-regressive models to generate image pixels sequentially. For instance, PixelRNN and PixelCNN [44] utilize LSTM [9] layers and masked convolutions to capture pixel inter-dependencies in a fixed order. Gated PixelCNN [32] describes a gated convolution to improve the generation quality with lower computational cost. However, deep auto-regressive models still struggle to generate high-fidelity images due to the limitation of sequential prediction of pixels. To address this issue, VQ-VAE [33] adapts an encoder-decoder structure to learns discrete latent representations for autoregressive modeling, which enables high fidelity image synthesis.

Leveraging their powerful attention mechanisms, transformers [45] allow to establish long-range dependencies effectively and have been adopted in various computer vision tasks. In image generation, Chen *et al.* [3] introduce a sequence Transformer to generate low-resolution images auto-regressively. Based on VQ-VAE [33], Esser *et al.* [6] propose a VQ-GAN to learn a discrete codebook and utilize the transformers to efficiently model sequence distributions for high-resolution images synthesis. Nevertheless, the aforementioned methods all neglect exposure bias which often introduces clear misalignment between the training and the inference. The proposed Gumbel sampling strategy introduces uncertainty in training stage which mitigates the misalignment greatly.

3   Proposed Method

3.1   Overall Framework

The framework of the proposed IQ-VAE is illustrated in Fig. 1. The IQ-VAE is first trained to learn discrete feature representations of the real image and conditional input with learnable codebook as shown in Fig. 2(a). With the learnt IQ-VAE and codebook, the conditional input and real image can be quantized into discrete sequences by IQ-VAE encoders E_x and E_c. The transformer then auto-regressively models the distribution of the image sequences with a given sequence of conditional input. With the sequence distributions predicted by the transformer, diverse sequences can be sampled and inversely quantized into feature vectors based on the learnt codebook. Finally, the inversely quantized feature vectors are concatenated with the conditional features and fed into the IQ-VAE decoder D_x to achieve diverse image generation. Details of IQ-VAE and auto-regressive transformer will be discussed in the ensuing subsections.

3.2   Integrated Quantization

For the task of conditional image generation, [6] employ two VQ-VAEs [33] to quantize the features of conditional inputs and real images independently. However, this naive quantization approach neglects the potential coupling between

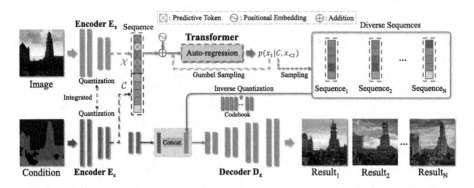

Fig. 1. The framework of the proposed auto-regressive image generation with integrated quantization: We design an integrated quantization VAE (IQ-VAE) with E_x and E_c to encode the *Image* and *Condition* into discrete representation sequences $\mathcal{X}$ and $\mathcal{C}$ concurrently. The distribution $p(x_t|\mathcal{C}, x_{<t})$ of sequence $\mathcal{X}$ conditioned on $\mathcal{C}$ is modeled by an auto-regressive *Transformer*. Finally, diverse sequences are sampled from the predicted distribution $p(x_t|\mathcal{C}, x_{<t})$ which are further inversely quantized and concatenated with the encoded condition features for diverse generation via the IQ-VAE decoder D_x.

conditional inputs and real images in the latent spaces. Intuitively, as conditional inputs imply certain information (e.g., edges) of the corresponding images, certain coupling or correlation should exist between their latent feature spaces. Explicitly regularizing such coupling between images and conditional inputs will be beneficial for the modeling of image distribution from the given conditional inputs.

We propose an integrated quantization scheme to regularize the discretization of the image and conditional input as illustrated in Fig. 2(a). Specially, two VQ-VAEs are employed to encode the image and conditional input to a pair of feature distributions as denoted by $\mathcal{X} = [x_1, x_2, \cdots, x_n]$ and $\mathcal{C} = [c_1, c_2, \cdots, c_n]$. An intuitive method to regularize the feature distributions is to employ KL divergence to measure and minimize their inter-domain discrepancy, namely $\mathrm{KL}(\mathcal{C}, \mathcal{X})$. However, this approach fails when a meaningful cost across the distributions cannot be defined. This is especially true for heterogeneous conditional inputs (e.g., texts and audios) that have incomparable latent spaces with respect to the image. Under such context, the KL divergence is ill-suited and inapplicable to capture the discrepancy between distributions. We thus design a novel variational regularizer that leverages the intra-domain variations of distributions to adaptively regularize their latent distributions.

Variational Regularizer. Inspired by the differential circuit which takes the variation of two signals as the valid input, we propose a variational regularizer that penalizes the inter-domain discrepancy via the intra-domain variations as illustrated in Fig. 2(b). Although the discrepancy between incomparable domain features $\mathcal{C} = [c_1, c_2, \cdots, c_n]$ and $\mathcal{X} = [x_1, x_2, \cdots, x_n]$ cannot be duly measured,

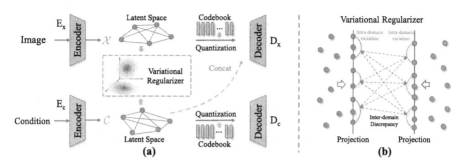

Fig. 2. (a) illustrates the framework of the proposed integrated quantization scheme. We introduce a variational regularizer to regularize their feature distributions in latent spaces. As shown in (b), the variational regularizer employs the intra-domain variations to penalize the structural inter-domain discrepancy, and it is optimized through a sliced projection.

the distance (or variation) among samples in the same domain can be effectively measured with some simple metric $\mathcal{M}$ (Euclidean distances is adopted in this work). We thus first compute the distances among intra-domain samples for the conditioned input and real image as denoted by $\mathcal{M}(c_i, c_k)$ and $\mathcal{M}(x_j, x_l)$. The discrepancy between intra-domain variations $\{\mathcal{M}(c_i, c_k)\}, i, k \in [1, n]$ and $\{\mathcal{M}(x_j, x_l)\}, j, l \in [1, n]$ can then serve as a proxy to indicate the inter-domain discrepancy between the conditional input and real image.

To regularize the structural difference between two latent distributions effectively, we adopt the discrete optimal transport (OT) [36, 41] with a 2th Euclidean distance cost as the discrepancy metric which naturally induces the intrinsic geometries of distributions and can measure the discrepancy between intra-domain variations as follows:

$$\mathrm{OT}(\mathcal{C}, \mathcal{X}) = \min_{\Gamma \in \prod(\alpha, \beta)} \sum_{i,j,k,l} \left| \mathcal{M}(c_i, c_k) - \mathcal{M}(x_j, x_l) \right|^2 \Gamma_{ij} \Gamma_{kl} \tag{1}$$

where Γ_{ij} and Γ_{kl} are entries of coupling matrice Γ, $\prod(\alpha, \beta) = \{\Gamma \in \mathbb{R}^{n \times n} | \Gamma \vec{1}_n = \alpha, \ \Gamma^T \vec{1}_n = \beta\}$, $\vec{1}_n$ is a n-dimensional all-one vector, $\alpha = \{\alpha_i\}$ and $\beta = \{\beta_j\}, i, j \in [1, n]$ are vectors of probability weights associated with c_i and x_j ($\alpha_i = 1/n$, $\beta_j = 1/n$). The formulation in Eq. (1) is often referred as Gromov Wasserstein (GW) distance [28] between distributions $\mathcal{C}$ and $\mathcal{X}$.

With GW distance as the metric in variational regularizer, we impose a constraint on the posterior distributions defined in different latent spaces which encourages structural similarity between them [51]. This regularizer helps avoid over-regularization as it does not enforce a shared latent distribution across different or heterogeneous domains. In addition, the GW distance is invariant to translations, permutations or rotations on both distributions when Euclidean distances are used, which allows to capture discrepancy between complex latent distributions effectively.

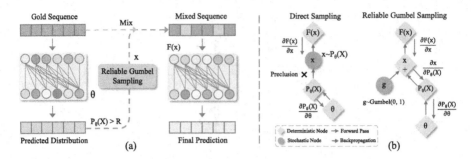

Fig. 3. (a) illustrates the framework of the proposed Gumbel sampling with twice executions. In the first forward pass, token distribution $P_\theta(X)$ is predicted from the gold sequence with network parameters θ. A sample x is sampled from $P_\theta(X)$ according to a reliability-based scheduling and is mixed with the gold sequence for the second pass (namely final pass). (b) compares the gradient flows of direct sampling and Gumbel sampling. The presence of stochastic node x in direct sampling precludes the back-propagation of gradient from x to $P_\theta(X)$. Gumbel sampling allows gradient flow from x to $P_\theta(X)$ through a reparameterization trick which transfers the stochasticity to a Gumbel distribution.

Optimization. The solution of the variational regularizer in Eq. (1) is a non-convex optimization problem. Grounded in the well-studied theory of Wasserstein distance [46], Eq. (1) can be solved through sliced Gromov Wasserstein (sliced GW) distance [46]. Specifically, the original metric measure spaces are projected to 1D spaces with random directions, and the sliced GW corresponds to the expectation of the GW distances in these projected 1D spaces. In this case, the sliced GW is approximated based on sample observations from the distributions shown in Fig. 2(b).

In particular, given $[c_1, c_2, \cdots, c_n]$ from $\mathcal{C}$ and $[x_1, x_2, \cdots, x_n]$ from $\mathcal{X}$ and L projection vectors $\{\gamma_m\}_{m=1}^{L}$, the empirical sliced GW can be formulated by:

$$\frac{1}{L} \sum_{m=1}^{L} \min_{\Gamma_{ij}, \Gamma_{kl} \in \prod(p,q)} \sum_{i,j,k,l} \left| \mathcal{M}(\langle c_i, \gamma_m \rangle, \langle c_j, \gamma_m \rangle) - \mathcal{M}(\langle x_k, \gamma_m \rangle, \langle x_l, \gamma_m \rangle) \right|^2 \Gamma_{ij} \Gamma_{kl}. \tag{2}$$

where $\langle c_i, \gamma_m \rangle$ denotes the projection of c_i on direction γ_m. Compared with direct computation via proximal gradient optimization [52], the sliced GW has much lower computational complexity of $\mathcal{O}(nd)$, where n and d denote the sample number and sample dimension, respectively.

Besides the loss of variational regularizer (namely sliced GW) as denoted by $\mathcal{L}_{reg}$ for the optimization of IQ-VAE, we also include reconstruction loss $\mathcal{L}_{recon}$ and quantization loss $\mathcal{L}_{quan}$ of the conditional input and real image. To further improve the image quality, a perceptual loss $\mathcal{L}_{perc}$ and discriminator loss $\mathcal{L}_{dis}$ are also included. Thus, the overall objective for the IQ-VAE network is:

$$\mathcal{L}_{IQ-VAE} = \lambda_1 \mathcal{L}_{reg} + \lambda_2 \mathcal{L}_{recon} + \lambda_3 \mathcal{L}_{quan} + \lambda_4 \mathcal{L}_{perc} + \lambda_5 \mathcal{L}_{dis}. \tag{3}$$

where λ balances the loss terms.

3.3   Auto-regression

Auto-regressive (AR) modeling is representative objective to accommodate sequence dependencies in a raster scan order. The probability of each position in the sequence is conditioned on all previously prediction and the joint distribution of sequences is modeled as the product of conditional distributions: $p(x) = \prod_{t=1}^{n} p(x_t|x_1, x_2, \cdots, x_{t-1}) = \prod_{t=1}^{n} p(x_t|x_{<t})$. Under the context of conditional image generation, a conditional auto-regression is actually adopted for the modeling of image distribution. For clarity, we still denote the discrete image sequence as $\mathcal{X} = [x_1, x_2, \ldots, x_n]$, the conditional sequence as $\mathcal{C} = [c_1, c_2, \ldots, c_n]$. Then the joint distribution of image sequence conditioned on $\mathcal{C}$ can be formulated as:

$$p(x|\mathcal{C}) = \prod_{t=1}^{n} p(x_t|c_1, c_2, \cdots, c_n, x_1, x_2, \cdots, x_{t-1}) = \prod_{t=1}^{n} p(x_t|\mathcal{C}, x_{<t}). \quad (4)$$

Auto-regressive models factorize the predicted tokens with chain rule of probability, which establishes the output dependency effectively for yielding better predictions. During inference, each token is predicted auto-regressively in a raster-scan order. A top-k (k is 100 in this work) sampling strategy is adopted to randomly sample from the k most likely next tokens, which naturally enables diverse sampling results. The predicted tokens are then concatenated with the previous sequence as conditions for the prediction of next token. This process repeats iteratively until all the tokens are sampled.

Gumbel Sampling. Auto-regressive models are trained using the ground truth sequence (i.e., gold sequence). This framework leads to quick convergence during training, but it is misaligned with the inference stage where gold sequence is not available and decisions are purely conditioned on previous predictions. This phenomenon is typically referred as exposure bias [40]. Intuitively, this problem can be tackled by using the previous predictions as conditions with certain probability in training stage as mentioned in [30].

Specially, in order to conduct sampling from previous predictions, the auto-regression process is executed twice in training stage as illustrated in Fig. 3. In the first execution, the predictions are conditioned on the gold sequence and yield discrete distribution $p_\theta(X) = [p_1, \cdots, p_l]$ for each token (θ is network parameter, l is the number of codebook embedding). In the second execution, we aim to sample tokens according to the discrete distributions. However, direct sampling from a distribution will preclude the gradient backpropagation as shown in Fig. 3(b). A Gumbel sampling strategy is thus introduced with a reparameterization trick [13] to enable gradient backpropagation in discrete distribution sampling. Specially, the sampling operation is conducted on a Gumbel-softmax distribution [13] which is defined by: $\text{softmax}(1/\tau(p_\theta(X) + g))$, where $g \sim \text{Gumbel}(0, 1) = -\log(-\log U)$, $U \sim \text{Uniform}(0, 1)$. A sample x_i drawn from the Gumbel-softmax distribution can be denoted by:

$$x_i = \frac{\exp((\log(p_i)) + g_i)/\tau)}{\sum_{j=1}^{n} \exp((\log(p_j) + g_j)/\tau)} \quad \text{for } i = 1, 2, \cdots, n.$$

where τ is an annealing parameter. The sampling from a Gumbel-softmax distribution exactly approximates the sampling from the categorical distribution $p_\theta(X)$ as proved in [27]. In forward pass of network training, sampling is actually conducted on the Gumbel(0, 1) distribution which is independent of the network parameter θ. In backpropagation, the sampling operation is not involved in the gradient flow, which means that the stochasticity of sampling operation is transferred from $p_\theta(X)$ to the Gumbel(0, 1) distribution.

To schedule the sampling in accordance with the training process, we design a *Gumbel sampling* strategy based on the prediction reliability. Considering sampled tokens are more difficult to learn than the ground truth especially at the early training stage, we only sample tokens for positions with high prediction reliability as denoted by R [20]. For a ground truth embedding γ_i and predicted distributions $[p_1, \cdots , p_l]$ associated with normalized codebook embeddings $[\gamma_1, \ldots , \gamma_l]$, the prediction reliability R_i can be quantified by the weighted summation of the inner products of embeddings:

$$R_i = \sum_{j=1}^{l} p_j * \gamma_j \cdot \gamma_i, \quad i \in [1, n]. \tag{5}$$

$R_i \in [0, 1]$ accurately indicates the similarity between the predicted token distribution and the ground truth token, and measures whether the prediction reliability reaches the threshold (0.9 by default) to conduct token sampling.

After obtaining a sequence representing the model prediction for each position, we mix the gold tokens and predicted tokens with a given probability which is a function of the training step and is calculated with a selected schedule. We then pass the new mixed sequence to the transformer for the second execution to yield the final predictions. Note that only the gradient of the second execution is backpropagated in model training.

Computational Cost. Twice execution for Gumbel sampling will increase the training time, which can be mitigated by reducing the frequency of applying Gumbel sampling. In our implementation, the Gumbel sampling is applied for every 4 iterations by default. The average speed of our model with Gumbel sampling is 2.8 iteration/s, and the model speed without Gumbel sampling is 3.0 iteration/s. Therefore, the increase of computational cost is very limited.

4   Experiments

4.1   Experimental Settings

Datasets. We benchmark our method over multiple public datasets in conditional image generation.

- ADE20k [66] has 20k training images associated with a 150-class segmentation mask. We use its semantic segmentation as conditional inputs in experiments.

Table 1. Comparing IQ-VAE with state-of-the-art image generation methods over four conditional image generation tasks. The adopted evaluation metrics include FID, SWD and LPIPS.

Methods	ADE20K			CelebA-HQ (edge)			DeepFashion		
	FID ↓	SWD ↓	LPIPS ↑	FID ↓	SWD ↓	LPIPS ↑	FID ↓	SWD ↓	LPIPS ↑
Pix2pixHD [48]	61.08	28.47	N/A	42.70	33.30	N/A	25.20	**16.40**	N/A
Pix2pixSC [47]	56.23	24.52	0.378	49.39	33.20	0.193	28.49	21.13	0.172
BicycleGAN [68]	62.52	33.27	0.405	44.63	31.96	0.224	29.82	22.74	0.251
StarGAN v2 [4]	98.72	65.47	**0.451**	48.63	41.96	0.214	43.29	30.87	0.296
DRIT++ [16]	105.1	81.82	0.432	50.31	47.21	0.313	52.67	42.34	0.281
SPADE [35]	33.90	19.70	0.344	31.50	26.90	0.207	36.20	27.80	0.231
SMIS [70]	42.17	22.67	0.416	23.71	22.23	0.201	26.23	23.73	0.240
VQ-GAN [6]	35.50	21.50	0.421	16.23	23.33	0.330	16.49	21.20	0.314
IQ-VAE	**29.77**	**17.44**	0.447	**14.71**	**19.74**	**0.344**	**11.15**	19.01	**0.320**

Fig. 4. Qualitative illustration of IQ-VAE and state-of-the-art image generation methods over four types of generation tasks. IQ-VAE is able to generate faithful images with high fidelity.

- CelebA-HQ [22] has 30,000 high quality face images whose semantic maps and edges serve as the condition for image generation.
- DeepFashion [21] has 52,712 person images of different appearances and poses. We use the key points of the person images as conditional inputs in experiments.
- COCO-Stuff [1] augments COCO [19] with pixel-level stuff annotations. We use its layout as condition for image generation.
- CUB-200 [49] has 200 bird species with attribute labels and we use it for text-to-image generation.
- Sub-URMP [2] is a subset of URMP [17] and we use it for audio-to-image generation.

Evaluation Metrics. We evaluate the proposed IQ-VAE on the tasks of semantic-to-image, edge-to-image and keypoint-to-image generation, as these tasks have rich prior studies for comprehensive yet fair benchmarking. We

Table 2. The parameter setting in the proposed transformer and IQ-VAE.

Transformer		IQ-VAE	
Parameters	Setting	Parameters	Setting
Learning rate	1.5e−4	Learning rate	1.5e−4
Batch size	32	Batch size	32
Epoch	50	Epoch	100
Vocabulary size	1024	Codebook embedding number	1024
Embedding number	1024	Codebook embedding dimension	256
Sequence length	512	Feature number	256
Number of transformer block	24		

Fig. 5. Illustration of diverse image generation by the proposed IQ-VAE: faithful yet diverse images are successfully generated with different types of conditional inputs such as semantic maps, edge maps, key points, layout maps, as well as heterogeneous conditions like texts and audios.

assess the compared methods with several widely adopted evaluation metrics. Specifically, *Fréchet Inception Score (FID)* [8] and *Sliced Wasserstein distance (SWD)* [14] are employed to evaluate the quality of generated images. *Learned Perceptual Image Patch Similarity (LPIPS)* [64] measures the distance between image patches, which is employed to evaluate the diversity of generated images and reconstruction performance of auto-encoder.

Implementation Details. The proposed model is optimized with a learning rate of 1.5e−4. The auto-regressive transformer is implemented based on the GPT2 architecture [38] with a input size of 256. AdamW [23] solver is adopted with $\beta_1 = 0.9$ and $\beta_2 = 0.95$. All experiments are conducted on 4 T V100 GPUs with a batch size of 32. The size of generated images is 256×256 for all evaluated generation tasks. The transformer is implemented based on minGPT[1]. Table 2 shows parameter setting in the transformer and IQ-VAE.

---

[1] https://github.com/karpathy/minGPT.

4.2 Quantitative Results

We compare the proposed IQ-VAE with several state-of-the-art conditional image generation methods including 1) Pix2pixHD [48]; 2) Pix2pixSC [47]; 3) BicycleGAN [68]; 4) StarGAN v2 [4]; 5) DRIT++ [16]; 6) SPADE [35]; 7) SMIS [70]; 8) Taming Transformer [6].

Table 3. Ablation study of IQ-VAE on ADE20k. VR and None denote the proposed variational regularizer and no regularization, respectively. GS denotes the proposed Gumbel sampling.

Models	FID ↓	SWD ↓	LPIPS ↑
VQ-GAN	35.50	21.50	0.421
IQ-VAE(None)	31.88	19.14	0.441
IQ-VAE(VR)	31.41	18.71	**0.450**
IQ-VAE(VR) + GS	**29.77**	**17.44**	0.447

In the quantitative experiments, all compared methods generate diverse images except Pix2PixHD [48] which does not support diverse generation. Table 1 shows experimental results in FID, SWD and LPIPS. It can be observed that IQ-VAE outperforms all compared methods across most metrics and tasks consistently. DRIT++ [16] and StarGAN v2 [4] achieve relatively high LPIPS scores by sacrificing the image quality as measured by FID and SWD, while SPADE [35] and SMIS [70] achieve decent FID and SWD scores with degraded LPIPS scores. The proposed IQ-VAE employs powerful variational auto-encoders to achieve high-fidelity image synthesis and a auto-regressive model for faithful image diversity modeling, thus achieving superior performance in terms of image quality and diversity. Compared with Taming transformer [6], the proposed IQ-VAE allows to quantize the image sequences and conditional sequence jointly and boosts the auto-regressive modeling for better FID and SWD scores. In addition, the proposed Gumbel sampling introduces uncertainty of distribution sampling into the training process which mitigates the exposure bias and improves the inference performance clearly. As the mixed sequence serves as certain extra data augmentation, the Gumbel sampling also helps to alleviate the over-fitting of auto-regressive model effectively.

4.3 Qualitative Evaluation

We perform qualitative comparisons as shown in Fig. 4. The experiments are conducted over six datasets including ADE20k [66], CelebA-HQ [22], Deep-Fashion [21], COCO-Stuff [1], CUB-200 [49], and Sub-URMP [2]. The splits of training and testing sets on all above datasets follow the default split settings. In addition, the data used in the experiments do not contains person identity related information or offensive contents. It can be seen that IQ-VAE achieves the best visual quality and presents remarkable coherence with the condition. SPADE [35] and SMIS [70] adopt VAE to constraint the distribution of encoded features which cannot capture the complex distributions of real images. Star-GAN v2 [4] and DRIT++ [16] adopt single latent code to encode image styles, which tends to capture global styles but misses local details.

IQ-VAE also generalizes well and demonstrates superior synthesis quality and diversity in various generation tasks as illustrated in Fig. 5. It can be observed that IQ-VAE is capable of synthesizing high-fidelity images with various conditional inputs such as semantic maps, edge maps, keypoints, layout maps as well as heterogeneous conditions such as texts and audios.

4.4   Ablation Study

We conduct extensive ablation studies to evaluate IQ-VAE as shown in Table 3. The baseline is selected as VQ-GAN (namely Taming Transformer [6]). Replacing VQ-GAN with the proposed IQ-VAE without any regularization in IQ-VAE(None) brings in marginal improvement. The proposed variational regularizer with adaptive weights in IQ-VAE(VR) improves the generation performance, demonstrating the effectiveness of adaptive weights learning. Finally, including the Gumbel sampling remarkably boosts the performance as indicated in IQ-VAE(VR)+GS.

We study the effect of feature sizes for discrete representation in IQ-VAE and Fig. 6 shows experimental results on the CelebaHQ dataset. As Fig. 6 shows, we specify the size of representation features in terms of a factor F where Fx denotes a feature size of $x \times x$. Note the input size of transformer is always fixed at 16×16. The horizontal axis of the graph shows reconstruction error as measured by LPIPS [64] which indicates the upper bound of generation quality (lower is better), while the vertical axis shows negative log-likelihood from the transformer which indicates

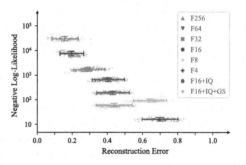

Fig. 6. Trade-off between negative log-likelihood and reconstruction error with different sizes of encoded features on CelebaHQ [22].

the performance of the auto-regressive modeling (lower is better). We can see that there is a trade-off between the negative log-likelihood and reconstruction error. Though an encoded feature of small size allows the transformer to better model the image distribution, the reconstruction deteriorates severely after a certain value (F16 in this case). The proposed integrated quantization and Gumbel sampling instead improve the negative log-likelihood remarkably without sacrificing the reconstruction performance clearly.

4.5   User Study

We conduct crowdsourcing user study to evaluate the quality of generated images as shown in Fig. 7. Specifically, 100 pairs of images generated by all compared methods are shown to 10 users who selected the image with the best visual quality. As shown in Fig. 7, we compared the proposed IQ-VAE with several state-of-the-art generation methods including BicycleGAN [68], SPADE [35], SMIS [70], and Taming Transformer [6]. The images generated by the proposed IQ-VAE are much more realistic according to the user feedback.

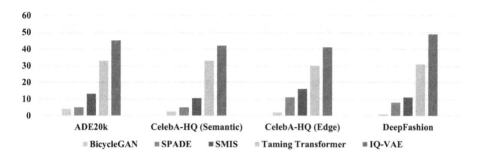

Fig. 7. User study over four datasets ADE20K [66], CelebA-HQ [22] (semantic), CelebA-HQ [22] (edge), DeepFashion [21]. The bars show the number of images that AMT users ranked with the best visual quality.

5   Conclusions

This paper presents IQ-VAE, an auto-regressive framework with integrated quantization for conditional image synthesis. We propose a novel variational regularizer to regularize the feature distribution structures of conditional inputs and real images, which boosts the auto-regressive modeling clearly. To mitigate the misalignment between training and inference of auto-regressive model, a Gumbel sampling strategy with a reliability-based scheduling is included in the training stage and improves the inference performance by a large margin. Quantitative and qualitative experiments show that IQ-VAE is capable of generating diverse yet high-fidelity images with multifarious conditional inputs.

Limitations. As auto-regression is adopted in the model to predict image sequence, the inference speed is inevitably constrained which may limit the application of the proposed model in time-critical tasks. Although some works [31,50] have been proposed to speed up the autoregressive sampling, the acceleration for the inference of auto-regressive model is still an open challenge.

Potential Negative Societal Impacts. This work aims to synthesize diverse yet high-fidelity images with given conditional inputs. It could have negative impacts if it is used for certain illegal purpose such as image forgery and manipulation.

Acknowledgement. This study is supported under the RIE2020 Industry Alignment Fund - Industry Collaboration Projects (IAF-ICP) Funding Initiative, as well as cash and in-kind contribution from the industry partner(s).

References

1. Caesar, H., Uijlings, J., Ferrari, V.: COCO-Stuff: thing and stuff classes in context. In: Proceedings of the IEEE Conference on Computer Vision and Pattern Recognition, pp. 1209–1218 (2018)
2. Chen, L., Srivastava, S., Duan, Z., Xu, C.: Deep cross-modal audio-visual generation. In: Proceedings of the on Thematic Workshops of ACM Multimedia 2017, pp. 349–357 (2017)
3. Chen, M., et al.: Generative pretraining from pixels. In: International Conference on Machine Learning, pp. 1691–1703. PMLR (2020)
4. Choi, Y., Uh, Y., Yoo, J., Ha, J.W.: StarGAN v2: diverse image synthesis for multiple domains. In: Proceedings of the IEEE/CVF Conference on Computer Vision and Pattern Recognition, pp. 8188–8197 (2020)
5. Doersch, C.: Tutorial on variational autoencoders. arXiv preprint arXiv:1606.05908 (2016)
6. Esser, P., Rombach, R., Ommer, B.: Taming transformers for high-resolution image synthesis. In: Proceedings of the IEEE Conference on Computer Vision and Pattern Recognition (2021)
7. Goodfellow, I., et al.: Generative adversarial nets. In: Advances in Neural Information Processing Systems, vol. 27 (2014)
8. Heusel, M., Ramsauer, H., Unterthiner, T., Nessler, B., Hochreiter, S.: GANs trained by a two time-scale update rule converge to a local Nash equilibrium. In: Advances in Neural Information Processing Systems, pp. 6626–6637 (2017)
9. Hochreiter, S., Schmidhuber, J.: Long short-term memory. Neural Comput. **9**(8), 1735–1780 (1997)
10. Huang, X., Belongie, S.: Arbitrary style transfer in real-time with adaptive instance normalization. In: Proceedings of the IEEE International Conference on Computer Vision, pp. 1501–1510 (2017)
11. Huang, X., Liu, M.-Y., Belongie, S., Kautz, J.: Multimodal unsupervised image-to-image translation. In: Ferrari, V., Hebert, M., Sminchisescu, C., Weiss, Y. (eds.) ECCV 2018. LNCS, vol. 11207, pp. 179–196. Springer, Cham (2018). https://doi.org/10.1007/978-3-030-01219-9_11
12. Isola, P., Zhu, J.Y., Zhou, T., Efros, A.A.: Image-to-image translation with conditional adversarial networks. In: Proceedings of the IEEE Conference on Computer Vision and Pattern Recognition, pp. 1125–1134 (2017)
13. Jang, E., Gu, S., Poole, B.: Categorical reparameterization with gumbel-softmax. arXiv preprint arXiv:1611.01144 (2016)
14. Karras, T., Aila, T., Laine, S., Lehtinen, J.: Progressive growing of GANs for improved quality, stability, and variation. arXiv preprint arXiv:1710.10196 (2017)
15. Kingma, D.P., Welling, M.: Auto-encoding variational bayes. arXiv preprint arXiv:1312.6114 (2013)
16. Lee, H.Y., et al.: DRIT++: diverse image-to-image translation via disentangled representations. Int. J. Comput. Vis. **128**(10), 2402–2417 (2020). https://doi.org/10.1007/s11263-019-01284-z
17. Li, B., Liu, X., Dinesh, K., Duan, Z., Sharma, G.: Creating a multitrack classical music performance dataset for multimodal music analysis: challenges, insights, and applications. IEEE Trans. Multimed. **21**(2), 522–535 (2018)
18. Li, Y., Cheng, Y., Gan, Z., Yu, L., Wang, L., Liu, J.: BachGAN: high-resolution image synthesis from salient object layout. In: Proceedings of the IEEE Conference on Computer Vision and Pattern Recognition (2020)

19. Lin, T.-Y., et al.: Microsoft COCO: common objects in context. In: Fleet, D., Pajdla, T., Schiele, B., Tuytelaars, T. (eds.) ECCV 2014. LNCS, vol. 8693, pp. 740–755. Springer, Cham (2014). https://doi.org/10.1007/978-3-319-10602-1_48
20. Liu, Y., Meng, F., Chen, Y., Xu, J., Zhou, J.: Confidence-aware scheduled sampling for neural machine translation. arXiv preprint arXiv:2107.10427 (2021)
21. Liu, Z., Luo, P., Qiu, S., Wang, X., Tang, X.: DeepFashion: powering robust clothes recognition and retrieval with rich annotations. In: Proceedings of the IEEE Conference on Computer Vision and Pattern Recognition, pp. 1096–1104 (2016)
22. Liu, Z., Luo, P., Wang, X., Tang, X.: Deep learning face attributes in the wild. In: Proceedings of the IEEE International Conference on Computer Vision, pp. 3730–3738 (2015)
23. Loshchilov, I., Hutter, F.: Decoupled weight decay regularization. arXiv preprint arXiv:1711.05101 (2017)
24. Lucas, J., Tucker, G., Grosse, R., Norouzi, M.: Don't blame the ELBO! A linear VAE perspective on posterior collapse. arXiv preprint arXiv:1911.02469 (2019)
25. Ma, L., Jia, X., Georgoulis, S., Tuytelaars, T., Van Gool, L.: Exemplar guided unsupervised image-to-image translation with semantic consistency. In: International Conference on Learning Representations (2018)
26. Ma, L., Jia, X., Sun, Q., Schiele, B., Tuytelaars, T., Van Gool, L.: Pose guided person image generation. In: Advances in Neural Information Processing Systems, pp. 406–416 (2017)
27. Maddison, C.J., Tarlow, D., Minka, T.: A* sampling. In: NIPS (2014)
28. Mémoli, F.: Gromov-Wasserstein distances and the metric approach to object matching. Found. Comput. Math. **11**(4), 417–487 (2011). https://doi.org/10.1007/s10208-011-9093-5
29. Men, Y., Mao, Y., Jiang, Y., Ma, W.Y., Lian, Z.: Controllable person image synthesis with attribute-decomposed GAN. In: Proceedings of the IEEE/CVF Conference on Computer Vision and Pattern Recognition, pp. 5084–5093 (2020)
30. Mihaylova, T., Martins, A.F.: Scheduled sampling for transformers. arXiv preprint arXiv:1906.07651 (2019)
31. Oord, A., et al.: Parallel WaveNet: fast high-fidelity speech synthesis. In: International Conference on Machine Learning, pp. 3918–3926. PMLR (2018)
32. van den Oord, A., Kalchbrenner, N., Vinyals, O., Espeholt, L., Graves, A., Kavukcuoglu, K.: Conditional image generation with PixelCNN decoders. arXiv preprint arXiv:1606.05328 (2016)
33. van den Oord, A., Vinyals, O., Kavukcuoglu, K.: Neural discrete representation learning. arXiv preprint arXiv:1711.00937 (2017)
34. Park, T., Efros, A.A., Zhang, R., Zhu, J.-Y.: Contrastive learning for unpaired image-to-image translation. In: Vedaldi, A., Bischof, H., Brox, T., Frahm, J.-M. (eds.) ECCV 2020. LNCS, vol. 12354, pp. 319–345. Springer, Cham (2020). https://doi.org/10.1007/978-3-030-58545-7_19
35. Park, T., Liu, M.Y., Wang, T.C., Zhu, J.Y.: Semantic image synthesis with spatially-adaptive normalization. In: Proceedings of the IEEE Conference on Computer Vision and Pattern Recognition, pp. 2337–2346 (2019)
36. Peyré, G., Cuturi, M., et al.: Computational optimal transport: with applications to data science. Found. Trends® Mach. Learn. **11**(5–6), 355–607 (2019)
37. Qiao, T., Zhang, J., Xu, D., Tao, D.: MirrorGAN: learning text-to-image generation by redescription. In: Proceedings of the IEEE/CVF Conference on Computer Vision and Pattern Recognition, pp. 1505–1514 (2019)
38. Radford, A., Wu, J., Child, R., Luan, D., Amodei, D., Sutskever, I., et al.: Language models are unsupervised multitask learners. OpenAI Blog **1**(8), 9 (2019)

39. Ramesh, A., et al.: Zero-shot text-to-image generation. arXiv preprint arXiv:2102.12092 (2021)
40. Schmidt, F.: Generalization in generation: a closer look at exposure bias. arXiv preprint arXiv:1910.00292 (2019)
41. Solomon, J.: Optimal transport on discrete domains. AMS Short Course on Discrete Differential Geometry (2018)
42. Sun, W., Wu, T.: Image synthesis from reconfigurable layout and style. In: Proceedings of the IEEE International Conference on Computer Vision, pp. 10531–10540 (2019)
43. Tang, H., Xu, D., Liu, G., Wang, W., Sebe, N., Yan, Y.: Cycle in cycle generative adversarial networks for keypoint-guided image generation. In: Proceedings of the 27th ACM International Conference on Multimedia, pp. 2052–2060 (2019)
44. Van Oord, A., Kalchbrenner, N., Kavukcuoglu, K.: Pixel recurrent neural networks. In: International Conference on Machine Learning, pp. 1747–1756. PMLR (2016)
45. Vaswani, A., et al.: Attention is all you need. arXiv preprint arXiv:1706.03762 (2017)
46. Vayer, T., Flamary, R., Tavenard, R., Chapel, L., Courty, N.: Sliced Gromov-Wasserstein. arXiv preprint arXiv:1905.10124 (2019)
47. Wang, M., et al.: Example-guided style-consistent image synthesis from semantic labeling. In: Proceedings of the IEEE Conference on Computer Vision and Pattern Recognition, pp. 1495–1504 (2019)
48. Wang, T.C., Liu, M.Y., Zhu, J.Y., Tao, A., Kautz, J., Catanzaro, B.: High-resolution image synthesis and semantic manipulation with conditional GANs. In: Proceedings of the IEEE Conference on Computer Vision and Pattern Recognition, pp. 8798–8807 (2018)
49. Welinder, P., et al.: Caltech-UCSD birds 200. California Institute of Technology (2010)
50. Wiggers, A., Hoogeboom, E.: Predictive sampling with forecasting autoregressive models. In: International Conference on Machine Learning, pp. 10260–10269. PMLR (2020)
51. Xu, H., Luo, D., Henao, R., Shah, S., Carin, L.: Learning autoencoders with relational regularization. In: International Conference on Machine Learning, pp. 10576–10586. PMLR (2020)
52. Xu, H., Luo, D., Zha, H., Duke, L.C.: Gromov-Wasserstein learning for graph matching and node embedding. In: International Conference on Machine Learning, pp. 6932–6941. PMLR (2019)
53. Yu, Y., et al.: Towards counterfactual image manipulation via clip. arXiv preprint arXiv:2207.02812 (2022)
54. Zhan, F., Xue, C., Lu, S.: GA-DAN: geometry-aware domain adaptation network for scene text detection and recognition. In: Proceedings of the IEEE International Conference on Computer Vision, pp. 9105–9115 (2019)
55. Zhan, F., et al.: Unbalanced feature transport for exemplar-based image translation. In: Proceedings of the IEEE Conference on Computer Vision and Pattern Recognition (2021)
56. Zhan, F., et al.: Bi-level feature alignment for versatile image translation and manipulation. arXiv preprint arXiv:2107.03021 (2021)
57. Zhan, F., et al.: GMLight: lighting estimation via geometric distribution approximation. arXiv preprint arXiv:2102.10244 (2021)
58. Zhan, F., Yu, Y., Wu, R., Zhang, J., Lu, S.: Multimodal image synthesis and editing: a survey. arXiv preprint arXiv:2112.13592 (2021)

59. Zhan, F., Yu, Y., Wu, R., Zhang, J., Lu, S., Zhang, C.: Marginal contrastive correspondence for guided image generation. In: Proceedings of the IEEE/CVF Conference on Computer Vision and Pattern Recognition, pp. 10663–10672 (2022)

60. Zhan, F., Zhang, C.: Spatial-aware GAN for unsupervised person re-identification. In: 2020 25th International Conference on Pattern Recognition (ICPR), pp. 6889–6896. IEEE (2021)

61. Zhan, F., et al.: EMLight: lighting estimation via spherical distribution approximation. In: Proceedings of the AAAI Conference on Artificial Intelligence, vol. 35, pp. 3287–3295 (2021)

62. Zhan, F., Zhang, J., Yu, Y., Wu, R., Lu, S.: Modulated contrast for versatile image synthesis. In: Proceedings of the IEEE/CVF Conference on Computer Vision and Pattern Recognition, pp. 18280–18290 (2022)

63. Zhang, P., Zhang, B., Chen, D., Yuan, L., Wen, F.: Cross-domain correspondence learning for exemplar-based image translation. In: Proceedings of the IEEE/CVF Conference on Computer Vision and Pattern Recognition, pp. 5143–5153 (2020)

64. Zhang, R., Isola, P., Efros, A.A., Shechtman, E., Wang, O.: The unreasonable effectiveness of deep features as a perceptual metric. In: Proceedings of the IEEE Conference on Computer Vision and Pattern Recognition, pp. 586–595 (2018)

65. Zhao, B., Meng, L., Yin, W., Sigal, L.: Image generation from layout. In: Proceedings of the IEEE Conference on Computer Vision and Pattern Recognition, pp. 8584–8593 (2019)

66. Zhou, B., Zhao, H., Puig, X., Fidler, S., Barriuso, A., Torralba, A.: Scene parsing through ade20k dataset. In: Proceedings of the IEEE Conference on Computer Vision and Pattern Recognition, pp. 633–641 (2017)

67. Zhu, J.Y., Park, T., Isola, P., Efros, A.A.: Unpaired image-to-image translation using cycle-consistent adversarial networks. In: Proceedings of the IEEE International Conference on Computer Vision, pp. 2223–2232 (2017)

68. Zhu, J.Y., et al.: Toward multimodal image-to-image translation. In: Advances in Neural Information Processing Systems 2017, pp. 466–477 (2017)

69. Zhu, P., Abdal, R., Qin, Y., Wonka, P.: SEAN: image synthesis with semantic region-adaptive normalization. In: Proceedings of the IEEE/CVF Conference on Computer Vision and Pattern Recognition, pp. 5104–5113 (2020)

70. Zhu, Z., Xu, Z., You, A., Bai, X.: Semantically multi-modal image synthesis. In: Proceedings of the IEEE/CVF Conference on Computer Vision and Pattern Recognition, pp. 5467–5476 (2020)

JoJoGAN: One Shot Face Stylization

Min Jin Chong[✉][iD] and David Forsyth[iD]

University of Illinois at Urbana-Champaign, Champaign, IL 61820, USA
mchong6@illinois.edu, daf@illinois.edu

Abstract. A *style mapper* applies some fixed style to its input images (so, for example, taking faces to cartoons). This paper describes a simple procedure – JoJoGAN – to learn a style mapper from a single example of the style. JoJoGAN uses a GAN inversion procedure and StyleGAN's style-mixing property to produce a substantial paired dataset from a single example style. The paired dataset is then used to fine-tune a Style-GAN. An image can then be style mapped by GAN-inversion followed by the fine-tuned StyleGAN. JoJoGAN needs just one reference and as little as 30 s of training time. JoJoGAN can use extreme style references (say, animal faces) successfully. Furthermore, one can control what aspects of the style are used and how much of the style is applied. Qualitative and quantitative evaluation show that JoJoGAN produces high quality high resolution images that vastly outperform the current state-of-the-art.

Keywords: Generative models · One-shot stylization · StyleGAN · Style transfer

1 Introduction

A *style mapper* applies some fixed style to its input images (so, for example, taking faces to cartoons). This paper describes a simple procedure to learn a style mapper from a single example of the style. Our procedure allows, for example, an unsophisticated user to provide a style example, and then apply that style to their choice of image. Because stylizing face images – make me look like JoJo – is so desirable to unsophisticated users, we describe our method in the context of face images, but the method applies to anything (Fig. 1).

To be useful, a procedure for learning a style mapper should: be easy to use; produce compelling and high quality results; require only one style reference, but accept and benefit from more; allow users to control how much style to transfer; and allow more sophisticated users to control what aspects of the style get transferred. We demonstrate with qualitative and quantitative evidence that our method meets these goals.

Learning a style mapper is hard because the natural method – use paired or unpaired image translation [4,13,40] – isn't really practical. Collecting a new

---

Supplementary Information The online version contains supplementary material available at https://doi.org/10.1007/978-3-031-19787-1_8.

S. Avidan et al. (Eds.): ECCV 2022, LNCS 13676, pp. 128–152, 2022.
https://doi.org/10.1007/978-3-031-19787-1_8

Fig. 1. JoJoGAN accepts a single style reference image (**top row**) and very quickly produces a style mapper that accepts an input (**left column**) and applies the style to that input. JoJoGAN can use extreme style references (**OOD stylizations**; the cat faces are JoJoGAN outputs for the human inputs above. Furthermore, JoJoGAN can apply styles to different extents (**Continuous stylization**); each row shows input; lightly stylized output; and strongly stylized output.

dataset per style is clumsy, and for many styles – Lucien Freud portraits, say – there may not be all that many examples. One might use few-shot learning techniques to fine-tune a StyleGAN [16] by adjusting the discriminator (as in [20, 23, 24, 29]). But these methods do not have detailed supervision from pixel-level losses and so mostly fail to capture distinct style details and diversity.

In contrast, JoJoGAN (our procedure) takes a reference image (or images – but one image is enough) and makes a paired dataset using GAN inversion and StyleGAN's style-mixing property. This paired dataset is used to fine-tune Style-GAN using a novel direct pixel-level loss. The mechanics are straightforward: we can obtain a mapper (and so a rich supply of stylized portraits) from a single reference image in under a minute. JoJoGAN can use extreme style references (say, animal faces) successfully. Natural procedures control what aspects of the style are used and how much of the style is applied. Qualitative examples show that the resulting images look much better than alternative methods produce. Quantitative evidence strongly supports our method. Training and demo code is available at https://github.com/mchong6/JoJoGAN.

2   Related Work

Style transfer methods likely start with [7,10]; these are one shot methods, but do not result in style mappers in any natural way. Neural style transfer (NST) methods start with [9]; Johnson *et al.* offer a learned mapper, trained with a large dataset and Gatys *et al.*'s procedure to stylize [14]. In contrast, our method uses much lesser data and produces much higher resolution images. A rich literature has followed, but general style transfer methods (for example [12,19,26]) cannot benefit from the detailed semantic and structural information captured by a GAN. Style transfer evaluation is mostly qualitative, but see [38]. Deformable Style Transfer (DST) [17] corrects structural errors by estimating spatial warps, then performing traditional neural style transfer; DST achieves impressive one-shot stylization, but warp estimation errors have significant effects and are hard to avoid (Fig. 10).

StyleGAN [15,16] remains the state-of-the-art unconditional generative model due to its unique style-based architecture. StyleGAN's AdaIN modulation layers (originally from [12]) have been shown to be disentangled and exhibit impressive editability [2,31,32]. StyleGAN has also been used as a prior for numerous tasks such as superresolution [22] and face restoration [35]. Pinkney *et al.* [27] first showed that finetuning the StyleGAN on a new dataset and performing layer swapping allows the StyleGAN to learn image to image translation with a relatively small dataset. But even obtaining a small paired dataset is hard: collection is difficult and expensive; one needs a new dataset for each new style; and in some cases (for example, Lucien Freud portrait style) there won't be many style images in the first place. In contrast, JoJoGAN creates a paired dataset from a single style reference by manipulating a pretrained StyleGAN2 [16] and a GAN inversion procedure, then finetunes using the created dataset.

One shot learning covers many applications (detection; classification; image synthesis), and methods remain specialized to their application. This paper focuses on one-shot image stylization, with a particular emphasis on faces.

One shot face stylization is now established. Learning a style mapper from very few examples results in overfitting problems. To control overfitting, [20,25] introduce regularization terms while [23,29] enforces constraints in the network's weights. These methods need tens to hundreds of style example images; in contrast, JoJoGAN works with one. Furthermore, these methods have difficulty capturing small style details, likely because they rely on an adversarial loss. BlendGAN [21] introduced a VGG-based style encoder and a weight blending module to learn arbitrary face stylization over a large styled faces dataset. As our comparisons show, this method fails to capture small but pertinent style details in face images. StyleGAN-NADA [8] uses CLIP [28] to perform zero/one shot image stylization based on text/image prompts, resulting in very strong generalization. However, as our comparisons show, StyleGAN-NADA fails to capture minute facial details that are important for face stylization.

Most similar to JoJoGAN is work by Zhu *et al.* [41] (detailed experimental comparison in Fig. 11 and Appendix); this also uses GAN inversion to find a corresponding real face from a reference, so creating a paired datapoint. Zhu *et al.* use this simple datapoint and a number of CLIP-based losses (from [28]). In contrast, JoJoGAN creates a large dataset of paired datapoints from a single one, and so needs only a simple pixel loss (with an optional identity loss). Zhu *et al.* use gradient descent inversion II2S (from [42]), which is slow but more accurate. In contrast, JoJoGAN uses feed forward inversion based on a simple encoder. Complex losses and slow inversion procedures mean Zhu *et al.* require some 15 min to train on a Titan XP; in contrast, JoJoGAN requires 1.

3 Methodology

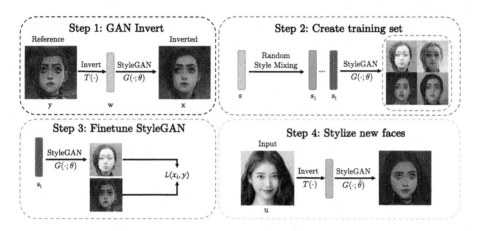

Fig. 2. Workflow: JoJoGAN's steps are: **GAN Inversion** to obtain a code s from the style reference; creating a **training set** $\mathcal{S}$ of similar s_i via random style mixing; **finetuning** a StyleGAN to obtain $\hat{\theta}$ so that $G(w_i; \hat{\theta}) \approx y$ using our perceptual loss; and **inference** by computing $G(T(u); \hat{\theta})$ for input u.

Write T for GAN inversion, G for StyleGAN, s for style parameters in StyleGAN's $\mathcal{S}$-space (notation after [36]; mixing in $\mathcal{S}$-space works better, see Appendix A.3), and θ for the parameters of the vanilla StyleGAN. JoJoGAN uses four steps (Fig. 2):

1. **GAN inversion:** We GAN invert the reference style image y to obtain a style code $w = T(y)$ and from that a set of s parameters $s(w)$.
2. **Training set:** We use s to find a set of style codes $\mathcal{S}$ that are "close" to s. Pairs (s_i, y) for $s_i \in \mathcal{S}$ will be our paired training set.
3. **Finetuning:** We finetune the StyleGAN to obtain $\hat{\theta}$ such that $G(s_i; \hat{\theta}) \approx y$.
4. **Inference:** For input u, our stylized face is $G(s(T(u)); \hat{\theta})$ (so $G \circ s \circ T$ is our style mapper).

Step 1: GAN Inversion: Remarkably, for any but extreme face style references y, $G(s(T(y)); \theta)$ is able to produce a fairly realistic unstylized face image (e.g., Fig. 2, step 1). Different GAN inverters recover different face images (Appendix Fig. 14) and we explore the control this can bring in Sect. 5.1.

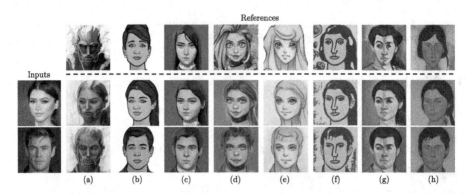

Fig. 3. JoJoGAN takes a single style reference image and produces a style mapper (reference images on top row; inputs far left). Note: clear following of input gender; subtle style details transferred (chin dimples in c; lip specularities in d); style lighting preserved (c); strong style effects in output, even from difficult styles (f, g, h); style idiosyncracies preserved (muscle fiber in a; bent nose in h; earrings in b).

Step 2: Training Set: We must find a set of style codes $\mathcal{S}$ that are "close" to $s(w)$. We use StyleGAN's style mixing mechanic. We use a 1024 resolution StyleGAN2 with 26 style modulation layers, so $s \in \mathbb{R}^{26 \times 512}$. Write $M \in \{0, 1\}^{26}$ for a fixed mask, FC for the style mapping layer of the StyleGAN and $z_i \sim \mathcal{N}(0, I)$. We produce new style codes using

$$s_i = M \cdot s + (1 - M) \cdot s(FC(z_i)) \tag{1}$$

(and do so per batch). Different M result in different stylization effects (Sect. 4).

Step 3: Finetuning StyleGAN: We now assume that a properly trained style mapper will map $s_i \in \mathcal{S}$ to y. This assumption certainly works, and is reasonable when the style mapper "reduces information" – so, for example, mapping faces with slightly different eye sizes or hair textures to the same reference image. We finetune StyleGAN to obtain

$$\hat{\theta} = \underset{\theta}{\text{argmin}} \, \text{loss}(\theta) = \underset{\theta}{\text{argmin}} \, \frac{1}{N} \sum_i^N \mathcal{L}(G(s_i; \theta), y) \tag{2}$$

where $\mathcal{L}$ is a novel perceptual loss (this choice is important; Sect. 3.1).

Step 4: Inference: For input u, our stylized face is $G(s(T(u)); \hat{\theta})$ (so $G \circ s \circ T$ is our style mapper). We could also generate random stylized samples by sampling random noise and generating with our finetuned StyleGAN (Fig. 3).

3.1   Perceptual Loss

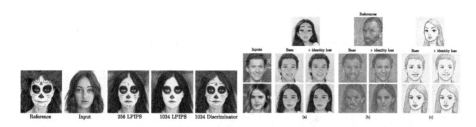

Fig. 4. Left: The choice of loss is important (this example is typical). For the style reference and the face shown, we train JoJoGAN using different losses. LPIPS at resolution 256 resolution leads to a loss of detail due to downsampling. LPIPS at 1024 does not control details, as the VGG filters (trained at 224) are not adapted to this scale. We match activations at layers of the pretrained discriminator from FFHQ-trained StyleGAN to compute a perceptual loss that preserves detail better. **Right:** some style inputs can result in outputs that lose identity (beard in b, for example). A straightforward identity loss can successfully control this effect, details in text.

The choice of loss in Eq. 2 is important (Fig. 4). While LPIPS [39] is a natural choice, it produces methods that lose detail. LPIPS is built on a VGG [33] backbone trained at a 224×224 resolution, but StyleGAN produces 1024×1024 images. The standard way to handle this mismatch is to downsample the images to 256×256 before computing LPIPS [1, 16, 34]. But this downsampling means we cannot control fine-grained details, which are mostly lost. Similarly, computing LPIPS at the native 1024 resolution leads to a complete loss of fine-grained detail as the VGG filters are not adapted to this resolution.

The pretrained StyleGAN discriminator is trained at the same resolution as the generator. The training process means that the discriminator computes features that do not ignore details (otherwise the generator could produce low-detail images). Discriminator features are known to stabilize GAN training when averaged over batches [30]. We choose to use the difference in discriminator activations at particular layers, per image (details in Appendix A.5). Write $D(\cdot)$ for the activations; then $\mathcal{L}(G(s_i; \theta), y) = \| D(G(s_i; \theta)) - D(y) \|_1$. A version of this loss is used in GPEN [37] but to our knowledge, we are the first to compare it with others and show how effective it is.

4   Variants

Controlling Identity: Some style references distort the original identity of the inputs (Fig. 4). In such cases, writing sim for cosine similarity and F for a pretrained face embedding network (we use ArcFace [6]), we use

$$\mathcal{L}_{id} = 1 - sim(F(G(s_i; \theta)), F(G(s_i; \hat{\theta}))) \tag{3}$$

to compel the finetuned network to preserve identity. We use identity loss only in Fig. 4(b) for references that severely distort the identity. We do not use identity loss for any other figures.

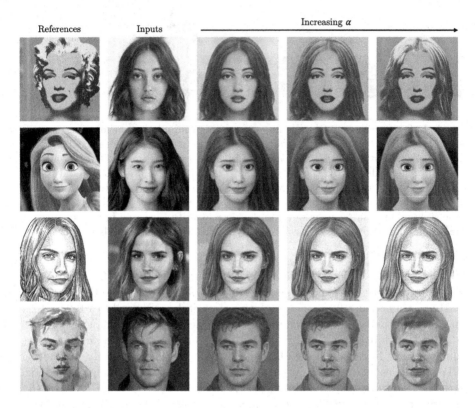

Fig. 5. Feature interpolation allows a user to control style intensity. As α increases, the results take the style of the reference more strongly.

Controlling Style Intensity by Feature Interpolation: Feature interpolation [5] allows us to vary the intensity of the style. Let f_i^A be intermediate feature maps of layer i from the original StyleGAN and f_i^B from JoJoGAN; we get continuous face stylization by using $f = (1 - \alpha)f_i^A + \alpha f_i^B$ where α is the interpolation factor. Increasing α results in stronger style intensity (Fig. 5).

Extreme Style References: For JoJoGAN to work, $\mathcal{S}$ has to consist of s_i that produce sensible responses from the StyleGAN. If the style reference is (roughly) a human face, there are no problems. An extreme style reference image is one where GAN inversion produces s that is out of distribution for the StyleGAN, for example, an image of an animal face. We are not aware of any test (other than trying) to distinguish between extreme and standard style references, but Fig. 19 in the Appendix demonstrates that using s from GAN inversion on animal

Mean Face   Inputs

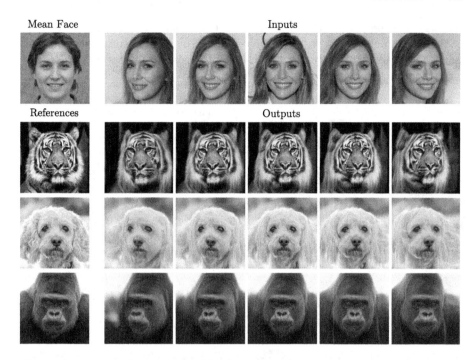

References   Outputs

Fig. 6. OOD references and using $\overline{w}$: JoJoGAN is able to handle OOD references that do not invert well by using mean style code $\overline{w}$. Even on animal faces which are semantically very different the human faces StyleGAN was trained on, JoJoGAN can generate realistic animal faces with poses that matches the input.

faces results in poor style transfer. For extreme style references y, rather than use $s(T(y))$ to construct $\mathcal{S}$, we use the mean style code $\overline{s} = \sum_1^{10000} s(FC(z \sim \mathcal{N}(0, I)))$ (note this style code is the best possible estimate of $s(T(y))$ for an image y *that one does not have*). With this modification, JoJoGAN works well on extreme style references (Fig. 6; note how the animal head poses are controlled by the input images) (Fig. 7).

Multi-shot Stylization: JoJoGAN extends to multi-shot stylization in a natural way (use each reference to construct a $\mathcal{S}_k$ for each reference y_k; now finetune using

$$\frac{1}{M * N} \sum_j^M \sum_i^N \mathcal{L}(G(s_{ij}; \theta), y_j).$$

Using more than one reference produces small but useful qualitative improvements in the style mapper (Fig. 12)

5   Controlling Aspects of Style

Style transfer is intrinsically ambiguous. The output should be "like" the reference in style, and "like" the input in content, but the distinction between content and style is vague. JoJoGAN offers methods to choose whether (say) the output should have exaggerated eyes (like the reference) or more natural eyes (like the input). Simple control is obtained by choice of mask and by loss. More detailed control follows by careful attention to the GAN inversion.

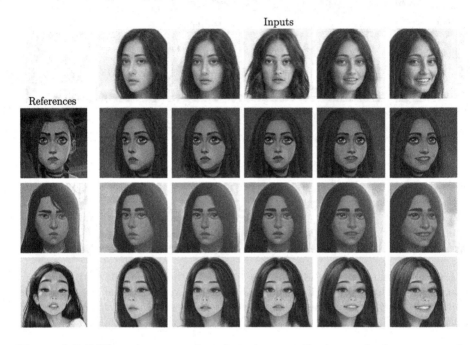

Fig. 7. JoJoGAN produces smooth and consistent stylization as the face moves and changes expressions.

Controlling Aspects of Style by Mask Choice and by Loss: Different choices of M will produce significant differences in $\mathcal{S}$, and so in results. Replacing too many elements of s with random numbers may result in a JoJoGAN that maps every face to the style reference; replacing too few means finetuning sees too few examples. Furthermore, replacing elements at locations corresponding to different StyleGAN layers controls different effects (see [16]). Figure 8 demonstrates this choice has significant effects by displaying results from two different M. The first gives dataset $\mathcal{X}$, the second $\mathcal{C}$. Both masks are chosen to maintain the input face pose and hairstyles while allowing features such as eye sizes and textures to vary, so the mask has ones in locations known to correspond to pose and zeros in those known to correspond to eye sizes, see [2]. But $\mathcal{C}$ is chosen so that the color of the input is preserved (so ones in relevant locations); and $\mathcal{X}$ so that color is driven by the style example. To ensure that the color of the input is preserved for the $\mathcal{C}$ case, we apply the loss in Eq. (2) to grayscale versions of the relevant images. This means the StyleGAN is finetuned to obtain the spatial appearance of the style target, but not its colors (variants in Appendix A.4)

Input   References

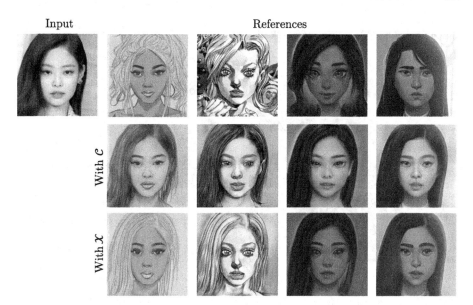

Fig. 8. The aspects of style that are transferred can be chosen using M. Section 5 describes our procedures to create two different datasets $\mathcal{C}$ and $\mathcal{X}$ from different choices of M that yield different stylizations. Finetuning using $\mathcal{C}$ mostly preserves the colors of the input; finetuning using $\mathcal{X}$ mostly reproduces the colors of the style example.

5.1   Control by Manipulating GAN Inversion

The choice of GAN inverter matters. If the GAN inverter produces an extremely realistic face from the reference, JoJoGAN will be trained to map s_i that represent highly realistic faces to the style reference, and so will tend to produce aggressively stylized faces. By the same argument, if the GAN inverter produces a somewhat stylized face from the reference, JoJoGAN will tend to produce lightly stylized faces and preserve the features of the input face (so an input with small eyes will result in an output with small eyes, say – example in Appendix Fig. 14). This effect can be used to control how much and what style is transferred by blending inverted codes.

Using two GAN inverters is clumsy in practice, but recall the mean style code is the best possible estimate of $s(T(y))$ for an image y *that one does not have*). It can thus represent the output of a (rather bad, but very fast) GAN inverter. We produce a virtual inverter $V(y)$ by blending the code produced by our standard inverter with the mean, using the procedure of Sect. 3 (but a different mask M). The blend is adjusted so that $G(s(V(y)); \theta)$ has desirable properties (so, for example, to preserve the eyes of the reference, $G(s(V(y)); \theta)$ should have realistic eyes). We then apply the JoJoGAN pipeline using V rather than T to generate training data. Using V rather than T in training changes the pairs (s_i, y) used in finetuning, and so the behavior of G. At inference, we compute $G(s(T(u); \hat{\theta})$ as before. Figure 9 demonstrates the extent of our style

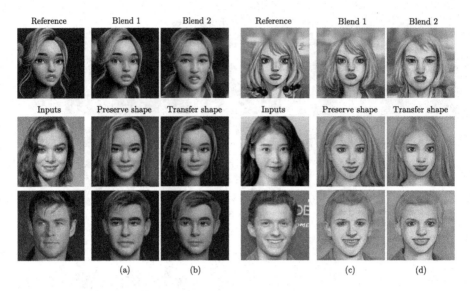

Fig. 9. How style is transferred can be controlled by blending the codes from two GAN inverters, then applying the JoJoGAN pipeline. (for these examples, one inverter just produces the mean code). For each reference, **top** shows $G(s(V(y)); \theta)$ for different blends. Notice how blending the inverter codes produces substantial changes in the inversion (eg **left** reference). By choice of blend, we can produce style mappers that tend to **preserve** the shape of input eye, nose and face or to **transfer** shapes from the reference. So, for example, (a) and (c) have eyes more like the input; but (b) and (d) have larger eyes, more like the reference. (b) has significantly smaller faces than (a).

control. In Fig. 9(b), using the blended inversion gives us larger eyes and thicker lips compared to using the accurate inversion (a). Further detail on blending the inverter is in Appendix A.1.

6   Experiments

Setup: For GAN inversion, we use ReStyle [1]. We finetune JoJoGAN for 200 to 500 iterations depending on the reference with Adam optimizer [18] at a learning rate of 2×10^{-3}. Finetuning on an Nvidia A40 takes about 30 to 60 s.

Qualitative Evaluation: A style mapper should: produce good looking outputs; faithfully transfer features from the style reference; and preserve the identity of the input. Qualitative evaluation shows JoJoGAN has these properties and vastly outperforms current methods.

Comparisons: Figure 10 shows comparisons of JoJoGAN to the state-of-the-art one/few shot stylization methods StyleGAN-NADA [8], BlendGAN [21], Ojha *et al.* [24] and DST [17]. JoJoGAN captures small style-defining details well while preserving the input identity. JoJoGAN results are typically improved when there are multiple consistent style references. Figure 12 compares several

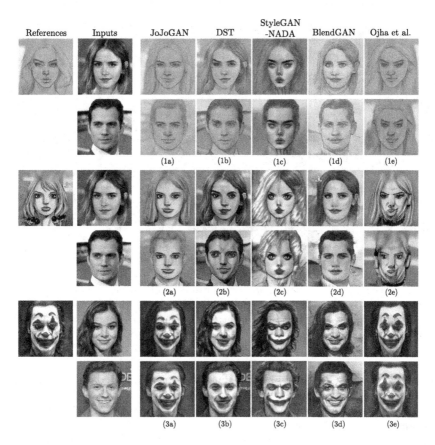

Fig. 10. JoJoGAN offers visible qualitative improvements over current SOTA methods for one shot face stylization. JoJoGAN captures the distinctive rendering style of the reference while preserving input pose, expression and identity. Note: excessive contrast (1b); color errors (1c, 2b, 3d); distorted facial layout (d, e); chin shape (b).

one-shot stylizations of individual references along with a multi-shot stylization using all references. Notice that the one-shot stylizations copy effects from the style reference aggressively (as it must), whereas when there are multiple style examples, JoJoGAN is able to blend details to hew more closely to the input.

Figure 11 shows a comparison with [41] (two examples in figure; others – except 2, for which we cannot find source – in supplementary). Note we can use only references shown in their paper, as the method is not open sourced.

Quantitative Evaluation by User Study: We proceed in two stages to reduce choice fatigue for users. From Fig. 10, DST gives good results in most cases while other methods produce examples with severe problems. We therefore compare JoJoGAN to non-DST methods in a first study, and to DST in a second. In each, users see a style reference, an input face, and stylizations from the methods and are asked to choose the stylization that best captures the style reference

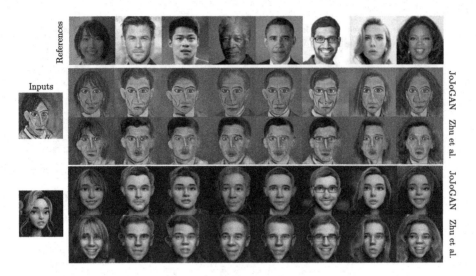

Fig. 11. We compare with Zhu *et al.* [41] on two examples for references used in their paper and described as hard cases there (others in supplementary). For each reference, the top row is JoJoGAN while the second row is Zhu *et al.* Note how their method distorts chin shape, while JoJoGAN produces strong outputs.

while preserving the original identity. The first study resulted in a total of 186 responses from 31 participants who overwhelmingly prefer JoJoGAN to other methods at 80.6%. The second study gathered 96 responses from 16 participants who prefer JoJoGAN to DST at 74%.

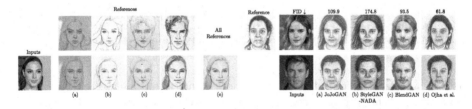

Fig. 12. Left: JoJoGAN's method extends cleanly to deal with multiple style references, if they are available. The figure compares one-shot stylizations of a reference with a multi-shot stylization for one input (more in supplementary). Note aggressive copying in the case of a single reference, including: noses in (a); lips in (b) and (c); and chin dimples in (b) and (d). This effect is notably muted when more references are available (and JoJoGAN can blend details from references), so (e) mouth and chin follow the input more closely. **Right:** JoJoGAN's FID score on the sketches dataset [24] is significantly larger than that of the best comparison. BlendGAN gets a better FID, but does not capture reference style well (note strong shading gradients, absent from the style reference); Ojha *et al.* get the best FID, but impose strong distortions on the input face (note other comparisons in Fig. 10).

Quantitative Evaluation by FID: FID [11] is a metric that is widely used to evaluate the quality and diversity of generated images by comparing population statistics. FID can be used to evaluate style mappers as follows [25]. Randomly select a reference from the style dataset and perform one shot stylization with it; now stylize a set of face images and compute the FID between the result and the original style dataset. To compute FID, we perform one shot stylization using the sketches dataset [24] and compute FID using the test set. JoJoGAN scores well behind SOTA on this metric. We report FID for JoJoGAN for candor and show FID for SOTA comparisons in Fig. 12, but point out that FID is a poor metric for style mappers. The procedure described cannot measure the fidelity with which the mapper preserves the input (for example, the FID for the completely ineffectual mapper that just produces a random sample from the style dataset would be close to zero). Furthermore, a perfect style mapper might produce a high FID with the protocol described, because its stylized images should be biased toward the input (for example, a perfect mapper with only male input images should produce a population of sketches that is not close to the original set of sketches). Finally, the datasets used for stylization are often very small (290 in the case of the sketches dataset), and computing FID for a small dataset is dangerous due to large biases [3].

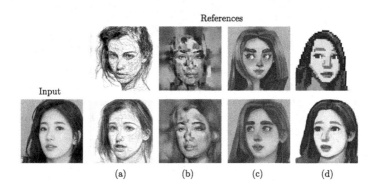

Fig. 13. Some style references are hard for JoJoGAN, likely a result of complicated structures in the style reference that are unfamiliar to StyleGAN. Note: loops in (a) mapped to strokes in the output; structure of brush strokes in (b) being broken up in output; gaze direction in (c) controlled by style reference rather than by input; high frequency pixel grids in (d) map to smooth strokes.

Failures: Using too small a $\mathcal{S}$ leads to problems (Appendix Fig. 17), typically artifacts and missing style details. As JoJoGAN only sees a single style reference, it does not always work for all style references. One common issue JoJoGAN has is that the eye gaze direction is often driven by the reference image rather than the input. The intended behavior is to preserve the gaze direction of the original input, yet JoJoGAN copies the reference instead. Figure 13 shows results on very difficult references, illustrating visual failure modes.

A   Appendix

A.1   Choice of GAN Inversion

Reference   e4e   II2S   ReStyle

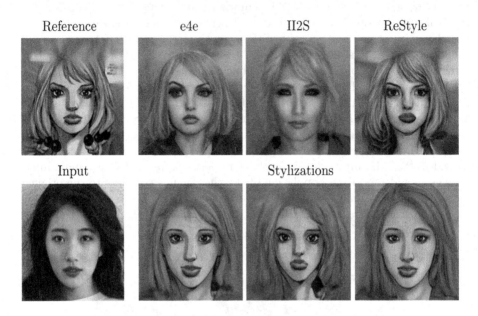

Input   Stylizations

Fig. 14. The choice of GAN inversion matters. We compare JoJoGAN trained on e4e [34], II2S [42], and ReStyle [1] inversions. II2S gives the most realistic inversions leading to stylizations that preserves shapes and proportions of the reference. ReStyle gives the most accurate reconstruction leading to stylization that better preserves the features and proportions of the input.

JoJoGAN relies on GAN inversion to create a paired dataset. We investigate the effect of using 3 different GAN inversion methods, e4e [34], II2S [42], and ReStyle [1] in Fig. 14.

Using e4e fails to accurately recreate the style reference and conveniently gives us a corresponding real face. On the other hand, ReStyle more accurately inverts the reference, giving a non-realistic face. II2S is a gradient-descent based method with a regularization term that allows us to map the style code to a higher density region in the latent space. The regularization term results in very realistic faces that are somewhat inaccurate to the reference.

The different inversions give us different JoJoGAN results. Training with ReStyle leads to clean stylization that accurately preserves the features and proportions of the input face. Training with II2S on the other hand leads to heavy stylization that borrows the shapes and proportions from the reference. However, this also leads to pretty heavy semantic changes from the input face and artifacts (note the change of identity and artifacts along the neck).

Reference   M1   M2   M3

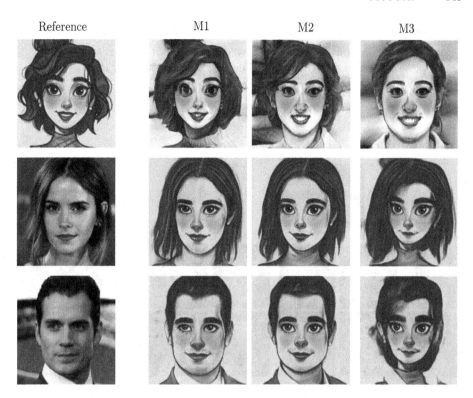

Fig. 15. The choice of M matters. M controls the blend between the inverted style with the mean style. $M1$ is the closest to the reference, leading to smaller features (e.g., eyes). $M3$ is the closest to a real face, leading to exaggerated features more like reference and also significant artifacts.

In practice, we blend the style codes from ReStyle and the mean face. For M, we borrow the style code from the mean face at layers 7, 9, and 11. This borrows the facial features of the mean face to the inversion. However, it is impossible to only affect the proportions of the features by simply blending coarsely at a layer level. For example, naively blending the mean face can change the expression of the inversion, e.g. from neutral to smiling or introduce artifacts. We thus have to blend at a finer scale, which we are able to do so by isolating specific facial features in the style space using RIS [2]. Figure 15 compares the results of using different M for blending. Note that when the blended image is more face-like ($M3$), the exaggerated features of the reference is transferred. However, significant artifacts are introduced, see $M3$ row 2. By carefully selecting M, we can transfer the exaggerated features while avoiding artifacts, see $M2$.

A.2   Identity Loss

Before computing identity loss, we grayscale the input images to prevent the identity loss from affecting the colors. The weight of the identity loss is reference-dependent, but we typically choose between 2×10^3 to 5×10^3.

A.3   Choice of Style Mixing Space

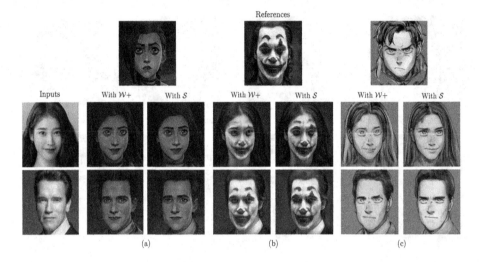

Fig. 16. We study how the choice of latent space to do style mixing affects JoJoGAN. Style mixing in $\mathcal{S}$ space gives more accurate color reproduction in (a) and (b) and better stylization effect (note the eyes) in (c).

Style mixing in Eq. (1) allows us to generate more paired datapoints. It is reasonable to map faces with slight differences in textures and colors to the same reference. As such it is pertinent that while we style mix to generate different faces, we need certain features such as identity, face pose, etc. to remain the same. We study how the choice of latent space to do style mixing affects the stylization. In Fig. 16 we see that style mixing in $\mathcal{S}$ gives better color reproduction and overall stylization effect. This is because $\mathcal{S}$ is more disentangled [36] and allows us to more aggressively style mix without changing the features we want intact.

A.4   Varying Dataset

Using $\mathcal{C}$ and $\mathcal{X}$ gives different stylization effects. Finetuning with $\mathcal{X}$ accurately reproduces the color profile of the reference while $\mathcal{C}$ tries to preserve the input color profile. However, this is insufficient to fully preserve the colors as we see in Fig. 17. Grayscaling the images before computing the loss in Eq. (2) in addition to finetuning with $\mathcal{C}$ gives us stylization effects without altering the color profile. We show that it is necessary to use both $\mathcal{C}$ and grayscaling to achieve this effect and using $\mathcal{X}$ and grayscaling is insufficient.

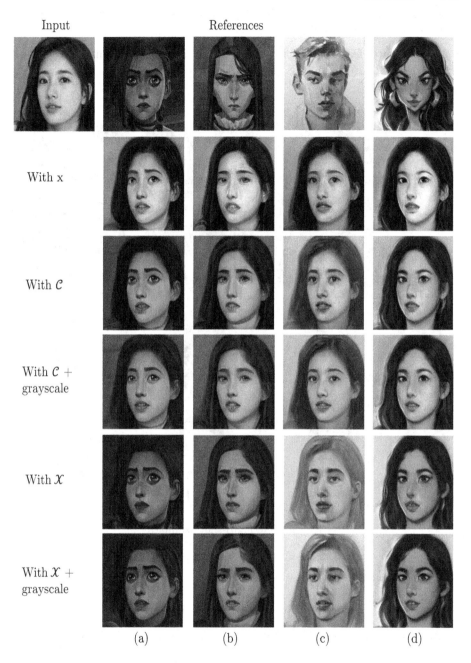

Fig. 17. The choice of training data has an effect. **First row:** when there is just one example in $\mathcal{W}$, JoJoGAN transfers relatively little style, likely because it is trained to map "few" images to the stylized example. **Second row:** same training procedure as in Fig. 8, using $\mathcal{C}$. **Third row:** same training procedure as second row but with grayscale images for Eq. (2). **Fourth row:** same training procedure as in Fig. 8, using $\mathcal{X}$. **Fifth row:** same training procedure as Fourth row but with grayscale images for Eq. (2).

A.5   Feature Matching Loss

For discriminator feature matching loss, we compute the intermediate activations after resblock $2, 4, 5, 6$ (Figs. 18, 20 and 21).

Fig. 18. More multi-shot examples

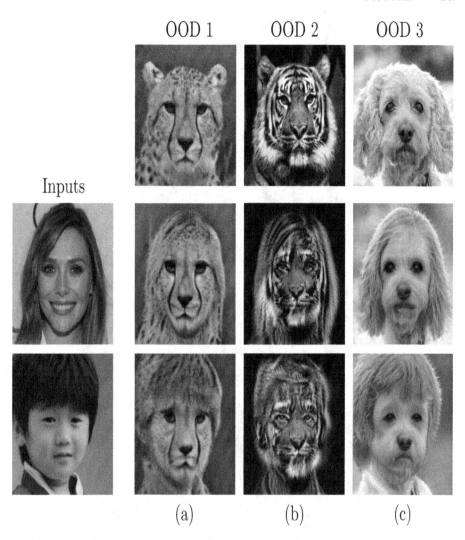

Fig. 19. JoJoGAN produces unsatisfactory style transfers on OOD cases, producing human-animal hybrids.

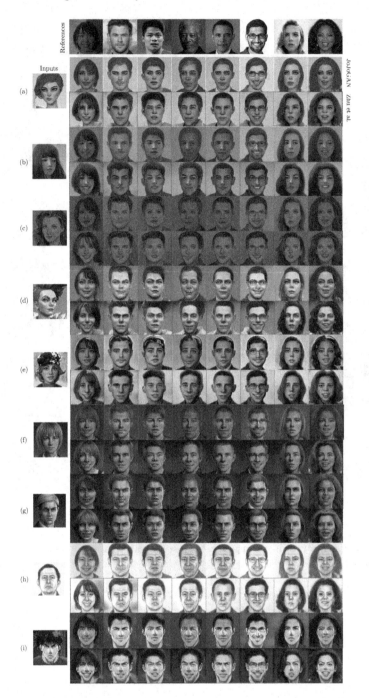

Fig. 20. We compare with Zhu *et al.* [41] on all examples for references used in their paper and described as hard cases there. For each reference, the top row is JoJoGAN while the second row is Zhu *et al.* Note how their method distorts chin shape, while JoJoGAN produces strong outputs.

References

Inputs

Fig. 21. JoJoGAN is a method to benefit from what a StyleGAN knows, and so should apply to other domains where a well-trained StyleGAN is available. Here we demonstrate JoJoGAN applied to LSUN-Churches.

References

1. Alaluf, Y., Patashnik, O., Cohen-Or, D.: Restyle: a residual-based StyleGAN encoder via iterative refinement. In: Proceedings of the IEEE/CVF International Conference on Computer Vision (ICCV), October 2021
2. Chong, M.J., Chu, W.S., Kumar, A., Forsyth, D.: Retrieve in style: unsupervised facial feature transfer and retrieval. In: Proceedings of the IEEE/CVF International Conference on Computer Vision (ICCV), pp. 3887–3896, October 2021

3. Chong, M.J., Forsyth, D.: Effectively unbiased fid and inception score and where to find them. In: Proceedings of the IEEE/CVF Conference on Computer Vision and Pattern Recognition, pp. 6070–6079 (2020)
4. Chong, M.J., Forsyth, D.: GANs N' roses: stable, controllable, diverse image to image translation (works for videos too!) (2021)
5. Chong, M.J., Lee, H.Y., Forsyth, D.: StyleGAN of all trades: image manipulation with only pretrained stylegan. arXiv preprint arXiv:2111.01619 (2021)
6. Deng, J., Guo, J., Xue, N., Zafeiriou, S.: Arcface: additive angular margin loss for deep face recognition. In: Proceedings of the IEEE/CVF Conference on Computer Vision and Pattern Recognition, pp. 4690–4699 (2019)
7. Efros, A.A., Freeman, W.T.: Image quilting for texture synthesis and transfer. In: Proceedings of SIGGRAPH 2001, pp. 341–346 (2001)
8. Gal, R., Patashnik, O., Maron, H., Chechik, G., Cohen-Or, D.: StyleGAN-nada: clip-guided domain adaptation of image generators (2021)
9. Gatys, L.A., Ecker, A.S., Bethge, M.: Image style transfer using convolutional neural networks. In: Proceedings of the IEEE Conference on Computer Vision and Pattern Recognition (CVPR), June 2016
10. Hertzmann, A., Jacobs, C.E., Oliver, N., Curless, B., Salesin, D.H.: Image analogies. In: Proceedings of SIGGRAPH 2001, pp. 327–340 (2001)
11. Heusel, M., Ramsauer, H., Unterthiner, T., Nessler, B., Hochreiter, S.: GANs trained by a two time-scale update rule converge to a local nash equilibrium. In: Advances in Neural Information Processing Systems, vol. 30 (2017)
12. Huang, X., Belongie, S.: Arbitrary style transfer in real-time with adaptive instance normalization. In: Proceedings of the IEEE International Conference on Computer Vision, pp. 1501–1510 (2017)
13. Huang, X., Liu, M.Y., Belongie, S., Kautz, J.: Multimodal unsupervised image-to-image translation. In: ECCV (2018)
14. Johnson, J., Alahi, A., Fei-Fei, L.: Perceptual losses for real-time style transfer and super-resolution. In: European Conference on Computer Vision (2016)
15. Karras, T., Laine, S., Aila, T.: A style-based generator architecture for generative adversarial networks. In: Proceedings of the IEEE/CVF Conference on Computer Vision and Pattern Recognition, pp. 4401–4410 (2019)
16. Karras, T., Laine, S., Aittala, M., Hellsten, J., Lehtinen, J., Aila, T.: Analyzing and improving the image quality of StyleGAN. In: Proceedings of CVPR (2020)
17. Kim, S.S.Y., Kolkin, N., Salavon, J., Shakhnarovich, G.: Deformable style transfer. In: Vedaldi, A., Bischof, H., Brox, T., Frahm, J.-M. (eds.) ECCV 2020. LNCS, vol. 12371, pp. 246–261. Springer, Cham (2020). https://doi.org/10.1007/978-3-030-58574-7_15
18. Kingma, D.P., Ba, J.: Adam: a method for stochastic optimization. CoRR abs/1412.6980 (2015)
19. Li, Y., Fang, C., Yang, J., Wang, Z., Lu, X., Yang, M.H.: Universal style transfer via feature transforms. In: Advances in Neural Information Processing Systems (2017)
20. Li, Y., Zhang, R., Lu, J.C., Shechtman, E.: Few-shot image generation with elastic weight consolidation. In: Advances in Neural Information Processing Systems (2020)
21. Liu, M., Li, Q., Qin, Z., Zhang, G., Wan, P., Zheng, W.: BlendGAN: implicitly GAN blending for arbitrary stylized face generation. In: Advances in Neural Information Processing Systems (2021)

22. Menon, S., Damian, A., Hu, S., Ravi, N., Rudin, C.: Pulse: self-supervised photo upsampling via latent space exploration of generative models. In: Proceedings of the IEEE/CVF Conference on Computer Vision and Pattern Recognition, pp. 2437–2445 (2020)
23. Mo, S., Cho, M., Shin, J.: Freeze the discriminator: a simple baseline for fine-tuning GANs. In: CVPR AI for Content Creation Workshop (2020)
24. Ojha, U., et al.: Few-shot image generation via cross-domain correspondence. In: CVPR (2021)
25. Ojha, U., et al.: Few-shot image generation via cross-domain correspondence. In: Proceedings of the IEEE/CVF Conference on Computer Vision and Pattern Recognition, pp. 10743–10752 (2021)
26. Park, D.Y., Lee, K.H.: Arbitrary style transfer with style-attentional networks. In: Proceedings of the IEEE/CVF Conference on Computer Vision and Pattern Recognition, pp. 5880–5888 (2019)
27. Pinkney, J.N., Adler, D.: Resolution dependent GAN interpolation for controllable image synthesis between domains. arXiv preprint arXiv:2010.05334 (2020)
28. Radford, A., et al.: Learning transferable visual models from natural language supervision. arXiv preprint arXiv:2103.00020 (2021)
29. Robb, E., Chu, W.S., Kumar, A., Huang, J.B.: Few-shot adaptation of generative adversarial networks. arXiv preprint arXiv:2010.11943 (2020)
30. Salimans, T., Goodfellow, I., Zaremba, W., Cheung, V., Radford, A., Chen, X.: Improved techniques for training GANs. Adv. Neural. Inf. Process. Syst. **29**, 2234–2242 (2016)
31. Shen, Y., Yang, C., Tang, X., Zhou, B.: InterFaceGAN: interpreting the disentangled face representation learned by GANs. IEEE Trans. Pattern Anal. Mach. Intell. (2020)
32. Shen, Y., Zhou, B.: Closed-form factorization of latent semantics in GANs. In: Proceedings of the IEEE/CVF Conference on Computer Vision and Pattern Recognition, pp. 1532–1540 (2021)
33. Simonyan, K., Zisserman, A.: Very deep convolutional networks for large-scale image recognition. In: International Conference on Learning Representations (2015)
34. Tov, O., Alaluf, Y., Nitzan, Y., Patashnik, O., Cohen-Or, D.: Designing an encoder for StyleGAN image manipulation. arXiv preprint arXiv:2102.02766 (2021)
35. Wang, X., Li, Y., Zhang, H., Shan, Y.: Towards real-world blind face restoration with generative facial prior. In: Proceedings of the IEEE/CVF Conference on Computer Vision and Pattern Recognition, pp. 9168–9178 (2021)
36. Wu, Z., Lischinski, D., Shechtman, E.: Stylespace analysis: disentangled controls for StyleGAN image generation. In: Proceedings of the IEEE/CVF Conference on Computer Vision and Pattern Recognition, pp. 12863–12872 (2021)
37. Yang, T., Ren, P., Xie, X., Zhang, L.: GAN prior embedded network for blind face restoration in the wild. In: IEEE Conference on Computer Vision and Pattern Recognition (CVPR) (2021)
38. Yeh, M.C., Tang, S., Bhattad, A., Zou, C., Forsyth, D.: Improving style transfer with calibrated metrics. In: Proceedings of the IEEE/CVF Winter Conference on Applications of Computer Vision (WACV), March 2020
39. Zhang, R., Isola, P., Efros, A.A., Shechtman, E., Wang, O.: The unreasonable effectiveness of deep features as a perceptual metric. In: CVPR (2018)
40. Zhu, J.Y., Park, T., Isola, P., Efros, A.A.: Unpaired image-to-image translation using cycle-consistent adversarial networks. In: 2017 IEEE International Conference on Computer Vision (ICCV) (2017)

41. Zhu, P., Abdal, R., Femiani, J., Wonka, P.: Mind the gap: domain gap control for single shot domain adaptation for generative adversarial networks. In: International Conference on Learning Representations (2022). https://openreview.net/forum?id=vqGi8Kp0wM
42. Zhu, P., Abdal, R., Qin, Y., Femiani, J., Wonka, P.: Improved StyleGAN embedding: where are the good latents? arXiv preprint arXiv:2012.09036 (2020)

VecGAN: Image-to-Image Translation with Interpretable Latent Directions

Yusuf Dalva[(⊠)], Said Fahri Altındiş, and Aysegul Dundar

Bilkent University, Ankara, Turkey
{yusuf.dalva,fahri.altindis}@bilkent.edu.tr, adundar@cs.bilkent.edu.tr

Abstract. We propose VecGAN, an image-to-image translation framework for facial attribute editing with interpretable latent directions. Facial attribute editing task faces the challenges of precise attribute editing with controllable strength and preservation of the other attributes of an image. For this goal, we design the attribute editing by latent space factorization and for each attribute, we learn a linear direction that is orthogonal to the others. The other component is the controllable strength of the change, a scalar value. In our framework, this scalar can be either sampled or encoded from a reference image by projection. Our work is inspired by the latent space factorization works of fixed pre-trained GANs. However, while those models cannot be trained end-to-end and struggle to edit encoded images precisely, VecGAN is end-to-end trained for image translation task and successful at editing an attribute while preserving the others. Our extensive experiments show that Vec-GAN achieves significant improvements over state-of-the-arts for both local and global edits.

Keywords: Image translation · Generative adversarial networks · Latent space manipulation · Face attribute editing

1 Introduction

There has been a significant progress in image-to-image translation methods [6,13,20,21,23,24,35,37] especially for facial attribute editing [5,19,25,33,38] powered with generative adversarial networks (GANs). A main challenge of facial attribute editing methods is to be able to change only one attribute of an image without affecting others such as global lighting parameters of the images, identity of the persons, background, or their other attributes. The other challenge is the interpretability of the style codes so that one can control the attribute intensity of the edit, e.g. increase the intensity of smile or aging.

To achieve the targeted attribute editing while preserving the others, many works set a separate style encoder and an image editing network where modified styles are injected into it [5,19]. During image-to-image translation, a style

Supplementary Information The online version contains supplementary material available at https://doi.org/10.1007/978-3-031-19787-1_9.

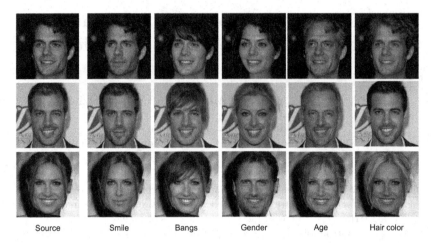

| Source | Smile | Bangs | Gender | Age | Hair color |

Fig. 1. Attribute editing results of VecGAN. The first column shows the source images, and other columns show the results of editing a specific attribute. Each edited image has an attribute value opposite to that of the source one. For hair color, sources are translated to brown, black, and blonde hair, respectively.

encoded from another image or a newly sampled style latent code can be used to output diverse images. To disentangle attributes, works focus on style encoding and progress from a shared style code, SDIT [30], to mixed style codes, Star-GANv2 [5], to hierarchical disentangled styles, HiSD [19]. Among these works, HiSD independently learn styles of each attribute, bangs, hair color, glasses and introduces a local translator which uses attention masks to avoid global manipulations. HiSD showcases successes on those three local attribute editing task and is not tested for global attribute editing, e.g. age, smile. Furthermore, one limitation of these works is the uninterpretablity of style codes as one cannot control the intensity of attribute (e.g. blondness) in a straight-forward manner.

To overcome the challenges of facial attribute editing task, we propose a novel framework, VecGAN, and image-to-image translation framework with interpretable latent directions. Our framework does not require a separate style encoder as in the previous works since we achieve the translation in the encoded latent space directly. The attribute editing directions are learned in the latent space and regularized to be orthogonal to each other for style disentanglement. The other component of our framework is the controllable strength of the change, a scalar value. This scalar can be either sampled from a distribution or encoded from a reference image by projection in the latent space. Our framework not only achieves significant improvements over state-of-the-arts for both local and global edits but also provides a knob to control the editing attribute intensity via its design.

VecGAN is encouraged by the findings that well-trained generative models organize their latent space as disentangled representations with meaningful directions in a completely unsupervised way. Exploring these interpretable directions in latent codes has emerged as an important research endeavor on the fixed pretrained GANs [9,26–28,32]. These works show that images can be mapped to the GANs latent space and edits can be achieved by manipulations in the latent space. However, since these models are not trained end-to-end, the results are sub-optimal as will also be shown in our experiments.

To enable VecGAN, different than previous works of image-to-image translation networks, we use a deeper neural network architecture. Image-to-image translation methods, such as state-of-the-art HiSD [19] uses a network with small receptive fields that decreases the image resolution only by four times in the encoder. However, we want an organization in a latent space such that we can take meaningful linear directions. Therefore, images should be encoded to a spatially smaller feature space and a network should have a full understanding of an image. For that reason, we set a deep encoder and decoder network architecture but then this network faces the challenges of reconstructing all the details from the input image. To solve this problem, we use a skip connection between the encoder and decoder but only at lower resolution to find the optimal equilibrium of the information flow between with and without dimensionality reduction bottleneck. In summary, our main contributions are:

– We propose VecGAN, a novel image-to-image translation network that is trained end to end with interpretable latent directions. Our framework does not employ a separate style network as in the previous works and translations are achieved with a single deep encoder-decoder architecture.
– VecGAN enables both reference attribute copy and attribute strength manipulation. Reference style encoding is designed in a novel way by using the same encoder from the translation pipeline. First, encoder is used to obtain latent codes of a reference image and it is followed by the projection of the codes into learned latent directions for different attributes.
– We conduct extensive experiments to show the effectiveness of our framework and achieve significant improvements over state-of-the-art for both local and global edits. Qualitative results of our framework can be seen in Fig. 1.

2   Related Works

Image to Image Translation. Image-to-image translation algorithms aim at preserving a given content while changing targeted attributes. Examples range from translating semantic maps into RGB images [29], to translating summer images into winter images [13], to portrait drawing [36] and very popularly to editing faces [2,5,7,11,19,25,31,33,38]. These algorithms powered with GAN loss [8] set an encoder-decoder architecture. In models that learn a deterministic mapping from one domain to the other, images are processed with encoder and decoder to output translated images [24,29]. In multi-modal image-to-image translation methods, style is encoded separately from an another image or sampled from a distribution [5,12]. In the generator, style and content are either combined with concatenation [41], or combined with a mask [19] or fed separately through instance normalization blocks [12,42]. The generator also uses an encoder-decoder architecture [19,34] that is separate than the style encoder. In our work, we are interested in designing the attribute as a learnable linear direction in the latent space and we do not employ a separate style encoder which results in a more intuitive framework.

Learning Interpretable Latent Directions. In another line of research, it is shown that GANs that are trained to synthesize faces can also be used for face attribute manipulations [3,15,16]. Initially, these networks are not designed or trained to translate images but rather to synthesize high fidelity images. However, it is shown that one can embed existing images into the GAN's embedding space [1] and further one can find latent directions to edit those images [9,26–28,32]. These directions are explored in supervised [26] and unsupervised ways [9,27,28,32]. It is quite remarkable when the generative network is only taught to synthesize realistic images, it organizes the use of latent space such that linear shifts on them change a specific attribute. Inspired by these findings, we design our image to image translation such that a linear shift in the encoded features is expected to change a single attribute of an image. Different than previous works, our framework is trained end-to-end for translation task and allows for reference guided attribute manipulation via projection.

3   Method

We follow the hierarchical labels defined by [19]. For a single image, its attribute for tag $i \in \{1, 2, ..., N\}$ can be defined as $j \in \{1, 2, ..., M_i\}$, where N is the number of tags and M_i is the number of attributes for tag i. For example i can be tag of hair color, and attribute j can take the value of black, brown, or blond.

Our framework has two main objectives. As the main task, we aim to be able to perform the image-to-image translation task in a feature (tag) specific manner. While performing this translation, as the second objective, we also want to obtain an interpretable feature space which allows us to perform tag-specific feature interpolation.

3.1   Generator Architecture

For image to image translation task, we set an encoder-decoder based architecture and latent space translation in the middle as given in Fig. 2. We perform the translation in the encoded latent space, e, which is obtained by $e = E(x)$ where E refers to the encoder. The encoded features go through a transformation T which is discussed in the next section. The transformed features are then decoded by G to reconstruct the translated images. The image generation pipeline following feature encoding is described in Eq. 1.

$$e' = T(e, \alpha, i)$$
$$x' = G(e') \tag{1}$$

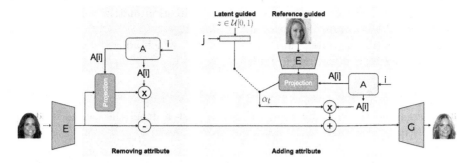

Fig. 2. VecGAN pipeline. Our translator is built on the idea of interpretable latent directions. We encode images with an Encoder to a latent representation from which we change a selected tag (i), e.g. hair color with a learnable direction A_i and a scale α. To calculate the scale, we subtract the target style scale from the source style. This operation corresponds to removing an attribute and adding an attribute. To remove the image's attribute, source style is encoded and projected from the source image. To add the target attribute, target style scale is sampled from a distribution mapped for the given attribute (j), e.g. blonde, brown or encoded and projected from a reference image.

Previous image-to-image translation networks [5,19,34] set a shallow encoder decoder architecture to translate an image and a separate deep network for style encoding. In most cases, the style encoder includes separate branches for each tag. The shallow architecture that is used to translate images prevents the model from making drastic changes in the images and this helps preserving the identity of the persons. Our framework is different as we do not employ a separate style encoder and instead have a deep encoder-decoder architecture for translation. That is because to be able to organize the latent space in an interpretable way, our framework requires a full understanding of the image and therefore a larger receptive field; deeper network architecture. A deep architecture with decreasing size of feature size, on the other hand, faces the challenges of reconstructing all the fine details from the input image.

With the motivation of helping the network to preserve tag independent features such as the fine details from background, we use skip connections between our encoder and decoder. However, we observe that the flow of information should be limited to force the encoder-decoder architecture learn facial attributes and well-organized latent representations. Because of that reason, we only allow skip connection at low resolution. This design is extensively justified in our Ablation Studies.

3.2 Translation Module

To achieve a style transformation, we perform the tag-based feature manipulation in a linear fashion in the latent space. First, we set a feature direction matrix A which contains learnable feature directions for each tag. In our formulation A_i

denotes the learned feature direction for tag i. Direction matrix A is randomly initialized and learned during the training process.

Our translation module is formulated in Eq. 2, which adds the desired shift on top of the encoded features e similar to [28].

$$T(e, \alpha, i) = e + \alpha \times A_i \tag{2}$$

We compute the shift by subtracting target style from the source style as given in Eq. 3.

$$\alpha = \alpha_t - \alpha_s \tag{3}$$

Since the attributes are designed as linear steps in the learnable directions, we find the style shift by subtracting the target attribute scale from source attribute scale. This way the same target attribute α_t can have the same impact on the translated images no matter what the attributes were of the original images. For example, if our target scale corresponds to brown hair, the source scale can be coming from an image with blonde or back hair but since we take a step for difference of the scales, they can be both translated to an image with the same shade of brown hair.

To extract the target shifting scale for feature (tag) i, α_t, there are two alternative pathways. The first pathway, named as latent-guided path, samples a $z \in \mathcal{U}[0, 1)$ and applies a linear transformation $\alpha_t = w_{i,j} \cdot z + b_{i,j}$, where α_t denotes sampled shifting scale for tag i and attribute j. Here tag i can be hair color and attribute j can be blonde, brown, or back hair. For each attribute we learn a different transformation module which is denoted as $M_{i,j}(z)$. Since we learn a single direction for every tag for example for hair color, this transformation module can put the initially sampled z's into correct scale in the linear line based on the target hair color attribute. As the other alternative pathway, we encode the scalar value α_t in a reference-guided manner. We extract α_t for tag i from a provided reference image by first encoding it into the latent space, e_r, and projecting e_r via by A_i as given in Eq. 4.

$$\alpha_t = P(e_r, A_i) = \frac{e_r \cdot A_i}{||A_i||} \tag{4}$$

In the reference guidance set-up, we do not use the information of attribute j, since it is encoded by the tag i features of the image.

The source scale, α_s, is obtained by the same way we obtain α_t from reference image. We perform the projection for the corresponding tag we want to manipulate, i, by $P(e, A_i)$. We formulate our framework with the intuition that the scale controls the amount of feature to be added. Therefore, especially when the attribute is copied over from a reference image, the amount of features that will be added will be different based on the source image. It is for this reason, we find the amount of shift by subtraction as given in Eq. 3. Our framework is intuitive and relies on a single encoder-decoder architecture. Figure 2 shows the overall pipeline.

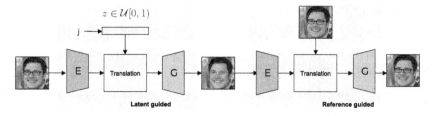

Fig. 3. Overview of cycle translation path.

3.3   Training Pathways

Modifying the translation paths defined by [19], we train our network using two different paths. For each iteration to optimize our model, we sample a tag i for shift direction, a source attribute j as the current attribute and a target attribute $\hat{j}$.

Non-translation Path. To ensure that the encoder-decoder structure preserves details of the images, we perform a reconstruction of the input image without applying any style shifts. The resulting image is denoted as x_n as given in Eq. 5.

$$x_n = G(E(x)) \tag{5}$$

Cycle-Translation Path. We apply a cyclic translation to ensure that we get a reversible translation from a latent guided scale. In this path, as shown in Fig. 3, we first apply a style shift by sampling $z \in \mathcal{U}[0,1)$ and obtaining target α_t with $M_{i,\hat{j}}(z)$ for target attribute $\hat{j}$. The translation uses α that is obtained by subtracting α_t from the source style. Decoder generates an image, x_t as given in Eq. 6 where e is encoded features from input image x, $e = E(x)$. x_t refers to the image without glasses in Fig. 3.

$$x_t = G(T(e, M_{i,j}(z) - P(e,i), i)) \tag{6}$$

Then by using the original image, x, as a reference image, we aim to reconstruct the original image by translating x_t. Overall, this path attempts to reverse a latent-guided style shift with a reference-guided shift. The second translation is given in Eq. 7 where $e_t = E(x_t)$.

$$x_c = G(T(e_t, P(e,i) - P(e_t,i), i)) \tag{7}$$

In our learning objectives, we use x_n and x_c for reconstruction and x_t and x_c for adversarial losses, and $M_{i,j}(z)$ for the shift reconstruction loss. Details about the learning objectives are given in the next section.

3.4   Learning Objectives

Given an input image $x_{i,j} \in \mathcal{X}_{i,j}$, where i is the tag to manipulate and j is the current attribute of the image, we optimize our model with the following objectives. In our equations, $x_{i,j}$ is shown as x.

Adversarial Objective. During training, our generator performs a style-shift either in a latent-guided way or a reference-guided way, which results in a translated image. In our adversarial loss, we receive feedback from the two steps of cycle-translation path. As the first component of the adversarial loss, we feed a real image x with tag i and attribute j to the discriminator as the real example. To give adversarial feedback to latent-guided path, we use the intermediate image generated in cycle-translation path, x_t. Finally, to provide adversarial feedback to reference-guided path, we use the final outcome of the cycle-translation path x_c. Only x acts as real image, both x_t and x_c are translated images, and they are treated as fake images with different attributes. The discriminator aims at classifying whether an image, given its tag and attribute, is real or not. The objective is given in Eq. 8.

$$\mathcal{L}_{adv} = 2log(D_{i,j}(x)) + log(1 - D_{i,\hat{j}}(x_t)) + log(1 - D_{i,j}(x_c)) \tag{8}$$

Shift Reconstruction Objective. As the cycle-consistency loss performs reference-guided generation followed by latent-guided generation, we utilize a loss function to make these two methods consistent with each other [12,17–19]. Specifically, we would like to obtain the same target scale, α_t, both from the mapping and from the encoded reference image generated by the mapped α_t. The loss function is given in Eq. 9.

$$\mathcal{L}_{shift} = ||M_{i,j}(z) - P(e_t, i)||_1 \tag{9}$$

Those parameters, $M_{i,j}(z)$ and $P(e_t, i)$, are calculated for the cycle-translation path as given in Eq. 6 and 7.

Image Reconstruction Objective. In all of our training paths, the purpose it to be able to re-generate the original image again. To supervise this desired behavior, we use L_1 loss for reconstruction loss. In our formulation x_n and x_c are outputs of non-translation path and cycle-translation path, respectively. Formulation of this objective is provided in Eq. 10.

$$\mathcal{L}_{rec} = ||x_n - x||_1 + ||x_c - x||_1 \tag{10}$$

Orthogonality Objective. To encourage the orthogonality between directions, we use soft orthogonality regularization based on Frobenius norm, which is given in Eq. 11. This orthogonality further encourages a disentanglement in the learned style directions.

$$\mathcal{L}_{ortho} = ||A^T A - I||_F \tag{11}$$

Full Objective. Combining all of the loss components described, we reach to the overall objective for optimization as given in Eq. 12. We additionally add L1 loss on the matrix A parameters to encourage its sparsity.

$$\min_{E,G,M,A} \max_D \lambda_a \mathcal{L}_{adv} + \lambda_s \mathcal{L}_{shift} + \lambda_r \mathcal{L}_{rec} + \lambda_o \mathcal{L}_{ortho} + \lambda_{sp} \mathcal{L}_{sparse} \quad (12)$$

To control the dominance of each loss component, we use $\lambda_a, \lambda_s, \lambda_r, \lambda_o,$ and λ_{sp} hyperparameters. These hyperparameter values and training details are given in Supplementary.

4   Experiments

4.1   Dataset and Settings

We train our model on CelebA-HQ dataset [22] which contains 30,000 face images. To extensively compare with state-of-the-arts, we follow two training-evaluation protocols as follows:

Setting A. In our first setting, we follow the set-up from HiSD [19]. Following HiSD, we use the first 3000 images of CelebA-HQ dataset as the test set and 27000 as the training set. These images include annotations for different attributes from which we use hair color, presence of glass, and bangs attributes for translation task in this setting. Hair color attribute includes 3 tags, black, brown, and blonde whereas the other attributes are binary. The images are resized to 128×128. Following the evaluation protocol proposed by HiSD [19], we compute FID scores on bangs addition task. For each test image without bangs, we translate them to images with bangs with latent and reference guidance. In latent guidance, 5 images are generated for each test image by randomly sampling scale from a uniform distribution. Then this generated set of images are compared with images that have attribute bangs in terms of their FIDs. FIDs are calculated for these 5 sets and averaged. For reference guidance, we randomly pick 5 references images to extract the style scale. FIDs are calculated for these 5 sets separately and averaged.

Setting B. In this setting, we follow the set-up from L2M-GAN [34]. The training/test split is obtained by re-indexing each image in CelebA-HQ back to the original CelebA and following the standard split of CelebA. This results in 27,176 training and 2,824 test images. Models are trained for hair color, presence of glasses, bangs, age, smiling, and gender attributes. Images are resized to 256×256 resolution. For evaluation, smiling attribute is used following L2M-GAN [34]. It is noted that smiling is one of the most challenging among the CelebA facial attributes because adding/removing a smile requires high-level understanding of the input face image for modifying multiple facial components simultaneously. FIDs are calculated for adding and removing the smile attribute.

Table 1. Comparisons with state-of-the-art competing methods. Please refer to Sect. 4 for details on training and evaluation protocol of Setting A and B.

Method	Lat.	Ref.
SDIT [30]	33.73	33.12
StarGANv2 [5]	26.04	25.49
Elegant [33]	-	22.96
HiSD [19]	21.37	21.49
VecGAN (Ours)	**20.17**	**20.72**

(a) Quantitative results for Setting A. Lat: Latent guided, Ref: Reference guided. FID scores are given. Lower is better.

Method	FID (+)	FID (−)	Avg
StarGAN [4]	32.6	38.6	35.6
CycleGAN [40]	22.5	24.4	23.5
Elegant [33]	39.7	42.9	41.3
PA-GAN [10]	20.5	21.4	21.0
InterFaceGAN [26]	24.8	24.9	24.9
L2M-GAN [34]	17.9	23.3	20.6
VecGAN (Ours)	**17.7**	**20.3**	**19.0**

(b) Quantitative results for Setting B. FID (+) (or FID (-)) denotes the FID score for adding (or removing) a smile.

Table 2. User study results conducted with smiling attribute. Smiling (+) denotes the results of adding a smile, Smiling (−) refers to the results of removing a smile, and Smiling (avg) denotes the average of Smiling (+) and Smiling (−). Percentages show the preference rates of our method versus the other competing method.

Comparisons	Smiling (+) Quality	Smiling (+) Fidelity	Smiling (−) Quality	Smiling (−) Fidelity	Smiling (Avg) Quality	Smiling (Avg) Fidelity
VecGAN (Ours) vs L2M-GAN	57.96%	70.94%	60.93%	77.50%	59.45%	74.22%
VecGAN (Ours) vs InterFaceGAN	88.13%	91.56%	77.50%	90.62%	82.82%	91.09%

4.2 Results

We extensively compare our results with other competing methods in Table 1. In Setting A, as given in Table 1a, we compare with SDIT [30], StarGANv2 [5], Elegant [33], and HiSD [19] models. Among these methods, HiSD learns a hierarchical style disentanglement whereas StarGANv2 learns a mixed style code. Therefore, StarGANv2 when translating images also does other unnecessary manipulations and does not strictly preserve the identity. Our work is most similar to HiSD as we also learn disentangled style directions. However, HiSD learns feature based local translators which is an approach known to be successful on local edits, e.g. bangs. Ours results show that VecGAN achieves significantly better quantitative results than HiSD both in latent guided and reference guided evaluations even though they are compared on a local edit task.

Figure 4 shows reference guided results of our model versus HiSD. We compare with HiSD since it provides with the best results after ours. As can be seen from Fig. 4, both methods achieve attribute disentanglement, they do not change any other attribute of the image than the bangs tag. However, HiSD outputs artifacts especially for the reference image from the last column. On the other hand, VecGAN outputs higher quality results. As the second example, we pick a very challenging example to compare these methods. Even though, our results can be further improved to look more realistic, it achieves significantly better outputs with no artifacts compared to HiSD.

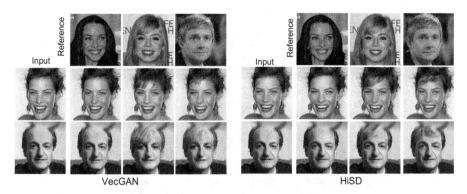

Fig. 4. Qualitative results of bangs attribute of our model (VecGAN) and HiSD. In the second example, we provide a very challenging sample where VecGAN even though not perfect achieves significantly better results than HiSD.

In our second set-up of evaluation, we compare our method with many state-of-the-art methods as given in Table 1b. We compare with StarGAN [4], Cycle-GAN [40], Elegant [33], PA-GAN [10], InterFaceGAN [26], and L2M-GAN [34]. For InterFaceGAN, we use the GAN Inversion [39] as the encoder and pre-trained StyleGAN [15] as the generator backbone. As can be seen from Table 1b, we achieve significantly better scores on both settings and in average.

In our visual comparisons, we mainly focus on L2M-GAN and InterFaceGAN since L2M-GAN is the second best model after ours and InterFaceGAN shares the same intuition with our model and performs edits by latent code manipulation. The results are shown in Fig. 5 where the first four examples show smile addition and the other four examples show smile removal manipulations. The most prominent limitation of L2M-GAN and InterFaceGAN is that they do not preserve the other attributes of images, especially on the background whereas VecGAN does a very good job at that. Smile attribute addition and removal of L2M-GAN is better than InterFaceGAN, however, worse than ours. VecGAN is the only method among them that can produce manipulated images with high fidelity to the originals with only targeted attribute manipulated in a natural and realistic way.

We also conduct a user study on the first 64 images of validation set among 10 users. We set an A/B test and provide users with input images and translations obtained by VecGAN and other competing methods. The left-right order is randomized to ensure fair comparisons. We perform two separate tests. 1) Quality: We ask users to select the best result according to i) whether the smile attribute is correctly added, ii) whether irrelevant facial attributes preserved, iii) and over-all whether the output image looks realistic and high quality. 2) Fidelity: We ask users to pay attention if details from the input image is preserved in addition to the quality. When only asked for quality, users pay attention to facial attributes and do not pay much attention to the background, ornament, details of hair of the image, and so on. In this test, we remind the users to pay attention to those

Fig. 5. Qualitative results of smile attribute of our model (VecGAN), L2M-GAN, and InterFaceGAN. The first four examples show smile addition and the other four shows smile removal manipulations.

as well. Table 2 shows the results of the user study. Users preferred our method as opposed to L2M-GAN 59.45% of the time (50% is tie), and as opposed to InterFaceGAN 82.82% of the time for the quality measure in average of smile addition and removal results. When users asked to pay attention to non-facial attributes as well, they preferred our method as opposed to L2M-GAN 74.22% of the time, and as opposed to InterFaceGAN 91.09% of the time in average.

4.3   Ablation Study

We conduct ablation studies for network architecture and loss objectives as given in Table 3. We first experiment with a shallower architecture where encoder decreases the input dimension of 128×128 to a spatial dimension of 8×8. This version gives reasonable scores, however, we are interested in a better latent space organization. For that, we use a deeper encoder-decoder architecture where encoded latent space goes as low as 1×1 which we refer as deep architecture. Deep architecture without skip connections is not able to minimize the reconstruction objective and results in a high FID. On the other hand, deep architecture with a skip connection at each resolution from encoder to decoder can minimize the reconstruction loss however the latent space is not well organized since the model tends to pass all the information from the encoder which instabilizes the training. Our architecture with single skip layer at resolution 32×32 provides a good balance between the information flow from encoder-decoder and the latent space bottleneck.

Fig. 6. Qualitative results of ablation study of orthogonality loss. Bangs tag transferred from the reference image.

Table 3. FID results of ablation study with Setting A. Lat: Latent guided, Ref: Reference guided.

Method	Lat.	Ref.
Shallow	21.30	20.94
Deep w/o skip	88.62	127.65
Deep all skip	273.80	273.97
Ours	**20.17**	**20.72**
w/o Orthogonality	21.98	22.50
w/o Sparsity	24.07	22.43

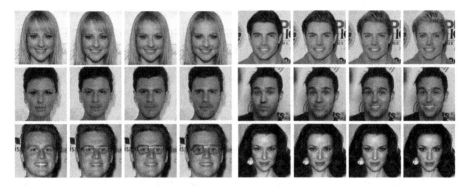

Fig. 7. Results of changing the strength of a manipulation gradually. Each example shows a different attribute manipulation. Rows show bangs, hair color, gender, smile, glasses, and age manipulations in this order.

Next, we experiment the effect of loss functions. First, we remove the orthogonality loss of A directions. This results in worse FID scores but more importantly we observe that the styles are not disentangled, e.g. changing bangs attribute changes the gender as can be seen in Fig. 6. Even without this loss function, we observe that during training the orthogonality loss of A decreases but to a higher value than when this loss is added to the final objective. That is because the framework and other loss objectives also encourage the disentanglement of attribute manipulations and it shows in the orthogonality of direction vectors. This also shows the importance of orthogonality in style disentanglement and this targeted loss helps improve that significantly. We also observe that sparsity loss applied on the directional vectors stabilizes the training and without that FIDs are much higher.

4.4   Other Capabilities of VecGAN

Gradually Increased Scale. We translate images with gradually increased attribute strength as shown in Fig. 7. We plot the manipulation results on six

(a) Multi-attribute editing results.   (b) Generalization results.

Fig. 8. Results of multi-attribute editing and cross-dataset generalization results of VecGAN.

different attributes. These results show that attributes that are designed as linear transformations are disentangled, and changing one attribute does not affect the other components. In these results, as scales are gradually increased, the strength of the tag smoothly increases with the identity of the person preserved.

Multi-tag Edits. We additionally experiment with multi-tag manipulation. To change two attributes, instead of encoding and decoding the image twice with a translation in between each time, we perform two translation operations in the latent code simultaneously. That is we apply Eq. 2 twice for two different i. Figure 8a shows results of the multi-tag edits. In the first row, we consider gender and smile tags, and first edit those attributes individually. In the last coloumn, we edit the image with these two tags simultaneously. The second row shows a similar experiment with smile and age tags. We observe that VecGAN provides with disentangled tag control and can successfully edit tags independently.

Generalization to Other Domains. We apply VecGAN model to MetFace dataset [14] without any retraining. The results are provided in Fig. 8b. The first row shows source images, and the second row shows outputs of our model. In the first two examples, we increase the smile attribute, and in the other two, we decrease it. The results show that VecGAN has a good generalization ability and works reasonably well across datasets.

5   Conclusion

This paper introduces VecGAN, an image-to-image translation framework with interpretable latent directions. This framework includes a deep encoder and decoder architecture with latent space manipulation in between. Latent space manipulation is designed as vector arithmetic where for each attribute, a linear direction is learned. This design is encouraged by the finding that well-trained generative models organize their latent space as disentangled representations with meaningful directions in a completely unsupervised way. Each change in the architecture and loss functions is extensively studied and compared with state-of-the-arts. Experiments show the effectiveness of our framework.

References

1. Abdal, R., Qin, Y., Wonka, P.: Image2StyleGAN: how to embed images into the StyleGAN latent space? In: Proceedings of the IEEE/CVF International Conference on Computer Vision, pp. 4432–4441 (2019)
2. Abdal, R., Zhu, P., Mitra, N.J., Wonka, P.: StyleFlow: attribute-conditioned exploration of StyleGAN-generated images using conditional continuous normalizing flows. ACM Trans. Graph. (ToG) **40**(3), 1–21 (2021)
3. Brock, A., Donahue, J., Simonyan, K.: Large scale GAN training for high fidelity natural image synthesis. arXiv preprint arXiv:1809.11096 (2018)
4. Choi, Y., Choi, M., Kim, M., Ha, J.W., Kim, S., Choo, J.: StarGAN: unified generative adversarial networks for multi-domain image-to-image translation. In: Proceedings of the IEEE Conference on Computer Vision and Pattern Recognition, pp. 8789–8797 (2018)
5. Choi, Y., Uh, Y., Yoo, J., Ha, J.W.: StarGAN v2: diverse image synthesis for multiple domains. In: Proceedings of the IEEE Conference on Computer Vision and Pattern Recognition (2020)
6. Dundar, A., Sapra, K., Liu, G., Tao, A., Catanzaro, B.: Panoptic-based image synthesis. In: Proceedings of the IEEE/CVF Conference on Computer Vision and Pattern Recognition, pp. 8070–8079 (2020)
7. Gao, Y., et al.: High-fidelity and arbitrary face editing. In: Proceedings of the IEEE/CVF Conference on Computer Vision and Pattern Recognition, pp. 16115–16124 (2021)
8. Goodfellow, I., et al.: Generative adversarial nets. In: Advances in Neural Information Processing Systems, vol. 27 (2014)
9. Härkönen, E., Hertzmann, A., Lehtinen, J., Paris, S.: GANSpace: discovering interpretable GAN controls. In: Advances in Neural Information Processing Systems, vol. 33 (2020)
10. He, Z., Kan, M., Zhang, J., Shan, S.: Pa-GAN: progressive attention generative adversarial network for facial attribute editing. arXiv preprint arXiv:2007.05892 (2020)
11. Hou, X., Zhang, X., Liang, H., Shen, L., Lai, Z., Wan, J.: GuidedStyle: attribute knowledge guided style manipulation for semantic face editing. Neural Netw. **145**, 209–220 (2022)
12. Huang, X., Liu, M.Y., Belongie, S., Kautz, J.: Multimodal unsupervised image-to-image translation. In: European Conference on Computer Vision (2018)
13. Isola, P., Zhu, J.Y., Zhou, T., Efros, A.A.: Image-to-image translation with conditional adversarial networks. In: IEEE Conference on Computer Vision and Pattern Recognition (2017)
14. Karras, T., Aittala, M., Hellsten, J., Laine, S., Lehtinen, J., Aila, T.: Training generative adversarial networks with limited data. Adv. Neural. Inf. Process. Syst. **33**, 12104–12114 (2020)
15. Karras, T., Laine, S., Aila, T.: A style-based generator architecture for generative adversarial networks. In: Proceedings of the IEEE/CVF Conference on Computer Vision and Pattern Recognition, pp. 4401–4410 (2019)
16. Karras, T., Laine, S., Aittala, M., Hellsten, J., Lehtinen, J., Aila, T.: Analyzing and improving the image quality of StyleGAN. In: Proceedings of the IEEE/CVF Conference on Computer Vision and Pattern Recognition, pp. 8110–8119 (2020)
17. Lee, H.Y., Tseng, H.Y., Huang, J.B., Singh, M., Yang, M.H.: Diverse image-to-image translation via disentangled representations. In: Proceedings of the European Conference on Computer Vision (ECCV), pp. 35–51 (2018)

18. Li, X., et al.: Attribute guided unpaired image-to-image translation with semi-supervised learning. arXiv preprint arXiv:1904.12428 (2019)
19. Li, X., et al.: Image-to-image translation via hierarchical style disentanglement. In: Proceedings of the IEEE/CVF Conference on Computer Vision and Pattern Recognition, pp. 8639–8648 (2021)
20. Liu, M.Y., Breuel, T., Kautz, J.: Unsupervised image-to-image translation networks. In: Advances in Neural Information Processing Systems (2017)
21. Liu, M.Y., Tuzel, O.: Coupled generative adversarial networks. Adv. Neural. Inf. Process. Syst. **29**, 469–477 (2016)
22. Liu, Z., Luo, P., Wang, X., Tang, X.: Deep learning face attributes in the wild. In: Proceedings of International Conference on Computer Vision (ICCV), December 2015
23. Mardani, M., Liu, G., Dundar, A., Liu, S., Tao, A., Catanzaro, B.: Neural FFTS for universal texture image synthesis. Adv. Neural. Inf. Process. Syst. **33**, 14081–14092 (2020)
24. Park, T., Liu, M.Y., Wang, T.C., Zhu, J.Y.: Semantic image synthesis with spatially-adaptive normalization. In: Proceedings of the IEEE Conference on Computer Vision and Pattern Recognition, pp. 2337–2346 (2019)
25. Shen, W., Liu, R.: Learning residual images for face attribute manipulation. In: Proceedings of the IEEE Conference on Computer Vision and Pattern Recognition, pp. 4030–4038 (2017)
26. Shen, Y., Gu, J., Tang, X., Zhou, B.: Interpreting the latent space of GANs for semantic face editing. In: Proceedings of the IEEE/CVF Conference on Computer Vision and Pattern Recognition, pp. 9243–9252 (2020)
27. Shen, Y., Zhou, B.: Closed-form factorization of latent semantics in GANs. In: Proceedings of the IEEE/CVF Conference on Computer Vision and Pattern Recognition, pp. 1532–1540 (2021)
28. Voynov, A., Babenko, A.: Unsupervised discovery of interpretable directions in the GAN latent space. In: International Conference on Machine Learning, pp. 9786–9796. PMLR (2020)
29. Wang, T.C., Liu, M.Y., Zhu, J.Y., Tao, A., Kautz, J., Catanzaro, B.: High-resolution image synthesis and semantic manipulation with conditional GANs. In: Proceedings of the IEEE Conference on Computer Vision and Pattern Recognition, pp. 8798–8807 (2018)
30. Wang, Y., Gonzalez-Garcia, A., van de Weijer, J., Herranz, L.: SDIT: scalable and diverse cross-domain image translation. In: Proceedings of the 27th ACM International Conference on Multimedia, pp. 1267–1276 (2019)
31. Wu, P.W., Lin, Y.J., Chang, C.H., Chang, E.Y., Liao, S.W.: RelGAN: multi-domain image-to-image translation via relative attributes. In: Proceedings of the IEEE/CVF International Conference on Computer Vision, pp. 5914–5922 (2019)
32. Wu, Z., Lischinski, D., Shechtman, E.: StyleSpace analysis: disentangled controls for StyleGAN image generation. In: Proceedings of the IEEE/CVF Conference on Computer Vision and Pattern Recognition, pp. 12863–12872 (2021)
33. Xiao, T., Hong, J., Ma, J.: ELEGANT: exchanging latent encodings with GAN for transferring multiple face attributes. In: Proceedings of the European Conference on Computer Vision (ECCV), pp. 168–184 (2018)
34. Yang, G., Fei, N., Ding, M., Liu, G., Lu, Z., Xiang, T.: L2M-GAN: learning to manipulate latent space semantics for facial attribute editing. In: Proceedings of the IEEE/CVF Conference on Computer Vision and Pattern Recognition, pp. 2951–2960 (2021)

35. Yi, R., Liu, Y.J., Lai, Y.K., Rosin, P.L.: APDrawingGAN: generating artistic portrait drawings from face photos with hierarchical GANs. In: Proceedings of the IEEE/CVF Conference on Computer Vision and Pattern Recognition, pp. 10743–10752 (2019)
36. Yi, R., Liu, Y.J., Lai, Y.K., Rosin, P.L.: Unpaired portrait drawing generation via asymmetric cycle mapping. In: Proceedings of the IEEE/CVF Conference on Computer Vision and Pattern Recognition, pp. 8217–8225 (2020)
37. Yi, Z., Zhang, H., Tan, P., Gong, M.: DualGAN: unsupervised dual learning for image-to-image translation. In: International Conference on Computer Vision (2017)
38. Zhang, G., Kan, M., Shan, S., Chen, X.: Generative Adversarial Network with Spatial Attention for Face Attribute Editing. In: Proceedings of the European Conference on Computer Vision (ECCV), pp. 417–432 (2018)
39. Zhu, J., Shen, Y., Zhao, D., Zhou, B.: In-domain GAN inversion for real image editing. In: Vedaldi, A., Bischof, H., Brox, T., Frahm, J.-M. (eds.) ECCV 2020. LNCS, vol. 12362, pp. 592–608. Springer, Cham (2020). https://doi.org/10.1007/978-3-030-58520-4_35
40. Zhu, J.Y., Park, T., Isola, P., Efros, A.A.: Unpaired image-to-image translation using cycle-consistent adversarial networks. In: International Conference on Computer Vision (2017)
41. Zhu, J.Y., et al.: Multimodal image-to-image translation by enforcing bi-cycle consistency. In: Advances in Neural Information Processing Systems, pp. 465–476 (2017)
42. Zhu, P., Abdal, R., Qin, Y., Wonka, P.: SEAN: image synthesis with semantic region-adaptive normalization. In: Proceedings of the IEEE/CVF Conference on Computer Vision and Pattern Recognition, pp. 5104–5113 (2020)

Any-Resolution Training
for High-Resolution Image Synthesis

Lucy Chai[1]([✉]), Michaël Gharbi[2], Eli Shechtman[2], Phillip Isola[1],
and Richard Zhang[2]

[1] MIT, Cambridge, USA
lrchai@mit.edu
[2] Adobe Research, San Jose, USA

Abstract. Generative models operate at fixed resolution, even though
natural images come in a variety of sizes. As high-resolution details are
downsampled away and low-resolution images are discarded altogether,
precious supervision is lost. We argue that every pixel matters and create datasets with variable-size images, collected at their native resolutions. To take advantage of varied-size data, we introduce *continuous-
scale* training, a process that samples *patches* at random scales to train
a new generator with variable output resolutions. First, conditioning the
generator on a target scale allows us to generate higher resolution images
than previously possible, without adding layers to the model. Second, by
conditioning on continuous coordinates, we can sample patches that still
obey a consistent global layout, which also allows for scalable training at
higher resolutions. Controlled FFHQ experiments show that our method
can take advantage of multi-resolution training data better than discrete
multi-scale approaches, achieving better FID scores and cleaner high-
frequency details. We also train on other natural image domains including churches, mountains, and birds, and demonstrate arbitrary scale synthesis with both coherent global layouts and realistic local details, going
beyond 2K resolution in our experiments. Our project page is available
at: https://chail.github.io/anyres-gan/.

Keywords: Unconditional image synthesis · Generative adversarial
networks · Continuous coordinate functions · Multi-scale learning

1 Introduction

The first step of typical generative modeling pipelines is to build a dataset with
a fixed, target resolution. Images above the target resolution are downsampled,
removing high-frequency details, and data of insufficient resolution is omitted,
discarding structural information about low frequencies. Our insight is that this
process wastes potentially learnable information. We propose to embrace the natural diversity of image sizes, processing them at their *native* resolution (Fig. 1).

Supplementary Information The online version contains supplementary material
available at https://doi.org/10.1007/978-3-031-19787-1_10.

Fig. 1. Trained on a dataset of varied-size images, our unconditional generator learns to synthesize patches at continuous scales to match the distribution of real patches. Here, we render crops of the image at different resolutions, indicating the target resolution for each. We indicate the region of each crop in the top-left panel, which is the image directly sampled without scale input.

Relaxing the fixed-dataset assumption offers new, previously unexplored opportunities. One can potentially simultaneously learn global structure – for which large sets of readily-available low-resolution images suffice – and fine-scale details – where even a handful of high-resolution images may be adequate, especially given their internal recurrence [48]. This enables generating images at higher resolutions than previously possible, by adding in higher-resolution images to currently available fixed-size datasets.

This problem setting offers unique challenges, both in regards to modeling and scaling. First, one must generate images across multiple scales in order to compare with the target distribution. Naïvely downsampling the full-resolution image is suboptimal, as it is common to even have images of 8× difference in scale in the dataset. Secondly, generating and processing high-resolution images offers scaling challenges. Modern training architectures at 1024 resolution already push the current hardware to the limits in memory and training time, and are unable to fully make use of images above that resolution.

To bypass these issues, we design a generator to synthesize image crops at arbitrary scales, hence, performing *any-resolution* training. We modify the state-of-the-art StyleGAN3 [26] architecture to take a grid of continuous coordinates, defined on a bounded domain, as well as a target scale, inspired by recent work in coordinate-based conditioning [2,7,32,37]. By keeping the latent code constant and varying the crop coordinates and scale, our generator can output patches of the same image [31,44], but at various scales. This allows us to (1) efficiently generate at arbitrary scale, so that a discriminator critic can compare generations to a multi-resolution dataset, and (2) decouple high-resolution synthesis from increasing architecture size and prohibitive memory requirements.

We first experiment on FFHQ images [24] as a controlled setting, showing efficient data usage by comparing our training on image subsets to using the entire full resolution dataset. We find minimal degradation (FIDs varying by 0.3), even at highly skewed distributions – 98% of the dataset at 4× lower resolution, and just 2% at a higher, mixed resolutions. Practically, this means we can leverage large-scale (>100k) lower-resolution datasets, such as LSUN Churches [65], Flickr Mountains [41], and Flickr Birds collected by us, and add a relatively small amount of high-resolution images (∼ 6000), for continuous resolution synthesis beyond the 1024 resolution limit of current generators. To summarize, we:

- propose to train on mixed-resolution datasets from images in-the-wild.
- modify the generator to be amenable to such data, sampling patches at arbitrary scales during our *any-resolution* training procedure.
- demonstrate successful generations beyond 1024 × 1024, with fine details and coherent global structure, without a larger and more expensive generator.
- introduce a variant of the FID metric that captures image statistics at multiple scales, thus accounting for the details of high resolutions.

2   Related Works

Unconditional Image Synthesis. Recent generative models including GANs [3,16,26,27], Variational Autoencoders [29], diffusion models [11,17,38,50,51], and autoregressive models [30,40,54] such as transformers [6,14,55] are rapidly improving in quality. Of these, we focus on GANs, which offer state-of-the-art performance along with efficient inference and effective editing properties. A key innovation in GANs has been multi-resolution supervision during training. Works such as LapGAN [10], the Progressive/StyleGAN family [24,26–28], MSG-GAN [23], and AnyCost-GAN [33] have demonstrated stable training by growing the generator with additional layers that increase resolution by factors of two. Such a strategy even works for single-image GANs [44,47], based on the observation that images share statistics across scales. While several works [25,69,70] show data augmentations, such as small jitters in scale, can help stabilize training, they are processing the same, underlying fixed-resolution dataset. We draw upon the insights in these works for stable training, and seek to unlock training on an any-resolution dataset. Importantly, our generator does not use additional layers and can synthesize images at continuous scales, not only powers of two.

Coordinate-Based Functions. Coordinate-based encodings enable spatial conditioning and provide an inductive bias towards natural images [52]. Recent methods use point-based neural mappings to transform 2D or 3D coordinates to a color value for the purposes of unconditional generation [1,9,26,31,32,49], conditional generation [45], 3D view synthesis [5,37,43], or fitting arbitrary signals [7,36]. By oversampling the coordinate grid, one can generate a larger image at inference time. However, because these models keep the same fixed-scale dataset assumption during training, the outputs struggle to offer additional high-frequency details without a high-frequency training signal. We draw upon the innovations in coordinate-based functions to sample patches at different scales and locations, enabling us to efficiently train on multiple scales. MS-PIE [62] and MS-PE [9] add positional encodings for multi-scale synthesis, but retain a global image discriminator at smaller resolution. Concurrently, ScaleParty [39] also samples patches, but their goal is to generate at multiple scales with cross-scale consistency while we focus on training with arbitrary size real images.

Extrapolation. One method of generating "infinite" resolution is extrapolating an image. Early texture synthesis works [12,13,60] focus on stationary textures. Recent advances [72] explore non-stationary textures, with large-scale structures and inhomogeneous patterns. Similar approaches operate by outpainting images, extending images beyond their boundaries in a conditional setting [35,53,64,72]. Recent generative models synthesize large scenes [8,32], typically casting synthesis as an outpainting problem [35,64]. These methods are most effective for signals with a strong stationary component, such as landscapes, although extrapolation of structured scenes can be achieved in some domains [59]. Unlike textures, the images we wish to synthesize typically have a strong global structure. In a sense, we seek to extrapolate by "zooming in" or out, rather than "panning" beyond an image's boundaries.

Super-Resolution. An alternative approach to generating high-resolution imagery would be to start with an off-the-shelf generative model and feed its outputs to a super-resolution method [7,18,57,58], possibly exploiting the self-similarity properties of images [15,19,46,48]. Applying super-resolution models is challenging, in part because of the specific blur kernels super-resolution models are trained on [66]. Furthermore, though generations continue to improve, there remains a persistent domain shift between synthesized and real images [4,56]. Finally, super-resolution is a local, conditional problem where the global structure is dictated by the low-resolution input, and optionally an additional high-resolution reference image [22,34,61,63,71]. We synthesize plausible images unconditionally, leveraging a *set* of high-resolution images to produce both realistic global structure and coherent fine details.

3   Methods

In standard GAN training, all training images share a common fixed resolution, which matches the generator's output size. We seek to exploit the variety of image resolutions available in the wild, learning from pixels that are usually discarded, to enable high- and continuously-variable resolution synthesis. We achieve this by switching from the common fixed-resolution thinking,

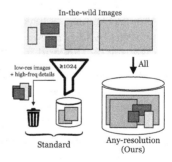

Fig. 2. Data comes at a variety of scales. Traditional dataset construction filters out low-resolution images and downsamples high-resolution images to a fixed, training resolution. We aim to keep images at their original resolution.

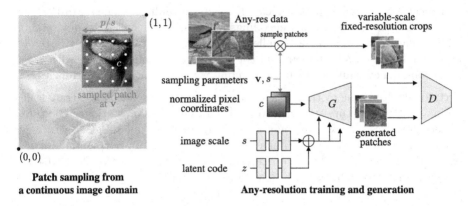

Fig. 3. Overview. (Left) We parameterize images (real or synthetic) as continuous functions over a normalized domain and extract random patches at various scales s, but constant resolution p. (Right) To train, we sample crops at random scales and offsets $\mathbf{v}$ from full-size real images. The same crops are sampled from the generator, by passing it a grid of the desired coordinates $c_{\mathbf{v},s}$, and injecting the image scale s through modulation, in addition to a global latent code z.

to a novel 'any-resolution' approach, where the original size of each training image is preserved (Fig. 2). We introduce a new class of GAN generators that learn from this multi-resolution signal to synthesize images at any resolution (Sect. 3.1), and show how to train them by sampling patches at multiple scales to jointly supervise the global-structure and fine image details (Sect. 3.2).

3.1 Multi-resolution GAN

We design our approach to leverage state-of-the-art GANs. We keep the architecture of the discriminator unchanged. Since the discriminator operates at fixed resolution, we modify the generator to synthesize images at any resolution and receive the discriminator's fixed-resolution supervision. Our implementation builds on the StyleGAN3 framework [26], which is conditioned on a fixed coordinate grid. We modify this grid for any-resolution and patch-based synthesis.

Continuous-Resolution Generator. We treat each image as a continuous function defined on a bounded normalized coordinate domain $[0, 1] \times [0, 1]$. The generator G always generates patches at a fixed pixel resolution $p \times p$, but each patch implicitly corresponds to a square sub-region, centered at $\mathbf{v} \in [0, 1]^2$, of the larger image. Denoting the resolution of the larger image as $s \times s$, we have that the patch size is p/s in normalized coordinates (see Fig. 3, left). During training, we sample patches from images at multiple scales s, either from the generator or from the multi-resolution dataset, before passing them to the fixed-resolution discriminator D. Formally, our generator takes three inputs: a regular grid of normalized continuous pixel coordinates $c_{\mathbf{v},s} \in \mathbb{R}^{p \times p \times 2}$, the resolution $s \in \mathbb{N}$ of the (implicit) larger image the patch is extracted from, and z, the latent code representing this larger image. It synthesizes the patch's pixel values at the sampled coordinates as:

$$G(z, c_{\mathbf{v},s}, s) = G(F(c_{\mathbf{v},s}); M(z, s)), \tag{1}$$

where F is a Fourier embedding of the continuous coordinates [26], and M is an auxiliary function that maps the latent code and sampling resolution into a set of modulation parameters for the StyleGAN3 generator (see Sect. 3.3 for details). Our method therefore modifies two components from StyleGAN3. First, we replace the fixed coordinate grid with *patch-dependent* coordinates to train on variable-resolution images; these coordinates are adjusted to account for upsampling in StyleGAN3. Second, we append an auxiliary branch M to inject scale information throughout the generator.

At test time, we can generate images at arbitrarily high resolutions by sampling the full continuous domain $[0, 1] \times [0, 1]$ at the desired sampling rate. Theoretically, the maximum resolution is infinite, but in practice the amount of detail that the model can generate is determined by generator resolution p and the resolutions of the training images.

3.2   Two-Phase Training

We train our generator in two phases. In the first, we want the generator to learn to generate globally-coherent images. For this, we disable the patch sampling mechanism and pretrain the generator at a fixed scale, corresponding to the full continuous image domain. That is, we fix $s = p$, and $\mathbf{v} = (0.5, 0.5)$, which is equivalent to standard fixed-resolution GAN training. Both the coordinate tensor $c_{\mathbf{v},s}$ and the scale conditioning variable s are constant in this phase, so we simply refer to the image generated as $G_{\text{fixed}}(z)$ and follow the training procedure of StyleGAN3. In the second phase, we enable patch sampling for both the real and synthetic images and continue training the generator using variable-scale patches, so it learns to synthesize fine details at any resolution. We found that using a copy of the pretrained fixed-scale generator G_{fixed} as a teacher model helps stabilize training in this phase.

Global Fixed-Resolution Pretraining. During the pretraing phase, we effectively resample all the training images to a fixed resolution $p \times p$, as in standard GAN training. Let $x \sim \mathcal{D}_{\text{fixed}}$ denote an image sampled from this fixed size dataset. We optimize a standard GAN objective with non-saturating logistic loss and R_1 regularization on the discriminator:

$$V(D, G(z), x) = D(x) - D(G(z)), \quad R_1(D, x) = ||\nabla D(x)||^2,$$

$$G_{\text{fixed}} = \arg \min_{G} \max_{D} \mathbb{E}_{z, x \sim \mathcal{D}_{\text{fixed}}} V(D, G(z), x) - \frac{\lambda_{R_1}}{2} R_1(D, x). \tag{2}$$

We follow the recommended values for λ_{R_1}, depending on the generator resolution p [26].

Mixed-Resolution Patch-Based Training. In the second phase, we enable multi-resolution sampling, alternating between extracting random crops from our any-resolution dataset and generating them with our continuous generator.

For synthetic patches, we sample a patch location $\mathbf{v}$ uniformly in the continuous domain $[0, 1] \times [0, 1]$; and an arbitrary image resolution $s \geq p$, corresponding to the implicit full image around the square patch. From those, we derive the sampling coordinate grid $c_{\mathbf{v},s}$, and synthesize the patch image $G(z, c_{\mathbf{v},s}, s)$, as described earlier.

For 'real' patches, we sample an image from our dataset. Because this image can have any resolution $s_{\text{im}} \geq p$, we crop it to a random square matching its smallest dimension, then Lanczos downsample this square to a random resolution $s \times s$ with $s_{\text{im}} \geq s \geq p$. Finally, we extract a random $p \times p$ crop from the downsampled image, recording its center $\mathbf{v}$. To preserve the generator's global coherence and continuous generation ability, we sample at global scale $s = p$ and $\mathbf{v} = (0.5, 0.5)$ (similar to the pretraining step) with probability 50%. We found that image quality at global resolution $s = p$ degrades otherwise, and we refer to these generated full images of size $p \times p$ as "base images." Our any-resolution GAN optimizes the following objective during this phase:

$$
G^* = \arg\min_G \min_D \ \mathbb{E}_{z,\{x,s,\mathbf{v}\}\sim\mathcal{D}} \ V(D, G(z, c_{\mathbf{v},s}, s), x)
$$
$$
+ \lambda_{\text{teacher}}\mathcal{L}_{\text{teacher}}(G, G_{\text{fixed}}, z) - \frac{\lambda_{R_1}}{2} R_1(D, x). \tag{3}
$$

We use $\lambda_{\text{teacher}} = 5$; other values offer slight tradeoffs between similarity to the base teacher model G_{fixed}, and FID score (see supplemental). $\mathcal{L}_{\text{teacher}}$ is an auxiliary loss to encourage faithfulness to the pretrained fixed-resolution generator G_{fixed}. The architecture of D remains the same as in the pretraining step; we found that modifying the discriminator setup did not further improve results (see supplemental).

Teacher Model. For the second training phase above, we initialize G with the pretrained weights of G_{fixed}. Weights for discriminator D are also kept for fine-tuning. We keep a separate copy of G_{fixed} with frozen weights, the teacher, for additional supervision. We design a loss function that encourages the generated patch (at any resolution), to match the teacher's fixed-resolution output in the region corresponding to the patch, after downsampling and proper alignment [21]. Formally, this loss is given by:

$$
\mathcal{L}_{\text{teacher}}(G, G_{\text{fixed}}, z) = d\left(m \odot w_{\mathbf{v},s}(G(z, c_{\mathbf{v},s}, s)), \ m \odot G_{\text{fixed}}(z)\right), \tag{4}
$$

where d is the sum of a pixel-wise ℓ_1 loss, and the LPIPS perceptual distance [20,67]. The warp function $w_{\mathbf{v},s}$ transforms and resamples the generated high-resolution patch using a band-limited Lanczos kernel, to project it in the coordinate frames of the low-resolution, global image $G_{\text{fixed}}(z)$. Because the warped patch does not cover the entire image domain, we multiply it with a binary mask m to indicate the valid pixels, prior to computing loss d.

3.3   Implementation Details

Scale-Conditioning. In addition to the pixel location $c_{\mathbf{v},s}$, we also pass the global resolution information s to the generator. Knowledge of the global image scale is important to enable continuous scale variations and proper anti-aliasing [2,7]. We found it beneficial to explicitly inject this information into all intermediate layers of the generator. To achieve this, we use a dual modulation approach [68], embedding the latent code z and scale s separately using two independent sub-networks (we use the same mapping network architecture for each). The two outputs are summed to obtain a set of modulation parameters $M(z,s)$, used to modulate the main generator features. Architectural details of the generator G and mapping network M, can be found in the supplemental.

Synthesizing Large Images. Our fully-convolutional generator can render image at arbitrary resolutions. But images larger than 1024×1024 require significant GPU memory. Equivalently, we can render non-overlapping tiles that we assemble into a larger image. Our patch-based multi-resolution training and the Fourier encoding of the spatial coordinates make the tile junctions seamless.

4   Experiments

We introduce a modified image quality metric that computes FID over multi-scale image patches without downsampling, which is largely correlated to the standard FID metric when ground truth high-resolution images are available (FFHQ), yet more sensitive to the quality of larger resolutions. We then compare our model to alternative approaches for variable scale synthesis and super-resolution on other natural image domains (Sect. 4.1). Finally, we investigate variations of our model and training procedure to validate our design decisions (Sect. 4.2).

Data. Our method is general and can work on collections of any-resolution data. As such, when targeting high-resolution generation, rather than starting over, we can add additional high-resolution (HR) images to existing, fixed-size, low-resolution (LR) datasets. Our datasets and their statistics are listed in Table 1. Figure 4 shows the resolution distribution in each dataset.

We begin initial experiments with a controlled setting of FFHQ, which contains 70K images at 1024 resolution. From these, we construct a varied-size dataset by (1) using 256 resolution for all images (2) downsampling a 5K subset between 512–1024 (uniformly distributed) and (3) add 1k subset at full 1024. The last step enables us to judiciously compare to methods that are limited to synthesizing images at strict powers of 2. We refer to this mixture as FFHQ6k. We use the full 1024 dataset as ground-truth for evaluation metrics.

In the remaining domains, we push current generation results to higher resolution by scraping HR images from Flickr. In cases where a standard fixed-size dataset is available (LSUN Churches [65] and Mountains [41]), we select the additional HR images to approximately match the LR domain. Our final generators synthesize realistic details despite the majority of the training set being LR. For

Table 1. Any-resolution datasets and generator settings. We build upon low-resolution (LR) datasets, and use it for fixed-size dataset pre-training. We add additional high-resolution (HR) images, of mixed resolutions. Note that the number of HR images is small (∼2–8% of LR size). Patches of size p are sampled from both subsets during training, with average sampled scale $\mathbb{E}[s]$.

Domain	Dataset							Generator		
	Source	# Imgs		Resolutions				Config	Resolutions	
		LR	HR	LR	HR_{min}	HR_{med}	HR_{max}		p	$\mathbb{E}[s]$
Faces	FFHQ	70,000	6000	256	512	819	1024	R	256	458
Churches	LSUN & Flickr	126,227	6253	256	1024	2836	18,000	T	256	1061
Birds	Flickr	112,266	7625	256	512	1365	2048	T	256	585
Mountains	Flickr	507,495	9361	1024	2049	3168	12,431	T	1024	1823

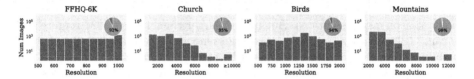

Fig. 4. Training set size distributions. Histogram shows the size distribution of the HR images (y-axis in log scale); pie chart indicates the proportion of LR to HR images.

Birds and Churches, > 92% of the training set is at 256 resolution but our model maintains quality beyond 1024; for Mountains > 98% of training set at 1024 but our model can generate beyond 2048. These categories cover a range between objects (but without the strong alignment of FFHQ) and outdoor scenes.

4.1 Continuous Multi-scale Image Synthesis

Qualitative Examples. We show qualitative examples in Fig. 5. Our generated images preserve the fine details of HR structures, such as bricks, rocky slopes, feathers, or hair. Pushing the inference resolution towards and beyond the higher resolutions of training images, we find that textured surfaces typically deteriorate first before edge boundaries deteriorate eventually (Fig. 6).

Patch-FID Metric. Standard FID evaluates *global structure* by first downsampling all images to a common size of 299. By design, this ignores high-resolution details (and itself can cause artifacts [42]). Therefore, we propose a modification, which we dub 'patch-FID', to specifically evaluate *texture* synthesized at higher resolutions. Our patch-FID randomly resizes and crops patches from the HR dataset, and computes FID on real and generated patches, sampled at corresponding scales and locations. We use 50k patches, matching standard FID. By avoiding downsampling, our patch-FID is more sensitive to blurriness or artifacts at higher resolutions, resulting in larger absolute difference compared to standard FID at 1024 resolution. As a sanity check, when a full HR ground-truth is available, we find it is largely correlated to standard FID (see Table 2).

Fig. 5. The inset shows the entire generated, high-resolution images (between 1000–3000 resolution), with enlarged regions of interest outlined in the white box. Note that our model can render the image (or any sub-region) at any resolution.

Fig. 6. Extrapolation limits. We test the extrapolation capabilities of our model by specifying the inference scale s. Typically, textures such as bricks and feathers deteriorate first before edges degrade. The dotted line indicates when generation starts to exceed the average scale sampled in training $\mathbb{E}[s]$ (which is 585 for birds and 1061 for churches).

To summarize, to evaluate structure, we compute standard FID on images generated at specified resolutions, e.g., FID (256). To evaluate texture, we sample

Table 2. Varied-size training and inference. Random-resize MS-PE [9] performs varied-size synthesis, but assumes a fixed-size dataset. AnyCost-GAN handles varied training at powers of 2. Our method directly utilizes training images at any size, achieving better results by FID. († = copied from paper)

FFHQ6K	FID			pFID
	256	512	1024	Random
MS-PE [9]†	6.75	30.41	–	–
Anycost [33]	4.24	5.94	6.47	18.39
Ours	3.34	3.71	4.06	2.96

Table 3. Comparison to super-resolution using patch-FID (pFID). For each domain, we compare our model to continuous-scale (LIIF) and fixed-scale super-resolution (Real-ESRGAN) models. Lower pFID suggests that our model can generate realistic details at high resolutions, not achievable with super-resolution alone.

	pFID (random)			
	FFHQ6K	Church	Bird	Mountain
LIIF [7]	22.93	83.88	30.19	23.10
Real-ESRGAN [57]	16.92	23.04	16.10	19.05
Ours	2.96	9.89	6.52	7.99

patches at random scales and locations and measure our patch-FID, which we denote as pFID (random). Lower numbers are better in both cases.

Alternative Methods of Multi-size Training and Generation. Our generator is encouraged to synthesize realistic high-resolution textures at training, even when the discriminator does not get to see the full image. While MS-PE [9] also enables continuous resolution synthesis, the discriminator learns only at a single resolution and the generator is not trained patch-wise. We find that this downsampling for the discriminator is detrimental to image quality at higher resolutions. Anycost-GAN [33] performs image synthesis at powers-of-two resolutions by adding additional synthesis blocks. For comparison, we modify it to handle a multi-size dataset by downsampling images to the nearest power of two and training each layer only on the valid image subset. Compared to Anycost-GAN, our model is more data-efficient, due to weight sharing for generation at multiple scales. Anycost-GAN learns a separate module for each increase in resolution, creating artifacts at higher resolutions when fewer HR training images are available and higher FID scores (Table 2). Additionally, Anycost-GAN increases the generator and discriminator size for synthesis at higher resolutions, whereas our model incurs a constant training cost, regardless of the inference scale.

Comparison to Super-Resolution. Most super-resolution methods require LR/HR image pairs, whereas there is no ground-truth HR counterpart to a LR image synthesized by our fixed-scale generator G_{fixed}. The teacher regularization encourages similarity between G_{fixed} and G's outputs, but unlike super-resolution, this supervision occurs at low-resolution, allowing variations in fine details. Figure 7 compares our model to super-resolution methods applied to the output of G_{fixed}. Our method produces much sharper details than LIIF [7], a recent continuous-scale super-resolution technique, and cleaner images than the state-of-the-art Real-ESRGAN [57]. The latter is a fixed-resolution model, so we run it iteratively until exceeding a target resolution, and then Lanczos downsample the result to the target size. Real-ESRGAN's outputs are either overly

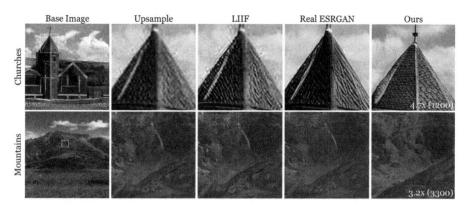

Fig. 7. Super-resolution Comparisons. Qualitative comparisons of Lanczos upsampling a patch from the base image (upsample), continuous (LIIF [7]) and fixed-factor (Real ESRGAN [57]) super-resolution models, and our trained model. LIIF tends to amplify artifacts from the base image (e.g. the JPEG artifacts around the church). While Real-ESRGAN is better at suppressing artifacts, it tends to overly smooth surfaces or synthesize grid-like textures (mountain). Our model is not a super-resolution model; it can add additional details to the low-resolution image but tolerates slight distortions in structure which are regularized with the teacher weight.

smooth, or exhibit grid-like artifacts. Our method generates realistic textures based on the low-resolution output of G_{fixed} and reaches a lower pFID (Table 3).

4.2   Model Variations

Using the full high-resolution FFHQ dataset as a benchmark, we investigate individual components of our architecture and training process. We train each model variation for 5M images and record metrics from the best FID@1024 checkpoint. We only report quantitative metrics in the main paper and refer to the supplemental for further evaluations and visual comparisons.

Impact of Teacher Regularization. Our full model uses an "inverse" teacher regularizer to encourages a downsampled HR patch to match the low-resolution teacher as described in Sect. 3.2. We also explored a variant with a "forward" teacher loss, in which the generated patch is encouraged to match the *upsampled* teacher output. This variation is qualitatively inferior and blurs details; it has worse FID at higher resolutions (see supplemental for details and visuals). Removing the teacher altogether improves pFID but degrades FID. Qualitatively, the generated patches diverge significantly from the fixed-size global image. We hypothesize that the global change in structure negatively impacts overall image quality, causing global FID metrics to increase, but this cannot be captured from evaluating patches alone. We found $\lambda_{\text{teacher}} = 5$ to provide the best balance between global and local image quality, but we observe minimal differences in FID and pFID for other values, evidence that the model can tolerate a range

Table 4. Multisize Training. Downsampling or upsampling all images to a common size, or using only the subset of the largest images, worsens FID compared to our training strategy. (*) indicates our default setting.

	FID			pFID
	256	512	1024	Random
Resize down to 512	3.31	4.11	19.18	26.83
Resize up to 1024	3.46	13.43	4.86	6.65
Train 1024 subset	3.46	12.41	4.67	5.43
Multisize training (*)	3.37	4.41	4.47	4.28

Table 5. Number of HR images. Our method is robust to a wide range of HR images, even when only 1K images at HR are available (<2% of the full ground-truth dataset).

	FID			pFID
	256	512	1024	Random
1k (1.4%)	3.43	5.13	4.38	3.73
5k (7.1%) (*)	3.36	4.97	4.54	3.48
10k (14.2%)	3.46	4.96	4.65	3.54
70k (100%)	3.42	4.88	4.52	3.42

of values for this parameter. See supplemental for a parameter sweep with full scores.

Removing Scale Conditioning Degrades Quality. We inject the scale information to intermediate layers of the generator through scale-conditioning. Adding this improves FID@1024 from 4.88 to 4.47, and pFID from 4.67 to 4.28.

Multi-size Training Improves Fidelity at all Scales. Our multi-size data pipeline lets our model learn to synthesize at continuous scales, which is a strictly more challenging than learning at a fixed scale. In Table 4, we investigate to what extent learning from images of varied sizes offers benefits over fixed-scale training on a smaller dataset. Visual comparisons can be found in supplemental. In a first alternative, using the same FFHQ-6K dataset, we resize all images down to 512 and train the model to generate patches at 512×512. In this case, the model performs well up to 512 scale, but does not generalize beyond (e.g., 1024) since it cannot exploit the information lost in downsampling. In two other variants, we train models for 1024 resolution in the first case by upsampling all images up to 1024, in the second by keeping only the 1K subset of images at 1024 resolution. Both variants are trained to output images specifically at 1024 resolution. The former approach (upsampling) increases blurriness. In the latter, FID@1024 remains worse than that of our multi-size training, which can take advantage of more data despite most of it being *smaller* than 1024.

Impact of Number of HR Images. Due to the patch-based training procedure, we find that our model can be trained with a small fraction of HR images, compared to the 70k LR images in the dataset. In Table 5, we use progressively larger subsets of HR images: 1k, 5k, 10k. We found that the FID scores are largely similar (within 0.3) to using the entire 70k HR images. However, training with 1k or fewer HR images shows evidence of divergence during training (see supplemental), but stabilizes by 5k HR images, For the remaining domains, we collect roughly 5K-10K images to construct the HR dataset.

4.3   Properties of Multi-scale Generation

Correcting Artifacts from Low Resolution. Because our model is not directly trained with corresponding LR and HR image pairs, we find that there can be small distortions between the upsampled base image and the HR generation from the same latent code. In some cases, this can be a desirable property (Fig. 8). For instance, the base generator on the birds dataset can struggle in synthesizing the eye of the bird, which is less apparent at low resolutions, but more salient at high resolution. Consequently, our HR generation will add the missing eye, and also synthesizes additional feather and beak details. In the churches domain, because the LR and HR datasets are collected separately, we find that the synthesized watermarks and JPEG artifacts at the base resolution disappear at higher resolution, because the HR dataset we used is of higher quality and does not have any watermark. The similarity between the LR and HR generations can be tuned using λ_{teacher} during training.

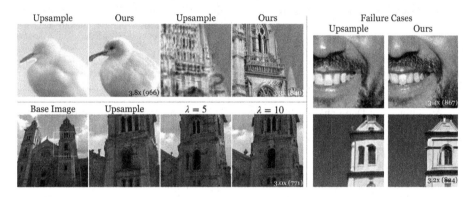

Fig. 8. Model Properties and Failure Cases. As fine details can be more difficult to learn at low resolution, our model is capable of adding corrections when generating at higher resolutions. In the case of inconsistencies between the LR and HR data sources, the model deletes patterns that are not present in the HR dataset (*e.g.* watermarks and compression artifacts), influenced by the teacher regularization weight. Failure cases include biases towards circular or ring-like structures.

Failure Cases. Our model tends to inherit the artifacts from StyleGAN3, such as a centered front tooth in FFHQ. In instances in which the base resolution image contains uneven surfaces, the model may fail to fully mitigate them at higher resolutions. These artifacts are often subtle at the low resolution, but become more apparent when upsampling the base image or generating at a larger target scale. In some cases, our model also has a tendency to generate "watery" circular or ring-like artifacts (Fig. 8).

5 Conclusion

We propose an image synthesis approach that can train on images of varied resolution and perform inference at continuous resolutions. This lifts the fixed-resolution requirement of prior generative models, which discard higher-resolution details. To do this, we train a generator jointly on a low-resolution dataset to learn global structure, and on patches from the varied-size dataset to learn details. At inference time, we can synthesize an image at any resolution by supplying the appropriate coordinate grid and scale factor to the generator. By using training images at their native resolutions and a single model for continuous-resolution synthesis, our method can efficiently leverage information present in only a handful of high-resolution images to complement a large set of low-resolution images. This approach enables high-resolution synthesis without a larger generator or large dataset of fixed-size, high-resolution images.

Acknowledgements. We thank Assaf Shocher for feedback and Taesung Park for dataset collection advice. LC is supported by the NSF Graduate Research Fellowship under Grant No. 1745302 and Adobe Research Fellowship. This work was started while LC was an intern at Adobe Research.

References

1. Anokhin, I., Demochkin, K., Khakhulin, T., Sterkin, G., Lempitsky, V., Korzhenkov, D.: Image generators with conditionally-independent pixel synthesis. In: IEEE Conference on Computer Vision and Pattern Recognition, pp. 14278–14287 (2021)
2. Barron, J.T., Mildenhall, B., Tancik, M., Hedman, P., Martin-Brualla, R., Srinivasan, P.P.: Mip-NeRF: a multiscale representation for anti-aliasing neural radiance fields. In: International Conference on Computer Vision, pp. 5855–5864 (2021)
3. Brock, A., Donahue, J., Simonyan, K.: Large scale GAN training for high fidelity natural image synthesis. In: International Conference on Learning Representations (2018)
4. Chai, L., Bau, D., Lim, S.-N., Isola, P.: What makes fake images detectable? Understanding properties that generalize. In: Vedaldi, A., Bischof, H., Brox, T., Frahm, J.-M. (eds.) ECCV 2020. LNCS, vol. 12371, pp. 103–120. Springer, Cham (2020). https://doi.org/10.1007/978-3-030-58574-7_7
5. Chan, E.R., Monteiro, M., Kellnhofer, P., Wu, J., Wetzstein, G.: pi-GAN: periodic implicit generative adversarial networks for 3D-aware image synthesis. In: IEEE Conference on Computer Vision and Pattern Recognition, pp. 5799–5809 (2021)
6. Chen, M., et al.: Generative pretraining from pixels. In: International Conference on Machine Learning, pp. 1691–1703. PMLR (2020)
7. Chen, Y., Liu, S., Wang, X.: Learning continuous image representation with local implicit image function. In: IEEE Conference on Computer Vision and Pattern Recognition, pp. 8628–8638 (2021)
8. Cheng, Y.C., Lin, C.H., Lee, H.Y., Ren, J., Tulyakov, S., Yang, M.H.: In&out: diverse image outpainting via GAN inversion. arXiv preprint arXiv:2104.00675 (2021)

9. Choi, J., Lee, J., Jeong, Y., Yoon, S.: Toward spatially unbiased generative models. In: International Conference on Computer Vision (2021)

10. Denton, E., Chintala, S., Szlam, A., Fergus, R.: Deep generative image models using a laplacian pyramid of adversarial networks. In: Advances in Neural Information Processing Systems (2015)

11. Dhariwal, P., Nichol, A.: Diffusion models beat GANs on image synthesis, vol. 34 (2021)

12. Efros, A.A., Freeman, W.T.: Image quilting for texture synthesis and transfer. In: Proceedings of the 28th Annual Conference on Computer Graphics and Interactive Techniques, pp. 341–346 (2001)

13. Efros, A.A., Leung, T.K.: Texture synthesis by non-parametric sampling. In: International Conference on Computer Vision, vol. 2, pp. 1033–1038. IEEE (1999)

14. Esser, P., Rombach, R., Ommer, B.: Taming transformers for high-resolution image synthesis. In: IEEE conference on Computer Vision and Pattern Recognition, pp. 12873–12883 (2021)

15. Glasner, D., Bagon, S., Irani, M.: Super-resolution from a single image. In: International Conference on Computer Vision, pp. 349–356. IEEE (2009)

16. Goodfellow, I.J., Shlens, J., Szegedy, C.: Explaining and harnessing adversarial examples. In: International Conference on Learning Representations (2015)

17. Ho, J., Jain, A., Abbeel, P.: Denoising diffusion probabilistic models. In: Advances in Neural Information Processing Systems, vol. 33, pp. 6840–6851 (2020)

18. Hu, X., Mu, H., Zhang, X., Wang, Z., Tan, T., Sun, J.: Meta-SR: a magnification-arbitrary network for super-resolution. In: IEEE Conference on Computer Vision and Pattern Recognition, pp. 1575–1584 (2019)

19. Huang, J.B., Singh, A., Ahuja, N.: Single image super-resolution from transformed self-exemplars. In: IEEE Conference on Computer Vision and Pattern Recognition, pp. 5197–5206 (2015)

20. Huh, M., Zhang, R., Zhu, J.-Y., Paris, S., Hertzmann, A.: Transforming and projecting images into class-conditional generative networks. In: Vedaldi, A., Bischof, H., Brox, T., Frahm, J.-M. (eds.) ECCV 2020. LNCS, vol. 12347, pp. 17–34. Springer, Cham (2020). https://doi.org/10.1007/978-3-030-58536-5_2

21. Irani, M., Peleg, S.: Improving resolution by image registration. CVGIP Graph. Models Image Process. 53(3), 231–239 (1991)

22. Jiang, Y., Chan, K.C., Wang, X., Loy, C.C., Liu, Z.: Robust reference-based super-resolution via C2-matching. In: IEEE Conference on Computer Vision and Pattern Recognition, pp. 2103–2112 (2021)

23. Karnewar, A., Wang, O.: MSG-GAN: multi-scale gradients for generative adversarial networks. In: IEEE Conference on Computer Vision and Pattern Recognition, pp. 7799–7808 (2020)

24. Karras, T., Aila, T., Laine, S., Lehtinen, J.: Progressive growing of GANs for improved quality, stability, and variation. In: International Conference on Learning Representations (2018)

25. Karras, T., Aittala, M., Hellsten, J., Laine, S., Lehtinen, J., Aila, T.: Training generative adversarial networks with limited data. In: Advances in Neural Information Processing Systems (2020)

26. Karras, T., et al.: Alias-free generative adversarial networks. In: Advances in Neural Information Processing Systems, vol. 34 (2021)

27. Karras, T., Laine, S., Aila, T.: A style-based generator architecture for generative adversarial networks. In: IEEE Conference on Computer Vision and Pattern Recognition (2019)

28. Karras, T., Laine, S., Aittala, M., Hellsten, J., Lehtinen, J., Aila, T.: Analyzing and improving the image quality of StyleGAN. In: IEEE Conference on Computer Vision and Pattern Recognition (2020)
29. Kingma, D.P., Welling, M.: Auto-encoding variational bayes. arXiv preprint arXiv:1312.6114 (2013)
30. Larochelle, H., Murray, I.: The neural autoregressive distribution estimator. In: Proceedings of the Fourteenth International Conference on Artificial Intelligence and Statistics, pp. 29–37. JMLR Workshop and Conference Proceedings (2011)
31. Lin, C.H., Chang, C.C., Chen, Y.S., Juan, D.C., Wei, W., Chen, H.T.: Coco-GAN: generation by parts via conditional coordinating. In: International Conference on Computer Vision, pp. 4512–4521 (2019)
32. Lin, C.H., Lee, H.Y., Cheng, Y.C., Tulyakov, S., Yang, M.H.: InfinityGAN: towards infinite-resolution image synthesis. In: International Conference on Learning Representations (2021)
33. Lin, J., Zhang, R., Ganz, F., Han, S., Zhu, J.Y.: Anycost GANs for interactive image synthesis and editing. In: IEEE Conference on Computer Vision and Pattern Recognition, pp. 14986–14996 (2021)
34. Lu, L., Li, W., Tao, X., Lu, J., Jia, J.: Masa-SR: matching acceleration and spatial adaptation for reference-based image super-resolution. In: IEEE Conference on Computer Vision and Pattern Recognition, pp. 6368–6377 (2021)
35. Ma, Y., et al.: Boosting image outpainting with semantic layout prediction. arXiv preprint arXiv:2110.09267 (2021)
36. Mehta, I., Gharbi, M., Barnes, C., Shechtman, E., Ramamoorthi, R., Chandraker, M.: Modulated periodic activations for generalizable local functional representations. In: International Conference on Computer Vision, pp. 14214–14223 (2021)
37. Mildenhall, B., Srinivasan, P.P., Tancik, M., Barron, J.T., Ramamoorthi, R., Ng, R.: NeRF: representing scenes as neural radiance fields for view synthesis. In: Vedaldi, A., Bischof, H., Brox, T., Frahm, J.-M. (eds.) ECCV 2020. LNCS, vol. 12346, pp. 405–421. Springer, Cham (2020). https://doi.org/10.1007/978-3-030-58452-8_24
38. Nichol, A.Q., Dhariwal, P.: Improved denoising diffusion probabilistic models. In: International Conference on Machine Learning, pp. 8162–8171. PMLR (2021)
39. Ntavelis, E., Shahbazi, M., Kastanis, I., Timofte, R., Danelljan, M., Van Gool, L.: Arbitrary-scale image synthesis. In: IEEE Conference on Computer Vision and Pattern Recognition (2022)
40. Van den Oord, A., Kalchbrenner, N., Espeholt, L., Vinyals, O., Graves, A., et al.: Conditional image generation with PixelCNN decoders. In: Advances in Neural Information Processing Systems, vol. 29 (2016)
41. Park, T., et al.: Swapping autoencoder for deep image manipulation. In: Advances in Neural Information Processing Systems (2020)
42. Parmar, G., Zhang, R., Zhu, J.Y.: On aliased resizing and surprising subtleties in GAN evaluation. In: IEEE Conference on Computer Vision and Pattern Recognition (2022)
43. Schwarz, K., Liao, Y., Niemeyer, M., Geiger, A.: GRAF: generative radiance fields for 3D-aware image synthesis. In: Advances in Neural Information Processing Systems, vol. 33, pp. 20154–20166 (2020)
44. Shaham, T.R., Dekel, T., Michaeli, T.: SinGAN: learning a generative model from a single natural image. In: International Conference on Computer Vision, pp. 4570–4580 (2019)

45. Shaham, T.R., Gharbi, M., Zhang, R., Shechtman, E., Michaeli, T.: Spatially-adaptive pixelwise networks for fast image translation. In: IEEE Conference on Computer Vision and Pattern Recognition, pp. 14882–14891 (2021)
46. Shechtman, E., Irani, M.: Matching local self-similarities across images and videos. In: IEEE Conference on Computer Vision and Pattern Recognition, pp. 1–8. IEEE (2007)
47. Shocher, A., Bagon, S., Isola, P., Irani, M.: InGAN: capturing and retargeting the "DNA" of a natural image. In: International Conference on Computer Vision, pp. 4492–4501 (2019)
48. Shocher, A., Cohen, N., Irani, M.: "Zero-shot" super-resolution using deep internal learning. In: IEEE Conference on Computer Vision and Pattern Recognition, pp. 3118–3126 (2018)
49. Skorokhodov, I., Ignatyev, S., Elhoseiny, M.: Adversarial generation of continuous images. In: IEEE Conference on Computer Vision and Pattern Recognition, pp. 10753–10764 (2021)
50. Song, J., Meng, C., Ermon, S.: Denoising diffusion implicit models. In: International Conference on Learning Representations (2021)
51. Song, Y., Ermon, S.: Improved techniques for training score-based generative models. In: Advances in Neural Information Processing Systems, vol. 33, pp. 12438–12448 (2020)
52. Tancik, M., et al.: Fourier features let networks learn high frequency functions in low dimensional domains. In: Advances in Neural Information Processing Systems, vol. 33, pp. 7537–7547 (2020)
53. Teterwak, P., et al.: Boundless: generative adversarial networks for image extension. In: International Conference on Computer Vision, pp. 10521–10530 (2019)
54. Van Oord, A., Kalchbrenner, N., Kavukcuoglu, K.: Pixel recurrent neural networks. In: International Conference on Machine Learning, pp. 1747–1756. PMLR (2016)
55. Vaswani, A., et al.: Attention is all you need. In: Advances in Neural Information Processing Systems, vol. 30 (2017)
56. Wang, S.Y., Wang, O., Zhang, R., Owens, A., Efros, A.A.: CNN-generated images are surprisingly easy to spot... for now. In: IEEE Conference on Computer Vision and Pattern Recognition (2020)
57. Wang, X., Xie, L., Dong, C., Shan, Y.: Real-ESRGAN: training real-world blind super-resolution with pure synthetic data. In: International Conference on Computer Vision, pp. 1905–1914 (2021)
58. Wang, X., et al.: ESRGAN: enhanced super-resolution generative adversarial networks. In: Proceedings of the European Conference on Computer Vision (ECCV) Workshops (2018)
59. Wang, Y., Tao, X., Shen, X., Jia, J.: Wide-context semantic image extrapolation. In: IEEE Conference on Computer Vision and Pattern Recognition, pp. 1399–1408 (2019)
60. Wexler, Y., Shechtman, E., Irani, M.: Space-time completion of video. IEEE Trans. Pattern Anal. Mach. Intell. **29**(3), 463–476 (2007)
61. Xia, B., Tian, Y., Hang, Y., Yang, W., Liao, Q., Zhou, J.: Coarse-to-fine embedded patchmatch and multi-scale dynamic aggregation for reference-based super-resolution. arXiv preprint arXiv:2201.04358 (2022)
62. Xu, R., Wang, X., Chen, K., Zhou, B., Loy, C.C.: Positional encoding as spatial inductive bias in GANs. In: IEEE Conference on Computer Vision and Pattern Recognition, pp. 13569–13578 (2021)

63. Yang, F., Yang, H., Fu, J., Lu, H., Guo, B.: Learning texture transformer network for image super-resolution. In: IEEE Conference on Computer Vision and Pattern Recognition, pp. 5791–5800 (2020)

64. Yang, Z., Dong, J., Liu, P., Yang, Y., Yan, S.: Very long natural scenery image prediction by outpainting. In: International Conference on Computer Vision, pp. 10561–10570 (2019)

65. Yu, F., Zhang, Y., Song, S., Seff, A., Xiao, J.: LSUN: construction of a large-scale image dataset using deep learning with humans in the loop. arXiv preprint arXiv:1506.03365 (2015)

66. Zhang, K., et al.: AIM 2020 challenge on efficient super-resolution: methods and results. In: Bartoli, A., Fusiello, A. (eds.) ECCV 2020. LNCS, vol. 12537, pp. 5–40. Springer, Cham (2020). https://doi.org/10.1007/978-3-030-67070-2_1

67. Zhang, R., Isola, P., Efros, A.A., Shechtman, E., Wang, O.: The unreasonable effectiveness of deep features as a perceptual metric. In: IEEE Conference on Computer Vision and Pattern Recognition (2018)

68. Zhao, S., et al.: Large scale image completion via co-modulated generative adversarial networks. In: International Conference on Learning Representations (2021)

69. Zhao, S., Liu, Z., Lin, J., Zhu, J.Y., Han, S.: Differentiable augmentation for data-efficient GAN training. In: Advances in Neural Information Processing Systems, vol. 33, pp. 7559–7570 (2020)

70. Zhao, Z., Zhang, Z., Chen, T., Singh, S., Zhang, H.: Image augmentations for GAN training. arXiv preprint arXiv:2006.02595 (2020)

71. Zheng, H., Ji, M., Wang, H., Liu, Y., Fang, L.: CrossNet: an end-to-end reference-based super resolution network using cross-scale warping. In: European Conference on Computer Vision, pp. 88–104 (2018)

72. Zhou, Y., Zhu, Z., Bai, X., Lischinski, D., Cohen-Or, D., Huang, H.: Non-stationary texture synthesis by adversarial expansion. ACM Trans. Graph. (2018)

CCPL: Contrastive Coherence Preserving Loss for Versatile Style Transfer

Zijie Wu[1], Zhen Zhu[2], Junping Du[3], and Xiang Bai[1(✉)]

[1] Huazhong University of Science and Technology, Wuhan, China
{zijiewu,xbai}@hust.edu.cn
[2] University of Illinois at Urbana-Champaign, Champaign, USA
zhenzhu4@illinois.edu
[3] Beijing University of Posts and Telecommunications, Beijing, China

Abstract. In this paper, we aim to devise a universally versatile style transfer method capable of performing artistic, photo-realistic, and video style transfer jointly, without seeing videos during training. Previous single-frame methods assume a strong constraint on the whole image to maintain temporal consistency, which could be violated in many cases. Instead, we make a mild and reasonable assumption that global inconsistency is dominated by local inconsistencies and devise a generic Contrastive Coherence Preserving Loss (CCPL) applied to local patches. CCPL can preserve the coherence of the content source during style transfer without degrading stylization. Moreover, it owns a neighbor-regulating mechanism, resulting in a vast reduction of local distortions and considerable visual quality improvement. Aside from its superior performance on versatile style transfer, it can be easily extended to other tasks, such as image-to-image translation. Besides, to better fuse content and style features, we propose Simple Covariance Transformation (SCT) to effectively align second-order statistics of the content feature with the style feature. Experiments demonstrate the effectiveness of the resulting model for versatile style transfer, when armed with CCPL.

Keywords: Image style transfer · Video style transfer · Temporal consistency · Contrastive learning · Image-to-image translation

1 Introduction

Over the past years, much progress has been made on style transfer to make the result exceptionally pleasant and artistically valuable. In this work, we are interested in *versatile style transfer*. Apart from artistic style transfer and photo-realistic style transfer, our derived method is versatile in performing video style transfer well without explicitly training with videos. The code is available at https://github.com/JarrentWu1031/CCPL.

Z. Wu and Z. Zhu—Equal contribution.

Supplementary Information The online version contains supplementary material available at https://doi.org/10.1007/978-3-031-19787-1_11.

S. Avidan et al. (Eds.): ECCV 2022, LNCS 13676, pp. 189–206, 2022.
https://doi.org/10.1007/978-3-031-19787-1_11

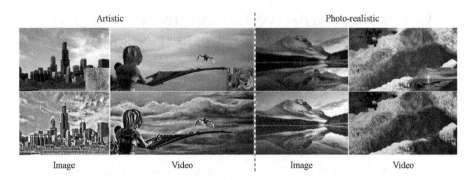

Artistic		Photo-realistic	
Image	Video	Image	Video

Fig. 1. Our algorithm can perform versatile style transfer. From left to right are examples of artistic image/video style transfer, photo-realistic image/video style transfer. Adobe Acrobat Reader is recommended to see the animations.

One naive solution to produce a stylized video is to independently transfer the style of successive frames with the same style reference. Since no temporal consistency constraint is enforced, the generated video usually has obvious flicker artifacts and incoherence between two consecutive frames. To combat this problem, former methods [4,16,19,22,39,40] used optical flow as guidance to restore the estimated motions of the original videos. However, estimating optical flow requires much computation, and the accuracy of estimated motions tightly constrains the quality of the stylized video. Recently, some algorithms [14,29,33] tried to improve the temporal consistency of the video outputs with single-frame regularizations. They attempted to ensure a linear transformation from the content feature to the fused feature. The underlying idea is to encourage preserving the dense pairwise relations within the content source. However, without explicit guidance, the linearity is largely affected by the global style optimization. Therefore, their video results are still not that temporally consistent. We notice that most video results show good structure rigidity to their content video inputs, but the local noise escalates the impression of inconsistency. So instead of considering a global constraint that could be easily violated, we start by thinking about a more relaxed constraint defined on local patches.

As shown in Fig. 2, our idea is simple: the change between patches denoted by R'_A and R'_B of the same location in the stylized images should be similar to patches R_A and R_B of two adjacent content frames. If the two consecutive content frames are shot within a short period, it is likely to find a similar patch to R_B in the neighboring area, which is denoted by R_C (in the blue box). In other words, we can treat two nearby patches in the same image as patches of the same location in consecutive frames. Therefore, we can apply the constraint even when we only have single-frame images. However, forcing these patch differences to be the same is unreliable since it will encourage the outputs to be the same as the content images. Then no style transfer effects would appear in the results. Inspired by recent advances in contrastive learning [8,35,37], we use the InfoNCE loss [35] to maximize the mutual information between the positive pair (from the

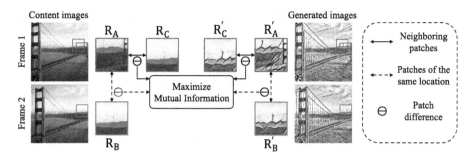

Fig. 2. Intuition of the contrastive coherence preserving loss. The regions denoted with red boxes from the first frame (R_A or R'_A) have the same location with corresponding patches in the second frame wrapped with brown box (R_B or R'_B). R_C and R'_C (in the blue boxes) are cropped from the first frames but their semantics align with R_B and R'_B. The difference between two patches is denoted as $\mathcal{D}$ (*e.g.*, $\mathcal{D}(R_A, R_B)$). The mutual information between $\mathcal{D}(R_A, R_C)$ and $\mathcal{D}(R'_A, R'_C)$ ($\mathcal{D}(R_A, R_B)$ and $\mathcal{D}(R'_A, R'_B)$) is encouraged to be maximized to preserve the coherence of the content source. (Color figure online)

same region) of patch differences relative to other negative pairs (from different regions). By sampling a sufficient number of negative pairs, the loss encourages the positive pair to be close while keeping away from negative samples. We call the derived loss as **Contrastive Coherence Preserving Loss (CCPL)**.

After applying CCPL, we note that the temporal consistency of the video outputs improves substantially while the stylization remains satisfying (see Fig. 5 and Table 1). Besides, due to the neighbor-regulating strategy of the CCPL, the local patches of the generated image are constrained by their neighboring patches, which reduces local distortions significantly, thus leading to better visual quality. Our proposed CCPL does not require video inputs and is not bound to specific network architecture. Therefore we can apply it to any existing image style transfer networks during training to improve their performance on images and videos (see Fig. 9 and Table 1). The significant improvement in visual quality and its flexibility empowers CCPL for photo-realistic style transfer with minor modifications, marking it a vital tool towards versatile style transfer (see Fig. 1).

With CCPL, we now aspire to fuse content and style features both efficiently and effectively. To realize this, we propose an efficient network for versatile style transfer, called **SCTNet**. The critical element of SCTNet is the **Simple Covariance Transformation (SCT)** module to fuse style features and content features. It computes the covariance of the style feature and directly multiplies the feature covariance with the normalized content features. Compared to the fusing operations in AdaIN [23] and Linear [29], our SCT is simple and can capture precise style information at the same time.

To summarize, our contributions are three-fold:

1. We propose Contrastive Coherence Preserving Loss (CCPL) for versatile style transfer. It encourages consistency between the content image and gener-

ated image in terms of the difference of an image patch with its neighboring patches. It is effective and transferable to other style transfer methods.

2. We propose Simple Covariance Transformation (SCT) to align second-order statistics of content and style features effectively. The resulted SCTNet is structurally simple and remains efficient (about 25 frames per second at the scale of 512×512), which is of great potential for practical use.

3. We apply our CCPL to other tasks, such as image-to-image translation, and improve the temporal consistency and visual quality of results without further modifications, demonstrating the flexibility of CCPL.

2   Related Works

Image Style Transfer. These algorithms aim at generating an image with the structure of one image and the style of another. Gatys *et al.* first pioneered Neural Style Transfer (NST) [17]. For acceleration, some algorithms [25,45] approximated the iterative optimization procedure as feed-forward networks and achieved style transfer with a fast forward pass. For broader applications, several algorithms tried to transfer multiple styles within a single model [5,15]. Nevertheless, these models have limitations on the number of learnt styles. Since then, various methods have been designed to transfer style from random images.

Style-swap methods [7,42] swapped each content patch with its closest style patch before reconstructing the image. WCT [30] utilized singular value decomposition to whiten and then re-color images. AdaIN [23] replaced the feature means and standard deviations with those from the style source. Recently, many attention-based algorithms came forth. For example, Li *et al.* [29] devised a linear transformation to align second-order statistics between the fused feature and the style feature. Deng *et al.* [14] improved it with multi-channel correlating. SANet [36] re-arranged style features utilizing spatial correlations with content features. AdaAttN [33] combined AdaIN [23] and SANet [36] to balance global and local style effects. Cheng *et al.* [11] proposed style-aware normalized loss to balance stylization. Another branch aims to transfer photo-realistic style onto images. Luan *et al.* [34] designed a color transformation network inspired by the Matting Laplacian [28]. Li *et al.* [31] replaced the upsampling layers of WCT [30] with unpooling layers and added max-pooling masks to alleviate detail losses. Yoo *et al.* [47] introduced the wavelet transform to preserve structural information. An *et al.* [2] used neural architecture search algorithms to find the appropriate decoder design for better performance.

Video Style Transfer. Existing video style transfer algorithms can be roughly divided into two categories according to whether to use the optical flow or not.

One line of work leverages optical flow when producing the video output. These algorithms try to estimate the motion of the original video and restore it in the generated video. Ruder *et al.* [39] proposed a temporal loss to regulate the current frame with the warped previous frame to extend the image style transfer algorithm [17] to videos. Chen *et al.* [4] designed an RNN structure baseline and performed the warping operation in the feature domain. Gupta *et al.* [19]

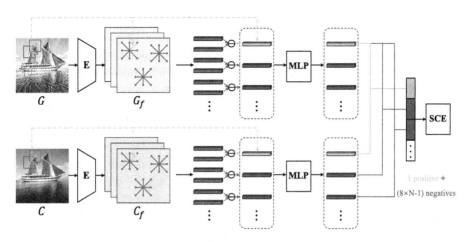

Fig. 3. Diagram of the proposed CCPL. C_f and G_f represent the encoded features from a specific layer of the encoder E. $\ominus$ denotes vector subtraction, and SCE means softmax cross-entropy. The yellow dashed lines illustrate how the positive pair is produced. (Color figure online)

concatenated the former stylized frame with the current content frame before rendering and formed a flow loss as a constraint. Huang *et al.* [22] tried to integrate temporal coherence into the stylization network with a hybrid loss. Ruder *et al.* [40] extended their previous work [39] with new initializations and loss functions to improve robustness against large motions and strong occlusions. Temporal consistency can be improved with these optical flow constraints. However, optical flow estimation is not perfectly accurate, resulting in artifacts in the video results. Besides, it is computationally expensive, especially when the image size scales up. Considering these, another line of work tries to maintain the coherence of content inputs without using optical flow.

Li *et al.* [29] and Deng *et al.* [14] devised linear transformations for content features to preserve structure affinity. Liu *et al.* [33] used L1 normalization to replace the softmax operation of SANet [36] to get a more flat attention score distribution. Wang *et al.* [46] proposed compound temporal regularization to enhance the robustness of the network to motions and illumination changes. Compared to these approaches, our proposed CCPL poses no requirements for the network architecture, making it exceptionally adaptive to other networks. With our SCTNet, the temporal consistency of video outputs surpasses SOTAs while the stylization remains satisfying. We also apply CCPL to other networks. The results show similar improvements in video stability (see Table 1).

Contrastive Learning. The original purpose of contrastive learning algorithms is to learn a good feature representation in a self-supervised scenario. A rich family of methods tried to achieve this by maximizing the mutual information of positive feature pairs while minimizing it in negative pairs [8–10,18,20,35]. Recent works extended contrastive learning to the field of image-to-image translation [37] and image style transfer [6]. Our work is most relevant to CUT [37] in using patch-based InfoNCE loss [35]. But CUT [37] utilized the correspondence of patches at the same locations for the image-to-image (Im2Im) translation task. However, our CCPL incorporates a neighbor-regulating scheme to preserve the correlations among neighboring patches, making it suitable for image and video generation. Besides, our experiment illustrates the effectiveness of CCPL on top of CUT [37] in the Im2Im translation task, as depicted in Sect. 4.4.

3   Methods

3.1   Contrastive Coherence Preserving Loss

Given two frames C_t and $C_{t+\Delta t}$ where Δt is the time interval in between, we assume the difference between the corresponding generated images G_t and $G_{t+\Delta t}$ is linearly dependent on the difference between C_t and $C_{t+\Delta t}$, when Δt is small:

$$\lim_{\Delta t \to 0} \mathcal{D}(C_{t+\Delta t}, C_t) \simeq \mathcal{D}(G_{t+\Delta t}, G_t), \tag{1}$$

where $\mathcal{D}(a, b)$ represents the difference between a and b. This constraint is probably too strict to hold for the whole image but technically sound for local patches where usually only simple image transformations, e.g., translation or rotation, can occur. Under this assumption, we propose a generic Contrastive Coherence Preserving Loss (CCPL) applied to local patches to enforce this constraint. We show in Sect. 1 that our loss applied on neighboring patches is equivalent to that on corresponding patches of two frames, assuming Δt is small. Operating on a single frame frees us from processing multiple frames of a video source, saving computation budget.

To apply CCPL, first, we send the generated image G and its content input C to the fixed image encoder E to get feature maps of a specific layer, denoted as G_f and C_f (shown in Fig. 3). Second, we randomly sample N vectors[1] from G_f (red dots in Fig. 3), denoted as G_a^x where $x = 1, \cdots, N$. Third, we sample the *eight* nearest neighboring vectors of each G_a^x (blue dots in Fig. 3), denoted by $G_n^{x,y}$ where $y = 1, \cdots, 8$ is the neighbor index. Then, we accordingly sample from C_f at the same locations to get C_a^x and $C_n^{x,y}$, respectively. The differences between a vector and its neighboring vectors are measured by:

$$d_g^{x,y} = G_a^x \ominus G_n^{x,y}, \ d_c^{x,y} = C_a^x \ominus C_n^{x,y}, \tag{2}$$

where $\ominus$ represents vector subtraction. In order to realize Eq. 1, one simple thought is to enforce d_g equal to d_c. But in this case, an easy workaround of the network is to encourage G similar to C, meaning that this constraint would contradict the purpose of style transfer. Inspired by the recent progress in contrastive learning [8,20,35], we instead try to maximize the mutual information

---

[1] As encoded features are spatially decreased, each vector in the feature level corresponds to an image patch in the image level.

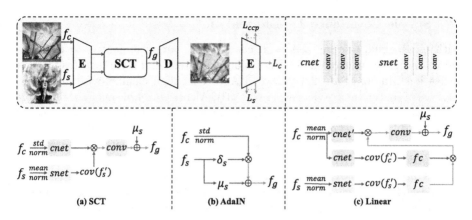

Fig. 4. Details of the proposed SCT module and its comparison with similar algorithms (AdaIN [23], Linear [29]). Here *conv* represents a convolutional layer, and the yellow lines in *cnet* and *snet* denote *relu* layers. Besides, *std norm* represents normalizing features by the means and standard deviations of channels, while *mean norm* normalizes features by the means of its channels. (Color figure online)

between "positive" difference vector pairs. A pair is only defined between a difference vector from C_f and G_f. Namely, the difference vectors of the same locations are defined as positive pairs between d_g and d_c, otherwise negative. The underlying intuition is also straightforward: the difference vectors of the same location should be most relevant in the latent space compared to other random pairs.

We follow the design of [8] to build a two-layer MLP (multi-layer perceptron) to map the difference vectors and normalize them onto a unit sphere before computing InfoNCE loss [35]. Mathematically:

$$L_{\text{ccp}} = \sum_{m=1}^{8 \times N} -\log\left[\frac{\exp(d_g^m \cdot d_c^m / \tau)}{\exp(d_g^m \cdot d_c^m / \tau) + \sum_{n=1, n \neq m}^{8 \times N} \exp(d_g^m \cdot d_c^n / \tau)}\right], \quad (3)$$

where τ stands for a temperature hyper-parameter set to 0.07 by default. With this setting, the temporal consistency of video outputs improves significantly (see Fig. 5 and Table 1) while the stylization remains satisfying or even gets better (see Fig. 6, Fig. 9, dirty texture disappears with our CCPL).

This loss avoids direct contradiction with style losses used to ensure style coherence between the generated and style image. Meanwhile, it can improve the temporal consistency of the generated video even without leveraging information from other frames of the input video. The complexity of CCPL is $\mathcal{O}(8 \times N)^2$, where $8 \times N$ represents the number of sampled difference vectors. It is computationally affordable during training and has zero influence on inference speed (shown in Fig. 8a). CCPL can even work as a simple plugin to extend methods of other *image generation* tasks to produce videos with much better temporal consistency, as shown in Sect. 4.4.

3.2 Simple Covariance Transformation

With CCPL guaranteeing temporal consistency, our next goal is to design a simple and effective module for the fusion of content and style features for rich stylization. Huang *et al.* [23] proposed AdaIN to align channel-wise mean and variance of content and style features directly. Although simple enough, the inter-channel correlations are ignored, which are verified to be effective in the latter literature [14,29]. Li *et al.* [29] devised a channel-attention mechanism to transfer second-order statistics of style features onto corresponding content features. But we empirically find that the structure of Linear [29] can be simplified.

To combine the advantages of AdaIN [23] and Linear [29], we design a **Simple Covariance Transformation (SCT)** module to fuse style and content features. As shown in Fig. 4, first, we normalize the content feature f_c by the means and deviations of its channels [23] and the style feature f_s by the means of its channels [29] to get $\bar{f}_c$ and $\bar{f}_s$. To reduce computation costs, we send $\bar{f}_c$ and $\bar{f}_s$ to *cnet* and *snet* (*cnet* and *snet* both contain three convolutional layers, and two *relu* layers in between) to gradually reduce the dimension of channels $(512 \rightarrow 32)$, and get f_c' and f_s'. Then we flatten f_s' and calculate its covariance matrix $cov(f_s')$ to find out the channel-wise correlations. After that, we simply fuse the features by performing a matrix multiplication between $cov(f_s')$ and f_c' to get f_g. Finally, we use a single convolutional layer (denoted as *conv* in Fig. 4) to restore the channel dimension of f_g back to normal $(32 \rightarrow 512)$ and add channel means of the original style feature before sending it to the decoder.

Combined with a symmetric encoder-decoder module, we name the whole network as **SCTNet**. The encoder is a VGG-19 network [43] pre-trained on ImageNet [13] to extract features from the content and style images, while the symmetric decoder needs to convert the fused feature back to images. Experiments suggest that our SCTNet is comparable to Linear [29] in stylization effects (see Fig. 6 and Table 1), while being lighter and faster (see Table 3).

3.3 Loss Function

Apart from the proposed CCPL, we adopt two commonly used losses [1,14,23,33] for style transfer. The overall training loss is a weighted sum of these three losses:

$$L_{\text{totoal}} = \lambda_c \cdot L_c + \lambda_s \cdot L_s + \lambda_{ccp} \cdot L_{ccp}. \tag{4}$$

The content loss L_c (the style loss L_s) is measured by the Frobenius norm of the differences between (means $\mu(\cdot)$ and standard deviations $\sigma(\cdot)$ of) the generated features and the content (style) features:

$$L_c = \|\phi_l(I_g) - \phi_l(I_c)\|_F, \tag{5}$$

$$L_s = \sum_l (\|\mu(\phi_l(I_g)) - \mu(\phi_l(I_s))\|_F + \|\sigma(\phi_l(I_g)) - \sigma(\phi_l(I_s))\|_F), \tag{6}$$

where $\phi_l(\cdot)$ denotes the feature map from the l-th layer of the encoder. For artistic style transfer, we use the features from $\{relu4\_1\}$, $\{relu1\_1, relu2\_1,$

$relu3\_1$, $relu4\_1\}$, $\{relu2\_1$, $relu3\_1$, $relu4\_1\}$ to calculate the content loss, style loss, and CCPL, respectively. As for photo-realistic style transfer, we set the loss layers to $\{relu3\_1\}$, $\{relu1\_1$, $relu2\_1$, $relu3\_1\}$, $\{relu1\_1$, $relu2\_1$, $relu3\_1\}$ for the above losses. The loss weights are set to $\lambda_c = 1.0$, $\lambda_s = 10.0$, $\lambda_{ccp} = 5.0$ by default. Please check Sect. 4.3 for details about how we find these configurations.

4   Experiments

4.1   Experimental Settings

Implementation Details. We adopt content images from MS-COCO [32] data-set and style images from Wikiart [38] data-set to train our network. Both data-sets contain approximately 80,000 images. We use the Adam optimizer [26] with a learning rate of 1e-4 and the batch size of 8 to train the model for 160k iterations by default. During training, we first resize the smaller dimension of images to 512. Then we randomly crop 256 × 256 patches from images as the final input. For CCPL, we only treat difference vectors within the same content image as negative samples. More details are provided in the supplemental file.

Metrics. To comprehensively evaluate the performance of different algorithms and make the comparison fair, we adopt several metrics to assess the results'

Table 1. Quantitative comparison of video and artistic style transfer. Here i stands for the interval of frames, and Pre. stands for *human preference score*. We show the *human preference score* of both artistic image style transfer (Art) and video style transfer (Vid) in the table. The results of *temporal loss* are magnified 100 times. We show the **first-place** score in bold and the second-place score with underlining.

Methods	SIFID (↓)	LPIPS(↓)		Temporal loss (↓)		Pre. (↑)	
		$i=1$	$i=10$	$i=1$	$i=10$	Art	Vid
AdaIN [23]	2.44	0.184	0.444	5.16	7.92	0.028	0.028
AdaIN [23]+L_{ccp}	2.58	0.163	0.408	4.21	6.72	0.054	0.054
SANet [36]	2.40	0.227	0.478	6.31	13.72	0.062	0.046
SANet [36]+L_{ccp}	2.60	0.167	0.390	4.42	7.09	0.084	0.086
Linear [29]	2.38	0.160	0.417	4.25	7.61	0.076	0.080
Linear [29]+L_{ccp}	2.47	0.147	0.370	4.01	6.96	0.082	0.088
MCCNet [14]	2.34	0.162	0.424	4.21	7.64	0.088	0.106
AdaAttN [33]	2.48	0.207	0.419	4.87	6.49	0.098	0.094
DSTN [21]	2.83	0.234	0.450	5.72	10.76	0.070	0.038
IE [6]	2.99	0.182	0.379	4.35	6.76	0.054	0.058
ReReVST [46]	2.78	**0.137**	**0.359**	**2.97**	5.19	0.046	0.062
SCTNet	**2.29**	0.187	0.446	4.82	12.22	0.066	0.060
SCTNet+L_{nor} [11]	2.31	0.191	0.439	5.07	11.54	0.070	0.062
SCTNet+L_{ccp}	2.43	0.144	0.367	3.45	**5.08**	**0.122**	**0.138**

Table 2. Quantitative comparison of photo-realistic style transfer.

Metrics	Linear [29]	WCT^2 [47]	StyleNAS [2]	DSTN [21]	SCTNet	SCTNet+L_{ccp}
SIFID (↓)	1.82	1.86	2.37	3.35	**1.65**	2.14
LPIPS (↓)	0.395	0.419	0.379	0.464	0.427	**0.351**
Pre. (↑)	0.176	0.186	0.180	0.068	0.128	**0.262**

stylization effects and temporal consistency. To evaluate stylization effects, we compute *SIFID* [41] between the generated image and its style input to measure their style distribution distance. Lower SIFID represents closer style distributions of a pair. To evaluate the visual quality and temporal consistency, we opt to *LPIPS* [48], which is originally used to measure the diversity of the generated images [12,24,27]. In our cases, small LPIPS represents few local distortions of the photo-realistic results or minor changes between two stylized video frames. Nonetheless, LPIPS only considers the correlations between stylized video frames while ignoring the changes between the original frames. As a supplement, we also adopt the *temporal loss* defined in [46] to measure temporal consistency. It is done by utilizing the optical flow between two frames to warp one stylized result and compute the Frobenius difference with another. We evaluate short-term (two adjacent frames) and long-term (9 frames in between) consistency for video style transfer. For short-term consistency, we directly use the ground-truth optical flow from the MPI Sintel data-set [3]. Otherwise, we use PWC-Net [44] to estimate the optical flow between two frames. The lower temporal loss represents better preservation of coherence between two frames.

For image style transfer comparison, we randomly choose 10 content images and 10 style images to synthesize 100 stylized images for each method and calcu-

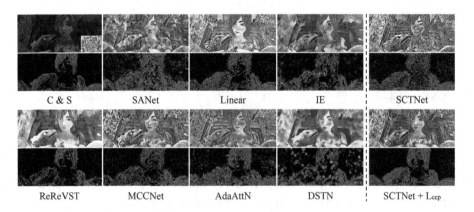

Fig. 5. Qualitative comparison of short-term temporal consistency. We compare our method with seven algorithms: SANet [36], Linear [29], IE [6], ReReVST [46], MCC-Net [14], AdaAttN [33], DSTN [21]. The odd rows show the previous frames. The even rows show the heat-maps of differences between consecutive frames.

late their mean SIFID as the stylization metric. Besides, we compute the mean LPIPS to measure the visual quality of photo-realistic results. As for temporal consistency, we randomly select 10 video clips (50 frames, 12 FPS each) from the MPI Sintel dataset [3] and use 10 style images to transfer these videos, respectively. Then we compute the mean LPIPS and temporal loss as the temporal consistency metrics. We also include human evaluation, which is more representative in image generation tasks. To do so, we invite 50 participants to choose their favorite stylized image/video from each image/video-style pair considering the visual quality, stylization effect, and temporal consistency. These participants come from different backgrounds, making the evaluation less biased towards a certain group of people. Overall, we get 500 votes for images and videos, respectively. Then we calculate the percentage of votes as the *human preference score*. All the evaluations are shown in Table 1 and Table 2.

4.2   Comparison with Former Methods

For video and artistic image style transfer, we compare our method with nine algorithms: AdaIN [23], SANet [36], DSTN [21], ReReVST [46], Linear [29], MCCNet [14], AdaAttN [33], IE [6], L_{nor} [11], which are the SOTAs of artistic image style transfer. Among these methods, [6,14,29,33] are also the most advanced single-frame-based video style transfer methods while ReReVST [46] is the SOTA multi-frames-based method. As for photo-realistic image style transfer, we compare our method with four SOTAs: Linear [29], WCT^2 [47], Style-NAS [2], DSTN [21]. Note that among all these mentioned algorithms, Linear [29] and DSTN [21] are most relevant to our method, since both of them are capable of transferring artistic and photo-realistic style onto images. We obtain all the test results from the official codes these methods provide.

Video Style Transfer. As shown in Table 1, our original SCTNet scores the best in SIFID, indicating its superiority in obtaining correct styles. Also, we can see the proposed CCPL improves the temporal consistency a lot with a minor decrease of the SIFID score, when the loss is applied to different methods. And our full model (with CCPL) exceeds all the single-frame methods [6,14,21,29,33,36] in both short-term and long-term temporal consistency, which are measured by LPIPS [48] and temporal loss, and performs on par with the SOTA multi-frame method: ReReVST [46]. However, our SIFID score exceeds ReReVST [46] significantly, which is consistent with the results shown in the qualitative comparison (See Fig. 6). The qualitative comparisons also show the advantage of our CCPL in maintaining short-term (Fig. 5) temporal consistency of the original video as our heat-map difference is mostly similar to ground-truth. We have another figure in the supplemental file to show the comparison of long-term temporal consistency. In terms of human preference score, our full model also ranks the best, further validating the effectiveness of our CCPL.

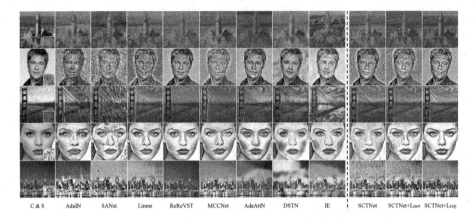

Fig. 6. Qualitative comparison of artistic style transfer. We compare our method with nine algorithms: AdaIN [23], SANet [36], Linear [29], ReReVST [46], MCCNet [14], AdaAttN [33], DSTN [21], IE [6], L_{nor} [11].

Artistic Style Transfer. As shown in Fig. 6, AdaIN [23] generates results with severe shape distortion (*e.g.*, house in the 1st and bridge in the 3rd row) and disarranged texture patterns (4th, 5th rows). SANet [36] also has shape distortion and misses some structural details in its results ($1^{st} \rightarrow 3^{rd}$ rows). Linear [29] and MCCNet [14] have relatively quite clean outputs. However, Linear [29] loses some content details ($1^{st}, 3^{rd}$ rows), and some results of MCCNet [14] have checkerboard artifacts in local regions (around collar in the 2nd row and corner of mouth in the 4th row). ReReVST [46] shows obvious color distortion ($2^{nd} \rightarrow 5^{th}$ rows). AdaAttN [33] is effective in reducing messy textures but the stylization effect seems to degenerate in some cases (1st row). The results of DSTN [21] have severe obvious distortion ($3^{rd}, 4^{th}$ rows). And the results of IE [6] are less similar to the original style ($1^{st}, 3^{rd}, 5^{th}$ rows). Our original SCTNet captures accurate style ($2^{nd}, 3^{rd}$ rows), but there are some messy regions in the generated images as well ($4^{th}, 5^{th}$ rows). When adding L_{nor} [11], some results are even messier ($4^{th}, 5^{th}$ rows). However, with CCPL, the generated results of our full model maintain well the structures of their content sources with vivid and appealing colorization. Besides, this effect is reinforced by its multi-level scheme. Therefore, irregular textures and local color distortions are decreased significantly. It even helps to improve stylization with better preservation of the semantic information of the content sources (as shown in Fig. 9).

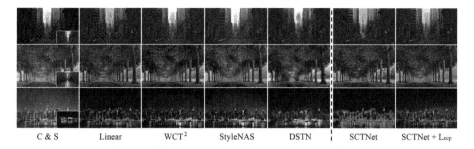

| C & S | Linear | WCT² | StyleNAS | DSTN | SCTNet | SCTNet + L_ccp |

Fig. 7. Qualitative comparison of photo-realistic style transfer. We compare our method with four algorithms: Linear [29], WCT² [47], StyleNAS [2] and DSTN [21].

Photo-Realistic Style Transfer. Since CCPL can preserve the semantic information of the content source and significantly reduce local distortions, it is well-suited for the task of photo-realistic style transfer. We make slight changes to SCTNet to enable it for this task: build a shallower encoder by throwing off layers beyond $relu3\_1$, then use feature maps from all three layers to calculate CCPL. As shown in Fig. 7, Linear [29] and DSTN [21] generates results with detail losses (vanished windows in the 3rd row). As for WCT² [47] and StyleNAS [2], some results of them show unreasonable color distribution (red road in the 2nd row). In comparison, our full model generates results comparable or even better than those SOTAs, with high visual quality and appropriate stylization, which is consistent with the quantitative comparison shown in Table 2.

Efficiency Analysis. Our model is quite efficient due to the simple feed-forward architecture of the network and the efficient feature fusion module SCT. We use a single 12 GB Titan XP GPU with no other ongoing programs to compare its running speed with other algorithms. Table 3 shows the average running speed (over 100 independent runs) of different methods on three input image scales. The result suggests that SCTNet surpasses the SOTAs in efficiency at different scales (comparisons for photo-realistic style transfer methods are provided in the supplemental file), indicating the feasibility of our algorithm for real-time use.

4.3 Ablation Studies

There are several factors relevant to the performance induced by the CCPL: 1) layers to apply the loss; 2) the number of difference vectors sampled each layer; 3) the loss weight ratio with the style loss. Therefore, we conduct several experiments by enumerating the number of CCPL layers from 0 to 4 (start from the deepest layer) and choosing from [16, 32, 64, 128] as the number of sampled combinations to show the impacts of the first two factors. Then we adjust the loss weight ratio between the CCPL and the style loss to manifest which ratio gives the best trade-off between style effects and temporal coherence. To be noted, the stylization score here represents the SIFID score, and the temporal consistency is measured by: $(20-10\times LPIPS -$ temporal loss) to show the escalating trend.

Table 3. Execution speed comparison (unit: FPS). We use a single 12 GB Titan XP GPU for all the execution time testing. OOM denotes the Out-Of-Memory error.

Artistic	Ad [23]	SA [36]	LT [29]	Re [46]	MC [14]	AN [33]	DN [21]	IE [6]	Ours
256 × 256	40.0	34.5	66.7	37.0	22.2	15.6	15.9	31.3	**77.0**
512 × 512	12.5	14.3	18.9	13.7	8.1	12.5	4.2	13.0	**21.7**
1024 × 1024	2.7	2.7	4.6	2.8	1.9	2.1	1.2	2.6	**5.0**

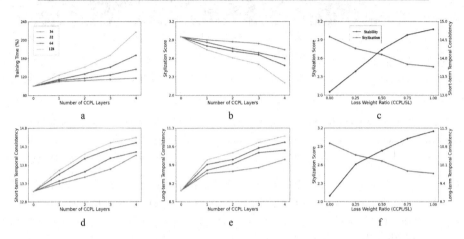

Fig. 8. Ablation studies on three factors of the CCPL: 1) layers to apply the loss; 2) the number of vectors sampled each layer; 3) the loss weight ratio with style loss. (Color figure online)

From the sub-figures, we can see that, as the number of CCPL layers increases, the short-term (Fig. 8d) and long-term (Fig. 8e) temporal consistency increases with the reduction of stylization score (Fig. 8b) and greater computation (Fig. 8a). And when the number of CCPL layers increases from 3 to 4, the changes of temporal consistency are minor. In contrast, the computation costs increase significantly, and the stylization effects are much weaker. Therefore, we choose 3 as the default setting for the number of CCPL layers.

As for the number of sampled difference vectors (per layer), the blue lines (64 sampled vectors) in Fig. 8d & e are near the yellow lines (128 sampled vectors), which means the performance of these two settings are close on improving temporal consistency. However, sampling 128 difference vectors per layer brings a significantly heavier computation burden and style degeneration. So we sample 64 difference vectors per layer by default.

The loss weight ratio can also be regarded as a handle to adjust temporal consistency and stylization. Figure 8c & f show the trade-off between temporal consistency and stylization when the loss weight ratio changes. We find 0.5 a good choice for the weight ratio because it gives a good trade-off between temporal consistency improvement and stylization score reduction. We show the qualitative results of ablation studies on CCPL in the supplemental file and more analysis, such as different sampling strategies in CCPL.

Fig. 9. CCPL can be easily applied to other methods, such as AdaIN [23], SANet [36] and Linear [29], to improve visual quality.

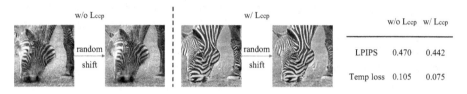

Fig. 10. Comparison of applying the CCPL on CUT [37] with its original model.

4.4 Applications

CCPL on Existing Methods. CCPL is highly flexible and can be plugged into other methods with minor modifications. We apply the proposed CCPL on three typical former methods: AdaIN [23], SANet [36], Linear [29]. All these methods achieve consistent improvements in temporal consistency with only a slight decrease on the SIFID score (see Table 1 and Fig. 9). The result reveals the effectiveness and flexibility of the CCPL.

Image-to-Image Translation. CCPL can be easily added to other generation tasks like image-to-image translation. We apply our CCPL on a recent image-to-image translation method CUT [37] and then train the model with the same horse2zebra dataset. The results in Fig. 10 demonstrate that our CCPL improves both the visual quality and temporal consistency. Please refer to the supplemental file for more applications.

5 Conclusions

In this work, we propose CCPL to preserve content coherence during style transfer. By contrasting the feature differences of image patches, the loss encourages the difference of patches of the same location in the content and generated images to be similar. Models trained with CCPL achieve a good trade-off between temporal consistency and style effects. We also propose a simple and effective module for aligning second-order statistics of the content feature with style feature.

Combining these two techniques, our full model is light and fast while generating satisfying image and video results. Besides, we demonstrate the effectiveness of the proposed loss on other models and tasks, such as image-to-image style transfer, which shows the vast potential of our loss for broader applications.

Acknowledgements. This work was supported by the National Natural Science Foundation of China 62192784.

References

1. An, J., Huang, S., Song, Y., Dou, D., Liu, W., Luo, J.: Artflow: unbiased image style transfer via reversible neural flows. In: Proceedings of the IEEE/CVF Conference on Computer Vision and Pattern Recognition, pp. 862–871 (2021)
2. An, J., Xiong, H., Ma, J., Luo, J., Huan, J.: Stylenas: an empirical study of neural architecture search to uncover surprisingly fast end-to-end universal style transfer networks. arXiv preprint arXiv:1906.02470 (2019)
3. Butler, D.J., Wulff, J., Stanley, G.B., Black, M.J.: A naturalistic open source movie for optical flow evaluation. In: Fitzgibbon, A., Lazebnik, S., Perona, P., Sato, Y., Schmid, C. (eds.) ECCV 2012. LNCS, vol. 7577, pp. 611–625. Springer, Heidelberg (2012). https://doi.org/10.1007/978-3-642-33783-3_44
4. Chen, D., Liao, J., Yuan, L., Yu, N., Hua, G.: Coherent online video style transfer. In: Proceedings of the IEEE International Conference on Computer Vision, pp. 1105–1114 (2017)
5. Chen, D., Yuan, L., Liao, J., Yu, N., Hua, G.: Stylebank: an explicit representation for neural image style transfer. In: Proceedings of the IEEE Conference on Computer Vision and Pattern Recognition, pp. 1897–1906 (2017)
6. Chen, H., et al.: Artistic style transfer with internal-external learning and contrastive learning. In: Advances in Neural Information Processing Systems, vol. 34 (2021)
7. Chen, T.Q., Schmidt, M.: Fast patch-based style transfer of arbitrary style. arXiv preprint arXiv:1612.04337 (2016)
8. Chen, T., Kornblith, S., Norouzi, M., Hinton, G.: A simple framework for contrastive learning of visual representations. In: International Conference on Machine Learning, pp. 1597–1607. PMLR (2020)
9. Chen, X., Fan, H., Girshick, R., He, K.: Improved baselines with momentum contrastive learning. arXiv preprint arXiv:2003.04297 (2020)
10. Chen, X., He, K.: Exploring simple siamese representation learning. In: Proceedings of the IEEE/CVF Conference on Computer Vision and Pattern Recognition, pp. 15750–15758 (2021)
11. Cheng, J., Jaiswal, A., Wu, Y., Natarajan, P., Natarajan, P.: Style-aware normalized loss for improving arbitrary style transfer. In: Proceedings of the IEEE/CVF Conference on Computer Vision and Pattern Recognition, pp. 134–143 (2021)
12. Choi, Y., Uh, Y., Yoo, J., Ha, J.W.: Stargan v2: diverse image synthesis for multiple domains. In: Proceedings of the IEEE/CVF Conference on Computer Vision and Pattern Recognition, pp. 8188–8197 (2020)
13. Deng, J., Dong, W., Socher, R., Li, L.J., Li, K., Fei-Fei, L.: Imagenet: a large-scale hierarchical image database. In: 2009 IEEE Conference on Computer Vision and Pattern Recognition, pp. 248–255. IEEE (2009)

14. Deng, Y., Tang, F., Dong, W., Huang, H., Ma, C., Xu, C.: Arbitrary video style transfer via multi-channel correlation. arXiv preprint arXiv:2009.08003 (2020)
15. Dumoulin, V., Shlens, J., Kudlur, M.: A learned representation for artistic style. arXiv preprint arXiv:1610.07629 (2016)
16. Gao, C., Gu, D., Zhang, F., Yu, Y.: ReCoNet: real-time coherent video style transfer network. In: Jawahar, C.V., Li, H., Mori, G., Schindler, K. (eds.) ACCV 2018. LNCS, vol. 11366, pp. 637–653. Springer, Cham (2019). https://doi.org/10.1007/978-3-030-20876-9_40
17. Gatys, L.A., Ecker, A.S., Bethge, M.: Image style transfer using convolutional neural networks. In: Proceedings of the IEEE Conference on Computer Vision and Pattern Recognition, pp. 2414–2423 (2016)
18. Grill, J.B., et al.: Bootstrap your own latent: a new approach to self-supervised learning. arXiv preprint arXiv:2006.07733 (2020)
19. Gupta, A., Johnson, J., Alahi, A., Fei-Fei, L.: Characterizing and improving stability in neural style transfer. In: Proceedings of the IEEE International Conference on Computer Vision, pp. 4067–4076 (2017)
20. He, K., Fan, H., Wu, Y., Xie, S., Girshick, R.: Momentum contrast for unsupervised visual representation learning. In: Proceedings of the IEEE/CVF Conference on Computer Vision and Pattern Recognition, pp. 9729–9738 (2020)
21. Hong, K., Jeon, S., Yang, H., Fu, J., Byun, H.: Domain-aware universal style transfer. In: Proceedings of the IEEE/CVF International Conference on Computer Vision, pp. 14609–14617 (2021)
22. Huang, H., et al.: Real-time neural style transfer for videos. In: Proceedings of the IEEE Conference on Computer Vision and Pattern Recognition, pp. 783–791 (2017)
23. Huang, X., Belongie, S.: Arbitrary style transfer in real-time with adaptive instance normalization. In: Proceedings of the IEEE International Conference on Computer Vision, pp. 1501–1510 (2017)
24. Huang, X., Liu, M.Y., Belongie, S., Kautz, J.: Multimodal unsupervised image-to-image translation. In: Proceedings of the European Conference on Computer Vision (ECCV), pp. 172–189 (2018)
25. Johnson, J., Alahi, A., Fei-Fei, L.: Perceptual losses for real-time style transfer and super-resolution. In: Leibe, B., Matas, J., Sebe, N., Welling, M. (eds.) ECCV 2016. LNCS, vol. 9906, pp. 694–711. Springer, Cham (2016). https://doi.org/10.1007/978-3-319-46475-6_43
26. Kingma, D.P., Ba, J.: Adam: a method for stochastic optimization. arXiv preprint arXiv:1412.6980 (2014)
27. Lee, H.Y., Tseng, H.Y., Huang, J.B., Singh, M., Yang, M.H.: Diverse image-to-image translation via disentangled representations. In: Proceedings of the European Conference on Computer Vision (ECCV), pp. 35–51 (2018)
28. Levin, A., Lischinski, D., Weiss, Y.: A closed-form solution to natural image matting. IEEE Trans. Pattern Anal. Mach. Intell. $30(2)$, 228–242 (2007)
29. Li, X., Liu, S., Kautz, J., Yang, M.H.: Learning linear transformations for fast image and video style transfer. In: Proceedings of the IEEE/CVF Conference on Computer Vision and Pattern Recognition, pp. 3809–3817 (2019)
30. Li, Y., Fang, C., Yang, J., Wang, Z., Lu, X., Yang, M.H.: Universal style transfer via feature transforms. arXiv preprint arXiv:1705.08086 (2017)
31. Li, Y., Liu, M.Y., Li, X., Yang, M.H., Kautz, J.: A closed-form solution to photorealistic image stylization. In: Proceedings of the European Conference on Computer Vision (ECCV), pp. 453–468 (2018)

32. Lin, T.-Y., et al.: Microsoft COCO: common objects in context. In: Fleet, D., Pajdla, T., Schiele, B., Tuytelaars, T. (eds.) ECCV 2014. LNCS, vol. 8693, pp. 740–755. Springer, Cham (2014). https://doi.org/10.1007/978-3-319-10602-1_48

33. Liu, S., et al.: Adaattn: revisit attention mechanism in arbitrary neural style transfer. In: Proceedings of the IEEE/CVF International Conference on Computer Vision, pp. 6649–6658 (2021)

34. Luan, F., Paris, S., Shechtman, E., Bala, K.: Deep photo style transfer. In: Proceedings of the IEEE Conference on Computer Vision and Pattern Recognition, pp. 4990–4998 (2017)

35. Oord, A.V.D., Li, Y., Vinyals, O.: Representation learning with contrastive predictive coding. arXiv preprint arXiv:1807.03748 (2018)

36. Park, D.Y., Lee, K.H.: Arbitrary style transfer with style-attentional networks. In: Proceedings of the IEEE/CVF Conference on Computer Vision and Pattern Recognition, pp. 5880–5888 (2019)

37. Park, T., Efros, A.A., Zhang, R., Zhu, J.-Y.: Contrastive learning for unpaired image-to-image translation. In: Vedaldi, A., Bischof, H., Brox, T., Frahm, J.-M. (eds.) ECCV 2020. LNCS, vol. 12354, pp. 319–345. Springer, Cham (2020). https://doi.org/10.1007/978-3-030-58545-7_19

38. Phillips, F., Mackintosh, B.: Wiki art gallery, Inc.: a case for critical thinking. Issues Account. Educ. **26**(3), 593–608 (2011)

39. Ruder, M., Dosovitskiy, A., Brox, T.: Artistic style transfer for videos. In: Rosenhahn, B., Andres, B. (eds.) GCPR 2016. LNCS, vol. 9796, pp. 26–36. Springer, Cham (2016). https://doi.org/10.1007/978-3-319-45886-1_3

40. Ruder, M., Dosovitskiy, A., Brox, T.: Artistic style transfer for videos and spherical images. Int. J. Comput. Vision **126**(11), 1199–1219 (2018)

41. Shaham, T.R., Dekel, T., Michaeli, T.: Singan: learning a generative model from a single natural image. In: Proceedings of the IEEE/CVF International Conference on Computer Vision, pp. 4570–4580 (2019)

42. Sheng, L., Lin, Z., Shao, J., Wang, X.: Avatar-net: multi-scale zero-shot style transfer by feature decoration. In: Proceedings of the IEEE Conference on Computer Vision and Pattern Recognition, pp. 8242–8250 (2018)

43. Simonyan, K., Zisserman, A.: Very deep convolutional networks for large-scale image recognition. arXiv preprint arXiv:1409.1556 (2014)

44. Sun, D., Yang, X., Liu, M.Y., Kautz, J.: PWC-net: CNNs for optical flow using pyramid, warping, and cost volume. In: Proceedings of the IEEE Conference on Computer Vision and Pattern Recognition, pp. 8934–8943 (2018)

45. Ulyanov, D., Lebedev, V., Vedaldi, A., Lempitsky, V.S.: Texture networks: feedforward synthesis of textures and stylized images. In: ICML, vol. 1, p. 4 (2016)

46. Wang, W., Yang, S., Xu, J., Liu, J.: Consistent video style transfer via relaxation and regularization. IEEE Trans. Image Process. **29**, 9125–9139 (2020)

47. Yoo, J., Uh, Y., Chun, S., Kang, B., Ha, J.W.: Photorealistic style transfer via wavelet transforms. In: Proceedings of the IEEE/CVF International Conference on Computer Vision, pp. 9036–9045 (2019)

48. Zhang, R., Isola, P., Efros, A.A., Shechtman, E., Wang, O.: The unreasonable effectiveness of deep features as a perceptual metric. In: Proceedings of the IEEE Conference on Computer Vision and Pattern Recognition, pp. 586–595 (2018)

CANF-VC: Conditional Augmented Normalizing Flows for Video Compression

Yung-Han Ho[1], Chih-Peng Chang[1], Peng-Yu Chen[1], Alessandro Gnutti[2],
and Wen-Hsiao Peng[1(✉)]

[1] Department of Computer Science, National Yang Ming Chiao Tung University,
Taipei, Taiwan
wpeng@cs.nctu.edu.tw
[2] Department of Information Engineering, CNIT - University of Brescia,
Brescia, Italy
alessandro.gnutti@unibs.it

Abstract. This paper presents an end-to-end learning-based video compression system, termed CANF-VC, based on conditional augmented normalizing flows (CANF). Most learned video compression systems adopt the same hybrid-based coding architecture as the traditional codecs. Recent research on conditional coding has shown the suboptimality of the hybrid-based coding and opens up opportunities for deep generative models to take a key role in creating new coding frameworks. CANF-VC represents a new attempt that leverages the conditional ANF to learn a video generative model for conditional inter-frame coding. We choose ANF because it is a special type of generative model, which includes variational autoencoder as a special case and is able to achieve better expressiveness. CANF-VC also extends the idea of conditional coding to motion coding, forming a purely conditional coding framework. Extensive experimental results on commonly used datasets confirm the superiority of CANF-VC to the state-of-the-art methods. The source code of CANF-VC is available at https://github.com/NYCU-MAPL/CANF-VC.

1 Introduction

Video compression is an active research area. The video traffic continues to grow exponentially due to an increased demand for various emerging video applications, particularly on social media platforms and mobile devices. The traditional video codecs, such as HEVC [33] and VVC [7], are still thriving towards being more efficient, hardware-friendly, and versatile. However, their backbones follow the hybrid-based coding framework–namely, spatial/temporal predictive coding plus transform-based residual coding–which has not changed since decades ago.

Supplementary Information The online version contains supplementary material available at https://doi.org/10.1007/978-3-031-19787-1_12.

The arrival of deep learning spurs a new wave of developments in end-to-end learned image and video compression [9,15,23,28,30,32]. The seminal work [4] by Ballé et $al.$ connects for the first time the learning of an image compression system to learning a variational generative model, known as the variational autoencoder (VAE) [19]. VAE involves learning the autoencoder network jointly with the prior distribution network by maximizing the variational lower bound (ELBO) on the image likelihood $p(x)$. Many follow-up works have been centered around enhancing the autoencoder network [8,9] and/or improving the prior modeling [9,30]. Lately, there have been few attempts at introducing normalizing flow models [13,28] to learned image compression

Inspired by the success of learned image compression, research on learned video compression is catching up quickly. However, most end-to-end learned video compression systems [14,24,26,27,32] were developed based primarily on the traditional, hybrid-based coding architecture, replacing key components, such as inter-frame prediction and residual coding, with neural networks. The idea of residual coding is to encode a target frame x_t by coding the prediction residual $r_t = x_t - x_c$ between x_t and its motion-compensated reference frame x_c. The recent revisit of residual coding as a problem of conditional coding in [21–23] opens up a new dimension of thinking. Arguably, the entropy $H(x_t - x_c)$ of the residual between the coding frame x_t and its motion-compensated reference frame x_c is greater than or equal to the conditional entropy $H(x_t|x_c)$, i.e. $H(x_t - x_c) \geq H(x_t|x_c)$. How to learn $p(x_t|x_c)$ is apparently the key to the success of conditional coding.

In this paper, we present a conditional augmented normalizing flow-based video compression (CANF-VC) system, which is inspired partly by the ANF-based image compression (ANFIC) [13]. However, while ANFIC [13] adopts ANF to learn the (unconditional) image distribution $p(x)$ for image compression, we address video compression from the perspective of learning a video generative model by maximizing the conditional likelihood $p(x_t|x_c)$. We choose the conditional augmented normalizing flow (CANF) to learn $p(x_t|x_c)$, because ANF is a special type of generative model, which includes VAE as a special case and is able to achieve superior expressiveness to VAE.

Our work has three main contributions: (1) CANF-VC is the first normalizing flow-based video compression system that leverages CANF to learn a video generative model for conditional inter-frame coding; (2) CANF-VC extends the idea of conditional inter-frame coding to conditional motion coding, forming a purely conditional coding framework; and (3) extensive experimental results confirm the superiority of CANF-VC to the state-of-the-art methods.

2   Related Work

2.1   Learned Video Compression

End-to-end learned video compression is a hot research area. DVC [26] presents the first end-to-end learned video coding framework based on temporal predictive coding. Since then, there have been several improvements on learning-based

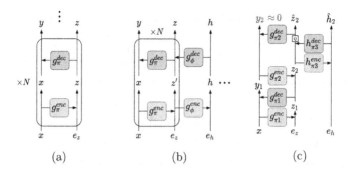

Fig. 1. The architectures of ANF: (a) N-step ANF, (b) N-step hierarchical ANF, and (c) ANF for image compression (ANFIC).

motion-compensated prediction. Agustsson *et al.* [2] estimate the uncertainty about the flow map in forming a frame predictor, with a scale index sent for each pixel to determine a spatially-varying Gaussian kernel for blurring the reference frame. Liu *et al.* [25] perform feature-domain warping in a coarse-to-fine manner. Hu *et al.* [15] adopt deformable convolution for feature warping. Lin *et al.* [24] and Yang *et al.* [37] form a multi-hypothesis prediction from multiple reference frames. To reduce motion overhead, Lin *et al.* [24] use predictive motion coding by extrapolating a flow map predictor from the decoded flow maps. Rippel *et al.* [32] use the flow map predictor for motion compensation and signal an incremental flow map between the resulting motion-compensated frame and the target frame. Hu *et al.* [14] adapt, either locally or globally, the resolution of the flow map features. Most learned video codecs encode the residual frame or the residual flow map by a variational autoencoder (VAE)-based image coder [4]. Some additionally leverage a recurrent neural network to propagate causal, temporal information in forming a temporal prior for entropy coding [25,38].

2.2   Conditional Coding

The idea of encoding the residual signal has recently been revisited from the information-theoretic perspective. Ladune *et al.* [21] show that coding a video frame x_t conditionally based on its motion-compensated reference frame x_c can achieve a lower entropy rate than coding the residual signal $x_t - x_c$ unconditionally. The fact motivates their converting the VAE-based residual coder into a conditional VAE by concatenating x_c and x_t for encoding, and their latent representations for decoding. The idea was extended in [22] for conditional motion coding, which encodes motion latents in an implicit, one-stage manner. However, Fabian *et al.* [6] show that these conditional VAE-based approaches [21,22] may suffer from the *bottleneck* issue; that is, the latent representation of x_c produced by a neural network for conditional decoding may not capture all the information of x_c, which serves as a condition for encoding x_t. Such information loss and asymmetry can harm the efficiency of conditional coding. Li *et al.* [23] improve the work in [21] by ensuring that the same information-rich latent representation of x_c is utilized for both conditional encoding and decoding. Likewise, the work in [12] creates the same coding context for conditional encoding and decoding via a feedback recurrent module that aggregates the past latent information. In common, these approaches do not evaluate any residual signal explicitly.

2.3   Augmented Normalizing Flows (ANF)

To learn properly the conditional distribution $p(x_t|x_c)$ for conditional coding, we turn to augmented normalizing flows (ANF), a special type of generative model able to achieve superior expressiveness to VAE. Different from the vanilla flow models [10,18,20], ANF [16] augments the input x with an independent noise e_z (Fig. 1a), allowing the augmented noise to induce a complex marginal on x [16]. ANF contains the autoencoding transform g_π as a basic building block, where the encoding transformation g_π^{enc} from (x, e_z) to (x, z) and the decoding transformation g_π^{dec} from (x, z) to (y, z) are specified by

$$g_\pi^{enc}(x, e_z) = (x, s_\pi^{enc}(x) \odot e_z + m_\pi^{enc}(x)) = (x, z), \tag{1}$$

$$g_\pi^{dec}(x, z) = ((x - \mu_\pi^{dec}(z))/\sigma_\pi^{dec}(z), z) = (y, z), \tag{2}$$

respectively. s_π^{enc}, m_π^{enc}, μ_π^{dec} and σ_π^{dec} are element-wise affine transform parameters, and they are driven by neural networks parameterized by π.

The training of ANF aims to maximize the *augmented data likelihood*—namely, $\arg\max_\pi p_\pi(x, e_z) = p(g_\pi(x, e_z))|det(\partial g_\pi(x, e_z)/\partial(x, e_z))|$. Performing one autoencoding transformation $(y, z) = g_\pi(x, e_z)$ (known as the one-step ANF) is equivalent to training a VAE by maximizing the evidence lower bound (ELBO) on the log-marginal $\log p_\pi(x)$ [19]. As such, the learned image compression with the factorized prior [3] can be viewed as an one-step ANF that adopts a purely additive autoencoding transform (i.e. $s_\pi^{enc}(x) = \sigma_\pi^{dec}(z) = 1$) and an augmented noise $e_z \sim \mathcal{U}(-0.5, 0.5)$ modeling the uniform quantization. In this case, the latents y, z follow the standard Normal $\mathcal{N}(0, I)$ and the learned factorized prior, respectively. In particular, the hyperprior extension [4] has a similar structure to the *hierarchical ANF* (Fig. 1b), an enhanced form of ANF [16] with $g_\phi^{enc}, g_\phi^{dec}$ playing a similar role to the hyper codec. For better expressiveness, one can stack multiple one-step ANF's as the *multi-step ANF*. In [13], Ho et al. introduce the first ANF-based image compression (Fig. 1c), which combines the multi-step and the hierarchical ANF's.

3   Proposed Method

3.1   Problem Statement

In this section, we formally define our task and objective. Let $x_{1:T} \in \mathbb{R}^{T \times 3 \times H \times W}$ denote a (RGB) video sequence of width W and height H to be encoded, and $\hat{x}_{1:T}$ the decoded video. The video compression task is to strike a good balance between the distortion $d(\hat{x}_{1:T}, x_{1:T})$ of the decoded video $\hat{x}_{1:T}$ and the rate

$r(\hat{x}_{1:T})$ needed to represent it. When $T = 1$, the task reduces to image compression, of which the problem is cast as learning a VAE by maximizing the ELBO on the log-likelihood $\log p(x)$ in [4]. The same perspective is applicable to video compression yet with the aim of learning a VAE that maximizes the joint log-likelihood $\log p(x_{1:T})$. Because $p(x_{1:T})$ factorizes as $\prod_{t=1}^{T} p(x_t|x_{<t})$, with $x_{<t}$ representing collectively the video frames up to time instance $t - 1$, video compression is often done frame-by-frame by learning the conditional distribution $p(x_t|x_{<t})$. In our task, the decoded frames $\hat{x}_{<t}$ are used in place of $x_{<t}$.

With the traditional predictive coding framework, the ELBO on $\log p(x_t|\hat{x}_{<t})$ has a form of

$$E_{q(\hat{f}_t,\hat{r}_t|x_t,\hat{x}_{<t})} \log p(x_t|\hat{f}_t,\hat{r}_t,\hat{x}_{<t}) - D_{KL}(q(\hat{f}_t,\hat{r}_t|x_t,\hat{x}_{<t})\|p(\hat{f}_t,\hat{r}_t|\hat{x}_{<t})), \quad (3)$$

where the latents $\hat{f}_t \in \mathbb{R}^{2 \times H \times W}, \hat{r}_t \in \mathbb{R}^{3 \times H \times W}$ represent the (quantized) optical flow map and the (quantized) motion-compensated residual frame associated with $x_t \in R^{3 \times H \times W}$, respectively. The encoding distribution $q(\hat{f}_t,\hat{r}_t|x_t,\hat{x}_{<t}) = q(\hat{f}_t|x_t,\hat{x}_{<t})q(\hat{r}_t|\hat{f}_t,x_t,\hat{x}_{<t})$ specifies the generation of $\hat{f}_t,\hat{r}_t$, while the decoding distribution $p(x_t|\hat{f}_t,\hat{r}_t,\hat{x}_{<t}) = \mathcal{N}(\hat{r}_t + warp(\hat{x}_{t-1};\hat{f}_t),\frac{1}{2\lambda}I)$ models the reconstruction process of x_t, with $warp(\hat{x}_{t-1};\hat{f}_t)$ denoting the backward warping of $\hat{x}_{t-1}$ based on $\hat{f}_t$, and $\frac{1}{2\lambda}$ being the variance of the Gaussian. Assuming the use of uniform quantization function for obtaining $\hat{f}_t,\hat{r}_t$, the Kullback-Leibler (KL) divergence $D_{KL}(\cdot\|\cdot)$ evaluates to the rate costs associated with their transmission:

$$D_{KL}(q(\hat{f}_t,\hat{r}_t|x_t,\hat{x}_{<t})\|p(\hat{f}_t,\hat{r}_t|\hat{x}_{<t})) \quad (4)$$
$$= E_{q(\hat{f}_t,\hat{r}_t|x_t,\hat{x}_{<t})}(-\log p(\hat{f}_t|\hat{x}_{<t}) - \log p(\hat{r}_t|\hat{f}_t,\hat{x}_{<t})).$$

Substituting Eq. (4) into Eq. (3) and applying the law of total expectation yields

$$E_{q(\hat{f}_t|x_t,\hat{x}_{<t})}(RD_r(x_t|\hat{f}_t,\hat{x}_{<t}) + \log p(\hat{f}_t|\hat{x}_{<t})), \quad (5)$$

where

$$RD_r(x_t|\hat{f}_t,\hat{x}_{<t}) = E_{q(\hat{r}_t|\hat{f}_t,x_t,\hat{x}_{<t})}(\log p(x_t|\hat{f}_t,\hat{r}_t,\hat{x}_{<t}) + \log p(\hat{r}_t|\hat{f}_t,\hat{x}_{<t})), \quad (6)$$

which bears the interpretation of the ELBO on $\log p(x_t|\hat{f}_t,\hat{x}_{<t})$, with the latent being the quantized residual frame $\hat{r}_t$.

From Eqs. (5) and (6), we see that the traditional predictive coding of a video frame x_t includes (1) encoding the residual frame $\hat{r}_t$ based on $\hat{f}_t,\hat{x}_{<t}$ in order to maximize the log-likelihood $\log p(x_t|\hat{f}_t,\hat{x}_{<t})$ and (2) encoding the flow map $\hat{f}_t$ in a way that strikes a good balance between the maximization of $\log p(x_t|\hat{f}_t,\hat{x}_{<t})$ and the (negative) rate $E_{q(\hat{f}_t|x_t,\hat{x}_{<t})} \log p(\hat{f}_t|\hat{x}_{x<t})$ needed to signal $\hat{f}_t$.

In this work, we propose to turn the maximization of the log-likelihood $\log p(x_t|\hat{f}_t,\hat{x}_{<t})$, i.e. Eq. (6), into a problem of conditional coding, where $\hat{f}_t,\hat{x}_{<t}$ are utilized to formulate the motion-compensated frame $x_c \in \mathbb{R}^{3 \times H \times W}$ as a condition. Unlike the existing works [12,21–23], which adopt the conditional VAE, our conditional coder is constructed based on multi-step CANF in modeling $p(x_t|x_c)$ for its better expressiveness.

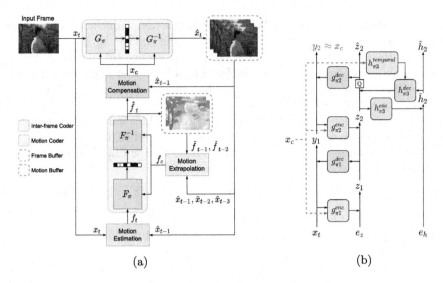

Fig. 2. Illustration of (a) the proposed CANF-VC framework and (b) the CANF-based inter-frame coder $\{G_\pi, G_\pi^{-1}\}$. The CANF-based motion coder $\{F_\pi, F_\pi^{-1}\}$ follows the same design as the inter-frame coder, with x_t, x_c replaced by f_t, f_c, respectively.

3.2 System Overview

Figure 2a depicts our CANF-based video compression system, abbreviated as CANF-VC. It includes two major components: (1) the CANF-based inter-frame coder $\{G_\pi, G_\pi^{-1}\}$ and (2) the CANF-based motion coder $\{F_\pi, F_\pi^{-1}\}$. The inter-frame coder encodes a video frame x_t conditionally, given the motion-compensated frame x_c. It departs from the conventional residual coding by maximizing the conditional log-likelihood $p(x_t|x_c)$ with CANF model (Sect. 3.3). The motion coder shares a similar architecture to the inter-frame coder. It extends conditional coding to motion coding, in order to signal the flow map f_t, which characterizes the motion between x_t and its reference frame $\hat{x}_{t-1}$. In our work, f_t is estimated by PWC-Net [34]. The compressed flow map $\hat{f}_t$ serves to warp the reference frame $\hat{x}_{t-1}$, with the warped result enhanced further by a motion compensation network to arrive at x_c. To formulate a condition for conditional motion coding, we introduce a flow extrapolation network to extrapolate a flow map f_c from three previously decoded frames $\hat{x}_{t-1}, \hat{x}_{t-2}, \hat{x}_{t-3}$ and two decoded flow maps $\hat{f}_{t-1}, \hat{f}_{t-2}$. Note that we expand the condition of $p(x_t|\hat{x}_{<t})$ from previously decoded frames $\{\hat{x}_{<t}\}$ to include also previously decoded flows $\{\hat{f}_{<t}\}$.

3.3 CANF-Based Inter-frame Coder

Figure 2b presents the architecture of our CANF-based inter-frame coder, which aims to learn the conditional distribution $p(x_t|x_c)$ of the coding frame x_t given the motion-compensated frame x_c. This is achieved by maximizing the augmented likelihood $p(x_t, e_z, e_h|x_c)$ in the CANF framework, where $e_z \in \mathbb{R}^{C \times \frac{H}{16} \times \frac{W}{16}}$, $e_h \in \mathbb{R}^{C \times \frac{H}{64} \times \frac{W}{64}}$ are the two augmented noise inputs. It is shown in [16] that maximizing $p(x_t, e_z, e_h|x_c)$ is equivalent to maximizing a lower bound on the marginal likelihood $p(x_t|x_c)$.

Architecture: Motivated by [13], our conditional inter-frame coder is a hybrid of the two-step and the hierarchical ANF's. The two autoencoding transforms $\{g_{\pi_1}^{enc}, g_{\pi_1}^{dec}\}, \{g_{\pi_2}^{enc}, g_{\pi_2}^{dec}\}$ convert x_t, e_z into their latents y_2, z_2, respectively, while the hierarchical autoencoding transform $\{h_{\pi_3}^{enc}, h_{\pi_3}^{dec}\}$ acts as the hyperprior codec, encoding the latent z_2 into the hyperprior representation $\hat{h}_2$. The volume preserving property of CANF requires that the latents y_2, z_2 (or $\hat{z}_2$), $\hat{h}_2$ have the same dimensions as their respective inputs x_t, e_z, e_h. One notable distinction between CANF and ANFIC [13] is the incorporation of the condition x_c into the autoencoding transforms and the prior distribution, as will be detailed next.

Conditional Encoding: The core idea of our conditional coding is to let the latent y_2, which represents a transformed version of the target frame x_t, approximate the condition x_c, with the latents $z_2, \hat{h}_2$ encoding the information necessary for instructing the transformation. Specifically, the two autoencoding transforms operate similarly and successively. Taking $\{g_{\pi_1}^{enc}, g_{\pi_1}^{dec}\}$ as an example (see Fig. 2b), we have

$$g_{\pi_1}^{enc}(x_t, e_z|x_c) = (x_t, e_z + m_{\pi_1}^{enc}(x_t, x_c)) = (x_t, z_1), \tag{7}$$

$$g_{\pi_1}^{dec}(x_t, z_1) = (x_t - \mu_{\pi_1}^{dec}(z_1), z_1) = (y_1, z_1). \tag{8}$$

That is, the one-step transformation from x_t to y_1 is done by subtracting the decoder output $\mu_{\pi_1}^{dec}(z_1)$ from x_t. Note that $\mu_{\pi_1}^{dec}(z_1)$ decodes the latent z_1, which aggregates the information of x_t, x_c, and the augmented noise e_z. We remark that the encoding process is made conditional on x_c by concatenating x_c and x_t to form the encoder input. Intuitively, supplying x_c as an auxiliary signal should ease the transformation from x_t to x_c. This process is repeated by taking y_1 and z_1 as the inputs to the next autoencoding transform $\{g_{\pi_2}^{enc}, g_{\pi_2}^{dec}\}$. In fact, the number of the autoencoding transforms is flexible. In comparison with Eqs. (1) and (2), our autoencoding transform is purely additive (i.e. $s_\pi^{enc}, \sigma_\pi^{dec}$ in Eqs. (1) and (2) are set to 1), which is found beneficial in terms of training stability.

The hierarchical autoencoding transform $\{h_{\pi_3}^{enc}, h_{\pi_3}^{dec}\}$ serves to estimate the probability distribution of z_2 for entropy coding. It operates according to

$$h_{\pi_3}^{enc}(z_2, e_h) = (z_2, e_h + m_{\pi_3}^{enc}(z_2)) = (z_2, \hat{h}_2), \tag{9}$$

$$h_{\pi_3}^{dec}(z_2, \hat{h}_2|x_c) = (\lfloor z_2 - \mu_{\pi_3}^{dec}(\hat{h}_2, h_{\pi_3}^{temporal}(x_c)) \rceil, \hat{h}_2) = (\hat{z}_2, \hat{h}_2), \tag{10}$$

where $\lfloor \cdot \rceil$ (depicted as Q in Fig. 2b) denotes the nearest-integer rounding, which is needed to express z_2 in fixed-point representation for lossy compression. At training time, the rounding effect is modeled by additive quantization noise. It is worth noting that x_c is provided as an auxiliary input to $\mu_{\pi_3}^{dec}$ to exert a combined effect of the hyperprior and the temporal prior ($h_{\pi_3}^{temporal}$ in Fig. 2b).

Conditional Decoding: The decoding process of our inter-frame coder updates the motion-compensated frame x_c successively to reconstruct x_t. It starts by entropy decoding the latents $\hat{z}_2, \hat{h}_2$, and substituting x_c for y_2. The quantized z_2 will then be recovered and decoded to reconstruct $\mu_{\pi_2}^{dec}(z_2)$, which updates x_c as $y_1 = x_c + \mu_{\pi_2}^{dec}(z_2)$. Subsequently, y_1 will be encoded conditionally based on x_c using $m_{\pi_2}^{enc}(y_1, x_c)$ in order to update the latent z_2 as $z_1 = z_2 - m_{\pi_2}^{enc}(y_1, x_c)$. Finally, z_1 is decoded by $\mu_{\pi_1}^{dec}(z_1)$ to update y_1 as the reconstructed version $\hat{x}_t = y_1 + \mu_{\pi_1}^{dec}(z_1)$ of x_t. In a sense, the reconstruction of x_t is achieved by passing the latent $\hat{z}_2$ through the composition of the decoding and encoding transforms to update x_c.

Conditional Prior Distribution: Another strategy we adopt to learn $p(x_t, e_z, e_h | x_c)$ is to introduce a conditional prior distribution $p(y_2, \hat{z}_2, \hat{h}_2 | x_c)$. Specifically, we assume that it factorizes as follows:

$$p(y_2, \hat{z}_2, \hat{h}_2 | x_c) = p(y_2 | x_c) p(\hat{z}_2 | \hat{h}_2, x_c) p(\hat{h}_2). \tag{11}$$

Because we require y_2 to approximate x_c, it is natural to choose $p(y_2 | x_c)$ to be $\mathcal{N}(x_c, \frac{1}{2\lambda_1} I)$ with a small variance $\frac{1}{2\lambda_1}$. Moreover, following the hyperprior [4], $p(\hat{z}_2 | \hat{h}_2, x_c)$ and $p(\hat{h}_2)$ are modeled by

$$p(\hat{z}_2 | \hat{h}_2, x_c) = \mathcal{N}(0, (\sigma_{\pi_3}^{dec}(\hat{h}_2, h_{\pi_3}^{temporal}(x_c)))^2 I) * \mathcal{U}(-0.5, 0.5)$$

$$p(\hat{h}_2) = \mathcal{P}_{\hat{h}_2 | \psi} * \mathcal{U}(-0.5, 0.5),$$

where $*$ denotes convolution and $\mathcal{P}_{\hat{h}_2 | \psi}$ is a factorized prior parameterized by ψ. The use of the motion-compensated frame x_c along with $\hat{h}_2$ in estimating the distribution of $\hat{z}_2$ combines temporal prior ($h_{\pi_3}^{temporal}$ in Fig. 2b) and hyperprior.

Augmented Noises e_z, e_h**:** In the theory of ANF [16], the augmented noises are meant to induce a complex marginal on the input x. For the compression task, we fix e_z at 0 during training and test, in order not to increase the entropy rate at $\hat{z}_2$. For training, the quantization Q in Fig. 2b is simulated by additive noise. In contrast, we draw $e_h \sim \mathcal{U}(-0.5, 0.5)$ for simulating the quantization of the hyperprior at training time, and set it to zero at test time when the hyperprior is actually rounded.

Extension to Conditional Motion Coding: The CANF-based motion coder follows the same design as the CANF-based inter-frame coder. The coding frame x_t is replaced with the optical flow map f_t and the motion-compensated frame x_c with the extrapolated flow map f_c. In addition, the temporal prior takes the extrapolated frame $warp(\hat{x}_{t-1}; f_c)$ as input. To perform the flow map extrapolation, we adopt a U-Net-based network (see supplementary document).

In the supplementary document, we provide another CANF implementation, which additionally accepts x_c as input to the decoding transforms. We choose the current implementation due to its comparable performance and simpler design.

3.4   Training Objective

We train the conditional inter-frame and motion coders end-to-end. Inspired by Eq. (5), we first turn the maximization of the ELBO (i.e. RD_r in Eq. (6)) on $\log p(x_t | \hat{f}_t, \hat{x}_{<t})$ into maximizing $\log p(x_t, e_z, e_h | x_c)$. That is, to minimize

$$-\log p(x_t, e_z, e_h | x_c) = -\log p(\hat{h}_2) - \log p(\hat{z}_2 | \hat{h}_2, x_c)$$
$$+\lambda_1 \|y_2 - x_c\|^2 - \log \left| \det \frac{\partial G_\pi(x_t, e_z, e_h | x_c)}{\partial (x_t, e_z, e_h)} \right|.$$

To ensure the reconstruction quality, we follow [13] to replace the negative log-determinant of the Jacobian with a weighted reconstruction loss $\lambda_2 d(x_t, \hat{x}_t)$, arriving at

$$-\log p(x_t, e_z, e_h | x_c) \approx \underbrace{-\log p(\hat{h}_2) - \log p(\hat{z}_2 | \hat{h}_2, x_c)}_{R} + \lambda_1 \|y_2 - x_c\|^2 + \underbrace{\lambda_2 d(x_t, \hat{x}_t)}_{D},$$

which includes the rate R needed to signal the transformation between x_t and x_c, the regularization term requiring y_2 to approximate x_c, and the distortion D of $\hat{x}_t$. To complete the loss function, we also follow the second term in Eq. (5) to include the conditional motion rate used to signal f_t given f_c, which leads to

$$\mathcal{L} = -\log p(\hat{h}_2) - \log p(\hat{z}_2 | \hat{h}_2, x_c) + \lambda_1 \|y_2 - x_c\|^2 \tag{12}$$
$$-\log p_f(\hat{h}_2) - \log p_f(\hat{z}_2 | \hat{h}_2, f_c) + \lambda_2 d(x_t, \hat{x}_t),$$

where p_f (respectively, p) describes the prior distribution over the motion (respectively, inter-frame) latents.

3.5   Comparison with ANFIC and Other VAE-Based Schemes

Our CANF-VC is based on ANF, as well as ANFIC, a learned image compression system proposed in [13]. However, they significantly differ from each other, not only because they refer to different applications. ANFIC [13] adopts an *unconditional* ANF to learn the image distribution $p(x)$ for image compression, whereas CANF-VC uses two *conditional* ANF's (CANF's) to learn the conditional distribution $p(x_t | x_c)$ for inter-frame coding and the conditional rate needed to signal the motion part, respectively. As a result, CANF-VC is a complete video coding framework. Note that how the conditional information x_c and f_c are both incorporated in the respective autoencoding transforms and in the respective prior distributions is first proposed in this work.

CANF-VC is also distinct from conditional VAE-based frameworks, such as DCVC [23] and [22]. CANF-VC bases the compression backbone on CANF, which is a flow-based model and includes VAE as a special case. As compared with DCVC [23], CANF-VC additionally features conditional motion coding. Although conditional motion coding also appears in [22], their VAE-based approach does not explicitly estimate a flow map prior to conditional coding, and may suffer from the bottleneck issue [6] (Sect. 2.2). In contrast, CANF-VC takes an explicit approach and avoids the bottleneck issue by using the same x_c symmetrically in the encoder and the decoder due to its invertible property.

Table 1. BD-rate comparison with GOP size 10/12. The anchor is x265 in veryslow mode. The best performer is marked in red and the second best in blue.

	Intra coder (PSNR-RGB/MS-SSIM-RGB)	BD-rate (%) PSNR-RGB			BD-rate (%) MS-SSIM-RGB			Size
		UVG	MCL-JCV	HEVC-B	UVG	MCL-JCV	HEVC-B	
DVC_Pro [27]	-/-	−3.0	–	−13.1	−5.2	–	−20.8	29M
M-LVC [24]	BPG/BPG	−15.3	18.8	−38.6	−0.2	4.7	−37.9	–
RaFC [14]	hyperprior/ hyperprior	−11.1	4.4	−9.3	−25.5	−27.9	−37.2	–
FVC [15]	BPG/BPG	−16.9	−3.8	−17.8	−45.0	−46.1	−54.3	26M
HM (LDP, 4 refs)	-/-	−29.4	−13.9	−29.6	−18.9	−13.8	−17.1	–
CANF-VC*	BPG/BPG	−35.5	**−14.6**	−35.4	−46.6	**−46.7**	−53.2	31M
DCVC [23]	cheng2020-anchor/hyperprior	−23.8	−14.4	−34.9	−43.9	−44.9	−50.7	8M
DCVC (ANFIC)	ANFIC/ANFIC	−24.8	−13.6	−34.0	−41.9	−43.7	−51.1	8M
CANF-VC Lite	ANFIC/ANFIC	**−37.3**	−14.3	**−39.8**	**−47.6**	−44.2	−56.8	15M
CANF-VC	ANFIC/ANFIC	−42.5	−21.0	−40.1	−51.4	−47.6	**−54.7**	31M

4 Experiments

4.1 Settings and Implementation Details

Training Details: We train our model on Vimeo-90k [36] dataset, which contains 91,701 7-frame sequences with resolution 448×256. We randomly crop these video clips into 256×256 for training. We adopt the Adam [17] optimizer with the learning rate 10^{-4} and the batch size 32. Separate models are trained to optimize first the mean-square error with $\lambda_2 = \{256, 512, 1024, 2048\}$ and $\lambda_1 = 0.01 * \lambda_2$ (see Eq. (12)). We then fine-tune these models for Multi-scale Structural Similarity Index (MS-SSIM), with λ_2 set to $\{4, 8, 16, 32, 64\}$. All the low-rate models are adapted from the one trained for the highest rate point.

Evaluation Methodologies: We evaluate our models on commonly used datasets, including UVG [29], MCL-JCV [35], and HEVC Class B [11]. We follow common test protocols to provide results in Table 1 for 100-frame encoding with GOP[1] size 10 on HEVC Class B, and full-sequence encoding with GOP size 12 on the other datasets. Additionally, we present results for GOP size 32 in Table 2, to underline the contributions of our inter-frame and motion coders. For this additional setting, all the learned codecs use ANFIC [13] as the intra-frame coder and encode only the first 96 frames in every test sequence. To evaluate the rate-distortion performance, the bit rates are measured in bits per pixel (bpp), and the quality in PSNR-RGB and MS-SSIM-RGB. Moreover, we use x265 in veryslow mode as the anchor for reporting BD-rates.

Baseline Methods: The baseline methods for comparison include x265, HEVC Test Model (HM) [1] and several recent publications, including DVC_Pro [27],

---

[1] GOP refers to Group-of-Pictures and is often used interchangeably with the intra period in papers on learned video codecs.

Table 2. BD-rate comparison with GOP size 32. All the competing methods (except HM) use ANFIC [13] as the intra-frame coder. The anchor is x265 in veryslow mode. The best performer is marked in red and the second best in **blue**.

	Intra coder (PSNR-RGB/MS-SSIM-RGB)	BD-rate (%) PSNR-RGB			BD-rate (%) MS-SSIM-RGB		
		UVG	MCL-JCV	HEVC-B	UVG	MCL-JCV	HEVC-B
M-LVC (ANFIC)	ANFIC/ANFIC	−12.1	−5.3	−9.7	−7.5	−8.4	−18.8
DCVC (ANFIC)	ANFIC/ANFIC	−16.3	−21.3	−10.5	**−38.8**	**−48.9**	−39.3
CANF-VC Lite	ANFIC/ANFIC	**−36.1**	−26.5	**−30.3**	−37.9	−47.8	−44.2
CANF-VC⁻	ANFIC/ANFIC	−31.1	−29.5	−23.6	−36.2	−47.8	−38.0
CANF-VC	ANFIC/ANFIC	−35.9	**−32.0**	−27.7	−40.3	−49.6	**−41.3**
HM (LDP, 4 refs)	-/-	−41.6	−38.6	−32.1	−34.3	−32.0	−31.0

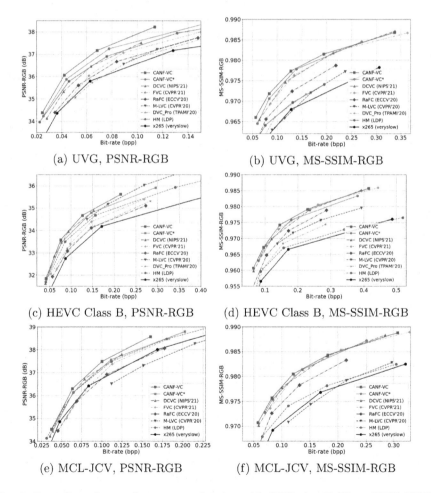

(a) UVG, PSNR-RGB

(b) UVG, MS-SSIM-RGB

(c) HEVC Class B, PSNR-RGB

(d) HEVC Class B, MS-SSIM-RGB

(e) MCL-JCV, PSNR-RGB

(f) MCL-JCV, MS-SSIM-RGB

Fig. 3. Rate-distortion performance evaluation with GOP size 10/12 on UVG, HEVC Class B, and MCL-JCV datasets for both PSNR-RGB and MS-SSIM-RGB.

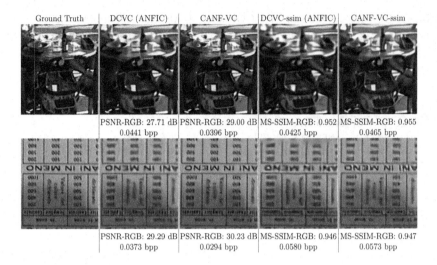

| Ground Truth | DCVC (ANFIC) | CANF-VC | DCVC-ssim (ANFIC) | CANF-VC-ssim |

| | PSNR-RGB: 27.71 dB | PSNR-RGB: 29.00 dB | MS-SSIM-RGB: 0.952 | MS-SSIM-RGB: 0.955 |
| | 0.0441 bpp | 0.0396 bpp | 0.0425 bpp | 0.0465 bpp |

| | PSNR-RGB: 29.29 dB | PSNR-RGB: 30.23 dB | MS-SSIM-RGB: 0.946 | MS-SSIM-RGB: 0.947 |
| | 0.0373 bpp | 0.0294 bpp | 0.0580 bpp | 0.0573 bpp |

Fig. 4. Subjective quality comparison between CANF-VC and DCVC (ANFIC).

M-LVC [24], RaFC [14], FVC [15] and DCVC [23]. Because these baseline methods adopt different intra-frame coders (see the second column of Table 1), which are critical to the overall rate-distortion performance, we provide results with ANFIC [13] (CANF-VC) and BPG (CANF-VC*) as the intra-frame coders to ease comparison. Note that ANFIC [13] shows comparable performance to cheng2020-anchor [5]. It is to be noted that x265, HM [1], DVC_Pro [27], and M-LVC [24] use the same model optimized for PSNR to report PSNR-RGB and MS-SSIM-RGB results. While the other methods train separate models in reporting these results. We also present CANF-VC$^-$ and CANF-VC Lite as two additional variants of CANF-VC. CANF-VC$^-$ disables conditional motion coding while CANF-VC Lite implements a lightweight version of CANF-VC by reducing the channels in the autoencoding and the hyperprior transforms, and adopting SPyNet [31] as the flow estimation network.

4.2   Rate-Distortion and Subjective Quality Comparison

Rate-Distortion Comparison: The upper part of Table 1 compares the competing methods with their intra-frame coders, e.g. hyperprior [3], performing comparably to BPG. We see that our CANF-VC* (with BPG as the intra-frame coder) outperforms most of these baselines across different datasets in terms of PSNR-RGB. Its slight rate inflation (3%) as compared to M-LVC [24] on HEVC-B class may be attributed to the not-fully-aligned rate range in which the BD-rate is measured (see Fig. 3). Note that M-LVC [24] is initially trained for GOP size 100. With no access to its training software, a rate shift occurs when its test code is re-run for GOP size 10/12. Another observation is that CANF-VC* shows similar MS-SSIM-RGB results to FVC [15], while surpassing the others considerably. The lower part of Table 1 further shows that in terms of both quality metrics, our full model CANF-VC performs consistently better than both DCVC variants, where one uses ANFIC [13] and the other adopts cheng2020-anchor [5] as their respective intra-frame coders. The same observation can be made with CANF-VC$^-$ and CANF-VC Lite, except that they perform similarly to DCVC [23] on MCL-JCV.

Under the long GOP setting (Table 2), the gain of our schemes (all three variants) over DCVC (ANFIC) and M-LVC (ANFIC) becomes more significant in terms of PSNR-RGB, while CANF-VC$^-$ and CANF-VC Lite show comparable or better MS-SSIM-RGB results than DCVC (ANFIC). Interestingly, the gap in PSNR-RGB between the more capable HM and the learned coders is still considerable, although the latter outperform HM in terms of MS-SSIM-RGB.

Subjective Quality Comparison: Figure 4 presents a subjective comparison between our CANF-VC and DCVC (ANFIC). Both schemes are trained for PSNR-RGB and MS-SSIM-RGB, use ANFIC as the intra-frame coder, and set GOP size to 32. Our CANF-VC is seen to preserve better the shape of the objects and has no color bias, as compared to DCVC (ANFIC).

4.3   Ablation Experiments

In this section, unless otherwise stated, all the experiments are conducted on UVG dataset [29], with the BD-rates reported against x265 in veryslow mode.

Conditional Inter-frame Coding vs. Residual Coding: To single out the gain of conditional inter-frame coding over residual coding, Table 3a presents a breakdown analysis in terms of BD-rate savings. In this ablation experiment, the conditional motion coding is disabled and replaced with the motion coder from DVC [26]. Besides, the residual coding schemes adopt ANFIC [13] for coding the residual frame $x_t - x_c$ as an intra image. The variants with the temporal prior additionally involve the motion-compensated frame x_c in estimating the coding probabilities of the latent code (i.e. $\hat{z}_2$ in Fig. 2b). As seen in the table, the conditional inter-frame coding outperforms the residual coding significantly, whether the temporal prior is enabled or not. This suggests that a direct application of ANFIC to residual coding is unable to achieve the same level of gain as our CANF-based inter-frame coding. The temporal prior additionally improves the rate savings of both schemes by 2.5% to 4.2%.

Conditional Motion Coding vs. Predictive/Intra Motion Coding: This ablation experiment addresses the benefits of conditional motion coding. To this end, different competing settings employ the same conditional inter-frame coding, but change the way the flow map f_t is coded. The baseline settings use ANFIC [13] to code f_t as an intra image or the flow map residual $f_t - f_c$ without any condition. For the conditional motion coding, we additionally present results by simply using the previously decoded flow map $\hat{f}_{t-1}$ as the condition. Separate models are trained for each test case. From Table 3b, our conditional motion coding (i.e. coding f_t based on f_c) achieves the best performance. In terms of rate savings, its gain over the two unconditional variants, i.e. coding f_t or $f_t - f_c$

Table 3. (a) Comparison of conditional inter-frame coding and residual coding under the settings with and without the temporal prior. (b) Comparison of the conditional motion coding, predictive motion coding, and intra motion coding. (c)(d) Comparisons of the conditional motion and inter-frame coders with a varied number of autoencoding transforms. The rows with blue color are our proposed full model.

Cond. Inter-frame Coding	Residual Coding	Temporal Prior	BD-Rate
	✓		−21.8%
	✓	✓	−24.3%
✓			−28.9%
✓		✓	−33.1%

(a)

Input of Motion Coder	Cond.	BD-Rate
f_t	-	−33.4%
$f_t - f_c$	-	−35.3%
f_t	$\hat{f}_{t-1}$	−35.2%
f_t	f_c	−42.5%

(b)

Motion Coder	Inter-frame Coder	BD-Rate
DVC [26]	1-step CANF	−31.4%
DVC [26]	2-step CANF	−33.1%
1-step CANF	2-step CANF	−38.1%
2-step CANF	2-step CANF	−42.5%

(c)

Motion Coder	Inter-frame Coder	BD-Rate
1-step CANF	1-step CANF	35.3%
2-step CANF	2-step CANF	−42.5%
3-step CANF	3-step CANF	−31.4%

(d)

unconditionally, is quite significant. This result corroborates the superiority of our conditional motion coding to predictive motion coding (i.e. coding $f_t - f_c$). As expected, the quality of the condition has a crucial effect on compression performance. The trivial use of the previously decoded flow $\hat{f}_{t-1}$ does not show much gain as compared to unconditional coding. The fact substantiates the effectiveness of our extrapolation network.

The Number of Autoencoding Transforms: Table 3c explores the effect of the number of autoencoding transforms on compression performance. The 1-step models are obtained by skipping the autoencoding transform $\{g_{\pi_1}^{enc}, g_{\pi_1}^{dec}\}$ in Fig. 2b. To have the model size compatible with the 2-step models, the 1-step models have more channels in every autoencoding transform. We first experiment with the conditional inter-frame coding, with the motion coder from DVC [26]. In this case, the 2-step model improves the rate saving of the 1-step model by 1.7%. Given the 2-step inter-frame coder, it is further seen that the 2-step motion coder also improves the rate saving of the 1-step motion coder by 4.4%. This suggests that with a similar model size, the 2-step model is superior to the 1-step model in both inter-frame and motion coding.

Table 3d complements Table 3c to present results for 1-, 2- and 3-step CANF when applied to both the motion and inter-frame codecs. 3-step CANF extends straightforwardly the 2-step CANF by incorporating one additional autoencoding transform. Despite a larger capacity, the 3-step CANF performs worse than the 2-step CANF and comparably to the 1-step CANF. From Fig. 2b, the quantization error introduced to the latent code z_2 and the approximation error between x_c (used for decoding) and y_2 (generated during encoding) are propagated and accumulated (from top to bottom in Fig. 2b) during decoding. The cascading effect, compounded by temporal error propagation, may outweigh the benefits of having more autoencoding transforms.

5   Conclusion

This work introduces CANF-VC for conditional inter-frame and motion coding. CANF-VC achieves the state-of-the-art video compression performance. Our major findings include: (1) the CANF-based inter-frame coding outperforms residual coding; (2) likewise, our conditional motion coding outperforms predictive motion coding at the cost of additional buffer requirements; (3) the quality of the conditioning variable is critical to compression performance; (4) our 2-step CANF performs better than 1-step CANF, justifying the use of multi-step CANF. Lastly, we note that CANF-VC does not use auto-regressive models in inter-frame and motion coding. Its operations are parallelizable.

Acknowledgements. This work was supported by MediaTek, National Center for High-Performance Computing, Taiwan, Ministry of Science and Technology, Taiwan under Grand Application 110-2221-E-A49-065-MY3 and 110-2634-F-A49-006-, and Italian Ministry of University and Research under Grant Application PRIN 2022N25TSZ.

References

1. HM reference software for HEVC. https://vcgit.hhi.fraunhofer.de/jvet/HM/-/tree/HM-16.20. Accessed 03 Mar 2022
2. Agustsson, E., Minnen, D., Johnston, N., Balle, J., Hwang, S.J., Toderici, G.: Scale-space flow for end-to-end optimized video compression. In: Proceedings of the IEEE/CVF Conference on Computer Vision and Pattern Recognition, pp. 8503–8512 (2020)
3. Ballé, J., Laparra, V., Simoncelli, E.P.: End-to-end optimized image compression. In: International Conference for Learning Representations (2017)
4. Ballé, J., Minnen, D., Singh, S., Hwang, S.J., Johnston, N.: Variational image compression with a scale hyperprior. In: International Conference on Learning Representations (2018)
5. Bégaint, J., Racapé, F., Feltman, S., Pushparaja, A.: Compressai: a pytorch library and evaluation platform for end-to-end compression research. arXiv preprint arXiv:2011.03029 (2020)
6. Brand, F., Seiler, J., Kaup, A.: Generalized difference coder: a novel conditional autoencoder structure for video compression. arXiv:2112.08011 (2021)
7. Bross, B., et al.: Overview of the versatile video coding (VVC) standard and its applications. IEEE Trans. Circuits Syst. Video Technol. **31**(10), 3736–3764 (2021)
8. Chen, T., Liu, H., Ma, Z., Shen, Q., Cao, X., Wang, Y.: End-to-end learnt image compression via non-local attention optimization and improved context modeling. IEEE Trans. Image Process. **30**, 3179–3191 (2021)

9. Cheng, Z., Sun, H., Takeuchi, M., Katto, J.: Learned image compression with discretized gaussian mixture likelihoods and attention modules. In: Proceedings of the IEEE/CVF Conference on Computer Vision and Pattern Recognition, pp. 7939–7948 (2020)
10. Dinh, L., Sohl-Dickstein, J., Bengio, S.: Density estimation using real NVP. Computing Research Repository (CoRR) (2016)
11. Frank, B.: Common test conditions and software reference configurations. JCTVC-L1100 **12**(7) (2013)
12. Golinski, A., Pourreza, R., Yang, Y., Sautiere, G., Cohen, T.S.: Feedback recurrent autoencoder for video compression. In: Proceedings of the Asian Conference on Computer Vision (2020)
13. Ho, Y.H., Chan, C.C., Peng, W.H., Hang, H.M., Domański, M.: ANFIC: image compression using augmented normalizing flows. IEEE Open J. Circuits Syst. **2**, 613–626 (2021)
14. Hu, Z., Chen, Z., Xu, D., Lu, G., Ouyang, W., Gu, S.: Improving deep video compression by resolution-adaptive flow coding. In: Vedaldi, A., Bischof, H., Brox, T., Frahm, J.-M. (eds.) ECCV 2020. LNCS, vol. 12347, pp. 193–209. Springer, Cham (2020). https://doi.org/10.1007/978-3-030-58536-5_12
15. Hu, Z., Lu, G., Xu, D.: FVC: a new framework towards deep video compression in feature space. In: Proceedings of the IEEE/CVF Conference on Computer Vision and Pattern Recognition, pp. 1502–1511 (2021)
16. Huang, C.W., Dinh, L., Courville, A.: Augmented normalizing flows: bridging the gap between generative flows and latent variable models. arXiv preprint arXiv:2002.07101 (2020)
17. Kingma, D.P., Ba, J.: Adam: a method for stochastic optimization. In: International Conference for Learning Representations (2015)
18. Kingma, D.P., Dhariwal, P.: Glow: generative flow with invertible 1x1 convolutions. arXiv:1807.03039 (2018)
19. Kingma, D.P., Welling, M.: Auto-encoding variational bayes. arXiv preprint arXiv:1312.6114 (2013)
20. Kobyzev, I., Prince, S.J., Brubaker, M.A.: Normalizing flows: an introduction and review of current methods. IEEE Trans. Pattern Anal. Mach. Intell. **43**(11), 3964–3979 (2020)
21. Ladune, T., Philippe, P., Hamidouche, W., Zhang, L., Déforges, O.: Optical flow and mode selection for learning-based video coding. In: 2020 IEEE 22nd International Workshop on Multimedia Signal Processing (MMSP), pp. 1–6. IEEE (2020)
22. Ladune, T., Philippe, P., Hamidouche, W., Zhang, L., Déforges, O.: Conditional coding for flexible learned video compression. In: Neural Compression: From Information Theory to Applications-Workshop@ ICLR 2021 (2021)
23. Li, J., Li, B., Lu, Y.: Deep contextual video compression. In: Advances in Neural Information Processing Systems (2021)
24. Lin, J., Liu, D., Li, H., Wu, F.: M-LVC: multiple frames prediction for learned video compression. In: Proceedings of the IEEE/CVF Conference on Computer Vision and Pattern Recognition, pp. 3546–3554 (2020)
25. Liu, H., et al.: Neural video coding using multiscale motion compensation and spatiotemporal context model. IEEE Trans. Circuits Syst. Video Technol. **31**(8), 3182–3196 (2020)
26. Lu, G., Ouyang, W., Xu, D., Zhang, X., Cai, C., Gao, Z.: DVC: an end-to-end deep video compression framework. In: Proceedings of the IEEE/CVF Conference on Computer Vision and Pattern Recognition, pp. 11006–11015 (2019)

27. Lu, G., Zhang, X., Ouyang, W., Chen, L., Gao, Z., Xu, D.: An end-to-end learning framework for video compression. IEEE Trans. Pattern Anal. Mach. Intell. **43**(10), 3292–3308 (2020)

28. Ma, H., Liu, D., Yan, N., Li, H., Wu, F.: End-to-end optimized versatile image compression with wavelet-like transform. IEEE Trans. Pattern Anal. Mach. Intell. (2020)

29. Mercat, A., Viitanen, M., Vanne, J.: UVG dataset: 50/120fps 4K sequences for video codec analysis and development. In: Proceedings of the 11th ACM Multimedia Systems Conference, pp. 297–302 (2020)

30. Minnen, D., Ballé, J., Toderici, G.D.: Joint autoregressive and hierarchical priors for learned image compression. Adv. Neural. Inf. Process. Syst. **31**, 10771–10780 (2018)

31. Ranjan, A., Black, M.J.: Optical flow estimation using a spatial pyramid network. In: Proceedings of the IEEE Conference on Computer Vision and Pattern Recognition, pp. 4161–4170 (2017)

32. Rippel, O., Anderson, A.G., Tatwawadi, K., Nair, S., Lytle, C., Bourdev, L.: ELF-VC: efficient learned flexible-rate video coding. In: Proceedings of the IEEE/CVF International Conference on Computer Vision (ICCV), pp. 14479–14488, October 2021

33. Sullivan, G.J., Ohm, J.R., Han, W.J., Wiegand, T.: Overview of the high efficiency video coding (HEVC) standard. IEEE Trans. Circuits Syst. Video Technol. **22**(12), 1649–1668 (2012)

34. Sun, D., Yang, X., Liu, M.Y., Kautz, J.: PWC-net: CNNs for optical flow using pyramid, warping, and cost volume. In: Proceedings of the IEEE Conference on Computer Vision and Pattern Recognition, pp. 8934–8943 (2018)

35. Wang, H., et al.: MCL-JCV: a JND-based H.264/AVC video quality assessment dataset. In: 2016 IEEE International Conference on Image Processing (ICIP), pp. 1509–1513. IEEE (2016)

36. Xue, T., Chen, B., Wu, J., Wei, D., Freeman, W.T.: Video enhancement with task-oriented flow. Int. J. Comput. Vision **127**(8), 1106–1125 (2019)

37. Yang, R., Mentzer, F., Gool, L.V., Timofte, R.: Learning for video compression with hierarchical quality and recurrent enhancement. In: Proceedings of the IEEE/CVF Conference on Computer Vision and Pattern Recognition, pp. 6628–6637 (2020)

38. Yang, R., Mentzer, F., Van Gool, L., Timofte, R.: Learning for video compression with recurrent auto-encoder and recurrent probability model. IEEE J. Sel. Top. Signal Process. **15**(2), 388–401 (2020)

Bi-level Feature Alignment for Versatile Image Translation and Manipulation

Fangneng Zhan[1], Yingchen Yu[2], Rongliang Wu[2], Jiahui Zhang[2],
Kaiwen Cui[2], Aoran Xiao[2], Shijian Lu[2]([✉]), and Chunyan Miao[2]

[1] Max Planck Institute for Informatics, Saarbrücken, Germany
fzhan@mpi-inf.mpg.de
[2] Nanyang Technological University, Singapore, Singapore
{yingchen001,ronglian001,jiahui003,kaiwen001,aoran.xiao}@e.ntu.edu.sg,
{shijian.lu,ascymiao}@ntu.edu.sg

Abstract. Generative adversarial networks (GANs) have achieved great success in image translation and manipulation. However, high-fidelity image generation with faithful style control remains a grand challenge in computer vision. This paper presents a versatile image translation and manipulation framework that achieves accurate semantic and style guidance in image generation by explicitly building a correspondence. To handle the quadratic complexity incurred by building the dense correspondences, we introduce a bi-level feature alignment strategy that adopts a top-k operation to rank block-wise features followed by dense attention between block features which reduces memory cost substantially. As the top-k operation involves index swapping which precludes the gradient propagation, we approximate the non-differentiable top-k operation with a regularized earth mover's problem so that its gradient can be effectively back-propagated. In addition, we design a novel semantic position encoding mechanism that builds up coordinate for each individual semantic region to preserve texture structures while building correspondences. Further, we design a novel confidence feature injection module which mitigates mismatch problem by fusing features adaptively according to the reliability of built correspondences. Extensive experiments show that our method achieves superior performance qualitatively and quantitatively as compared with the state-of-the-art.

1 Introduction

Image translation and manipulation aim to generate and edit photo-realistic images conditioning on certain inputs such as semantic segmentation [32,42], key points [5,38] and layout [19]. It has been studied intensively in recent years thanks to its wide spectrum of applications in various tasks [30,35,41]. However, achieving high fidelity image translation and manipulation with faithful style control remains a grand challenge due to the high complexity of natural image styles. A typical approach to control image styles is to encode image features

---

F. Zhan and Y. Yu—Equal contribution.

S. Avidan et al. (Eds.): ECCV 2022, LNCS 13676, pp. 224–241, 2022.
https://doi.org/10.1007/978-3-031-19787-1_13

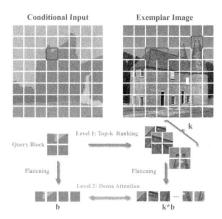

Fig. 1. Bi-level feature alignment via ranking and attention scheme: With a query block from the *Conditional Input*, we first retrieve the top-k most similar blocks from the *Exemplar Image* through a differentiable ranking operation, and then compute dense attention between features in query block and features in retrieved top-k blocks. Such bi-level alignment reduces the computational cost greatly, and allows to build high-resolution correspondences.

into a latent space with certain regularization (e.g., Gaussian distribution) on the latent feature distribution. For example, Park *et al.* [32] utilize VAE [4] to regularize the distribution of encoded features for faithful style control. However, VAE struggles to encode the complex distribution of natural image styles and often suffers from *posterior collapse* [25] which leads to degraded style control performance. Another strategy is to encode reference images into style codes [3, 64] to provide style guidance in image generation, while style codes often capture the global or regional style without an explicit style guidance for generating texture details.

To achieve more accurate style guidance and preserve details from exemplar, Zhang *et al.* [58] explore to build cross-domain correspondences with Cosine similarity to achieve exemplar-based image translation. Zhou *et al.* [62] propose a GRU-assisted Patch-Match [1] method to build high-resolution correspondences efficiently. Since the textures within a semantic region share identical semantic information, the existing methods tend to build correspondences based on the semantic coherence without considering the structure coherence within each semantic region. Warping exemplars with such pure semantic correspondence may cause destroyed texture patterns in the warped exemplars, and consequently result in inaccurate guidance for image generation.

This paper presents **RABIT**, a **R**anking and **A**ttention scheme with **Bi**-level feature alignment for versatile **I**mage **T**ranslation and manipulation. To mitigate the quadratic computational complexity issue of building the dense correspondence between conditional inputs (semantic guidance) and exemplars (style guidance), we design a bi-level alignment strategy with a Ranking and Attention Scheme (RAS) which builds feature correspondences efficiently at two levels: 1) a top-k ranking operation for dynamically generating block-wise ranking matrices; 2) a dense attention module that achieves dense correspondences between features within blocks as illustrated in Fig. 1. RAS enables to build high-resolution correspondences and reduces the memory cost from $\mathcal{O}(L^2)$ to $\mathcal{O}(N^2 + b^2)$ (L is the number of features for alignment, b is block size, and $N = \frac{L}{b}$). However, the top-k operation involves index swapping whose gradient cannot be back-propagated in networks. To address this issue, we approximate the top-k ranking operation with a regularized Earth Mover's problem [34] which enables gradient back-propagation effectively.

As in [58,62], building correspondences based on semantic information only often leads to the losing of texture structures and patterns in warped exemplars. Thus, spatial information should also be incorporated to preserve the texture structures and patterns and yield more accurate feature correspondences. A vanilla method to encode the position information is concatenating the semantic features with the corresponding feature coordinates via coordconv [22]. However, the vanilla position encoding builds a single coordinate system for the whole image which ignores the position information within each semantic region. Instead, we design a semantic position encoding (SPE) mechanism that builds a dedicated coordinate system for each semantic region which outperforms the vanilla position encoding significantly.

In addition, conditional inputs and exemplars are seldom perfectly matched, e.g., conditional inputs could contain several semantic classes that do not exist in exemplar images. Under such circumstances, the built correspondences often contain errors which lead to inaccurate exemplar warping and further deteriorated image generation. We tackle this problem by designing a CONfidence Feature Injection (CONFI) module that fuses the features of conditional inputs and warped exemplars according to the reliability of the built correspondences. Although the warped exemplar may not be reliable, the conditional input always provides accurate semantic guidance in image generation. The CONFI module thus assigns higher weights to the conditional input when the built correspondence (or warped exemplar) is unreliable. Experiments show that CONFI helps to generate faithful yet high-fidelity images consistently.

The contributions of this work can be summarized in three aspects. First, we propose a versatile image translation and manipulation framework which introduces a ranking and attention Scheme for bi-level feature alignment that greatly reduces the memory cost while building the correspondence between conditional inputs and exemplars. Second, we introduce a semantic position encoding mechanism that encodes region-level position information to preserve texture structures and patterns. Third, we design a confidence feature injection module that provides reliable feature guidance in image translation and manipulation.

2 Related Work

2.1 Image-to-Image Translation

Image translation has achieved remarkable progress in learning the mapping between images of different domains. It could be applied in different tasks such as style transfer [7,10,20], image super-resolution [15,16,21,57], domain adaptation [8,30,35,40,48], image composition [47,50,53,54,56] etc. To achieve high-fidelity and flexible translation, existing work uses different conditional inputs such as semantic segmentation [12,32,42,52,55], scene layouts [19,37,51,59], key points [5,27,29,49], edge maps [6,12], etc. However, effective style control remains a challenging task in image translation.

Style control has attracted increasing attention in image translation and generation. Earlier works such as [14] regularize the latent feature distribution to control the generation outcome. However, they struggle to capture the complex textures of natural images. Style encoding has been studied to address this issue. For example, [11] and [26] transfer style codes from exemplars to source images via adaptive instance normalization (AdaIN) [10]. [3] employs a style encoder for style consistency between exemplars and translated images. [64] designs semantic region-adaptive normalization (SEAN) to control the style of each semantic region individually. However, encoding style exemplars tends to capture the overall image style and ignores the texture details in local regions. To achieve accurate style guidance for each local region, Zhang et al. [58] build dense semantic correspondences between conditional inputs and exemplars with Cosine similarity to capture accurate exemplar details. To mitigate the quadratic complexity issue and enable high-resolution correspondence building, Zhou et al. [62] introduce the GRU-assisted Patch-Match to efficiently establish the high-resolution correspondence.

2.2 Semantic Image Editing

The arise of generative adversarial network (GANs) brings revolutionary advances to image editing [2,9,31,33,43–45,63]. As one of the most intuitive representation in image editing, semantic information has been extensively investigated in conditional image synthesis. For example, Park et al. [32] introduce spatially-adaptive normalization (SPADE) to inject guided features in image generation. MaskGAN [17] exploits a dual-editing consistency as auxiliary supervision for robust face image manipulation. Instead of directly learning a label-to-pixel mapping, Hong et al. [9] propose a semantic manipulation framework HIM that generates images guided by a predicted semantic layout. Upon this work, Ntavelis et al. [31] propose SESAME which requires only local semantic maps to achieve image manipulation. However, the aforementioned methods either only learn a global feature without local focus (e.g., MaskGAN [17]) or ignore the features in the editing regions of the original image (e.g., HIM [9], SESAME [31]). To better utilize the fine features in the original image, Zheng et al. [60] adapt exemplar-based image synthesis framework CoCosNet [58] for semantic image manipulation by building a high-resolution correspondence between the original image and the edited semantic map.

3 Proposed Method

The proposed RABIT consists of an alignment network and a generation network that are inter-connected as shown in Fig. 2. The alignment network learns

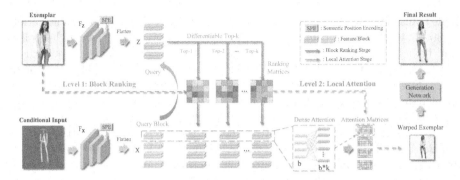

Fig. 2. The framework of the proposed RABIT: *Conditional Input* and *Exemplar* are fed to feature extractors F_X and F_Z to extract feature vectors X and Z where b local features form a feature block. In the first level, each block from the conditional input serves as the query to retrieve top-k similar blocks from the exemplar through a differentiable ranking operation. In the second level, *Dense Attention* is then built between the b features in query block and $b * k$ features in the retrieved blocks. The built *Ranking Matrices* and *Attention Matrices* are combined to warp the exemplar to be aligned with the conditional input as in *Warped Exemplar*, which serves as a style guidance to generate the final result through a generation network.

the correspondence between a conditional input and an exemplar for warping the exemplar to be aligned with the conditional input. The generation network produces the final generation under the guidance of the warped exemplar and the conditional input. RABIT is typically applicable in the task of conditional image translation with extra exemplar as style guidance. It is also applicable to the task of image manipulation by treating the exemplars as the original images for editing and the conditional inputs as the edited semantic. The detailed loss functions can be found in the supplementary materials.

3.1   Alignment Network

The alignment network aims to build the correspondence between conditional inputs and exemplars, and accordingly provide accurate style guidance by warping the exemplars to be aligned with the conditional inputs. As shown in Fig. 2, conditional input and exemplar are fed to feature extractors F_X and F_Z to extract two sets of feature vectors $X = [x_1, \cdots, x_L] \in \mathbb{R}^d$ and $Z = [z_1, \cdots, z_L] \in \mathbb{R}^d$, where L and d denote the number and dimension of feature vectors, respectively. Then X and Z can be aligned by building a $L \times L$ dense correspondence matrix where each entry denotes the Cosine similarity between the corresponding feature vectors in X and Z.

Semantic Position Encoding. Existing works [58,62] mainly rely on semantic features to establish the correspondences. However, as textures within a semantic region share the same semantic feature, the pure semantic correspondence fails to preserve the texture structures or patterns within each semantic region. Thus, the position information of features can be facilitated to preserve the texture structures and patterns. A vanilla method to encode the position information is employing a simple coordconv [22] to build a global coordinate for the full image. However, this vanilla position encoding mechanism builds a single coordinate system for the whole image, ignoring region-wise semantic differences. To preserve the fine texture pattern within each semantic region, we design a semantic position encoding (SPE) mechanism that builds a dedicated coordinate for each semantic region as shown in Fig. 3. Specifically, SPE treats the center of each semantic region as the origin of coordinate, and the coordinates within each semantic region are normalized to $[-1, 1]$. The proposed SPE outperforms the vanilla position encoding significantly as shown in Fig. 6 and to be evaluated in experiments.

Bi-level Feature Alignment. On the other hand, building correspondence has quadratic complexity which incurs large memory and computation costs. Most existing studies thus work with low-resolution exemplar images (e.g. 64×64 in CoCosNet [58]) which often struggle in generating realistic images with fine texture details. In this work, we propose a bi-level alignment strategy via a novel ranking and attention scheme (RAS) that greatly reduces computational costs and allows to build correspondences with high-resolution images as shown in Fig. 6. Instead of building correspondences between features directly, the bi-level alignment strategy builds correspondences at two levels, including the first level that introduces top-k ranking to generate block-wise ranking matrices dynamically and the second level that achieves dense attention between the features within blocks. As Fig. 2 shows, b local features are grouped into a block, thus the features of conditional input and exemplar are partitioned into N blocks ($N = L/b$) as denoted by $X = [X_1, \cdots, X_N] \in \mathbb{R}^{bd}$ and $Z = [Z_1, \cdots, Z_N] \in \mathbb{R}^{bd}$. In the first level of top-k ranking, each block feature of the conditional input serves as a query to retrieve top-k block features from the exemplar according to the Cosine similarity between blocks. In the second level of local attention, the features in each query block further attends to the features in the top-k retrieved blocks to build up local attention matrices within the block features. The correspondence between the exemplar and conditional input can thus be built much more efficiently by combining such inter-block ranking and inner-block attention.

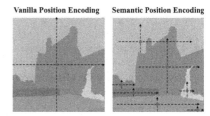

Vanilla Position Encoding   Semantic Position Encoding

Fig. 3. The comparison of vanilla position encoding and the proposed semantic position encoding (SPE). Red dots denote the coordinate origin. (Color figure online)

The ranking and attention scheme employs a top-k operation that ranks the correlative blocks. However, the original top-k operation involves index swapping whose gradient cannot be computed and so cannot be integrated into end-to-end network training. Inspired by Xie *et al.* [46], we tackle this issue by formulating the top-k ranking as a regularized earth mover's problem which allows

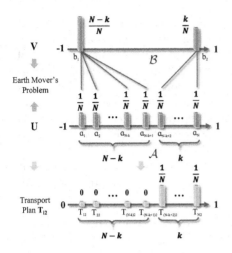

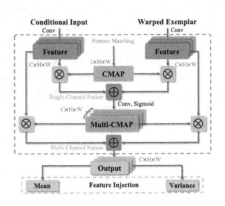

Fig. 4. Illustration of the earth mover's problem in top-k retrieval. Earth mover's problem is conducted between distributions U and V which is defined on supports $\mathcal{A} = [a_1, \cdots, a_N]$ and $\mathcal{B} = [b_1, b_2]$. Transport Plan T_{i2} indicates the retrieved top-k elements.

Fig. 5. Illustration of confidence feature injection: Conditional input and warped exemplar are initially fused with a confidence map (CMAP) of size $1 \times H \times W$. A multi-channel confidence map (Multi-CMAP) of size $C \times H \times W$ is then obtained from the initial fusion which further fuses the conditional input and warped exemplar in multiple channels.

gradient computation via implicit differentiation. Earth mover's problem aims to find a transport plan that minimizes the total cost to transform one distribution to another. Consider two discrete distributions $U = [\mu_1, \ldots, \mu_N]^\top$ and $V = [\nu_1, \ldots, \nu_M]^\top$ defined on supports $\mathcal{A} = [a_1, \cdots, a_N]$ and $\mathcal{B} = [b_1, \cdots, b_M]$, with probability (or amount of earth) $\mathbb{P}(a_i) = \mu_i$ and $\mathbb{P}(b_j) = \nu_j$. We define $C \in \mathbb{R}^{N \times M}$ as the cost matrix where C_{ij} denotes the cost of transportation between a_i and b_j, and T as a transport plan where T_{ij} denotes the amount of earth transported between μ_i and ν_j. An earth mover's (EM) problem can be formulated by: $EM = \min_T \langle C, T \rangle$, s.t. $T\mathbf{1}_M = U$, $T^\top \mathbf{1}_N = V$ where $\mathbf{1}$ denotes a vector of ones, $\langle \rangle$ denotes inner product. By treating a correlation scores between a query block and N key blocks as $\mathcal{A} = [a_1, \cdots, a_N], a_i \in [-1, 1]$ and defining $\mathcal{B} = \{-1, 1\}$, $U = [\mu_1, \cdots, \mu_N]$ and $V = [\nu_1, \nu_2]$, it can be proved that solving the Earth Mover's problem is equivalent to select the largest K elements from $\mathcal{A} = [a_1, \cdots, a_N]$. The detailed proof and optimization of the earth mover's problem is provided in supplementary material. Figure 4 illustrates the earth mover's problem and transport plan T which indicates the top-k elements.

Complexity Analysis. The vanilla dense correspondence has a self-attention memory complexity $\mathcal{O}(L^2)$ where L is the input sequence length. For our bi-level alignment strategy, the memory complexity of building block ranking matrices and local attention matrices are $\mathcal{O}(N^2)$ and $\mathcal{O}(b * (kb))$, where b, N ($N = L/b$)

| Conditional Input | Exemplar | CoCosNet (64) | CoCosNet v2 (64) | CoCosNet v2 (128) | Baseline (64) | SPE (64) | RAS (128) | SPE+RAS (128) |

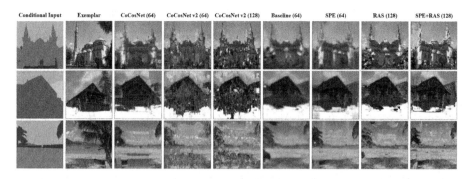

Fig. 6. Warped exemplars with different methods: '64' and '128' mean to build correspondences at resolutions 64×64 and 128×128. CoCosNet [58] tends to lose texture details and structures, while CoCosNet v2 [62] tends to generate messy warping. The *Baseline* denotes building correspondences with Cosine similarity, which tends to lose textures details and structures. The proposed ranking and attention scheme (RAS) allows efficient image warping at high resolutions, the proposed semantic position encoding (SPE) can better preserve texture structures. The combination of the two as denoted by SPE+RAS achieves the best warping performance with high resolution and preserved texture structures.

and k are block size, block number and the number of top-k selection. Thus, the overall memory complexity is $\mathcal{O}(N^2 + b * (kb))$.

3.2 Generation Network

The generation network aims to synthesize images under the semantic guidance of conditional inputs and style guidance of exemplars. The overall architecture of the generation network is similar to SPADE [32]. Please refer to supplementary material for details of the network structure.

State-of-the-art approach [58] simply concatenates the warped exemplar and conditional input to guide the image generation process. However, the warped input image and edited semantic map could be structurally aligned but semantically different especially when they have severe semantic discrepancy. Such unreliably warped exemplars could serve as false guidance and heavily deteriorate the generation performance. Therefore, a mechanism is required to identify the semantic reliability of warped exemplar to provide reliable guidance for the generation network. To this end, we propose a CONfidence Feature Injection (CONFI) module that adaptively weights the features of conditional input and warped exemplar according to the reliability of feature matching.

Confidence Feature Injection. Intuitively, in the case of lower reliability of the feature correspondence, we should assign a relatively lower weight to the warped exemplar which provides unreliable style guidance and a higher weight to the conditional input which consistently provides accurate semantic guidance.

As illustrated in Fig. 5, the proposed CONFI fuses the features of the conditional input and warped exemplar based on a confidence map (CMAP) that captures the reliability of the feature correspondence. To derive the confidence map, we first obtain a block-wise correlation map of size $N \times N$ by computing element-wise Cosine distance between $X = [X_i, \cdots, X_N]$ and $Z = [Z_i, \cdots, Z_N]$. For a block X_i, the correlation score with Z is denoted by $\mathcal{A} = [a_1, \cdots, a_N]$. As higher correlation scores indicate more reliable feature matching, we treat the peak value of $\mathcal{A}$ as the confidence score of X_i. Similar for other blocks, we can obtain the confidence map (CMAP) of size $1 \times H \times W$ ($N = H * W$) which captures the semantic reliability of all blocks. The features of the conditional input and exemplar (both of size $C \times H \times W$ after passing through convolution layers) can thus be fused via weighted sum based on the confidence map CMAP: $F = X*(1-\text{CMAP})+(T \cdot Z)*\text{CMAP}$ where T is the built correspondence matrix. As the confidence map contains only one channel ($1 \times H \times W$), the above feature fusion is conducted in $H \times W$ but ignores that in C channel. To achieve thorough feature fusion in all channels, we feed the initial fusion F to convolution layers to generate a multi-channel confidence map (Multi-CMAP) of size $C \times H \times W$. The conditional input and warped exemplar are then thoroughly fused via a full channel-weighted summation according to the Multi-CMAP. The final fused feature is further injected to the generation process via spatial de-normalization [32] to provide accurate semantic and style guidance.

4   Loss Functions

The alignment network and generation network are jointly optimized. For clarity, we still denote the conditional input and exemplar as X and Z, the ground truth as X', the generated image as Y, the feature extractors for conditional input and exemplar as E_X and E_Z, the generator and discriminator in the generation network as G and D.

Alignment Network. First, the warping should be cycle consistent, i.e. the exemplar should be recoverable from the warped warped. We thus employ a cycle-consistency loss as follows:

$$\mathcal{L}_{cyc} = ||T^\top \cdot T \cdot Z - Z||_1$$

where T is the correspondence matrix. The feature extractors F_X and F_Z aim to extract invariant semantic information across domains, i.e. the extracted features from X and X' should be consistent. A feature consistency loss can thus be formulated as follows:

$$\mathcal{L}_{cst} = ||F_X(X) - F_Z(X')||_1$$

Generation Network. The generation network employs several losses for high-fidelity synthesis with consistent style with the exemplar and consistent semantic with the conditional input. As the generated image Y should be semantically consistent with the ground truth X', we employ a perceptual loss $\mathcal{L}_{perc}$ [13] to penalize their semantic discrepancy as below:

$$\mathcal{L}_{perc} = ||\phi_l(Y) - \phi(X')||_1 \tag{1}$$

where ϕ_l is the activation of layer l in pre-trained VGG-19 [36] model. To ensure the statistical consistency between the generated image Y and the exemplar Z, a contextual loss [28] is adopted:

$$\mathcal{L}_{cxt} = -\log(\sum_i \max_j CX_{ij}(\phi_l^i(Z), \phi_l^j(Y))) \tag{2}$$

where i and j are the indexes of the feature map in layer ϕ_l. Besides, a pseudo pairs loss $\mathcal{L}_{pse}$ as described in [58] is included in training.

The discriminator D is employed to drive adversarial generation with an adversarial loss $\mathcal{L}_{adv}$ [12]. The full network is thus optimized with the following objective:

$$\mathcal{L} = \min_{F_X, F_Z, G} \max_D (\lambda_1 \mathcal{L}_{cyc} + \lambda_2 \mathcal{L}_{cst} + \lambda_3 \mathcal{L}_{perc} \\ + \lambda_4 \mathcal{L}_{cxt} + \lambda_5 \mathcal{L}_{pse} + \lambda_6 \mathcal{L}_{adv}) \tag{3}$$

where the weights λ balance the losses in the objective.

5 Experiments

5.1 Experimental Settings

Datasets. We evaluate and benchmark our method over multiple datasets for image translation & manipulation tasks.

- ADE20K [61] is adopted for image translation conditioned on semantic segmentation. For image manipulation, we apply object-level affine transformations on the test set to acquire paired data (150 images) for evaluations as in [60].
- CelebA-HQ [24] is used for two translation tasks by using face semantics and face edges as conditional inputs. We use 2993 face images for translation evaluations as in [58], and manually edit 100 semantic maps which is randomly selected for image manipulation evaluations.
- DeepFashion [23] is used for image translation conditioned key points.

Implementation Details: The default size for our correspondence computation is 128×128 with a block size of 2×2. The number k in top-k ranking is set at 3 by default in our experiments. The default size of generated images is 256×256.

5.2 Image Translation Experiments

We compare RABIT with several state-of-the-art image translation methods.

Quantitative Results. In quantitative experiments, all methods translate images with the same exemplars except Pix2PixHD [42] which doesn't support style injection from exemplars. LPIPS is calculated by comparing the generated images with randomly selected exemplars. All compared methods adopt

Table 1. Comparing RABIT with state-of-the-art image translation methods over four translation tasks with FID, SWD and LPIPS as the evaluation metrics.

Methods	ADE20K			CelebA-HQ (Semantic)			DeepFashion			CelebA-HQ (Edge)		
	FID ↓	SWD ↓	LPIPS ↑	FID ↓	SWD ↓	LPIPS ↑	FID ↓	SWD ↓	LPIPS ↑	FID ↓	SWD ↓	LPIPS ↑
Pix2pixHD [42]	81.80	35.70	N/A	43.69	34.82	N/A	25.20	16.40	N/A	42.70	33.30	N/A
StarGAN v2 [3]	98.72	65.47	0.551	53.20	41.87	**0.324**	43.29	30.87	**0.296**	48.63	41.96	**0.214**
SPADE [32]	33.90	19.70	0.344	39.17	29.78	0.254	36.20	27.80	0.231	31.50	26.90	0.207
SelectionGAN [39]	35.10	21.82	0.382	42.41	30.32	0.277	38.31	28.21	0.223	34.67	27.34	0.191
SMIS [65]	42.17	22.67	0.476	28.21	24.65	0.301	22.23	23.73	0.240	23.71	22.23	0.201
SEAN [64]	24.84	10.42	0.499	**17.66**	14.13	0.285	16.28	17.52	0.251	16.84	14.94	0.203
CoCosNet [58]	26.40	10.50	**0.580**	21.83	12.13	0.292	14.40	17.20	0.272	14.30	15.30	0.208
RABIT	**24.35**	**9.893**	**0.571**	20.44	11.18	0.307	**12.58**	**16.03**	0.284	**11.67**	14.22	0.209

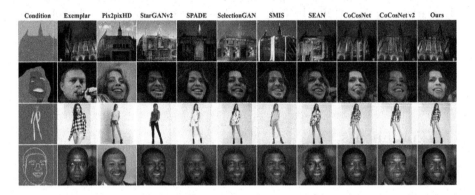

Fig. 7. Qualitative comparison of the proposed RABIT and state-of-the-art methods over four types of conditional image translation tasks.

three exemplars for each conditional input and the final LPIPS is obtained by averaging the LPIPS between any two generated images.

Table 1 shows experimental results. It can be seen that RABIT outperforms all compared methods over most metrics and tasks consistently. By building explicit yet accurate correspondences between conditional inputs and exemplars, RABIT enables direct and accurate guidance from the exemplar and achieves better translation quality (in FID and SWD) and diversity (in LPIPS) as compared with the regularization-based methods such as SPADE [32] and SMIS [65], and style-encoding methods such as StarGAN v2 [3] and SEAN [64]. Compared with correspondence-based method CoCosNet [58], the proposed bi-level alignment allows RABIT to build correspondences and warp exemplars at higher resolutions (e.g. 128×128) which offers more detailed guidance in the generation process and helps to achieve better FID and SWD. While compared with CoCosNet v2 [62], the proposed semantic position encoding enables to preserve the texture structures and patterns, thus yielding more accurate warped exemplars as guidance. Besides generation quality, RABIT achieves the best generation diversity in LPIPS except StarGAN v2 [3] which sacrifices the generation quality with much lower FID and SWD.

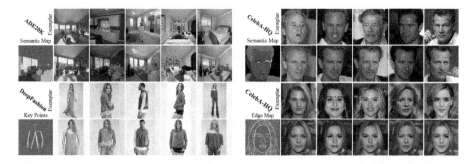

Fig. 8. Illustration of generation diversity of RABIT: With the same conditional input, RABIT can generate a variety of images that have consistent styles with the provided exemplars. It works for different types of conditional inputs consistently.

Table 2. Comparing RABIT with state-of-the art image manipulation methods on ADE20K [61] and CelebA-HQ [24].

ADE20K [61]				CelebA-HQ [24]			
Models	FID ↓	PSNR ↑	SSIM ↑	Models	FID ↓	SWD ↓	LPIPS ↓
SPADE [32]	120.2	13.11	0.334	**SPADE** [32]	105.1	41.90	0.376
HIM [9]	59.89	18.23	0.667	**SEAN** [64]	96.31	35.90	0.351
SESAME [31]	52.51	18.67	0.691	**MaskGAN** [17]	80.89	23.86	0.271
CoCosNet [58]	41.03	20.30	0.744	**CoCosNet** [58]	68.70	22.90	0.224
RABIT	**26.61**	**23.08**	**0.823**	**RABIT**	**60.87**	**21.07**	**0.176**

Qualitative Evaluations. Figure 7 shows qualitative comparisons on various conditional image translation tasks. It can be seen that RABIT achieves the best visual quality with faithful styles as exemplars. RABIT also demonstrates superior diversity in image translation as illustrated in Fig. 8.

5.3 Image Manipulation Experiment

RABIT manipulates images by treating input images as exemplars and edited semantic guidance as conditional inputs. We compare RABIT with several state-of-the-art image manipulation methods including 1) SPADE [32], 2) SEAN [64], 3) MaskGAN [18], 4) Hierarchical Image Manipulation (HIM) [9], 5) SESAME [31], 6) CoCosNet [58].

Quantitative Results: In quantitative experiments, all compared methods manipulate images with the same input image and edited semantic label map. Left side of Table 2 shows experimental results over the synthesized test set of ADE20K [61]. It can be observed that RABIT outperforms state-of-the-art methods over all evaluation metrics consistently. Right side of Table 2 shows experimental results over the CelebA-HQ dataset with manually edited semantic maps. It can be observed that RABIT outperforms the state-of-the-art methods by large margins in all perceptual quality metrics.

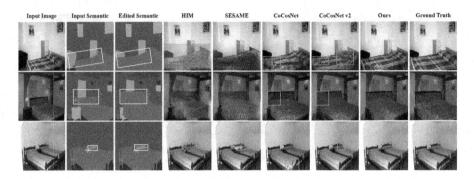

Fig. 9. Qualitative illustration of RABIT and SOTA image manipulation methods on the augmented test set of ADE20K with ground truth as described in [60].

Fig. 10. Various image editing by the proposed RABIT: With input images as the exemplars and edited semantic maps as the conditional input, RABIT generates new images with faithful semantics and high-fidelity textures with little artifacts.

Qualitative Evaluation: Figure 9 shows visual comparisons with state-of-art manipulation methods on ADE20K. Figure 10 shows the editing capacity of RABIT with various types of manipulation on semantic labels. We also compare RABIT with MaskGAN [17] on CelebA-HQ [18] in Fig. 12.

6   User Study

We conduct crowdsourcing user studies through Amazon Mechanical Turk (AMT) to evaluate the image translation & manipulation in terms of generation quality and style consistency. Specifically, each compared method generates 100 images with the same conditional inputs and exemplars. Then the generated images together with the conditional inputs and exemplars were presented to 10 users for assessment. For the evaluation of image quality, the users were instructed to pick the best-quality images. For the evaluation of style consistency, the users were instructed to select the images with best style relevance to the exemplar. The final AMT score is the averaged number of the methods to be selected as the best quality and the best style relevance.

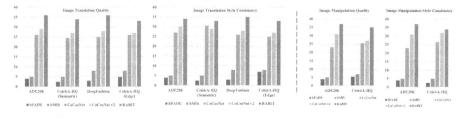

Fig. 11. AMT (Amazon Mechanical Turk) user studies of different image translation and image manipulation methods in terms of the visual quality and style consistency of the generated images.

Figure 11 shows AMT results on multiple datasets. It can be observed that RABIT outperforms state-of-the-art methods consistently in image quality and style consistency on both image translation & image manipulation tasks.

7 Conclusions

This paper presents RABIT, a versatile conditional image translation & manipulation framework that adopts a novel bi-level alignment strategy with a ranking and attention scheme (RAS) to align the features between conditional inputs and exemplars efficiently. A semantic position encoding mechanism is designed to facilitate semantic-level position information and preserve the texture patterns in the exemplars. To handle the semantic mismatching between the conditional inputs and warped exemplars, a novel confidence feature injection module is proposed to achieve multichannel feature fusion based on the matching reliability of warped exemplars. Quantitative and qualitative experiments over multiple datasets show that RABIT is capable of achieving high-fidelity image translation and manipulation while preserving consistent semantics with the conditional input and faithful styles with the exemplar.

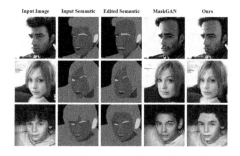

Fig. 12. The comparison of image manipulation by MaskGAN [17] and the proposed RABIT over dataset CelebA-HQ [24].

Acknowledgement. This study is supported under the RIE2020 Industry Alignment Fund - Industry Collaboration Projects (IAF-ICP) Funding Initiative, as well as cash and in-kind contribution from the industry partner(s).

References

1. Barnes, C., Shechtman, E., Finkelstein, A., Goldman, D.B.: PatchMatch: a randomized correspondence algorithm for structural image editing. ACM Trans. Graph. **28**(3), 24 (2009)
2. Choi, Y., Choi, M., Kim, M., Ha, J.W., Kim, S., Choo, J.: StarGAN: unified generative adversarial networks for multi-domain image-to-image translation. In: Proceedings of the IEEE Conference on Computer Vision and Pattern Recognition, pp. 8789–8797 (2018)
3. Choi, Y., Uh, Y., Yoo, J., Ha, J.W.: StarGAN V2: diverse image synthesis for multiple domains. In: Proceedings of the IEEE/CVF Conference on Computer Vision and Pattern Recognition, pp. 8188–8197 (2020)
4. Doersch, C.: Tutorial on variational autoencoders. arXiv preprint arXiv:1606.05908 (2016)
5. Dong, H., Liang, X., Gong, K., Lai, H., Zhu, J., Yin, J.: Soft-gated warping-GAN for pose-guided person image synthesis. arXiv preprint arXiv:1810.11610 (2018)
6. Fu, Y., Ma, J., Ma, L., Guo, X.: EDIT: exemplar-domain aware image-to-image translation. arXiv preprint arXiv:1911.10520 (2019)
7. Gatys, L.A., Ecker, A.S., Bethge, M.: Image style transfer using convolutional neural networks. In: Proceedings of the IEEE Conference on Computer Vision and Pattern Recognition, pp. 2414–2423 (2016)
8. Hoffman, J., et al.: CYCADA: cycle-consistent adversarial domain adaptation. In: International Conference on Machine Learning, pp. 1989–1998. PMLR (2018)
9. Hong, S., Yan, X., Huang, T., Lee, H.: Learning hierarchical semantic image manipulation through structured representations. In: NIPS (2018)
10. Huang, X., Belongie, S.: Arbitrary style transfer in real-time with adaptive instance normalization. In: Proceedings of the IEEE International Conference on Computer Vision, pp. 1501–1510 (2017)
11. Huang, X., Liu, M.-Y., Belongie, S., Kautz, J.: Multimodal unsupervised image-to-image translation. In: Ferrari, V., Hebert, M., Sminchisescu, C., Weiss, Y. (eds.) ECCV 2018. LNCS, vol. 11207, pp. 179–196. Springer, Cham (2018). https://doi.org/10.1007/978-3-030-01219-9_11
12. Isola, P., Zhu, J.Y., Zhou, T., Efros, A.A.: Image-to-image translation with conditional adversarial networks. In: Proceedings of the IEEE Conference on Computer Vision and Pattern Recognition, pp. 1125–1134 (2017)
13. Johnson, J., Alahi, A., Fei-Fei, L.: Perceptual losses for real-time style transfer and super-resolution. In: Leibe, B., Matas, J., Sebe, N., Welling, M. (eds.) ECCV 2016. LNCS, vol. 9906, pp. 694–711. Springer, Cham (2016). https://doi.org/10.1007/978-3-319-46475-6_43
14. Kingma, D.P., Welling, M.: Auto-encoding variational Bayes. arXiv preprint arXiv:1312.6114 (2013)
15. Lai, W.S., Huang, J.B., Ahuja, N., Yang, M.H.: Deep Laplacian pyramid networks for fast and accurate super-resolution. In: Proceedings of the IEEE Conference on Computer Vision and Pattern Recognition, pp. 624–632 (2017)
16. Ledig, C., et al.: Photo-realistic single image super-resolution using a generative adversarial network. In: Proceedings of the IEEE Conference on Computer Vision and Pattern Recognition, pp. 4681–4690 (2017)
17. Lee, C.H., Liu, Z., Wu, L., Luo, P.: MaskGAN: towards diverse and interactive facial image manipulation. In: CVPR, pp. 5549–5558 (2020)

18. Lee, C.H., Liu, Z., Wu, L., Luo, P.: MaskGAN: towards diverse and interactive facial image manipulation. In: IEEE Conference on Computer Vision and Pattern Recognition (CVPR) (2020)
19. Li, Y., Cheng, Y., Gan, Z., Yu, L., Wang, L., Liu, J.: BachGAN: high-resolution image synthesis from salient object layout. In: Proceedings of the IEEE Conference on Computer Vision and Pattern Recognition (2020)
20. Li, Y., Fang, C., Yang, J., Wang, Z., Lu, X., Yang, M.H.: Universal style transfer via feature transforms. arXiv preprint arXiv:1705.08086 (2017)
21. Lim, B., Son, S., Kim, H., Nah, S., Mu Lee, K.: Enhanced deep residual networks for single image super-resolution. In: Proceedings of the IEEE Conference on Computer Vision and Pattern Recognition Workshops, pp. 136–144 (2017)
22. Liu, R., et al.: An intriguing failing of convolutional neural networks and the Coord-Conv solution. In: Advances in Neural Information Processing Systems (2018)
23. Liu, Z., Luo, P., Qiu, S., Wang, X., Tang, X.: DeepFashion: powering robust clothes recognition and retrieval with rich annotations. In: Proceedings of the IEEE Conference on Computer Vision and Pattern Recognition, pp. 1096–1104 (2016)
24. Liu, Z., Luo, P., Wang, X., Tang, X.: Deep learning face attributes in the wild. In: Proceedings of the IEEE International Conference on Computer Vision, pp. 3730–3738 (2015)
25. Lucas, J., Tucker, G., Grosse, R., Norouzi, M.: Don't blame the ELBO! A linear VAE perspective on posterior collapse. arXiv preprint arXiv:1911.02469 (2019)
26. Ma, L., Jia, X., Georgoulis, S., Tuytelaars, T., Van Gool, L.: Exemplar guided unsupervised image-to-image translation with semantic consistency. In: International Conference on Learning Representations (2018)
27. Ma, L., Jia, X., Sun, Q., Schiele, B., Tuytelaars, T., Van Gool, L.: Pose guided person image generation. In: Advances in Neural Information Processing Systems, pp. 406–416 (2017)
28. Mechrez, R., Talmi, I., Zelnik-Manor, L.: The contextual loss for image transformation with non-aligned data. In: Ferrari, V., Hebert, M., Sminchisescu, C., Weiss, Y. (eds.) Computer Vision – ECCV 2018. LNCS, vol. 11218, pp. 800–815. Springer, Cham (2018). https://doi.org/10.1007/978-3-030-01264-9_47
29. Men, Y., Mao, Y., Jiang, Y., Ma, W.Y., Lian, Z.: Controllable person image synthesis with attribute-decomposed GAN. In: Proceedings of the IEEE/CVF Conference on Computer Vision and Pattern Recognition, pp. 5084–5093 (2020)
30. Murez, Z., Kolouri, S., Kriegman, D., Ramamoorthi, R., Kim, K.: Image to image translation for domain adaptation. In: Proceedings of the IEEE Conference on Computer Vision and Pattern Recognition, pp. 4500–4509 (2018)
31. Ntavelis, E., Romero, A., Kastanis, I., Van Gool, L., Timofte, R.: SESAME: semantic editing of scenes by adding, manipulating or erasing objects. In: Vedaldi, A., Bischof, H., Brox, T., Frahm, J.-M. (eds.) ECCV 2020. LNCS, vol. 12367, pp. 394–411. Springer, Cham (2020). https://doi.org/10.1007/978-3-030-58542-6_24
32. Park, T., Liu, M.Y., Wang, T.C., Zhu, J.Y.: Semantic image synthesis with spatially-adaptive normalization. In: Proceedings of the IEEE Conference on Computer Vision and Pattern Recognition, pp. 2337–2346 (2019)
33. Pumarola, A., Agudo, A., Martinez, A.M., Sanfeliu, A., Moreno-Noguer, F.: GAN-imation: anatomically-aware facial animation from a single image. In: Ferrari, V., Hebert, M., Sminchisescu, C., Weiss, Y. (eds.) ECCV 2018. LNCS, vol. 11214, pp. 835–851. Springer, Cham (2018). https://doi.org/10.1007/978-3-030-01249-6_50
34. Rubner, Y., Tomasi, C., Guibas, L.J.: The earth mover's distance as a metric for image retrieval. IJCV **40**, 99–121 (2000)

35. Shrivastava, A., Pfister, T., Tuzel, O., Susskind, J., Wang, W., Webb, R.: Learning from simulated and unsupervised images through adversarial training. In: Proceedings of the IEEE Conference on Computer Vision and Pattern Recognition, pp. 2107–2116 (2017)
36. Simonyan, K., Zisserman, A.: Very deep convolutional networks for large-scale image recognition. arXiv preprint arXiv:1409.1556 (2014)
37. Sun, W., Wu, T.: Image synthesis from reconfigurable layout and style. In: Proceedings of the IEEE International Conference on Computer Vision, pp. 10531–10540 (2019)
38. Tang, H., Xu, D., Liu, G., Wang, W., Sebe, N., Yan, Y.: Cycle in cycle generative adversarial networks for keypoint-guided image generation. In: Proceedings of the 27th ACM International Conference on Multimedia, pp. 2052–2060 (2019)
39. Tang, H., Xu, D., Sebe, N., Wang, Y., Corso, J.J., Yan, Y.: Multi-channel attention selection GAN with cascaded semantic guidance for cross-view image translation. In: Proceedings of the IEEE Conference on Computer Vision and Pattern Recognition, pp. 2417–2426 (2019)
40. Tsai, Y.H., Hung, W.C., Schulter, S., Sohn, K., Yang, M.H., Chandraker, M.: Learning to adapt structured output space for semantic segmentation. In: Proceedings of the IEEE Conference on Computer Vision and Pattern Recognition, pp. 7472–7481 (2018)
41. Wan, Z., et al.: Bringing old photos back to life. In: Proceedings of the IEEE/CVF Conference on Computer Vision and Pattern Recognition, pp. 2747–2757 (2020)
42. Wang, T.C., Liu, M.Y., Zhu, J.Y., Tao, A., Kautz, J., Catanzaro, B.: High-resolution image synthesis and semantic manipulation with conditional GANs. In: Proceedings of the IEEE Conference on Computer Vision and Pattern Recognition, pp. 8798–8807 (2018)
43. Wu, R., Lu, S.: LEED: label-free expression editing via disentanglement. In: Vedaldi, A., Bischof, H., Brox, T., Frahm, J.-M. (eds.) ECCV 2020. LNCS, vol. 12357, pp. 781–798. Springer, Cham (2020). https://doi.org/10.1007/978-3-030-58610-2_46
44. Wu, R., Zhang, G., Lu, S., Chen, T.: Cascade EF-GAN: progressive facial expression editing with local focuses. In: Proceedings of the IEEE/CVF Conference on Computer Vision and Pattern Recognition, pp. 5021–5030 (2020)
45. Xia, W., Yang, Y., Xue, J.H., Wu, B.: TediGAN: text-guided diverse face image generation and manipulation. In: Proceedings of the IEEE/CVF Conference on Computer Vision and Pattern Recognition, pp. 2256–2265 (2021)
46. Xie, Y., et al.: Differentiable top-k with optimal transport. In: Advances in Neural Information Processing Systems 33 (2020)
47. Zhan, F., Lu, S., Zhang, C., Ma, F., Xie, X.: Adversarial image composition with auxiliary illumination. In: Ishikawa, H., Liu, C.-L., Pajdla, T., Shi, J. (eds.) ACCV 2020. LNCS, vol. 12623, pp. 234–250. Springer, Cham (2021). https://doi.org/10.1007/978-3-030-69532-3_15
48. Zhan, F., Xue, C., Lu, S.: GA-DAN: geometry-aware domain adaptation network for scene text detection and recognition. In: Proceedings of the IEEE International Conference on Computer Vision, pp. 9105–9115 (2019)
49. Zhan, F., et al.: Unbalanced feature transport for exemplar-based image translation. In: Proceedings of the IEEE Conference on Computer Vision and Pattern Recognition (2021)
50. Zhan, F., et al.: GMLight: lighting estimation via geometric distribution approximation. arXiv preprint arXiv:2102.10244 (2021)

51. Zhan, F., Yu, Y., Wu, R., Zhang, J., Lu, S.: Multimodal image synthesis and editing: a survey. arXiv preprint arXiv:2112.13592 (2021)
52. Zhan, F., Yu, Y., Wu, R., Zhang, J., Lu, S., Zhang, C.: Marginal contrastive correspondence for guided image generation. In: Proceedings of the IEEE/CVF Conference on Computer Vision and Pattern Recognition, pp. 10663–10672 (2022)
53. Zhan, F., et al.: Sparse Needlets for lighting estimation with spherical transport loss. arXiv preprint arXiv:2106.13090 (2021)
54. Zhan, F., et al.: EMLight: lighting estimation via spherical distribution approximation. In: Proceedings of the AAAI Conference on Artificial Intelligence, vol. 35, pp. 3287–3295 (2021)
55. Zhan, F., Zhang, J., Yu, Y., Wu, R., Lu, S.: Modulated contrast for versatile image synthesis. In: Proceedings of the IEEE/CVF Conference on Computer Vision and Pattern Recognition, pp. 18280–18290 (2022)
56. Zhan, F., Zhu, H., Lu, S.: Spatial fusion GAN for image synthesis. In: Proceedings of the IEEE Conference on Computer Vision and Pattern Recognition, pp. 3653–3662 (2019)
57. Zhang, J., Lu, S., Zhan, F., Yu, Y.: Blind image super-resolution via contrastive representation learning. arXiv preprint arXiv:2107.00708 (2021)
58. Zhang, P., Zhang, B., Chen, D., Yuan, L., Wen, F.: Cross-domain correspondence learning for exemplar-based image translation. In: Proceedings of the IEEE/CVF Conference on Computer Vision and Pattern Recognition, pp. 5143–5153 (2020)
59. Zhao, B., Meng, L., Yin, W., Sigal, L.: Image generation from layout. In: Proceedings of the IEEE Conference on Computer Vision and Pattern Recognition, pp. 8584–8593 (2019)
60. Zheng, H., et al.: Semantic layout manipulation with high-resolution sparse attention. arXiv preprint arXiv:2012.07288 (2020)
61. Zhou, B., Zhao, H., Puig, X., Fidler, S., Barriuso, A., Torralba, A.: Scene parsing through ade20k dataset. In: Proceedings of the IEEE Conference on Computer Vision and Pattern Recognition, pp. 633–641 (2017)
62. Zhou, X., et al.: CoCosNet V2: full-resolution correspondence learning for image translation. In: CVPR (2021)
63. Zhu, J.-Y., Krähenbühl, P., Shechtman, E., Efros, A.A.: Generative visual manipulation on the natural image manifold. In: Leibe, B., Matas, J., Sebe, N., Welling, M. (eds.) ECCV 2016. LNCS, vol. 9909, pp. 597–613. Springer, Cham (2016). https://doi.org/10.1007/978-3-319-46454-1_36
64. Zhu, P., Abdal, R., Qin, Y., Wonka, P.: SEAN: image synthesis with semantic region-adaptive normalization. In: Proceedings of the IEEE/CVF Conference on Computer Vision and Pattern Recognition, pp. 5104–5113 (2020)
65. Zhu, Z., Xu, Z., You, A., Bai, X.: Semantically multi-modal image synthesis. In: Proceedings of the IEEE/CVF Conference on Computer Vision and Pattern Recognition, pp. 5467–5476 (2020)

High-Fidelity Image Inpainting
with GAN Inversion

Yongsheng Yu[1,2], Libo Zhang[1,2,3(✉)], Heng Fan[4], and Tiejian Luo[2]

[1] Institute of Software, Chinese Academy of Sciences, Beijing, China
`yuyongsheng19@mails.ucas.ac.cn`, `libo@iscas.ac.cn`
[2] University of Chinese Academy of Sciences, Beijing, China
`tjluo@ucas.ac.cn`
[3] Nanjing Institute of Software Technology, Nanjing, China
[4] Department of Computer Science and Engineering, University of North Texas,
Denton, USA
`heng.fan@unt.edu`

Abstract. Image inpainting seeks a semantically consistent way to recover the corrupted image in the light of its unmasked content. Previous approaches usually reuse the well-trained GAN as effective prior to generate realistic patches for missing holes with GAN inversion. Nevertheless, the ignorance of hard constraint in these algorithms may yield the gap between GAN inversion and image inpainting. Addressing this problem, in this paper we devise a novel GAN inversion model for image inpainting, dubbed *InvertFill*, mainly consisting of an encoder with a pre-modulation module and a GAN generator with $\mathcal{F}\&\mathcal{W}^+$ latent space. Within the encoder, the pre-modulation network leverages multi-scale structures to encode more discriminative semantic into style vectors. In order to bridge the gap between GAN inversion and image inpainting, $\mathcal{F}\&\mathcal{W}^+$ latent space is proposed to eliminate glaring color discrepancy and semantic inconsistency. To reconstruct faithful and photorealistic images, a simple yet effective Soft-update Mean Latent module is designed to capture more diverse in-domain patterns that synthesize high-fidelity textures for large corruptions. Comprehensive experiments on four challenging dataset, including Places2, CelebA-HQ, MetFaces, and Scenery, demonstrate that our InvertFill outperforms the advanced approaches qualitatively and quantitatively and supports the completion of out-of-domain images well. The code is available at https://github.com/yeates/InvertFill.

Keywords: Image inpainting · GAN inversion · $\mathcal{F}\&\mathcal{W}^+$ latent space

1 Introduction

Image inpainting is an ill-posed problem that requires to recover the missing or corrupted content based on incomplete images with masks. It has been

Supplementary Information The online version contains supplementary material available at https://doi.org/10.1007/978-3-031-19787-1_14.

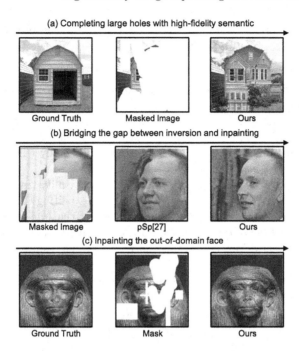

Fig. 1. Visual results of our contributions. Image (a) shows the high-fidelity inpainting results for large corruptions, image (b) exhibits the improvement of our method for the "gapping" issue over previous inversion-based inpainting method pSp [28], and image (c) demonstrates the semantically consistent results by our model for the out-of-domain masked image. *Best viewed in color for all figures throughout the paper.* (Color figure online)

widely adopted for manipulating photographs, such as corrupted image repairing, unwanted object removal, or object position modification [3,30,31].

The mainstream approaches [20,27] often employ an encoder-decoder architecture in UNet style [29] for image inpainting, and have demonstrated promising results in dealing with narrow holes or removing small objects. To apply to more complicated cases, later works have been focused on improving the performance with various discriminators [16,43,45], contextual attention mechanisms [16,45,47], and auxiliary information [8,21,26,46]. Nevertheless, limited by their model capacity, it remains challenging for these UNet-like methods to fill large corruptions with visually realistic patches.

Recently, generative adversarial network (GAN) models [11,13,14] have been verified to successfully produce high-resolution photorealistic images. In these models, GAN inversion [38,51] plays an important role. Specifically, when simply fed with stochastic vectors of latent space, GAN is not applicable to any image-to-image translation. To handle this problem, GAN inversion method uses a pre-trained GAN as prior, and encodes the given images into stochastic vectors that represent the target images, resulting to high-fidelity translation results. Inspired by this, several approaches [2, 6, 28] have made great efforts to introduced GAN inversion for image inpainting. Despite excellent performance, existing methods may suffer from following issues:

- **Distortion for extreme image inpainting.** Due to large corruptions, current methods (*e.g.*, [8, 16, 47]) may become degenerated because these models are *not* able to effectively extract correlation from inadequate knowledge in extremely degraded images. Such correlation information is crucial in eliminating the ambiguity of large continuous holes, especially where far from the boundary.
- **Inconsistency caused by hard constraint.** Unlike in regular conditional translation (*e.g.*, super-resolution [6], face editing [50] and label-to-image [28]), image inpainting has a hard constraint that *the unmasked regions in the input and the output should be the same.* Current inversion-based algorithms [2, 6, 28], however, ignore this constraint, which results in color discrepancy and semantic inconsistency as displayed in Fig. 1(b) and may require additional post-processing such as image blending [2]. We call this problem "gapping" in the following sections.
- **Robustness for out-of-domain inputs.** In order to reconstruct faithful images, the key is to find an in-domain latent code that can align with the domain of a well-trained GAN model [50]. Unfortunately, the encoder fails to invert out-of-domain inputs to produce accurate results. For example, the pSp [28] is hard to tackle the corrupted images with contents or masks from unseen domains, which is harmful to the applicability of GAN inversion.

To solve the above issues, we introduce a novel InvertFill network for image inpainting. It follows the encoder-based inversion fashion architecture [28] that consists of an encoder and a GAN generator. We first develop a new latent space $\mathcal{F\&W}^+$ (as explained later) that encodes the original images into style vector to enable the accessibility of the generator backbone to inputs, decreasing color discrepancy and semantic inconsistency. Besides, to make full use of the encoder, we present pre-modulation networks to amplify the reconstruction signals of the style vector based on the predicted multi-scale structures, further enhancing the discriminative semantic. Then, we propose a simple yet effective soft-update mean latent technique to sample a dynamic in-domain code for the generator. Compared to using a fixed code, our method is able to facilitate diverse downstream goals while reconstructing faithfully and photo-realistically, even in the task of unseen domain. To verify the superiority of our method, we conduct extensive experiments on four datasets, including CelebA-HQ [11], Places2 [49], MetFaces [12], and Scenery [41]. The results demonstrate that our method achieves favorable performance, especially for images with large corruptions. Furthermore, our approach can handle images and masks from unseen domains by optimizing a lightweight encoder without retraining the GAN generator on a large-scale dataset. Figure 1 shows several visual results of our approach.

The contributions of our work are summarized as three-fold: (**1**) We introduce a novel $\mathcal{F\&W}^+$ latent space to resolve the problems of color discrepancy and semantic inconsistency and thus bridge the gap between image inpainting and GAN inversion. (**2**) We propose (a) pre-modulation networks to encode more discriminative semantic from compact multi-scale structures and (b) soft-update mean latent to synthesize more semantically reasonable and visually realistic patches by leveraging diverse patterns. (**3**) Extensive experiments on CelebA-HQ [11], Places2 [49], MetFaces [12], and Scenery [41] show that the proposed approach outperforms current state-of-the-arts, evidencing its effectiveness.

2   Related Work

Image Inpainting. Image inpainting could be treated as a conditional translation task with hard constraint. The seminal learning-based work by Pathak *et al.* [27] integrates UNet [29] and GAN discriminator [5] for image inpainting, and subsequently derives many variants that effectively deal with narrow holes or remove small objects. More recently, several works have attempted to extending the idea in [27] to more complicated cases. Roughly speaking, these methods can be categorized into three types. The first one is to explicitly dispose of invalid signals at masked regions [20,23,43,44]. Among them, Liu *et al.* [20] attach heuristic mask update step to standard convolution and Yu *et al.* [43] formally replace the mask update process with a learnable convolution layer. The second type is called valuable signals shifting that is inspired by the traditional exemplar-based approach [3], which presently tends to model contextual attention to achieve [16,22,40,42,46]. In particular, RFR [16] applies multiple iterations at the bottleneck while sharing the attention scores to guide a patch-swap process. ProFill [47] iteratively performs inpainting based on the confidence map calculated by spatial attention. CRFill [46] yields a contextual reconstruction objective function that learns query-reference feature similarity. The third branch is to adopt auxiliary labels, which generate intermediate structures to assist with more accurate semantic [8,17,21,26]. In specific, EC [26] introduces canny edge to deliver finer inpainting structures. MEDFE [21] jointly learns to represent structures and textures and utilizes spatial and channel equalization to ensure consistency. CTSDG [8] couples texture and structure through parallel pathways and then fuses them by bidirectional gated layers. In addition to the above methods, there also exist other approaches. One notable example is Score-SDE [32] which proposes a scoring model that saves the gradient computation of energy-based models for efficient sampling.

Inpainting with GAN Inversion. StyleGAN [13] implicitly learns hierarchical latent styles $w \in \mathbb{R}^{1 \times 512}$ instead of the initial stochastic vector z, which provides control over the style of outputs at coarse-to-fine levels of detail by style-modulation modules [10]. StyleGAN2 [14] further proposes weight demodulation, path length regularization, and generator redesign for improved image quality. They are adept in the generation without any given images, but requires specialized networks [24] or regularization [7,25] and paired training data. GAN inversion [51] is a common practice that takes advantage of the intrinsic statistics of well-trained large-scale GAN as prior for generic applications [1,50].

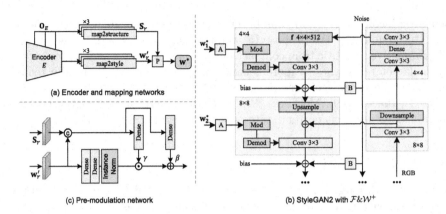

Fig. 2. The main components of InvertFill, including a feature pyramid-based encoder (image (a)), mapping network with pre-modulation network (image (c)) and a Style-GAN2 generator with the proposed $\mathcal{F}\&\mathcal{W}^+$ latent space (image (b)).

Existing GAN inversion approaches could be roughly divided as optimized-based [2,6,34,37] and encoder-based [28,39,50]. Among these methods, mGAN-prior [6] utilizes multiple latent codes and adaptive channel importance for faithful reconstruction and shows applications in different tasks including inpainting. pSp [28] synthesizes images with the mapping network to extract style vectors $w^+ \in \mathbb{R}^{18 \times 512}$ of latent space $\mathcal{W}^+$ [1] separately for corresponding 18 style-modulation layers of the StyleGAN. Nevertheless, these approaches ignore the "gapping" issue, resulting in color inconsistency and semantic misalignment.

Difference with Previous Studies. In this paper we focus on encoder-based GAN inversion to improve generation fidelity for image inpainting. The proposed InvertFill is related to but significantly different from previous studies. In specific, InvertFill is relevant to the methods in [28,39,50] where encoder-based architecture is adopted. However, differing from them, we introduce a new $\mathcal{F}\&\mathcal{W}^+$ latent space to explicitly handle the "gapping" issue which is ignored in previous algorithms. Our method also shares similar spirit with the works of [46,47] that adopt GAN for image inpainting. The difference is that these approaches may suffer from ambiguity when filling the large corruptions, while the proposed InvertFill exploits the priors of a large-scale generator and can achieve image inpainting with high-fidelity semantic.

3   The Proposed Method

Given an original image $\mathbf{I}$ and its corrupted image $\mathbf{I}_m = \mathbf{I} \odot (1 - \mathbf{M})$, where $\mathbf{M}$ is a binary mask and $\odot$ denotes element-wise product. The value of pixels in masked region $\mathbf{M}$ equal to 1 indicates invisible. We aim to produce a visually realistic reconstructed image $\mathbf{O}$ with the input of corrupted images $\mathbf{I}_m$.

3.1 $\mathcal{F}\&\mathcal{W}^+$ Latent Space

Our architecture mainly consists of three components: (i) A feature pyramid-based [19] encoder E that extracts input images and provides hierarchical reconstructed RGB images, (ii) the mapping networks with pre-modulation module, and (iii) a StyleGAN2 generator that takes in the style vectors as well as the input image $\mathbf{I}_m$ to generate a image. The details of InvertFill are shown in Fig. 2.

Specifically, we attach three RGB heads to the encoder E for generating reconstructed RGB images $\mathbf{O}_E^r = \{\mathbf{O}_E^1, \mathbf{O}_E^2, \mathbf{O}_E^3\}$ in correspondence to three different scale. We follow the *map2style* [28] for the mapping network and cut down the network number from 18 to 3, each of which corresponds to the disentanglement level of image representation (i.e., coarse, middle and fine [13,28]). Three *map2style* networks encode the output feature map of the encoder into the intermediate code $\mathbf{w}' \in \mathbb{R}^{3 \times 512}$. Similarly, we replicate *map2style* as *map2structure* to project reconstructed RGB images $\mathbf{O}_E^r$ gradually into structure vector $\mathbf{S}_r = \{\mathbf{S}_1, \mathbf{S}_2, \mathbf{S}_3\}$.

Before executing the style modulation in the generator, we perform L pre-modulation networks to project the semantic structure $\mathbf{S}_r$ into the style vector $\mathbf{w}^*$ in latent space $\mathcal{F}\&\mathcal{W}^+$, i.e., $\mathbf{w}^* = E(\mathbf{I}_m), \mathbf{w}^* \in \mathbb{R}^{L \times 512}$. $L = \log_2(s) \cdot 2 - 2$ denotes number of style-modulation layers of StyleGAN2 generator, and is adjusted by the image resolution s on the generator side. As Fig. 2(c) demonstrates, we adopt Instance Normalization (IN) [33] to regularize the $\mathbf{w}'$ latent code, then carry out denormalization according to multi-scale structure vector $\mathbf{S}_r$,

$$\mathbf{w}_l^* = \gamma \odot \mathrm{IN}\left(\mathbf{w}_r'\right) + \beta, \tag{1}$$

where $l \in \{1, 2, \ldots, L\}$ denotes the index of style vectors, $r \in [1, 3]$ indicates three vectors $\mathbf{w}'$ correspond to level of coarse to fine, (γ, β) is a pair of the affine transformation parameters learned by networks shown in Fig. 2(c). Different than previous methods in only using intermediate latent code from a network, the proposed pre-modulation module is a lightweight network and novel in applying more discriminative multi-scale features to help latent code perceive uncorrupted prior and better guide image generation.

The GAN is initially fed with a stochastic vector $z \in \mathcal{Z}$, and previous works [1,6,28,50] invert the source images into the intermediate latent space $\mathcal{W}$ or $\mathcal{W}^+$, which is a less entangled representation than latent space $\mathcal{Z}$. The style vectors $w \in \mathcal{W}$ or $w^+ \in \mathcal{W}^+$ are sent to the style-modulation layers of pre-trained StyleGAN2 to synthesize target images. These approaches can be formulated mathematically as follows,

$$\mathbf{O}_G = G(E(\mathbf{I}_m)), E(\mathbf{I}_m) \sim W^+, \tag{2}$$

where $E(\cdot)$ and $G(\cdot)$ represent the encoder that maps source images into latent space and the pre-trained GAN generator, respectively.

Nevertheless, the above formulation in Eq. (2) may encounter the "gapping" issue in image translation tasks with hard constraint, *e.g.*, image inpainting. The hard constraint requires that parts of the source and recovered image remain the same. We formally defined the hard constraint in image inpainting as $\mathbf{I} \odot (1 - \mathbf{M}) \equiv \mathbf{O} \odot (1 - \mathbf{M})$. Intuitively, we argue that the "gapping" issue is caused by that the GAN model cannot directly access pixels of the input image but the intermediate latent code. To avoid the semantic inconsistency and color discrepancy caused by this problem, we utilize the corrupted image $\mathbf{I}_m$ as one of the inputs to assist with the GAN generator inspired by skip connection of U-Net [29]. In detail, $\mathbf{I}_m$ is fed into the RGB branch as shown in Fig. 2(b), the feature map between RGB branch and the generator are connected by element-wise addition. Hence, the previous formulation in Eq. (2) is updated as:

$$\mathbf{O}_G = G(E(\mathbf{I}_m), \mathbf{I}_m), E(\mathbf{I}_m) \sim \mathcal{F} \& \mathcal{W}^+ \tag{3}$$

.

3.2 Soft-update Mean Latent

Pixels closer to the mask boundary are more accessible to inpainting, but conversely the model is hard to predict specific content missing. We find that the encoder learns a trick to averaging textures to reconstruct the region away from unmasked region. It causes blurring or mosaic in some areas of the output image, mainly located away from the mask borders, as shown in Fig. 7. Drawing inspiration from L2 regularization and motivated by the intuition that fitting diverse domains works better than fitting a preset static domain, a feasible solution is to make style code $\mathbf{w}^*$ be bounded by the mean latent code of pre-trained GAN.

The mean latent code is obtained from abundant random samples that restrict the encoder outputs to the average style hence lossy the diversity of output distribution of encoder. In addition, it introduces additional hyperparameters and a static mean latent code that requires loading when training the model.

We adopt dynamic mean latent code instead of static one by stochastically fluctuating the mean latent code while training. Further, we smooth the effect of fluctuating variance for convergence inspired by a reinforcement learning [18]. For initialization, target mean latent code $\overline{\mathbf{w}}_t$ and online mean latent code $\overline{\mathbf{w}}_o$ are sampled. $\overline{\mathbf{w}}_o$ is used in image generation instead of $\overline{\mathbf{w}}_t$, which is fixed until $\overline{\mathbf{w}}_o = \overline{\mathbf{w}}_t$ and then resampled. Between two successive sampled mean latent codes, $\overline{\mathbf{w}}_o$ is updated by $\overline{\mathbf{w}}_t \leftarrow \tau \overline{\mathbf{w}}_o + (1 - \tau)$ per iteration during training, where τ denotes updating factor and $\overline{\mathbf{w}}_t$ for soft updating target mean latent code. The soft-update mean latent degraded to static mean latent [28] when the parameter τ of soft-update mean latent approaching zero.

3.3 Optimization

Following prior work in inpainting [16,20], our architecture is supervised by regular inpainting loss $\mathcal{L}_{\text{ipt}}$, which consists of the pixel-wise Euclidean norm of valid and hole regions, the perceptual loss perc, the style loss style, and the total variation loss tv:

$$\mathcal{L}_{\text{ipt}} = \mathcal{L}_{\text{valid}} + \mathcal{L}_{\text{hole}} + \mathcal{L}_{\text{perc}} + \mathcal{L}_{\text{style}} + \mathcal{L}_{\text{tv}}, \tag{4}$$

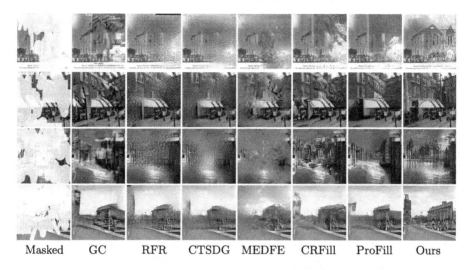

| Masked | GC | RFR | CTSDG | MEDFE | CRFill | ProFill | Ours |

Fig. 3. Qualitative results on Places2 dataset.

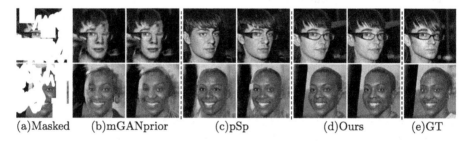

(a)Masked   (b)mGANprior   (c)pSp   (d)Ours   (e)GT

Fig. 4. Qualitative results on CelebA-HQ dataset. Two columns of (b-d) show the original model output and composition output, from left to right, respectively. The output of the GAN-inversion-based method (pSp [28] and mGANprior [6]) is inconsistency at the edge of the mask. Zoom-in to see the details.

where all above distance are calculated between $\mathbf{I}$ and $\mathbf{O}_G$. $\mathcal{L}_{\text{valid}}$ and $\mathcal{L}_{\text{hole}}$ are ℓ_1 norm on the known and masked region respectively. The perceptual loss $\mathcal{L}_{\text{perc}}$ and the style loss $\mathcal{L}_{\text{style}}$ are based on a pre-trained VGG-16 network. More details can be found in [16].

To directly optimize our encoder, the multi-scale reconstruction loss $\mathcal{L}_{\text{msr}}$ is utilized to penalize the deviation of $\mathbf{O}_E^r$ at each scale:

$$\mathcal{L}_{\text{msr}} = \sum_{r=1}^{3} (\mathcal{L}_{\text{perc}}^r + \mathcal{L}_{\text{style}}^r + \mathcal{L}_{\text{rec}}^r), \tag{5}$$

where $\mathcal{L}_{\text{rec}}$ is represented as mean-squared loss between $\mathbf{I}$ and $\mathbf{O}_E$. The multi-scale reconstruction loss $\mathcal{L}_{\text{msr}}$ contains three different losses including perceptual ($\mathcal{L}_{\text{perc}}^r$) [4], style ($\mathcal{L}_{\text{style}}^r$) [20] and mean-square ($\mathcal{L}_{\text{rec}}^r$) losses. The role of $\mathcal{L}_{\text{msr}}$ is

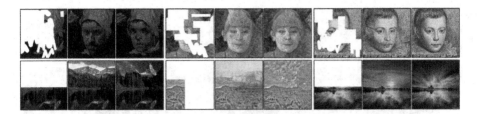

Fig. 5. The visual effect of our method for processing input images from unseen domain. The 1st row shows inpainting results of Metfaces, and the 2nd row shows outpainting results of Scenery. Each instance of results is laid out as the masked image, the model output, and the original image.

to supervise the generated image from decoder and make final generation close to the original image.

The soft-update mean latent is utilized to prevent the encoder from falling into the trick way. We adopt the following fidelity loss $\mathcal{L}_{\text{fid}}$ for improving the quality and diversity of output images:

$$\mathcal{L}_{\text{fid}} = \|\mathbf{w}^* - \overline{\mathbf{w}}\|_2, \ (\mathbf{w}^*, \overline{\mathbf{w}}) \in \mathbb{R}^{L \times 512}. \tag{6}$$

The fidelity loss $\mathcal{L}_{\text{fid}}$ is designed as a mean squared loss of style vectors $\mathbf{w}^*$ and mean latent code $\overline{\mathbf{w}}$. Its role is to improve the quality and diversity of the output images.

Overall, the loss of our networks is defined as the weighted sum of the inpainting loss, the multi-scale reconstruction loss, and the fidelity loss.

$$\mathcal{L} = \mathcal{L}_{\text{ipt}} + \lambda_{\text{msr}}\mathcal{L}_{\text{msr}} + \lambda_{\text{fid}}\mathcal{L}_{\text{fid}}, \tag{7}$$

where λ_{msr} and λ_{fid} are the balancing factors for the multi-scale reconstruction loss and the fidelity loss, respectively.

4   Experiments

We perform extensive validating experiments aiming to answer the following research questions:

- **RQ1:** How does our approach perform, compared with existing methods, especially the fidelity when the input is large-scale masked images.
- **RQ2:** Can our approach resolve the "gapping" issue?
- **RQ3:** Can our approach handle input from unseen domain by reusing the well-trained generator while only retraining a lightweight encoder?
- **RQ4:** How do different components (e.g., soft-update mean latent, pre-modulation) affect our approach?

Table 1. Quantitative comparison with the mainstream inpainting approaches on Places2 and CelebA-HQ datasets. *Hard, Extreme, All* masks denote the mask with coverage ratio of 50%–60%, 70%–90%, and 10%–90%, respectively. ↑ Higher is better, and ↓ lower is better. Best and second best results are **highlighted**.

			GC	RFR	MEDFE	ProFill	CTSDG	CRFill	Ours
Places2	*hard*	SSIM↑	0.624	0.645	0.598	**0.664**	0.651	0.629	0.641
		FID↓	22.05	27.77	44.38	21.49	35.77	22.46	**12.44**
		LPIPS↓	0.246	0.235	0.294	0.240	0.272	0.250	**0.232**
	extreme	SSIM↑	0.363	0.382	0.323	**0.409**	0.393	0.360	0.366
		FID↓	51.35	71.19	111.85	46.44	95.50	51.26	**21.08**
		LPIPS↓	0.407	0.395	0.495	0.402	0.438	0.413	**0.386**
	all	SSIM↑	0.734	0.750	0.714	**0.764**	0.755	0.738	0.761
		FID↓	14.19	16.26	26.15	13.81	21.36	14.44	**9.29**
		LPIPS↓	0.178	0.170	0.217	0.173	0.199	0.182	**0.155**
CelebA-HQ	*hard*	SSIM↑	0.790	**0.825**	0.781	–	0.818	0.810	0.812
		FID↓	17.38	9.98	21.97	–	15.13	13.78	**9.89**
		LPIPS↓	0.170	0.128	0.192	–	0.151	0.139	**0.121**
	extreme	SSIM↑	0.589	0.641	0.552	–	0.616	0.639	**0.652**
		FID↓	41.70	22.07	55.52	–	33.89	30.19	**13.21**
		LPIPS↓	0.297	0.241	0.359	–	0.281	0.275	**0.214**
	all	SSIM↑	0.852	**0.878**	0.846	–	0.875	0.859	0.867
		FID↓	11.78	7.96	15.52	–	10.32	11.94	**7.71**
		LPIPS↓	0.128	0.092	0.142	–	0.110	0.114	**0.089**

4.1 Experimental Settings

Datasets. Experiments for RQ1, RQ2 and RQ4 are conducted on two datasets, Places2 [49] and CelebA-HQ [11]. CelebA-HQ contains 30,000 high-resolution celebrity faces, and we follow [42,43] to split this dataset for training and testing. Places2 contains real-world photos, including more significant objects, such as streets, cars, houses, which is better suited for verifying models on large-scale masks than CelebA-HQ. Based on the official *train/val/test* split, we train the model on *train* plus *test* about 200,000 images, evaluate the model on first 5,000 images of *val*. With regard to RQ3, we utilize two datasets Scenery [41] and MetFaces [12]. Scenery dataset is a common benchmark for recent image outpainting tasks and contains 6,040 landscape photographs. We follow [41] to use about 5,000 images as training set and the remaining 1,000 images as test set. MetFaces consists of 1,336 human faces extracted from works of art, and we randomly select 1,000 images as training set and other images as test set. Our model and all baselines adopt the same training and test strategies to ensure experimental fairness.

Evaluation Metrics. We use three metrics following prior works to measure the quality and fidelity of inpainting results. SSIM [35] modeling image distortion by structure, luminance, and contrast, is a pixel-level objective metric similar to PSNR, and their drawbacks cause inconsistent evaluation results with the human eye. Despite that, they are classical metrics for image evaluation, one of which SSIM we selected for quantitative comparison. FID [9] is a deep metric

Masked   pSp   pSp+Blend   Ours   Masked   pSp   pSp+Blend   Ours

Fig. 6. Comparison with pSp [28] and pSp+Blend [36] that post-processing by image blending. The 1st row shows the color discrepancy that image blending is sufficient to resolve satisfactorily. The 2nd row shows that the semantic inconsistency is still reserved, except for our method. (Color figure online)

and closer to human perception. It measures the distribution distance with a pre-trained inception model, which better captures distortions. LPIPS [48] is another learned perceptual metric and commonly used to score the intra-conditioning diversity of models output. Following previous works [16,26,43], We calculate these quantitative metrics on original images $\mathbf{I}$ and composition images $\mathbf{I} \odot (1 - \mathbf{M}) + \mathbf{O}_G \odot \mathbf{M}$.

Baselines. We carefully select baseline methods mainly from two perspectives: UNet style methods and Inversion style methods to demonstrate our approach's characteristics and superiority. First, for the sake of validating the ability of InvertFill in filling images under large-scale masks, we compare it with the previous approaches including EC [26], GC [42], RFR [16], MEDFE [21], ProFill [47], CTSDG [8] and CRFill [46]. Second, we compare with the latest GAN inversion-based inpainting methods mGANprior [6] and pSp [28].

4.2 Implementation Details

We utilize eight A100 GPUs for pre-training the GAN generator, and one TITAN RTX GPU for optimizing the encoder and other experiments. Following [16], we scale the image size of all datasets to 256×256 as the input. In the light of the mask coverage, we classify the test masks into three difficulty levels: *Hard/Extreme/All*, indicates the mask with coverage ratio of 50%–60%, 70%–90%, 10%–90%, respectively. During testing, for a fair comparison, we use the same image-mask pair for all approaches. More details of implementation are shown in the supplementary.

4.3 Result Analysis

RQ1. We reproduce all the above baselines by utilizing their official implementations. Concerning Places2 dataset, we utilize the pre-trained weights officially released by the baselines. On CelebA-HQ dataset, EC [26], GC [43], mGANprior [6] offer pre-trained weights, we thereby carefully retrain other baselines through the official source codes. Because ProFill only offers Web API on Places2, we use placeholder '–' in Table 1 for ProFill on CelebA-HQ.

From Table 1, our method achieves the best or comparable performance among advanced inpainting approaches. In terms of the FID metric, our method

Table 2. Comparison with previous GAN inversion-based and diffusion-based approaches on CelebA-HQ dataset.

	FID↓	LPIPS↓	SSIM↑
Score-SDE [32]	24.76	0.337	0.428
mGANprior [6]	29.57	0.273	0.608
pSp [28]	25.61	0.248	0.594
pSp + Blend [36]	21.96	0.240	0.602
Ours	**13.21**	**0.214**	**0.652**

Table 3. Comparison with previous outpainting approaches and inpainting baselines on Scenery dataset.

	FID↓	LPIPS↓	SSIM↑
RFR [16]	138.31	0.455	0.376
pSp [28]	49.62	0.379	0.392
Boundless [15]	45.05	0.368	0.413
NS-outpaint [41]	38.95	0.342	0.410
Ours	**20.90**	**0.294**	**0.439**

Table 4. Comparison with previous inpainting methods on Metfaces. In this experimental setting, the model/generator is only trained on CelebA-HQ.

Method	Easy			Extreme		
	SSIM↑	FID↓	LPIPS↓	SSIM↑	FID↓	LPIPS↓
RFR	0.93	18.89	0.069	0.52	58.24	0.315
CRFill	0.95	13.67	0.042	0.54	50.93	0.278
pSp	0.95	14.91	0.040	0.49	65.04	0.341
Ours	**0.97**	**8.64**	**0.033**	**0.60**	**38.85**	**0.227**

at most produces a notable margin of 54.60% and 40.14% on Places2 and CelebA-HQ datasets, respectively. And our method also outperforms the second-best approach 11.2% and 10.4% improvements on another perceptual metric LPIPS, which validates the superiority of our design.

Figure 3 and 4 provide several visual inpainting results on Places2 and CelebA-HQ datasets. Figure 3 reveals that the prior works still struggle to generate refined texture if the input image with large corruptions, while our approach has been able to create semantically rich objects such as windows, towers, and woods. In Fig. 4, mGANprior [6] progressively erases the color discrepancy rely on optimized-based inversion but is unable to bypass semantic inconsistency. The encoder-based inversion method pSp [28] could synthesize realistic pixels for corrupted regions based on the well-trained model, though it still has not resolved the "gapping" issue. The results indicate that our method produces consistent output while generating high-fidelity texture compared to existing methods.

RQ2. The "gapping" causes color discrepancy and semantic inconsistency, and we are counting on image post-processing to tackle this issue at the beginning of this study. Specifically, we adopt image blending [36], which is effective in eliminating the color discrepancy but helpful in remedying semantic inconsistency.

To further demonstrate the superiority of our method, we construct the pSp+Blend variant that introduces an image blending [36] method after generating output images. In Fig. 6, the first row shows the distinct gap at the stitching boundary in pSp output, and pSp+Blend fixes this color discrepancy problem. Even so, the second row shows pSp+Blend unable to assist with semantic inconsistency problem given the glasses are still incomplete. Compared with the vanilla pSp and pSp+Blend, output images of our approach no longer suffer from color discrepancy or semantic inconsistency.

We conduct a comparison experiment on CelebA-HQ dataset with the *Extreme* level masks. As demonstrated in Table 2, our method performs better than a recent diffusion-based approach Score-SDE [32] w.r.t to FID, LPIPS and SSIM metrics. The results in Table 2 also show that our method performs best among the existing inversion-based inpainting approaches after resolving the "gapping" issue. Notably, our method does not require any image post-processing.

Masked   w/o SML   w/ SML   GT   Masked   w/o SML   w/ SML   GT

Fig. 7. The importance of soft-update mean latent.

RQ3. Concerning validating that our approach can reuse the pre-trained GAN generator as priors to tackle image from unseen domain, we conduct two extended tests that introduced images or masks from unseen domains and only required optimizing the lightweight encoder. The first is archaic photograph inpainting, and we use MetFaces [12] for optimizing the encoder, and remain the pre-trained weights of GAN generator of CelebA-HQ dataset. For the second one, we perform our approach with outpainting masks [15] on Scenery dataset. Similarly, the generator did not retrain on the Scenery dataset rather than remaining the weights for Places2.

The 1st row of Fig. 5 shows the inpainting results of archaic photograph inpainting. It demonstrates that our method enables the generator to synthesize semantically consistent style and patches, even in an unseen domain. From the 2nd row of Fig. 5, the outpainting results on the Scenery dataset show our approach still can synthesize realistic texture and significant objects, *e.g.*trees, mountains. To ensure the masks are unseen for the GAN generator, we only use the outpainting masks to train the encoder, not the GAN generator.

Furthermore, we quantitatively compare mainstream outpainting approaches as well as adopt RFR [16] and pSp [28] as additional baselines. As shown in Table 3, our model considerably outperforms the best outpainting baselines [15, 41] with respect to FID, LPIPS, and SSIM. Similarly, we conduct experiments compared with inpainting baselines on Metfaces, as show in Table 4. In summary, the results indicate that our proposed method is robust and extends to other tasks with out-of-domain inputs.

Due to limited space, please kindly refer to the supplementary material for more results.

Table 5. Ablation study comparison on Places2 dataset under *Extreme* mask setting.

$\mathcal{F\&W}^+$	SML	PM	$\mathcal{W}^+$	FID	LPIPS	SSIM	PSNR
✓				35.37	0.395	0.357	13.85
✓	✓			24.73	0.389	0.358	13.99
✓	✓	✓		21.08	0.386	0.366	14.62
	✓	✓	✓	42.85	0.392	0.361	14.25

Masked   Original   Ours   Masked   Original   Ours

Fig. 8. Illustration of two failure cases of the proposed method.

4.4 Ablation Study (RQ4)

The ablation experiments are carried out on the Places2 dataset under the *Extreme* mask setting. In Table 5, we construct three variants to verify the contribution of proposed modules, in which PM and SML denote pre-modulation and soft-update mean latent. By learning from these modules, our method considerably outperforms the most naive variant w.r.t FID, LPIPS, SSIM, and PSNR.

The soft-update mean latent is motivated by the intuition that fitting diverse domains works better than fitting a preset static domain, especially when the training dataset contains various scenarios such as street and landscape. As shown in Fig. 7, when we use SML code that dynamically fluctuates during training, the masked region far away from the mask border tends to be reconstructed by explicitly learned semantics instead of repetitive patterns. Notably, 'w/o SML' represents using regular static mean latent code.

4.5 Failure Cases and Discussion

Figure 8 shows two failure cases. Even if the model can recognize the corrupted objects (our method tends to recover the human face in the left case of Fig. 8), it mistakenly locates them and produces severe artifacts. When lacking sufficient prior knowledge, our method fails to reconstruct details. This demonstrates that these situations are challenging for image inpainting and need further study.

5 Conclusion

In this paper, we propose an encoder-based GAN inversion method InvertFill for image inpainting. The encoder projects corrupted images into a latent space

$\mathcal{F}\&\mathcal{W}^+$ with pre-modulation for learning more discriminative representation. The novel latent space $\mathcal{F}\&\mathcal{W}^+$ resolves the "gapping" issue when applied to GAN inversion in image inpainting. In addition, the soft-update mean latent dynamically samples diverse in-domain patterns, leading to more realistic textures. Extensive quantitative and qualitative comparisons demonstrate the superiority of our model over previous approaches and can cheaply support the semantically consistent completion of images or masks from unseen domains.

Acknowledgment. This work was supported by the Key Research Program of Frontier Sciences, CAS, Grant No. ZDBS-LY-JSC038. Libo Zhang was supported by the CAAI-Huawei MindSpore Open Fund and Youth Innovation Promotion Association, CAS (2020111). Heng Fan and his employer received no financial support for the research, authorship, and/or publication of this article.

References

1. Abdal, R., Qin, Y., Wonka, P.: Image2styleGAN: how to embed images into the styleGAN latent space? In: ICCV, pp. 4431–4440 (2019)
2. Cheng, Y., Lin, C.H., Lee, H., Ren, J., Tulyakov, S., Yang, M.: In&out: diverse image outpainting via GAN inversion. CoRR abs/2104.00675 (2021)
3. Criminisi, A., Pérez, P., Toyama, K.: Object removal by exemplar-based inpainting. In: CVPR, pp. 721–728 (2003)
4. Gatys, L.A., Ecker, A.S., Bethge, M.: A neural algorithm of artistic style. CoRR abs/1508.06576 (2015)
5. Goodfellow, I.J., et al.: Generative adversarial nets. In: NeurIPS, pp. 2672–2680 (2014)
6. Gu, J., Shen, Y., Zhou, B.: Image processing using multi-code GAN prior. In: CVPR, pp. 3009–3018 (2020)
7. Gulrajani, I., Ahmed, F., Arjovsky, M., Dumoulin, V., Courville, A.C.: Improved training of Wasserstein GANs. In: NeurIPS, pp. 5767–5777 (2017)
8. Guo, X., Yang, H., Huang, D.: Image inpainting via conditional texture and structure dual generation. In: ICCV, pp. 14134–14143 (2021)
9. Heusel, M., Ramsauer, H., Unterthiner, T., Nessler, B., Hochreiter, S.: GANs trained by a two time-scale update rule converge to a local Nash equilibrium. In: NeurIPS, pp. 6626–6637 (2017)
10. Huang, X., Belongie, S.J.: Arbitrary style transfer in real-time with adaptive instance normalization. In: ICCV, pp. 1510–1519 (2017)
11. Karras, T., Aila, T., Laine, S., Lehtinen, J.: Progressive growing of GANs for improved quality, stability, and variation. In: ICLR (2018)
12. Karras, T., Aittala, M., Hellsten, J., Laine, S., Lehtinen, J., Aila, T.: Training generative adversarial networks with limited data. In: NeurIPS (2020)
13. Karras, T., Laine, S., Aila, T.: A style-based generator architecture for generative adversarial networks. In: CVPR, pp. 4401–4410 (2019)
14. Karras, T., Laine, S., Aittala, M., Hellsten, J., Lehtinen, J., Aila, T.: Analyzing and improving the image quality of styleGAN. In: CVPR, pp. 8107–8116 (2020)
15. Krishnan, D., et al.: Boundless: generative adversarial networks for image extension. In: ICCV, pp. 10520–10529 (2019)
16. Li, J., Wang, N., Zhang, L., Du, B., Tao, D.: Recurrent feature reasoning for image inpainting. In: CVPR, pp. 7757–7765 (2020)

17. Liao, L., Xiao, J., Wang, Z., Lin, C., Satoh, S.: Image inpainting guided by coherence priors of semantics and textures. In: CVPR, pp. 6539–6548 (2021)
18. Lillicrap, T.P., et al.: Continuous control with deep reinforcement learning. In: ICLR (2016)
19. Lin, T., Dollár, P., Girshick, R.B., He, K., Hariharan, B., Belongie, S.J.: Feature pyramid networks for object detection. In: CVPR, pp. 936–944 (2017)
20. Liu, G., Reda, F.A., Shih, K.J., Wang, T.-C., Tao, A., Catanzaro, B.: Image inpainting for irregular holes using partial convolutions. In: Ferrari, V., Hebert, M., Sminchisescu, C., Weiss, Y. (eds.) ECCV 2018. LNCS, vol. 11215, pp. 89–105. Springer, Cham (2018). https://doi.org/10.1007/978-3-030-01252-6_6
21. Liu, H., Jiang, B., Song, Y., Huang, W., Yang, C.: Rethinking image inpainting via a mutual encoder-decoder with feature equalizations. In: Vedaldi, A., Bischof, H., Brox, T., Frahm, J.-M. (eds.) ECCV 2020. LNCS, vol. 12347, pp. 725–741. Springer, Cham (2020). https://doi.org/10.1007/978-3-030-58536-5_43
22. Liu, H., Jiang, B., Xiao, Y., Yang, C.: Coherent semantic attention for image inpainting. In: ICCV, pp. 4169–4178 (2019)
23. Ma, Y., Liu, X., Bai, S., Wang, L., He, D., Liu, A.: Coarse-to-fine image inpainting via region-wise convolutions and non-local correlation. In: IJCAI, pp. 3123–3129 (2019)
24. Mao, X., Li, Q., Xie, H., Lau, R.Y.K., Wang, Z., Smolley, S.P.: Least squares generative adversarial networks. In: ICCV, pp. 2813–2821 (2017)
25. Miyato, T., Kataoka, T., Koyama, M., Yoshida, Y.: Spectral normalization for generative adversarial networks. In: ICLR (2018)
26. Nazeri, K., Ng, E., Joseph, T., Qureshi, F.Z., Ebrahimi, M.: EdgeConnect: structure guided image inpainting using edge prediction. In: ICCVW, pp. 3265–3274 (2019)
27. Pathak, D., Krähenbühl, P., Donahue, J., Darrell, T., Efros, A.A.: Context encoders: feature learning by inpainting. In: CVPR, pp. 2536–2544 (2016)
28. Richardson, E., et al.: Encoding in style: a styleGAN encoder for image-to-image translation. In: CVPR, pp. 2287–2296 (2021)
29. Ronneberger, O., Fischer, P., Brox, T.: U-Net: convolutional networks for biomedical image segmentation. In: Navab, N., Hornegger, J., Wells, W.M., Frangi, A.F. (eds.) MICCAI 2015. LNCS, vol. 9351, pp. 234–241. Springer, Cham (2015). https://doi.org/10.1007/978-3-319-24574-4_28
30. Shetty, R., Fritz, M., Schiele, B.: Adversarial scene editing: automatic object removal from weak supervision. In: NeurIPS, pp. 7717–7727 (2018)
31. Song, L., Cao, J., Song, L., Hu, Y., He, R.: Geometry-aware face completion and editing. In: AAAI, pp. 2506–2513 (2019)
32. Song, Y., Sohl-Dickstein, J., Kingma, D.P., Kumar, A., Ermon, S., Poole, B.: Score-based generative modeling through stochastic differential equations. In: ICLR (2021)
33. Ulyanov, D., Vedaldi, A., Lempitsky, V.S.: Instance normalization: the missing ingredient for fast stylization. CoRR abs/1607.08022 (2016)
34. Wang, H.P., Yu, N., Fritz, M.: Hijack-GAN: unintended-use of pretrained, blackbox GANs. In: Proceedings of the IEEE/CVF Conference on Computer Vision and Pattern Recognition (CVPR), pp. 7872–7881, June 2021
35. Wang, Z., Bovik, A.C., Sheikh, H.R., Simoncelli, E.P.: Image quality assessment: from error visibility to structural similarity. TIP **13**, 600–612 (2004)
36. Wu, H., Zheng, S., Zhang, J., Huang, K.: GP-GAN: towards realistic high-resolution image blending. In: MM, pp. 2487–2495 (2019)

37. Wu, Z., Lischinski, D., Shechtman, E.: Stylespace analysis: disentangled controls for styleGAN image generation. In: Proceedings of the IEEE/CVF Conference on Computer Vision and Pattern Recognition (CVPR), pp. 12863–12872, June 2021

38. Xia, W., Zhang, Y., Yang, Y., Xue, J., Zhou, B., Yang, M.: GAN inversion: a survey. CoRR abs/2101.05278 (2021)

39. Xu, Y., Shen, Y., Zhu, J., Yang, C., Zhou, B.: Generative hierarchical features from synthesizing images. In: Proceedings of the IEEE/CVF Conference on Computer Vision and Pattern Recognition (CVPR), pp. 4432–4442, June 2021

40. Yan, Z., Li, X., Li, M., Zuo, W., Shan, S.: Shift-Net: image inpainting via deep feature rearrangement. In: Ferrari, V., Hebert, M., Sminchisescu, C., Weiss, Y. (eds.) Computer Vision – ECCV 2018. LNCS, vol. 11218, pp. 3–19. Springer, Cham (2018). https://doi.org/10.1007/978-3-030-01264-9_1

41. Yang, Z., Dong, J., Liu, P., Yang, Y., Yan, S.: Very long natural scenery image prediction by outpainting. In: ICCV, pp. 10560–10569 (2019)

42. Yu, J., Lin, Z., Yang, J., Shen, X., Lu, X., Huang, T.S.: Generative image inpainting with contextual attention. In: CVPR, pp. 5505–5514 (2018)

43. Yu, J., Lin, Z., Yang, J., Shen, X., Lu, X., Huang, T.S.: Free-form image inpainting with gated convolution. In: ICCV, pp. 4470–4479 (2019)

44. Yu, T., et al.: Region normalization for image inpainting. In: AAAI, pp. 12733–12740 (2020)

45. Zeng, Y., Fu, J., Chao, H., Guo, B.: Aggregated contextual transformations for high-resolution image inpainting. CoRR abs/2104.01431 (2021)

46. Zeng, Y., Lin, Z., Lu, H., Patel, V.M.: CR-Fill: generative image inpainting with auxiliary contextual reconstruction. In: ICCV, pp. 14164–14173, October 2021

47. Zeng, Yu., Lin, Z., Yang, J., Zhang, J., Shechtman, E., Lu, H.: High-resolution image inpainting with iterative confidence feedback and guided upsampling. In: Vedaldi, A., Bischof, H., Brox, T., Frahm, J.-M. (eds.) ECCV 2020. LNCS, vol. 12364, pp. 1–17. Springer, Cham (2020). https://doi.org/10.1007/978-3-030-58529-7_1

48. Zhang, R., Isola, P., Efros, A.A., Shechtman, E., Wang, O.: The unreasonable effectiveness of deep features as a perceptual metric. In: CVPR, pp. 586–595 (2018)

49. Zhou, B., Lapedriza, À., Khosla, A., Oliva, A., Torralba, A.: Places: a 10 million image database for scene recognition. TPAMI 40, 1452–1464 (2018)

50. Zhu, J., Shen, Y., Zhao, D., Zhou, B.: In-domain GAN inversion for real image editing. In: Vedaldi, A., Bischof, H., Brox, T., Frahm, J.-M. (eds.) ECCV 2020. LNCS, vol. 12362, pp. 592–608. Springer, Cham (2020). https://doi.org/10.1007/978-3-030-58520-4_35

51. Zhu, J.-Y., Krähenbühl, P., Shechtman, E., Efros, A.A.: Generative visual manipulation on the natural image manifold. In: Leibe, B., Matas, J., Sebe, N., Welling, M. (eds.) ECCV 2016. LNCS, vol. 9909, pp. 597–613. Springer, Cham (2016). https://doi.org/10.1007/978-3-319-46454-1_36

DeltaGAN: Towards Diverse Few-Shot Image Generation with Sample-Specific Delta

Yan Hong⬤, Li Niu⁽✉⁾⬤, Jianfu Zhang⬤, and Liqing Zhang⬤

MoE Key Lab of Artificial Intelligence, Shanghai Jiao Tong University,
Shanghai, China
{ustcnewly,c.sis}@sjtu.edu.cn, zhang-lq@cs.sjtu.edu.cn

Abstract. Learning to generate new images for a novel category based on only a few images, named as few-shot image generation, has attracted increasing research interest. Several state-of-the-art works have yielded impressive results, but the diversity is still limited. In this work, we propose a novel Delta Generative Adversarial Network (DeltaGAN), which consists of a reconstruction subnetwork and a generation subnetwork. The reconstruction subnetwork captures intra-category transformation, *i.e.*, "delta", between same-category pairs. The generation subnetwork generates sample-specific "delta" for an input image, which is combined with this input image to generate a new image within the same category. Besides, an adversarial delta matching loss is designed to link the above two subnetworks together. Extensive experiments on six benchmark datasets demonstrate the effectiveness of our proposed method. Our code is available at https://github.com/bcmi/DeltaGAN-Few-Shot-Image-Generation.

1 Introduction

With the great success of deep learning, existing deep image generation models [5,6,10,18,28,32,33,43,45,46] based on Variational Auto-Encoder (VAE) [35] or Generative Adversarial Network (GAN) [22] have made a significant leap forward for generating diverse and realistic images for a given category. These methods generally require amounts of training images to generate new images for a given category. For the long-tail or newly emerging categories with only a few images, directly training or finetuning on limited data may cause overfitting issue [19,71]. Besides, it is very tedious to finetune the model for each unseen category. Therefore, given a few images from an unseen category, it is necessary to consider how to generate new realistic and diverse images for this category instantly. This task is called few-shot image generation in previous literature [2,4,29,30]. In this paper, following [2,4,29,30], we target at achieving

Supplementary Information The online version contains supplementary material available at https://doi.org/10.1007/978-3-031-19787-1_15.

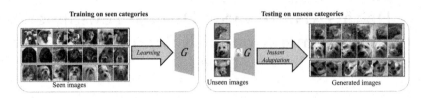

Fig. 1. The illustration of few-shot image generation task. We train a generative model on multiple seen categories. The learned generative model can be instantly applied to generate new images for unseen categories at test time. Each color indicates one category (Color figure online)

instant adaptation from multiple seen categories to unseen categories without finetuning as shown in Fig. 1, which can benefit a lot of downstream tasks like low-data classification and few-shot classification.

The abovementioned few-shot image generation methods [2,4,29,30] resort to seen categories with sufficient training images to train a generative model, which can be used to generate new images for an unseen category with only a few images, which are dubbed as conditional images. For brevity, we refer to the images from seen (*resp.*, unseen) categories as seen (*resp.*, unseen) images. We classify the few-shot image generation methods into fusion-based methods [4,23,29,30] and transformation-based method [2]. However, those fusion-based methods can only produce images similar to conditional images and cannot be applied to one-shot image generation. Although transformation-based method could produce new images based on one conditional image, however, it fails to produce diverse images.

Following the research line of transformation-based methods, we propose a novel Delta Generative Adversarial Network (DeltaGAN), which can generate new images based on one conditional image by sampling random vectors. Our DeltaGAN is inspired by few-shot feature generation method Delta-encoder [57], in which intra-category transformation (*i.e.*, the difference between two images within the same category) is called "delta". The main idea of Delta-encoder is shown in Fig. 2(a). In the training stage, Delta-encoder learns to extract delta $\boldsymbol{\Delta}^r$ from same-category feature pair $\{\boldsymbol{f}_{x_1}, \boldsymbol{f}_{x_2}\}$ of image pair $\{\boldsymbol{x}_1, \boldsymbol{x}_2\}$ from seen categories, in which $\boldsymbol{\Delta}^r$ is the additional information required to reconstruct $\boldsymbol{f}_{x_2}$ from $\boldsymbol{f}_{x_1}$. We refer to $\boldsymbol{x}_1$ as conditional (source) sample and $\boldsymbol{x}_2$ as target sample. In the testing stage, these extracted deltas are applied to a conditional feature $\boldsymbol{f}_y$ of image $\boldsymbol{y}$ from an unseen category to generate new feature $\tilde{\boldsymbol{f}}_y$ for this unseen category. However, Delta-encoder is a few-shot feature generation method, which cannot be directly applied to image generation. Besides, Delta-encoder relies on the deltas extracted from same-category training pairs, which does not support stochastic sampling (*i.e.*, sampling random vectors) to generate new samples in the testing stage.

In this paper, we aim to extend Delta-encoder to few-shot image generation method DeltaGAN, which supports producing diverse deltas based on random vectors. In this way, we can sample random vectors to generate diverse images without reaching training data in the testing stage. Considering that the

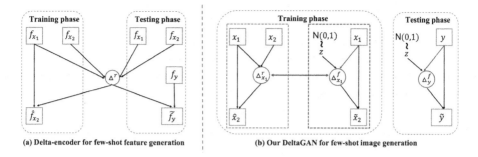

(a) Delta-encoder for few-shot feature generation   (b) Our DeltaGAN for few-shot image generation

Fig. 2. The illustration of evolving from Delta-encoder to our DeltaGAN. $\{x_1, x_2\}$ (*resp.*, $\{f_{x_1}, f_{x_2}\}$) is a same-category seen image pair (*resp.*, feature pair). y (*resp.*, f_y) is a conditional image (*resp.*, feature) from an unseen category. $\{\hat{x}_2, \tilde{x}_2, \tilde{y}\}$ (*resp.*, $\{\hat{f}_{x_2}, \tilde{f}_y\}$) are generated images (*resp.*, features). z is a random vector. $\{\Delta^r, \Delta^r_{x_1}\}$ (*resp.*, $\{\Delta^f_{x_1}, \Delta^f_y\}$) means real (*resp.*, fake) deltas. Red arrows indicate using adversarial delta matching loss to bridge the gap between real and fake delta. In (b), the green (*resp.*, blue) box encloses the reconstruction (*resp.*, generation) subnetwork, and pink arrows indicate the process of generating sample-specific delta (Color figure online)

plausibility of delta may depend on the conditional image [1], that is, a plausible delta for one conditional image may be unsuitable for another conditional image (see Sect. 4.3), we aim to produce sample-specific delta. In particular, we take in a random vector and a conditional image to generate sample-specific delta, which represents the transformation from this conditional image to another possible image from the same category. We conjecture that the ability of generating sample-specific delta can be transferred from seen categories to unseen categories. To this end, we develop our DeltaGAN according to Fig. 2(b). In the training phase, we use a reconstruction subnetwork to reconstruct x_2 from x_1 with the delta $\Delta^r_{x_1}$ (real delta) extracted from $\{x_1, x_2\}$. We also use a generation subnetwork to generate sample-specific delta $\Delta^f_{x_1}$ (fake delta) and produce new image $\tilde{x}_2$. To ensure that fake deltas function similarly to real ones, we introduce a novel adversarial delta matching loss by using a delta matching discriminator to judge whether an input-output image pair matches the corresponding delta. Besides, we employ a variant of mode seeking loss [44] to alleviate the mode collapse issue. We also employ typical adversarial loss and classification loss to make the generated images realistic and category-preserving. In the testing stage, given a conditional unseen image y, we can obtain its sample-specific delta Δ^f_y by sampling random vector z for producing new image $\tilde{y}$ from the category of y. Because each delta represents one possible intra-category transformation, given a conditional unseen image, different deltas can produce realistic and diverse images from the same unseen category. Extensive experiments on six benchmark datasets demonstrate the effectiveness of our proposed method. Our contributions can be summarized as follows:

– We propose a novel delta-based few-shot image generation method, which has never been explored before.

– Technically, we extend few-shot feature generation method Delta-encoder to few-shot image generation with stochastic sampling and sample-specific delta. We also design a novel adversarial delta matching loss.
– Our method can produce diverse and realistic images for each unseen category based on a single conditional image, surpassing existing few-shot image generation methods by a large margin.

2   Related Work

Data Augmentation: Data augmentation targets at augmenting training data with new samples. Traditional data augmentation tricks (*e.g.*, crop, flip, color jittering) only have limited diversity. Also, there are some methods [15,27,39,60] proposed to learn optimal augmentation strategies to improve the accuracy of classifiers. Similarly, neural augmentation [7,31,51,53,70] allowed a network to learn augmentations. As another research line, deep generative models can generate more diverse samples to augment training data, which can be categorized into feature-based augmentation methods [2] and image-based augmentation methods [57]. Feature-based augmentation methods [12,24] focused on generating more diverse deep features to augment the feature space of training data, while image-based augmentation methods [11,29,30,62] targeted at exploiting the distribution of training images and generating more diverse images.

Few-Shot Feature Generation. In existing few-shot feature generation methods, the semantic knowledge learned from the seen categories is transferred to compensate unseen categories in [17,24]. cCov-GAN [21] proposed a covariance-preserving adversarial augmentation network to generate more features for unseen categories. In [66], a generator subnetwork was added to a classification network to generate new examples. Intra-category diversity learned from seen categories was transferred to unseen categories to generate new features in [40,57]. Dual TriNet [12] proposed to synthesize instance features by leveraging semantics using a novel auto-encoder network for unseen categories. DTN [9] learned to transfer latent diversities from seen categories and composite them with support features to generate diverse features for unseen categories.

Few-Shot Image Generation. Compared with few-shot feature generation, few-shot image generation is a more challenging problem. Early methods can only be applied to generate new images for simple concepts, such as Bayesian program learning in [36], Bayesian reasoning in [55], and neural attention in [54]. Recently, several more advanced methods have been proposed to generate new real-world images in few-shot setting. To name a few, fusion-based method GMN [4] (*resp.*, MatchingGAN [29]) combined Matching Network [64] with Variational Auto-Encoder [52] (*resp.*, Generative Adversarial Network [22]) to generate new images without finetuning in the test phase. F2GAN [30] was designed to enhance the fusion ability of model by filling the details borrowed from conditional images. Transformation-based method DAGAN [2] proposed to produce new images by injecting random vectors into the generator conditioned on a

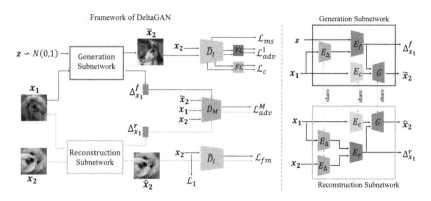

Fig. 3. Our DeltaGAN mainly consists of a reconstruction subnetwork and a generation subnetwork. Generation subnetwork learns to generate new image $\tilde{x}_2$ based on conditional image x_1 and random vector z. Reconstruction subnetwork learns to produce reconstructed target image $\hat{x}_2$ based on image pair $\{x_1, x_2\}$. Best viewed in color (Color figure online)

single image. Apart from fusion-based and transformation-based methods, there also exist optimization-based methods. For example, FIGR [13] (*resp.*, DAWSON [38]) combined adversarial learning with meta-learning method Reptile [48] (*resp.*, MAML [20]) to generate new images. However, they need to fine-tune the trained model with unseen category. Moreover, they can hardly produce sharp and realistic images. In this work, we propose a new transformation-based few-shot image generation method, which can produce more diverse images than previous methods based on a single image.

Note that some more recent works [37,50,56,65] are also called few-shot image generation. However, these works focus on adapting the generative model pretrained on a large dataset to a small dataset with a few examples, whose setting is quite different from ours. Firstly, these methods target at adapting from one source domain to another target domain, whereas our method adapts from multiple seen categories to unseen categories. Secondly, the models of these works need to be finetuned for each unseen domain, which is very tedious. Instead, the model of our method can be instantly applied to unseen categories without finetuning.

3   Our Method

We split all categories into seen categories and unseen categories, which have no overlap. Our DeltaGAN mainly consists of a reconstruction subnetwork and a generation subnetwork as shown in Fig. 3. The detailed architecture of each encoder/decoder is reported in Supplementary. In the training stage, given a same-category seen image pair $\{x_1, x_2\}$ where x_1 is the conditional image and x_2 is the target image, the reconstruction subnetwork extracts real delta $\Delta_{x_1}^r$ from this pair, and reconstructs the target image x_2 based on x_1 and $\Delta_{x_1}^r$. In the generation subnetwork, a random vector z and the conditional image x_1 are used to obtain fake sample-specific delta $\Delta_{x_1}^f$, which collaborates with x_1 to generate a new image $\tilde{x}_2$. Moreover, we design an adversarial delta matching loss to bridge the gap between real delta and fake delta. In the testing stage, given an unseen image y, only generation subnetwork is used to produce diverse and realistic images $\{\tilde{y}_k\}$ belonging to the same category of y.

3.1   Reconstruction Subnetwork

In the reconstruction subnetwork (see Fig. 3), there are three encoders E_Δ, E_c, E_r and a decoder G. Given a same-category seen image pair $\{x_1, x_2\}$, we use E_Δ to extract paired features $\{E_\Delta(x_1), E_\Delta(x_2)\} \in \mathcal{R}^{W \times H \times C}$, where $W \times H$ denotes the feature map size and C denotes the channel number. Then, we calculate the difference between $E_\Delta(x_2)$ and $E_\Delta(x_1)$, which is fed into E_r to obtain real delta $\Delta_{x_1}^r \in \mathcal{R}^{W \times H \times C}$:

$$\Delta_{x_1}^r = E_r(E_\Delta(x_2) - E_\Delta(x_1)), \tag{1}$$

where $\Delta_{x_1}^r$ contains the additional information needed to reconstruct x_2 from x_1. We do not restrict our delta features to be linear offsets, which enables the delta features to learn more complex transformations. Then, $\Delta_{x_1}^r$ is concatenated with $E_c(x_1) \in \mathcal{R}^{W \times H \times C}$ and fed into G to obtain the reconstructed image $\hat{x}_2$:

$$\hat{x}_2 = G(\Delta_{x_1}^r, E_c(x_1)). \tag{2}$$

We employ a reconstruction loss $\mathcal{L}_1$ to ensure that $\hat{x}_2$ is close to x_2:

$$\mathcal{L}_1 = ||\hat{x}_2 - x_2||_1. \tag{3}$$

Considering the instability issue of early training stage, we use a feature matching loss [3] by matching the discriminative feature of $\hat{x}_2$ with that of x_2. In detail, we use a feature extractor $\hat{D}_I$ to extract the discriminative features of $\hat{x}_2$ and x_2 in each layer to calculate the feature matching loss:

$$\mathcal{L}_{fm} = \frac{1}{L} \sum_{l=1}^{L} ||\hat{D}_I^l(x_2) - \hat{D}_I^l(\hat{x}_2)||_1, \tag{4}$$

where L is the layer number of $\hat{D}_I$.

To support stochastic sampling for generation, we design another generation subnetwork in parallel with the reconstruction subnetwork (see Fig. 3). Two subnetworks share two encoders E_Δ, E_c and the decoder G. Besides, a new encoder E_f is introduced to obtain fake sample-specific delta. In our generation subnetwork, we concatenate a random vector z sampled from unit Gaussian distribution and the feature of conditional image $E_\Delta(x_1) \in \mathcal{R}^{W \times H \times C}$, which is fed into E_f to obtain sample-specific delta $\Delta_{x_1}^f \in \mathcal{R}^{W \times H \times C}$:

$$\Delta_{x_1}^f = E_f(z, E_\Delta(x_1)), \tag{5}$$

where $\boldsymbol{\Delta}_{x_1}^f$ contains the additional information needed to transform conditional image x_1 to another possible image within the same category. Then, analogous to the reconstruction subnetwork, $\boldsymbol{\Delta}_{x_1}^f$ is concatenated with $E_c(x_1)$ and fed into G to produce a new image $\tilde{x}_2$ belonging to the category of x_1:

$$\tilde{x}_2 = G(\boldsymbol{\Delta}_{x_1}^f, E_c(x_1)), \qquad (6)$$

in which $\tilde{x}_2$ is the transformed result after applying delta $\boldsymbol{\Delta}_{x_1}^f$ to x_1.

3.2   Generation Subnetwork

Adversarial Loss: To make the generated image $\tilde{x}_2$ close to real images, we employ a standard adversarial loss using the discriminator D_I. D_I contains the feature extractor $\hat{D}_I$ mentioned in Sect. 3.1 and a fully-connected (FC) layer. We adopt the hinge adversarial loss proposed in [47]:

$$\begin{aligned}
\mathcal{L}_{adv,D}^I &= \mathbb{E}_{x_2}[\max(0, 1 - D_I(x_2))] + \mathbb{E}_{\tilde{x}_2}[\max(0, 1 + D_I(\tilde{x}_2))], \\
\mathcal{L}_{adv,G}^I &= -\mathbb{E}_{\tilde{x}_2}[D_I(\tilde{x}_2)].
\end{aligned} \qquad (7)$$

The discriminator D_I tends to distinguish fake images from real images by minimizing $\mathcal{L}_{adv,D}^I$, while the generator tends to generate realistic images to fool the discriminator by minimizing $\mathcal{L}_{adv,G}^I$.

Classification Loss: To ensure that $\tilde{x}_2$ belongs to the expected category, we construct a classifier by replacing the last FC layer of D_I with another FC layer (the number of outputs is the number of seen categories). Then, the images from different categories can be distinguished by a cross-entropy classification loss:

$$\mathcal{L}_c = -\log p(c(x)|x), \qquad (8)$$

where $c(x)$ is the category label of x. We train the classifier by minimizing $\mathcal{L}_{c,D} = -\log p(c(x_2)|x_2)$ of the target image x_2. We also expect the generated image $\tilde{x}_2$ to be classified as the same category of target image x_2. Thus, we minimize $\mathcal{L}_{c,G} = -\log p(c(x_2)|\tilde{x}_2)$ when updating the generator.

Adversarial Delta Matching Loss: To ensure that the generated sample-specific deltas function similarly to real deltas and encode the intra-category transformation, we design a novel adversarial delta matching loss to bridge the gap between real deltas and fake deltas. This goal is accomplished by a delta matching discriminator D_M, which takes a triplet (conditional image, output image, the delta between them) as input as shown in Fig. 3. Our delta matching discriminator D_M is constructed by feature extractor $\hat{D}_I$ and four FC layers following global average pooling. In delta matching discriminator D_M, we extract the features of paired images $\{\hat{D}_I(x_1), \hat{D}_I(x_2)\}$ (*resp.*, $\{\hat{D}_I(x_1), \hat{D}_I(\tilde{x}_2)\}$), which are concatenated with sample-specific delta $\boldsymbol{\Delta}_{x_1}^r$ (*resp.*, $\boldsymbol{\Delta}_{x_1}^f$) to form a real (*resp.*, fake) triplet. Then, the real triplet and fake triplet are fed into the four FC layers to judge whether this conditional-output image pair matches the corresponding delta, in other words, whether the delta is the additional information required to transform the conditional image to the output image. In adversarial learning, the discriminator D_M needs to distinguish the real triplet $\{x_1, x_2, \Delta_{x_1}^r\}$ from the fake triplet $\{x_1, \tilde{x}_2, \Delta_{x_1}^f\}$, while the generator aims to synthesize realistic fake triplet to fool the discriminator. The delta matching adversarial loss is also in the form of hinge adversarial loss [47], which can be written as

$$
\begin{aligned}
\mathcal{L}_{adv,D}^M &= \mathbb{E}_{x_1,x_2,\Delta_{x_1}^r}[\max(0, 1 - D_M(x_1, x_2, \Delta_{x_1}^r))] \\
&+ \mathbb{E}_{x_1,\tilde{x}_2,\Delta_{x_1}^f}[\max(0, 1 + D_M(x_1, \tilde{x}_2, \Delta_{x_1}^f))], \\
\mathcal{L}_{adv,G}^M &= -\mathbb{E}_{x_1,\tilde{x}_2,\Delta_{x_1}^f}[D_M(x_1, \tilde{x}_2, \Delta_{x_1}^f)],
\end{aligned}
\tag{9}
$$

where $\mathcal{L}_{adv,D}^M$ (resp., $\mathcal{L}_{adv,G}^M$) is optimized for updating $\{\hat{D}_I, D_M\}$ (resp., the generator).

Mode Seeking Loss: We observe that by varying random vector z, the generated images may collapse into a few modes, which is referred to as mode collapse [44]. Therefore, we use a variant of mode seeking loss [44] to seek for more modes to enhance the diversity of generated images. Different from [44], we apply mode seeking loss to multi-layer features extracted by $\hat{D}_I$. In particular, we minimize the ratio of the distance between z_1 and z_2 over the distance between $\hat{D}_I^l(\tilde{x}_2^1)$ and $\hat{D}_I^l(\tilde{x}_2^2)$ at the l-th layer of $\hat{D}_I$:

$$
\mathcal{L}_{ms} = \frac{1}{L} \sum_{l=1}^{L} \frac{||z_1 - z_2||_1}{||\hat{D}_I^l(\tilde{x}_2^1) - \hat{D}_I^l(\tilde{x}_2^2)||_1}.
\tag{10}
$$

Intuitively, when $||z_1 - z_2||_1$ is large, we expect $\hat{D}_I^l(\tilde{x}_2^1)$ and $\hat{D}_I^l(\tilde{x}_2^2)$ to be considerably different, which can push the generator to search more modes to produce diverse images. In our experiments (see Sect. 4.3), we find that mode seeking loss is critical for diversity. However, without the guidance of reconstruction subnetwork and adversarial delta matching loss, solely using mode seeking loss cannot generate meaningful deltas, with both diversity and realism significantly downgraded.

3.3 Optimization

We use θ_G to denote the model parameters of $\{E_\Delta, E_r, E_c, E_f, G\}$, while θ_D is used to denote the model parameters of $\{D_I, D_M\}$. The total loss function of our method can be written as

$$
\mathcal{L} = \mathcal{L}_{adv}^I + \mathcal{L}_{adv}^M + \lambda_1 \mathcal{L}_1 + \mathcal{L}_c + \lambda_{fm} \mathcal{L}_{fm} + \lambda_{ms} \mathcal{L}_{ms},
\tag{11}
$$

in which λ_1, λ_{fm}, and λ_{ms} are trade-off parameters. $\mathcal{L}_{adv}^I$ represents $\mathcal{L}_{adv,G}^I$ (resp., $\mathcal{L}_{adv,D}^I$) when updating the model parameters θ_G (resp., θ_D). Similarly, $\mathcal{L}_{adv}^M$ represents $\mathcal{L}_{adv,G}^M$ (resp., $\mathcal{L}_{adv,D}^M$) when updating the model parameters θ_G (resp., θ_D).

Table 1. FID ($\downarrow$) and LPIPS ($\uparrow$) of images generated by different methods for unseen categories on four datasets in 1/3-shot setting

Method	Shot	VGGFace		Flowers		Animal Faces		NABirds	
		FID	LPIPS	FID	LPIPS	FID	LPIPS	FID	LPIPS
FIGR [13]	3	139.83	0.0834	190.12	0.0634	211.54	0.0756	210.75	0.0918
DAWSON [38]	3	137.82	0.0769	188.96	0.0583	208.68	0.0642	181.97	0.1105
GMN [4]	3	136.21	0.0902	200.11	0.0743	220.45	0.0868	208.74	0.0923
DAGAN [2]	3	128.34	0.0913	151.21	0.0812	155.29	0.0892	159.69	0.1405
DAGAN [2]	1	*134.28*	*0.0608*	*179.59*	*0.0496*	*185.54*	*0.0687*	*183.57*	*0.0967*
MatchingGAN [29]	3	118.62	0.1695	143.35	0.1627	148.52	0.1514	142.52	0.1915
F2GAN [30]	3	109.16	0.2125	120.48	0.2172	117.74	0.1831	126.15	0.2015
LoFGAN [23]	3	106.24	0.2096	112.55	0.2687	116.45	0.1756	124.56	0.2041
DeltaGAN	3	**78.35**	**0.3487**	**104.62**	**0.4281**	**87.04**	**0.4642**	**95.97**	**0.5136**
DeltaGAN	1	*80.12*	*0.3146*	*109.78*	*0.3912*	*89.81*	*0.4418*	*96.79*	*0.5069*

θ_G and θ_D are optimized using related loss terms in an alternating fashion. In particular, θ_D is optimized by minimizing $\mathcal{L}_{adv,D}^I + \mathcal{L}_{adv,D}^M + \mathcal{L}_{c,D}$. θ_G is optimized by minimizing $\mathcal{L}_{adv,G}^I + \mathcal{L}_{adv,G}^M + \lambda_1 \mathcal{L}_1 + \mathcal{L}_{c,G} + \lambda_{fm}\mathcal{L}_{fm} + \lambda_{ms}\mathcal{L}_{ms}$, in which $\mathcal{L}_{c,D}$ and $\mathcal{L}_{c,G}$ are defined below Eq. 8.

4   Experiments

We conduct experiments on six few-shot image datasets: EMNIST [14], VGGFace [8], Flowers [49], Animal Faces [16], NABirds [63], and Foods [34]. Following MatchingGAN and FUNIT, we split all categories into seen categories and unseen categories. After having a few trials, we set $\lambda_1 = 10$, $\lambda_{fm} = 0.1$, and $\lambda_{ms} = 10$ by observing the quality of generated images during training. We adopt Adam optimizer with learning rate of $1e-4$. The batch size is set to 16 and our model is trained for 200 epochs. The details of datasets and implementation are reported in Supplementary.

4.1   Evaluation of Generated Images

To evaluate the quality of images generated by different methods, we calculate Fréchet Inception Distance (FID) [26] and Learned Perceptual Image Patch Similarity (LPIPS) [69] on four datasets. FID is used to measure the distance between the extracted features of generated unseen images and those of real unseen images. LPIPS is used to measure the diversity of generated unseen images. For each unseen category, the average of pairwise distances among generated images is calculated, and then the average of all unseen categories is calculated as the final LPIPS score. Since the number of conditional images in fusion-based methods GMN [4], MatchingGAN [29], F2GAN [30], and LoF-GAN [23]) is a tunable hyper-parameter, we use 3 conditional images in each training and testing episode. In the testing stage, if K images are provided for each unseen category, we refer to this setting as K-shot setting. We report the

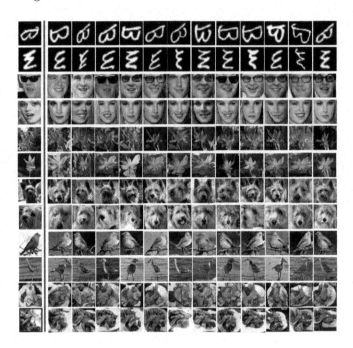

Fig. 4. Images generated by our DeltaGAN in 1-shot setting on four datasets (from top to bottom:EMNIST, VGGFace, Flowers, Animal Faces, NABirds, and Foods). The conditional images are in the leftmost column

3-shot results for all methods and 1-shot results for the methods which only require one conditional image.

In either setting, following [23, 30], we use each method to generate 128 images for each unseen category, which are used to calculate FID and LPIPS. For DeltaGAN and DAGAN which are applicable to both 1-shot and 3-shot settings, we generate 128 images based on one conditional image in 1-shot setting and generate 128 images by randomly sampling one conditional image each time in 3-shot setting. The results are summarized in Table 1, we can observe that our method achieves the lowest FID and highest LPIPS in the 3-shot setting, which demonstrates that our method could generate more diverse and realistic images compared with baseline methods. Besides, our method in 1-shot setting also achieves competitive results, which are even better than other baselines in 3-shot setting. We also compare our DeltaGAN with other few-shot image generation method [50] in Supplementary.

We show some example images generated by our DeltaGAN on six datasets in Fig. 4. We exhibit 12 generated images based on one conditional unseen image by sampling different random vectors. On EMNIST dataset, we can see that generated images maintain the concepts of conditional images and have remarkable diversity. On natural datasets VGGFace, Flowers, Animal Faces, NABirds, and Foods, our DeltaGAN can generate diverse images with high fidelity.

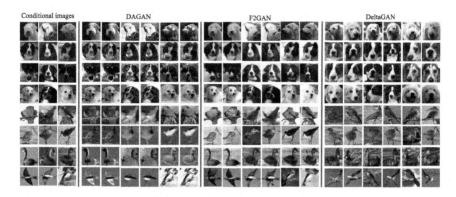

Fig. 5. Images generated by DAGAN, F2GAN, and our DeltaGAN in 3-shot setting on two datasets (from top to bottom: Animal Faces and NABirds). The conditional images are in the left three columns

For comparison, we also show some example images generated by DAGAN and F2GAN in Fig. 5. For DAGAN, we arrange the results according to the conditional image. It can be seen that the structures of images produced by DAGAN are almost the same as the conditional image. For F2GAN, the generated images are still close to one of the conditional images and may have unreasonable shapes when fusing conditional images. Apparently, our DeltaGAN can produce images of higher quality and more diversity.

4.2   Few-Shot Classification

In this section, we demonstrate that the new images generated by our DeltaGAN can greatly benefit few-shot classification. The experiments for low-data classification and comparison with traditional data augmentation methods can be found in Supplementary. Following the N-way C-shot setting in few-shot classification [20], in which evaluation episodes are created and the averaged accuracy over multiple evaluation episodes is calculated for evaluation. In each evaluation episode, N categories from unseen categories are randomly selected and C images from each of N categories are randomly selected. These selected $N \times C$ images are used as training set while the remaining unseen images from N unseen categories are used as test set. We pretrain ResNet18 [25] on the seen images and remove the last FC layer as the feature extractor, which is used to extract features for unseen images. In each evaluation episode in N-way C-shot setting, our DeltaGAN generates 512 new images to augment each of N categories. Based on the extracted features, we train a linear classifier with $N \times (C + 512)$ training images, which is then applied to the test set. we train a linear classifier to evaluate the few-shot generation ability of our DeltaGAN. Besides $N \times C$ training images, our generator can generate 512 images to augment each of N categories in the training set.

Table 2. Accuracy(%) of different methods on three datasets in few-shot classification setting (10-way 1/5-shot). Note that fusion-based methods MatchingGAN, F2GAN, and LoFGAN are not applicable in 1-shot setting

Method	VGGFace		Flowers		Animal Faces	
	1-shot	5-shot	1-shot	5-shot	1-shot	5-shot
MatchingNets [64]	33.68	48.67	40.96	56.12	36.54	50.12
MAML [20]	32.16	47.89	42.95	58.01	35.98	49.89
RelationNets [59]	39.95	54.12	48.18	61.03	45.32	58.12
MTL [58]	51.45	68.95	54.34	73.24	52.54	70.91
MatchingNet-LFT [61]	54.34	69.92	58.41	74.32	56.83	71.62
DPGN [67]	54.83	70.27	58.95	74.56	57.18	72.02
DeepEMD [68]	54.15	70.35	59.12	73.97	58.01	72.71
GCNET [41]	53.73	71.68	57.61	72.47	56.64	71.53
Delta-encoder [57]	53.19	67.57	56.05	72.84	56.38	71.29
MatchingGAN [29]	–	70.94	–	74.09	–	70.89
F2GAN [30]	–	72.31	–	75.02	–	73.19
LoFGAN [23]	–	73.01	–	75.86	–	73.43
DeltaGAN	**56.85**	**75.71**	**61.23**	**77.09**	**60.31**	**74.59**

We compare our DeltaGAN with existing few-shot classification methods, including the representative methods MatchingNets [64], RelationNets [59], MAML [20] as well as the state-of-the-art methods MTL [58], MatchingNet-LFT [61], DPGN [67], DeepEMD [68], and GCNET [41]. For these baselines, no augmented images are added to the training set in each evaluation episode. Instead, the images from seen categories are used to train those few-shot classifiers by strictly following their original training procedure. We also compare our DeltaGAN with few-shot image generation methods MatchingGAN and F2GAN as well as few-shot feature generation method Delta-encoder. We adopt the same augmentation strategy as our DeltaGAN in each evaluation episode. Besides, we compare our DeltaGAN with few-shot image translation method FUNIT [42] in Supplementary. By taking 10-way 1-shot/5-shot as examples, we report the averaged accuracy over 10 episodes on three datasets in Table 2. Our method achieves the best performance on all datasets compared with few-shot classification and few-shot generation baselines, which demonstrates the high quality of generated images by our DeltaGAN.

4.3   Ablation Studies

We analyze the impact of each loss and alternative network designs on Animal Faces dataset in 1-shot setting. For each ablated method, FID, LPIPS, and the accuracy of 10-way 1-shot classification augmented with generated images are reported in Table 3.

Table 3. Ablation studies of our loss terms and alternative network designs on Animal Faces dataset

Setting	Accuracy (%) ↑	FID ↓	LPIPS ↑
w/o $\mathcal{L}_1$	58.68	100.21	0.4191
w/o $\mathcal{L}_{ms}$	50.08	121.74	0.2976
w/o $\mathcal{L}_{fm}$	59.17	95.82	0.4324
w/o $\mathcal{L}_c$	42.21	196.18	0.4119
w/o $\mathcal{L}_{adv}^I$	52.18	139.46	0.3912
w/o $\mathcal{L}_{adv}^M$	57.12	115.11	0.4153
w/o real delta	53.03	128.69	0.3838
Global delta	58.96	94.51	0.4311
SC delta	56.11	101.05	0.4162
DC delta	55.29	105.91	0.4021
Simple D_1	54.53	129.17	0.3012
Simple D_2	58.01	109.54	0.4401
Simple D_3	59.51	94.12	0.4392
Linear delta	53.89	122.71	0.4091
DeltaGAN	**60.31**	**89.81**	**0.4418**

Loss Terms: In our method, we employ a reconstruction loss $\mathcal{L}_1$, a mode seeking loss $\mathcal{L}_{ms}$, a feature matching loss $\mathcal{L}_{fm}$, a classification loss $\mathcal{L}_c$, and an adversarial loss $\mathcal{L}_{adv}^I$. To investigate the impact of each loss term, we conduct ablation studies on Animal Faces dataset by removing each loss term from the final objective in Eq. 11 separately. The results are summarized in Table 3, which shows that the diversity and fidelity of generated images are compromised when removing $\mathcal{L}_1$. By removing mode seeking loss $\mathcal{L}_{ms}$, we can see that all metrics become much worse, which implies the mode collapse issue after removing $\mathcal{L}_{ms}$. Another observation is that ablating $\mathcal{L}_{fm}$ leads to slight degradation of generated images. Removing $\mathcal{L}_c$ results in severe degradation of generated images, since the generated images may not belong to the category of conditional image. When $\mathcal{L}_{adv}^I$ is removed from the final objective, the worse quality of generated images indicates that typical adversarial loss can ensure the fidelity of generated images. To investigate the impact of our adversarial delta matching loss $\mathcal{L}_{adv}^M$ in Eq. 9, We remove $\mathcal{L}_{adv}^M$ from the final objective in Eq. 11, which is referred to as "w/o $\mathcal{L}_{adv}^M$" in Table 3. We can see that the diversity and fidelity of generated images are compromised without $\mathcal{L}_{adv}^M$, because $\mathcal{L}_{adv}^M$ can bridge the gap between real delta and fake delta.

Without Real Delta: To investigate the necessity of enforcing generated fake deltas to be close to real deltas, we cut off the links between real delta and fake delta by removing the reconstruction subnetwork and adversarial delta matching loss (*i.e.*, removing $\{\mathcal{L}_{adv}^M, \mathcal{L}_1, \mathcal{L}_{fm}\}$), which is referred to as "w/o real delta" in

Table 3. Compared with DeltaGAN, both diversity and realism are significantly degraded, because generation subnetwork fails to generate meaningful deltas without the guidance of reconstruction subnetwork and adversarial delta matching loss. Thus, we conclude that mode seeking loss needs to cooperate with our framework to produce realistic and diverse images. Another observation is that "w/o $\mathcal{L}_{adv}^{M}$" is better than "w/o real delta", which can be explained as follows. Even without using adversarial delta matching loss, since the reconstruction subnetwork and the generation subnetwork share the same E_c and G, generated fake deltas have been implicitly pulled close to real deltas.

Sample-Specific Delta: To corroborate the superiority of sample-specific delta, we directly use random vectors to generate deltas, which is referred to as "Global delta" in Table 3. It can be seen that our design of sample-specific deltas can benefit the quality of generated images. Besides, with our trained DeltaGAN model, we exchange sample-specific deltas within images from the same category (*resp.*, across different categories) to generate new images, which is referred to as "SC delta" (*resp.*, "DC delta") in Table 3. Compared with "SC delta" and "DC delta", our DeltaGAN achieves the best performance on all metrics, which verifies our assumption that delta is sample-specific and exchangeable use of deltas may lead to performance drop. We also visualize some examples generated by "SC delta" (*resp.*, "DC delta") in Supplementary.

Delta Matching Discriminator: In Sect. 3.2, we use conditional image, sample-specific delta, and output image as input triplet $\{\hat{D}_I(\boldsymbol{x}_1), \boldsymbol{\Delta}_{x_1}, \hat{D}_I(\boldsymbol{x}_2)\}$ for our delta matching discriminator D_M, which judges whether the conditional-output image pair matches the corresponding sample-specific delta. To evaluate the effectiveness and necessity of this input format, we explore different types of inputs for delta matching discriminator. As shown in Table 3, we use $\{\boldsymbol{\Delta}_{x_1}\}$ (*resp.*, $\{\hat{D}_I(\boldsymbol{x}_1), \boldsymbol{\Delta}_{x_1}\}$, $\{\hat{D}_I(\boldsymbol{x}_2), \boldsymbol{\Delta}_{x_1}\}$) as inputs of D_M, which is referred to as "Simple D_1" (*resp.*, "Simple D_2", "Simple D_3"). We can see that "Simple D_1" is the worst, which demonstrates that only employing adversarial loss on delta does not work well. Besides, both "Simple D_2" and "Simple D_3" are worse than our DeltaGAN, which demonstrates the effectiveness of matching conditional-output image pair with the corresponding sample-specific delta.

Linear Offset Delta: To evaluate the effect of the learned non-linear "delta", we replace the non-linear "delta" with linear "delta", which is referred to as "Linear delta" in Table 3. In the reconstruction subnetwork, we set $\boldsymbol{\Delta}_{x_1}^r = E_\Delta(\boldsymbol{x}_2) - E_\Delta(\boldsymbol{x}_1)$, and $\hat{\boldsymbol{x}}_2 = G(\boldsymbol{\Delta}_{x_1}^r + E_c(\boldsymbol{x}_1))$, which means that we simply learn linear offset "delta" from same-class pairs of training data. In the generation subnetwork, we apply the generated fake "delta" $\boldsymbol{\Delta}_{x_1}^f$ to conditional image $\boldsymbol{x}_1$ to generate new image $\tilde{\boldsymbol{x}}_2 = G(\boldsymbol{\Delta}_{x_1}^f + E_c(\boldsymbol{x}_1))$. Based on Table 3, the FID gap between "Linear delta" and "DeltaGAN" indicates that complex transformations of intra-category pairs cannot be well captured by linear offset.

5   Conclusion

In this paper, we have explored applying sample-specific deltas to a conditional image to generate new images. Specifically, we have proposed a novel few-shot generation method DeltaGAN composed of a reconstruction subnetwork and a generation subnetwork, which are bridged by an adversarial delta matching loss. The experimental results on six datasets have shown that our DeltaGAN can substantially improve the quality and diversity of generated images compared with existing few-shot image generation methods.

Acknowledgement. The work is supported by Shanghai Municipal Science and Technology Key Project (Grant No. 20511100300), Shanghai Municipal Science and Technology Major Project, China (2021SHZDZX0102), and National Science Foundation of China (Grant No. 61902247).

References

1. Almahairi, A., Rajeswar, S., Sordoni, A., Bachman, P., Courville, A.C.: Augmented cycleGAN: learning many-to-many mappings from unpaired data. In: ICML (2018)
2. Antoniou, A., Storkey, A., Edwards, H.: Data augmentation generative adversarial networks. arXiv preprint arXiv:1711.04340 (2017)
3. Bao, J., Chen, D., Wen, F., Li, H., Hua, G.: CVAE-GAN: fine-grained image generation through asymmetric training. In: ICCV (2017)
4. Bartunov, S., Vetrov, D.: Few-shot generative modelling with generative matching networks. In: AISTATS (2018)
5. Binkowski, M., Sutherland, D.J., Arbel, M., Gretton, A.: Demystifying mmd GANs. In: ICLR (2018)
6. Brock, A., Donahue, J., Simonyan, K.: Large scale GAN training for high fidelity natural image synthesis. In: ICLR (2018)
7. Cao, B., Wang, N., Li, J., Gao, X.: Data augmentation-based joint learning for heterogeneous face recognition. IEEE Trans. Neural Netw. Learn. Syst. (TNNLS) **30**, 1731–1743 (2019)
8. Cao, Q., Shen, L., Xie, W., Parkhi, O.M., Zisserman, A.: VGGFace2: a dataset for recognising faces across pose and age. In: FG (2018)
9. Chen, M., et al.: Diversity transfer network for few-shot learning. In: AAAI (2020)
10. Chen, T., Zhai, X., Ritter, M., Lucic, M., Houlsby, N.: Self-supervised GANs via auxiliary rotation loss. In: CVPR (2019)
11. Chen, Z., Fu, Y., Wang, Y.X., Ma, L., Liu, W., Hebert, M.: Image deformation meta-networks for one-shot learning. In: CVPR (2019)
12. Chen, Z., Fu, Y., Zhang, Y., Jiang, Y.G., Xue, X., Sigal, L.: Multi-level semantic feature augmentation for one-shot learning. TIP **28**(9), 4594–4605 (2019)
13. Clouâtre, L., Demers, M.: FIGR: few-shot image generation with reptile. arXiv preprint arXiv:1901.02199 (2019)
14. Cohen, G., Afshar, S., Tapson, J., van Schaik, A.: EMNIST: an extension of MNIST to handwritten letters. In: IJCNN (2017)
15. Cubuk, E.D., Zoph, B., Mane, D., Vasudevan, V.K., Le, Q.V.: Autoaugment: learning augmentation strategies from data. In: CVPR (2019)
16. Deng, J., Dong, W., Socher, R., Li, L.J., Li, K., Fei-Fei, L.: ImageNet: a large-scale hierarchical image database. In: CVPR (2009)

17. Dixit, M., Kwitt, R., Niethammer, M., Vasconcelos, N.: AGA: attribute-guided augmentation. In: CVPR (2017)
18. Donahue, J., Simonyan, K.: Large scale adversarial representation learning. In: Advances in Neural Information Processing Systems (2019)
19. Feng, R., Gu, J., Qiao, Y., Dong, C.: Suppressing model overfitting for image super-resolution networks. In: CVPR (2019)
20. Finn, C., Abbeel, P., Levine, S.: Model-agnostic meta-learning for fast adaptation of deep networks. In: ICML (2017)
21. Gao, H., Shou, Z., Zareian, A., Zhang, H., Chang, S.: Low-shot learning via covariance-preserving adversarial augmentation networks. In: NeurIPS (2018)
22. Goodfellow, I., et al.: Generative adversarial nets. In: NeurIPS (2014)
23. Gu, Z., Li, W., Huo, J., Wang, L., Gao, Y.: LoFGAN: fusing local representations for few-shot image generation. In: ICCV (2021)
24. Hariharan, B., Girshick, R.B.: Low-shot visual recognition by shrinking and hallucinating features. In: ICCV (2017)
25. He, K., Zhang, X., Ren, S., Sun, J.: Deep residual learning for image recognition. In: CVPR (2016)
26. Heusel, M., Ramsauer, H., Unterthiner, T., Nessler, B., Hochreiter, S.: GANs trained by a two time-scale update rule converge to a local Nash equilibrium. In: NeurIPS (2017)
27. Ho, D., Liang, E., Chen, X., Stoica, I., Abbeel, P.: Population based augmentation: efficient learning of augmentation policy schedules. In: ICML (2019)
28. Hoffman, J., et al.: CyCADA: cycle-consistent adversarial domain adaptation. In: ICML (2018)
29. Hong, Y., Niu, L., Zhang, J., Zhang, L.: MatchingGAN: matching-based few-shot image generation. In: ICME (2020)
30. Hong, Y., Niu, L., Zhang, J., Zhao, W., Fu, C., Zhang, L.: F2GAN: fusing-and-filling GAN for few-shot image generation. In: ACM MM (2020)
31. Jo, H.J., Min, C.H., Song, J.B.: Bin picking system using object recognition based on automated synthetic dataset generation. In: 2018 15th International Conference on Ubiquitous Robots (UR) (2018)
32. Karras, T., Laine, S., Aila, T.: A style-based generator architecture for generative adversarial networks. In: CVPR (2019)
33. Karras, T., Laine, S., Aittala, M., Hellsten, J., Lehtinen, J., Aila, T.: Analyzing and improving the image quality of StyleGAN. arXiv preprint arXiv:1912.04958 (2019)
34. Kawano, Y., Yanai, K.: Automatic expansion of a food image dataset leveraging existing categories with domain adaptation. In: Agapito, L., Bronstein, M.M., Rother, C. (eds.) ECCV 2014. LNCS, vol. 8927, pp. 3–17. Springer, Cham (2015). https://doi.org/10.1007/978-3-319-16199-0_1
35. Kingma, D.P., Welling, M.: Auto-encoding variational Bayes. In: ICLR (2014)
36. Lake, B., Salakhutdinov, R., Gross, J., Tenenbaum, J.: One shot learning of simple visual concepts. In: CogSci (2011)
37. Li, Y., Zhang, R., Lu, J., Shechtman, E.: Few-shot image generation with elastic weight consolidation. In: NeurIPS (2020)
38. Liang, W., Liu, Z., Liu, C.: DAWSON: a domain adaptive few shot generation framework. arXiv preprint arXiv:2001.00576 (2020)
39. Lim, S., Kim, I., Kim, T., Kim, C., Kim, S.: Fast autoaugment. In: NeurIPS (2019)
40. Liu, J., Sun, Y., Han, C., Dou, Z., Li, W.: Deep representation learning on long-tailed data: a learnable embedding augmentation perspective. In: CVPR (2020)

41. Liu, L., et al.: GenDet: meta learning to generate detectors from few shots. TNNLS (2021)
42. Liu, M., et al.: Few-shot unsupervised image-to-image translation. In: ICCV (2019)
43. Makhzani, A., Frey, B.J.: PixelGAN autoencoders. In: Advances in Neural Information Processing Systems (2017)
44. Mao, Q., Lee, H.Y., Tseng, H.Y., Ma, S., Yang, M.H.: Mode seeking generative adversarial networks for diverse image synthesis. In: CVPR (2019)
45. Mao, X., Li, Q., Xie, H., Lau, R.Y., Wang, Z., Paul Smolley, S.: Least squares generative adversarial networks. In: ICCV (2017)
46. Miyato, T., Kataoka, T., Koyama, M., Yoshida, Y.: Spectral normalization for generative adversarial networks. In: Proceedings of 6th International Conference on Learning Representations (ICLR) (2018)
47. Miyato, T., Koyama, M.: CGANs with projection discriminator. In: ICLR (2018)
48. Nichol, A., Achiam, J., Schulman, J.: On first-order meta-learning algorithms. arXiv preprint arXiv:1803.02999 (2018)
49. Nilsback, M.E., Zisserman, A.: Automated flower classification over a large number of classes. In: CVGIP (2008)
50. Ojha, U., et al.: Few-shot image generation via cross-domain correspondence. In: CVPR (2021)
51. Perez, L., Wang, J.: The effectiveness of data augmentation in image classification using deep learning. In: CVPR (2017)
52. Pu, Y., Zhe, G., Henao, R., Xin, Y., Carin, L.: Variational autoencoder for deep learning of images, labels and captions. In: NeurIPS (2016)
53. Ratner, A.J., Ehrenberg, H., Hussain, Z., Dunnmon, J., Ré, C.: Learning to compose domain-specific transformations for data augmentation. In: NeurIPS (2017)
54. Reed, S., et al.: Few-shot autoregressive density estimation: towards learning to learn distributions. In: ICLR (2018)
55. Rezende, D.J., Mohamed, S., Danihelka, I., Gregor, K., Wierstra, D.: One-shot generalization in deep generative models. In: ICML (2016)
56. Robb, E., Chu, W.S., Kumar, A., Huang, J.B.: Few-shot adaptation of generative adversarial networks. arXiv preprint arXiv:2010.11943 (2020)
57. Schwartz, E., et al.: Delta-encoder: an effective sample synthesis method for few-shot object recognition. In: NeurIPS (2018)
58. Sun, Q., Liu, Y., Chua, T.S., Schiele, B.: Meta-transfer learning for few-shot learning. In: CVPR (2019)
59. Sung, F., Yang, Y., Zhang, L., Xiang, T., Torr, P.H., Hospedales, T.M.: Learning to compare: relation network for few-shot learning. In: CVPR (2018)
60. Tian, K., Lin, C., Sun, M., Zhou, L., Yan, J., Ouyang, W.: Improving auto-augment via augmentation-wise weight sharing. In: NeurIPS (2020)
61. Tseng, H., Lee, H., Huang, J., Yang, M.: Cross-domain few-shot classification via learned feature-wise transformation. In: Proceedings of 8th International Conference on Learning Representations (ICLR) (2020)
62. Tsutsui, S., Fu, Y., Crandall, D.: Meta-reinforced synthetic data for one-shot fine-grained visual recognition. In: NeurIPS (2019)
63. Van Horn, G., et al.: Building a bird recognition app and large scale dataset with citizen scientists: the fine print in fine-grained dataset collection. In: CVPR (2015)
64. Vinyals, O., Blundell, C., Lillicrap, T., Wierstra, D., et al.: Matching networks for one shot learning. In: NeurIPS (2016)
65. Wang, Y., Gonzalez-Garcia, A., Berga, D., Herranz, L., Khan, F.S., van de Weijer, J.: MineGAN: effective knowledge transfer from GANs to target domains with few images. In: CVPR (2020)

66. Wang, Y., Girshick, R.B., Hebert, M., Hariharan, B.: Low-shot learning from imaginary data. In: CVPR (2018)
67. Yang, L., Li, L., Zhang, Z., Zhou, X., Zhou, E., Liu, Y.: DPGN: distribution propagation graph network for few-shot learning. In: CVPR (2020)
68. Zhang, C., Cai, Y., Lin, G., Shen, C.: DeepEMD: few-shot image classification with differentiable earth mover's distance and structured classifiers. In: CVPR (2020)
69. Zhang, R., Isola, P., Efros, A.A., Shechtman, E., Wang, O.: The unreasonable effectiveness of deep features as a perceptual metric. In: CVPR (2018)
70. Zhang, X., Wang, Z., Liu, D., Ling, Q.: DADA: deep adversarial data augmentation for extremely low data regime classification. In: Proceedings of International Conference on Acoustics, Speech and Signal Processing (ICASSP) (2019)
71. Zhao, M., Cong, Y., Carin, L.: On leveraging pretrained GANs for generation with limited data. In: ICML (2020)

Image Inpainting with Cascaded Modulation GAN and Object-Aware Training

Haitian Zheng[1,2(✉)], Zhe Lin[2], Jingwan Lu[2], Scott Cohen[2], Eli Shechtman[2],
Connelly Barnes[2], Jianming Zhang[2], Ning Xu[2], Sohrab Amirghodsi[2],
and Jiebo Luo[1]

[1] University of Rochester, Rochester, USA
[2] Adobe Research, Bangalore, India
zheng.ht.ustc@gmail.com

Abstract. Recent image inpainting methods have made great progress but often struggle to generate plausible image structures when dealing with large holes in complex images. This is partially due to the lack of effective network structures that can capture both the long-range dependency and high-level semantics of an image. We propose cascaded modulation GAN (CM-GAN), a new network design consisting of an encoder with Fourier convolution blocks that extract multi-scale feature representations from the input image with holes and a dual-stream decoder with a novel cascaded global-spatial modulation block at each scale level . In each decoder block, global modulation is first applied to perform coarse and semantic-aware structure synthesis, followed by spatial modulation to further adjust the feature map in a spatially adaptive fashion. In addition, we design an object-aware training scheme to prevent the network from hallucinating new objects inside holes, fulfilling the needs of object removal tasks in real-world scenarios. Extensive experiments are conducted to show that our method significantly outperforms existing methods in both quantitative and qualitative evaluation . Please refer to the project page: https://github.com/htzheng/CM-GAN-Inpainting.

Keywords: Image inpainting · Generative adversarial networks

1 Introduction

Image inpainting refers to the task of completing missing regions of an image as shown in Fig. 1. It is one of the fundamental tasks in computer vision and has many practical applications, such as object removal [55,57] and manipulation [33, 34], image re-targeting [4,41,46], image compositing [7], and 3D photo effects [24, 32].

Supplementary Information The online version contains supplementary material available at https://doi.org/10.1007/978-3-031-19787-1_16.

Input   ProFill   LaMa   CoModGAN   CM-GAN (ours)

Fig. 1. Results of CM-GAN in comparison to state of the art methods: ProFill [57], LaMa [44] and CoModGAN [59]. CM-GAN generates more plausible and realistic results for the distractor removal scenario (1st row) and large holes (2nd row).

Early inpainting methods leverage patch-based synthesis [4,9,25] or color diffusion [3,6,10,42] to fill the holes by propagating repeating textures and patterns from the visible regions. To facilitate the completion of more complex image structures, recent research efforts have shifted to adopting a data-driven scheme where deep generative networks are learned to predict visual content and appearance [44,54,55,57,59]. By training on a large corpus of images and with assist of both reconstruction and adversarial losses, generative inpainting models have shown to produce more visually appealing results on various types of input data including natural images and faces.

While existing works have shown promising results on completing simple image structures, generating complex holistic structures and image contents with high-fidelity details remains a huge challenge, especially when the holes are large. Essentially, how to 1) *accurately propagate global context into the incomplete region* while 2) *synthesizing realistic local details that are coherent to the global clue* is the key question for image inpainting. To tackle global context propagation, existing networks leverage encoder-decoder structure [18,36,37], dilated convolution [53,55], contextual attention [52,54,55], or Fourier convolution [44] to incorporate long range feature dependency for expanding the effective receptive field [28]. Furthermore, two-stage approaches [31,49,50,54,55] and iterative hole filling [57] predict a coarse result such as a smoothed image, edge/semantic maps or partial completion to enhance the global structure. However, those models lack a mechanism to capture high-level semantics in the unmasked region and effectively propagate them into the hole to synthesize a holistic global structure. With the typical shallow bottleneck designs, the designed feature propagation layer is less aware of global semantics and more prone to generating incoherent local details potentially leading to visual artifacts.

More recently, feature modulation-based methods [20, 21, 36] have shown very promising results on controlling image generation with a global style code. Benefiting from a global code that captures the context of the entire image, CoModGAN [59] attempts to inject global context into the generator for filling in very large holes [59]. However, due to the lack of spatial adaptation, global modulation is sensitive to the corrupted encoding feature inside the inpainting region (shown in Fig. 3). Their results show that passing global information in this way is insufficient for synthesizing high quality global structures which may lead to severe artifacts such as the large unseen color blobs [59] or inconsistent visual appearances such as distorted structures, cf. Fig. 5.

To seek a better way to inject global context into the missing region in inpainting, we investigate a new modulation scheme by cascading global and spatial modulations. We propose **C**ascaded Modulation **GAN** (**CM-GAN**), a new generative network that can synthesize better holistic structure and local details, cf. Figs 1 and 5. Different from [59] that attempts to globally modulates the partially invalid encoder feature as shown Fig. 3, our spatially-adaptive modulation scheme cascaded after a global modulation is much more effective in processing invalid features inside the hole. Although several spatially-adaptive modulation schemes [22, 35] have been proposed in the past to tackle inpainting, our cascaded modulation approach is significantly different from those works in that: i) our spatial code comes from the *decoding stage* rather than the encoding stage to avoid modulating the decoder with invalid encoding feature (cf. Fig. 3), 2) we incorporate the global code into spatial modulation for enforcing the global-local consistency, 3) we introduce a dual-branch design that decouples global and local features for structure-details separation, and 4) we design spatially-aware demodulation instead of instance or batch normalization to avoid potential 'droplet artifact" [21]. Furthermore, on the encoder side, we inject the Fast Fourier convolution [8] at each stage of the encoder network to expand the receptive field of the encoder at early stages, allowing the network to capture long-range correlations across the image.

Another design aspect to consider is how to generate synthetic masks used during inpainting training. We need to design the masks tailored for real-world inpainting use cases, such as object removal and partial object completion. Previous methods generate training data by randomly locating rectangular [18, 37] or irregularly-shaped [55, 59] masks. However, the masks users draw in common use cases likely have a shape of an existing object, or part of an object or other relatively simple shapes such as a scribble or a blob. Moreover, users usually expect the removed objects to be filled by background textures and structures, thus new objects are not expected to appear inside the holes. However, models trained with randomly located masks tend to have the color-bleeding effect across object boundaries and generate object-like artifact blobs inside the hole [59]. In a better attempt, a recent work [57] leverages saliency annotation to simulate the holes left by distracting objects occluded by the foreground salient objects. However, saliency annotations only capture large dominant foreground objects, thus resulting in the algorithm constructing large holes that include most of the other less salient objects. This is different from the real use cases where only a few distracting objects need to be removed.

We found that the tendency to generate spurious objects, blobs and color-bleeding effects across object boundaries can be addressed by a proper and more carefully designed object-aware training scheme. In terms of mask sampling for training, different from random mask [27,39,44,55,59] or saliency-based mask [57], we leverage an instance-level panoptic segmentation model [26] to generate object-aware masks that better simulate real distractors or clutter removal use cases. To avoid generating distorted objects or color blobs inside the hole, we filter out cases where the entire object or a large part of the object is covered by the mask. Furthermore, the panoptic segmentation provides precise object boundaries and thus prevents the trained model from leaking colors at object boundaries.

Finally, for the training losses, we propose a masked R_1 regularization specifically designed for inpainting and augment the adversarial loss with a perceptual loss extracted by segmentation model for improving robustness. The new regularization avoids penalizing the model outside the mask and thus imposes a better separation of input condition from the generated contents. Consequently, the new regularization eliminates the potential harmful impact of computing regularization on the background. Our contributions are four-fold:

- Cascaded Modulation GAN, a new inpainting network architecture formed by a masked image encoder with Fourier convolution blocks and a cascaded global-spatial modulation-based decoder.
- An object-aware mask generation scheme preventing the model from generating new objects inside holes and mimicking realistic inpainting use cases.
- A masked R_1 regularization loss to stabilize the adversarial training for the inpainting tasks.
- State-of-the-art results on the Places2 dataset for various types of masks.

2   Related Works

2.1   Image Inpainting

Image inpainting has a long standing history. Traditionally, the patch-based methods [4,9,11,25] search and copy-paste patches from known region to progressively fill in the target hole. Meanwhile, the diffusion-based methods [3,6,10,42] describe and solve the color propagation inside the hole via partial differential equation. The above methods can produce high-quality stationary textures while completing simple shapes, but they lack the mechanisms to model the high-level semantics for completing new semantic structure inside the hole. Recently, inpainting methods have shifted to a data-driven scheme where deep generative models are learned to directly predict the filled in content inside the hole in an end-to-end fashion. By training deep generative models via adversarial training [12], the learned model can capture higher-level representation of images and hence can generate more visually plausible results. Specifically, Pathak *et al.* [37] first leverage an encoder-decoder network with a bottleneck layer to predict the missing structures for hole filling. Iizuka *et al.* [18] propose a two discriminator design to encourage the global and local consistency separately. To make the generator better at capturing the global context, several attempts have been made. Motivated by the structure-texture decomposition principle [2,5], two-stage networks predict an intermediate representation of image with smoothed image [15,47,52,54,57], edge [31,49], gradient [51] or segmentation map [43] for enhancing the final output. Yu *et al.* [54] design contextual attention to explicitly let the network borrow patch features at a global scale. Aiming at expanding the receptive field of the network, Iizuka *et al.* [18] Yu *et al.* [54] incorporate dilated convolutions to the generator. Likewise, Suvorov *et al.* [44] leverage Fourier convolution [8] to acquire a global receptive field. Furthermore, feature gating such as partial convolution [27] and gated convolution [55] is proposed to handle invalid features inside the hole. To enhance global prediction capacity, Zhao *et al.* [59] propose an encoder-decoder network that leverages style code modulation for global-level structure inpainting. To augment the adversarial loss and to suppress artifacts, existing works [31,44,54,55,57] often train the generator with additional reconstruction objectives such as ℓ_1, perceptual [19] or contextual [37] loss. Recently, Suvorov *et al.* [44] propose to use segmentation networks to compute perceptual loss which achieve better performance.

2.2 Feature Modulation for Image Synthesis

Originating from style transfer [16], feature modulation [17,20,21] has been widely adopted to incorporate input conditions for controlled generation [1,35, 36,45,48,59,60]. Existing modulation methods usually leverage batch normalization or instance normalization to normalize the input feature. The modulation is then achieved by scaling and shifting the normalized activation according to affine parameters predicted from input conditions. Recently, Karras *et al.* [21] find that normalization would cause the "droplet artifact" as the network can create a strong activation spike to sneak signal through normalization layer. Consequently, StyleGAN2 [21] replaces the feature normalization with a proposed demodulation step [21] for better image synthesis. To modulate the input feature according to a spatially-varying feature map, spatial modulation [1,22,35,48] are proposed. Essentially, those methods leverage convolutional layers to predict 2d affine parameters for spatially-controlled modulation. However, feature normalization makes the existing approaches [1,48] less consistent with the design principle of StyleGAN2.

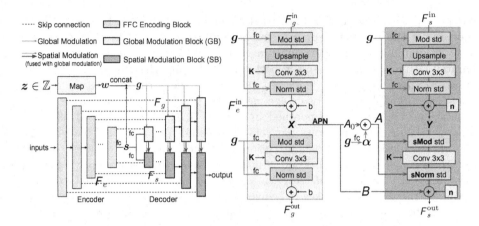

Fig. 2. Left: The CM-GAN architecture, which consists of an encoder with FFC blocks and a two-stream decoder with a cascade of global modulation block (GB) and subsequent spatial modulation block (SB). This cascaded modulation scheme extracts spatial style codes from the globally modulated feature map (instead of from the encoder feature map used in the previous work) to make spatial modulation more effective for inpainting. **Right**: Cascaded modulation at each scale. GB and SB take F_g^{in} and F_s^{in}, respectively, as inputs and produce the upsampled feature F_g^{out} and F_s^{out}. Specifically, we apply joint global-spatial modulation to ensure the generation consistency both at the global and local scales. (Color figure online)

2.3 Regularization for Adversarial Training

Adversarial training is known to be challenging [29] as it is hard for the adversarial networks to reach global Nash equilibrium [14]. Consequently, various regularization are proposed to stabilize the GAN training. In particular, weight normalization [40] and spectrum normalization [30] are proposed to constrain the the Lipschitz continuity of the discriminator. Likewise, Gulrajani *et al.* [13] propose a gradient penalty to impose a K-Lipschitz constraint to the discriminator. Mescheder *et al.* [29] propose R_1 regularization to penalize the discriminator gradient on real data, which is later used by [21,44,59]. Karras *et al.* [21] propose perceptual path length regularization on the generator to ensure smoothness mappings and lazy regularization to optimize the training efficiency.

3   Method

3.1   Cascaded Modulation GAN

To better model the global context for image completion [44,52,54–57,59], we propose a novel mechanism that *cascades global code modulation with spatial code modulation* to facilitate the processing of the partially invalid feature while better injecting the global context into the spatial region. It leads to a new architecture named Cascaded Modulation GAN (CM-GAN), which can synthesize holistic structures and local details surprisingly well as shown in Fig. 1.

Network Overview. As shown in Fig. 2 (left), CM-GAN is based on an encoder branch and two parallel cascaded decoder branches to generate visual output. Specifically, it starts with an encoder that takes the partial image and the mask as inputs to produce multi-scale feature maps $F_e^{(1)}, \cdots, F_e^{(L)}$ at each scale $1 \leq i \leq L$ (L is the highest level with the smallest spatial size). Unlike most encoder-decoder methods [44,54] and to facilitate the completion of holistic structure, we extract a global style code s from the highest level feature $F_e^{(L)}$ with a fully connected layer followed by a ℓ_2 normalization. Furthermore, an MLP-based mapping network [21] is used to generate a style code w from noise, simulating the stochasticity of image generation. The code w is joined with s to produce a global code $g = [s; w]$ for the consequent decoding steps.

Global-Spatial Cascaded Modulation. To better bridge the global context at the decoding stage, we propose *global-spatial* **C**ascaded **M**odulation (CM). As shown in Fig. 2 (right), the decoding stage is based on two branches of Global Modulation Block (GB) and Spatial Modulation Block (SB) to respectively upsample global feature F_g and local features F_s in parallel. Different from existing approaches [44,47,55,57,59], the CM design introduces a new way to inject the global context into the hole region. At a conceptual level, it consists of a cascade of global and spatial modulations between features at each scale and naturally integrates three compensating mechanisms for global context modeling: 1) *feature upsampling* allows both GB and SB to utilize the global context from the low-resolution features generated by both of the previous blocks; 2) the *global modulation* (cyan arrows of Fig. 2) allows both GB and SB to leverage the global code g for generating better global structure; and 3) *spatial modulation* (blue arrows of Fig. 2) leverages spatial code (intermediate feature output of GB) to further inject fine-grained visual details to SB.

More specifically, as shown in Fig. 2 (right), CM at each level of the decoder consists of the paralleled GB block (yellow) and SB block (blue) bridged by spatial modulation. Such parallel blocks takes F_g^{in} and F_s^{in} as input and output F_g^{out} and F_s^{out}. In particular, GB leverages an initial upsampling layer following a convolution layer to generate the intermediate feature X and global output F_g^{out}, respectively. Both layers are modulated by the global code g [21] to capture the global context.

Due to the limited expressive power of the global code g to represent a 2-d scene, and the noisy invalid features inside the inpainting hole [27,55], the global modulation alone generates distorted features inconsistent with the context as shown in Fig. 3 and leads to visual artifacts such as large color blobs and incorrect structure as demonstrated in Fig. 7. To address this critical issue, we cascade GB with an SB to correct invalid features while further injecting spatial details. SB also takes the global code g to synthesize local details while respecting global context. Specifically, taking the spatial feature F_s^{in} as input, SB first produces an initial upsampled feature Y with an upsampling layer modulated by global code g. Next, Y is jointly modulated by X and g in a spatially adaptive fashion following the *modulation-convolution-demodulation* principle [21]:

- *Global-spatial feature modulation.* A spatial tensor $A_0 = \text{APN}(X)$ is produced from feature X by a 2-layer convolutional affine parameter network (APN). Meanwhile, a global vector $\alpha = \text{fc}(g)$ is produced from the global code g with a fully connected layer (fc) to incorporate the global context. Finally, a fused spatial tensor $A = A_0 + \alpha$ leverages both global and spatial information extracted from g and X, respectively, to scale the intermediate feature Y with element-wise product $\odot$:

$$\bar{Y} = Y \odot A. \tag{1}$$

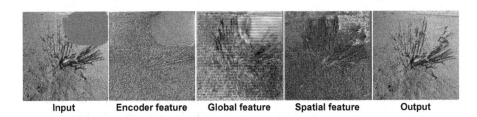

Input   **Encoder feature**   **Global feature**   **Spatial feature**   **Output**

Fig. 3. Visualization of the intermediate features for inpainting. From left to right are the incomplete image, encoded feature, our globally modulated feature and spatially modulated features at the 256×256 layer and the output image. (Color figure online)

- *Convolution.* The modulated tensor $\bar{Y}$ is then convolved with a 3×3 learnable kernel K, resulting in $\hat{Y}$

$$\hat{Y} = \bar{Y} * K. \tag{2}$$

- *Spatially-aware demodulation.* Different from existing spatial modulation methods [1,22,35], we discard instance or batch normalization to avoid the known "water droplet" artifact [21] and propose a spatially-aware demodulation step to produce normalized output $\widetilde{Y}$. Specifically, we assume that the input features Y are independent random variables with unit variance and after the modulation, the expected variance of the output does not change, i.e. $\mathbb{E}_{y \in \widetilde{Y}}[\text{Var}(y)] = 1$. This assumption gives the demodulation computation:

$$\widetilde{Y} = \hat{Y} \odot D, \tag{3}$$

where $D = 1/\sqrt{K^2 \odot \mathbb{E}_{a \in A}[a^2]}$ is the demodulation coefficient. Equation (3) is implemented with standard tensor operations as elaborated in the supplementary material.
- *Adding spatial bias and broadcast noise.* To introduce further spatial variation from feature X, the normalized feature $\widetilde{Y}$ is added to a shifting tensor $B =$

APN(X) produced by another affine parameter network from feature X along with the broadcast noise n to generate the new local feature F_s^{out}:

$$F_s^{\text{out}} = \widetilde{Y} + B + n. \tag{4}$$

As shown in the 4th column of Fig. 3, the cascaded SB block helps generate fine-grained visual details and improves the consistency of feature values inside and outside the hole.

Expanding the Receptive Field at Early Stages. The fully convolutional models suffer from slow growth of the effective receptive field [28], especially at early stages of the network. For this reason, an encoder based on strided convolution usually generates invalid features inside the hole region, making the feature correction at the decoding stage more challenging. A recent work [44] shows that fast Fourier convolution (FFC) [8] can help early layers achieve large receptive fields that cover the entire image. However, the work [44] stacks FFC at the bottleneck layer and is computationally demanding. Moreover, like many other works [55], due to the shallow bottleneck layers, [44] cannot capture global semantics effectively, limiting its ability to handle large holes. We propose to replace every convolutional blocks of the CNN encoder with FFC. By adopting FFC at all scale levels, we enable the encoder to propagate features at early stages and thus address the issue of generating invalid features inside the holes, helping improve the results as shown in the ablation study in Table 2.

Real Requests   **Masks of CoModGAN**   **Our Mask**

Fig. 4. Examples of our object-aware masks generated for training (right) in comparison to the real inpainting requests (left) and the masks generated by CoModGAN [59] (middle). Note that our masks are more consistent with real user requests.

3.2 Object-aware Training

The algorithm to generate masks for training is crucial. In essence, the sampled masks should be similar to the masks drawn in realistic use-cases. Moreover, the masks should avoid covering an entire object or most of any new object to discourage model from generating object-like patterns. Previous works generate mask with square-shaped masks [18,37] or use random strokes [27] or a mixture of both [55,59] for training. The oversimplified mask schemes may cause casual artifacts such as suspicious objects or color blobs.

To better support realistic object removal use cases while preventing the model from trying to synthesize new objects inside the holes, we propose an object-aware training scheme that generate more realistic masks during training as shown in Fig. 4. Specifically, we first pass the training images to PanopticFCN [26] to generate highly accurate instance-level segmentation annotations. Next, we sample a mixture of free-form holes [59] and object holes as the initial mask. Finally, we compute the overlapping ratio between a hole and each instance from an image. If the overlapping ratio is larger than a threshold, we exclude the foreground instance from the hole. Otherwise, the hole is unchanged to mimic object completion. We set the threshold to 0.5. We dilate and translate the object masks randomly to avoid overfitting. We also dilate the hole on the instance segmentation boundary to avoid leaking background pixels near the hole into the inpainting region.

3.3 Training Objective and Masked-R_1 Regularization

Our model is trained with a combination of adversarial loss [59] and segmentation-based perceptual loss [44]. The experiments show that our method can also achieve good results when purely using the adversarial loss, but adding the perceptual loss can further improve the performance. In addition, we propose a masked-R_1 regularization tailored to stabilize the adversarial training for the inpainting task. Different from [44,59] that naively apply R_1 regularization [29], we leverage the mask m to avoid computing the gradient penalty outside the mask, specifically:

$$\bar{R}_1 = \frac{\gamma}{2}\mathbb{E}_{p_{\text{data}}}[\|m \odot \nabla D(x)\|^2], \tag{5}$$

where m is the mask indicating the hole region and γ is a balancing weight. The new loss eliminates the potential harmful impact of computing gradient on real pixels, and therefore stabilizes the training.

Table 1. Quantitative evaluation of inpainting on the Places evaluation set. We report FID [14], LPIPS [58], U-IDS [59] and P-IDS [59] scores.

Methods	FID↓	LPIPS↓	U-IDS↑	P-IDS↑	Methods	FID↓	LPIPS↓	U-IDS↑	P-IDS↑
CM-GAN	**1.628**	**0.189**	**37.42**	**20.96**	DS [38]	16.003	0.399	10.72	0.47
CoModGAN [59]	3.724	0.229	32.38	14.68	EC [31]	12.086	0.414	9.21	0.28
Lama [44]	3.864	0.195	29.57	10.08	ICT [47]	16.405	0.424	8.12	0.25
ProFill [57]	7.700	0.230	21.19	3.87	HiFill [52]	37.484	0.336	9.86	0.96
CRFill [56]	9.657	0.233	22.90	5.53	SF [39]	28.252	0.489	6.05	0.13
DeepFillv2 [55]	13.597	0.371	14.23	1.67	MEDFE [15]	35.454	0.445	7.10	0.35

4   Experiments

Implementation Details. We conduct the image inpainting experiment at resolution 512×512 on the Places2 dataset [61]. Our model is trained with Adam optimizer [23]. The learning rate and batch size are set to 0.001 and 32, respectively. Our network takes the resized image as input, so that the model can predict the global structure of an image. We apply flip augmentation to increase the training samples.

Evaluation Metrics. We report the numerical metrics on the validation set of Places2 which contains 36.5k images. For the numerical evaluation, we compute *Frchet Inception Distance* (FID) [14] and *Perceptual Image Patch Similarity Distance* (LPIPS) [58]. We also adopt the *Paired/Unpaired Inception Discriminative Score* (P-IDS/U-IDS) [59] for evaluation.

4.1 Comparisons to Existing Methods

We set channel numbers of our network to have a similar model capacity as CoModGAN and LaMa as shown in Table 4.

Quantitative Evaluation. Table 1 presents the comparison of our method against a number of recent methods using our masks. Results show that our method significantly outperforms all other methods in terms of FID, LPIPS, U-IDS and P-IDS. We notice that with the assist of perceptual loss, LaMa [44] and our CM-GAN achieve significantly better LPIPS score than CoModGAN and other methods, attributing to additional semantic guidance provided of the pre-trained perceptual model. Compared to LaMa, our CM-GAN reduces FID by over 50% from 3.864 to 1.628, which can be explained by the typically blurry results of LaMa versus ours which tend to be sharper.

We evaluate generalization of CM-GAN to other types of masks including the wide mask [44] and the mask of CoModGAN [59]. We also fine-tune CM-GAN with masks of [44] and [59] (denoted by CM-GAN†) and report the results. As shown in Table 3, our models with and without fine-tuning achieve clear performance gain and demonstrate its generalization ability. Notably, CM-GAN trained on our object-aware masks outperforms CoModGAN on the CoModGAN mask, confirming the better generation capacity of CM-GAN. The strong capacity of CM-GAN brings further performance gain after fine-tuning.

Qualitative Evaluation. Figures 5, 6 and 8 presents visual comparisons of our method with state-of-the-art methods on our synthesized masks and other types of mask introduced by [44] and [59], respectively. ProFill [57] generates incoherent global structures such as the smoothed building and tends to bleed color to the background for the object-removal case. CoModGAN [59] produces structural artifacts and color blobs. LaMa [44] is superior on repeating structures, but tends to generate blurry results on large holes, especially on nature scenes. In contrast, our method produces more coherent semantic structures and hallucinates cleaner textures for various types of scenarios.

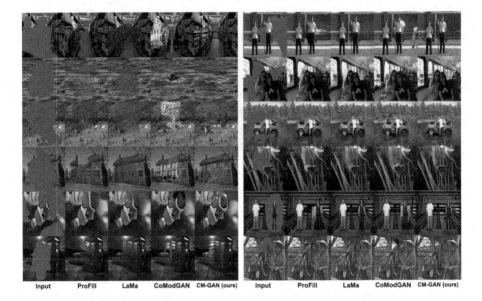

| Input | ProFill | LaMa | CoModGAN | CM-GAN (ours) | Input | ProFill | LaMa | CoModGAN | CM-GAN (ours) |

Fig. 5. Visual comparisons on Places2 with our synthesized masks including large random masks (left) and masks for the distractor removal scenario (right). We show the input images and the results of ProFill [57], Lama [44], CoModGAN [59] and CM-GAN (ours). Best viewed by zoom-in on screen. (Color figure online)

4.2 Ablation Study

We perform a set of ablation experiments to show the importance of each component of our model. All ablated models are trained and evaluated on the Places2 dataset. Results of the ablations are shown in Table 2. Below we describe the ablations from the following aspects:

Masked-R_1 Regularization. We start from a `baseline` with a simple encoder-decoder structure based on global-vector modulation [59] and skip connection. We compare the baseline trained with R_1 regularization with the model trained with masked R_1 regularization regularization (`baseline+m`R_1). From the result, the masked R_1 regularization improves the numerical metrics as the designed loss avoids computing gradient at the fixed input region.

Cascaded Modulation. We next evaluate the cascaded modulation design on top of the baseline network and mR_1 loss. Specifically, we evaluate a baseline model with spatial modulation instead of the global code modulation, i.e. `baseline + s`. The performance improvement verify the effectiveness of the spatial adaptation introduced by spatial modulation. Next, we cascade global and spatial modulation on the baseline to get the main CM model `baseline+CM (g-s)` ours which improves all numerical metrics. To better understand CM, we visually compare `baseline` with `baseline+CM (g-s)` ours in Fig. 7 and find that CM significantly improves the synthesized color, texture and global semantic

Fig. 6. Visual comparisons on Places2 with the masks proposed by Lama [44]. (Color figure online)

| Input | Baseline | Baseline + CM | Input | Baseline | Baseline + CM |

Fig. 7. Compared to a global modulation baseline, CM significantly improves the coherence of the synthesized color, textures, global structures and objects. (Color figure online)

and corrects the color blob artifact [59], which confirms the effectiveness of CM on correcting the incoherent feature in a global-semantic aware fashion.

Choices of Second-stage Modulation. We evaluate other variant of the second-stage modulation choices: i) we replace our StyleGAN2-compatible spatial with skip connection, resulting in a model that cascades global modulation twice, i.e. `baseline+CM(g-g)`, ii) we test the CM baseline with the spatial modulation of [22], i.e. `baseline+CM(g- [22])` and iii) we drop the demodulation step (Eq. (3)) from the spatial modulation step, resulting a model with a plain spatial modulation operation, i.e. `Baseline + CM (g-s) plain`. From the results, our spatial modulation outperforms the global modulation version as we modulate feature using both global and spatial code. We found [22] does not improve CM as the instance normalization of [22] is designed for StyleGAN and is less compatible with our baseline. Furthermore, demodulation seems crucial to the model as it regularizes the intermediate feature activation. Finally, for the same reason as [22], we found SPADE [35] not compatible with our baseline due to the use of batch normalization.

Perceptual Loss. The perceptual loss (perc.) provides additional semantic supervision to the network and can significantly improve the FID metrics. However, it slightly decreases the discriminative U/P-IDS scores as perceptual loss may lead to certain visual patterns that are imperceptible to human.

Fast Fourier Convolutions Encoder. The Fast Fourier Convolution (FFC) encoder further brings significant performance gains on top of the cascaded modulation and perceptual loss as shown in the table, which validates the importance of more effective encoder with wider receptive fields in early encoding stages.

Object-Aware Training. To study the effect of object-aware training (OT), we retrain LaMa [44] and CoModGAN [59] on our object-aware masks. The results show that object-aware training improves the performance of both of these models consistently. However, our full model still outperforms these retrained models significantly. Notably, our model reduces FID of the retrained CoModGAN (with OT) by 40% from 2.599 of to 1.628.

Table 2. Ablation study of our model design (architecture, loss, training scheme) including masked-R_1 loss (mR_1), cascaded modulation option (CM), Fourier convolution (FFC), perceptual loss (perc.) and object-aware training (OT). We report FID [14], LPIPS [58], U-IDS [59] and P-IDS [59] scores.

Ablations	Methods	FID↓	LPIPS↓	U-IDS↑	P-IDS↑
Masked-R_1 (mR_1)	Baseline	2.530	0.221	**36.59**	21.10
	Baseline + mR_1	2.475	0.221	36.58	**21.55**
Cascaded Modulation (CM)	Baseline + s	2.398	0.218	36.72	21.94
	Baseline + CM (g-g)	2.247	0.226	37.12	22.70
	Baseline + CM (g-[22])	2.915	**0.221**	35.72	20.44
	Baseline + CM (g-s) plain	2.392	0.224	37.23	22.67
	Baseline + CM (g-s)	**2.187**	0.225	**37.76**	**23.86**
FFC and perc.	Baseline + CM + perc	1.730	0.195	36.12	19.73
	Baseline + CM + perc. + FFC	**1.628**	**0.189**	**37.42**	**20.96**
Object-aware Training (OT)	CoModGAN [59]	3.724	0.229	32.38	14.68
	CoModGAN [59] + OT	2.599	0.222	35.45	20.08
	Lama [44]	3.864	0.195	29.57	10.08
	Lama [44] + OT	2.884	0.192	31.32	15.10
	CM-GAN (full)	**1.628**	**0.189**	**37.42**	**20.96**

Table 3. Generalization evaluation on other types of mask including wide masks [44] and CoMoGAN masks [59]. CMGAN† are models fine-tuned on the two types of mask.

Wide masks [44]					CoModGAN masks [59]				
Methods	FID↓	LPIPS↓	U-IDS ↑	P-IDS ↑	Methods	FID↓	LPIPS↓	U-IDS ↑	P-IDS ↑
CM-GAN	1.521	0.129	39.24	23.24	CM-GAN	6.811	0.313	26.13	10.84
CM-GAN†	1.329	0.126	40.20	25.59	CM-GAN†	5.863	0.310	27.92	12.45
LaMa [44]	1.838	0.123	35.00	15.12	CoModGAN [59]	7.790	0.344	24.87	10.47
CoModGAN [59]	1.964	0.140	37.69	21.42	LaMa [44]	12.442	0.316	18.71	4.36
ProFill [57]	3.333	0.142	29.12	8.39	ProFill [57]	20.314	0.352	11.21	1.23

Table 4. The inference complexities. Our model has a similar number of parameters and FLOPs as other recent models.

Models	#Params of $\mathcal{G}$	#Params of $\mathcal{D}$	FLOPs
CoModGAN [59]	79.79 M	28.98 M	345.54 G
LaMa [44]	51.25 M	9.258 M	395.30 G
CM-GAN (ours)	75.28 M	28.98 M	373.60 G

4.3   User Study

We conduct a user study to better evaluate the visual quality of our method. Specifically, we generate samples for evaluation using the Places2 evaluation set and three types of masks: our object-aware mask, wide mask [44] and mask from real user request. The former data class contains 30 samples while the latter contains 13 real inpainting requests online following [56]. Each input image with the region to be removed and the results of different methods are presented to online users who are asked to select the best result. Finally, we collect votes from all users. Results in Table 5 shows that our method receives the majority of votes on both the synthetic data and realistic object removal requests.

Table 5. The user study. For each mask type, we show the number of votes and percentages for different methods (ProFill, LaMa, CoModGAN, and ours)

Masks	ProFill [57]	Lama [44]	CoModGAN [59]	CM-GAN
Our mask	20 (5%)	68 (17%)	83 (20%)	**234 (58%)**
Wide mask	45 (10%)	107 (25%)	94 (22%)	**186 (43%)**
User mask	15 (5%)	101 (35%)	55 (19%)	**120 (41%)**

| Input | ProFill | LaMa | CoModGAN | CM-GAN (ours) |

Fig. 8. Visual comparison on the mask of CoModGAN [59]. Best viewed by zoom-in on screen. (Color figure online)

5   Conclusion

In this paper, we present a new approach tailored to real-world image inpainting. Our method is based on a new modulation block that cascades global modulation with spatial modulation for better pass global context into the hole region.

We further propose a training scheme based on object-aware mask sampling to improve generalization to real use cases. Finally, we propose specifically designed masked R_1 regularization to stabilize the adversarial training of image inpainting networks. Our method achieves the new state-of-the-art performance on the Places2 dataset and better visual quality.

Currently, our model is still limited in synthesizing large objects like humans or animals. One possible solution is to train a specialist inpainting model for specific types of objects. Another direction is to leverage depth and semantic segmentation for more precise structure-aware inpainting.

Acknowledgement. We would like to thank Qing Liu for the efforts and help on the demo interface and other experiments.

References

1. AlBahar, B., Lu, J., Yang, J., Shu, Z., Shechtman, E., Huang, J.B.: Pose with style: detail-preserving pose-guided image synthesis with conditional StyleGAN. ACM Trans. Graph. **40**, 1 (2021)
2. Aujol, J.F., Gilboa, G., Chan, T., Osher, S.: Structure-texture image decomposition-modeling, algorithms, and parameter selection. Int. J. Comput. Vision **67**(1), 111–136 (2006)
3. Ballester, C., Bertalmio, M., Caselles, V., Sapiro, G., Verdera, J.: Filling-in by joint interpolation of vector fields and gray levels. IEEE Trans. Image Process. **10**(8), 1200–1211 (2001)
4. Barnes, C., Shechtman, E., Finkelstein, A., Goldman, D.B.: PatchMatch: a randomized correspondence algorithm for structural image editing. ACM Trans. Graph. **28**(3), 24 (2009)
5. Bertalmio, M., Vese, L., Sapiro, G., Osher, S.: Simultaneous structure and texture image inpainting. IEEE Trans. Image Process. **12**(8), 882–889 (2003)
6. Chan, T.F., Shen, J.: Nontexture inpainting by curvature-driven diffusions. J. Vis. Commun. Image Represent. **12**(4), 436–449 (2001)
7. Chen, B.C., Kae, A.: Toward realistic image compositing with adversarial learning. In: Proceedings of the IEEE/CVF Conference on Computer Vision and Pattern Recognition, pp. 8415–8424 (2019)
8. Chi, L., Jiang, B., Mu, Y.: Fast Fourier convolution. Adv. Neural Inf. Process. Syst. **33**, 4479–4488 (2020)
9. Cho, T.S., Butman, M., Avidan, S., Freeman, W.T.: The patch transform and its applications to image editing. In: 2008 IEEE Conference on Computer Vision and Pattern Recognition, pp. 1–8. IEEE (2008)
10. Criminisi, A., Pérez, P., Toyama, K.: Region filling and object removal by exemplar-based image inpainting. IEEE Trans. Image Process. **13**(9), 1200–1212 (2004)
11. Efros, A.A., Freeman, W.T.: Image quilting for texture synthesis and transfer. In: Proceedings of the 28th Annual Conference on Computer Graphics and Interactive Techniques, pp. 341–346. ACM (2001)
12. Goodfellow, I., et al.: Generative adversarial nets. In: Advances in Neural Information Processing Systems, pp. 2672–2680 (2014)
13. Gulrajani, I., Ahmed, F., Arjovsky, M., Dumoulin, V., Courville, A.: Improved training of Wasserstein GANs. arXiv preprint arXiv:1704.00028 (2017)

14. Heusel, M., Ramsauer, H., Unterthiner, T., Nessler, B., Hochreiter, S.: GANs trained by a two time-scale update rule converge to a local Nash equilibrium. In: Advances in Neural Information Processing Systems, pp. 6626–6637 (2017)
15. Liu, H., Jiang, B., Song, Y., Huang, W., Chao, Y.: Rethinking image inpainting via a mutual encoder-decoder with feature equalizations. In: Proceedings of the European Conference on Computer Vision (2020)
16. Huang, X., Belongie, S.: Arbitrary style transfer in real-time with adaptive instance normalization. In: Proceedings of the IEEE International Conference on Computer Vision, pp. 1501–1510 (2017)
17. Huang, X., Liu, M.-Y., Belongie, S., Kautz, J.: Multimodal unsupervised image-to-image translation. In: Ferrari, V., Hebert, M., Sminchisescu, C., Weiss, Y. (eds.) ECCV 2018. LNCS, vol. 11207, pp. 179–196. Springer, Cham (2018). https://doi.org/10.1007/978-3-030-01219-9_11
18. Iizuka, S., Simo-Serra, E., Ishikawa, H.: Globally and locally consistent image completion. ACM Trans. Graph. (ToG) **36**(4), 1–14 (2017)
19. Johnson, J., Alahi, A., Fei-Fei, L.: Perceptual losses for real-time style transfer and super-resolution. In: Leibe, B., Matas, J., Sebe, N., Welling, M. (eds.) ECCV 2016. LNCS, vol. 9906, pp. 694–711. Springer, Cham (2016). https://doi.org/10.1007/978-3-319-46475-6_43
20. Karras, T., Laine, S., Aila, T.: A style-based generator architecture for generative adversarial networks. In: Proceedings of the IEEE/CVF Conference on Computer Vision and Pattern Recognition, pp. 4401–4410 (2019)
21. Karras, T., Laine, S., Aittala, M., Hellsten, J., Lehtinen, J., Aila, T.: Analyzing and improving the image quality of StyleGAN. In: Proceedings of CVPR (2020)
22. Kim, H., Choi, Y., Kim, J., Yoo, S., Uh, Y.: Exploiting spatial dimensions of latent in GAN for real-time image editing. In: Proceedings of the IEEE Conference on Computer Vision and Pattern Recognition (2021)
23. Kingma, D.P., Ba, J.: Adam: a method for stochastic optimization. arXiv preprint arXiv:1412.6980 (2014)
24. Kopf, J., et al.: One shot 3d photography. Trans. Graph. **39**(4), 76–81 (2020)
25. Kwatra, V., Essa, I., Bobick, A., Kwatra, N.: Texture optimization for example-based synthesis. In: ACM SIGGRAPH 2005 Papers, pp. 795–802 (2005)
26. Li, Y., et al.: Fully convolutional networks for panoptic segmentation. In: Proceedings of the IEEE/CVF Conference on Computer Vision and Pattern Recognition, pp. 214–223 (2021)
27. Liu, G., Reda, F.A., Shih, K.J., Wang, T.-C., Tao, A., Catanzaro, B.: Image inpainting for irregular holes using partial convolutions. In: Ferrari, V., Hebert, M., Sminchisescu, C., Weiss, Y. (eds.) ECCV 2018. LNCS, vol. 11215, pp. 89–105. Springer, Cham (2018). https://doi.org/10.1007/978-3-030-01252-6_6
28. Luo, W., Li, Y., Urtasun, R., Zemel, R.: Understanding the effective receptive field in deep convolutional neural networks. In: Proceedings of the 30th International Conference on Neural Information Processing Systems, pp. 4905–4913 (2016)
29. Mescheder, L., Geiger, A., Nowozin, S.: Which training methods for GANs do actually converge? In: International Conference on Machine Learning, pp. 3481–3490. PMLR (2018)
30. Miyato, T., Kataoka, T., Koyama, M., Yoshida, Y.: Spectral normalization for generative adversarial networks. arXiv preprint arXiv:1802.05957 (2018)
31. Nazeri, K., Ng, E., Joseph, T., Qureshi, F.Z., Ebrahimi, M.: Edgeconnect: generative image inpainting with adversarial edge learning. arXiv preprint arXiv:1901.00212 (2019)

32. Niklaus, S., Mai, L., Yang, J., Liu, F.: 3D Ken burns effect from a single image. ACM Trans. Graph. **38**(6), 184:1–184:15 (2019)
33. Ntavelis, E., Romero, A., Kastanis, I., Van Gool, L., Timofte, R.: SESAME: semantic editing of scenes by adding, manipulating or erasing objects. In: Vedaldi, A., Bischof, H., Brox, T., Frahm, J.-M. (eds.) ECCV 2020. LNCS, vol. 12367, pp. 394–411. Springer, Cham (2020). https://doi.org/10.1007/978-3-030-58542-6_24
34. Oh, B.M., Chen, M., Dorsey, J., Durand, F.: Image-based modeling and photo editing. In: Proceedings of the 28th Annual Conference on Computer Graphics and Interactive Techniques, pp. 433–442 (2001)
35. Park, T., Liu, M.Y., Wang, T.C., Zhu, J.Y.: Semantic image synthesis with spatially-adaptive normalization. In: Proceedings of the IEEE Conference on Computer Vision and Pattern Recognition (2019)
36. Park, T., et al.: Swapping autoencoder for deep image manipulation. arXiv preprint arXiv:2007.00653 (2020)
37. Pathak, D., Krahenbuhl, P., Donahue, J., Darrell, T., Efros, A.A.: Context encoders: Feature learning by inpainting. In: Proceedings of the IEEE Conference on Computer Vision and Pattern Recognition, pp. 2536–2544 (2016)
38. Peng, J., Liu, D., Xu, S., Li, H.: Generating diverse structure for image inpainting with hierarchical VQ-VAE. In: Proceedings of the IEEE/CVF Conference on Computer Vision and Pattern Recognition (CVPR), pp. 10775–10784 (2021)
39. Ren, Y., Yu, X., Zhang, R., Li, T.H., Liu, S., Li, G.: StructureFlow: image inpainting via structure-aware appearance flow. In: IEEE International Conference on Computer Vision (ICCV) (2019)
40. Salimans, T., Kingma, D.P.: Weight normalization: a simple reparameterization to accelerate training of deep neural networks. Adv. Neural. Inf. Process. Syst. **29**, 901–909 (2016)
41. Setlur, V., Takagi, S., Raskar, R., Gleicher, M., Gooch, B.: Automatic image retargeting. In: Proceedings of the 4th International Conference on Mobile and Ubiquitous Multimedia, pp. 59–68 (2005)
42. Shen, J., Chan, T.F.: Mathematical models for local nontexture inpaintings. SIAM J. Appl. Math. **62**(3), 1019–1043 (2002)
43. Song, Y., Yang, C., Shen, Y., Wang, P., Huang, Q., Kuo, C.C.J.: SPG-Net: segmentation prediction and guidance network for image inpainting. arXiv preprint arXiv:1805.03356 (2018)
44. Suvorov, R., et al.: Resolution-robust large mask inpainting with Fourier convolutions. arXiv preprint arXiv:2109.07161 (2021)
45. Tan, Z., et al.: Semantic image synthesis via efficient class-adaptive normalization. arXiv preprint arXiv:2012.04644 (2020)
46. Vaquero, D., Turk, M., Pulli, K., Tico, M., Gelfand, N.: A survey of image retargeting techniques. In: Applications of Digital Image Processing XXXIII, vol. 7798, pp. 328–342. SPIE (2010)
47. Wan, Z., Zhang, J., Chen, D., Liao, J.: High-fidelity pluralistic image completion with transformers. arXiv preprint arXiv:2103.14031 (2021)
48. Wang, X., Yu, K., Dong, C., Loy, C.C.: Recovering realistic texture in image super-resolution by deep spatial feature transform. In: Proceedings of the IEEE Conference on Computer Vision and Pattern Recognition, pp. 606–615 (2018)
49. Xiong, W., et al.: Foreground-aware image inpainting. In: Proceedings of the IEEE/CVF Conference on Computer Vision and Pattern Recognition, pp. 5840–5848 (2019)

50. Yang, C., Lu, X., Lin, Z., Shechtman, E., Wang, O., Li, H.: High-resolution image inpainting using multi-scale neural patch synthesis. In: 2017 IEEE Conference on Computer Vision and Pattern Recognition, CVPR 2017, Honolulu, HI, USA, 21–26 July 2017, pp. 4076–4084 (2017)
51. Yang, J., Qi, Z., Shi, Y.: Learning to incorporate structure knowledge for image inpainting. In: Proceedings of the AAAI Conference on Artificial Intelligence, vol. 34, pp. 12605–12612 (2020)
52. Yi, Z., Tang, Q., Azizi, S., Jang, D., Xu, Z.: Contextual residual aggregation for ultra high-resolution image inpainting. In: Proceedings of the IEEE/CVF Conference on Computer Vision and Pattern Recognition, pp. 7508–7517 (2020)
53. Yu, F., Koltun, V.: Multi-scale context aggregation by dilated convolutions. arXiv preprint arXiv:1511.07122 (2015)
54. Yu, J., Lin, Z., Yang, J., Shen, X., Lu, X., Huang, T.S.: Generative image inpainting with contextual attention. In: Proceedings of the IEEE Conference on Computer Vision and Pattern Recognition, pp. 5505–5514 (2018)
55. Yu, J., Lin, Z., Yang, J., Shen, X., Lu, X., Huang, T.S.: Free-form image inpainting with gated convolution. In: Proceedings of the IEEE International Conference on Computer Vision, pp. 4471–4480 (2019)
56. Zeng, Y., Lin, Z., Lu, H., Patel, V.M.: CR-FILL: generative image inpainting with auxiliary contextual reconstruction. In: Proceedings of the IEEE International Conference on Computer Vision (2021)
57. Zeng, Y., et al.: High-resolution image inpainting with iterative confidence feedback and guided upsampling. arXiv preprint arXiv:2005.11742 (2020)
58. Zhang, R., Isola, P., Efros, A.A., Shechtman, E., Wang, O.: The unreasonable effectiveness of deep features as a perceptual metric. In: Proceedings of the IEEE Conference on Computer Vision and Pattern Recognition, pp. 586–595 (2018)
59. Zhao, S., et al.:: Large scale image completion via co-modulated generative adversarial networks. arXiv preprint arXiv:2103.10428 (2021)
60. Zheng, H., Liao, H., Chen, L., Xiong, W., Chen, T., Luo, J.: Example-guided image synthesis using masked spatial-channel attention and self-supervision. In: Vedaldi, A., Bischof, H., Brox, T., Frahm, J.-M. (eds.) ECCV 2020. LNCS, vol. 12359, pp. 422–439. Springer, Cham (2020). https://doi.org/10.1007/978-3-030-58568-6_25
61. Zhou, B., Lapedriza, A., Khosla, A., Oliva, A., Torralba, A.: Places: a 10 million image database for scene recognition. IEEE Trans. Pattern Anal. Mach. Intell. **40**(6), 1452–1464 (2017)

StyleFace: Towards Identity-Disentangled Face Generation on Megapixels

Yuchen Luo[1], Junwei Zhu[2], Keke He[2], Wenqing Chu[2], Ying Tai[2], Chengjie Wang[2], and Junchi Yan[1(✉)]

[1] Department of CSE and MoE Key Lab of Artificial Intelligence, Shanghai Jiao Tong University, Shanghai, China
{592mcavoy,yanjunchi}@sjtu.edu.cn
[2] Youtu Lab, Tencent, Shanghai, China
{junweizhu,katehe,yingtai,jasoncjwang}@tencent.com

Abstract. Identity swapping and de-identification are two essential applications of identity-disentangled face image generation. Although sharing a similar problem definition, the two tasks have been long studied separately, and identity-disentangled face generation on megapixels is still under exploration. In this work, we propose StyleFace, a unified framework for 1024^2 resolution high-fidelity identity swapping and de-identification. To encode real identity while supporting virtual identity generation, we represent identity as a latent variable and further utilize contrastive learning for latent space regularization. Besides, we utilize StyleGAN2 to improve the generation quality on megapixels and devise an Adaptive Attribute Extractor, which adaptively preserves the identity-irrelevant attributes in a simple yet effective way. Extensive experiments demonstrate the state-of-the-art performance of StyleFace in high-resolution identity swapping and de-identification.

1 Introduction

A face image can be semantically separated into two parts, including the *identity* that contains the identifiable characteristics, and the identity-irrelevant *attributes*, such as pose, expression, background, *etc.*. Although many works are devoted to disentangling and editing the facial attributes, identity-disentangled face generation is still not well investigated.

Identity-disentangled face generation constrains the generation randomness on the identity property with conditioned attributes. It has two important applications, including *Identity Swapping* and *De-identification* (De-ID). Identity swapping changes the identity in the original image to that of a specific person, while De-ID changes it to a nonexistent one. Most identity swapping methods [5,9,22,38,42] learn by maximizing the identity similarity to a specific identity and maintaining the attributes in the original image, which formulates an

---

Supplementary Information The online version contains supplementary material available at https://doi.org/10.1007/978-3-031-19787-1_17.

Fig. 1. This paper presents *StyleFace*, the first unified framework for high-fidelity identity swapping (col 3) and de-identification (col 4) on megapixels. Both the embedded and sampled identities are visually realistic, and the attributes (*e.g.*, lighting, occlusion, expression, *etc.*.) are faithfully preserved.

effective semi-supervised learning scheme. On the contrary, due to the ambiguity of anonymization, De-ID methods modify the original identity by feature-level repulsion [8], representation manipulation [3], or indirect identity guidance [24]. These methods often suffer from poor anonymization diversity [8] and lack supervision on attribute retention, leading to unrealistic visual effect [3,8,24].

Identity swapping and De-ID have been treated as two independent tasks for a long time. Nevertheless, both tasks require the generated faces to faithfully preserve the identity-irrelevant attributes in the original image, differing only on the generated identity. Therefore, we wonder if it is possible to *unify the two tasks in one framework* and promote the De-ID performance with the supervision signals in the identity swapping scheme.

In addition, *high-fidelity identity-disentangled face generation on megapixels* is still an unresolved problem, albeit the rapid improvement of generative techniques [17,18]. Existing De-ID methods [8,10,24,27,39] are mostly cursed with limited resolution and poor fidelity. Early identity swapping methods utilize feature matching [5] or self-supervised refinement [21,22] to improve the fidelity, but mainly focus on 256^2 resolution generation or need extra super-resolution [21,38]. Currently, MegaFS [42] and InfoSwap [9] make early attempts on megapixel-level identity swapping, but they produce visible artifacts and have difficulty in detailed attributes recovery.

In this paper, we propose a novel framework *StyleFace*, which unifies identity swapping and de-identification in one model and renders identity-disentangled face images on mega-pixels. To bridge the gap between identity swapping and De-ID, we first design a Variational Auto-Encoder (VAE)-based projector, which encodes the identity priors from the face recognition model as a latent variable. On this basis, we can *embed the identity of a real person for identity swapping* and *sample virtual identities for de-identification*. We apply a hierarchical augmentation on the identity latent space to improve the effectiveness on different scales. Moreover, we introduce contrastive learning [11] to promote the uniformity in the intermediate latent space and improve the quality of de-identification.

With the unified framework design, we can train the model by the identity swapping objectives but directly apply it to de-identification at test time.

Thanks to the effective supervision signals on attribute retention and identity distinctiveness from the identity swapping scheme, the fidelity and realism of de-identified faces are largely promoted. Also, the diversity of de-identified faces is improved with the infinite sampling power of the latent identity space.

Next, to achieve megapixel-level generation, we utilize StyleGAN2 [18] as the generator. We formulate the projected identity as style and the identity-irrelevant attributes extracted by a carefully devised Adaptive Attribute Extractor (AAE) as noise input. Unlike [9] that erases redundant information with mutual information compression, which is time-consuming and often fails to maintain the details, AAE adaptively preserves desired attributes under simple constrain. We show that AAE is explainable and can effectively preserve even the fine-level facial details (*e.g.*, hair and wrinkles).

With the disentangled latent representation of identity, *StyleFace* unifies identity swapping and de-identification in one framework and conducts megapixel-level generation (see Fig. 1). Experiments on high-resolution identity swapping show the superiority of our model in synthesizing high-fidelity face images with precise identity control. Besides, *StyleFace* achieves state-of-the-art performance for de-identification and the de-identified faces are realistic and diverse.

In summary, our main contributions include:

- We represent identity as a latent variable and introduce contrastive learning for latent regularization. In this way, we propose *StyleFace*, to our best knowledge, the first unified high-fidelity face generation framework for both identity swapping and de-identification.
- We devise an attribute extractor to cooperate with a powerful generator (StyleGAN2) and achieve high-fidelity generation on megapixels.
- Extensive experiments show that the proposed model can generate visually appealing results with both real and virtual identities, achieving state-of-the-art performance in identity swapping and de-identification, respectively.

2   Related Works

Identity Swapping. Identity swapping is a long researched task. Early methods [5,22] mainly focus on 256-res face swapping or need extra super resolution [21,38], therefore cannot meet the requirement in a real-world application. Recently, InfoSwap [9] leverages the information-bottleneck principles and proposes an identity contrastive loss to promote the disentanglement. MegaFS [42] proposes a Face Transfer Module to modify identity by latent code manipulation. These methods achieve precise identity control with the carefully designed supervision objectives but are not satisfactory in maintaining the attributes.

Following this line, we train the model with the identity swapping objective but have mainly two differences. Firstly, we do not directly use the deterministic identity representations of the FR model but embed the identity priors into a latent distribution. Secondly, we preserve the attributes and details with a carefully designed feature extractor. Therefore, our model can generate new identities and have better fidelity on megapixels.

Face De-identification. Conventional de-identification methods use pixelation, blurring, or masking to conceal identifiable characteristics, which harm the original facial attributes. Current methods attach more importance to the quality and realism of the anonymized face but are still far from satisfactory. The faces generated by [8] and [27] are not natural enough and lack diversity. CIAGAN [24] shows better diversity but cannot handle complicated facial attributes and has poor fidelity. Currently, [3,10,39] focus on recoverable de-identification, but still produce visible artifacts. In this work, we attempt to increase the fidelity of anonymized faces with the help of identity swapping supervisions.

Face Identity Embedding with GANs. GAN has been widely used in face image manipulation. [1,2,33] propose to enlarge the latent space for better editing or transferring. For manipulating facial semantics, most works focus on changing the attributes [14,32,36]. [23] provides inspirational findings in learning identity-distilled and identity-dispelled features, but it focuses more on attribute editing and is usually applied at regular resolution. For editing the identity, SD-GAN [7] trains with a pair of images of the same identity and disentangles identity and attributes with the specialized discriminator. Recently, DiscoFace-GAN [6] embeds the 3D prior into adversarial learning with several VAE-based encoders. These methods conduct identity-specific generation but do not precisely disentangle identity from the other image content (*e.g.*, background, haircut, *etc.*.). In this work, we follow [6] to embed identity to the latent space but focus on the fine-grained control of the identifiable characteristics.

High-Resolution Face Generation. The image quality of generative methods, particularly Generative Adversarial Networks (GAN), have improved rapidly. StyleGAN [17] adopts a novel intermediate latent space and a *style ingcontrol* mechanism, which allows more disentangled and scale-specific control. Besides, it facilitates stochastic variation by providing additional random noise maps and further improves the generation fidelity. Recently, StyleGAN2 [18] and StyleGAN3 [16] fix some characteristics artifacts in [17] and yield the state-of-the-art generation quality. In this work, we utilize the powerful StyleGAN2 network and devise a new attribute extractor to improve the image fidelity.

3   Approach

We train StyleFace with the identity swapping scheme: Given the source image X_s and the target image X_t, we change the identity of X_t to that of X_s while preserving all the identity-irrelevant attributes, thus producing image $Y_{s,t}$. Once trained, we can anonymize the identity in image X_t by directly sampling a virtual identity. As presented in Fig. 2 (a), we utilize the StyleGAN2 model for high-resolution generation, with the identity as style and attribute as noise input. To construct the latent identity space and improve the generation fidelity, we devise the *Identity Projector* and the *Adaptive Attribute Extractor* (AAE). Next, we will explain the proposed framework in detail.

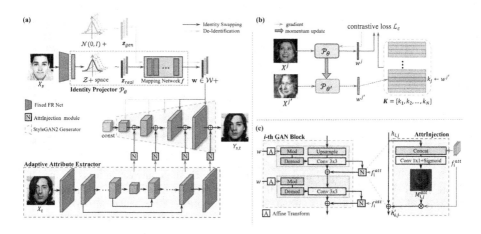

Fig. 2. The architecture of *StyleFace*. (a) The generation pipeline for identity swapping (with embedded z_{real}) and de-identification (with sampled z_{gen}). (b) Illustration of the contrastive loss. (c) Detailed structures of the GAN Block and the *AttrInjection* module. Please refer to [18] for details of the GAN block.

3.1 Identity Projector

To unify identity swapping and de-identification in one framework, we desire to represent identity as a latent variable to embed real identity and generate virtual identities. Inspired by DiscoFaceGAN [6], we devise a Variational Auto-Encoder (VAE)-based projector to project the identity prior from a pretrained Face Recognition (FR) network to the latent space of the StyleGAN2 model.

$\mathcal{Z}+$ **Space.** The original latent space $\mathcal{Z}$ of StyleGAN is a standard Gaussian distribution. Inspired by [1,2,33] that enlarge the latent space to increase the model's expressiveness, we expand $\mathcal{Z}$ by adopting three different latent codes from $\mathcal{Z}$ in the low (4^2–16^2), middle (32^2–128^2), and high-level (256^2–1024^2) layers. This is equivalent to sampling from a $\mathcal{Z}+$ space that consists of three versions of $\mathcal{Z}$. We empirically find that $\mathcal{Z}+$ provides a hierarchical identity control and increases the distinctiveness of identity change (Sect. 4.4).

To embed identity to the $\mathcal{Z}+$ space, we regard deep features from the pretrained FR model [13] as identity priors and devise a simple VAE-based projector. The projector is implemented as a one-layer MLP, which maps the features to the means and covariances of the $\mathcal{Z}+$ space. We regularize the latent space by the Kullback-Leibler divergence loss $\mathcal{L}_{kl}$:

$$\mathcal{L}_{kl} = \frac{1}{2}\sum_i(\mu_i^2 + \sigma_i^2 - \log\sigma_i^2 - 1), \tag{1}$$

where $\mu_i, \sigma_i \in \mathbb{R}^{1\times512}$ and $i \in \{l, m, h\}$. For simplicity, we use l, m, h to denote the low, middle, high-level layers, respectively. We do not add an extra reconstruction task as the typical VAE does but use the identity preserving objective

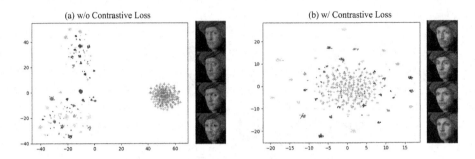

Fig. 3. Distributions of $\mathbf{w}_{real}$ and $\mathbf{w}_{gen}$ without/with the contrastive loss. '.' and '+' denote $\mathbf{w}_{real}$ and $\mathbf{w}_{gen}$, respectively. Different identities are marked by different colors. Examples of generated identity are shown on the right. (Color figure online)

to guarantee that the valid identity information is not lost. In the following sections, $\mathbf{z} = \{\mathbf{z}_i\}, i \in \{l, m, h\}$ is used to represent the latent code in $\mathcal{Z}+$ space.

Non-uniformity in the $\mathcal{W}+$ Space. There are two latent spaces in StyleGAN: the original latent space $\mathcal{Z}$, and the less entangled intermediate latent space $\mathcal{W}$. $\mathcal{W}$ is produced from $\mathcal{Z}$ by a non-linear mapping f. With identity embedded to the $\mathcal{Z}+$ space, we subsequently map $\mathbf{z}$ to $\mathbf{w}$ and compose the intermediate latent space $\mathcal{W}+$. The $\mathbf{w}$ vectors modulate the weights of corresponding convolution layers in the generator to control the generated identity.

Denoting the latent identity code embedded from a real image as $\mathbf{z}_{real}$ and that randomly sampled from the Gaussian distribution $\mathcal{N}(0, \mathbf{I})$ as $\mathbf{z}_{gen}$, we find that $\mathbf{z}_{real}$ can recover the corresponding identity for identity swapping, but $\mathbf{z}_{gen}$ fails to produce a feasible identity for de-identification. To figure out the reason, we use t-SNE to visualize the distribution of $\mathbf{w}_{real}$ and $\mathbf{w}_{gen}$ vectors. We randomly pick 100 identities from the training dataset and get 626 $\mathbf{w}_{real}$ vectors. Then we randomly sample 200 $\mathbf{z}_{gen}$ vectors from $\mathcal{N}(0, \mathbf{I})$ and map them to $\mathbf{w}_{gen}$.

As shown in Fig. 3 (a), the $\mathbf{w}_{real}$ codes are clustered to different centers and there is no overlap between $\mathbf{w}_{gen}$ and $\mathbf{w}_{real}$. Accordingly, the generated virtual identities are not feasible, indicating the $\mathbf{w}_{gen}$ codes are lying out of the reasonable intermediate space. We think the reason for the non-uniformity of $\mathcal{W}+$ is mainly two-fold: i) $\mathcal{W}+$ space is a complex non-Gaussian distribution that has no constrain on the uniformity. ii) The amount of different IDs in the training dataset is limited, thus $\mathbf{w}_{real}$ may not span the whole $\mathcal{W}+$ space but only a small subspace. Hence, we try to resolve this issue via *contrastive learning*.

Contrastive Constrain. Currently, [37] points out that the contrastive loss optimizes the alignment of features from positive pairs and the uniformity of the induced distribution. To generate reasonable identities from randomly sampled latent code, we desire the intermediate identity representation $\mathbf{w}$ to meet the following requirements: i) The $\mathbf{w}$ codes for samples of the same identity gather together. ii) All the **w** codes distribute uniformly in the $\mathcal{W}+$ space. To this end, we introduce a contrastive constrain on **w**.

We parameterize the process of embedding an image **X** to the intermediate identity representation **w** as an Identity Projector $\mathcal{P}_\theta$:

$$\mathbf{w} = \mathcal{P}_\theta(\mathbf{X}) = f(\phi(\omega(FR(\mathbf{X})))), \tag{2}$$

where FR is the fixed FR net, ω is the VAE-based projector, ϕ is the reparamterization process, f is the non-linear mapping from $\mathcal{Z}+$ to $\mathcal{W}+$, and θ denotes the learnable parameters in ω and f. Inspired by MoCo [11] that facilitates unsupervised contrastive learning with a large and consistent queue and a moving-averaged encoder, we build a dynamic list $\mathbf{K} = [\mathbf{k}_i]_{i=1}^N, \mathbf{k}_i \in \mathcal{R}^{1\times512}$, where N is the amount of distinct identity in the training set. Note that we build a dynamic list $\mathbf{K}_i$ for each $\mathbf{w}_i \in \mathbf{w}, i \in \{l, m, h\}$ and omit the subscript i for simplicity.

Figure 2 (b) illustrates the contrastive constrain. We create another encoder $\mathcal{P}_{\theta'}$, which has the same structure with $\mathcal{P}_\theta$. Given image $\mathbf{X}^j$ with the identity label j, we randomly pick another image $\mathbf{X}^{j'}$ of the same person to compose a positive pair. Then, we update $\mathcal{P}_{\theta'}$ by a momentum-based moving average of $\mathcal{P}_\theta$ so that $\theta' \leftarrow m\theta' + (1-m)\theta$, where $m \in [0,1)$ is the momentum coefficient. We encode $\mathbf{X}^j$ and $\mathbf{X}^{j'}$ with the projector $\mathcal{P}_\theta$ and the moving-averaged projector $\mathcal{P}_{\theta'}$, respectively. In this way, we get $\mathbf{w}^j = \mathcal{P}_\theta(\mathbf{X}^j)$ and $\mathbf{w}^{j'} = \mathcal{P}_{\theta'}(\mathbf{X}^{j'})$.

Unlike MoCo, which updates the dynamic queue by replacing the oldest sample in an unsupervised manner, we update the j-th item in $\mathbf{K}$ by $\mathbf{K}[j] \leftarrow \mathbf{w}^{j'}$. With $\mathbf{w}^j$ as the query, we regard $\mathbf{K}[j]$ as the positive key and the other items in $\mathbf{K}$ as the negative keys. Then, we normalize all the vectors to the unit space and measure the similarity between the query $\mathbf{w}^j$ and the dynamic list $\mathbf{K}$ by the InfoNCE [25] loss. In this way, the contrastive constrain $\mathcal{L}_c$ is formulated as:

$$\mathcal{L}_c = -\log \frac{\exp(\mathbf{w}^j \cdot \mathbf{K}[j]/\tau)}{\sum_{k=1}^N \exp(\mathbf{w}^j \cdot \mathbf{K}[k]/\tau)}, \tag{3}$$

where τ is the temperature. $\mathcal{L}_c$ encourages the **w** codes of the same identity to be similar to each other and dissimilar to those of other identities. Note that we empirically set $m = 0.999$ and $\tau = 0.07$.

As shown in Fig. 3 (b), with the intermediate representations constrained by the contrastive loss $\mathcal{L}_c$, the $\mathbf{w}_{real}$ codes uniformly distribute throughout the whole space and overlap with the $\mathbf{w}_{gen}$ codes. In this way, we can generate realistic identities with the randomly sampled latent codes.

3.2 Adaptive Attribute Extractor

A critical issue in megapixel-level face generation is to faithfully preserve the identity-irrelevant attributes and facial details, which is crucial for face realism and image quality. In this subsection, we introduce the Adaptive Attribute Extractor (AAE), which adaptively preserves the necessary information.

Multi-level Attribute Encoding. The attributes of a face image often span a large range of spatial resolution, such as the global-level position, the middle-level expression, and the fine-level details. Early work [22] demonstrates that multi-level features better preserve the image details than compressed single vectors. Therefore, we devise a lightweight U-shape DNN to extract features in various resolutions. Inspired by [40], we carefully design the DNN so that the feature map $\mathbf{f}_i^{att}$ in the i-th layer has the same shape as that in the i-th GAN block. Unlike [22] that injects attributes in a SPADE [26]-like design, we treat $\mathbf{f}_i^{att}$ in the i-th layer as the noise input of the corresponding i-th GAN block. In this way, we name the module as *AttrInjection*.

Adaptive Attribute Disentangle. The extracted multi-level features contain redundant information, such as the identity information of the target image. We desire to preserve just the least sufficient information of the target attributes. Therefore, we predict a control mask $\mathbf{M}_{i,j} \in [0,1]$ for the corresponding *AttrInjection* module, where $\mathbf{M}_{i,j}$ has the same shape as $\mathbf{f}_i^{att}$:

$$\mathbf{M}_{i,j} = \sigma(Conv(\mathbf{h}_{i,j} \circ \mathbf{f}_i^{att})). \tag{4}$$

$\mathbf{h}_{i,j}$ is the output of the j-th modulated convolution in i-th GAN block, σ is the *Sigmoid*($\cdot$) function and $\circ$ is the channel-wise concatenation. We compress the extracted attribute features $\mathbf{f}_i^{att}$ with the control mask $\mathbf{M}_{i,j}$ by

$$\mathbf{h}'_{i,j} = \mathbf{h}_{i,j} + \mathbf{M}_{i,j} \times \mathbf{f}_i^{att}. \tag{5}$$

As shown in Fig. 2 (c), we incorporate the distilled attribute information into the generation process without modifying the GAN structure.

Recent work [9] supervises the information compression by mutual information, but we observe that the imperfect information compression harms both the identity and attributes. Differently, we simply constrain the control mask by minimizing the activation in $\mathbf{M}_{i,j}$:

$$\mathcal{L}_{mask} = \sum_{i,j} ||\mathbf{M}_{i,j}||_1. \tag{6}$$

In this way, the multi-level information and spatial correspondence in the target image are maintained, and the redundant information is filtered.

3.3   Loss Function

Attribute Preserving Loss. When the source image $\mathbf{X}_s$ and the target image $\mathbf{X}_t$ have the same identity, we expect the output $\mathbf{Y}_{s,t}$ to be identical with $\mathbf{X}_t$, thus define the pixel-wise reconstruction loss as,

$$\mathcal{L}_{rec} = ||\mathbf{Y}_{s,t} - \mathbf{X}_t||_1 \quad if \, ID(\mathbf{X}_t) = ID(\mathbf{X}_s). \tag{7}$$

Following [5], we define the feature matching loss by minimizing the $L2$ distance between the multi-level features from the discriminator D for $\mathbf{X}_t$ and $\mathbf{Y}_{s,t}$. To eliminate the ghosting artifacts, we use a background mask $\mathbf{M}_{bg}$ from the segmentation model [34] in the shadow layers:

$$\mathcal{L}_{FM}^{low} = \sum_{i=1}^{m} \mathbf{M}_{bg} \cdot ||D^{(i)}(\mathbf{X}_t) - D^{(i)}(\mathbf{Y}_{s,t})||_2. \tag{8}$$

In deep layers, we match the features in the whole image:

$$\mathcal{L}_{FM}^{high} = \sum_{i=m}^{M} ||D^{(i)}(\mathbf{X}_t) - D^{(i)}(\mathbf{Y}_{s,t})||_2. \tag{9}$$

We define the total feature matching objective as the equally weighted sum:

$$\mathcal{L}_{FM} = \mathcal{L}_{FM}^{low} + \mathcal{L}_{FM}^{high}. \tag{10}$$

Identity Preserving Loss. To encourage the swapped identity to be more distinctive, we adopt the Identity Contrastive Loss (ICL) in [9]:

$$\begin{aligned} \mathcal{L}_{ICL} =& 1 - \cos < z_{id}(\mathbf{Y}_{s,t}), z_{id}(\mathbf{X}_s) > \\ &+ (\cos < z_{id}(\mathbf{Y}_{s,t}), z_{id}(\mathbf{X}_t) > - \cos < z_{id}(\mathbf{X}_s), z_{id}(\mathbf{X}_t) >)^2, \end{aligned} \tag{11}$$

where z_{id} is the 512-dim vector extracted by the FR net.

Overall Loss. We adopt the same GAN loss $\mathcal{L}_{GAN}$ for adversarial training as StyleGAN2, and the total objective is formulated as:

$$\begin{aligned} \mathcal{L}_{total} =& L_{GAN} + \mathcal{L}_c + \mathcal{L}_{mask} + \mathcal{L}_{FM} \\ &+ \lambda_{rec}\mathcal{L}_{rec} + \lambda_{ICL}\mathcal{L}_{ICL} + \lambda_{KL}\mathcal{L}_{KL}, \end{aligned} \tag{12}$$

where the contrastive loss $\mathcal{L}_c$ is defined in Eq. (3), the mask loss $\mathcal{L}_{mask}$ is defined in Eq. (6), and the KL-divergence loss is defined in Eq. (1), We train the whole model end-to-end with $\mathcal{L}_{total}$. Once the training is finished, the model can be directly used for de-identification (see Fig. 2 (a)).

4 Experiments

4.1 Implementation Details and Protocols

We train the model on a combination of FFHQ [17], VGGFace2 [4], and CelebAHQ [15], with all images aligned and cropped to 1024×1024. We devise the AAE (Sect. 3.2) to have the same spatial resolution as the StyleGAN2 model, but only one layer in each resolution and 1/8 the channel dimension. We use a 1×1 Conv to adjust the channel dimension of the DNN feature map before sending it to the *AttrInjection* module. The VAE-liked projector is a one-layer MLP. More details of the model's architecture are provided in the supplementary.

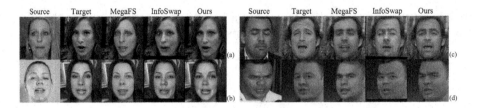

Fig. 4. Qualitative comparison with MegaFS [42] and InfoSwap [9] on FF++ [28]. Our model maintains the (a) eye color and (b) face shape of source identity, and better preserves target attributes, such as (c) expression and (d) skin color. (Color figure online)

Fig. 5. Qualitative comparison about identity swapping on the CelebAMaskHQ dataset. The first row shows source-target image pairs, and the last three rows show the results of MegaFS, InfoSwap, and *StyleFace* (ours) from top to bottom.

At the start of training, we set the ratio of source-target pairs with the same identity to 100% for a warm-up and linearly decrease it to 50%. Adam [19] is used with $\beta_1 = 0$, $\beta_2 = 0.99$. We first pretrain the generator on FFHQ for 20K steps and then train the whole model end-to-end. The learning rates of AAE and the generator are $1e-4$, while that of the identity projector is $1e-6$. For Eq. (12), we set $\lambda_{rec} = 10$, $\lambda_{ICL} = 5$, and $\lambda_{KL} = 1e-4$. The 1024^2-res model is trained using 4 A100 GPUs for 2 days with a batch size of 4.

4.2   High-Resolution Identity Swapping

In this section, we compare *StyleFace* with state-of-the-art high-resolution identity swapping methods, including MegaFS [42] and InfoSwap [9]. The public model and processing scripts are used in the following experiments.

Table 1. Quantitative test w.r.t.identity preserving, attribute preserving, and image quality. Values underlined are from [42] due to lack of ground-truth segmentation of FF++ [28]. Inference speed is tested on one V100 GPU over 1,000 independent runs. For MegaFS, the time for segmentation is excluded.

Model	ID↑	Shape↓	Pose↓	Exp.↓	FID ↓	Inference Speed (ms/1024^2 Image)	User Study
MegaFS [42]	90.83	–	2.64	–	16.64	91 ± 1.7	6.7%
InfoSwap [9]	**98.70**	0.57	3.11	0.31	3.39	329 ± 0.7	18.7%
Ours	96.34	**0.50**	**2.52**	**0.28**	**2.04**	**86 ± 1.8**	**74.6%**

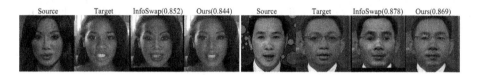

Fig. 6. Numbers in the bracket denotes the cosine similarity between the generated identity and the source identity. Our model produces more natural and perceptually similar faces than InfoSwap, though it has numerically lower scores.

Qualitative Comparison. We first compare on the FaceForensics++ (FF++) [28] dataset. As shown n Fig. 4, MegaFS produces distinct face contour and ignores the source face shape. Besides, InfoSwap produces visible skin artifacts and inconsistent eye color. Both two methods cannot preserve the target skin color. In contrast, our model maintains identity-level characteristics such as face shape and eye color and faithfully preserves the target attributes like pose and expression. In addition, the face images rendered by our model have distinctly better quality and are more visually appealing.

For megapixel-level identity swapping, we randomly compose 30K pairs of the source-target images from CelebAMaskHQ [20] and generate 1024^2 resolution results in Fig. 5. We observe that MegaFS [42] occasionally fails (col 5) due to unstable GAN-inversion. InfoSwap produces twisted hair (col 4), while our model preserves the detailed hair strands. Our model can better retain the lighting (col 1), expression (col 2), the source face shape (col 3/4), and better handle the occlusion (col 6/7). Besides, it produces fewer artifacts and maintains the image details, showing superior fidelity on megapixels.

Quantitative Comparison. Following [5, 22, 38], we conduct quantitative comparison on the FF++ [28] dataset on the following metrics: *ID retrieval, pose error, face shape error*, and *expression error*. For the ID retrieval rate, we use [35] to extract the identity embedding and report the Top-1 matching rate of the swapped image and the source image. We estimate the 3D pose by [29],

Fig. 7. (a) Examples of de-identified faces. (b) Comparison with CIAGAN [24]. Our method better preserves the original attributes (*e.g.*, lighting, expression, and occlusion), and generates identities that are more realistic and diverse.

Table 2. Comparison with recent de-identification methods on LFW [12].

Method	VGGFace2 [4]↓	CASIA [41]↓
Original	0.986 ± 0.010	0.965 ± 0.016
Gafni *et al.* [8]	0.038 ± 0.015	0.035 ± 0.011
CIAGAN [24]	0.034 ± 0.016	0.019 ± 0.008
Ours	**0.013 ± 0.006**	**0.012 ± 0.008**

and the expression and shape by [30]. We report the $L2$ distance between the regressed coefficients of swapped image and the ground truth for these three metrics. To further evaluate image fidelity and model efficiency on high-resolution generation, we compute the Fréchet Inception Distance (FID) score on the Celeb-MaskHQ dataset and the inference speed of generating one 1024^2-res image.

As shown in Table 1, our model has the lowest pose and expression error, indicating that the target attributes are well maintained. Besides, the FID scores imply that images generated by our model have high quality and fewer artifacts. Figure 6 shows that our model produces more natural and perceptually similar faces, although it has a slightly lower ID retrieval rate than InfoSwap, Besides, our model has the fastest inference speed, showing good efficiency. Moreover, we conduct a user study among 20 users on 50 source-target pairs from the CelebAHQ dataset, and each user selects the best one from three methods. As reported in Table 1, our method significantly outperforms the other methods.

4.3   Face De-identification

Qualitative Comparison. In Sect. 3.1, we design an Identity Projector to construct a latent identity space. Thus we can sample infinite virtual identities

Fig. 8. Identity swapping results with $\mathcal{Z}$ and $\mathcal{Z}+$ spaces.

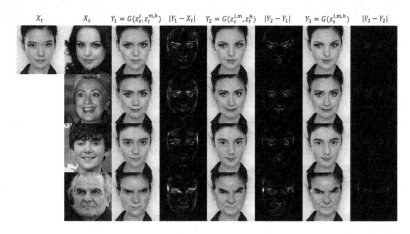

Fig. 9. Visualization of the hierarchical identity control of the $\mathcal{Z}+$ space. The brighter pixel indicates a larger difference.

from $\mathcal{Z}+$ space for face de-identification. Here we present some examples of the de-identified faces in Fig. 7 (a) and the qualitative comparison with the current state-of-the-art method CIAGAN [24] in Fig. 7 (b). It can be observed that faces de-identified by our model are more diverse and realistic, showing better preservation of the original attributes and better image quality.

Quantitative Comparison. Following [24], we anonymize the second image of each positive pair in the LFW [12] dataset. We utilize two FaceNet [31] models, which are pretrained on VGGFace2 [4] and CASIA-WebFace [41], respectively. The true acceptance rate is reported in Table 2 that lower value indicates better anonymization. We compare with the state-of-the-art De-ID methods, including Gafni *et al.* [8] and CIAGAN [24]. As presented in Table 2, when the face is anonymized by our model, the identification rate is lower than the other two methods, showing better de-identification ability.

4.4 Analysis

$\mathcal{Z}+$ **Space.** To verify the effectiveness of the $\mathcal{Z}+$ space (Sect. 3.1), we train another model with the original $\mathcal{Z}$ space. As shown in Fig. 8, the model with $\mathcal{Z}$ space has lower identity similarity and fails to recover the gaze direction.

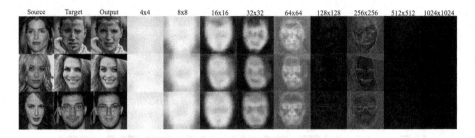

Fig. 10. Visualizing the control masks in AAE at different resolutions. The lighter pixel indicates a higher weight for attribute preserving.

Furthermore, we analyze the hierarchical identity control of the $\mathcal{Z}+$ space. With X_t as the attribute reference, we change the identity code from $\mathbf{z}_t$ to $\mathbf{z}_s$ by gradually replacing the low, middle, and high-level component of $\mathbf{z}_t$ with those of $\mathbf{z}_s$, generating $\mathbf{Y}_1$, $\mathbf{Y}_2$ and $\mathbf{Y}_3$. Then, we compute the differences to show the impact of each component in Fig. 9. The low-level code $\mathbf{z}^l$ affects the coarse-level attributes, such as the shape of face and eyebrow. The middle-level code $\mathbf{z}^m$ primarily affects the perceptual similarity, with the most facial attributes (*e.g.*, eye color, lips, *etc.*.) changed. Finally, the high-level codes $\mathbf{z}^h$ further strengthen some facial details, and the identity completely changes to $\mathbf{X}_s$.

Control Mask Visualization. In Sect. 3.2, we predict a control mask $M_{i,j}$ (Eq. (4)) to select the identity-irrelevant information from feature maps. Here we visualize the mean value of control masks at each resolution in Fig. 10. The mask highlights the whole face region in the low-level layers, indicating that the model learns to recover the global pose and facial layout. As the resolution increases, it focuses more on the background and facial decorations (*e.g.*, makeup and glasses). In the highest layers, the activation becomes sparser that only some edges and details are highlighted. The visualization shows that the AAE adaptively preserves the desired attributes at different resolutions.

5   Conclusion

In this paper, we have proposed a novel framework *StyleFace*, which unifies identity swapping and de-identification in one model and achieves high-fidelity face rendering on megapixels. To bridge the gap between identity swapping and de-identification, we embed identity prior into the latent space and introduce a contrastive constrain for further regularization. We utilize the StyleGAN2 for megapixel-level generation and devise an adaptive attribute extractor to preserve the identity-irrelevant information. We show that the proposed model can generate high-fidelity results with both embedded real identities and sampled virtual identities. Extensive experiments demonstrate the state-of-the-art performance of StyleFace in identity swapping and de-identification.

Acknowledgements. This work was partly supported by the Shanghai Municipal Science and Technology Major Project (2021SHZDZX0102) and the National Science of Foundation China (61972250, 72061127003).

References

1. Abdal, R., Qin, Y., Wonka, P.: Image2stylegan: How to embed images into the stylegan latent space? In: ICCV, Oct 2019
2. Abdal, R., Qin, Y., Wonka, P.: Image2stylegan++: How to edit the embedded images? In: CVPR, Jun 2020
3. Cao, J., Liu, B., Wen, Y., Xie, R., Song, L.: Personalized and invertible face de-identification by disentangled identity information manipulation. In: Proceedings of the IEEE/CVF International Conference on Computer Vision (ICCV), pp. 3334–3342, Oct 2021
4. Cao, Q., Shen, L., Xie, W., Parkhi, O.M., Zisserman, A.: Vggface2: a dataset for recognising faces across pose and age. In: FG 2018, pp. 67–74 (2018)
5. Chen, R., Chen, X., Ni, B., Ge, Y.: Simswap: an efficient framework for high fidelity face swapping. In: ACM MM, pp. 2003–2011 (2020)
6. Deng, Y., Yang, J., Chen, D., Wen, F., Tong, X.: Disentangled and controllable face image generation via 3d imitative-contrastive learning. In: CVPR (2020)
7. Donahue, C., Lipton, Z.C., Balsubramani, A., McAuley, J.J.: Semantically decomposing the latent spaces of generative adversarial networks. In: ICLR (2018)
8. Gafni, O., Wolf, L., Taigman, Y.: Live face de-identification in video. In: ICCV, pp. 9377–9386 (2019)
9. Gao, G., Huang, H., Fu, C., Li, Z., He, R.: Information bottleneck disentanglement for identity swapping. In: CVPR, pp. 3404–3413, Jun 2021
10. Gu, X., Luo, W., Ryoo, M.S., Lee, Y.J.: Password-conditioned anonymization and deanonymization with face identity transformers. In: Vedaldi, A., Bischof, H., Brox, T., Frahm, J.-M. (eds.) ECCV 2020. LNCS, vol. 12368, pp. 727–743. Springer, Cham (2020). https://doi.org/10.1007/978-3-030-58592-1_43
11. He, K., Fan, H., Wu, Y., Xie, S., Girshick, R.: Momentum contrast for unsupervised visual representation learning. In: Proceedings of the IEEE/CVF Conference on Computer Vision and Pattern Recognition (CVPR), Jun 2020
12. Huang, G.B., Ramesh, M., Berg, T., Learned-Miller, E.: Labeled faces in the wild: A database for studying face recognition in unconstrained environments. Tech. Rep. 07–49, University of Massachusetts, Amherst, Oct 2007
13. Huang, Y., et al.: Curricularface: adaptive curriculum learning loss for deep face recognition. In: CVPR, pp. 1–8 (2020)
14. Härk"onen, E., Hertzmann, A., Lehtinen, J., Paris, S.: Ganspace: discovering interpretable gan controls. In: Proceedings of NeurIPS (2020)
15. Karras, T., Aila, T., Laine, S., Lehtinen, J.: Progressive growing of GANs for improved quality, stability, and variation. In: ICLR (2018). https://openreview.net/forum?id=Hk99zCeAb
16. Karras, T., et al.: Alias-free generative adversarial networks. In: NIPS (2021)
17. Karras, T., Laine, S., Aila, T.: A style-based generator architecture for generative adversarial networks. In: CVPR, pp. 4401–4410 (2019)
18. Karras, T., Laine, S., Aittala, M., Hellsten, J., Lehtinen, J., Aila, T.: Analyzing and improving the image quality of StyleGAN. In: CVPR (2020)
19. Kingma, D.P., Ba, J.: Adam: a method for stochastic optimization. arXiv preprint arXiv:1412.6980 (2014)

20. Lee, C.H., Liu, Z., Wu, L., Luo, P.: Maskgan: towards diverse and interactive facial image manipulation. In: CVPR (2020)
21. Li, J., Li, Z., Cao, J., Song, X., He, R.: Faceinpainter: high fidelity face adaptation to heterogeneous domains. In: CVPR, pp. 5089–5098, Jun 2021
22. Li, L., Bao, J., Yang, H., Chen, D., Wen, F.: Advancing high fidelity identity swapping for forgery detection. In: CVPR, Jun 2020
23. Liu, Y., Wei, F., Shao, J., Sheng, L., Yan, J., Wang, X.: Exploring disentangled feature representation beyond face identification. In: CVPR (2018)
24. Maximov, M., Elezi, I., Leal-Taixé, L.: CIAGAN: conditional identity anonymization generative adversarial networks. In: CVPR, pp. 5446–5455 (2020)
25. Oord, A.v.d., Li, Y., Vinyals, O.: Representation learning with contrastive predictive coding. arXiv preprint arXiv:1807.03748 (2018)
26. Park, T., Liu, M.Y., Wang, T.C., Zhu, J.Y.: Semantic image synthesis with spatially-adaptive normalization. In: CVPR (2019)
27. Ren, Z., Lee, Y.J., Ryoo, M.S.: Learning to anonymize faces for privacy preserving action detection. In: Ferrari, V., Hebert, M., Sminchisescu, C., Weiss, Y. (eds.) ECCV 2018. LNCS, vol. 11205, pp. 639–655. Springer, Cham (2018). https://doi.org/10.1007/978-3-030-01246-5_38
28. Rössler, A., Cozzolino, D., Verdoliva, L., Riess, C., Thies, J., Nießner, M.: Faceforensics++: learning to detect manipulated facial images. In: ICCV (2019)
29. Ruiz, N., Chong, E., Rehg, J.M.: Fine-grained head pose estimation without keypoints. In: CVPRW, pp. 2074–2083 (2018)
30. Sanyal, S., Bolkart, T., Feng, H., Black, M.: Learning to regress 3d face shape and expression from an image without 3d supervision. In: CVPR (2019)
31. Schroff, F., Kalenichenko, D., Philbin, J.: Facenet: A unified embedding for face recognition and clustering. In: CVPR, pp. 815–823 (2015)
32. Shen, Y., Yang, C., Tang, X., Zhou, B.: Interfacegan: interpreting the disentangled face representation learned by gans. In: TPAMI (2020)
33. Song, G., et al.: Agilegan: stylizing portraits by inversion-consistent transfer learning. In: SIGGRAPH, Jul 2021
34. Sun, K., et al.: High-resolution representations for labeling pixels and regions. arXiv (2019)
35. Wang, H., et al.: Cosface: Large margin cosine loss for deep face recognition. In: CVPR, pp. 5265–5274 (2018)
36. Wang, T., Zhang, Y., Fan, Y., Wang, J., Chen, Q.: High-fidelity gan inversion for image attribute editing. arxiv:2109.06590 (2021)
37. Wang, T., Isola, P.: Understanding contrastive representation learning through alignment and uniformity on the hypersphere. In: ICML, pp. 9929–9939 (2020)
38. Wang, Y., et al.: Hififace: 3d shape and semantic prior guided high fidelity face swapping. In: IJCAI-2021, pp. 1136–1142 (8 2021)
39. Yamaç, M., Ahishali, M., Passalis, N., Raitoharju, J., Sankur, B., Gabbouj, M.: Reversible privacy preservation using multi-level encryption and compressive sensing. In: EUSIPCO, pp. 1–5 (2019)
40. Yang, T., Ren, P., Xie, X., Zhang, L.: Gan prior embedded network for blind face restoration in the wild. In: CVPR (2021)
41. Yi, D., Lei, Z., Liao, S., Li, S.Z.: Learning face representation from scratch. arXiv: 1411.7923 (2014)
42. Zhu, Y., Li, Q., Wang, J., Xu, C., Sun, Z.: One shot face swapping on megapixels. In: CVPR, pp. 4834–4844, Jun 2021

Video Extrapolation in Space and Time

Yunzhi Zhang$^{(\boxtimes)}$ and Jiajun Wu

Stanford University, Stanford, USA
{yzzhang,jiajunwu}@cs.stanford.edu

Abstract. Novel view synthesis (NVS) and video prediction (VP) are
typically considered disjoint tasks in computer vision. However, they can
both be seen as ways to observe the spatial-temporal world: NVS aims
to synthesize a scene from a new point of view, while VP aims to see
a scene from a new point of time. These two tasks provide complementary signals to obtain a scene representation, as viewpoint changes from
spatial observations inform depth, and temporal observations inform the
motion of cameras and individual objects. Inspired by these observations,
we propose to study the problem of Video Extrapolation in Space and
Time (VEST). We propose a model that leverages the self-supervision
and the complementary cues from both tasks, while existing methods
can only solve one of them. Experiments show that our method achieves
performance better than or comparable to several state-of-the-art NVS
and VP methods on indoor and outdoor real-world datasets. (Project
page: https://cs.stanford.edu/~yzzhang/projects/vest/.)

1 Introduction

Novel view synthesis (NVS) and video prediction (VP) are both widely studied
computer vision tasks. The former extrapolates a scene to a different camera
viewpoint, while the latter extrapolates to a future timestamp. NVS focuses
on the scene geometry revealed from discrete camera positions, whereas VP
extracts information from the moving trajectories of both cameras and objects.
The self-supervision signals from these two tasks can be jointly used to extract
a scene representation. To this end, we call attention to the problem of Video
Extrapolation in Space and Time (VEST) that considers both tasks.

We solve the problem of VEST by first developing a representation that incorporates both the spatial and temporal consistency from video data as inductive
biases. We generalize Multiplane Images (MPIs) [44], which is originally a layered representation that decomposes images into RGBA planes, by additionally
parameterizing the flow field of each plane in order to model the temporal dynamics. Images from a future timestamp and a novel viewpoint can be rendered from
flow-based and homography-based warping, respectively.

Compared with previous MPI-based NVS approaches [4,29,32,44], our generalized MPI representation leverages learning signals derived from both the
spatial and temporal coherence in video data, while previous NVS methods utilize frame tuples randomly sampled from the training video sequences without

S. Avidan et al. (Eds.): ECCV 2022, LNCS 13676, pp. 313–333, 2022.
https://doi.org/10.1007/978-3-031-19787-1_18

(a) Model inference   (b) Input frame I_{11} (c) Input frame I_{22} (d) Predicted frame I_{33} (e) Synthesized view I_{23}

Fig. 1. We propose the task of Video Extrapolation in Space and Time (VEST) that exploits both spatial and temporal consistency in video data. (a) During inference time, the model takes in two consecutive frames as inputs, colored in orange, and outputs images extrapolated along the temporal axis for future frame prediction and along the spatial axis for novel view synthesis, colored in green. The black curve denotes the camera trajectory. (b) and (c) are examples of model inputs; (d) and (e) are model outputs. Images from the bottom row are zoom-in views. (Color figure online)

considering the temporal information. Compared with other optical-flow-based VP methods [5], our method predicts the motion field individually for each MPI plane instead of the full scene. Since each MPI plane captures a relatively simple structure, we can effectively estimate the motion field of each plane with affine transformations [35].

We instantiate a model that performs the spatial-temporal extrapolation with the generalized MPI representation. As shown in Fig. 1, our model predicts MPI planes from monocular inputs and leverages historical frames for motion inference. Experiments show that our problem formulation and model are generally applicable to diverse scenarios: indoor and outdoor scenes, and videos taken by a single and multiple static or moving camera(s). Our method achieves favorable performance on *both* tasks compared to baselines designed for either.

Our main contributions are as follows:

- We propose VEST as a self-supervision task for learning a generalized MPI representation from video data.
- We instantiate a model that learns the proposed representation for simultaneous video extrapolation in space and time.
- Our VEST model produces realistic results in both space and time extrapolation on a diverse range of datasets.

2   Related Works

Novel view synthesis. Synthesizing novel views based on 2D images is a challenging problem, as it requires the reasoning of the 3D structure of the perceptual world. As shown in Fig. 2(a) and Fig. 2(b), monocular view synthesis approaches [10,32,38] take I_{11} as inputs and outputs the synthesized image for a query view $j \neq 1$. Stereo view synthesis approaches [44] are similar but take in two input views I_{11}, I_{12}. Most of these approaches synthesize novel views based

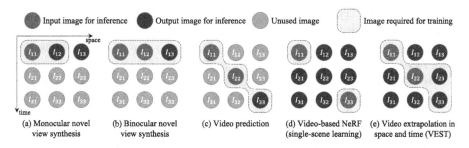

Fig. 2. Overview of vision tasks related to our work. I_{ij} denotes the RGB frame for the i-th timestamp in a video sequence taken from the j-th camera viewpoint. We restrict $i, j \in \{1, 2, 3\}$ for illustration. In (a) and (b), monocular or binocular NVS methods only extrapolate in space, while in (c), VP methods only extrapolate in time. Video-based NeRF methods in (d) train on a collection of images for each scene, and perform interpolation instead of extrapolation for inference. Finally, (e) illustrates the problem we focus on in this work. Given historical frames, the task consists of both predicting the future and extrapolating to a novel viewpoint. During inference, novel views I_{21}, I_{23} can be synthesized the same way as training time, while I_{22}, I_{23} can be inferred by duplicating I_{11} as input (VEST-single from Sect. 4.1).

on camera parameters estimated from Structure-from-Motion (SfM) techniques, except for Lai et al. [10] which uses camera parameters predicted by the model. These methods focus exclusively on spatial extrapolation and do not take the temporal axis into account. Previous works [27,33,44,45] use layered representation and apply per-layer warping to obtain the novel view based on camera poses. SynSin [38] and Worldsheet [8] predict the depth from a monocular image input and warps a 3D representation of the scene in the features space and pixel space, respectively, to synthesize novel views. In comparison, our proposed VEST encapsulates NVS but also has the ability to perform temporal extrapolation.

Closer to ours, Lin et al. [13] and Yoon et al. [41] tackle the problem of NVS for dynamic scenes. Lin et al. [13] addresses the temporal inconsistency of the MPI representation when applied to scenes with moving objects by identifying the error-prone regions with a learned 3D mask volume. There are three key differences between our model and Lin et al.First, they require two images from synchronized stereo cameras as inputs during inference, while our method takes in two consecutive frames from a single camera. Second, they assume a static camera and static background while we do not. Third, they rely on the static background computed from the full video sequences from two source views, while ours directly predicts the future video sequence based on two input frames. Yoon et al. [41] proposes to fuse the depth estimated from a single view and multiple views with a depth fusion network, and predicts the novel view on top. It requires segmentation labels and optical flow inputs, while our method is fully self-supervised.

Video prediction. Video prediction methods typically take in two or more frames taken by one camera and only extrapolate along the temporal axis as shown in Fig. 2(c). Deep neural networks have been widely used in video pre-diction [5,22,26,30]. Previous works studying temporal extrapolation for videos include hallucination-based methods [16,34] and warping-based methods [15]. More recently, Wang et al. [36] propose PredRNN-V2, integrating convolutional recurrent units with a pair of decoupled memory cells. Several methods also pro-posed to decompose high-dimensional videos into object-centric [39] or semantic-aware [1] regions, and model the motion of each region independently. These methods consider only monocular input and often require models pre-trained on large datasets for semantic and depth information, while our method factor-izes videos into depth-aware components by using the geometric cue from view synthesis, and it requires no external supervision.

Dynamic Neural Radiance Fields. Neural Radiance Fields (NeRF) [19] shows impressive results in synthesizing novel views with high fidelity. Not surprisingly, many follow-up papers have attempted to extend it to video for image synthesis in novel space and time point [3]. Our method differs from these methods in two ways. First, though on static scenes, methods like PixelNeRF can already learn a neural scene representation conditioned on one or few input images [12,20, 21,31,40,42], all video-based NeRF methods still employ single-video learning and need to be re-trained for every new scene, while our method performs 4D synthesis conditioned on the input images. Second, NeRF-based methods require hundreds of images for training, and essentially perform interpolation among input time- and viewpoints as shown in Fig. 2(d), while our method focuses on video extrapolation.

Layered Representations. The idea of decomposing images into RGBA layers is effective in multiple problem domains. For view synthesis, Shade et al. [25] proposed layered depth images to represent a scene with multi-layer depth and color images. Viewers can then see the scene from different points in space. For video prediction, there has also been a line of work inspired by the classic research on layered motion representations [35]. A notable example is a recent work by Lu et al. [17,18] on the problem of video manipulation, where they decompose videos into semantic-aware RGBA layers. Our layered representation builds upon all these ideas and aims to tackle both NVS and VP.

3   Method

3.1   Multiplane Images

Before introducing the representation we use, we first review the classical Mul-tiplane Images (MPIs) [44]. An image $I \in \mathbb{R}^{H \times W \times 3}$ is represented by D RGBA planes, $\{(c_i, \alpha_i)\}_{i=1}^{D}$, where $c_i \in \mathbb{R}^{H \times W \times 3}$ are RGB values and $\alpha_i \in \mathbb{R}^{H \times W \times 1}$ are alpha values. Each plane corresponds to fixed depth d_i. The plane is fronto-parallel to the camera and can be written as $n^T x - d_i = 0$, where $n = [0, 0, 1]^T$ is the normal vector of the plane.

Let (R, t) be the rotation and translation matrix from target to source view, and K, K' be the camera intrinsics for source and target views. The transformation for the i-th plane from target to source view, denoted as $\mathcal{W}_i^{R,t,K,K'}$, is defined as

$$\begin{bmatrix} u \\ v \\ 1 \end{bmatrix} \sim \mathcal{W}_i^{R,t,K,K'} \begin{bmatrix} u' \\ v' \\ 1 \end{bmatrix} := K\left(R - \frac{tn^T}{d_i}\right)(K')^{-1} \begin{bmatrix} u' \\ v' \\ 1 \end{bmatrix}, \tag{1}$$

where (u, v) and (u', v') are the coordinates from the source and target views, respectively.

The MPI representation for the target view $\{(c_i', \alpha_i')\}_{i=1}^D$ is computed as

$$c_{u',v'}' = c_{u,v}, \tag{2}$$

$$\alpha_{u',v'}' = \alpha_{u,v}. \tag{3}$$

Here (u, v) are sampled according to Eq. (1). Similar to Zhou et al. [44], we apply bilinear sampling from the neighboring grid corners when (u, v) is not aligned with the coordinate grid.

Finally, an MPI renderer synthesizes the target-view image with

$$\hat{I}' = \sum_{i=1}^D c_i' \alpha_i' \prod_{j=i+1}^D (1 - \alpha_j'). \tag{4}$$

3.2 Generalized Multiplane Images

We generalize the MPI representation to model the motion from image $I = I_t$ to $I' = I_{t+1}$, the next frame in a video sequence. Note that different from Sect. 3.1, here we use I' to refer to the frame for timestamp $t + 1$ which is not necessarily from a different camera viewpoint.

Formally, the generalized MPI representation for image I is denoted as $\{(c_i, \alpha_i, \mathcal{T}_i)\}_{i=1}^D$, where $\mathcal{T} = \mathcal{T}_i$ is an operator defined by

$$\begin{bmatrix} u \\ v \end{bmatrix} \sim \mathcal{T} \begin{bmatrix} u' \\ v' \end{bmatrix} := \begin{bmatrix} u' + \Delta u' \\ v' + \Delta v' \end{bmatrix}. \tag{5}$$

Let θ be a parameterization for $\mathcal{T}$. When $\mathcal{T}$ is fully parameterized, $\theta \in \mathbb{R}^{H \times W \times 2}$ is the pixel displacement field, and $\Delta u' = \theta_{u',v',0}$, $\Delta v' = \theta_{u',v',1}$ correspond to the number of pixels to be shifted in the u, v coordinates, respectively. However, since each MPI plane has a relatively simple structure, in practice we restrict $\mathcal{T}$ to be in the class of affine transformations. $\mathcal{T}$ can now be parameterized by $\theta \in \mathbb{R}^{2 \times 3}$, and

$$\mathcal{T} \begin{bmatrix} u' \\ v' \end{bmatrix} = \begin{bmatrix} \theta_{1,1} & \theta_{1,2} & \theta_{1,3} \\ \theta_{2,1} & \theta_{2,2} & \theta_{2,3} \end{bmatrix} \begin{bmatrix} u' \\ v' \\ 1 \end{bmatrix}. \tag{6}$$

We denote the generalized MPI representation for image I as

$$\{(c_i, \alpha_i, \theta_i)\}_{i=1}^D. \tag{7}$$

Finally, we predict the next frame $\hat{I}'$ based on Eq. (2)–(4).

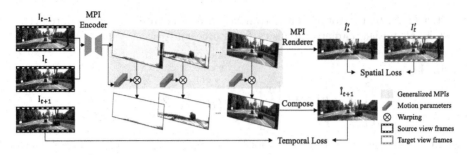

Fig. 3. Model architecture. The MPI encoder receives monocular frames I_{t-1} and I_t as inputs and outputs the generalized MPI representation for I_t, which consists of D RGBA planes and the motion parameters for each plane. Then the MPI renderer renders the target view image I'_t based on camera parameters and the RGBA planes. The next-frame prediction I_{t+1} is generated by first warping each RGBA plane with motion parameters, and then composing the planes. The training objective is to match $\hat{I}'_t$ and $\hat{I}_{t+1}$ with the ground truth. Images are licensed under CC BY-NC-SA 3.0.

3.3   Video Extrapolation in Space and Time

We now introduce our VEST model that predicts the generalized MPIs given monocular video frames, and performs spatial and temporal extrapolation. An overview of our model is shown in Fig. 3.

Training. The model takes in two consecutive frames from a video sequence, I_{t-1} and I_t, and outputs the generalized MPI representation (Eq. (7)) for I_t. The target-view image I'_t is synthesized following Sect. 3.1, and the next-frame prediction following Sect. 3.2.

Inference. During inference, with inputs I_{t-1} and I_t, the model can be queried to extrapolate to other space-time coordinates. Future frames with longer horizon $I_{t+2}, I_{t+3}, \cdots$ can be inferred by iteratively forwarding the model in an autoregressive manner.

Even when there is only one input frame I_t available, our model can synthesize the novel view I'_t by having I_t replicated twice as inputs and still produce realistic NVS results, corresponding to VEST-single from Table 1 and Table 2 as we will show later.

Losses. The training loss is the sum of spatial and temporal extrapolation errors, namely

$$\mathcal{L}^{\text{total}} = \mathcal{L}^{\text{space}}(\hat{I}'_t, I'_t) + \mathcal{L}^{\text{time}}(\hat{I}_{t+1}, I_{t+1}), \tag{8}$$

where

$$\mathcal{L}^{\text{space}} = \lambda_1^{\text{space}}\mathcal{L}_1 + \lambda_{\text{perc}}^{\text{space}}\mathcal{L}_{\text{perc}}, \tag{9}$$

$$\mathcal{L}^{\text{time}} = \lambda_1^{\text{time}}\mathcal{L}_1 + \lambda_{\text{perc}}^{\text{time}}\mathcal{L}_{\text{perc}}. \tag{10}$$

$\mathcal{L}_1$ is the ℓ_1 loss, and $\mathcal{L}_{\text{perc}}$ is the perceptual loss using pretrained VGG-19 [28] features.

3.4   Implementation Details

We adopt a model architecture similar to Tucker and Snavely [32], specified in Appendix A. The network outputs a tensor in $\mathbb{R}^{D \times H \times W \times 7}$, which is split channelwise into $f^\alpha \in \mathbb{R}^{D \times H \times W \times 1}$ for alpha values and $f^\theta \in \mathbb{R}^{D \times H \times W \times 6}$ for motion parameters. We set the RGB values for each plane (c_i from Eq. (7)) to be the RGB values of the source image. The final motion parameter θ for each MPI plane is computed as a weighted spatial average of f^θ:

$$w_i = \alpha_i \prod_{j=i+1}^{D} (1 - \alpha_j), \tag{11}$$

$$\theta_i = \mathrm{SpatialAverage}(w_i \otimes f^\theta), \tag{12}$$

where $\otimes$ denotes the element-wise multiplication. Note that w_i are the same weights for RGB values used in Eq. (4).

Camera parameters are estimated with SfM [23,24] and are used to render novel views. Since SfM models have ambiguous depth scales, we compute a depth scale factor σ similar to Tucker and Snavely [32] such that MPI planes are associated with scaled depth values $\{\sigma d_i\}_{i=1}^{D}$ instead of $\{d_i\}_{i=1}^{D}$. Here σ is computed to minimize the log-squared error of the predicted depth map $\hat{Z}$ and a set of sparse 3D points P_s,

$$\sigma = \exp\left[\frac{1}{|P_s|} \sum_{(u,v,z) \in P_s} (\ln z - \ln \hat{Z}_{u,v}) \right], \tag{13}$$

where the depth map is computed with Eq. (11) and

$$\hat{Z} = \sum_{i=1}^{D} d_i w_i. \tag{14}$$

For each element $(u, v, z) \in P_s$, (u, v) is the pixel coordinate in the source image, and z is the depth value of the corresponding 3D point. We run COLMAP [23] to obtain P_s for each video sequence.

4   Experiments

We conduct extensive experiments to validate that our method compares favorably to state-of-the-art methods on both monocular NVS and VP in diverse scenarios: during training, our model may learn from single- and multi-view videos, static and moving cameras, and indoor and outdoor scenes. In particular, our method is the only one that performs both tasks simultaneously. Finally in Sect. 4.5, we provide an analysis of intermediate model outputs for an intuitive understanding of our method.

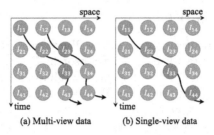

(a) Multi-view data   (b) Single-view data

Fig. 4. Training setups. Training inputs are colored in orange and outputs are in green. Black curves denote camera trajectories. Our method can be trained on multi-view or single-view datasets. I_{33} is the ground truth for future frame prediction. The NVS supervision comes from a stereo frame when available (I_{23} in (a)), or from a temporally nearby frame otherwise (I_{44} in (b)). (Color figure online)

4.1   Learning from Multi-View Videos with Moving Cameras

Setup. We first test our model on learning from videos with binocular views, as shown in Fig. 4(a). KITTI [6] is a benchmark dataset widely used for both NVS and VP. It contains street scenes captured by two stereo cameras mounted on a moving car. Following prior methods [11,32,33], we use the 28 city scenes from the KITTI-raw dataset and split them into 20 sequences for training, 4 for validation, and 4 for testing. This is denoted as LDI [33] split in Table 1. To be directly comparable with prior works on VP [1,39], we also evaluate on the split with 24 sequences for training and 4 for testing, denoted as FVS [39] split.

Evaluation. For NVS evaluation, images are cropped by 5% from the boundary of all sides and resized to 128×384. We report the similarity of synthesized images compared to the ground truth using LPIPS [43] computed with VGG features, SSIM [37] and PSNR. We report LPIPS with AlexNet features for VP to be directly comparable to prior methods [1,39].

Baselines. For NVS, we compare our results with Tucker and Snavely [32] and LDI [33]. LDI [33] derives a 2-layer representation from single-view inputs. We also compare with MINE [11], a recent work that extends Tucker and Snavely [32] to the continuous depth domain using implicit functions. MINE [11]-32 and MINE [11]-64 from Table 1 refer to two model variants with 32 and 64 planes, respectively, as reported in the original paper.

For VP, we compare our method with PredRNN-V2 [36], a recently proposed method using a pair of decoupled RNN memory cells. We retrain their non-action-conditioned model on the dataset with 2 input frames and a prediction length of 5. We also compare with two VP methods which decompose scenes for better dynamics modeling. FVS [39] predicts the next frame by decomposing frames in a video into object-centric layers and modeling the dynamics of each layer. Similar to ours, FVS also assumes the motion of each layer to be affine. Another baseline is SADM [1] which decomposes frames into semantically consistent regions and

Table 1. Results on KITTI [6] with three train-test splits used in previous methods. Our method achieves better performance than all NVS baselines, including those pretrained on ImageNet [2] (denoted by *). We also achieve competitive performance compared with VP baselines. LDI [33] and Tucker et al. [32] do not report LPIPS values, denoted by N/A. Baselines can do only one task while ours solves both, indicated by N/A.

Split	Method	Extrapolation in space			Extrapolation in time		
		LPIPS↓	SSIM↑	PSNR↑	LPIPS-AlexNet↓		
					$t+1$	$t+3$	$t+5$
LDI [33] split train 256 × 768 test 128 × 3840.	MINE [11]-32*	0.112	0.822	21.4	N/A		
	MINE [11]-64*	0.108	0.820	21.3	N/A		
	LDI [33]	N/A	0.572	16.5	N/A		
	Tucker et al. [32]	N/A	0.733	19.5	N/A		
	PredRNN-V2 [36]		N/A		0.3085	0.4573	0.5422
	VEST-no-perc	0.111	0.819	21.4	0.1129	0.2902	0.3943
	VEST-single	0.091	0.816	21.2	N/A		
	VEST (ours)	**0.085**	**0.825**	**21.6**	**0.1154**	**0.2881**	**0.3911**
LDI [33] split train 128 × 384 test 128 × 384	MINE [11]*	0.112	0.828	21.9	N/A		
	MINE [11]	0.129	0.812	**21.4**	N/A		
	PredRNN-V2 [36]		N/A		0.2153	0.3946	0.4984
	VEST-single	0.105	0.806	20.7	N/A		
	VEST (ours)	**0.097**	**0.818**	21.1	**0.0798**	**0.2348**	**0.3384**
FVS [39] split train 256 × 832 test 256 × 832	FVS [39]*		N/A		0.1848	0.2461	0.3049
	SADM [1]*		N/A		0.1441	0.2458	0.3116
	PredNet [16]		N/A		0.5535	0.5866	0.6295
	MCNet [34]		N/A		0.2405	**0.3171**	**0.3739**
	VoxelFlow [15]		N/A		0.3247	0.3743	0.4159
	VEST (ours)	**0.150**	**0.739**	**19.9**	**0.1560**	0.3441	0.4467

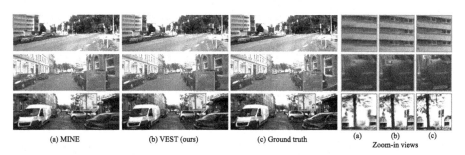

(a) MINE   (b) VEST (ours)   (c) Ground truth   (a)   (b)   (c)
Zoom-in views

Fig. 5. View synthesis results on KITTI. Left: (a) results from MINE [11]; (b) results from VEST (ours); (c) ground truth images. Right: zoom-in views. In the first two examples, our model produces fine details for small objects, such as windows and printed texts. The third example shows that our model performs more accurate extrapolation even for challenging, thin objects, such as trees with small distortion.

predicts the motion of each region. Both FVS and SADM require instance maps and optical flow inputs obtained from pretrained models.

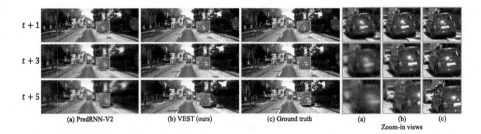

$t+1$

$t+3$

$t+5$

(a) PredRNN-V2   (b) VEST (ours)   (c) Ground truth   (a)   (b)   (c)
Zoom-in views

Fig. 6. Video prediction results on KITTI. Left: (a) predictions from PredRNN-V2 [36]; (b) predictions from VEST (ours); (c) ground truth images. Right: zoom-in views. Each row corresponds to frame prediction for $t+1$, $t+3$, $t+5$. Our VEST produces much sharper predictions compared to the baseline.

Results. Results are shown in Table 1. All our models are trained with the same configuration. For NVS, our model achieves a significant improvement across all metrics compared to the NVS baselines including variants of MINE [11] that are pre-trained on ImageNet [2]. We further show a qualitative comparison in Fig. 5 where MINE [11] uses ImageNet pretraining. Please refer to the project page for better visualization.

We additionally test our model with a different inference procedure (VEST-single as shown in Table 1). Instead of I_{t-1} and I_t, the model receives a single source frame I_t repeated twice as inputs to predict I'_t. In this way, our model receives no additional information from historical frames. VEST and VEST-single from Table 1 use the same checkpoint and both outperform all non-pretrained baselines across all metrics.

For VP, our approach leads to a significant improvement compared to PredRNN-V2 [36]. Figure 6 shows that our model makes prediction with sharper edges and higher fidelity. In these examples, street and cars are moving towards the camera, which is effectively a scale transformation in 2D. Because scale transformation is a subclass of affine transformation, our method can accurately capture such perpendicular motions in the scenes. All qualitative results from the KITTI dataset are used under license CC BY-NC-SA 3.0.

Ablation. We ablate the importance of using Perceptual loss for NVS. In Table 1, VEST-no-perc corresponds to a variant with $\lambda_{\text{perc}}^{\text{space}} = 0$, suggesting that perceptual loss in the spatial loss term largely improves performance. We also include an ablation study on the effect of the number of MPI planes in Sect. C.3.

4.2 Learning from Single-View Videos with a Moving Camera

Setup. We evaluate our method on video data with a single view, as shown in Fig. 4(b). RealEstate10K [44] is a standard NVS benchmark dataset consisting of 80K videos filming mostly indoor scenes. We follow the training-test split from Lai et al. [10], which uses a randomly sampled subset of the full dataset. It

Table 2. Results on RealEstate10K [44]. Our model achieves a better or comparable performance compared to the baselines, and only ours can predict future frames. SynSin [38] does not report LPIPS values, denoted by N/A. n specifies the number of frames between source and target images in the video sequence, and $n =$ rand means that n is uniformly sampled between 1 and 30. A larger n indicates larger range of spatial extrapolation. * indicates that the model requires pretraining.

Method	LPIPS↓		SSIM↑		PSNR↑		LPIPS	SSIM	PSNR
	$n = 5$	$n =$ rand	$n = 5$	$n =$ rand	$n = 5$	$n =$ rand	$t + 1$		
MINE [11]*	0.0986	0.1774	0.9018	0.8221	27.9837	24.3112	N/A		
SynSin [38]	N/A	N/A	N/A	0.7400	N/A	22.3100	N/A		
Tucker et al. [32]	0.0967	0.1761	0.8699	0.7851	27.0500	23.5200	N/A		
VEST-single	0.0944	0.1736	0.8700	0.7688	26.6599	24.6906	N/A		
Zhou et al. [44]	**0.0816**	0.1667	0.8943	**0.8014**	27.5788	24.1531	N/A		
VEST (ours)	0.0841	**0.1596**	**0.8987**	0.8003	**28.2078**	**25.7553**	0.0436	0.9334	31.6831

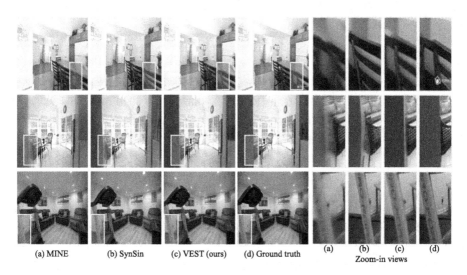

(a) MINE   (b) SynSin   (c) VEST (ours)   (d) Ground truth   (a) (b) (c) (d)
Zoom-in views

Fig. 7. Novel view synthesis results on RealEstate10K. (a)–(c) correspond to results from MINE [11], SynSin [38], and VEST (ours). (d) corresponds to the ground truth target image I_{22}. These examples show that VEST produces sharper details than MINE [11]. It also predicts object positions more accurately compared to SynSin [38].

includes 10K video sequences for training and 5K for testing. The training and evaluation resolution is 256×256.

Evaluation. Following MINE [11], for evaluation, we randomly sample 5 source frames from each testing sequence and sample target frames that are 5, 10, or at most 30 frames apart for each of the source frames. We evaluate on the intersection of the testing frames used in MINE and our testing split. This results in 372 testing sequences in total. The similarity scores are measured with LPIPS [43] with VGG features, SSIM [37], and PSNR.

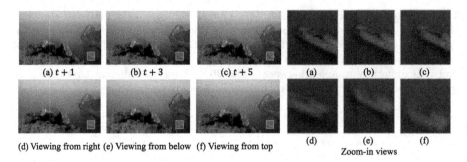

Fig. 8. Results on the ACID dataset filmed by a monocular moving camera. (a)–(c) are VP results and (d)–(f) NVS results.

Table 3. Results on the dataset from Lin et al. [13] filmed by multi-view static cameras. Our method achieves better performance than the baseline method under the same view setting.

Test-time Input	Method	LPIPS↓	SSIM↑	PSNR↑
Multi-view	Lin et al. [13]	0.1558	0.8667	21.1988
Single-view	Lin et al. [13]	0.3719	0.4929	16.7794
	VEST-single (ours)	0.2600	0.6242	21.6957
	VEST (ours)	**0.2591**	**0.6249**	**21.7664**

Baselines. We run the publicly released checkpoint from MINE [11] with 64 planes, which is pre-trained on ImageNet and trained on the full training split of RealEstate10K with a resolution of 384×256. We first run their model with resolution 384×256, and then downsample to 256×256 to compare with the ground truth. Both our model and MINE receive the same set of sparse points for scale-invariant analysis in Eq. (13). We also compare with Synsin [38] and Tucker and Snavely [32], both taking single-view inputs, and Zhou et al. [44], which takes in binocular inputs.

Results. We show quantitative results in Table 2. We obtain similar scores as reported in the original paper.[1] Results for Tucker and Snavely [32] and SynSin [38] are taken from the original paper for reference, and they are evaluated on a different test split. Our method outperforms MINE [11], a state-of-the-art single-view NVS method on this dataset, despite that the baseline uses ImageNet pre-training and trains on the full dataset. We show qualitative comparisons in Fig. 7.

On this dataset, modeling the motion is beneficial as our method uses the motion parallax from input frames by modeling the dynamics as opposed to explicit plane sweep volume construction as done in the stereo-based baselines [13,44]. Indeed, our method is comparable to the binocular baseline Zhou et al. [44].

Additionally, we train our model on Aerial Coastline Imagery Dataset (ACID) [14], a single-view dataset mostly filming outdoor natural scenes. Our method can synthesize images corresponding to the queried viewpoint and timestamp, as shown in Fig. 8. Both RealEstate10K and ACID datasets include YouTube videos under the Creative-Commons license.

4.3   Learning from Multi-View Videos with Static Cameras

Setup. We evaluate our method on learning from a multi-view dataset with static cameras. We use the dataset from Lin et al. [13], containing videos captured by 10 synchronized, static cameras, where scenes have a static background and human body movement in the foreground. The dataset is split into 86 scenes for training and 10 scenes for testing. The training and evaluation resolution is 256×256. We follow Lin et al. [13] and pretrain our model on RealEstate10K.

Results. We compare with Lin et al. [13], a state-of-the-art multi-view NVS method, which incorporates a learned 3D mask volume into the MPI representation to improve the temporal consistency of MPI planes.

As shown in Table 3, our model outperforms the baseline on novel view synthesis from a single view. Note that Lin et al. [13] was originally designed to perform view synthesis from stereo inputs and works well in their setup; in contrast, our model is designed to take single-view input during inference. We further show qualitative results in Fig. 9.

---

[1] The LPIPS scores of MINE [11] computed are slightly worse compared to the original paper due to a bug in the evaluation script in their public codebase, where tensors in range $[0, 1]$ are fed into an LPIPS package which expects inputs in range $[-1, 1]$.

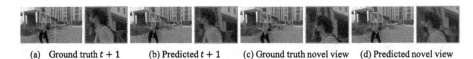

(a) Ground truth $t+1$ (b) Predicted $t+1$ (c) Ground truth novel view   (d) Predicted novel view

Fig. 9. Results on dataset from Lin et al. [13]. (a)–(b) VP results; (c)–(d) NVS results. Our method assigns the moving human figure and the background scene into different MPI layers.

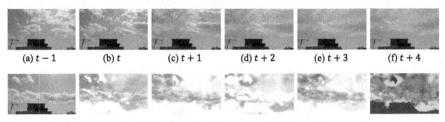

(a) $t-1$ (b) t (c) $t+1$ (d) $t+2$ (e) $t+3$ (f) $t+4$

(g) RGBA plane   (h) RGBA plane   (i) RGBA plane   (j) RGBA plane   (k) Predicted flow (l) Predicted depth

Fig. 10. Video prediction results on the cloud dataset. (a)–(b) are two input frames, and (c)–(f) are predicted future frames. (g)–(j) are RGBA planes ordered from far to near, and (k)–(l) are the flow and depth maps predicted by the model. In this example, regions of clouds and the building are assigned to different planes, corresponding to different motions.

4.4 Learning from Single-View Videos with a Static Camera

In the last camera setting, videos are filmed by a single, static camera, and NVS task becomes non-applicable due to a lack of viewpoint changes throughout a video sequence. While our method is motivated by leveraging the spatial and temporal cues from scenes, we evaluate our method under this camera setting for completeness, and show that our method can decompose scenes into layers based on different motion patterns.

We collect a dataset of 175 moving clouds videos from YouTube, which will be made publicly available. Our method produces realistic VP results as shown in Fig. 10. Since each affine transformation layer is weighted by the predicted alpha map, the overall scene dynamics is not restricted to be affine. We defer further analysis on modeling scene dynamics in such camera setting to Appendix C.4.

4.5 Qualitative Analysis of the Generalized MPIs

To understand how the model learns to solve NVS and VP simultaneously, we qualitatively analyze the intermediate model outputs as shown in Fig. 11. The depth map is computed from Eq. (14), and the flow map is composed from per-layer flow maps similarly. In the shown example, the camera is moving forward and objects closer to the camera tend to have more dominant motions. With the generalized MPI representation, the scene dynamics is decomposed into plane-wise motion fields. Since each plane is depth-aware, this representation helps

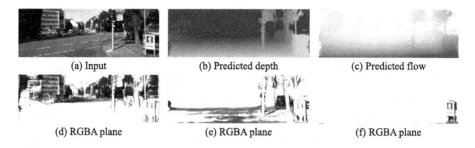

(a) Input   (b) Predicted depth   (c) Predicted flow

(d) RGBA plane   (e) RGBA plane   (f) RGBA plane

Fig. 11. Visualization of intermediate predictions on KITTI dataset. (a) The second input frame, which is the source image for MPI representations. (b) and (c) are depth and flow predictions. In (d–f), we show 3 out of 16 RGBA planes predicted by the model, ordered from far to near.

introduce an inductive bias for the model to learn the depth-motion correspondence which facilitates the learning of the two extrapolation tasks.

5   Conclusion

In this work, we view NVS and VP as extrapolation along two axes for the spatial-temporal coordinates of videos. NVS utilizes camera viewpoint changes in a video sequence to discover depth, while VP considers both camera and object motions. The two tasks can be jointly learned to develop a scene representation from video data with complementary learning signals coming from each of the tasks. We propose a generalized MPI representation to tackle both tasks, and develop a model that achieves superior or comparable performance compared to previous methods that tackle only one of the tasks, on natural datasets for indoor and outdoor scenes. Please see Appendix C.5 for more discussions.

Acknowledgement. We thank Angjoo Kanazawa, Hong-Xing (Koven) Yu, Huazhe (Harry) Xu, Noah Snavely, Ruohan Zhang, Ruohan Gao, and Shangzhe (Elliott) Wu for detailed feedback on the paper, and Kaidi Cao for collecting the cloud dataset. This work is in part supported by the Stanford Institute for Human-Centered AI (HAI), the Stanford Center for Integrated Facility Engineering (CIFE), the Samsung Global Research Outreach (GRO) Program, and Amazon, Autodesk, Meta, Google, Bosch, and Adobe.

A   Architecture Details

The architecture used for the MPI encoder is specified in Table 4.

Table 4. MP2 is max pooling with stride 2, Up2 is nearest-neighbor upsampling with scale 2, + is concatenation. Reshape transforms a tensor with $C \times D$ channels into C channels, and D is merged to the batch dimension, and ReshapeBack is the reverse operation. All layers up till up1b use ReLU activation and the layers for conv1, conv2 and conv3 use LeakyReLu with a negative slope 0.2. There is no activation following the very last layer. All layers use Instance Norm for activation normalization and Spectral Norm for weight normalization.

Input	k	c	Output	Input	k	c	Output
Concat(I_{t-1}, I_t)	7	32	down1	down1	7	32	down1b
MP2(down1b)	5	64	down2	down2	5	64	down2b
MP2(down2b)	3	128	down3	down3	3	128	down3b
MP2(down3b)	3	256	down4	down4	3	256	down4b
MP2(down4b)	3	512	down5	down5	3	512	down5b
MP2(down5b)	3	512	down6	down6	3	512	down6b
MP2(down6b)	3	512	mid1	mid1	3	512	mid2
Up2(mid2) + down6b	3	512	up6	up6	3	512	up6b
Up2(up6b) + down5b	3	512	up5	up5	3	512	up5b
Up2(up5b) + down4b	3	256	up4	up4	3	256	up4b
Up2(up4b) + down3b	3	128	up3	up3	3	128	up3b
Up2(up3b) + down2b	3	64	up2	up2	3	64	up2b
Up2(up2b) + down1b	3	64	post1	post1	3	64	post2
post2	3	64	up1	up1	3	64	up1b
up1b	3	64 x D	conv1	Reshape(conv1)	3	64	conv2
conv2	7	7	conv3	ReshapeBack(conv3)	-	-	output

B   Implementation Details

To have a better gradient flow, similar to Tucker et al. [32], we add a harmonious bias $1/i$ to the alpha channel prediction, so that w_i from Eq. (11) becomes uniformly $1/D$ during initialization. We also add an identity bias to f^θ such that each MPI plane is associated with zero motion during initialization.

In all experiments, we set the number of MPI planes to be $D = 16$. The depth values for MPI planes are linear in the inverse space, with $d_1 = 1000$ and $d_D = 1$.

Table 5. Ablation on the number of MPI planes D. Increasing the plane count improves the performance but also increases the training time. We adopt $D = 16$ in the main paper since further increasing D results in diminishing returns.

D	Extrapolation in Space			Extrapolation in Time		
	LPIPS↓	PSNR ↑	SSIM ↑	LPIPS↓	PSNR↑	SSIM↑
4	0.0987	19.3453	0.7180	0.0792	22.9415	0.7880
8	0.0874	20.5795	0.7881	0.0784	23.1073	0.7922
16	0.0786	21.1889	0.8188	0.0757	23.3812	0.7971
32	**0.0762**	**21.2279**	**0.8207**	**0.0726**	**23.7882**	**0.8083**

C   Training Details

C.1   KITTI

Since videos from KITTI are taken by stereo cameras with fixed relative poses, the depth scale is consistent across scenes and therefore we set it to be a constant $\sigma = 1$. We use $\lambda_1^{\text{space}} = 1000$, $\lambda_{\text{spec}}^{\text{space}} = 100$, $\lambda_1^{\text{time}} = 1000$, and $\lambda_{\text{perc}}^{\text{time}} = 10$. We use Adam Optimizer [9] with an initial learning rate 0.0002, which we exponentially decrease by a factor of 0.8 for every 5 epochs. We train our model for 200K iterations on two NVIDIA TITAN RTX GPUs for about two days. During training, we apply horizontal flip with 50% probability and apply color jittering as data augmentation.

C.2   RealEstate10K

We train our model for 200K iterations on one NVIDIA GeForce RTX 3090 GPU, which takes about one day. We use $\mathcal{L}_1^{\text{space}} = 10, \mathcal{L}_{\text{perc}}^{\text{space}} = 10, \mathcal{L}_1^{\text{time}} = 10, \mathcal{L}_{\text{perc}}^{\text{time}} = 0$. We use Adam Optimizer [9] with a constant learning rate 0.0002.

C.3   Ablations on the Number of MPI Planes

To study the effect of the number of MPI planes, we perform an ablation study on the KITTI [6] dataset with resolution 128×384. As shown in Table 5, a small number of MPI planes ($D = 4$ or 8) results in degraded model performance. Further increasing the number of planes from 16 to 32 results in marginal performance gain, with a cost of $2.1\times$ slower training time. Therefore, we use $D = 16$ for all other experiments.

C.4   Modeling Dynamic Scenes

To test whether our method is able to model more dynamic scenes, we test our method on CATER [7], a dataset of scenes with 5–10 individually moving objects. We show a quantitative comparison with a video prediction baseline

Table 6. Results of next-frame prediction on CATER [7]. Our model achieves better performance compared to PredRNN [36].

Method	LPIPS↓	PSNR↑	SSIM↑
PredRNN [36]	0.0600	37.02	0.9643
Ours	**0.0122**	**42.58**	**0.9762**

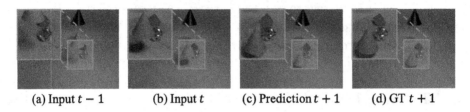

(a) Input $t-1$ (b) Input t (c) Prediction $t+1$ (d) GT $t+1$

Fig. 12. Model prediction on an example scene with occlusion. (a) and (b) are two historical frames as model inputs, (c) and (d) are the predicted and ground truth next frame, respectively. Top-left corners of subfigures are zoomed-in views for occluded regions.

PredRNN [36]. As shown in Table 6, our model achieves better performance across all three metrics.

Qualitatively, our method makes temporal prediction consistent with the ground truth object motions on this dataset. In Fig. 12, the model correctly recovers the purple object and the gold object occluded by the blue cone. Our model effectively handles object occlusions by warping from neighboring pixels with similar RGB values.

C.5 Discussions

While we focus on demonstrating the possibility of simultaneous extrapolation in both space and time, specific modules can be further optimized for each task. For example, it is possible to improve the dynamic scene representation to better handle video prediction with long horizons or highly complex motion, or to synthesize novel views with a large viewpoint change.

In the meantime, while our method is designed for natural scenes with many potential positive impacts such as interactive scene exploration for family entertainment, like all other visual content generation methods, our method might be exploited by malicious users with potential negative impacts. We expect such impacts to be minimal as our method is not designed to work with human videos. In our code release, we will explicitly specify allowable uses of our system with appropriate licenses. We will use techniques such as watermarking to label visual content generated by our system.

References

1. Bei, X., Yang, Y., Soatto, S.: Learning semantic-aware dynamics for video prediction. In: CVPR (2021)
2. Deng, J., Dong, W., Socher, R., Li, L.J., Li, K., Fei-Fei, L.: ImageNet: a large-scale hierarchical image database. In: CVPR (2009)
3. Du, Y., Zhang, Y., Yu, H.X., Tenenbaum, J.B., Wu, J.: Neural radiance flow for 4D view synthesis and video processing. In: ICCV (2021)
4. Flynn, J., et al.: DeepView: view synthesis with learned gradient descent. In: CVPR (2019)
5. Gao, H., Xu, H., Cai, Q.Z., Wang, R., Yu, F., Darrell, T.: Disentangling propagation and generation for video prediction. In: ICCV (2019)
6. Geiger, A., Lenz, P., Stiller, C., Urtasun, R.: Vision meets robotics: the KITTI dataset. Int. J. Robot. Res. **32**(11), 1231–1237 (2013)
7. Girdhar, R., Ramanan, D.: CATER: a diagnostic dataset for compositional actions and temporal reasoning. In: ICLR (2020)
8. Hu, R., Ravi, N., Berg, A.C., Pathak, D.: Worldsheet: wrapping the world in a 3D sheet for view synthesis from a single image. In: ICCV (2021)
9. Kingma, D.P., Ba, J.: Adam: a method for stochastic optimization. In: ICLR (2015)
10. Lai, Z., Liu, S., Efros, A.A., Wang, X.: Video autoencoder: self-supervised disentanglement of static 3D structure and motion. In: ICCV (2021)
11. Li, J., Feng, Z., She, Q., Ding, H., Wang, C., Lee, G.H.: MINE: towards continuous depth MPI with NeRF for novel view synthesis. In: ICCV (2021)
12. Li, Z., Niklaus, S., Snavely, N., Wang, O.: Neural scene flow fields for space-time view synthesis of dynamic scenes. In: CVPR (2021)
13. Lin, K.E., Xiao, L., Liu, F., Yang, G., Ramamoorthi, R.: Deep 3D mask volume for view synthesis of dynamic scenes. In: ICCV (2021)
14. Liu, A., Tucker, R., Jampani, V., Makadia, A., Snavely, N., Kanazawa, A.: Infinite nature: perpetual view generation of natural scenes from a single image. In: ICCV (2021)
15. Liu, Z., Yeh, R., Tang, X., Liu, Y., Agarwala, A.: Video frame synthesis using deep voxel flow. In: ICCV (2017)
16. Lotter, W., Kreiman, G., Cox, D.: Deep predictive coding networks for video prediction and unsupervised learning. In: ICLR (2017)
17. Lu, E., et al.: Layered neural rendering for retiming people in video. In: SIGGRAPH Asia (2020)
18. Lu, E., Cole, F., Dekel, T., Zisserman, A., Freeman, W.T., Rubinstein, M.: Omnimatte: associating objects and their effects in video. In: CVPR (2021)
19. Mildenhall, B., Srinivasan, P.P., Tancik, M., Barron, J.T., Ramamoorthi, R., Ng, R.: NeRF: representing scenes as neural radiance fields for view synthesis. In: Vedaldi, A., Bischof, H., Brox, T., Frahm, J.-M. (eds.) ECCV 2020. LNCS, vol. 12346, pp. 405–421. Springer, Cham (2020). https://doi.org/10.1007/978-3-030-58452-8_24
20. Park, K., et al.: Nerfies: deformable neural radiance fields. In: ICCV (2021)
21. Pumarola, A., Corona, E., Pons-Moll, G., Moreno-Noguer, F.: D-NeRF: neural radiance fields for dynamic scenes. In: CVPR (2021)
22. Ranzato, M., Szlam, A., Bruna, J., Mathieu, M., Collobert, R., Chopra, S.: Video (language) modeling: a baseline for generative models of natural videos. arXiv preprint arXiv:1412.6604 (2014)
23. Schönberger, J.L., Frahm, J.M.: Structure-from-motion revisited. In: CVPR (2016)

24. Schönberger, J.L., Zheng, E., Frahm, J.-M., Pollefeys, M.: Pixelwise view selection for unstructured multi-view stereo. In: Leibe, B., Matas, J., Sebe, N., Welling, M. (eds.) ECCV 2016. LNCS, vol. 9907, pp. 501–518. Springer, Cham (2016). https://doi.org/10.1007/978-3-319-46487-9_31
25. Shade, J., Gortler, S., He, L.w., Szeliski, R.: Layered depth images. In: SIGGRAPH (1998)
26. Shi, X., Chen, Z., Wang, H., Yeung, D.Y., Wong, W.K., Woo, W.C.: Convolutional LSTM network: a machine learning approach for precipitation nowcasting. In: NeurIPS (2015)
27. Shih, M.L., Su, S.Y., Kopf, J., Huang, J.B.: 3D photography using context-aware layered depth inpainting. In: CVPR (2020)
28. Simonyan, K., Zisserman, A.: Very deep convolutional networks for large-scale image recognition. In: ICLR (2015)
29. Srinivasan, P.P., Tucker, R., Barron, J.T., Ramamoorthi, R., Ng, R., Snavely, N.: Pushing the boundaries of view extrapolation with multiplane images. In: CVPR (2019)
30. Srivastava, N., Mansimov, E., Salakhutdinov, R.: Unsupervised learning of video representations using LSTMs. In: ICML (2015)
31. Tretschk, E., Tewari, A., Golyanik, V., Zollhöfer, M., Lassner, C., Theobalt, C.: Non-rigid neural radiance fields: reconstruction and novel view synthesis of a dynamic scene from monocular video. In: ICCV (2021)
32. Tucker, R., Snavely, N.: Single-view view synthesis with multiplane images. In: CVPR, pp. 551–560 (2020)
33. Tulsiani, S., Tucker, R., Snavely, N.: Layer-structured 3D scene inference via view synthesis. In: Ferrari, V., Hebert, M., Sminchisescu, C., Weiss, Y. (eds.) ECCV 2018. LNCS, vol. 11211, pp. 311–327. Springer, Cham (2018). https://doi.org/10.1007/978-3-030-01234-2_19
34. Villegas, R., Yang, J., Hong, S., Lin, X., Lee, H.: Decomposing motion and content for natural video sequence prediction. In: ICLR (2017)
35. Wang, J.Y.A., Adelson, E.H.: Layered representation for motion analysis. In: CVPR (1993)
36. Wang, Y., Wu, H., Zhang, J., Gao, Z., Wang, J., Yu, P., Long, M.: Predrnn: A recurrent neural network for spatiotemporal predictive learning. IEEE TPAMI (2022)
37. Wang, Z., Bovik, A.C., Sheikh, H.R., Simoncelli, E.P.: Image quality assessment: from error visibility to structural similarity. IEEE TIP 13(4), 600–612 (2004)
38. Wiles, O., Gkioxari, G., Szeliski, R., Johnson, J.: SynSin: end-to-end view synthesis from a single image. In: CVPR (2020)
39. Wu, Y., Gao, R., Park, J., Chen, Q.: Future video synthesis with object motion prediction. In: CVPR (2020)
40. Xian, W., Huang, J.B., Kopf, J., Kim, C.: Space-time neural irradiance fields for free-viewpoint video. In: CVPR (2021)
41. Yoon, J.S., Kim, K., Gallo, O., Park, H.S., Kautz, J.: Novel view synthesis of dynamic scenes with globally coherent depths from a monocular camera. In: CVPR, pp. 5336–5345 (2020)
42. Yu, A., Ye, V., Tancik, M., Kanazawa, A.: pixelNeRF: neural radiance fields from one or few images. In: CVPR (2021)
43. Zhang, R., Isola, P., Efros, A.A., Shechtman, E., Wang, O.: The unreasonable effectiveness of deep networks as a perceptual metric. In: CVPR (2018)

44. Zhou, T., Tucker, R., Flynn, J., Fyffe, G., Snavely, N.: Stereo magnification: learning view synthesis using multiplane images. In: SIGGRAPH (2018)
45. Zhou, T., Tulsiani, S., Sun, W., Malik, J., Efros, A.A.: View synthesis by appearance flow. In: Leibe, B., Matas, J., Sebe, N., Welling, M. (eds.) ECCV 2016. LNCS, vol. 9908, pp. 286–301. Springer, Cham (2016). https://doi.org/10.1007/978-3-319-46493-0_18

Contrastive Learning for Diverse Disentangled Foreground Generation

Yuheng Li[1,2(✉)], Yijun Li[2], Jingwan Lu[2], Eli Shechtman[2], Yong Jae Lee[1], and Krishna Kumar Singh[2]

[1] University of Wisconsin-Madison, Madison, USA
li2464@wisc.edu
[2] Adobe Research, Bangalore, India

Abstract. We introduce a new method for diverse foreground generation with explicit control over various factors. Existing image inpainting based foreground generation methods often struggle to generate diverse results and rarely allow users to explicitly control specific factors of variation (e.g., varying the facial identity or expression for face inpainting results). We leverage contrastive learning with latent codes to generate diverse foreground results for the same masked input. Specifically, we define two sets of latent codes, where one controls a pre-defined factor ("known"), and the other controls the remaining factors ("unknown"). The sampled latent codes from the two sets jointly bi-modulate the convolution kernels to guide the generator to synthesize diverse results. Experiments demonstrate the superiority of our method over state-of-the-arts in result diversity and generation controllability.

Keywords: Foreground generation · Diversity · Disentanglement

1 Introduction

Foreground object generation is the task of filling in the missing foreground region in a given context, such as generating human faces as shown in Fig. 1. This task is useful in practice, e.g., for privacy-related applications (anonymizing a person's face by generating a new identity) or replacing/adding objects in an image (replacing a car in a photo if one does not like the original one). It is a special case of image inpainting in which the entire foreground object is masked. In inpainting, when the missing region (hole) is small, there may only be one or few "correct" completions (e.g., if only one eye is masked, then it mostly can be inferred from the other eye), but as the hole gets bigger there should be more diversity in the generated completion, especially when an entire object is masked. As there can be many different plausible solutions for filling in the missing region, this task naturally demands learning a "one-to-many" mapping between the

---

Supplementary Information The online version contains supplementary material available at https://doi.org/10.1007/978-3-031-19787-1_19.

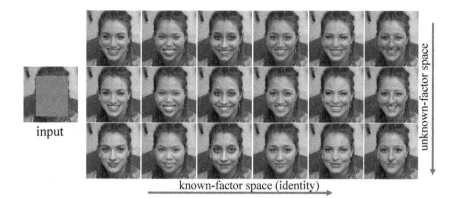

Fig. 1. Foreground generation on the same mask. We use contrastive learning to increase generation diversity. We also explicitly disentangle out an expected predefined factor (human identity here) to increase diversity and controllability.

input and outputs (e.g., Fig. 1). That is, a good method should 1) synthesize foreground objects that are both *realistic* and *semantically coherent* with the surrounding unmasked context; 2) have the capability to generate *diverse* results for the same missing region and context; and 3) provide *control* over different properties of the synthesized results. While tremendous progress has been made to obtain better realism and coherence [11,26,40,60], progress in diversity is still unsatisfactory and increasing controllability for the results is also relatively under explored.

Like the inpainting task [3,4,12,28,32,61], foreground generation needs to consider coherence between the given context and the generated object. Existing inpaiting work can generate good quality object/foreground, but it usually lacks diversity and controllability. Although there are many inpainting methods trying to generate diverse results [26,59–61], the results are still less satisfactory. These methods typically have an encoder-decoder architecture. To achieve diversity, different latent codes can be sampled and injected into these models. However, although the output is a function of both the masked image as well as the latent code, the spatial features from the encoder usually dominate the final results and prevent the latent codes from inducing large changes. For example, in [60], an encoder is used to extract 2D spatial features from the masked image, and skip connections are added to all levels of the encoder and decoder. The information from the latent code can be easily submerged by the large number of features from the encoder.

In this paper, we propose a novel approach for diverse and controllable foreground generation. As shown in Fig. 1, our method can generate diverse results for the same input. To synthesize diverse content, we condition the generation on both the masked image and the sampled latent codes, and apply contrastive learning [5] so that the latent codes that are close/far in code space result in corresponding synthesized images that are close/far in image space.

Besides diversity, controllability is another desired property in foreground generation. Thus, we also try to explicitly disentangle a predefined factor by using a pretrained classifier on this factor. For example, as shown in Fig. 1, one can disentangle human identity (rows) from other attributes (columns) for face images. We explicitly use two sets of latent codes, where one represents the predefined factor ("known"), and the other controls all the other factors ("unknown"). This allows us to change the unknown factors while keeping the known factor fixed (e.g.,, in Fig. 1, changing the facial attributes which are unknown during training while keeping the identity of the face intact). To inject these two codes, we propose a bi-modulated convolution module where the convolution kernels are modulated by the two latent codes from different spaces. We design each training batch to contain a mix of instances that share the same known latent code while differing in the unknown, and instances that share the same unknown latent code while differing in the known. We use a contrastive loss to ensure that known and unknown codes control their respective factors.

Contributions. (1) We propose a novel contrastive learning based approach for diverse foreground generation; (2) An explicit disentangled latent space for controllability via a novel bi-modulated convolution module; (3) More diverse results compared to existing state-of-the-art methods on three different datasets.

2   Related Work

Image Inpainting. This problem has been studied for decades due to its importance. Traditional methods [2–4,12,41] typically rely on low-level assumptions and image statistics, leading to over smoothing and results with limited visual semantics. Recently, deep learning methods [8,18,28,33,34,46,48,50,51,53,55, 57,59] dramatically boosted the quality, in terms of both visual quality and semantic coherence. [32] first uses an encoder-decoder architecture in inpainting with reconstruction loss and adversarial loss [10]. [28] and [55] proposes the use of partial and gated convolutions on irregular masks. However, these methods only generate deterministic results. Thus [61] proposes a VAE-based [23] method allowing pluralistic image completion. Recently proposed [26,60] use StyleGAN [20,21] architecture for inpainting. [60] combines encoded features from a masked image with a random latent code to co-modulate StyleGAN convolution kernels. [26] has a similar setting as ours as instead of traditional inpainting, they use a foreground model to synthesize high quality foreground objects conditioned on the background context. In both work, diverse images can be generated by sampling different latent codes injected into StyleGAN. But their diversity in the latent code space is restricted due to extra spatial features from the encoder which usually determine most of the aspects of the generation. [45,56] also try to use transformer to realize diversity in image inpainting. They both use bidirectional attention to predict missing tokens. However, their image quality suffers compared to styleGAN2-based architectures. Also, none of the existing pluralistic inpainting work enables user controllability in the results via latent code disentanglement.

Contrastive Learning. Contrastive learning [5,13,30,42,63] has shown great potential in representation learning. Among them, [5] proposes a simple framework for contrastive learning without requiring specialized architectures or memory bank. Recently [31] proposes to use contrastive learning in image translation task. Also, there are a few work [29,62] studying contrastive learning in the image inpainting task. Like most inpainting methods, [29,62] use encoder-decoder architecture. [62] encode more discriminative features using contrastive loss in different semantic sub-regions. [29] applies the contrastive loss to the output features of encoder by setting two identical images with different masks as positive pairs while different images as negative pairs. However, they are both deterministic inpainting methods which means they only produce a single result per input. Different from prior work, instead of learning a better intermediate feature using contrastive loss, we use contrastive learning to achieve disentanglement and diversity in the latent space, enabling us to produce diverse inpainting results for foreground generation in a controllable way.

Disentanglement Learning. For a generative model, it is desirable to disentangle the factors of variation. One way is to explicitly learn a disentangled latent space: having separate codes for different factors. A large number of work try to disentangle object shape/structure from appearance [7,27,37,39,49]. Searching for semantic directions in a pre-trained GAN latent space is another way to achieve disentanglement. This method is getting popular recently and both unsupervised methods [17,36,43] and supervised methods [19,35,52] are heavily explored. Despite the progress made in the field, few work explore it for the foreground generation task. We use contrastive learning to explicitly learn a disentangled latent space for controllable foreground generation.

3   Approach

Our goal is to propose a model (ContrasFill) which is able to generate diverse foreground objects for the same masked region while providing control over different factors of generated results. We encode spatial features corresponding to masked image and modulate them with randomly sampled latent codes to generate diverse results. However, without applying explicit training loss on diversity, different latent codes might introduce only minor changes as in [26,60]. Thus we use contrastive learning to encourage the model to synthesize diverse results by forcing latent codes closer in the latent space to produce images closer in the image space and vice versa. To gain explicit control over certain factor, we also try to disentangle the latent space into two spaces: a known factor space which corresponds to an expected factor, and an unknown factor space which controls the rest of other factors. Section 3.1 introduces the training details for achieving diversity and disentanglement using the contrastive loss. Section 3.2 talks about how do we inject two codes into our model using the proposed bi-modulation.

3.1   Contrastive Learning for Diversity and Disentanglement

Figure 2 shows the framework of our method. Our model takes a masked image I with context only and two latent codes as inputs: a known factor code k

Fig. 2. ContrasFill takes as input two sets of codes (squares on the left): known-factor code (e.g., identity) and unknown-factor code (non-identity factors) to synthesize images. Two encoders (E_k and E_u) embed images into different features (bars on the right, color between code and feature refers to correspondence). A contrastive loss forces features with same/different colors closer/further in the feature spaces.

drawn from a distribution ϕ_k and an unknown factor code u from a distribution ϕ_u to output the synthesized image S. If we define our model as G, then we have $S_{k,u,I} = G(k,u,I)$. Since the input context is independent of the following analysis, we will omit I in $S_{k,u,I}$ for simplicity and talk about the context later. One can refer to supp. for visual examples to help understand the following analysis.

Suppose we sample N codes from each latent space, we will have N^2 combinations between the code k and u in total. To enforce a contrastive loss in the known and unknown factor latent spaces, we first define an image pair as:

$$p_{(k,u),(k',u')} = (S_{k,u}, S_{k',u'}), \tag{1}$$

Note, we do not consider images sharing the same known and unknown codes as a valid pair. In order words, $k = k'$ and $u = u'$ cannot hold at the same time. We will next define the positive and negative pairs used in our contrastive learning scheme. To simplify the explanation, we first consider the case of the known space.

Contrastive Pairs in the Known Space. A positive image pair contains two images sharing the same known codes but different unknown codes (i.e., $k = k', u \neq u'$ in the Eq. 1). We define $P_{\mathbf{k},u}$ as a set of all positive pairs *associated* with the code combination (k, u) in the known space (bold indicates the space). For example, in Fig. 2 where we set the known factor as human identity, $P_{\mathbf{k}1,u1} = \{p_{(k1,u1),(k1,u2)}\}$. To ease explanation, we denote $P_{\mathbf{K}}$ as *all* positive pairs in the known space. Here we have $P_{\mathbf{K}} = \{p_{(k1,u1),(k1,u2)}, p_{(k2,u1),(k2,u2)}\}$.

For the negative pair, we define two images not sharing the same known codes (i.e., $k \neq k'$). In this case, we construct two types of negative pairs. The first case is the hard negative pair where two images sharing different known codes but the same unknown codes (i.e., $k \neq k', u = u'$, images in each column in Fig. 2). The reason is that these images share the same features (e.g., smile) in the unknown space which forces the learned known latent code to control different aspects of the face due to the use of contrastive loss which we will introduce later. Pairs of images sharing different known and unknown codes (i.e., $k \neq k', u \neq u'$) are easy negative pairs (diagonal image pairs in Fig. 2). Similarly, we define $N_{\mathbf{k},u}$ as all negative pairs *associated* with the code combination (k, u) in the known factor space. For example, for image $S_{k1,u1}$ in Fig. 2, $N_{\mathbf{k1},u1} = \{P_{(k1,u1),(k2,u1)}, P_{(k1,u1),(k2,u2)}\}$.

Contrastive Loss in the Known Space. The job of the known space is to control an expected factor during the generation. To push the model to learn this correspondence, we use contrastive learning. The intuition is to push images closer/further if they are positive/negative pairs. In order to measure the distance between two images, we define the similarity score f as:

$$f_{(k,u),(k',u')} = e^{\mathrm{sim}(z_{k,u}, z_{k',u'})/\tau}, \tag{2}$$

where $z_{k,u}$ is the extracted feature of the image $S_{k,u}$ from an encoder, $\mathrm{sim}(\cdot, \cdot)$ is the cosine similarity and τ denotes a temperature parameter. To force our known space to control the expected factor, we assume having access to a pretrained and fixed classifier. The encoder E_k for the known space will output the penultimate feature of the classifier. For example, in faces, we use a pretrained ArcFace [6] to extract identity features.

For an image $S_{k,u}$, and its positive pair $S_{k,u'}$ in the known space, the contrastive loss becomes:

$$\ell_{(k,u),(k,u')} = -\log \frac{f_{(k,u),(k,u')}}{f_{(k,u),(k,u')} + FN_{\mathbf{k},u}}, \tag{3}$$

where $FN_{\mathbf{k},u}$ is the sum of similarity scores of all negative pairs with respect to the image $S_{k,u}$. In other words, it is the summation of Eq. 2 over all elements in the $N_{\mathbf{k},u}$. Finally, the total loss for the known space becomes:

$$\mathcal{L}_{known} = \frac{1}{|P_K|} \sum \ell_{(k,u),(k,u')}, \tag{4}$$

where the summation is over all positive pairs P_K in the known space.

Contrastive Learning in the Unknown Space. The contrastive learning idea is similarly applied to the unknown latent space and we highlight the main difference below.

In the unknown space, positive pairs share the same unknown code (each column in Fig. 2) and negative pairs have different unknown codes. Similarly, we define P_U as *all* positive pairs in the unknown space. For example, in Fig. 2, $P_U = \{P_{(k1,u1),(k2,u1)}, P_{(k1,u2),(k2,u2)}\}$. For an image $S_{k,u}$, we define all negative pairs associated with it in the unknown space as $N_{k,\mathbf{u}}$ (bold indicates space). For example, $N_{k1,\mathbf{u1}} = \{P_{(k1,u1),(k1,u2)}, P_{(k1,u1),(k2,u2)}\}$ In the unknown space, the image feature $z_{k,u}$ for calculating image pair similarity in the Eq. 2 is extracted from an encoder E_u which is trained from scratch. This is because it is hard to define what factors can be controlled in the unknown space beforehand. Also, this avoids pre-training an additional feature extractor and simplifies our approach.

Then, for an image $S_{k,u}$, and its positive pair $S_{k',u}$, the counterpart of Eq. 3 in the unknown space becomes

$$\ell_{(k,u),(k',u)} = -\log \frac{f_{(k,u),(k',u)}}{f_{(k,u),(k',u)} + FN_{k,\mathbf{u}}}, \tag{5}$$

where $FN_{k,\mathbf{u}}$ is the sum of similarity score of all negative pairs with respect to image $S_{k,u}$ in the unknown space. The total loss for the unknown space becomes:

$$\mathcal{L}_{unknown} = \frac{1}{|P_U|} \sum \ell_{(k,u),(k',u)}, \tag{6}$$

where the summation is over all positive pairs in the P_U in the unknown space.

In this way the disentanglement can be learned because we use a pretrained encoder for the known-factor which only extracts expected features, thus the model will synthesize known-factors in images according to codes sampled from known space. For the unknown space, due to the existence of hard negative pair (sharing the same known factors), different unknown codes need to generate factors that are different from known factor to minimize the contrastive loss.

Overall we have the final loss $\mathcal{L}$ as

$$\mathcal{L} = \mathcal{L}_{gan} + \lambda_1 \mathcal{L}_{known} + \lambda_2 \mathcal{L}_{unknown}, \tag{7}$$

where $\mathcal{L}_{gan}$ is same as the one used in the StyleGAN2 [22]. $\mathcal{L}_{known}$ and $\mathcal{L}_{unknown}$ are two contrastive losses in known and unknown latent spaces. λ_1 and λ_2 are their weights. We sample different context (background) for different code combinations (e.g., N^2 in total) since we want to have the same context distribution for both the real and fake batches when training the discriminator. More details are presented in the supp.

3.2   Codes Injection with Bi-Modulated Convolution

Our model uses an encoder-decoder architecture (details in the supp). Inspired by StyleGAN2 that shows the effectiveness of modulation, we also use our latent codes to modulate convolution kernels. However, since we have two latent codes, we propose the bi-modulation, where the convolution kernel is modulated by two codes. We use this novel modulation scheme for all convolutions in our model.

Figure 3 shows the bi-modulation process. The two codes k and u first go to two separate fully connected layer to become scaling vectors s and t. The length of scaling vectors is the same as the number of input channels of a convolution kernel. Then the scaling vectors bi-modulate the convolution weight by: $w'_{ijk} = s_i \cdot t_i \cdot w_{ijk}$, where w and w' are the original and the bi-modulated weights. s_i and t_i are the scaling factors corresponding to the ith input feature map. j and k enumerate the output feature maps and the spatial footprint of the convolution.

4   Experiments

We perform quantitative and qualitative evaluations via comparing our proposed foreground generation model ContrasFill with prior arts.

Fig. 3. The proposed bi-modulation scheme, where convolution kernels are modulated by two disentangled latent codes.

Datasets. We conduct the evaluation on three different datasets: 1) **Face.** We use CelebAMask-HD [25] that includes 30,000 face images with segmentation masks. We follow the official training/testing split. To acquire more training data, we use a publicly available face parsing model [1] on FFHQ [21] as extra training data. We use a pretrained face recognition model [6] as our known factor feature extractor. 2) **Bird.** We use the *bird* category from LSUN dataset [54]. We choose images greater than certain resolution and run the pretrained MaskR-CNN [14,47] to remove bad images. In total, we have 34,969 images and we randomly select 10% (3,497) as test data. We train a fine-grained classification model [9] on the CUB dataset [44] as our known factor feature extractor. 3) **Car.** We use the *car* category from LSUN dataset [54] and same preprocessing steps to clean our data. In total, we have 77,840 images and we randomly select 10% (7,784) as test data. We train a shape classifier [9] on the Stanford car dataset [24] as our known factor feature extractor. To measure the extent of our ability to synthesize diverse results, we use the object bounding box as the missing region in our main study.

We train our model at 256×256 resolution on all datasets. Our unknown factor code is drawn from the normal distribution. Our known factor codes are drawn from one-hot distribution for the cars and birds; for faces, we choose to draw from a hypersphere which is the feature distribution of penultimate layer of ArcFace. We sample $N = 8$ different known and unknown codes in each training minibatch. Due to memory issue, we can not fit all 64 combinations, thus we subsample one hard negative pair for each code, resulting in a batch size of 16 during training. Please refer to supp for more dataset and implementation details.

Baselines. We mainly compare with: 1) **CollageGAN** [26], which generates foreground object conditioned on the background; 2) **CoModGAN** [60], a state-of-the-art image inpainting model. These two methods, built on top of StyleGAN2 [22], are able to generate multiple results via sampling codes in the latent space; 3) **BAT-fill** [56], a recently proposed two-stage inpainting model using transformer. It first autoregressively predicts missing tokens in a 32×32 image with bidirectional attention, and then use a convolutional network to perform upsampling to 256×256. It samples different plausible missing tokens in the 32×32 grid to achieve diversity. This work has demonstrated the benefits of adding autoregressive predication over other transformer-based inpainting work [45]. We train all baseline models with the same input mask setting as ours.

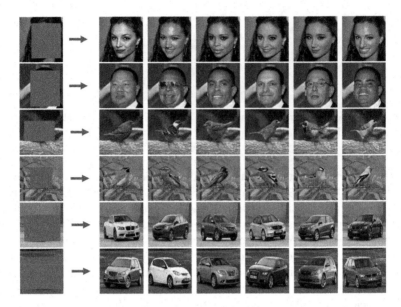

Fig. 4. We can achieve diverse samples from the same masked image by sampling in both known and unknown code spaces.

Evaluation. We use the following metrics: 1) **FID** [15] measures the quality and diversity by comparing distributions between the real and the generated images. 2) **LPIPS** [58] measures distance between two images in deep feature space. For each testing image, we compare pairs of inpainting results generated from the same input mask, we use this to measure diversity. 3) Known Factor Feature Angle (**KFFA**). To better understand how we can improve result diversity using the disentangled known factor, we sample 10 inpainting results for each input image. Then we compute deep features of these 10 results from a known factor classifier. We report average angle between all normalized feature pairs. For a fair comparison, we use feature extractors different from the one used in the training. For face, we use CurricularFace [16], and for bird and car, we train a new classifier using the VGG architecture [38]. Note that L1, SSIM and PSNR are also commonly used metrics for inpainting tasks. However, they all favor deterministic methods which aim to reproduce the single ground truth. As pointed out by [45], these metrics are more suitable for small mask cases where the synthesized contents are more likely to be similar to the ground truth. With large holes covering an entire semantic region or object, synthesized diverse contents might look plausible but different from the ground truth.

4.1 Qualitative Results

Figure 4 shows random samples from our model given the same masked input. Our method can generate diverse identities, facial attributes for faces and synthesize diverse shapes, poses and object appearances for birds and cars. Figure 5

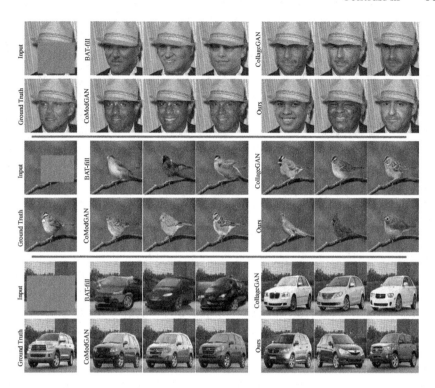

Fig. 5. Compared with the baselines, our method generates more diverse results. For faces, we have more variations in identity. For birds and cars, we have different object shapes and textures.

shows side-by-side comparisons between our method and other baselines. Our results on faces are more diverse compared to CoModGAN [60] and Collage-GAN [26]. For unaligned dataset (cars and birds), these two methods tends to generate results with the same shape. BAT-fill [56] results have better diversity, but lower image quality. It sometimes generates artifacts on faces or distorted geometries for cars.

We also evaluate how disentangled our results are in Fig. 6. Each column shares the same known factor and each row shares the same unknown factor. For faces, the known code controls identity and the unknown code controls facial attributes such as smile, glass and lighting condition. For cars, the known code controls car shape (e.g., sedan-like and wagon-like in the second and third column), and the unknown code changes color and orientation. For birds, the known factor changes species (color is associated with species) and the unknown factor changes pose and orientation. When one latent code changes, the image changes only along the direction it is supposed to, e.g., as the identities change, the same facial expression remains within each row in the face results.

Fig. 6. Known factor code is same for each column, unknown factor code is same for each row.

Table 1. Our method has comparable image quality with the state-of-the-art, but with more diversity.

	Face			Bird			Car		
	FID	LPIPS	KFFA	FID	LPIPS	KFFA	FID	LPIPS	KFFA
CoModGAN	8.88	0.045	52.41	11.35	0.090	52.45	6.59	0.183	44.34
CollageGAN	8.77	0.069	66.00	12.11	0.100	61.08	6.57	0.191	48.67
BAT-Fill	15.08	**0.102**	75.98	37.15	0.117	55.41	22.20	0.270	51.98
ContrasFill-1	8.40	0.072	74.71	**11.29**	0.151	66.06	**6.24**	0.310	63.09
ContrasFill (Ours)	**8.36**	0.075	**83.66**	11.97	**0.160**	**74.58**	6.46	**0.327**	**82.96**

4.2 Image Quality and Diversity

Besides our model ContrasFill, we also evaluate one variant of our approach, where we only have one latent space using contrastive learning without explicit latent disentanglement (denoted as "ContrasFill-1", see supp for details about this variant). This entangled latent space models all factors together for generation. It is used to show the effectiveness of the contrastive loss on diversity.

Table 1 shows comparison in terms of image quality (FID) and diversity (LPIPS for overall, KFFA for known factor). Overall, our model has comparable image quality with the state-of-the-art methods, and it performs favorably against all baselines in terms of known factor diversity, especially compared with CoModGAN [60] and CollageGAN [26]. We also have the highest overall diversity on bird and car datasets. Although, BAT-fill [56] has better LPIPS distance in face dataset, but their image quality is worse (Fig. 5) and they sometimes generate artifacts, which can often results in larger LPIPS difference. Our model also has better diversity, especially on the known factors, compared with our single code variant (ContrasFill-1).

We also compared with UCTGAN [59] which is designed for diverse hole filling. Due to code unavailablity, we only compare with it in CelebA-HQ dataset and use the same setting as theirs. Here we measure diversity LPIPS for full output and only mask region. We grab their numbers and notations (ours first): I_{out}: **0.036** vs 0.030; $I_{out(m)}$: **0.101** vs 0.092.

CoModGAN   CollageGAN

ContrasFill-1   ContrasFill (ours)

Fig. 7. Moving along the discovered identity direction causes changes in non-identity factors such as facial expressions and lighting for baselines. Our results only vary in identity.

4.3   Disentanglement Study

We compare with baselines to show that having two explicit latent spaces improves the disentanglement. Recent works [17, 36, 43] show that postprocessing can be applied to find disentangled latent directions in a pretrained GAN space. Thus we use a supervised method [35] to find known factor directions for CoModGAN, CollageGAN and ContrasFill-1. We use pretrained known factor classifiers to get labels for sampled latent codes and then train a linear regressor to find latent directions [35]. We do not compare with BAT-fill since they lack controllability.

Next, we generate images with different known factors. For baselines, given a masked image, we first randomly sample a latent code, and then move along the discovered known direction to generate 10 different results. For our approach, since we have a disentangled space, thus we directly sample 10 different codes in known factor space by fixing unknown code. We calculate KFFA for 1,000 different contexts and report the average number in the Table 2. Our model has the best KFFA scores. This demonstrates the benefit of having an explicitly disentangled latent space.

Since we has two spaces, we also conduct an experiment where we fix our known code and randomly sample 10 unknown codes. The last row of Table 2 shows the average KFFA numbers over 1,000 context images which indicates when unknown factor varies, our known factor is less influenced. Please refer to Fig. 6 for qualitative results.

We also visually examine the baselines in Fig. 7. By moving along the discovered known (identity) direction, those baselines not only change identity to some extent, but also alter other attributes, such as smile, skin tone, and gaze, whereas our sampled results maintain the attributes controlled by the unknown factor while the known factor changes. This means that certain factors such as human identity can not be easily disentangled during the latent direction discovery stage even with explicit supervision. We also show that this is true for vanilla unconditional StyleGAN [21, 22] in the supp. Note, we move large steps in two directions on purpose (leftmost and rightmost) to show the full effect of the discovered directions. Details about this study can be found in the supp.

Table 2. High KFFA shows our latent space is more diverse compared with discovered latent directions in baselines. The last row is a different setting, please refer the text.

	Face	Bird	Car
CoModGAN	56.00	47.55	44.25
CollageGAN	57.78	58.15	47.49
ContrasFill-1	67.40	61.14	52.01
ContrasFill	**82.03**	**75.20**	**83.71**
ContrasFill (known fixed)	26.15	21.69	35.47

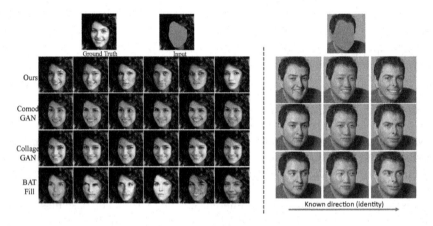

Fig. 8. Our model can generate diverse disentangled results in semantic mask case.

4.4 Ablation

Other Type of Masks. We also analyze our model's performance on the inpainting task where the input mask is of the shape of an object instead of a box. In this case, our model can no longer change the object shape and pose but can still generate diverse appearance in the mask. Figure 8 (left) shows that our results are more diverse than CoModGAN and CollageGAN, especially on human identity. BAT-fill has worse image quality. We can also achieve disentanglement (Fig. 8 right). We compare image quality and diversity (LPIPS for overall, KFFA for known factor), and report numbers in Table 3 (left) for the face dataset. We also study the case of arbitrarily-shaped masks that cover part of the object in random places (e.g., half of face is hidden); see supp for details.

Latent Codes Injection Method. To study the effectiveness of our bi-modulated convolutions (Fig. 3), we try the following alternative approaches: (1) We concatenate s with t and pass the result to fully-connected layers to output a single scale vector to modulate the convolutions (denoted as "concat"); (2) We use each code to modulate a different set of convolution kernels. And the two sets of modulated convolutions are sequentially applied to image features.

Table 3. (Left) Comparison in mask case. Our method has comparable image quality but with more diversity on faces when using face masks as inpainting regions. **(Right) Ablation.** results indicate the effectiveness of bi-modulation (the first three columns) and contrastive loss (4th column).

	CoMod	Collage	BAT	Ours		Concat	u-k	k-u	Repredict	Ours
FID	**5.73**	6.07	11.97	5.95	FID	8.64	**8.41**	8.42	14.41	**8.41**
LPIPS	0.029	0.029	**0.050**	0.048	LPIPS	0.048	0.052	0.056	**0.120**	0.075
KFFA	51.19	58.73	72.48	**83.39**	KFFA	48.99	78.19	77.58	**86.87**	83.06

Depending on the order, we denote them as "k-u" and "u-k" (k and u stand for known and unknown codes). The first three columns of Table 3 (right) indicate that, these alternative designs achieve similar image quality, but lower level of diversity. Our bi-modulation is a more direct way to inject information to the generator which makes the learning process easier compared with the "concat" alternative. The approach of applying two separate sets of convolutions results in poor diversity. We hypothesize that the later convolution set may undo what is learnt by the previous set as their objective functions are different (factors in two spaces should learn different things).

The Loss Choice. We use the contrastive loss to encourage diversity by forcing images with the same latent codes to have similar factors. Another way to learn this correspondence is to repredict the input code from the resulting image. For example, by sampling a code in the identity space, one can use ArcFace to repredict this code from the generated image. Table 3 (right 4th column) shows that replacing the contrastive loss with reprediction loss encourages more diversity, but at the cost of image quality. This is because a foreground generation model needs to consider the compatibility between the sampled latent code and the input context. For example, if the context contains light skin pixels on the neck, then latent codes that generate dark-skinned faces are not compatible. However, the reprediction loss forces the model to synthesize a dark-skinned face, which may not look real according to the discriminator. However if a contrastive loss is applied, which considers the *relative distance* in the feature space, then the model can adjust the input identity code based on the context information to synthesize a face that looks more plausible in the context.

5 Conclusion and Limitations

We propose ContrasFill, a novel approach for diverse and controllable foreground generation by contrastive learning. We demonstrate superior diversity and controllability over previous work. Our method has some limitations. We found that our model is sometimes sensitive to the pretrained classifier which may be biased due to the training data. For example, certain types of car are more common in certain color (e.g., van are usually white). Thus our model may also be biased.

Acknowledgement. This work was supported in part by Sony Focused Research Award, NSF CAREER IIS-2150012, Wisconsin Alumni Research Foundation, and NASA 80NSSC21K0295.

References

1. https://github.com/zllrunning/face-parsing.pytorch
2. Ballester, C., Bertalmio, M., Caselles, V., Sapiro, G., Verdera, J.: Filling-in by joint interpolation of vector fields and gray levels. IEEE Trans. Image Process. **10**(8), 1200–1211 (2001). https://doi.org/10.1109/83.935036
3. Barnes, C., Shechtman, E., Finkelstein, A., Goldman, D.B.: PatchMatch: a randomized correspondence algorithm for structural image editing. In: SIGGRAPH 2009 (2009)
4. Bertalmío, M., Sapiro, G., Caselles, V., Ballester, C.: Image inpainting, pp. 417–424 (2000)
5. Chen, T., Kornblith, S., Norouzi, M., Hinton, G.E.: A simple framework for contrastive learning of visual representations. arXiv:abs/2002.05709 (2020)
6. Deng, J., Guo, J., Zafeiriou, S.: ArcFace: additive angular margin loss for deep face recognition. In: 2019 IEEE/CVF Conference on Computer Vision and Pattern Recognition (CVPR), pp. 4685–4694 (2019)
7. Denton, E.L., Birodkar, V.: Unsupervised learning of disentangled representations from video. arXiv:abs/1705.10915 (2017)
8. Ding, D., Ram, S., Rodríguez, J.J.: Image inpainting using nonlocal texture matching and nonlinear filtering. IEEE Trans. Image Process. **28**, 1705–1719 (2019)
9. Du, R., et al.: Fine-grained visual classification via progressive multi-granularity training of jigsaw patches. In: Vedaldi, A., Bischof, H., Brox, T., Frahm, J.-M. (eds.) ECCV 2020. LNCS, vol. 12365, pp. 153–168. Springer, Cham (2020). https://doi.org/10.1007/978-3-030-58565-5_10
10. Goodfellow, I., et al.: Generative adversarial nets. In: NeurIPS (2014)
11. Guo, X., Yang, H., Huang, D.: Image inpainting via conditional texture and structure dual generation. In: Proceedings of the IEEE/CVF International Conference on Computer Vision (ICCV), pp. 14134–14143, October 2021
12. Hays, J., Efros, A.A.: Scene completion using millions of photographs. In: SIGGRAPH 2007 (2007)
13. He, K., Fan, H., Wu, Y., Xie, S., Girshick, R.B.: Momentum contrast for unsupervised visual representation learning. In: 2020 IEEE/CVF Conference on Computer Vision and Pattern Recognition (CVPR), pp. 9726–9735 (2020)
14. He, K., Gkioxari, G., Dollár, P., Girshick, R.B.: Mask R-CNN. IEEE Trans. Pattern Anal. Mach. Intell. **42**, 386–397 (2020)
15. Heusel, M., Ramsauer, H., Unterthiner, T., Nessler, B., Hochreiter, S.: GANs trained by a two time-scale update rule converge to a local Nash equilibrium. In: NIPS (2017)
16. Huang, Y., Wang, Y., Tai, Y., Liu, X., Shen, P., Li, S., Jilin Li, F.H.: CurricularFace: adaptive curriculum learning loss for deep face recognition, pp. 1–8 (2020)
17. Härkönen, E., Hertzmann, A., Lehtinen, J., Paris, S.: GANspace: Discovering interpretable GAN controls. In: Proceedings of NeurIPS (2020)
18. Iizuka, S., Simo-Serra, E., Ishikawa, H.: Globally and locally consistent image completion. ACM Trans. Graph. (TOG) **36**, 1–14 (2017)
19. Jahanian, A., Chai, L., Isola, P.: On the "steerability" of generative adversarial networks. In: International Conference on Learning Representations (2020)

20. Karras, T., Laine, S., Aila, T.: A style-based generator architecture for generative adversarial networks. In: CVPR (2018)
21. Karras, T., Laine, S., Aila, T.: A style-based generator architecture for generative adversarial networks. In: CVPR (2019)
22. Karras, T., Laine, S., Aittala, M., Hellsten, J., Lehtinen, J., Aila, T.: Analyzing and improving the image quality of StyleGAN. In: Proceedings of CVPR (2020)
23. Kingma, D.P., Welling, M.: Auto-encoding variational bayes. CoRR abs/1312.6114 (2014)
24. Krause, J., Stark, M., Deng, J., Fei-Fei, L.: 3D object representations for fine-grained categorization. In: 4th International IEEE Workshop on 3D Representation and Recognition (3dRR-2013), Sydney, Australia (2013)
25. Lee, C.H., Liu, Z., Wu, L., Luo, P.: MaskGAN: towards diverse and interactive facial image manipulation. In: IEEE Conference on Computer Vision and Pattern Recognition (CVPR) (2020)
26. Li, Y., Li, Y., Lu, J., Shechtman, E., Lee, Y.J., Singh, K.K.: Collaging class-specific GANs for semantic image synthesis. In: ICCV (2021)
27. Li, Y., Singh, K.K., Ojha, U., Lee, Y.J.: MixnMatch: multifactor disentanglement and encoding for conditional image generation. In: 2020 IEEE/CVF Conference on Computer Vision and Pattern Recognition (CVPR), pp. 8036–8045 (2020)
28. Liu, G., Reda, F.A., Shih, K.J., Wang, T.-C., Tao, A., Catanzaro, B.: Image inpainting for irregular holes using partial convolutions. In: Ferrari, V., Hebert, M., Sminchisescu, C., Weiss, Y. (eds.) ECCV 2018. LNCS, vol. 11215, pp. 89–105. Springer, Cham (2018). https://doi.org/10.1007/978-3-030-01252-6_6
29. Ma, X., Zhou, X., Huang, H., Chai, Z., Wei, X., He, R.: Free-form image inpainting via contrastive attention network. In: 2020 25th International Conference on Pattern Recognition (ICPR), pp. 9242–9249 (2021)
30. Misra, I., van der Maaten, L.: Self-supervised learning of pretext-invariant representations. In: 2020 IEEE/CVF Conference on Computer Vision and Pattern Recognition (CVPR), pp. 6706–6716 (2020)
31. Park, T., Efros, A.A., Zhang, R., Zhu, J.Y.: Contrastive learning for unpaired image-to-image translation. In: ECCV (2020)
32. Pathak, D., Krähenbühl, P., Donahue, J., Darrell, T., Efros, A.A.: Context encoders: Feature learning by inpainting. In: 2016 IEEE Conference on Computer Vision and Pattern Recognition (CVPR), pp. 2536–2544 (2016)
33. Ren, Y., Yu, X., Zhang, R., Li, T.H., Liu, S., Li, G.: Structureflow: image inpainting via structure-aware appearance flow. In: 2019 IEEE/CVF International Conference on Computer Vision (ICCV), pp. 181–190 (2019)
34. Sagong, M.C., Shin, Y.G., Kim, S.W., Park, S., Ko, S.: Pepsi : fast image inpainting with parallel decoding network. In: 2019 IEEE/CVF Conference on Computer Vision and Pattern Recognition (CVPR), pp. 11352–11360 (2019)
35. Shen, Y., Gu, J., Tang, X., Zhou, B.: Interpreting the latent space of GANs for semantic face editing. In: 2020 IEEE/CVF Conference on Computer Vision and Pattern Recognition (CVPR), pp. 9240–9249 (2020)
36. Shen, Y., Zhou, B.: Closed-form factorization of latent semantics in GANs. arXiv:abs/2007.06600 (2020)
37. Shu, Z., Sahasrabudhe, M., Alp Güler, R., Samaras, D., Paragios, N., Kokkinos, I.: Deforming autoencoders: unsupervised disentangling of shape and appearance. In: Ferrari, V., Hebert, M., Sminchisescu, C., Weiss, Y. (eds.) ECCV 2018. LNCS, vol. 11214, pp. 664–680. Springer, Cham (2018). https://doi.org/10.1007/978-3-030-01249-6_40

38. Simonyan, K., Zisserman, A.: Very deep convolutional networks for large-scale image recognition. CoRR abs/1409.1556 (2015)
39. Singh, K.K., Ojha, U., Lee, Y.J.: FineGAN: unsupervised hierarchical disentanglement for fine-grained object generation and discovery. In: 2019 IEEE/CVF Conference on Computer Vision and Pattern Recognition (CVPR), pp. 6483–6492 (2019)
40. Suin, M., Purohit, K., Rajagopalan, A.N.: Distillation-guided image inpainting. In: Proceedings of the IEEE/CVF International Conference on Computer Vision (ICCV), pp. 2481–2490, October 2021
41. Telea, A.: An image inpainting technique based on the fast marching method. J. Graph. Tools. **9**, 23–34 (004). https://doi.org/10.1080/10867651.2004.10487596
42. Tian, Y., Krishnan, D., Isola, P.: Contrastive multiview coding. In: Vedaldi, A., Bischof, H., Brox, T., Frahm, J.-M. (eds.) ECCV 2020. LNCS, vol. 12356, pp. 776–794. Springer, Cham (2020). https://doi.org/10.1007/978-3-030-58621-8_45
43. Voynov, A., Babenko, A.: Unsupervised discovery of interpretable directions in the GAN latent space. arXiv:abs/2002.03754 (2020)
44. Wah, C., Branson, S., Welinder, P., Perona, P., Belongie, S.: The Caltech-UCSD Birds-200-2011 Dataset. Technical report, CNS-TR-2011-001 (2011)
45. Wan, Z., Zhang, J., Chen, D., Liao, J.: High-fidelity pluralistic image completion with transformers. In: ICCV (2021)
46. Wang, Y., Tao, X., Qi, X., Shen, X., Jia, J.: Image inpainting via generative multi-column convolutional neural networks. In: NeurIPS (2018)
47. Wu, Y., Kirillov, A., Massa, F., Lo, W.Y., Girshick, R.: Detectron2 (2019). https://github.com/facebookresearch/detectron2
48. Xie, C., et al.: Image inpainting with learnable bidirectional attention maps. In: 2019 IEEE/CVF International Conference on Computer Vision (ICCV), pp. 8857–8866 (2019)
49. Xing, X., Gao, R., Han, T., Zhu, S.C., Wu, Y.N.: Deformable generator network: Unsupervised disentanglement of appearance and geometry. IEEE Trans. Pattern Anal. Mach. Intell. (2020)
50. Xiong, W., et al.: Foreground-aware image inpainting. In: 2019 IEEE/CVF Conference on Computer Vision and Pattern Recognition (CVPR), pp. 5833–5841 (2019)
51. Yan, Z., Li, X., Li, M., Zuo, W., Shan, S.: Shift-Net: image inpainting via deep feature rearrangement. arXiv:abs/1801.09392 (2018)
52. Yang, C., Shen, Y., Zhou, B.: Semantic hierarchy emerges in deep generative representations for scene synthesis. Int. J. Comput. Vis. **129**, 1451–1466 (2021)
53. Yang, C., Lu, X., Lin, Z.L., Shechtman, E., Wang, O., Li, H.: High-resolution image inpainting using multi-scale neural patch synthesis. In: 2017 IEEE Conference on Computer Vision and Pattern Recognition (CVPR), pp. 4076–4084 (2017)
54. Yu, F., Zhang, Y., Song, S., Seff, A., Xiao, J.: LSUN: construction of a large-scale image dataset using deep learning with humans in the loop. arXiv:abs/1506.03365 (2015)
55. Yu, J., Lin, Z.L., Yang, J., Shen, X., Lu, X., Huang, T.S.: Free-form image inpainting with gated convolution. In: 2019 IEEE/CVF International Conference on Computer Vision (ICCV), pp. 4470–4479 (2019)
56. Yu, Y., et al.: Diverse image inpainting with bidirectional and autoregressive transformers. In: Proceedings of the 29th ACM International Conference on Multimedia (2021)
57. Zeng, Y., Fu, J., Chao, H., Guo, B.: Learning pyramid-context encoder network for high-quality image inpainting. In: 2019 IEEE/CVF Conference on Computer Vision and Pattern Recognition (CVPR), pp. 1486–1494 (2019)

58. Zhang, R., Isola, P., Efros, A.A., Shechtman, E., Wang, O.: The unreasonable effectiveness of deep features as a perceptual metric. In: CVPR (2018)
59. Zhao, L., et al.: UCTGAN: diverse image inpainting based on unsupervised cross-space translation. In: 2020 IEEE/CVF Conference on Computer Vision and Pattern Recognition (CVPR), pp. 5740–5749 (2020)
60. Zhao, S., et al.: Large scale image completion via co-modulated generative adversarial networks. arXiv:abs/2103.10428 (2021)
61. Zheng, C., Cham, T., Cai, J.: Pluralistic image completion. In: 2019 IEEE/CVF Conference on Computer Vision and Pattern Recognition (CVPR), pp. 1438–1447 (2019)
62. Zhou, X., Li, J., Wang, Z., He, R., Tan, T.: Image inpainting with contrastive relation network. In: 2020 25th International Conference on Pattern Recognition (ICPR), pp. 4420–4427 (2021)
63. Zhuang, C., Zhai, A., Yamins, D.: Local aggregation for unsupervised learning of visual embeddings. In: 2019 IEEE/CVF International Conference on Computer Vision (ICCV), pp. 6001–6011 (2019)

BIPS: Bi-modal Indoor Panorama Synthesis via Residual Depth-Aided Adversarial Learning

Changgyoon Oh[1], Wonjune Cho[2], Yujeong Chae[1], Daehee Park[1], Lin Wang[3], and Kuk-Jin Yoon[1(✉)]

[1] Visual Intelligence Lab., KAIST, Daejeon, South Korea
{changgyoon,yujeong,bag2824,kjyoon}@kaist.ac.kr
[2] NAVER LABS, Los Angeles, USA
wonjune.cho@naverlabs.com
[3] AI Thrust, HKUST Guangzhou and Department of CSE, HKUST, Hong Kong, China
linwang@ust.hk

Abstract. Providing omnidirectional depth along with RGB information is important for numerous applications. However, as omnidirectional RGB-D data is not always available, synthesizing RGB-D panorama data from limited information of a scene can be useful. Therefore, some prior works tried to synthesize RGB panorama images from perspective RGB images; however, they suffer from limited image quality and can not be directly extended for RGB-D panorama synthesis. In this paper, we study a new problem: RGB-D panorama synthesis under the various configurations of cameras and depth sensors. Accordingly, we propose a novel bi-modal (RGB-D) panorama synthesis (BIPS) framework. Especially, we focus on indoor environments where the RGB-D panorama can provide a complete 3D model for many applications. We design a generator that fuses the bi-modal information and train it via residual depth-aided adversarial learning (RDAL). RDAL allows to synthesize realistic indoor layout structures and interiors by jointly inferring RGB panorama, layout depth, and residual depth. In addition, as there is no tailored evaluation metric for RGB-D panorama synthesis, we propose a novel metric (FAED) to effectively evaluate its perceptual quality. Extensive experiments show that our method synthesizes high-quality indoor RGB-D panoramas and provides more realistic 3D indoor models than prior methods. Code is available at https://github.com/chang9711/BIPS.

Keywords: RGB-D panorama synthesis · Indoor layout · GAN · VR/AR

C. Oh, W. Cho, Y. Chae and D. Park—Equal contribution.

Supplementary Information The online version contains supplementary material available at https://doi.org/10.1007/978-3-031-19787-1_20.

1 Introduction

Omnidirectional RGB-D data is important for numerous applications, *e.g.*, VR/ AR, yet it is not always available. Manually creating a 3D space from scratch is unrealistic and requires a huge effort, while capturing and restoring whole real-world requires high computational cost [35]. Synthesizing RGB-D panorama from limited input information can overcome the limitations and generate 3D virtual space with minimal time and effort. Even though prior works have tried to synthesize RGB panorama images from perspective RGB images [20,60], these methods show limited performance on synthesizing panoramas from small partial views and can not be directly extended for RGB-D panorama synthesis (Fig. 1).

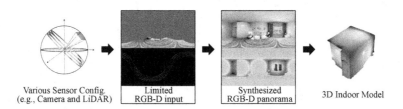

| Various Sensor Config. (e.g., Camera and LiDAR) | Limited RGB-D input | Synthesized RGB-D panorama | 3D Indoor Model |

Fig. 1. Overall scheme for our BIPS framework, which takes RGB-D input from cameras and depth sensors in various configurations and synthesizes an RGB-D panorama.

By contrast, *jointly learning to synthesize depth data along with the RGB images* accompanies two advantages: (1) Depth panorama, which is useful to plenty of applications, can be directly obtained without additional endeavors such as monocular depth estimation or depth completion. (2) The quality of generated RGB and depth panorama can be improved to complement each other. It is because they share the semantic correspondence of the scene, and this correspondence is learned during the joint learning. The extensive experiments in Sect. 4.2 demonstrates the mutual gain between RGB and depth panorama. Therefore, it is promising to synthesize RGB-D panorama from the cameras and depth sensors, such that we can synthesize realistic 3D indoor models.

In this paper, we consider a novel problem: *RGB-D panorama synthesis from limited input visual information of a scene.* To maximize usability, we consider the various configurations of cameras and depth sensors. To this end, we design the various sensor configurations by randomly sampling the number of sensors, their intrinsic parameters, and extrinsic parameters, assuming that the sensors are calibrated and aligned to each other. This enables to represent most of the possible combinations of cameras and depth sensors. Accordingly, our novel bi-modal panorama synthesis (BIPS) framework synthesizes RGB-D indoor panoramas from the camera and depth sensors in various configurations via adversarial learning (See Fig. 3). We thus design a generator that fuses the bi-modal (RGB and depth) features. Through the generator, multiple latent features from one branch can help the other by providing the relevant information of different modalities. For synthesizing the depth of *indoor* scenes, we rely on the fact that the overall layout is usually made of flat surfaces, while interior components have various structures. Thus, we propose to separate the depth of a scene I^d into two components: layout depth $I^{d,lay}$ and residual depth $I^{d,res}$. Here, $I^{d,lay}$ corresponds to the depth of planar surfaces, and $I^{d,res}$ corresponds to the depth of other objects, *e.g.*, furniture. With this relation, we propose a joint learning scheme called *Residual Depth-aided Adversarial Learning (RDAL)*. RDAL jointly trains RGB panorama, layout depth, and residual depth to synthesize more realistic RGB-D panoramas and 3D indoor models (Sect. 3.2).

Previously, some metrics [22,56] have been proposed to evaluate the outputs of generative models using latent feature distribution of a pre-trained classification network [62]. However, the input modality of utilizing an off-the-shelf network is only limited to perspective RGB images. For this reason, we propose a novel metric, called Fréchet Auto-Encoder Distance (FAED), to evaluate the perceptual quality for RGB-D panorama synthesis (Sect. 3.3). FAED adopts an auto-encoder to reconstruct the inputs from latent features with an unlabeled dataset. Then, the latent feature distribution of the trained auto-encoder is used to calculate the Fréchet distance between the synthesized and GT RGB-D data. Extensive experimental results demonstrate that our RGB-D panorama synthesis method significantly outperforms the extensions of the prior image inpainting [46,61,88], image outpainting [32,60], and image-guided depth synthesis methods [11,24,37,51] modified to synthesize RGB-D panorama from partial RGB-D inputs. Moreover, we show the validity of the proposed FAED by showing how well it captures the disturbance level [22] of synthesized RGB-D panorama.

In summary, our main contributions are three-fold: (I) We introduce a new problem of generating RGB-D panoramas from partial RGB-D inputs under various sensor configurations. (II) We propose a BIPS framework that synthesizes RGB-D panoramas via residual depth-aided adversarial learning. (III) We introduce a novel evaluation metric, FAED, for RGB-D panorama synthesis.

2   Related Works

Image Inpainting. Conventional approaches explore diffusion or patch matching [5–8,13,14,16]. However, they have limited ability inpainting largely missing regions. The learning-based methods often use generative adversarial networks (GANs) [26,38,79,89], optimized by the minimax loss [27]. Some works explored different convolution layers, *e.g.*, partial convolution [40] and gated convolution [50,80], to handle missing pixels. Moreover, attention mechanism [65,66] has also been applied to capture the contextual information [39,41,70,75,79]. Recently, research has been made to synthesize high-resolution outputs [52,59, 72] or semantically diverse outputs [42,87]. Although endeavors have tackled large completion problem [46,61,88], they often fail to synthesize visually pleasing panoramas due to only using perspective RGB inputs.

Image Outpainting. Conventional methods extend an input image to a larger seamless one; however, they require manual guidance [4,6,86] or image sets of the same scene category [29,58,69]. By contrast, learning-based methods synthesize large images with novel textures that do not exist in the input image [18,19, 30,31,34,47,55,74,82]. Some works focus on driving scenes [73,84] or synthesize panorama with iterative extension or multiple perspective images [20, 32, 60, 78]. Although performance has been greatly improved, the existing methods are still afflicted by the limited quality from the perspective images.

Image-guided Depth Synthesis. One line of research attempts to fuse the bi-modal information, *i.e.*, the RGB image and sparse depth. Some methods, *e.g.* [44], fuse the sparse depth and RGB image via early fusion while others [17, 25, 28, 37, 43, 63] utilize a late fusion scheme, or jointly utilize both [36, 64, 67]. Another line of research focuses on utilizing affinity or geometric information of the scene via surface normal, occlusion boundaries, and the geometric convolutional layer [10, 11, 24, 33, 51, 53, 76, 85]. However, these works only generate dense depth maps that have the same FoV with the input perspective RGB images.

Evaluation of Generative Models. Image quality assessment can be classified into three groups: full-reference (FR), reduced-reference (RR), and no-reference (NR). There exist many conventional FR metrics, *e.g.*, PSNR, MSE, and SSIM, and deep learning (DL)-based FR metrics, *e.g.*, LPIPS [83]. These metrics typically calculate either pixel-wise or patch-wise similarity to the ground truth images. By contrast, NR methods, *e.g.*, BRISQUE [48] and NIQE [49] do not require reference image. Among the DL-based NR metrics, Inception Score (IS) [56] and Fréchet Inception Distance (FID) [22] are widely used [2]. IS and FID scores are calculated from pretrained classification models to capture the high-level features. Unfortunately, these metrics are less applicable for RGB-D panorama evaluation because (1) they are trained only with perspective RGB images, and (2) there are no labeled panorama images for training. They are highly sensitive to the distortion of panoramas, making them hard to capture perceptual quality properly on panoramas. Furthermore, naively using them on RGB-D leads to an imprecise measure of the semantic correspondence between the two different modalities. Therefore, *we propose FAED, which aims to directly evaluate the RGB-D panorama quality. FAED can be adaptively applied to evaluate multi-modal data that lacks a labeled dataset.*

3   Proposed Methods

3.1   Problem Formulation

Previous works, *e.g.*, [20, 60], generate an equirectangular projection (ERP) image (ERP^{rgb}) from input RGB image(s) ($I_{\mathrm{in}}^{\mathrm{rgb}}$). Then, an RGB panorama $I_{\mathrm{out}}^{\mathrm{rgb}}$ can be created via a function G, mapping $I_{\mathrm{in}}^{\mathrm{rgb}}$ into a ERP^{rgb} [21], which can be formulated as $I_{\mathrm{out}}^{\mathrm{rgb}} = ERP^{\mathrm{rgb}} = G(I_{\mathrm{in}}^{\mathrm{rgb}})$.

However, as it is crucial to provide omnidirectional depth information [1, 54] in many applications, many studies tried to synthesize depth panoramas from input RGB panorama and partial depth measurements [23, 68]. One solution to synthesize an RGB-D panorama would be to sequentially synthesize RGB panorama from input RGB images, and then apply the depth synthesis methods to generate an omnidirectional depth map. However, such an approach is cumbersome and less effective, as shown in the experimental results (See Table 4). We solve this novel yet challenging problem by jointly utilizing the input RGB image

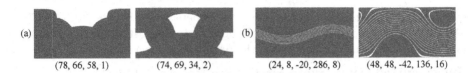

(78, 66, 58, 1)   (74, 69, 34, 2)   (24, 8, -20, 286, 8)   (48, 48, -42, 136, 16)

Fig. 2. Sampled input masks. (a) Input mask of cameras and perspective depth sensors with parameters $(\delta_H, \delta_V, \psi, n)$ and (b) mechanical LiDARs with $(\delta_L, \delta_U, \psi, \omega, \eta)$.

(I_{in}^{rgb}) and depth data (I_{in}^{d}). Our goal is to directly generate the RGB panorama (ERP^{rgb}) and depth panorama (ERP^d) simultaneously via a mapping function G, which can be described as $(I_{out}^{rgb}, I_{out}^d) = (ERP^{rgb}, ERP^d) = G(I_{in}^{rgb}, I_{in}^d)$. G can be formulated by learning a *single* network to synthesize ERP^{rgb} and ERP^d using I_{in}^{rgb} and I_{in}^d obtained in various sensor configurations. As the information in the left and right boundaries in ERP images should be connected, our designed G uses circular padding [57] before each convolutional operation.

Consequently, we design the various input configurations by randomly sampling the parameters of cameras and depth sensors to provide the input to the G during training. Since our model takes partial ERP input, data obtained from sensors should be projected to ERP image space. The masks of the sensor input area projected on the ERP can be represented as shown in Fig. 2. Therefore, the partial RGB-D input projected on the ERP space, I_{in}^{rgb} and I_{in}^d, can be obtained by multiplying sampled mask to the full ERP image. We also randomly choose whether to use cameras only, depth sensors only, or both, to handle the cases where only cameras or depth sensors are available.

Parameters of RGB Cameras. We denote the parameters of RGB cameras, horizontal FoV as δ_H, vertical FoV as δ_V, pitch angle as ψ, and the number of viewpoints as n. When $n > 1$, we arrange the viewpoints in a circle having the sampled pitch angle from the equator and at the same intervals. We do not consider roll and yaw, as they do not affect the results (*i.e.*, the output is equivariant to the horizontal shift of input) thanks to using circular padding. Considering general setting of cameras, we sample the parameters from $\{\delta_H, \delta_V\} \sim \mathcal{U}[60°, 90°]$, $\psi \sim \mathcal{U}[-90°, 90°]$, and $n \sim \mathcal{U}\{0, 1, 2, 3, 4\}$, where $\mathcal{U}[\cdot]$ represents uniform distribution.

Parameters of Depth Sensors. I_{in}^d can be obtained from mechanical LiDARs or perspective depth sensors thus we should generate various depth input masks for both. For the LiDARs, we denote lower FoV as δ_L, upper FoV as δ_U, pitch angle as ψ, yaw angle as ω, and the number of channels as η. The yaw angle is needed to consider the relative yaw motion to the camera arrangement. For the perspective depth sensors providing dense depth, we use the same sampled parameters with the cameras $(\delta_H, \delta_V, \psi, n)$. In practice, we first sample the parameters from $\psi \sim \mathcal{U}[-90°, 90°]$, $\omega \sim \mathcal{U}[0°, 360°]$, and $\eta \sim \mathcal{U}\{0, 2, 4, 8, 16\}$. Then, we sample δ_L and δ_U from $\mathcal{U}\{\eta, 2\eta, 3\eta\}$. Finally, our problem is formulated as

$$(I_{out}^{rgb}, I_{out}^d) = (ERP^{rgb}, ERP^d) \tag{1}$$

$$= G(I_{in}^{rgb}(\delta_H, \delta_V, \psi, n), I_{in}^d(\delta_L, \delta_U, \psi, \omega, \eta, \delta_H, \delta_V, n)) \tag{2}$$

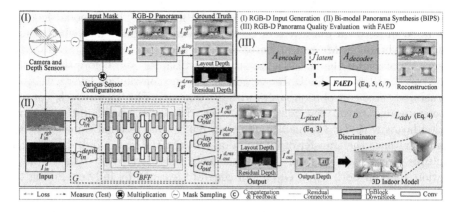

Fig. 3. Overall structure of our bi-modal indoor panorama synthesis (BIPS) framework. Our framework takes RGB-D input provided by various sensor configurations, integrates the bi-modal input data with BFF branch in the generator network, and jointly trains to synthesize layout depth and residual depth. Then, the perceptual quality of the synthesized RGB-D panorama is measured by proposed FAED metric.

3.2 RGB-D Panorama Synthesis Framework

Overview. An overview of the proposed BIPS framework is depicted in Fig. 3. BIPS consists of a generator G, and a discriminator D. G takes the partial RGB image I_{in}^{rgb} and depth I_{in}^{d} as inputs. We notice that the quality of the RGB-D panorama depends on both the overall (mostly rectangular) layout and how the furniture are arranged in the indoor scene. Inspired by [81], we separate the depth data I_{gt}^{d} into *layout depth* $I_{gt}^{d,lay}$, and *residual depth (interior components)* $I_{gt}^{d,res}$, which is defined as $(I_{gt}^{d} - I_{gt}^{d,lay})$. The generator G outputs the RGB panorama I_{out}^{rgb}, the layout depth panorama $I_{out}^{d,lay}$, and the residual depth panorama $I_{out}^{d,res}$ simultaneously. As these are jointly trained with adversarial loss, we call this learning scheme *Residual Depth-aided Adversarial Learning (RDAL)*.

Generator. G is composed of input branch G_{in}, bi-modal feature fusion (BFF) branch G_{BFF}, and output branch G_{out}, as shown in Fig. 3. G_{in} consists of two encoding branches: G_{in}^{rgb} and G_{in}^{depth}, which take I_{in}^{rgb} and I_{in}^{d} respectively, that independently extract RGB and depth features. Then, G_{BFF} fuses the highly correlated features to fully exploit the bi-modal information of the scene. Lastly, G_{out} have three decoding branches and each of them generates RGB panorama I_{out}^{rgb}, layout depth panorama $I_{out}^{d,lay}$, and residual depth panorama $I_{out}^{d,res}$, respectively. Since realistic indoor space comes with clean layout structure and detailed interiors, we design G_{out} to jointly synthesize the layout and residual depth of an indoor scene. The detailed structure can be found in suppl. material.

BFF Branch. G_{BFF} takes $G_{in}^{rgb}(I_{in}^{rgb})$ and $G_{in}^{depth}(I_{in}^{d})$ as inputs. Since I_{in}^{rgb} and I_{in}^{d} are captured in the same scene, their bi-modal information is highly correlated in a semantic manner. To utilize this correlation, G_{BFF} consists of two-stream encoder-decoder networks fusing the bi-modal features. The bi-modal features are fused in between the layers of G_{BFF} four times. In particular, the features from both branches are concatenated and fed back to each other. Overall, the fusion is done after the features pass two 'DownBlocks' and before the features pass two 'UpBlocks'. The 'UpBlock' consists of one convolution layer with 4×4 kernel and three convolution layers with 3×3 kernel. The 'DownBlock' consists of one upsample layer, one convolution layer with 4×4 kernel and three convolution layers with 3×3 kernel. In addition, multi-scale residual connections are used to vitalize the transfer of information between the layers and branches. As multiple latent features from one branch help the other by sharing the information apart in both ways, G_{BFF} can generate features by fully exploiting the information of the scene.

Discriminator. We use the multi-scale discriminator D from [71], but modify it to have five input channels (three for I^{rgb}, one for $I^{\mathrm{d,lay}}$, and one for $I^{\mathrm{d,res}}$). The detailed discriminator structure can be found in the suppl. material.

Loss Function. For training G, we use a weighted sum of the pixel-wise L1 loss and adversarial loss. The pixel-wise L1 loss between the GT and the output panorama, denoted as L_{pixel}, consists of three terms as the G has three outputs (RGB, layout depth, residual depth panorama):

$$L_{\mathrm{pixel}}^{\mathrm{total}} = L_{\mathrm{pixel}}^{\mathrm{rgb}} + L_{\mathrm{pixel}}^{\mathrm{d,lay}} + L_{\mathrm{pixel}}^{\mathrm{d,res}}. \tag{3}$$

For adversarial loss L_{adv}, we used LSGAN loss [45]: $L_{\mathrm{adv}} = \frac{1}{2}\,\mathbb{E}\left[(D(I_{\mathrm{out}}^{\mathrm{total}}) - 1)^2\right]$, where $I_{\mathrm{out}}^{\mathrm{total}}$ is the concatenation of generator outputs $I_{\mathrm{out}}^{\mathrm{rgb}}$, $I_{\mathrm{out}}^{\mathrm{d,lay}}$ and $I_{\mathrm{out}}^{\mathrm{d,res}}$, and D is a discriminator. By decomposing the total depth loss into $L^{\mathrm{d,lay}}$ and $L^{\mathrm{d,res}}$, our RDAL scheme allows G to synthesize RGB-D panorama that has a highly plausible interior. Finally, the total loss for the generator is:

$$L_{\mathrm{G}} = \lambda L_{\mathrm{pixel}}^{\mathrm{total}} + L_{\mathrm{adv}} \tag{4}$$

where λ is a weighting factor. For detailed loss terms, refer to suppl. material.

3.3 Fréchet Auto-Encoder Distance (FAED)

Auto-Encoder Network. Similar to the high-level features in a CNN trained with large-scale semantic labels, latent features f_{latent} in a trained auto-encoder also contain high-level information, as auto-encoder is trained to reconstruct the input from the latent features. Moreover, the auto-encoder has the advantage that it does not need any labels for training. Since there is no dataset including semantic labels for RGB-D panoramas, we propose to train an auto-encoder A to generate RGB-D panoramas and use its latent features to calculate the perceptual quality. The detailed structure of A is given in the suppl. material.

Calculation of FAED for RGB-D Panorama. We denote f_{latent} at c-th channel, h-th row, and w-th column as $f_{\mathrm{latent}}(c, h, w)$. Note that as we use ERP, the h and w have one-to-one relation to latitude and longitude.

Longitudinal Invariance. To evaluate the performance of G, we extract f_{latent} from generated samples using A_{encoder}. However, as we generate the upright ERP image, it is expected to have a distance metric that is invariant to the longitudinal shift. This is because an upright ERP panorama represents the same scene when it is cyclically shifted along the longitudinal direction. Therefore, to make the resulting distance metric invariant to the longitudinal shift, we take the mean along the longitudinal direction of f_{latent} as (Fig. 4):

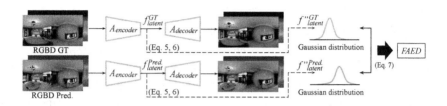

Fig. 4. The proposed FAED metric for RGB-D panorama quality evaluation. It measures the distance of the distributions of latent features extracted from the pre-trained auto-encoder network on RGB-D panorama.

$$f'_{\text{latent}}(c, h) = \frac{1}{W} \sum_{w} f_{\text{latent}}(c, h, w). \tag{5}$$

Latitudinal Equivariance. As ERP has varying sampling rates depending on the latitude ϕ, we apply different weights on f'_{latent} considering information density along latitude. Specifically, we multiply $\cos(\phi)$ to feature at the latitude ϕ, since in ERP, each pixel occupies $\cos(\phi)$ area in the spherical surface, compared with the pixels in the equator. The resulting feature is expressed as:

$$f''_{\text{latent}}(c, h) = \cos(\phi) \cdot f'_{\text{latent}}(c, h). \tag{6}$$

Fréchet Distance. We treat the resulting f''_{latent} as a vector and assume that it has a multi-dimensional Gaussian distribution. Then, we get the distribution of ground truths $\mathcal{N}(m, C)$ and that of generated samples $\mathcal{N}(\hat{m}, \hat{C})$, and calculate the Fréchet distance d between them as given by [15]:

$$d^2(\mathcal{N}(m, C), \mathcal{N}(\hat{m}, \hat{C})) = ||m - \hat{m}||_2^2 + Tr(C + \hat{C} - 2(C\hat{C})^{1/2}). \tag{7}$$

We use d^2 as a perceptual distance metric where m and C is mean and covariance.

4   Experimental Results

Synthetic Dataset. Structured3D dataset [90] provides various textures of indoor scenes with a 512×1024 resolution. We split the dataset into 17468 train, 2183 validation, and 2184 test data. Then we augmented the entire data with three random horizontal shifts then the number of the dataset has quadrupled. In addition, with the corner locations provided in the dataset, we manually

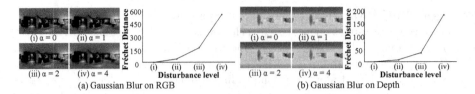

Fig. 5. Verification of FAED in Structured3D dataset. It can be seen that FAED correlates well with perceptual evaluation of humans, as FAED increases as the data becomes more corrupted. For more detailed results, please refer to the suppl. material.

generated layout depth maps of each 3D scene. The residual depth maps are obtained by subtracting the layout depth from the GT depth map. More details about GT layout and residual depth generation can be found in suppl. material.

Real Dataset. We used a combination of two datasets: Matterport3D [9] and 2D-3D-S dataset [3]. Both datasets provide real-world indoor RGB-D panorama captured with real sensors, so depth data in those datasets contain sensor noise or missing holes. However, since this dataset does not provide a sufficient number of annotated layouts, it is only used for test purpose.

Implementation Details. Please refer the suppl. material. for the details.

4.1   Verification of FAED

To show the effectiveness of FAED on measuring the perceptual quality of RGB-D panorama, we corrupt the Structured3D dataset [90] in two ways: corrupting RGB only or corrupting depth only. Following [22], we corrupt the dataset by applying various types of noise: Gaussian blur, Gaussian noise, uniform patches, swirl, and salt and pepper noise. Also, we utilized discrete cosine transform that causes blocking effects to show that our model is sensitive to GAN-like artifacts. Here, we plot the result of Gaussian blur in Fig. 5. Other results can be found in suppl. material. *Note that the evaluation is done for RGB-D panorama, neither for RGB image alone nor for depth map alone.* As shown in Fig. 5, the Fréchet distance for both RGB and depth panorama increases as the disturbance level is increased. We show that the same applies to the other five types of noises in the suppl. material due to the lack of space. *This demonstrates the perceptual quality of RGB-D panorama becomes poorer as the FAED increases.*

Moreover, we calculated FAED of paired and unpaired RGB-D panorama to verify that FAED is effective in considering semantic alignment between RGB and depth panorama. Unpaired RGB-D panorama consists of RGB panorma and randomly selected depth panorama not corresponded to RGB panorama, and its FAED score is 168.0. 3D indoor model from unpaired RGB-D panorama has inconsistent semantic information, *e.g.*, misaligned corner of indoor room and distorted furniture. The visual results of the inconsistency can be found in suppl. material. *Consequently, it indicates that the higher FAED score denotes poorer semantic alignment between RGB and depth panorama.*

Table 1. Quantitative results of RGB panorama synthesis on Structured3D dataset. As [60] needs 4 perspective RGB inputs, we report our results in the same setting. In other cases, 1~4 number of RGB inputs are randomly used. The depth input is not used to compare with image synthesis methods. For FAED calculation, GT full depth is used with synthesized RGB panorama. Best results in **bold**.

Category	Method	Input no. (n)		RGB metric			Layout metric	Proposed metric
		RGB	Depth	PSNR(↑)	SSIM(↑)	LPIPS(↓)	2D Corner error(↓)	FAED(↓)
Inpainting	BRGM [46]	1/2/3/4	0	14.00	0.5310	0.6192	72.52	442.3
	CoModGAN [88]			14.35	0.5837	0.4768	62.45	208.2
	LaMa [61]			13.74	0.5207	0.5658	51.12	379.2
Outpainting	Boundless [32]			13.74	0.5663	0.6144	74.47	429.4
	Ours			**16.21**	**0.6161**	**0.4549**	**39.63**	**162.3**
Panorama syn.	Sumantri *et. al.*[60]	4	0	**18.49**	**0.6680**	0.4190	50.76	443.4
	Ours			17.29	0.6510	**0.3975**	**34.68**	**103.1**

Masked GT RGB Image   Sumantri et. al.   Ours

Fig. 6. Qualitative comparison to Sumantri *et. al.* [60]. While the result from [60] is blurry, our result is sharp and realistic.

4.2 RGB-D Panorama Synthesis

Evaluation on RGB Panorama Synthesis. Table 1 shows the quantitative comparison with the inpainting and outpainting methods on the Structured3D dataset. Our model takes partial RGB inputs and no depth as inputs. We use PSNR, SSIM, and LPIPS to evaluate the quality of RGB panorama. We also measure 2D corner error, where the 2D GT corner points are compared with the estimated 2D corner points using DuLa-Net [77] on the synthesized RGB panorama. We also use the proposed FAED to evaluate the perceptual quality. Here, synthesized RGB panorama and GT depth are used to compute FAED.

As shown in Table 1, our method outperforms the image inpainting and outpainting methods: BRGM [46], CoModGAN [88], LaMa [61], and Boundless [32], by a large margin for all metrics. For instance, our method outperforms the best inpainting method, CoModGAN, by a 4.6% decrease in LPIPS score, 36.5% drop of 2D corner error, and 22% decline of FAED score. The effectiveness can also be visually verified in Fig. 7(a). Our method produces clearer RGB panorama images compared with LaMa producing blurry images. Although CoModGAN produces clear RGB outputs, it does not consider the indoor layout and semantic information of the furniture, *e.g.*, the electric cooker is combined with bookshelves, as shown in Fig. 7. The reason our model has higher performance than the existing SoTA inpainting/outpainting method is that RDAL helps the generator to learn distinguishing features of layout and residual during joint learning. Although the layout and the residual are separated only in the depth image, our joint learning framework induces learning of highly correlated features by

Table 2. Quantitative results of depth panorama synthesis on Structured3D dataset. Depth input type L/P means that we use LiDAR (L) and dense perspective depth sensor (P). The full GT RGB is used with synthesized depth panorama for FAED calculation. Best results in **bold**.

Category	Method	Input type		Depth metric		Layout metric	Proposed metric
		RGB	Depth	AbsREL($\downarrow$)	RMSE($\downarrow$)	2D IoU($\uparrow$)	FAED($\downarrow$)
Depth syn.	CSPN [11]	Full	L/P	0.0855	2214	0.8062	428.9
	NLSPN [51]			0.1268	2807	0.7333	836.1
	MSG-CHN [37]			0.1764	3296	0.6724	896.4
	PENet [24]			0.1740	3145	0.7033	906.0
	Ours			**0.0844**	**1942**	**0.8286**	**131.5**

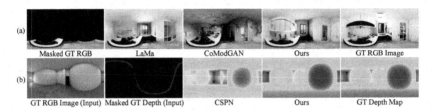

(a)
Masked GT RGB   LaMa   CoModGAN   Ours   GT RGB Image

(b)
GT RGB Image (Input) Masked GT Depth (Input)   CSPN   Ours   GT Depth Map

Fig. 7. (a) Visual results for RGB panorama synthesis on Structured3D dataset. Two methods, LaMa and CoMoGAN, are visualized for comparison. (b) Visual results for depth panorama synthesis on Structured3D dataset. CSPN is also visualized for comparison. More qualitative results can be found in suppl. material.

exchanging information between depth and RGB. Therefore, it is possible to create a very realistic indoor environment panorama even in RGB compared to other models that do not take this into account.

We also compare our model with the panorama synthesis method, Sumantri *et. al.* [60]. Our method shows slightly lower scores using the conventional metrics, PSNR and SSIM; however, it shows a much better LPIPS score, 2D corner error, and FAED score as shown Table 1. We argue that PSNR and SSIM merely measure local photometric similarity and thus fail to reflect the perceptual quality while FAED catches the perceptual quality. This can be visually verified in Fig. 6. Our method synthesizes better textures and shows much more visually plausible output. More visual results can be found in suppl. material.

Evaluation on Depth Panorama Synthesis. We compare our method with the image-guided depth synthesis methods: CSPN [11], NLSPN [51], MSG-CHN [37] and PENet [24]. AbsREL, RMSE, layout 2D IoU [12] and the proposed FAED are used for evaluation. The details of the metrics can be found in the suppl. material. Table 2 shows the quantitative comparison with the depth synthesis methods. In particular, our method outperforms one of the best depth synthesis method, CSPN, with all metrics. We also compared our model with the 360° monocular depth estimation method [67]. The visual result of [67] was not reasonable, and its FAED score was 1140. With the proposed RDAL scheme, our model understands the structure of the indoor scene and learns the relative

Table 3. FAED scores of our model according to the amount of RGB-D inputs.

FAED (The redder the cell, the lower the value)		Number of Depth Input							
		Perspective (num. of NFoVs)				LiDAR (num. of channels)			
		0	1	2	3	4	2	4	8	16
	0	-	2077	746.4	439.8	371.0	1345	1030	631.1	382.6
Number of	1	910.0	695.1	246.6	176.5	152.2	316.6	267.9	210.6	151.7
Perspective	2	461.8	295.4	202.3	152.2	132.7	229.4	207.8	174.6	134.4
RGB Input	3	365.8	233.7	154.0	128.3	107.5	189.9	171.4	141.7	101.4
	4	346.1	214.5	141.7	108.0	91.9	176.4	156.3	127.9	87.2

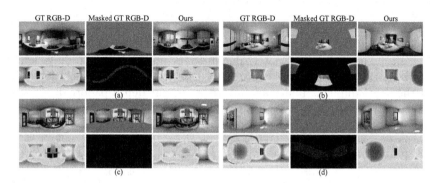

Fig. 8. Visualization of our synthesized RGB-D panorama results from RGB-D data in various configurations. (a) and (b) take both RGB and depth data, (c) takes only RGB, and (d) takes only depth data. More results are visualized in suppl. materials.

depth of interior components. Therefore, our method estimates the best layout depth, which is demonstrated by the highest layout 2D IoU.

Figure 7(b) shows the qualitative comparison with CSPN [11]. CSPN failed to synthesize valid layouts with non-planar output depth map on the walls and ceiling, which incurs unrealistic 3D indoor model. In contrast, our result shows undisturbed, clear layouts. More of these results can be found in suppl. material.

Evaluation on RGB-D Panorama Synthesis. To show the effectiveness of our model quantitatively, we compared our model with 'inpainting with depth synthesis' (IwDS). To be specific, an RGB panorama is first synthesized from partial RGB input using the image inpainting method. Then, depth panorama is synthesized by applying the depth synthesis method to the synthesized RGB panorama and partial depth input. We chose CoModGAN [88] and CSPN [11] for RGB and depth synthesis methods, which showed the highest FAED score in Table 1 and Table 2. In Table 4, it can be seen that IwDS leads lower 2D IoU score and a much higher FAED score than our method. Also, FAED score of IwDS with [60] and [11] was 722.1, even though [60] uses 1∼4 RGB inputs. These indicate that the two-stage, sequential synthesis of RGB-D panorama is less effective than our BIPS framework that fuses the bi-modal features, trained with one-stage, joint learning scheme. Also, IwDS fails to generate realistic 3D indoor models, with distorted indoor layouts and severe bumpy surfaces, as shown in Fig. 10.

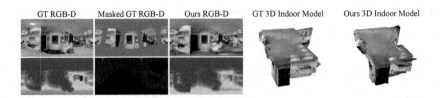

GT RGB-D   Masked GT RGB-D   Ours RGB-D   GT 3D Indoor Model   Ours 3D Indoor Model

Fig. 9. Visual results for RGB panorama synthesis on Matterport3D dataset.

Figure 8 shows the qualitative results of our model generated from the partial RGB and depth inputs, including RGB only or depth only cases. The mutual gain for using RGB and depth information together is visually demonstrated in Fig. 8(a). The upper shelf not visible in RGB input is successfully generated by utilizing the corresponding depth data. Likewise, the lower shelf not visible from the depth input is plausibly created in output depth panorama referring to RGB information. It means that the bi-modal information is exchanged in a bi-directional manner, which enables our model to understand the overall scene. *More results with 3D indoor models are visualized in suppl. materials.*

In Table 3, we quantitatively analyze the performance with FAED regarding the amount of input information. Overall, *using both types of input shows better panorama synthesis quality than that of using a single input.* For example, using 2 RGB and 2 depth inputs shows much better FAED score (202.3) than those of using 4 RGB or 4 depth inputs (FAED scores: 371.0, 346.1). The fusion between textural information from RGB and structural information from depth through BFF enables a more comprehensive understanding of the scene.

RGB-D Panorama Evaluation on Real Dataset. We evaluated our synthesized RGB-D panorama on real indoor scenes in Matterport3D and 2D-3D-S dataset. Figure 9 shows an output RGB-D panorama and its 3D indoor model. Overall, our method synthesizes high-quality RGB-D panorama on real indoor scenes, which are unseen during training. Our synthesized depth panorama shows the precise indoor layout and plausible residuals, generating a realistic 3D indoor model. For quantitative results, our method achieved a much better FAED score than IwDS (1123 vs. 5099). Since the domain of real dataset is different from the domain of synthetic dataset, FAED scores generally increased. More visual results can be found in suppl. material.

4.3   Ablation Study and Analysis

Impact of BFF. We studied the effectiveness of RGB-D panorama synthesis by removing the BFF branch in the generator. In details, G_{BFF} is replaced with a single branch network taking the concatenation of $G_{\text{in}}^{\text{rgb}}(I_{\text{in}}^{\text{rgb}})$ and $G_{\text{in}}^{\text{depth}}(I_{\text{in}}^{\text{d}})$. For fair comparison, we designed a single branch network to be of similar capacity to our final model (23,254 vs. 22,642 MB). As shown in Table 4, the 2D IoU drops and FAED increases without BFF. Figure 10 shows that the texture of the RGB-D output is not consistent with the given RGB-D input. This reflects that BFF

Table 4. Quantitative results of IwDS and ablation study of BIPS framework.

	IwDS ([88] + [11])	Ours w/o BFF	Ours w/o RDAL	Ours
2D IoU (↑)	0.7561	0.7859	0.7164	**0.8158**
FAED (↓)	640.9	381.4	329.0	**198.0**

Input   IwDS   Ours w/o BFF   Ours w/o RDAL   Ours   GT

Fig. 10. Visualization of IwDS and our ablation study results on Structured3D dataset.

encourages the information exchange of bi-modal information and significantly contributes to having minimal artifacts.

Impact of RDAL. We further compared the model without RDAL to validate its effectiveness. In detail, the number of output branches is reduced to two, and each is designed to learn RGB and total depth panorama. As shown in Table 4, the 2D IoU drops and FAED increases without RDAL. It shows that RDAL is critical for estimating precise indoor layouts. The impact of RDAL is visually verified in Fig. 10. The result without RDAL shows a distorted indoor layout while having fewer artifacts than ours without BFF. In summary, jointly learning layout and residual depth helps to synthesize a more structural 3D indoor model.

Analysis on Robustness of BIPS. We conducted various experiments: applying on noisy sensor inputs, comparison on different input data of the same 3D scene, and generalization on unseen input configurations. The results show that the proposed BIPS can synthesize visually pleasing RGB-D panorama under these scenarios, making it directly applicable to real-world applications. The implementation details, results, and discussion are included in suppl. material.

5   Conclusion

We tackled a novel problem of synthesizing RGB-D indoor panoramas from various configurations of RGB and depth inputs. Our method can synthesize high-quality RGB-D panoramas with the proposed BIPS framework by utilizing the BFF and jointly training through RDAL. Extensive experiments show that this bi-modal joint learning enables the generator to effectively understand the structure of indoor scene, so our model achieved the highest performance in indoor RGB-D panorama synthesis than conventional methods. Moreover, a label-free novel image quality assessment metric, FAED, was proposed, and its validity was demonstrated.

Acknowledgements. This work was supported by the National Research Foundation of Korea (NRF) grant funded by the Korea government (MSIT) (NRF2022R1A2B5B03002636). This research was conducted while Wonjune Cho and Lin Wang were with KAIST.

References

1. Alaee, G., Deasi, A.P., Pena-Castillo, L., Brown, E., Meruvia-Pastor, O.: A user study on augmented virtuality using depth sensing cameras for near-range awareness in immersive VR. In: IEEE VR's 4th Workshop on Everyday Virtual Reality (WEVR 2018). vol. 10 (2018)
2. Ali, B.: Pros and cons of Gan evaluation measures. Comput. Vis. Image Underst. **179**, 41–65 (2019)
3. Armeni, I., Sax, S., Zamir, A.R., Savarese, S.: Joint 2d–3d-semantic data for indoor scene understanding. arXiv preprint arXiv:1702.01105 (2017)
4. Avidan, S., Shamir, A.: Seam carving for content-aware image resizing. In: ACM SIGGRAPH 2007 Papers, p. 10. SIGGRAPH 2007, Association for Computing Machinery, New York, NY, USA (2007). https://doi.org/10.1145/1275808.1276390
5. Ballester, C., Bertalmio, M., Caselles, V., Sapiro, G., Verdera, J.: Filling-in by joint interpolation of vector fields and gray levels. IEEE Trans. Image Process. **10**(8), 1200–1211 (2001)
6. Barnes, C., Shechtman, E., Finkelstein, A., Goldman, D.B.: PatchMatch: a randomized correspondence algorithm for structural image editing. ACM Trans. Graph. **28**(3), 24 (2009)
7. Bertalmio, M., Sapiro, G., Caselles, V., Ballester, C.: Image inpainting. In: Proceedings of the 27th Annual Conference on Computer Graphics and Interactive Techniques, pp. 417–424 (2000)
8. Bertalmio, M., Vese, L., Sapiro, G., Osher, S.: Simultaneous structure and texture image inpainting. IEEE Trans. Image Process. **12**(8), 882–889 (2003)
9. Chang, A., et al.: Matterport3d: learning from RGB-D data in indoor environments. In: International Conference on 3D Vision (3DV) (2017)
10. Cheng, X., Wang, P., Chenye, G., Yang, R.: CSPN++: learning context and resource aware convolutional spatial propagation networks for depth completion. In: Proceedings of the AAAI Conference on Artificial Intelligence, vol. 34, 10615–10622 (2020). https://doi.org/10.1609/aaai.v34i07.6635
11. Cheng, X., Wang, P., Yang, R.: Learning depth with convolutional spatial propagation network. IEEE Trans. Pattern Anal. Mach. Intell. **42**(10), 2361–2379 (2020). https://doi.org/10.1109/TPAMI.2019.2947374
12. Choi, D.: 3d room layout estimation beyond the Manhattan world assumption. arXiv preprint arXiv:2009.02857 (2020)
13. Criminisi, A., Perez, P., Toyama, K.: Object removal by exemplar-based inpainting. In: Proceedings of 2003 IEEE Computer Society Conference on Computer Vision and Pattern Recognition, vol. 2, p. 2. IEEE (2003)
14. Criminisi, A., Pérez, P., Toyama, K.: Region filling and object removal by exemplar-based image inpainting. IEEE Trans. Image Process. **13**(9), 1200–1212 (2004)
15. Dowson, D., Landau, B.: The fréchet distance between multivariate normal distributions. J. Multivar. Anal. **12**(3), 450–455 (1982)
16. Efros, A.A., Leung, T.K.: Texture synthesis by non-parametric sampling. In: Proceedings of the seventh IEEE International Conference on Computer Vision. vol. 2, pp. 1033–1038. IEEE (1999)

17. Eldesokey, A., Felsberg, M., Khan, F.S.: Confidence propagation through CNNs for guided sparse depth regression. IEEE Trans. Pattern Anal. Mach. Intell. **42**(10), 2423–2436 (2020). https://doi.org/10.1109/tpami.2019.2929170

18. Guo, D., Feng, J., Zhou, B.: Structure-aware image expansion with global attention. In: SIGGRAPH Asia 2019 Technical Briefs, pp. 13–16. SA 2019, Association for Computing Machinery, New York, NY, USA (2019). https://doi.org/10.1145/3355088.3365161

19. Guo, D., et al.: Spiral generative network for image extrapolation. In: Vedaldi, A., Bischof, H., Brox, T., Frahm, J.-M. (eds.) ECCV 2020. LNCS, vol. 12364, pp. 701–717. Springer, Cham (2020). https://doi.org/10.1007/978-3-030-58529-7_41

20. Hara, T., Harada, T.: Spherical image generation from a single normal field of view image by considering scene symmetry. Proc. AAAI Conf. Artif. Intell. **35**(2), 1513–1521 (2021)

21. He, Y., Ye, Y., Hanhart, P., Xiu, X.: Geometry padding for motion compensated prediction in 360 video coding. In: 2017 Data Compression Conference (DCC), pp. 443–443. IEEE Computer Society (2017)

22. Heusel, M., Ramsauer, H., Unterthiner, T., Nessler, B., Hochreiter, S.: Gans trained by a two time-scale update rule converge to a local Nash equilibrium. In: Advances in Neural Information Processing Systems, pp. 6626–6637 (2017)

23. Hirose, N., Tahara, K.: Depth360: Monocular depth estimation using learnable axisymmetric camera model for spherical camera image. CoRR abs/2110.10415 (2021). https://arxiv.org/abs/2110.10415

24. Hu, M., Wang, S., Li, B., Ning, S., Fan, L., Gong, X.: PeNet: towards precise and efficient image guided depth completion. In: IEEE International Conference on Robotics and Automation, ICRA 2021, Xi'an, China, May 30 - June 5, 2021, pp. 13656–13662. IEEE (2021). https://doi.org/10.1109/ICRA48506.2021.9561035

25. Huang, Z., Fan, J., Cheng, S., Yi, S., Wang, X., Li, H.: HMS-Net: hierarchical multi-scale sparsity-invariant network for sparse depth completion. IEEE Trans. Image Process. **29**, 3429–3441 (2020). https://doi.org/10.1109/TIP.2019.2960589

26. Iizuka, S., Simo-Serra, E., Ishikawa, H.: Globally and locally consistent image completion. ACM Trans. Gr. (ToG) **36**(4), 1–14 (2017)

27. Isola, P., Zhu, J.Y., Zhou, T., Efros, A.A.: Image-to-image translation with conditional adversarial networks. In: Proceedings of the IEEE Conference on Computer Vision and Pattern Recognition, pp. 1125–1134 (2017)

28. Jaritz, M., De Charette, R., Wirbel, E., Perrotton, X., Nashashibi, F.: Sparse and dense data with CNNs: depth completion and semantic segmentation. In: 2018 International Conference on 3D Vision (3DV), pp. 52–60. IEEE (2018)

29. Kaneva, B., Sivic, J., Torralba, A., Avidan, S., Freeman, W.T.: Infinite images: creating and exploring a large photorealistic virtual space. Proc. IEEE **98**(8), 1391–1407 (2010). https://doi.org/10.1109/JPROC.2009.2031133

30. Kasaraneni, S.H., Mishra, A.: Image completion and extrapolation with contextual cycle consistency. In: 2020 IEEE International Conference on Image Processing (ICIP), pp. 1901–1905 (2020). https://doi.org/10.1109/ICIP40778.2020.9191339

31. Kim, K., Yun, Y., Kang, K.W., Kong, K., Lee, S., Kang, S.J.: Painting outside as inside: Edge guided image outpainting via bidirectional rearrangement with progressive step learning. 2021 IEEE Winter Conference on Applications of Computer Vision (WACV) pp. 2121–2129 (2021)

32. Krishnan, D., et al.: Boundless: generative adversarial networks for image extension. In: 2019 IEEE/CVF International Conference on Computer Vision (ICCV), pp. 10520–10529 (2019). https://doi.org/10.1109/ICCV.2019.01062

33. Lee, B., Jeon, H., Im, S., Kweon, I.S.: Depth completion with deep geometry and context guidance. In: 2019 International Conference on Robotics and Automation (ICRA), pp. 3281–3287 (2019). https://doi.org/10.1109/ICRA.2019.8794161
34. Lee, D., Yun, S., Choi, S., Yoo, H., Yang, M.-H., Oh, S.: Unsupervised holistic image generation from key local patches. In: Ferrari, V., Hebert, M., Sminchisescu, C., Weiss, Y. (eds.) ECCV 2018. LNCS, vol. 11209, pp. 21–37. Springer, Cham (2018). https://doi.org/10.1007/978-3-030-01228-1_2
35. Lee, J.K., Yea, J., Park, M.G., Yoon, K.J.: Joint layout estimation and global multi-view registration for indoor reconstruction. In: Proceedings of the IEEE International Conference on Computer Vision, pp. 162–171 (2017)
36. Lee, S., Lee, J., Kim, D., Kim, J.: Deep architecture with cross guidance between single image and sparse lidar data for depth completion. IEEE Access 8, 79801–79810 (2020). https://doi.org/10.1109/ACCESS.2020.2990212
37. Li, A., Yuan, Z., Ling, Y., Chi, W., Zhang, S., Zhang, C.: A multi-scale guided cascade hourglass network for depth completion. In: Proceedings of the IEEE/CVF Winter Conference on Applications of Computer Vision (WACV), March 2020
38. Li, Y., Liu, S., Yang, J., Yang, M.H.: Generative face completion. In: Proceedings of the IEEE Conference On Computer Vision and Pattern Recognition, pp. 3911–3919 (2017)
39. Liao, L., Xiao, J., Wang, Z., Lin, C.W., Satoh, S.: Image inpainting guided by coherence priors of semantics and textures. In: Proceedings of the IEEE/CVF Conference on Computer Vision and Pattern Recognition, pp. 6539–6548 (2021)
40. Liu, G., Reda, F.A., Shih, K.J., Wang, T.-C., Tao, A., Catanzaro, B.: Image inpainting for irregular holes using partial convolutions. In: Ferrari, V., Hebert, M., Sminchisescu, C., Weiss, Y. (eds.) ECCV 2018. LNCS, vol. 11215, pp. 89–105. Springer, Cham (2018). https://doi.org/10.1007/978-3-030-01252-6_6
41. Liu, H., Jiang, B., Xiao, Y., Yang, C.: Coherent semantic attention for image inpainting. In: Proceedings of the IEEE/CVF International Conference on Computer Vision, pp. 4170–4179 (2019)
42. Liu, H., Wan, Z., Huang, W., Song, Y., Han, X., Liao, J.: PD-GAN: Probabilistic diverse Gan for image inpainting. In: Proceedings of the IEEE/CVF Conference on Computer Vision and Pattern Recognition, pp. 9371–9381 (2021)
43. Ma, F., Cavalheiro, G.V., Karaman, S.: Self-supervised sparse-to-dense: Self-supervised depth completion from lidar and monocular camera. In: 2019 International Conference on Robotics and Automation (ICRA), pp. 3288–3295 (2019). https://doi.org/10.1109/ICRA.2019.8793637
44. Mal, F., Karaman, S.: Sparse-to-dense: depth prediction from sparse depth samples and a single image. In: 2018 IEEE International Conference on Robotics and Automation (ICRA), pp. 1–8. IEEE (2018)
45. Mao, X., Li, Q., Xie, H., Lau, R.Y., Wang, Z., Paul Smolley, S.: Least squares generative adversarial networks. In: Proceedings of the IEEE International Conference on Computer Vision, pp. 2794–2802 (2017)
46. Marinescu, R.V., Moyer, D., Golland, P.: Bayesian image reconstruction using deep generative models. In: Advances in Neural Information Processing Systems (2021)
47. Mastan, I.D., Raman, S.: DeepCFL: deep contextual features learning from a single image. In: 2021 IEEE Winter Conference on Applications of Computer Vision (WACV), pp. 2896–2905 (2021)
48. Mittal, A., Moorthy, A.K., Bovik, A.C.: No-reference image quality assessment in the spatial domain. IEEE Trans. Image Process. 21(12), 4695–4708 (2012)
49. Mittal, A., Soundararajan, R., Bovik, A.C.: Making a "completely blind" image quality analyzer. IEEE Signal Process. Lett. 20(3), 209–212 (2012)

50. Navasardyan, S., Ohanyan, M.: Image inpainting with onion convolutions. In: Proceedings of the Asian Conference on Computer Vision (2020)
51. Park, J., Joo, K., Hu, Z., Liu, C.-K., So Kweon, I.: Non-local spatial propagation network for depth completion. In: Vedaldi, A., Bischof, H., Brox, T., Frahm, J.-M. (eds.) ECCV 2020. LNCS, vol. 12358, pp. 120–136. Springer, Cham (2020). https://doi.org/10.1007/978-3-030-58601-0_8
52. Peng, J., Liu, D., Xu, S., Li, H.: Generating diverse structure for image inpainting with hierarchical VQ-VAE. In: Proceedings of the IEEE/CVF Conference on Computer Vision and Pattern Recognition, pp. 10775–10784 (2021)
53. Qiu, J., et al.: DeepLiDAR: deep surface normal guided depth prediction for outdoor scene from sparse lidar data and single color image. In: Proceedings of the IEEE/CVF Conference on Computer Vision and Pattern Recognition (CVPR), June 2019
54. Rosin, P.L., Lai, Y.-K., Shao, L., Liu, Y.: RGB-D Image Analysis and Processing. ACVPR, Springer, Cham (2019). https://doi.org/10.1007/978-3-030-28603-3
55. Sabini, M., Rusak, G.: Painting outside the box: image outpainting with GANs. arXiv:abs/1808.08483 (2018)
56. Salimans, T., Goodfellow, I., Zaremba, W., Cheung, V., Radford, A., Chen, X.: Improved techniques for training GANs. In: Advances in Neural Information Processing Systems, pp. 2234–2242 (2016)
57. Schubert, S., Neubert, P., Pöschmann, J., Pretzel, P.: Circular convolutional neural networks for panoramic images and laser data. In: 2019 IEEE Intelligent Vehicles Symposium (IV), pp. 653–660. IEEE (2019)
58. Shan, Q., Curless, B., Furukawa, Y., Hernandez, C., Seitz, S.M.: Photo Uncrop. In: Fleet, D., Pajdla, T., Schiele, B., Tuytelaars, T. (eds.) ECCV 2014. LNCS, vol. 8694, pp. 16–31. Springer, Cham (2014). https://doi.org/10.1007/978-3-319-10599-4_2
59. Suin, M., Purohit, K., Rajagopalan, A.: Distillation-guided image inpainting. In: Proceedings of the IEEE/CVF International Conference on Computer Vision, pp. 2481–2490 (2021)
60. Sumantri, J.S., Park, I.K.: 360 panorama synthesis from a sparse set of images on a low-power device. IEEE Trans. Comput. Imaging **6**, 1179–1193 (2020). https://doi.org/10.1109/TCI.2020.3011854
61. Suvorov, R., et al.: Resolution-robust large mask inpainting with Fourier convolutions. In: 2022 IEEE Winter Conference on Applications of Computer Vision (WACV) (2022)
62. Szegedy, C., Vanhoucke, V., Ioffe, S., Shlens, J., Wojna, Z.: Rethinking the inception architecture for computer vision. In: CVPR, pp. 2818–2826 (2016)
63. Tang, J., Tian, F.P., Feng, W., Li, J., Tan, P.: Learning guided convolutional network for depth completion. IEEE Trans. Image Process. **30**, 1116–1129 (2020)
64. Van Gansbeke, W., Neven, D., De Brabandere, B., Van Gool, L.: Sparse and noisy lidar completion with RGB guidance and uncertainty. In: 2019 16th International Conference on Machine Vision Applications (MVA), pp. 1–6 (2019). https://doi.org/10.23919/MVA.2019.8757939
65. Vaswani, A., et al.: Attention is all you need. In: Advances in Neural Information Processing Systems, pp. 5998–6008 (2017)
66. Wan, Z., Zhang, J., Chen, D., Liao, J.: High-fidelity pluralistic image completion with transformers. In: 2021 IEEE/CVF International Conference on Computer Vision (ICCV), pp. 4672–4681 (2021)
67. Wang, B., An, J.: FIS-Nets: full-image supervised networks for monocular depth estimation. CoRR abs/2001.11092 (2020). arXiv:2001.11092

68. Wang, F.E., Yeh, Y.H., Sun, M., Chiu, W.C., Tsai, Y.H.: BiFuse: monocular 360 depth estimation via bi-projection fusion. In: CVPR (2020)
69. Wang, M., Lai, Y.K., Liang, Y., Martin, R.R., Hu, S.M.: BiggerPicture: data-driven image extrapolation using graph matching. ACM Trans. Graph. **33**(6), 1–13 (2014). https://doi.org/10.1145/2661229.2661278
70. Wang, N., Li, J., Zhang, L., Du, B.: Musical: Multi-scale image contextual attention learning for inpainting. In: IJCAI, pp. 3748–3754 (2019)
71. Wang, T.C., Liu, M.Y., Zhu, J.Y., Tao, A., Kautz, J., Catanzaro, B.: High-resolution image synthesis and semantic manipulation with conditional GANs. In: Proceedings of the IEEE Conference on Computer Vision and Pattern Recognition, pp. 8798–8807 (2018)
72. Wang, W., Zhang, J., Niu, L., Ling, H., Yang, X., Zhang, L.: Parallel multi-resolution fusion network for image inpainting. In: Proceedings of the IEEE/CVF International Conference on Computer Vision, pp. 14559–14568 (2021)
73. Wang, Y., Tao, X., Shen, X., Jia, J.: Wide-context semantic image extrapolation. In: 2019 IEEE/CVF Conference on Computer Vision and Pattern Recognition (CVPR), pp. 1399–1408 (2019). https://doi.org/10.1109/CVPR.2019.00149
74. Wu, X., et al.: Deep portrait image completion and extrapolation. IEEE Trans. Image Process. **29**, 2344–2355 (2020). https://doi.org/10.1109/tip.2019.2945866
75. Xie, C., et al.: Image inpainting with learnable bidirectional attention maps. In: Proceedings of the IEEE/CVF International Conference on Computer Vision, pp. 8858–8867 (2019)
76. Xu, Y., Zhu, X., Shi, J., Zhang, G., Bao, H., Li, H.: Depth completion from sparse lidar data with depth-normal constraints. In: Proceedings of the IEEE/CVF International Conference on Computer Vision (ICCV), October 2019
77. Yang, S.T., Wang, F.E., Peng, C.H., Wonka, P., Sun, M., Chu, H.K.: DuLA-Net: a dual-projection network for estimating room layouts from a single RGB panorama. In: Proceedings of the IEEE/CVF Conference on Computer Vision and Pattern Recognition, pp. 3363–3372 (2019)
78. Yang, Z., Dong, J., Liu, P., Yang, Y., Yan, S.: Very long natural scenery image prediction by outpainting. In: Proceedings of the IEEE International Conference on Computer Vision, pp. 10561–10570 (2019)
79. Yu, J., Lin, Z., Yang, J., Shen, X., Lu, X., Huang, T.S.: Generative image inpainting with contextual attention. In: Proceedings of the IEEE Conference on Computer Vision and Pattern Recognition, pp. 5505–5514 (2018)
80. Yu, J., Lin, Z., Yang, J., Shen, X., Lu, X., Huang, T.S.: Free-form image inpainting with gated convolution. In: Proceedings of the IEEE/CVF International Conference on Computer Vision, pp. 4471–4480 (2019)
81. Zeng, W., Karaoglu, S., Gevers, T.: Joint 3D layout and depth prediction from a single indoor panorama image. In: Vedaldi, A., Bischof, H., Brox, T., Frahm, J.-M. (eds.) ECCV 2020. LNCS, vol. 12361, pp. 666–682. Springer, Cham (2020). https://doi.org/10.1007/978-3-030-58517-4_39
82. Zhang, L., Wang, J., Shi, J.: Multimodal image outpainting with regularized normalized diversification. In: 2020 IEEE Winter Conference on Applications of Computer Vision (WACV), pp. 3422–3431 (2020). https://doi.org/10.1109/WACV45572.2020.9093636
83. Zhang, R., Isola, P., Efros, A.A., Shechtman, E., Wang, O.: The unreasonable effectiveness of deep features as a perceptual metric. In: Proceedings of the IEEE Conference on Computer Vision and Pattern Recognition, pp. 586–595 (2018)

84. Zhang, X., Chen, F., Wang, C., Tao, M., Jiang, G.P.: SiENet: Siamese expansion network for image extrapolation. IEEE Signal Process. Lett. **27**, 1590-1594 (2020). https://doi.org/10.1109/LSP.2020.3019705

85. Zhang, Y., Funkhouser, T.: Deep depth completion of a single RGB-D image. In: Proceedings of the IEEE Conference on Computer Vision and Pattern Recognition (CVPR), June 2018

86. Zhang, Y., Xiao, J., Hays, J., Tan, P.: FrameBreak: dramatic image extrapolation by guided shift-maps. In: Proceedings of the IEEE Conference on Computer Vision and Pattern Recognition, pp. 1171–1178 (2013)

87. Zhao, L., et al.: UctGAN: diverse image inpainting based on unsupervised cross-space translation. In: Proceedings of the IEEE/CVF Conference on Computer Vision and Pattern Recognition, pp. 5741–5750 (2020)

88. Zhao, S., et al.: Large scale image completion via co-modulated generative adversarial networks. In: 9th International Conference on Learning Representations, ICLR 2021, Virtual Event, Austria, 3–7 May 2021. OpenReview.net (2021). https://openreview.net/forum?id=sSjqmfsk95O

89. Zheng, C., Cham, T.J., Cai, J.: Pluralistic image completion. In: Proceedings of the IEEE Conference on Computer Vision and Pattern Recognition, pp. 1438–1447 (2019)

90. Zheng, J., Zhang, J., Li, J., Tang, R., Gao, S., Zhou, Z.: Structured3D: a large photo-realistic dataset for structured 3d modeling. In: Vedaldi, A., Bischof, H., Brox, T., Frahm, J.-M. (eds.) ECCV 2020. LNCS, vol. 12354, pp. 519–535. Springer, Cham (2020). https://doi.org/10.1007/978-3-030-58545-7_30

Augmentation of rPPG Benchmark Datasets: Learning to Remove and Embed rPPG Signals via Double Cycle Consistent Learning from Unpaired Facial Videos

Cheng-Ju Hsieh, Wei-Hao Chung, and Chiou-Ting Hsu[✉]

National Tsing Hua University, Hsinchu, Taiwan
cthsu@cs.nthu.edu.tw

Abstract. Remote estimation of human physiological condition has attracted urgent attention during the pandemic of COVID-19. In this paper, we focus on the estimation of remote photoplethysmography (rPPG) from facial videos and address the deficiency issues of large-scale benchmarking datasets. We propose an end-to-end RErPPG-Net, including a Removal-Net and an Embedding-Net, to augment existing rPPG benchmark datasets. In the proposed augmentation scenario, the Removal-Net will first erase any inherent rPPG signals in the input video and then the Embedding-Net will embed another PPG signal into the video to generate an augmented video carrying the specified PPG signal. To train the model from unpaired videos, we propose a novel double-cycle consistent constraint to enforce the RErPPG-Net to learn to robustly and accurately remove and embed the delicate rPPG signals. The new benchmark "Aug-rPPG dataset" is augmented from UBFC-rPPG and PURE datasets and includes 5776 videos from 42 subjects with 76 different rPPG signals. Our experimental results show that existing rPPG estimators indeed benefit from the augmented dataset and achieve significant improvement when fine-tuned on the new benchmark. The code and dataset are available at https://github.com/nthumplab/RErPPGNet.

Keywords: Remote photoplethysmography · Data augmentation · Double-cycle consistency · Remote heart rate measurement

1 Introduction

Contactless and video-based methods for heart rate (HR) estimation have attracted enormous research interests. Especially, remote photoplethysmography (rPPG), which analyzes the subtle chrominance changes reflected on skin,

Supplementary Information The online version contains supplementary material available at https://doi.org/10.1007/978-3-031-19787-1_21.

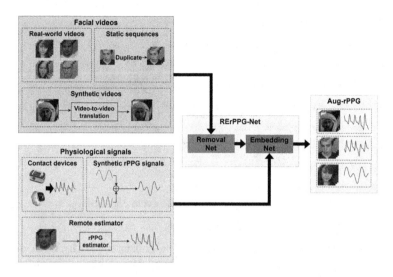

Fig. 1. The proposed scenario of rPPG data augmentation.

captures the heart rate related information [3,4]. Recent learning-based methods for rPPG estimation fall into two categories. The first category [8,10,17,22] involved a number of preprocessing steps, such as facial landmark detection or regions-of-interest detection, to obtain a spatial-temporal representation as the input to the CNN-based model. The second category [2,3,5,13,14,16,23] focused on training an end-to-end architecture to directly estimate either rPPG signals or HR from an input facial video.

There remain a number of challenges in developing robust rPPG or HR estimation. First, since the success of deep learning-based methods heavily relies on large-scale supervised datasets, there are unfortunately only few datasets publicly available for rPPG or HR estimation. In addition, the ground truth labels of these datasets are not always accurate and thus usually lead to unstable estimation. Finally, because of the lack of large-scale dataset, previous methods tend to overfit to a certain dataset but poorly generalize to others.

In this paper, to tackle the aforementioned issues, we propose the RErPPG-Net to augment existing rPPG datasets. As shown in Fig. 1, the RErPPG-Net consists of a Removal-Net and an Embedding-Net. We first use the Removal-Net to erase any possible rPPG-relevant signals in the input video and then use the Embedding-Net to embed a PPG signal into the resultant video.

However, training the Removal-Net and Embedding-Net is highly challenging, because no paired videos (i.e., facial videos from the same subject with and without PPG signals) are available for model training. Inspired by the success of cycle consistency learning [25] from unpaired data, we propose a novel double-cycle consistent constraint into our model training. We use Fig. 2 to illustrate the idea of single cycle consistent and the proposed double-cycle consistent learning. In Fig. 2, when given an input X and two translators T_1 and T_2, the original

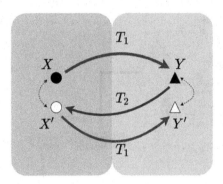

Fig. 2. Illustration of the double-cycle consistency.

single cycle consistency [25] enforces X' to be consistent with X. In our case, because the rPPG signals are extremely delicate in comparison with the facial content, we found this single cycle consistency between X' and X tends to focus on facial contents instead of the rPPG signals. Therefore, we adopt an additional cycle consistent constraint on Y and Y' to ensure that rPPG-related information in X' and X are well preserved. In addition, we also refer to the background signals of the input videos to guide the model training.

Our contributions are summarized below:

1) We propose the RErPPG-Net to generate an augmented rPPG estimation dataset: Aug-rPPG dataset, for public use on research of rPPG estimation.
2) We devise a novel double-cycle consistent constraint into the learning of RErPPG-Net and successfully generate high-quality videos with specified PPG signals.
3) Experimental results on UBFC-rPPG and PURE datasets show that existing rPPG estimators substantially improve the estimation accuracy and achieve state-of-the-art performance when fine-tuned on Aug-rPPG dataset.

2   Related Work

2.1   Remote Photoplethysmography Estimation

The goal of rPPG estimation is to remotely measure the heart rate from facial videos. Traditional approaches [4,7,12,18,20,21] focused on separate physiological signals from facial videos under different prior assumptions. These methods generally perform well on videos recorded under controlled environment but may not generalize well to other scenarios. Many learning-based methods [2,3,5,8,10,13,14,16,17,23] have also been developed for rPPG estimation. In [8], the authors proposed a Dual-GAN framework to learn a noise-resistant mapping from input spatial-temporal maps to ground truth blood volume pulse signals. In [3], the DeepPhys framework was proposed to simultaneously generate an attention mask for RoI detection and to recover rPPG signals using the convolutional attention network. In [23], the authors proposed a STVEN network to enhance hidden rPPG information from highly compressed videos and an rPPGNet to predict rPPG signals.

2.2 Data Augmentation

Data augmentation has been widely adopted to alleviate the shortage of well-labeled training data. Traditional augmentation methods include image flipping, rotating, cropping, scaling, shifting, and so on. With the success of Generative Adversarial Networks (GANs) [6,19,25] and autoencoder [16] in generating high fidelity data, many methods are proposed to use generators to automate the data augmentation. In [6], the authors used conditional GANs to achieve both age progression and regression. In [9], the authors utilized 3D avatars to synthesize facial videos with blood flow and breathing patterns. In [25], the authors proposed an unsupervised method with cycle-consistency to solve image-to-image translation from unpaired data. In [16], the authors proposed a multi-task framework to predict rPPG signals and to augment data simultaneously.

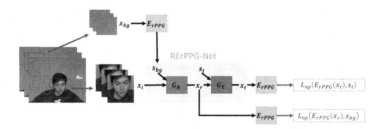

Fig. 3. The proposed RErPPG-Net, which consists of a Removal-Net G_R and an Embedding-Net G_E. The rPPG-removed video x_r is expected to carry no rPPG signal; whereas the rPPG-embedded video x_t is expected to carry the specified signal s_t.

3 Proposed Method

3.1 Overview

In this paper, we propose a RErPPG-Net to augment existing rPPG datasets by embedding ground-truth PPG signals into any existing facial videos. As shown in Fig. 3, the proposed RErPPG-Net consists of a Removal-Net G_R and an Embedding-Net G_E and aims to remove any inherent rPPG signals existing in the input videos and then to embed the specified PPG signals into the rPPG-removed videos. To train the model from unpaired videos, we propose a novel double-cycle consistent learning to enforce the Embedding-Net G_E and the Removal-Net G_R to learn to robustly and accurately embed and remove the delicate rPPG signals.

3.2   RErPPG-Net

Figure 3 illustrates the proposed RErPPG-Net. Let $x_i \in \mathbb{R}^{H \times W \times C \times T}$ be an input facial video; $s_t \in \mathbb{R}^T$ denote the specified PPG signal, and $x_t \in \mathbb{R}^{H \times W \times C \times T}$ denote the generated facial video, where H, W, C, and T denote the height, width, the number of channels, and the length of the video, respectively. E_{rPPG} denotes an off-the-shelf rPPG estimator.

Note that, the input facial video may come from different sources, e.g., broadcast videos, video-to-video translated results, spoof videos, or temporally duplication of static images. Therefore, we first need to ensure that its inherent rPPG signals (if any) are completely erased before we embed another PPG signal. However, because we do not know whether the input video x_i carries rPPG signals or not, training the Removal-Net G_R is not a trivial task. Hence, we assume that the background region of any input video should carry no rPPG information and propose to use the signal estimated from the background region as the pseudo ground truth of "no rPPG signal". As shown in Fig. 3, we crop the upper left corner of each video frame as the reference background and let $s_{bg} \in \mathbb{R}^T$ denote the signal predicted by the rPPG estimator E_{rPPG} from the background region. Given the input x_i, we refer to s_{bg} to remove the rPPG signals by the Removal-Net G_R:

$$x_r = G_R(x_i, s_{bg}), \tag{1}$$

where $x_r \in \mathbb{R}^{H \times W \times C \times T}$.

Next, we embed the specified PPG signal s_t into the rPPG-removed video x_r by the Embedding-Net G_E:

$$x_t = G_E(x_r, s_t). \tag{2}$$

To ensure that the rPPG-removed video x_r carries no rPPG signal and that the rPPG-embedded video x_t carries the signal s_t, we formulate the rPPG loss L_{rPPG}^{re} as:

$$L_{rPPG}^{re} = L_{np}(s_{bg}, E_{rPPG}(x_r)) + L_{np}(s_t, E_{rPPG}(x_t)), \tag{3}$$

in terms of the negative Pearson correlation loss L_{np} defined by

$$L_{np}(s, s') = 1 - \frac{(s - \bar{s})^t(s' - \bar{s'})}{\sqrt{(s - \bar{s})^t(s - \bar{s})}\sqrt{(s' - \bar{s'})^t(s' - \bar{s'})}}, \tag{4}$$

where $s \in \mathbb{R}^T$ and $s' \in \mathbb{R}^T$.

3.3   Double-Cycle Consistent Learning for Embedding-Net

Nevertheless, training RErPPG-Net in terms of only the rPPG loss L_{rPPG}^{re} is far from enough. Specifically, there is no guarantee that the output video x_t is perceptually satisfactory and that x_t carries only the specified signal s_t. Therefore, we devise a double-cycle consistent learning to constrain the Embedding-Net to learn to generate perceptually plausible results embedded with only the specified PPG signals.

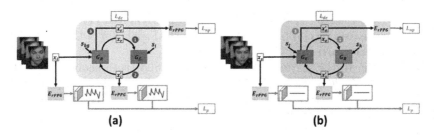

Fig. 4. (a) Double-cycle consistent learning for Embedding-Net G_E; and (b) double-cycle consistent learning for Removal-Net G_R.

As shown in Fig. 4 (a), we illustrate the double-cycle consistent learning with three stages. Here we assume the input video x_i and its ground truth PPG signal s_i are available during this training stage. First, we obtain the rPPG-removed video x_r by Eq. (1). Second, we embed the original PPG signal s_i of x_i back into x_r and obtain x'_i by

$$x'_i = G_E(x_r, s_i), \qquad (5)$$

where $x'_i \in \mathbb{R}^{H \times W \times C \times T}$.

Third, we input x'_i into the Removal-Net G_R to obtain its rPPG-removed video x'_r by

$$x'_r = G_R(x'_i, s_{bg}), \qquad (6)$$

where $x'_r \in \mathbb{R}^{H \times W \times C \times T}$.

From Eqs. (5) and (6), we expect that the rPPG-carrying videos x'_i should be consistent with x_i and that the rPPG-removed videos x'_r should be consistent with x_r. Thus, we formulate the double-cycle consistent loss L_{dc}^{embed} by

$$L_{dc}^{embed} = ||x_i - x'_i||_1 + ||x_r - x'_r||_1. \qquad (7)$$

In addition, to ensure the predicted signals from x'_i and x'_r are highly correlated with s_i and s_{bg}, respectively, we define an rPPG-embedding loss term by:

$$\begin{aligned} L_{rPPG}^{embed} = L_{np}(s_i, E_{rPPG}(x'_i)) \\ + L_{np}(s_{bg}, E_{rPPG}(x'_r)). \end{aligned} \qquad (8)$$

To constrain the perceptual consistency between x'_i and x_i in the feature space, we further include a multi-layer perception loss [24] by

$$L_p^{embed} = \sum_k ||E_{rPPG}^k(x_i) - E_{rPPG}^k(x'_i)||_1, \qquad (9)$$

where $E_{rPPG}^k(\cdot)$ is the feature of the k-th layer of the rPPG estimator E_{rPPG}.

3.4   Double-Cycle Consistent Learning for Removal-Net

To train the Removal-Net G_R, the major difficulty lies in the lack of a paired video without rPPG signals. Therefore, we randomly select one frame from the input video x_i and temporally duplicate this frame to create a static video $x_s \in \mathbb{R}^{H \times W \times C \times T}$ as the reference ground-truth of rPPG-removed video. In addition, because there exists no chrominance change on the facial skin of the static video x_s, we assume x_s carries only a flat DC signal s_h.

By referring to the static video x_s and its ground truth signal s_h, we then devise a double-cycle consistent learning to train the Removal-Net G_R. As shown in Fig. 4 (b), we illustrate the training with three stages. First, we embed the PPG signal s_t into the static video x_s by

$$x_e = G_E(x_s, s_t), \tag{10}$$

where $x_e \in \mathbb{R}^{H \times W \times C \times T}$.

In the second and third stages, similarly to the case in Sect. 3.3, we remove the rPPG signal from the rPPG-embedded video x_e to obtain x'_s and then again embed s_t back to x'_s to obtain an embedded video x'_e by:

$$x'_s = G_R(x_e, s_h), \tag{11}$$

and

$$x'_e = G_E(x'_s, s_t), \tag{12}$$

respectively.

Similar to Eq. (7), we again impose the double-cycle consistent constraints between x'_s and x_s and between x'_e and x_e and define the loss by

$$L_{dc}^{remove} = ||x_s - x'_s||_1 + ||x_e - x'_e||_1. \tag{13}$$

To ensure that x_e and x'_e indeed carry the signal s_t and that x'_s carries a flat signal, we define the rPPG-removal loss by

$$\begin{aligned} L_{rPPG}^{remove} = & L_{np}(s_t, E_{rPPG}(x_e)) \\ & + L_{np}(s_t, E_{rPPG}(x'_e)) \\ & + L_{var}(E_{rPPG}(x'_s)), \end{aligned} \tag{14}$$

where we use the signal variance L_{var} to measure whether x'_s carries a DC signal or not, because the negative Pearson correlation coefficient is inapplicable in this case. L_{var} is defined by

$$L_{var}(s) = (s - \bar{s})^t (s - \bar{s}). \tag{15}$$

Here, we again adopt the multi-layer perception loss to ensure the consistency between x'_s and x_s in the feature space by

$$L_p^{remove} = \sum_k ||E_{rPPG}^k(x_s) - E_{rPPG}^k(x'_s)||_1. \tag{16}$$

3.5   Loss Function

Finally, we include the rPPG loss, the double-cycle consistent loss, and the perceptual loss to define the total loss for training G_E and G_R by:

$$L_{total} = \lambda_1 L_{rPPG} + L_{dc} + L_p, \tag{17}$$

where

$$L_{rPPG} = L_{rPPG}^{re} + L_{rPPG}^{embed} + L_{rPPG}^{remove}, \tag{18}$$

$$L_{dc} = L_{dc}^{embed} + L_{dc}^{remove}, \text{and} \tag{19}$$

$$L_p = L_p^{embed} + L_p^{remove}, \tag{20}$$

and λ_1 is a hyper-parameter and is empirically set as 0.01 in all our experiments.

4   Experiments

4.1   Datasets

The UBFC-rPPG dataset [1] contains 42 RGB videos, each is recorded from a single individual. All the videos are recorded by Logitech C920 HD Pro with resolution of 640 × 480 pixels in uncompressed 8-bit format and 30 fps. CMS50E transmissive pulse oximeter is used to monitor the PPG signals and corresponding heart rates.

Because there is no pre-defined data split for training and testing on UBFC-rPPG dataset, previous methods did not all follow the same setting for evaluation. In [8,10,13], the training and the testing sets contain the first 30 subjects and the rest 12 subjects, respectively. In [16], the training and testing sets contain 28 and 14 subjects, respectively. In [2,5], no description about the data split is given. In our experiment, to have a balanced rPPG distribution within the training and testing sets, we include 35 subjects and the rest 7 subjects in the training and testing sets. In addition, to have a fair comparison with [8,10,13], we also conduct experiments using their setting with 30 and 12 subjects in training/testing sets. More detailed description and results are given in Sect. 4.5.

The PURE dataset [15] contains 10 subjects performing six different and controlled head motions in front of the camera. The six setups include: (1) sitting still, (2) talking, (3) slowly moving the head, (4) quickly moving the head, (5) rotating the head with 20° angles, and (6) rotating the head with 35° angles. All the videos are recorded by evo274CVGE camera with resolution of 640 × 480 pixels and 30 fps. Pulox CMS50E finger clip pulse oximeter is adopted to capture PPG signals with sampling rate 60 Hz. The PPG signals are reduced to 30 fps with linear interpolation to align with the videos. We follow [16] to split the dataset into the training and testing sets with videos from 7 and 3 subjects, respectively.

The VIPL-HR dataset [11] contains 2378 RGB videos of 107 subjects captured with 9 scenarios and recorded by 3 different devices. Because the sampling rates between videos and PPG signals are different, similar to [8], we additionally resample the PPG signals to the corresponding video frame rates by cubic spline interpolation.

4.2   Implementation Details

We develop the proposed Removal-Net G_R and Embedding-Net G_E using the generator proposed in [6]. As to the rPPG estimator E_{rPPG}, we adopt the rPPG model in [16] and then follow [23] to aggregate the features in the middle layer to predict the rPPG signals. We train the RErPPG-Net (i.e., G_R and G_E) and the rPPG estimator E_{rPPG} with Nvidia RTX 2080 and RTX 3080 for 900 and 500 epochs, respectively, and use Adam optimizer with the learning rate of 0.001. The RErPPG-Net is trained with batch size 1 and E_{rPPG} is trained with batch size 3. In each epoch, we randomly sample 60 consecutive frames from each training video to train RErPPG-Net and E_{rPPG}.

4.3   Evaluation Metrics

To assess how the proposed data augmentation improves the rPPG estimation, we follow [13] to derive heart rate (HR) from the predicted rPPG signals and then evaluate the results in terms of the following metrics: (1) Mean absolute error (MAE), (2) Root mean square error (RMSE), (3) Pearson correlation coefficient (R), (4) Peak signal-to-noise ratio (PSNR), and (5) Structural similarity (SSIM).

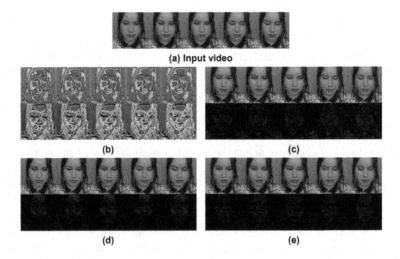

Fig. 5. Visualized examples of ablation study on UBFC-rPPG dataset, when training G_R and G_E using: (b) L_{rPPG}^{re}; (c) L_{rPPG}^{re} with double-cycle consistent loss and rPPG loss on Embedding-Net; (d) L_{rPPG}^{re} with double-cycle consistent loss and rPPG loss on Embedding-Net and Removal-Net; and (e) the proposed total loss.

Table 1. Ablation study on UBFC-rPPG dataset.

L^{re}_{rPPG}	L^{embed}_{dc}	L^{remove}_{dc}	L^{embed}_{rPPG}	L^{remove}_{rPPG}	L_p	MAE↓	RMSE↓	PSNR↑	SSIM↑
✓						44.79	48.53	5.11	0.1642
✓	✓		✓			4.14	9.51	49.72	0.9995
✓	✓	✓	✓	✓		2.30	5.63	51.08	0.9997
✓	✓	✓	✓	✓	✓	**0.71**	**1.48**	**52.71**	**0.9998**

4.4   Ablation Study

We conduct several ablation studies on UBFC-rPPG dataset and show the results in Table 1 and Fig. 5. First, to evaluate how the rPPG estimator E_{rPPG} may benefit from the proposed RErPPG-Net, we train E_{rPPG} using the augmented data from UBFC-rPPG training set and then test on the UBFC-rPPG testing set. Next, to evaluate the perceptual quality of the augmented videos, we embed the original PPG signals into the rPPG-removed videos and then measure the PSNR and SSIM between the augmented videos and the input videos.

In Table 1, the first column L^{re}_{rPPG} indicates that the RErPPG-Net is trained using only the rPPG loss L^{re}_{rPPG} but without the double-cycle consistent learning or the perceptual loss. We first evaluate the effectiveness of the double-cycle consistent loss. When including the loss terms L^{embed}_{dc} and L^{embed}_{rPPG}, we significantly reduce MAE by about 91% and RMSE by about 80%. As shown in Fig. 5 (b) and (c), the visual quality of Fig. 5 (c) is greatly improved with PSNR increased from 5.11 to 49.72 and SSIM from 0.1642 to 0.9995. This improvement comes from that the double-cycle consistent loss on the Embedding-Net effectively constrains the generated video to be consistent with the input on the pixel level. Next, when including the loss terms L^{remove}_{dc} and L^{remove}_{rPPG}, we further reduce MAE by about 44% and RMSE by about 41% and also increase PSNR and SSIM with 1.36 and 0.0002, respectively. These results again verify that double-cycle consistent learning effectively constrains the RErPPG-Net to learn to remove and embed rPPG signals as well as to generate photo-realistic facial videos. Finally, when including the perception loss L_p in the training stage, we enforce the augmented videos to be consistent with the input videos in the feature space and achieve the best performance among all settings.

In Fig. 5, we show the visual comparisons of videos generated by RErPPG-Net trained using different losses. In each setting, the first row shows the augmented video and the second row shows the residual between the augmented video and the input video. The residual results are intensity enhanced by gamma transformed with $\gamma = 3$ to highlight the differential areas. As shown in Fig. 5 (c), (d), and (e), the major differences in the residual videos locate around the face area; these results show that RErPPG-Net indeed learns to erase and embed the rPPG information on the facial regions.

In Table 2, we compare the performance of using single-cycle and double-cycle consistent learning. To train the single-cycle framework, we remove all the loss terms related to x'_r and x'_e in Eqs. (7), (8), (13) and (14). The results show

Table 2. Comparison of single-cycle and double-cycle framework on UBFC-rPPG dataset.

Method	MAE↓	RMSE↓	R↑
Ours (single-cycle)	2.38	4.73	0.84
Ours (double-cycle)	0.71	1.48	0.96

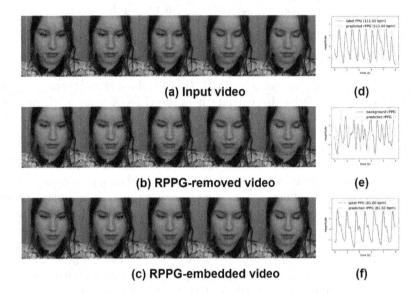

(a) Input video   (d)

(b) RPPG-removed video   (e)

(c) RPPG-embedded video   (f)

Fig. 6. (a) An input video x_i from UBFC-rPPG dataset; (b) The rPPG-removed video x_r; (c) The rPPG-embedded video x_t; (d) The ground truth PPG signal s_i (blue) and the predicted rPPG signal of x_i (orange); (e) The background signal s_{bg} (blue) and the predicted signal of x_r (orange); and (f) The specified PPG signal s_t (blue) and the predicted rPPG signal of x_t (orange). (Color figure online)

that double-cycle consistent learning significantly reduces MAE by about 70% and RMSE by about 69% over the single-cycle consistent constraint and verify its effectiveness for generating photo-realistic augmented videos.

Finally, in Fig. 6, we give an example to visualize the rPPG-removed and rPPG-embedded videos. As shown in Fig. 6 (b) and (c), both videos are visually indistinguishable from the input one. Moreover, in Fig. 6 (e) and (f), the estimated rPPG signals from the rPPG-removed and rPPG-embedded videos are highly correlated with the background signal and the ground truth signal, respectively. These results verify that the proposed RErPPG-Net successfully erases the rPPG signal from the input video and embeds the specified PPG signal into the rPPG-removed video.

Table 3. Comparison on UBFC-rPPG dataset.

Method	MAE↓	RMSE↓	R↑
Meta-rPPG [5]	5.97	7.42	0.53
SynRhythm [10]	5.59	6.82	0.72
3D CNN [2]	5.45	8.64	-
PulseGAN [13]	1.19	2.10	0.98
Multi-task [16]	0.47	2.09	-
Dual-GAN [8]	0.44	0.67	0.99
rPPGNet [23]	0.72	1.47	0.96
rPPGNet (All) [23]	0.56	0.73	0.991
Ours* (Original)	0.64	0.95	0.94
Ours* (Aug)	1.84	3.81	0.85
Ours* (All)	0.75	1.05	0.94
Ours (Original)	0.66	1.40	0.96
Ours (Aug)	0.71	1.48	0.96
Ours (All)	**0.41**	**0.56**	**0.994**

4.5 Results and Comparison

In Table 3, we compare the HR predictions using our rPPG estimator E_{rPPG} with other methods [2,5,8,10,13,16,23] on the UBFC-rPPG dataset. The settings "Original", "Aug", and "All" indicate that we train E_{rPPG} using (1) only the original training videos in UBFC-rPPG dataset, (2) only the augmented videos, and (3) both the original training data and the augmented videos. The methods: 3D CNN [2], Meta-rPPG [5], Multi-task [16], and rPPGNet [23] are developed with the end-to-end architecture; whereas the other three methods [8,10,13] need to compute the spatial-temporal maps before using convolutional neural networks. The result of "Ours (Aug)" shows that E_{rPPG}, even only trained with augmented data, performs pretty well. The setting "Ours (All)" achieves the best performance with MAE 0.41, RMSE 0.56, and Pearson correlation coefficient (R) 0.994.

To have a fair comparison with [8,10,13], we follow the setting in [13] to train our RErPPG-Net and rPPG estimator E_{rPPG} and mark the results with "*". Although we observe degraded performance in "Ours* (Aug)" and "Ours* (All)", we believe the reason comes from the imbalanced HR distribution between the training and testing sets adopted in [13]. As shown in Fig. 7 (a), the HR distributions between the training and testing sets are very different; therefore, this imbalanced issue may grow even worse in the augmented data. On the other hand, in Fig. 7 (b), the data distribution in our setting is more balanced and thus the augmented data serve a better training set.

Furthermore, to show that our augmented dataset can also boost the performance of other rPPG estimators, we re-implement the rPPGNet [23] and show

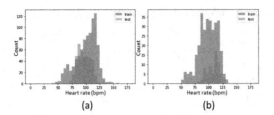

Fig. 7. Heart rate distribution of the training/testing data split in UBFC-rPPG dataset using (a) the setting in [13] and (b) our setting.

the results in Table 3. In comparison with "rPPGNet", the setting "rPPGNet (All)" has MAE reduced by about 22% and RMSE reduced by about 50% when training with both the original training data and the augmented videos. These results verify that existing rPPG estimators can indeed benefit from the proposed augmented dataset.

Comparisons with the methods [4,7,8,14,16,21] on PURE dataset are shown in Table 4, where 2SR [21], CHROME [4], and LiCVPR [7] are not learning-based methods. The setting "Ours (All)" again outperforms the other rPPG estimation methods with MAE 0.38 and RMSE 0.54.

Table 4. Comparison on PURE dataset.

Method	MAE↓	RMSE↓	R↑
LiCVPR [7]	28.22	30.96	−0.38
2SR [21]	2.44	3.06	0.98
CHROME [4]	2.07	2.50	**0.99**
HR-CNN [14]	1.84	2.37	0.98
Dual-GAN [8]	0.82	1.31	**0.99**
Multi-task [16]	0.40	1.07	0.92
Ours (Original)	0.69	1.24	0.96
Ours (All)	**0.38**	**0.54**	0.96

Finally, we conduct a cross-dataset validation to evaluate the generalization capability of the rPPG estimator when fine-tuning E_{rPPG} on our augmented videos. In Table 5, the settings "UBFC", "UBFC + Aug-U", "PURE", "PURE + Aug-P", "UBFC + PURE", and "UBFC + PURE + Aug-rPPG" indicate that the rPPG estimator E_{rPPG} is obtained by (1) training on UBFC-rPPG dataset, (2) training on the UBFC-rPPG dataset and then fine-tuning with the augmented videos of UBFC-rPPG dataset, (3) training on PURE dataset, (4) training on PURE dataset and then fine-tuning with the augmented videos of PURE dataset, (5) training on UBFC-rPPG and PURE datasets, and (6)

Table 5. Comparison of cross-dataset testing.

Training	Testing	MAE↓	RMSE↓	R↑
UBFC	PURE	14.18	20.20	0.34
UBFC+Aug-U	PURE	4.36	6.69	0.60
PURE	UBFC	3.78	6.69	0.71
PURE+Aug-P	UBFC	2.23	4.66	0.78
UBFC+PURE	VIPL-HR	28.94	33.73	0.18
UBFC+PURE+Aug-rPPG	VIPL-HR	25.40	31.14	0.15

training on both UBFC-rPPG and PURE dataset and then fine-tuning with our Aug-rPPG dataset. When involving the augmented data in the training stage, the results show that we reduce the MAE by about 69%, 41%, and 12% and RMSE by about 67%, 30%, and 8% in cross-dataset testing on PURE, UBFC-rPPG, and VIPL-HR datasets, respectively, Note that, because VIPL-HR is a much larger dataset and includes various head movements and illumination conditions, the cross-dataset testing on VIPL-HR usually results in poorer performance when training on small-scale datasets (e.g. UBFC-rPPG and PURE). Nevertheless, when fine-tuning E_{rPPG} on the proposed Aug-rPPG, we show that the cross-dataset testing on VIPL-HR indeed benefits from the proposed augmented data and reaches a better generalization capability.

4.6   Aug-rPPG Dataset

To generate the Aug-rPPG dataset, we use all the 76 training videos and the corresponding PPG signals from UBFC-rPPG and PURE datasets as the inputs to the proposed RErPPG-Net. The 76 input videos are from 42 subjects, where 35 subjects are from UBFC-rPPG training set and 7 subjects are from PURE training set. By running every possible combination of the videos and PPG signals, we generate $76^2 = 5776$ videos of resolution 200×200 pixels. Note that, because we only include the facial region of 200×200 pixels in the data augmentation, our generated videos are of the same quality as the two benchmark datasets.

5   Conclusion

In this paper, we propose the RErPPG-Net to augment existing rPPG benchmark datasets. The proposed RErPPG-Net includes (1) a Removal-Net to erase any inherent rPPG signals in facial videos and (2) an Embedding-Net to embed the specified PPG signals into the videos. To train the model from unpaired videos, we propose a novel double-cycle consistent constraint to enforce the Embedding-Net and the Removal-Net to learn to robustly and accurately embed and remove the delicate rPPG signals. The Aug-rPPG dataset is augmented from

UBFC-rPPG and PURE datasets and includes 5776 videos with the same resolution as the original datasets. Experimental results on UBFC-rPPG, PURE, and VIPL-HR datasets verify the effectiveness of the proposed RErPPG-Net and also show that the augmented data indeed improve the estimation accuracy and the generalization capability of existing rPPG estimators.

References

1. Bobbia, S., Macwan, R., Benezeth, Y., Mansouri, A., Dubois, J.: Unsupervised skin tissue segmentation for remote photoplethysmography. Pattern Recogn. Lett. **124**, 82–90 (2019)
2. Bousefsaf, F., Pruski, A., Maaoui, C.: 3d convolutional neural networks for remote pulse rate measurement and mapping from facial video. Appl. Sci. **9**(20), 4364 (2019)
3. Chen, W., McDuff, D.: DeepPhys: video-based physiological measurement using convolutional attention networks. In: Ferrari, V., Hebert, M., Sminchisescu, C., Weiss, Y. (eds.) ECCV 2018. LNCS, vol. 11206, pp. 356–373. Springer, Cham (2018). https://doi.org/10.1007/978-3-030-01216-8_22
4. De Haan, G., Jeanne, V.: Robust pulse rate from chrominance-based rPPG. IEEE Trans. Biomed. Eng. **60**(10), 2878–2886 (2013)
5. Lee, E., Chen, E., Lee, C.-Y.: Meta-rPPG: remote heart rate estimation using a transductive meta-learner. In: Vedaldi, A., Bischof, H., Brox, T., Frahm, J.-M. (eds.) ECCV 2020. LNCS, vol. 12372, pp. 392–409. Springer, Cham (2020). https://doi.org/10.1007/978-3-030-58583-9_24
6. Li, Q., Liu, Y., Sun, Z.: Age progression and regression with spatial attention modules. In: Proceedings of the AAAI Conference on Artificial Intelligence, pp. 11378–11385 (2020)
7. Li, X., Chen, J., Zhao, G., Pietikainen, M.: Remote heart rate measurement from face videos under realistic situations. In: Proceedings of the IEEE conference on computer vision and pattern recognition. pp. 4264–4271 (2014)
8. Lu, H., Han, H., Zhou, S.K.: Dual-GAN: joint BVP and noise modeling for remote physiological measurement. In: Proceedings of the IEEE/CVF Conference on Computer Vision and Pattern Recognition, pp. 12404–12413 (2021)
9. McDuff, D., Liu, X., Hernandez, J., Wood, E., Baltrusaitis, T.: Synthetic data for multi-parameter camera-based physiological sensing. In: 2021 43rd Annual International Conference of the IEEE Engineering in Medicine and Biology Society (EMBC), pp. 3742–3748. IEEE (2021)
10. Niu, X., Han, H., Shan, S., Chen, X.: SynRHythm: learning a deep heart rate estimator from general to specific. In: 2018 24th International Conference on Pattern Recognition (ICPR), pp. 3580–3585. IEEE (2018)
11. Niu, X., Han, H., Shan, S., Chen, X.: VIPL-HR: a multi-modal database for pulse estimation from less-constrained face video. In: Jawahar, C.V., Li, H., Mori, G., Schindler, K. (eds.) ACCV 2018. LNCS, vol. 11365, pp. 562–576. Springer, Cham (2019). https://doi.org/10.1007/978-3-030-20873-8_36
12. Poh, M.Z., McDuff, D.J., Picard, R.W.: Advancements in noncontact, multiparameter physiological measurements using a webcam. IEEE Trans. Biomed. Eng. **58**(1), 7–11 (2010)
13. Song, R., Chen, H., Cheng, J., Li, C., Liu, Y., Chen, X.: PulseGAN: learning to generate realistic pulse waveforms in remote photoplethysmography. IEEE J. Biomed. Health Inform. **25**(5), 1373–1384 (2021)

14. Špetlík, R., Franc, V., Matas, J.: Visual heart rate estimation with convolutional neural network. In: Proceedings of the British machine vision conference, Newcastle, UK, pp. 3–6 (2018)
15. Stricker, R., Müller, S., Gross, H.M.: Non-contact video-based pulse rate measurement on a mobile service robot. In: The 23rd IEEE International Symposium on Robot and Human Interactive Communication, pp. 1056–1062. IEEE (2014)
16. Tsou, Y.Y., Lee, Y.A., Hsu, C.T.: Multi-task learning for simultaneous video generation and remote photoplethysmography estimation. In: Proceedings of the Asian Conference on Computer Vision (2020)
17. Tsou, Y.Y., Lee, Y.A., Hsu, C.T., Chang, S.H.: Siamese-rPPG network: remote photoplethysmography signal estimation from face videos. In: Proceedings of the 35th Annual ACM Symposium on Applied Computing, pp. 2066–2073 (2020)
18. Verkruysse, W., Svaasand, L.O., Nelson, J.S.: Remote plethysmographic imaging using ambient light. Opt. Express **16**(26), 21434–21445 (2008)
19. Wang, T.C., et al.: Video-to-video synthesis. arXiv preprint arXiv:1808.06601 (2018)
20. Wang, W., den Brinker, A.C., Stuijk, S., De Haan, G.: Algorithmic principles of remote PPG. IEEE Trans. Biomed. Eng. **64**(7), 1479–1491 (2016)
21. Wang, W., Stuijk, S., De Haan, G.: A novel algorithm for remote photoplethysmography: spatial subspace rotation. IEEE Trans. Biomed. Eng. **63**(9), 1974–1984 (2015)
22. Wang, Z.K., Kao, Y., Hsu, C.T.: Vision-based heart rate estimation via a two-stream CNN. In: 2019 IEEE International Conference on Image Processing (ICIP), pp. 3327–3331. IEEE (2019)
23. Yu, Z., Peng, W., Li, X., Hong, X., Zhao, G.: Remote heart rate measurement from highly compressed facial videos: an end-to-end deep learning solution with video enhancement. In: Proceedings of the IEEE/CVF International Conference on Computer Vision, pp. 151–160 (2019)
24. Zhang, X., Ng, R., Chen, Q.: Single image reflection separation with perceptual losses. In: Proceedings of the IEEE Conference on Computer Vision and Pattern Recognition, pp. 4786–4794 (2018)
25. Zhu, J.Y., Park, T., Isola, P., Efros, A.A.: Unpaired image-to-image translation using cycle-consistent adversarial networks. In: Proceedings of the IEEE International Conference on Computer Vision, pp. 2223–2232 (2017)

Geometry-Aware Single-Image Full-Body Human Relighting

Chaonan Ji[1], Tao Yu[1], Kaiwen Guo[2], Jingxin Liu[3], and Yebin Liu[1(✉)]

[1] Department of Automation, Tsinghua University, Beijing, China
liuyebin@mail.tsinghua.edu.cn
[2] Meta Reality Labs, Redmond, USA
[3] Guangdong OPPO Mobile Telecommunications Corporation Limited,
Dongguan, China

Abstract. Single-image human relighting aims to relight a target human under new lighting conditions by decomposing the input image into albedo, shape and lighting. Although plausible relighting results can be achieved, previous methods suffer from both the entanglement between albedo and lighting and the lack of hard shadows, which significantly decrease the realism. To tackle these two problems, we propose a geometry-aware single-image human relighting framework that leverages single-image geometry reconstruction for joint deployment of traditional graphics rendering and neural rendering techniques. For the de-lighting, we explore the shortcomings of UNet architecture and propose a modified HRNet, achieving better disentanglement between albedo and lighting. For the relighting, we introduce a ray tracing-based per-pixel lighting representation that explicitly models high-frequency shadows and propose a learning-based shading refinement module to restore realistic shadows (including hard cast shadows) from the ray-traced shading maps. Our framework is able to generate photo-realistic high-frequency shadows such as cast shadows under challenging lighting conditions. Extensive experiments demonstrate that our proposed method outperforms previous methods on both synthetic and real images.

Keywords: Single-image human relighting · Single-image geometry reconstruction · Ray tracing · Neural rendering

1 Introduction

Human relighting aims to relight a target human with correct shadow effects under a desired illumination. Relighting can realize seamless replacement of the background while ensuring the light consistency of the foreground and the background and has a huge application prospect in film-making, video chat and

Supplementary Information The online version contains supplementary material available at https://doi.org/10.1007/978-3-031-19787-1_22.

Input   Albedo   Relit images   Relit images (Rotate lighting)

Fig. 1. Given a single human image and an arbitrary high dynamic range lighting environment, our framework estimates a high-quality albedo map and generates photorealistic relit images of the target human subject under the desired lighting conditions.

Virtual Reality. Single-image human relighting for a general human person is convenient and promising for amateur photographers but also more challenging because the difficulty of decoupling lighting, geometry, and surface material from a single image.

Most existing single-image human relighting methods focus on portrait relighting [5,16,18,22,32,33,37,38,40,43,44,46,47,50–52,58,59] and only a few works [11,25,28] focus on single-image full-body relighting. Since the style, texture and material of human clothes are widely varied and the geometry and poses of clothed humans are usually complex, decoupling the albedo and lighting is highly ill-posed. Moreover, mutual occlusion of limbs will produce self-shadows, which are not only difficult to remove for de-lighting tasks, but also challenging to generate under the target lighting conditions for relighting tasks. Previous full-body relighting methods [11,25,28] have attempted to address these issues, but the results still have the following drawbacks: (i) inability to disambiguate albedo and lighting (ii) inability to model high-frequency shadows due to reliance on Spherical harmonics representation of lighting.

We propose a novel geometry-aware single-image human relighting framework that leverages SOTA single-image geometry reconstruction [19,42] for fully leveraging the traditional graphics rendering and neural rendering techniques simultaneously. First of all, accurate intrinsic decomposition for albedo estimation (de-lighting) is the cornerstone of producing high-quality relighting results. Previous methods [11,25,28] have used the UNet to infer albedo and suffer from severe entanglement between lighting and albedo. We discover that the skip connections in UNet is the culprit and found that HRNet [49] performs much better for de-lighting. In addition we further improve the de-lighting performance by: i) removing the aggressive down-sampling operations in the early stage, ii) eliminating skip connections and iii) fusing multi-scale features while maintaining high-resolution representations in the HRNet, and finally achieve better disentanglement between lighting and albedo on both synthetic and real images.

More importantly, even with high-quality albedo, we still need photo-realistic shadows to produce realistic relit results. Limited by the expressive ability of the spherical harmonics (SH) lighting model and the lack of 3D geometry information, previous methods [11,25,28] can only produce low-frequency shadows.

Low-resolution environment maps [50,52] and pre-filtered environment maps [40] have also been explored and struggle to generate hard cast shadows. With the development of single-image 3D human reconstruction technology, obtaining high-quality mannequins from a single image is possible, which can provide more 3D prior information for human relighting. Hence, we propose a geometry-aware relighting method, which consists of a ray tracing-based per-pixel lighting representation that explicitly models high-frequency shadows and a learning-based shading refinement module to restore realistic shadows. The key idea of our method is that with the estimated 3D human model, we can render photo-realistic shading maps that encode full-band lighting information under target lighting conditions using ray tracing. This lighting representation is able to preserve high-frequency shadows and significantly improve the quality of the relit results which remains difficult for learning-based methods. However, the ray-traced shadows still suffer from artifacts due to the errors in geometry estimation, which prevents direct composition of the relit results. Thus, we further propose a learning-based refinement module that utilizes the ray-traced shading maps and the inferred ambient occlusion map as input to restore photo-realistic high-frequency shadows with rich local details and clear shadow boundaries. The proposed framework dramatically improves relighting performance and produces a photo-realistic relit image with high-frequency self-shadowing effects under the target lighting conditions.

To conclude, our main contributions are the following:

- A novel geometry-aware single-image human relighting framework that combines single-image 3D human reconstruction, ray tracing and neural rendering technologies.
- We demonstrate the UNet with skip connections is not suitable for the delighting task and propose a modified HRNet for achieving better disentanglement between lighting and albedo on both synthetic images and real photographs.
- We propose a ray tracing-based per-pixel lighting representation and a learning-based shading refinement module that utilizes the inferred ambient occlusion map as auxiliary input to restore photo-realistic shadows while preserving rich local details.

2   Related Work

2.1   Person-Specific Human Relighting

Turner Whitted [55] first used recursive ray tracing to simulate global illumination and render realistic images under target lighting conditions. However, the forward rendering technique requires a high-precision 3d model and corresponding PBR textures of the target object, which are unavailable or costly for real-world objects. Debevec et al. [13] proposed collecting OLAT (one-look-at-a-time) images using LightStage to synthesize specific person's face under novel illuminations with no 3D model required. Subsequent works improved LightStage to capture higher-quality images and a larger range of human bodies [8,12,14,20,21,53,54,57]. Guo et al. [20] attempted to combine image-based rendering technique with geometry and material esrimated by LightStage and achieved unprecedented quality and photorealism for free viewpoint videos. However, collecting OLAT data, gradient data and object geometry requires professional equipment and it is a time-consuming and complicated process for person-specific relighting. Li *et al.* [31] attempted to relight target human from multi-view video recorded under unknown illumination conditions and Imber *et al.* [23] extended this approach to scene relighting by introducing intrinsic textures. These two methods can produce high-quality relit results but multiview video must be collected for every target human or scene.

2.2 General Human Relighting

Deep learning makes general human relighting possible. Most existing general human relighting methods [5,16,18,22,32,33,35,37,38,40,43,44,46,47,50–52,58,59] focus on human portrait relighting. [40,50,57] achieved amazing single-image portrait relighting utilizing a dataset synthesized from OLAT images. For human full-body relighting, Meka *et al.* [36] combined traditional geometry pipelines with neural rendering to generate relit results using gradient images and estimated human geometry. Kanamori *et al.* [25] proposed an occlusion-aware single-image full-body human relighting method to infer albedo, geometry and illumination leveraging spherical harmonics (SH) lighting model. On the basis of [25], Lagunas *et al.* [28] added specular reflectance and light-dependent residual terms to explicitly handle highlights and Tajima *et al.* [11] used a residual network to restore neglected light effects. However, due to the limitation of the expressive ability of the SH lighting model, [11,25,28] could only produce low-frequency shadows and their method may fail under harsh illuminations.

2.3 Inverse Rendering

Ramachandran *et al.* [41] proposed the *shape-from-shading* method to estimate shape from shading given an input image with known lighting conditions. Subsequent works [10,34,39] assume a simple light source model such as directional and point light sources. Based on the Retinex theory [29], intrinsic images [4,6,7,15,27,30,45,56] aims to decompose an input image into reflectance and shading. Recent single-image human relighting studies [11,25,28] have drawn on the idea of intrinsic images to decompose an input image into albedo, shape and illumination using UNet. We have further improved the intrinsic images decomposition results in the de-lighting stage utilizing the modified HRNet [49] and achieved better disentanglement between albedo and lighting.

3 Overview

Following current work [52], our framework consists of two stages: de-lighting and relighting. Figure 2 shows the architecture of the whole framework. For the de-lighting stage, we first use two geometry networks (Geometry Module) to

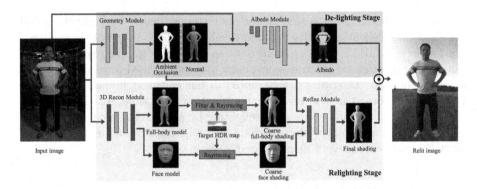

Fig. 2. Illustration of our framework architecture. There are two stages in our method: de-lighting and relighting. The de-lighting stage takes the input image and outputs estimated albedo $\widehat{A}$ (Sect. 4). For the relighting stage (Sect. 5), a full-body 3D model and a face 3D model are estimated by 3D Recon Module and then are sent to the renderer to render coarse shading maps (Sect. 5.1). The Refine Module takes the coarse shading maps and the inferred ambient occlusion map as input and produces the final shading map (Sect. 5.2).

infer the per-pixel normal map $\widehat{N}$ and ambient occlusion map $\widehat{AO}$ separately. Similar to [40], the inferred normal map and the input image are concatenated as the input of Albedo Module to infer albedo $\widehat{A}$. For the relighting stage, we first render the 3D human models estimated by [42] and [19] (3D Recon Module) using ray tracing to obtain the coarse full-body shading map S^{body}_{coarse} and coarse face shading map S^{face}_{coarse}. Then the Refine Module takes these coarse shading maps and the inferred ambient occlusion map as input to produce the refined full-body shading map $\widehat{S^{body}_{fine}}$ and refined face shading map $\widehat{S^{face}_{fine}}$, which are composited together to obtain the final shading map $\widehat{S}$. Details regarding the specific architecture and implementations of the networks are provided in the supplementary materials. Finally, the dot product of the inferred albedo and the final shading map is obtained to produce the relit image I_{relit} using the following formula:

$$I_{relit} = \widehat{A} \odot \widehat{S}$$

4   De-lighting

One of the most important factor for achieving better de-lighting results is the network architecture. Although the UNet architecture has been widely used in previous de-lighting networks [25,28,40,43,52], we find that it usually produces heavy lighting and albedo entangled results, possibly because of the fact that the lighting and albedo features remain coupled in the UNet architecture.

To guarantee that the lighting and albedo can be disentangled in the network architecture level, we modified the HRNet [49] architecture and achieved better delighting performance than UNet as shown in Fig. 3. Note that the vanilla

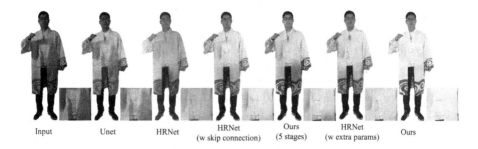

| Input | Unet | HRNet | HRNet (w skip connection) | Ours (5 stages) | HRNet (w extra params) | Ours |

Fig. 3. Estimated albedo of a real image. Unet comes from RH [25] and is retrained on our dataset. We zoom in the area in the red anchor and place it at the right of corresponding image. (Color figure online)

HRNet can only produce oversmoothed albedo map due to the aggressive downsampling operations in the beginning stage. Even with skip connections between the input downsampling features and output upsampling features, the results of vanilla HRNet remain poor and even make unremoved shadows appear in the inferred albedo map, especially in outdoor scenarios. We assume that skip connections have a negative impact on the final de-lighting results because high-resolution features from the input still contain environmental lighting information. Based on this assumption, we modify the vanilla HRNet by removing the downsampling operations directly (which also avoid skip connections) to fuse multiscale features while maintaining high-resolution representations. Moreover, we modify HRNet to output lighting prediction at the layer with the lowest resolution of the final stage. This strategy guarantees a better decomposition between lighting and albedo at both the architecture and feature representation level. Details regarding the specific architecture and implementations are provided in the supplementary materials.

To estimate the albedo of an input human image, we employ losses as follows:

$$L_{DL} = \lambda_{input} \left\| \widehat{A} - A \right\|_1 + \lambda_{vgg} Vgg(\widehat{A}, A)$$
$$+ \lambda_{light} \left\| w \odot log(1 + \widehat{I}_l) - log(1 + I_l) \right\|_2^2$$

where $\widehat{A}$ is the inferred albedo and A is the ground truth albedo. Similar to portrait relighting [50], we use the weighted log-L_2 loss. $\widehat{I}_l$ and I_l are latitude-longitude representation of environmental illumination and are used to describe the estimated lighting and ground truth lighting, respectively. The ground truth lighting map is downsampled to $16 \times 32 \times 3$ using Gaussian pyramid. w is the solid angle of each "pixel". The L_{DL} is the total loss for the de-lighting stage, and λ_{input}, λ_{vgg} and λ_{light} are the weight factors. Empirically, we find $\lambda_{input} = 500$, $\lambda_{vgg} = 100$, $\lambda_{light} = 0.025$ achieves the best performance.

5   Relighting

To render photorealistic shadows, we divide the relighting stage into two sub-stages: ray tracing and shading refinement. We first estimate 3D models of the target human because the classical ray tracing algorithm requires a 3D model of target object.

For geometry estimation, given a single human image, we use PIFuHD [42] to estimate a complete 3D human model. Then we crop the human face from the input image and align it with the dense 3D face model following [19] by regressing the 3DMM parameters. Both the full-body 3D model and face 3D model have no texture because we only need to render shading maps of the target human.

5.1   Ray Tracing

For photorealistic rendering, we use the Cycles rendering engine in Blender [1] and a Principled BSDF shader. We set the camera mode to orthographic to ensure that the rendered shading map is pixel-aligned with the input image. Due to the limitations of the PIFuHD [42] network's generation capacity and memory space, the surface of the estimated full-body 3D model is not smooth enough, which may produce checkered artifacts after raytracing. The schematic diagram of the artifacts is provided in the supplementary materials. To solve the problem, we adopt Laplacian smoothing [48] for the full-body 3D model and set smoothing steps to 10 with cotangent weight. Then the smoothed full-body model is sent to the renderer to render the full-body shading map under the target lighting environment using ray tracing. In addition, the generated 3D face model was directly sent to the renderer without smoothing for ray tracing.

In a word, we render a coarse full-body shading map S_{coarse}^{body} and coarse face shading map S_{coarse}^{face} under the target lighting environment using ray tracing. The rendered coarse shading maps are pixel-aligned with the input image.

5.2   Shading Refinement

After ray tracing under the target lighting environment, we obtained roughly correct shadows (especially for self-occlusions and hard shadows such as cast shadows) of the target human. However, the full-body 3D model reconstructed from a single image is not completely accurate and thus the ray-traced shading maps contain unnatural shadows and obvious geometry errors. To enhance the realism of the ray-traced full-body shading map, we introduce two refinement networks to compensate for shadow-rendering errors and restore a high-quality shading map. The first refinement network is designed to paint in the ray-traced full-body shading map and improve the overall quality of the shading details. The second refinement network is designed to refine facial shading details to ensure that we can fully leverage the geometry priors of human faces. Figure 4 shows the entire shading refinement process.

Full-Body Shading Refinement. The full-body refinement network takes the coarse full-body shading S_{coarse}^{body} and the inferred ambient occlusion map $\widehat{AO}$ as input and outputs the refined full-body shading residual. We choose an ambient occlusion map instead of a normal map for the following three reasons. First, the

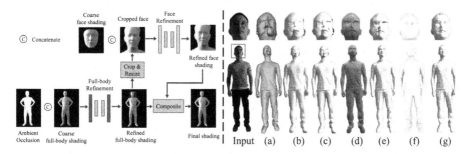

Fig. 4. Left: Illustration of Refine Module. A refined full-body shading map is inferred by the full-body refinement network. Then the cropped face from the refined full-body shading map and coarse face shading map is concatenated as the input of face refinement network, which outputs the refined face shading map. Finally, the refined face shading map and the refined full-body shading map are composited to generate the final relit shading map. **Right**: Refined shading maps. (a) 3D model estimated by PIFuHD [42] (b) Ray-traced shading map (c) Refined shading map without the inferred ambient occlusion map (d) Inferred normal map (e) Refined shading map with the inferred normal map (f) Inferred ambient occlusion map (g) Refined shading map with the inferred ambient occlusion map. Cropped faces of corresponding shading map are placed at the top.

ambient occlusion map can supplement part of the self-shadows lost due to geometry prediction errors. Second, compared with inferring a normal map, inferring an ambient occlusion map is more robust under various lighting environments even with extreme lighting distributions. Third, existing 3D human reconstruction methods such as [24,42] are highly dependent on the surface normal to predict 3D geometry surface details, which means that the normal prediction errors are consistent with the geometry errors and S_{coarse}^{body} cannot obtain extra correct geometry information from the normal map to compensate for existing shading errors. Figure 4 shows the refined results with the normal and ambient occlusion maps as auxiliary inputs, respectively.

The architecture of full-body refinement network is similar to MIMO-UNet [9], and we deepen the network to 4 downsampling operations. The coarse full-body shading map S_{coarse}^{body} and the inferred ambient occlusion map $\widehat{AO}$ are concatenated together as the original scale input and the downsampled $\widehat{AO}$ is used as multiscale input. We use multiscale output for supervision. We add adversarial loss in the final highest-resolution output layer to help the network generate plausible shading effects and use PatchGAN as the discriminator. The training loss consists of content loss, fft loss and PatchGAN loss and is defined as follows:

$$L_{fb} = \lambda_{content} \sum_{k=1}^{K} \left\| \widehat{S}_{fine}^{body^k} - S^k \right\|_1 + \left\| P(\widehat{S}_{fine}^{body}) - P(S) \right\|_2^2$$
$$+ \lambda_{fft} \sum_{k=1}^{K} \left\| F(\widehat{S}_{fine}^{body^k}) - F(S^k) \right\|_1$$

where K is the number of levels and $K = 5$; $\widehat{S}_{fine}^{body^k}$ is the k_{th} level output and S^k is the k_{th} downsampled ground truth shading map. P is the PatchGAN discriminator, and $F()$ denotes the fast Fourier transform (FFT). $\lambda_{content}$ and λ_{fft} are weight factors and we empirically set $\lambda_{content} = 10$ and $\lambda_{fft}=0.001$.

Face Shading Refinement. Although the above-mentioned refinement network produces realistic shading with rich details, the human eye is very sensitive to the details of the face and is able to distinguish small geometry and shadow errors. Therefore, we continue to refine the face region on the basis of the refined full-body shading map $\widehat{S}_{fine}^{body}$. The face refinement network takes S_{crop}^{face} cropped from $\widehat{S}_{fine}^{body}$ and S_{coarse}^{face} as the input and outputs the refined face shading residual. The training loss can be expressed as follows:

$$L_{ff} = \lambda_{face} \left\| \widehat{S_{fine}^{face}} - S^{face} \right\|_1 + \left\| L(\widehat{S_{fine}^{face}}) - L(S^{face}) \right\|_2^2$$

where L is the LSGAN discriminator, λ_{face} is the weight factor and is set to be $\lambda_{face} = 5$. S^{face} is the ground truth face shading map. We use UNet architecture as the backbone of face refinement network and details regarding the architecture are provided in the supplementary materials.

6   Implementation Details

We carefully select 811 scanned 3D human figures with good lighting conditions from Twindom [2], of which 700 figures are used for training and 111 figures for testing. The use of the dataset has been officially approved by Twindom. We collect 480 panoramic lighting environments sourced from www.HDRIHaven. com [3] and rotate them every 36°C to generate a total of 4800 HDR environment maps. We allocate 4600 HDR environment maps for training and 200 HDR environment maps for testing. To balance the amount of lighting in indoor and outdoor scenes, we add an extra 150 indoor HDR environment maps from the Laval Indoor Dataset [17] to the testing dataset. None of the test lighting conditions appear in the training dataset. Details regarding the specific data rendering, training and testing are provided in the supplementary materials.

7   Experiments

In this section, we first compare our method with previous state-of-the-art methods quantitatively and qualitatively to show that our method performs better on de-lighting task and produces more photorealistic relit results under challenging lighting conditions. Then we evaluate the key contributions of our proposed method and prove the effectiveness of the entire framework and each module. For quantitative comparisons, we adopt the metrics MSE, PSNR and SSIM to

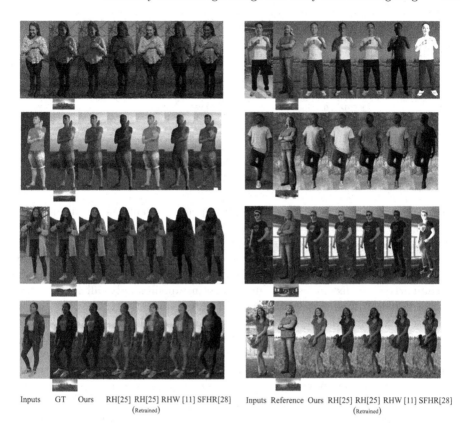

Inputs   GT   Ours   RH[25] RH[25] RHW [11] SFHR[28]   Inputs Reference Ours RH[25] RH[25] RHW [11] SFHR[28]
(Retrained)   (Retrained)

Fig. 5. Relit results on synthetic and real images. The first column: synthetic images from our testing dataset. The second column: real images. For real images, "Reference" are the rendered images of a virtual 3D human model under the target lighting conditions and are used to indicate the position of shadows. The target HDR environment map is placed under the ground truth image or reference image.

compare the inferred albedo, relit shading and relit images with the corresponding ground truth images in our testing dataset. For qualitative comparisons, we show some intrinsic images decomposition results and relit results under target lighting conditions using both real images and synthetic images.

7.1   Comparisons with SOTA Methods

We compare our method with the state-of-the-art single-image human relighting methods RH [25], SFHR [28] and RHW [11]. All of them are based on intrinsic images decomposition and require only a single human image and target lighting for relighting as in our approach. Details regarding the specific comparison settings are provided in the supplementary materials.

Table 1. Quantitative comparisons of our single-image human relighting framework against prior works. Shading indicates the estimated shading map under target lighting condition. The value of MSE is scaled by multiplying 10^3.

	Relight			Albedo			Shading		
	MSE ↓	SSIM ↑	PSNR ↑	MSE ↓	SSIM ↑	PSNR ↑	MSE ↓	SSIM ↑	PSNR ↑
RH [25]	8.501	0.9275	23.70	4.528	0.9521	24.69	18.96	0.9162	20.26
RH [25] (Re-trained)	3.781	0.9506	26.45	2.819	0.9559	26.73	8.333	0.9236	22.39
RHW [11]	10.90	0.9085	22.97	3.624	0.9586	25.28	26.18	0.9037	19.43
SFHR [28]	5.833	0.9349	24.66	4.185	0.9613	24.98	12.79	0.9155	21.41
Ours	**1.311**	**0.9638**	**30.20**	**1.468**	**0.9814**	**30.27**	**0.729**	**0.9626**	**32.86**

Comparison on Synthetic Data. We first perform quantitative and qualitative comparisons on the testing dataset where we have ground truth images as a comparison. Table 1 shows quantitative comparison of intrinsic images decomposition performance and single-image human relighting quality. Our method outperforms the competitive methods on every metric for both tasks. To limit the comparison to decomposition and relighting quality only, all metrics are computed only on the foreground region for all methods.

Qualitative comparisons for relighting are shown in Fig. 5. To improve the visual effects, we use MODNet [26] to infer alpha channel of the input image and change the background of the relit images to the corresponding part of the environment lighting map. Benefiting from the combination of classical forward rendering and deep learning, our method can produce photorealistic shadows under arbitrary lighting conditions. As a comparison, RH [25], SFHR [28] and RHW [11] can only produce low-frequency shadows. Moreover, they also fail to relight images under outdoor lighting conditions and produce overly bright or dark relit results. As shown in Fig. 6, our method can produce plausible hard cast shadows and achieve better disentanglement between albedo and shading, which remain challenging for previous methods. Since all methods rely on albedo estimation at first for relighting, we also compare the de-lighting performance. As shown in Fig. 7, compared with RH [25], SFHR [28] and RHW [11], our method achieves better disentanglement between lighting and albedo and is able to remove large areas of shadows (even including self-shadows from concave regions on the real-world 3D shape of the human body, e.g., armpits, crotch, neck under the chin or folds on the clothing.

Comparison on Real-World Images. Although our method is trained on synthetic datasets, it is generalizable to real data. The second column of Fig. 6 and Fig. 7 show shading estimation and de-lighting results on real images, respectively, and the second column of Fig. 5 shows the relit results of images photographed in the real world under arbitrary and complex illumination conditions. RH [25], SFHR [28] and RHW [11] still suffer from the entanglement of lighting and albedo and struggle to produce high-frequency shadows. By comparison, our method performs better at removing detailed shadows from the original images and generating photorealistic shadows on the relit images.

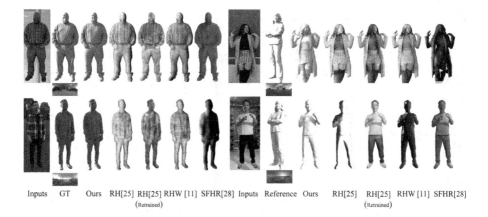

Inputs GT Ours RH[25] RH[25] RHW [11] SFHR[28] Inputs Reference Ours   RH[25]   RH[25] RHW [11] SFHR[28]
                       (Retrained)   (Retrained)

Fig. 6. Qualitative results for shading estimation under target lighting conditions. The first column: synthetic images from our testing dataset. The second column: real images.

Table 2. Quantitative results for ablation study of the full-body shading refinement.

	MSE $\downarrow$ ($\times 10^{-3}$)	SSIM $\uparrow$	PSNR $\uparrow$
Raytracing	1.492	0.9499	29.84
Ours (w/o ambient)	1.250	0.9524	30.89
Ours (w normal)	0.886	0.9561	32.29
Ours	**0.729**	**0.9626**	**32.86**

7.2 Ablation Study

To demonstrate the effectiveness of our delighting network, full-body refinement network and face refinement network, we conduct comprehensive ablation studies both quantitatively and qualitatively.

First, we compare our delighting network with vanilla UNet, vanilla HRNet and vanilla HRNet with skip connections. Table 3 shows the quantitative results of different networks and Fig. 3 presents the de-lighting results on a real image. The vanilla HRNet is HRNet-W32 [49] has two transposed convolution layers with stride 2 to ensure that the output size and input size are the same. Based on the vanilla HRNet, the vanilla HRNet with skip connections further adds skip connections between the downsampled features of the first stage and the transposed convolution features of the output. "HRNet(w extra params)" means vanilla HRNet with skip connections, extra stages, modules and blocks. "Ours(5 stages)" indicates that the architecture of network is similar to that of our de-lighting network but only contains 5 stages. The proposed network "Ours" outperforms other networks on all metrics and achieves better disentanglement between lighting and albedo. By contrast, the vanilla HRNet fails to produce high-resolution results and the removal of partial self-shadows such as those caused by clothing folds remains difficult, even with skip connections and

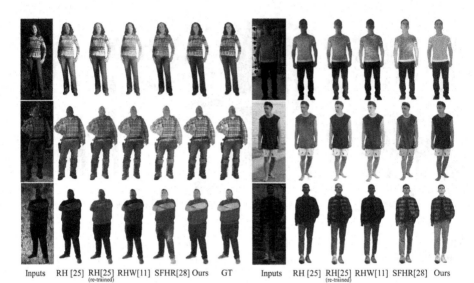

Inputs   RH [25]   RH[25]   RHW[11]   SFHR[28]   Ours   GT   Inputs   RH [25]   RH[25]   RHW[11]   SFHR[28]   Ours
          (re-trained)   (re-trained)

Fig. 7. De-lighting results on synthetic and real images. The first column: synthetic images from our testing dataset. The second column: real-world images.

Table 3. Quantitative results for ablation study of the de-lighting network. Vanilla Unet comes from RH [25] and is trained on our dataset.

	MSE ↓ ($\times 10^{-3}$)	SSIM ↑	PSNR ↑
Vanilla Unet	2.819	0.9559	26.73
Vanilla HRNet	2.470	0.9550	27.21
Vanilla HRNet (w skip connections)	2.150	0.9767	28.53
Ours (5 stages)	1.728	0.9808	28.91
Ours	**1.468**	**0.9814**	**30.27**

extra parameters. Compared with UNet, our method greatly improves PSNR (increasing by 3) and MSE (dropping to half of the UNet's).

Second, we verify the effectiveness of the ambient occlusion map used by the full-body refinement network. "Refinement Net(w/o ambient)" means that the refinement network only takes coarse full-body shading as input and removes SCM and FAM modules of MIMO-UNet [9]. "Refinement Net(w normal)" means that the refinement network takes the normal map as the auxiliary input rather than the ambient occlusion map. Table 2 shows quantitative results and Fig. 4 shows qualitative results. Without an ambient occlusion map, the refinement work cannot fill in missing geometry details and shadows. Moreover, when the input image is under extreme lighting conditions, the inference of normal map may fail around the boundary of the shadows. By contrast, the inferred ambient

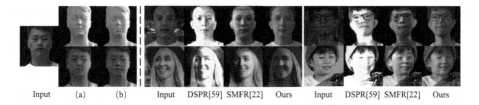

Input   (a)   (b)   Input   DSPR[59] SMFR[22]   Ours   Input   DSPR[59] SMFR[22]   Ours

Fig. 8. Left: Ablation of face refinement module. (a) Top: cropped face from the refined full-body shading map Bottom: corresponding relit result (b) Top: refined face shading map Bottom: corresponding relit result. **Right**: Comparison with existing portrait relighting methods.

occlusion map is unaffected by shadows and able to recover more geometry and occlusion details, thus restoring better shading maps.

Finally, we evaluate the face refinement network. To highlight the role of this module, we present the cropped face shading for comparison. Figure 8 shows the qualitative results for face refinement. The face without refinement contains unnatural facial details such as a twisted nose and asymmetric eyes. By contrast, thanks to the geometry priors provided by 3DMM templates, the refined face owns clearer and more natural facial features. Compared with DSPR [59] and SMFR [22], our method can generate plausible hard cast shadows, especially around the nose and neck, whereas SMFR [22] may produces patchy shadows and DSPR [59] may produces overexposed results.

8   Discussion

Conclusion. We propose a geometry-aware single-image human relighting framework that leverages 3D geometry prior information to produce higher-quality relit results. Our framework contains two stages: de-lighting and relighting. For the de-lighting stage, we use a modified HRNet as the de-lighting network and achieve better disentanglement between lighting and albedo. For the relighting stage, we use ray tracing to render the shading map of the target human, and further refine it utilizing learning-based refinement networks. The extensive results demonstrate that our framework can produce photorealistic high-frequency shadows with clear boundaries under challenging lighting conditions and outperforms the existing SOTA method on both synthetic images and real images.

Limitations. Due to the limitation of the dataset, we adopt the assumption of Lambertian materials for the clothed humans, which fails to produce specular reflectance in the relit results. For the same reason, our delighting network struggles to remove the highlights on the face. Moreover, inaccurate single-image geometry reconstruction may generate unnatural refined shading results.

Acknowledgement. This paper is supported by National Key R&D Program of China (2021ZD0113501) and the NSFC project No.62125107, No.62171255 and No.61827805.

References

1. https://www.blender.org/
2. https://web.twindom.com/
3. https://polyhaven.com/hdris
4. Barron, J.T., Malik, J.: Color constancy, intrinsic images, and shape estimation. In: Fitzgibbon, A., Lazebnik, S., Perona, P., Sato, Y., Schmid, C. (eds.) ECCV 2012. LNCS, vol. 7575, pp. 57–70. Springer, Heidelberg (2012). https://doi.org/10.1007/978-3-642-33765-9_5
5. Barron, J.T., Malik, J.: Shape, illumination, and reflectance from shading. IEEE Trans. Pattern Anal. Mach. Intell. **37**(8), 1670–1687 (2014)
6. Baslamisli, A.S., Le, H.A., Gevers, T.: CNN based learning using reflection and retinex models for intrinsic image decomposition. In: Proceedings of the IEEE Conference on Computer Vision and Pattern Recognition, pp. 6674–6683 (2018)
7. Bonneel, N., Kovacs, B., Paris, S., Bala, K.: Intrinsic decompositions for image editing. In: Computer Graphics Forum, vol. 36, pp. 593–609. Wiley Online Library (2017)
8. Chabert, C.F., et al.: Relighting human locomotion with flowed reflectance fields. In: ACM SIGGRAPH 2006 Sketches, p. 76 (2006)
9. Cho, S.J., Ji, S.W., Hong, J.P., Jung, S.W., Ko, S.J.: Rethinking coarse-to-fine approach in single image deblurring. In: Proceedings of the IEEE/CVF International Conference on Computer Vision, pp. 4641–4650 (2021)
10. Christou, C.G., Koenderink, J.J.: Light source dependence in shape from shading. Vision. Res. **37**(11), 1441–1449 (1997)
11. Tajima, D., Kanamori, Y., Endo, Y.: Relighting humans in the wild: monocular full-body human relighting with domain adaptation. Comput. Graph. Forum **40**(7), 205–216 (2021)
12. Debevec, P.: The light stages and their applications to photoreal digital actors. SIGGRAPH Asia **2**(4), 1–6 (2012)
13. Debevec, P., Hawkins, T., Tchou, C., Duiker, H.P., Sarokin, W., Sagar, M.: Acquiring the reflectance field of a human face. In: Proceedings of the 27th Annual Conference on Computer Graphics and Interactive Techniques, pp. 145–156 (2000)
14. Debevec, P., Wenger, A., Tchou, C., Gardner, A., Waese, J., Hawkins, T.: A lighting reproduction approach to live-action compositing. ACM Trans. Graphics (TOG) **21**(3), 547–556 (2002)
15. Ding, S., Sheng, B., Hou, X., Xie, Z., Ma, L.: Intrinsic image decomposition using multi-scale measurements and sparsity. In: Computer Graphics Forum, vol. 36, pp. 251–261. Wiley Online Library (2017)
16. Egger, B., et al.: Occlusion-aware 3d morphable models and an illumination prior for face image analysis. Int. J. Comput. Vision **126**(12), 1269–1287 (2018)
17. Gardner, M.A., et al.: Learning to predict indoor illumination from a single image. ACM Trans. Graph. (TOG) **36**(6), 1–14 (2017)
18. Genova, K., Cole, F., Maschinot, A., Sarna, A., Vlasic, D., Freeman, W.T.: Unsupervised training for 3D morphable model regression. In: Proceedings of the IEEE Conference on Computer Vision and Pattern Recognition, pp. 8377–8386 (2018)

19. Guo, J., Zhu, X., Yang, Y., Yang, F., Lei, Z., Li, S.Z.: Towards fast, accurate and stable 3D dense face alignment. In: Vedaldi, A., Bischof, H., Brox, T., Frahm, J.-M. (eds.) ECCV 2020. LNCS, vol. 12364, pp. 152–168. Springer, Cham (2020). https://doi.org/10.1007/978-3-030-58529-7_10

20. Guo, K., et al.: The relightables: volumetric performance capture of humans with realistic relighting. ACM Trans. Graph. (TOG) **38**(6), 1–19 (2019)

21. Hawkins, T., Cohen, J., Debevec, P.: A photometric approach to digitizing cultural artifacts. In: Proceedings of the 2001 Conference on Virtual Reality, Archeology, and Cultural Heritage, pp. 333–342 (2001)

22. Hou, A., Zhang, Z., Sarkis, M., Bi, N., Tong, Y., Liu, X.: Towards high fidelity face relighting with realistic shadows. In: Proceedings of the IEEE/CVF Conference on Computer Vision and Pattern Recognition, pp. 14719–14728 (2021)

23. Imber, J., Guillemaut, J.-Y., Hilton, A.: Intrinsic textures for relightable free-viewpoint video. In: Fleet, D., Pajdla, T., Schiele, B., Tuytelaars, T. (eds.) ECCV 2014. LNCS, vol. 8690, pp. 392–407. Springer, Cham (2014). https://doi.org/10.1007/978-3-319-10605-2_26

24. Jafarian, Y., Park, H.S.: Learning high fidelity depths of dressed humans by watching social media dance videos. In: Proceedings of the IEEE/CVF Conference on Computer Vision and Pattern Recognition, pp. 12753–12762 (2021)

25. Kanamori, Y., Endo, Y.: Relighting humans: occlusion-aware inverse rendering for full-body human images. ACM Trans. Graph. (TOG) **37**(6), 1–11 (2018)

26. Ke, Z., et al.: Is a green screen really necessary for real-time portrait matting? (2020)

27. Laffont, P.Y., Bazin, J.C.: Intrinsic decomposition of image sequences from local temporal variations. In: Proceedings of the IEEE International Conference on Computer Vision, pp. 433–441 (2015)

28. Lagunas, M., et al.: Single-image full-body human relighting. arXiv preprint arXiv:2107.07259 (2021)

29. Land, E.H., McCann, J.J.: Lightness and retinex theory. Josa **61**(1), 1–11 (1971)

30. Li, C., Zhou, K., Wu, H.T., Lin, S.: Physically-based simulation of cosmetics via intrinsic image decomposition with facial priors. IEEE Trans. Pattern Anal. Mach. Intell. **41**(6), 1455–1469 (2018)

31. Li, G., et al.: Capturing relightable human performances under general uncontrolled illumination. In: Comput. Graph. Forum, vol. 32, pp. 275–284. Wiley Online Library (2013)

32. Li, Y., Liu, M.-Y., Li, X., Yang, M.-H., Kautz, J.: A closed-form solution to photorealistic image stylization. In: Ferrari, V., Hebert, M., Sminchisescu, C., Weiss, Y. (eds.) ECCV 2018. LNCS, vol. 11207, pp. 468–483. Springer, Cham (2018). https://doi.org/10.1007/978-3-030-01219-9_28

33. Lin, J., Yuan, Y., Shao, T., Zhou, K.: Towards high-fidelity 3D face reconstruction from in-the-wild images using graph convolutional networks. In: Proceedings of the IEEE/CVF Conference on Computer Vision and Pattern Recognition, pp. 5891–5900 (2020)

34. Lopez-Moreno, J., Hadap, S., Reinhard, E., Gutierrez, D.: Light source detection in photographs. In: CEIG, pp. 161–167 (2009)

35. Meka, A., et al.: Deep reflectance fields: high-quality facial reflectance field inference from color gradient illumination. ACM Trans. Graph. (TOG) **38**(4), 1–12 (2019)

36. Meka, A., et al.: Deep relightable textures: volumetric performance capture with neural rendering. ACM Trans. Graph. (TOG) **39**(6), 1–21 (2020)

37. Nagano, K., et al.: Deep face normalization. ACM Trans. Graph. (TOG) **38**(6), 1–16 (2019)

38. Nestmeyer, T., Lalonde, J.F., Matthews, I., Lehrmann, A.: Learning physics-guided face relighting under directional light. In: Proceedings of the IEEE/CVF Conference on Computer Vision and Pattern Recognition, pp. 5124–5133 (2020)

39. Okatani, T., Deguchi, K.: Shape reconstruction from an endoscope image by shape from shading technique for a point light source at the projection center. Comput. Vis. Image Underst. **66**(2), 119–131 (1997)

40. Pandey, R., et al.: Total relighting: learning to relight portraits for background replacement. ACM Trans. Graph. (TOG) **40**(4), 1–21 (2021)

41. Ramachandran, V.S.: Perception of shape from shading. Nature **331**(6152), 163–166 (1988)

42. Saito, S., Simon, T., Saragih, J., Joo, H.: PIFuHD: multi-level pixel-aligned implicit function for high-resolution 3D human digitization. In: Proceedings of the IEEE/CVF Conference on Computer Vision and Pattern Recognition, pp. 84–93 (2020)

43. Sengupta, S., Kanazawa, A., Castillo, C.D., Jacobs, D.W.: SfSNet: learning shape, reflectance and illuminance of faces 'in the wild'. In: Proceedings of the IEEE conference on computer vision and pattern recognition, pp. 6296–6305 (2018)

44. Shahlaei, D., Blanz, V.: Realistic inverse lighting from a single 2D image of a face, taken under unknown and complex lighting. In: 2015 11th IEEE International Conference and Workshops on Automatic Face and Gesture Recognition (FG), vol. 1, pp. 1–8. IEEE (2015)

45. Sheng, B., Li, P., Jin, Y., Tan, P., Lee, T.Y.: Intrinsic image decomposition with step and drift shading separation. IEEE Trans. Visual Comput. Graphics **26**(2), 1332–1346 (2018)

46. Shu, Z., Hadap, S., Shechtman, E., Sunkavalli, K., Paris, S., Samaras, D.: Portrait lighting transfer using a mass transport approach. ACM Trans. Graph. (TOG) **36**(4), 1 (2017)

47. Shu, Z., Yumer, E., Hadap, S., Sunkavalli, K., Shechtman, E., Samaras, D.: Neural face editing with intrinsic image disentangling. In: Proceedings of the IEEE Conference on Computer Vision and Pattern Recognition, pp. 5541–5550 (2017)

48. Sorkine, O.: Laplacian mesh processing. In: Eurographics (State of the Art Reports), pp. 53–70. Citeseer (2005)

49. Sun, K., Xiao, B., Liu, D., Wang, J.: Deep high-resolution representation learning for human pose estimation. In: Proceedings of the IEEE/CVF Conference on Computer Vision and Pattern Recognition, pp. 5693–5703 (2019)

50. Sun, T., et al.: Single image portrait relighting. ACM Trans. Graph. **38**(4), 1–79 (2019)

51. Tewari, A., et al.: MoFA: model-based deep convolutional face autoencoder for unsupervised monocular reconstruction. In: Proceedings of the IEEE International Conference on Computer Vision Workshops, pp. 1274–1283 (2017)

52. Wang, Z., Yu, X., Lu, M., Wang, Q., Qian, C., Xu, F.: Single image portrait relighting via explicit multiple reflectance channel modeling. ACM Trans. Graph. (TOG) **39**(6), 1–13 (2020)

53. Wenger, A., Gardner, A., Tchou, C., Unger, J., Hawkins, T., Debevec, P.: Performance relighting and reflectance transformation with time-multiplexed illumination. ACM Trans. Graph. (TOG) **24**(3), 756–764 (2005)

54. Weyrich, T., et al.: Analysis of human faces using a measurement-based skin reflectance model. ACM Trans. Graph. (ToG) **25**(3), 1013–1024 (2006)

55. Whitted, T.: An improved illumination model for shaded display. In: Proceedings of the 6th annual conference on Computer graphics and interactive techniques, p. 14 (1979)
56. Ye, G., Garces, E., Liu, Y., Dai, Q., Gutierrez, D.: Intrinsic video and applications. ACM Trans. Graph. (ToG) **33**(4), 1–11 (2014)
57. Zhang, L., Zhang, Q., Wu, M., Yu, J., Xu, L.: Neural video portrait relighting in real-time via consistency modeling. arXiv preprint arXiv:2104.00484 (2021)
58. Zhang, X., et al.: Portrait shadow manipulation. ACM Trans. Graph. (TOG) **39**(4), 1–78 (2020)
59. Zhou, H., Hadap, S., Sunkavalli, K., Jacobs, D.W.: Deep single-image portrait relighting. In: Proceedings of the IEEE/CVF International Conference on Computer Vision, pp. 7194–7202 (2019)

3D-Aware Indoor Scene Synthesis
with Depth Priors

Zifan Shi[1], Yujun Shen[2], Jiapeng Zhu[1], Dit-Yan Yeung[1], and Qifeng Chen[1(✉)]

[1] The Hong Kong University of Science and Technology, Kowloon, China
{dyyeung,cqf}@cse.ust.hk
[2] ByteDance Inc., Beijing, China

Abstract. Despite the recent advancement of Generative Adversarial Networks (GANs) in learning 3D-aware image synthesis from 2D data, existing methods fail to model indoor scenes due to the large diversity of room layouts and the objects inside. We argue that indoor scenes do not have a shared intrinsic structure, and hence only using 2D images cannot adequately guide the model with the 3D geometry. In this work, we fill in this gap by introducing depth as a 3D prior (Depth is essentially a 2.5D prior, but in this paper we use 3D for simplicity). Compared with other 3D data formats, depth better fits the convolution-based generation mechanism and is more easily accessible in practice. Specifically, we propose a dual-path generator, where one path is responsible for depth generation, whose intermediate features are injected into the other path as the condition for appearance rendering. Such a design eases the 3D-aware synthesis with explicit geometry information. Meanwhile, we introduce a switchable discriminator both to differentiate real *v.s.* fake domains and to predict the depth from a given input. In this way, the discriminator can take the spatial arrangement into account and advise the generator to learn an appropriate depth condition. Extensive experimental results suggest that our approach is capable of synthesizing indoor scenes with impressively good quality and 3D consistency, significantly outperforming state-of-the-art alternatives. (Project page can be found here.)

Keywords: 3D-aware image synthesis · Scene synthesis · Depth priors

1 Introduction

Generative Adversarial Networks (GANs) [12] have enabled high-fidelity 2D image synthesis, but how to make a GAN model aware of 3D information remains unsolved. Along with the recent advent of Neural Radiance Field (NeRF) [22] for 3D scene reconstruction, some attempts [4,5,13,27,35,41] propose to incorporate NeRF into GANs to learn a 3D-aware image generator from a 2D image

Supplementary Information The online version contains supplementary material available at https://doi.org/10.1007/978-3-031-19787-1_23.

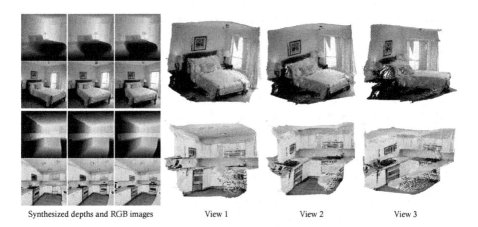

Synthesized depths and RGB images   View 1   View 2   View 3

Fig. 1. Photo-realistic 3D-aware synthesis results on bedrooms and kitchens. Left: Two sets of synthesized depth maps and their corresponding rendered images from three different viewpoints. Right: Visualization of the 3D reconstruction results (using [32] and [45]) from the synthesized samples

collection. Instead of using 2D convolutional layers, the generator is asked to learn a point-wise implicit function, which maps the 3D coordinates to volume densities and colors [22,35].

Although existing methods show promising results in learning 3D-aware object synthesis, such as human faces and cars, they exhibit severe performance degradation on indoor scene datasets, such as bedrooms and kitchens. There are mainly two reasons. First, objects normally have a shared intrinsic structure, which eases the difficulty of modeling 3D geometry from 2D images *only*. For instance, human heads share similar shapes, and each face consists of two eyes located at relatively defined positions. On the contrary, indoor scenes have much higher diversity, considering the complex room layout and the interior decoration [42]. Second, existing methods assume the distribution of camera poses [26,35]. Such an assumption is sound under the case of object synthesis because objects are commonly placed at the center of a 2D image. Indoor scene images are usually shot from far more diverse viewpoints, making it too challenging for the NeRF-based approaches to handle.

In this work, we propose a new paradigm for 3D-aware image synthesis by explicitly introducing a 3D prior into 2D GANs. Compared with the volume renderer equipped with Multi-Layer Perceptron (MLP) [4,27,35], GANs built on Convolutional Neural Network (CNN) achieve much more appealing synthesis performances [17–19], especially from the image quality and the image resolution perspectives. Among numerous 3D data formats, such as point cloud [33,34], voxel [2,7], and implicit surface [21,29], we choose depth as our prior as it is defined in the 2D domain and hence naturally suitable for the convolution-based generator. In addition, there are many publicly available depth datasets [8,20,23] and depth predictors [31,43], making depth data easily accessible in practice.

To sufficiently leverage the depth prior, we re-design the objectives of both the generator and the discriminator in a conventional GAN. For one thing, we ask the generator to synthesize a 2D image accompanied by its corresponding depth. To meet this goal, we carefully tailor a dual-path architecture, where the appearance-path takes the multi-level feature maps from the depth-path as the input conditions. Through such a design, we manage to explicitly inject the geometry information into the generator. For another, unlike the conventional discriminator that makes the real/fake decision from the 2D space, we learn a 3D-aware switchable discriminator. Specifically, it is asked to distinguish the real and synthesized samples based on the image-depth joint distribution and, simultaneously, predict the depth from an input image. The depth prediction is trained on real data and further used to supervise the fake data. In this way, the discriminator is able to gain more knowledge on the spatial layout and better guide the generator from the 3D perspective.

We evaluate our approach, termed as **DepthGAN**, on a couple of challenging indoor scene datasets. Both qualitative and quantitative results demonstrate the sufficient superiority of DepthGAN over existing methods. For example, we improve Fréchet Inception Distance (FID) [15] from 15.560 to 4.797 on the LSUN bedroom dataset [44] in 256×256 resolution. 3D visualization on a set of synthesized images is shown in Fig. 1.

2   Related Work

GAN-Based Image Synthesis. With the advent of Generative Adversarial Networks (GANs) [12], a large number of works have been proposed to generate high-quality photorealistic images [3,16,18,19]. To gain explicit control of the images, researchers study the disentanglement of different properties such as poses. Supervised methods [36] leverage off-the-shelf attribute classifiers or image transformations to annotate the synthesized data and use the labeled data to guide the subspace learning in the latent space. Unsupervised methods [14,37,46] learn the control by analyzing the statistics or the model weights. While these works can control the poses with the azimuth and elevation angles, the changes may violate the consistency in the 3D space since there is no such constraint.

3D-Aware Image Synthesis. Realizing that previous image synthesis methods do not consider 3D geometry, a large number of works have started to add 3D constraints for image synthesis. Voxel-based methods [24,25] learn low-dimensional 3D representations with deep voxels, followed by a learnable 3D-to-2D projection. Inspired by NeRF [22], some works [4,5,9,13,27,30,35,41] incorporate neural radiance fields for 3D-aware image generation and render more consistent images. RGBD-GAN [28] synthesizes RGBD images under two views and warps them to each other to ensure 3D consistency. In contrast, we synthesize RGB images conditioned on depth images with carefully designed architectures, which models the geometry-appearance relationship better. All the works mentioned above learn geometry and appearance from 2D RGB images alone. Due

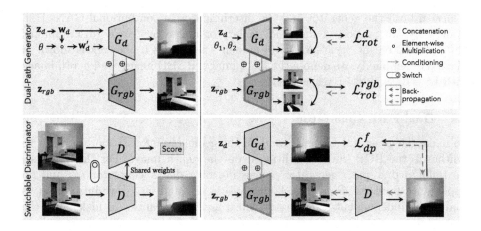

Fig. 2. Framework of DepthGAN, consisting of a *dual-path* generator that takes in two latent codes to generate the RGBD image with the appearance conditioned on the geometry, and a *switchable* discriminator that produces the realness score from an RGBD image and predicts the depth map from an RGB image. Black arrows indicate the forward computation, while dashed arrows under different colors stand for the back-propagation regarding different objective functions (Color figure online)

to the complexity of 3D geometry modeling and the lack of explicit 3D information, they target objects or well-aligned scenes and fail to generate high-quality 3D-aware images for complex scenes like bedrooms and kitchens. On the contrary, some other works utilize 3D prior knowledge to facilitate the learning of 3D consistency. Some researchers [6,40,47] select shape as the 3D prior and use the expensive 3D-conv-based GAN to learn the geometry information, which is costly and unable to model fine details of the shape. Others [1,6] utilize more than one 3D prior, such as albedo maps and normal maps, resulting in multiple 2D GANs to learn all the 3D attributes. Instead of generating objects only, S^2-GAN [39] synthesizes indoor scenes with the help of normal maps, but it adopts the two-stage training to learn geometry first and the appearance next. All these works either have separate 3D and 2D discriminators to learn geometry and texture distributions independently, or use 2D discriminators only to make the real/fake decision on one 3D attribute or the appearance. In contrast, our discriminator is endowed with 3D and 2D knowledge simultaneously. GSN [10] follows the NeRF rendering structure and adds another depth channel in the discriminator to incorporate 3D priors, but it fails to generate images with a large diversity and reasonable fidelity due to the complex rendering process, the special requirements of training data, and the inadequacy of its discriminator.

3   Method

In this work, we propose a new paradigm for 3D-aware image synthesis via introducing depth as a 3D prior into 2D GANs. To adequately use the depth prior, we re-design both the generator and the discriminator in conventional GANs [12]. Concretely, we propose a *dual-path* generator and a *switchable* discriminator based on the recent StyleGAN2 [19] model. The overall framework is shown in Fig. 2. For simplicity, we denote the RGB image, RGBD image, and depth image with $\mathbf{I}_{rgb}$, $\mathbf{I}_{rgbd}$, and $\mathbf{I}_d$, respectively.

3.1   Dual-Path Generator

To make the generator become aware of the geometry information, we ask it to synthesize the RGB image conditioned on the depth image. For this purpose, we tailor a dual-path generator, consisting of a depth generator G_d and an appearance renderer G_{rgb}. Two latent spaces, $\mathcal{Z}_d$ and $\mathcal{Z}_{rgb}$, are introduced to enable the independent sampling of depth and appearance. To make sure the appearance is properly rendered on top of the geometry, we feed the intermediate feature maps of G_d into G_{rgb} as the condition.

Depth Generator. To control the viewing point of the generated depth, we uniformly sample an angle θ from $[\theta_L, \theta_R]$. Since networks tend to learn better information from high-frequency signals [38], we encode θ with

$$\gamma(\theta, t) = h(sin(\theta), cos(\theta), ..., sin(t\theta), cos(t\theta)), \tag{1}$$

where t determines the maximum frequency. $h : \mathbb{R}^{2t} \rightarrow \mathbb{R}^m$ stands for a non-linear mapping, which is implemented by a two-layer fully-connection (FC) and a Leaky ReLU activation in between. Like StyleGAN2 [19], the raw depth latent code $\mathbf{z}_d \in \mathcal{Z}_d$ is projected into a more disentangled latent space, resulting in $\mathbf{w}_d \in \mathbb{R}^m$. The angle information is then injected to $\mathbf{w}_d$ through

$$\mathbf{w}'_d = \mathbf{w}_d \circ \gamma(\theta, t), \tag{2}$$

where $\circ$ denotes the element-wise multiplication. $\mathbf{w}'_d$ guides G_d on synthesizing the depth image, $\mathbf{I}^f_d$, via layer-wise style modulation [19]. Note that only the first two layers of G_d employ $\mathbf{w}'_d$ while the remaining layers still use $\mathbf{w}_d$, because only early layers correspond to the viewing point of the output image [42].

Depth-Conditioned Appearance Renderer. G_{rgb} shares a similar structure as G_d with three modifications. First, the number of output channels is 3 ($\mathbf{I}^f_{rgb}$) instead of 1 ($\mathbf{I}^f_d$). Second, G_{rgb} does not take the angle θ as the input. Third, most importantly, G_{rgb} takes the intermediate feature maps of G_d as the conditions to acquire the geometry information. Specifically, we first concatenate the per-layer feature $\mathbf{\Psi}_i$ of G_d with that of G_{rgb}, $\mathbf{\Phi}_i$. Here, i denotes the layer index. We then transform the concatenated result with

$$\mathbf{\Phi}'_i = f(\mathbf{\Psi}_i \oplus \mathbf{\Phi}_i), \tag{3}$$

where $\oplus$ stands for the concatenation operation, and f is implemented with a two-layer convolution. $\mathbf{\Phi}'_i$ has the same number of channels as $\mathbf{\Phi}_i$.

3.2 Switchable Discriminator

Unlike the discriminator in conventional GANs that simply differentiates the real and fake domains from the RGB image space, we propose a switchable discriminator to compete with the generator by taking the spatial arrangement into account. This is achieved from two aspects. On one hand, D makes the real/fake decision based on the joint distribution of RGB images and the corresponding depths. In other words, D takes an RGBD image as the input and outputs the realness score. On the other hand, to better capture the relationship between the image and the depth, we ask D to predict the depth from a given RGB image. Concretely, we introduce a separate branch on top of some intermediate feature maps of D for depth prediction. Detailed structure of the depth prediction branch can be found in the *Supplementary Material*.

To summarize, D switches between the 4-channel RGBD inputs (*i.e.*, for realness discrimination) and the 3-channel RGB inputs (*i.e.*, for depth prediction). To achieve this goal, we come up with a switchable input layer that adaptively adjusts the number of convolutional kernels.

3.3 Training Objectives

Adversarial Loss. We adopt the standard adversarial loss for GAN training:

$$\mathcal{L}_{adv}^{d} = -\mathbb{E}[\log(D(\mathbf{I}_{rgbd}^{r}))] - \mathbb{E}[\log(1 - D(\mathbf{I}_{rgbd}^{f}))], \tag{4}$$

$$\mathcal{L}_{adv}^{g} = -\mathbb{E}[\log(D(\mathbf{I}_{rgbd}^{f}))], \tag{5}$$

where $\mathbf{I}_{rgbd}^{r}$ represents the real RGBD data, and $\mathbf{I}_{rgbd}^{f}$ concatenates the generated RGB image $\mathbf{I}_{rgb}^{f}$ and the conditioned depth $\mathbf{I}_{d}^{f}$.

Rotation Consistency Loss. We design the rotation consistency loss [28] to enhance the consistency between the synthesis from different viewpoints, *i.e.*, θ. Specifically, two angles, θ_1 and θ_2, are randomly sampled, leading to two samples, $\mathbf{I}_{rgbd,1}^{f}$ and $\mathbf{I}_{rgbd,2}^{f}$ with the same latent codes, $\mathbf{z}_d$ and $\mathbf{z}_{rgb}$. We fix the camera and rotate the scene around its central axis. We assume an underlying camera intrinsic parameter, $\mathbf{K}$, which is fixed in the training process. After rotating $\mathbf{I}_{rgbd,1}^{f}$ from θ_1 to θ_2, we will get

$$P(\mathbf{I}_{rgbd,1}^{f,rot}) = \mathbf{K}R(\theta_1, \theta_2)\mathbf{K}^{-1}P(\mathbf{I}_{rgbd,1}^{f}), \tag{6}$$

where $R(\cdot, \cdot)$ denotes the rotation operation based on the depth image, $\mathbf{I}_{d,1}^{f}$, and $P(\cdot)$ represents the coordinates of the pixels. More details are available in the *Supplementary Material*.

The rotation consistency losses $\mathcal{L}_{rot}^{d}$ and $\mathcal{L}_{rot}^{rgb}$ for the dual-path generator are then defined as

$$\mathcal{L}_{rot}^{d} = \|\mathbf{I}_{d,1}^{f,rot} - \mathbf{I}_{d,2}^{f}\|_1, \tag{7}$$

$$\mathcal{L}_{rot}^{rgb} = \|\mathbf{I}_{rgb,1}^{f,rot} - \mathbf{I}_{rgb,2}^{f}\|_1, \tag{8}$$

where $\|\cdot\|_1$ denotes the ℓ_1 norm.

Depth Prediction Loss. As discussed above, besides differentiating real and fake data, our switchable discriminator is also asked to predict the depth from a given RGB image. Such a prediction is trained on real image-depth pairs, and further used to guide the synthesis. Following [11], our depth prediction is learned with a k-class classification. Thus, the depth prediction branch of D, $D_d(\cdot)$, produces a k-channel output map, indicating the class probability for each pixel. The loss function is formulated as

$$\mathcal{L}_{dp}^r = \mathcal{H}(D_d(\mathbf{I}_{rgb}^r), \mathbf{I}_d^r), \tag{9}$$

where $\mathcal{H}(\cdot, \cdot)$ denotes the pixel-wise cross-entropy loss. $\mathbf{I}_{rgb}^r$ and $\mathbf{I}_d^r$ stand for the ground-truth pair.

In order to help the generated appearance, $\mathbf{I}_{rgb}^f$, better fit the geometry, $\mathbf{I}_d^f$, we also predict the depth from the synthesized image to in turn guide the generator with

$$\mathcal{L}_{dp}^f = \mathcal{H}(D_d(\mathbf{I}_{rgb}^f), \mathbf{I}_d^f). \tag{10}$$

Full Objectives. In summary, the dual-path generator (*i.e.*, G_d and G_{rgb}) and the switchable discriminator (*i.e.*, D) are jointly optimized with

$$\mathcal{L}_{G_d} = \mathcal{L}_{adv}^g + \lambda_1 \mathcal{L}_{rot}^d, \tag{11}$$

$$\mathcal{L}_{G_{rgb}} = \mathcal{L}_{adv}^g + \lambda_2 \mathcal{L}_{rot}^{rgb} + \lambda_3 \mathcal{L}_{dp}^f, \tag{12}$$

$$\mathcal{L}_D = \mathcal{L}_{adv}^d + \lambda_4 \mathcal{L}_{dp}^r, \tag{13}$$

where $\{\lambda_i\}_{i=1}^4$ are loss weights to balance different terms.

4   Experiments

Datasets. We conduct experiments on LSUN bedroom and kitchen datasets [44]. Details are available in the *Supplementry Material*.

Metrics. We use the following metrics for evaluation: Fréchet Inception Distance (FID) [15], Chamfer Distance (CD), Rotation Precision (RP), and Rotation Consistency (RC). FID evaluates the quality of both the generated RGB images and depth images. FID for depth images is obtained by repeating the one-channel depth image to a three-channel image as input. CD measures the 3D consistency in 3D space, which computes the cross-view distance via warping point clouds. In addition, we propose another two metrics for evaluation. *(1) Rotation Precision (RP)* is aimed to measure the accuracy of the angle of rotation given two generated images from different views of the same scene. The formulation is the same as Eq. (7), and it evaluates depths in the range [0, 1]. *(2) Rotation Consistency (RC)* targets at the rotation consistency evaluation and has the same format as Eq. (8). It evaluates RGB images with pixel range

Table 1. Quantitative comparisons with existing 3D-aware image synthesis models on the LSUN bedroom and kitchen datasets [44] under both 128×128 and 256×256 resolutions. FID [15] regarding RGB images and depths, rotation precision (RP), rotation consistency (RC) and Chamfer distance (CD) are used as the metrics to evaluate the synthesis quality and the 3D controllability. CD is reported in the order of 10^{-3}. ↓ means lower value is better

(a) Evaluation on bedrooms

Method	128×128					256×256				
	FID↓	FID (D)↓	RP↓	RC↓	CD↓	FID↓	FID (D)↓	RP↓	RC↓	CD↓
2D-GAN SeFa [37]	8.650	-	0.572	1.027	-	7.190	-	0.401	1.110	-
RGBD-GAN [28]	28.694	370.526	-	-	-	59.026	360.858	-	-	-
GRAF [35]	63.940	184.379	0.218	1.149	44.321	66.856	188.368	0.219	0.880	66.702
GRAF(D) [35]	158.503	107.653	0.135	1.108	17.590	194.260	156.081	0.154	1.193	52.176
GIRAFFE [27]	48.412	422.634	-	-	-	44.232	420.681	-	-	-
π-GAN [4]	28.128	201.722	0.033	0.572	0.744	48.926	174.744	0.052	0.597	3.403
π-GAN(D) [4]	30.932	101.739	**0.022**	**0.420**	**0.355**	49.640	94.196	0.036	0.510	1.201
StyleNeRF [13]	13.675	284.088	0.140	1.026	0.841	15.560	288.379	0.159	1.055	2.352
VolumeGAN [41]	18.121	175.963	0.088	0.707	1.599	17.345	164.190	0.110	0.672	4.038
DepthGAN (Ours)	**4.040**	**18.874**	0.040	0.530	0.461	**4.797**	**17.140**	**0.025**	**0.456**	**0.339**

(b) Evaluation on kitchens

Method	128×128					256×256				
	FID↓	FID (D)↓	RP↓	RC↓	CD↓	FID↓	FID (D)↓	RP↓	RC↓	CD↓
2D-GAN SeFa [37]	11.530	-	0.748	1.115	-	10.850	-	0.480	1.163	-
RGBD-GAN [28]	33.425	267.036	-	-	-	51.044	392.126	-	-	-
GRAF [35]	86.920	239.657	0.224	1.326	48.265	94.095	204.050	0.227	0.928	72.472
GRAF(D) [35]	139.902	157.801	0.110	0.970	19.237	244.480	142.436	0.133	0.966	52.775
GIRAFFE [27]	42.923	307.233	-	-	-	50.256	370.760	-	-	-
π-GAN [4]	29.790	398.146	0.028	0.702	0.832	41.178	398.946	0.051	0.726	3.833
π-GAN(D) [4]	46.332	112.171	**0.025**	**0.482**	**0.258**	77.066	104.865	0.039	0.566	1.092
DepthGAN (Ours)	**5.068**	**17.655**	0.038	0.551	0.468	**6.051**	**25.335**	**0.028**	**0.502**	**0.329**

normalized to $[-1, 1]$. Since our discriminator can be the role of depth estimator, in ablation studies, we also report the depth prediction accuracy *DP (Real)* on real images from the test set. Besides, we predict the depth for the synthesized RGB image by the pre-trained depth prediction model [43], and compare it with the generated depth to form the evaluation metric *DP (Fake)*.

Baselines. [1] We compare against seven state-of-the-art methods for 3D-aware image synthesis: HoloGAN [24], RGBD-GAN [28], GRAF [35], GIRAFFE [27], π-GAN [4], StyleNeRF [13] and VolumeGAN [41]. For a fair comparison, we also incorporate depth information into existing methods by either employing another discriminator for depth learning or changing the input/output from RGB to RGBD image, which result in the two variants, GRAF(D) and π-GAN(D). Another baseline is a 2D-based approach named SeFa [37], which can rotate the scenes through interpolation in the latent space.

---

[1] More details of the baselines can be found in the *Supplementary Material*.

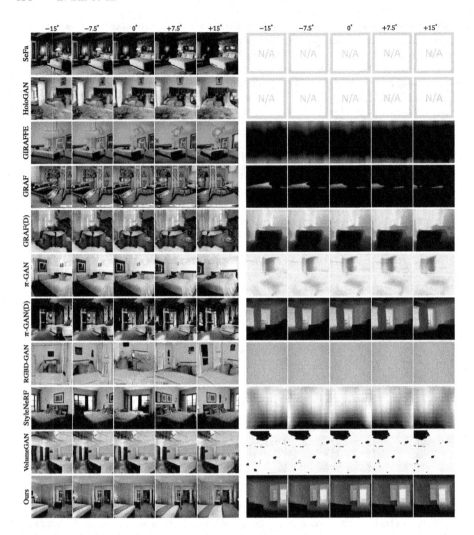

Fig. 3. Qualitative comparisons on LSUN bedroom [44] with existing 3D-aware image synthesis models. *Left:* RGB images. *Right:* The corresponding depths. Each scene is evenly rotated by 30 °C to generate five samples. Zoom in for details (Color figure online)

4.1   Quantitative Results

Table 1 reports the quantitative comparisons. Since RGBD-GAN does not learn geometry at all as shown in Figs. 3 and 4, it makes no sense to evaluate it with RP, RC, and CD. GIRAFFE fixes the background and rotates objects only, and thus we do not report RP, RC, and CD on it. We show significant improvement of image quality compared with 3D-aware image synthesis baselines in terms of FID scores on both RGB images and depth images. When 3D-aware image

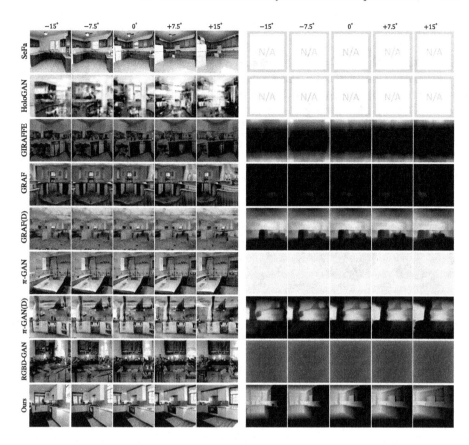

Fig. 4. Qualitative comparisons on LSUN kitchen [44] with existing 3D-aware image synthesis models. *Left:* RGB images. *Right:* The corresponding depths. Each scene is evenly rotated by 30 °C to generate five samples. Zoom in for details (Color figure online)

synthesis methods are given with the depth information for training, the quality of the geometry generally improves, but the quality of the appearance decreases. While maintaining the high quality of images, ours ensures the 3D consistency as well. Note that while π-GAN has lower RP, RC, and CD values than ours, it produces depth maps with simple geometry reflected by FID on depth images and the qualitative results, which makes it easier to maintain consistency.

4.2 Qualitative Results

The generated images from each baseline and our DepthGAN are shown in Figs. 3 and 4. 2D-GANs can generate RGB images of high quality. However, interpolation in the latent space does not guarantee 3D consistency, and thus both the geometry and appearance can be changed during rotation. Though 3D-aware

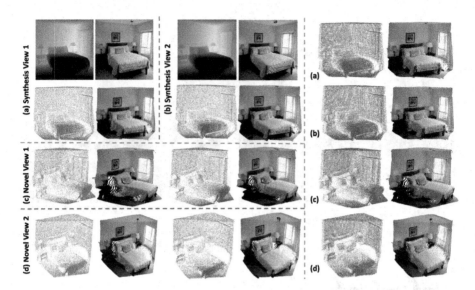

Fig. 5. Geometry visualization. *Left:* (a, b) Original syntheses (*i.e.*, depth and RGB on the top row) from two different views by our dual-path generator, as well as the extracted geometry (*i.e.*, point cloud and rendered appearance on the bottom row). (c, d) Two novel views through rotating (a, b). ***Right:*** Overlaying the two synthesized geometry and visualizing them from four different views (same as those on the *Left*). Zoom in for details (Color figure online)

image synthesis methods can synthesize RGB images of discernible scenes, they generally fail to learn a reasonable geometry unsupervisedly for both the bedrooms and the kitchens. This indicates that in the previous works, generating a visually-pleasing RGB image does not require a good understanding of the underlying 3D geometry. Besides, the image quality degrades significantly compared with that of 2D GANs. With the help of ground-truth depth information, GRAF(D) and π-GAN(D) can generate geometries of higher quality but sacrifice the quality of the appearance. In contrast, our DepthGAN can generate images with reasonable geometries and photo-realistic appearance simultaneously, which mitigates the gap between the 3D-aware image synthesis and 2D GANs and surpass the recent 3D-aware image synthesis methods on scene generation as well.

We also analyze the consistency by visualizing the geometry learnt by Depth-GAN in Fig. 5, where we drag point clouds from two RGBD images to the same view and overlay them. The overlapped point clouds on the right side demonstrate geometry rendered from two views are consistent with each other (see pillow, lamp, and edge of bed). Besides, they complement each other for a more complete point cloud (*i.e.*, fewer holes on the overlapped point clouds).

Table 2. Ablation study conducted on LSUN bedroom dataset [44] under 128×128 resolution. FID [15] regarding RGB images and depths, rotation precision (RP) and rotation consistency (RC), depth prediction (DP) on real and fake samples are used as the metrics to evaluate the synthesis quality and the 3D controllability. ↓ means lower value is better

	FID↓	FID (D)↓	RP↓	RC↓	DP (Real)↓	DP (Fake)↓
$w/o\ L_{dp}^r,\ L_{dp}^f$	4.882	26.518	0.067	0.711	-	0.317
$w/o\ L_{dp}^f$	5.441	24.633	0.066	0.683	1.334	0.318
$w/o\ L_{rgb}^{rot}$	5.038	24.834	0.067	0.716	1.303	0.315
$w/o\ L_d^{rot}$	4.504	**8.315**	0.152	1.196	1.279	0.311
w/o condition	24.062	119.917	0.097	1.443	1.242	0.343
w/o condition, rotation	**2.793**	21.225	-	-	1.205	0.312
Ours-full	4.040	18.874	**0.040**	**0.530**	**1.201**	**0.310**

Fig. 6. Diverse synthesis via varying the appearance latent code z_{rgb}, with the depth latent code z_d fixed. We can tell that all samples are with the same geometry, benefiting from our *dual-path* generator that conditions the appearance generation branch on the depth branch (Color figure online)

4.3 Ablation Study

We analyze the effectiveness of each component of DepthGAN. Evaluation results are shown in Table 2. There is a discrepancy between the generated depth and RGB images if there is no depth prediction loss. As such, the discriminator struggles to lead the generator to capture a coherent relationship between the geometry and the appearance, and all the metrics drop significantly. Without the rotation-consistency loss on RGB images, the RGB consistency completely depends on the conditioning and the discriminator, which forces the network to figure out the consistency on RGB images by itself. While the network is working hard to learn such consistency, it hinders other aspects of learning to some extent. To test the performance of DepthGAN without rotation-consistency loss on depth, we allow the rotation-consistency loss on RGB images to back-propagate the gradients to G_d, which is different from our original design. Without the consistency loss on the rotation of depth images, there are fewer constraints on the depth generation, resulting in a lower FID score on the generated

Fig. 7. Diverse geometries via varying the depth latent code z_d, with the appearance latent code z_{rgb} fixed. We can observe the appropriate alignment between the depth image and the corresponding appearance, where all RGB images are rendered with the same style (Color figure online)

depth images. However, the rotation precision and consistency measurements experience a significant drop due to the lack of explicit supervision on the depth rotation. We also report the result without conditioning appearance features on depth features. For discriminator, the depth prediction from a real image is preserved to enhance the 3D knowledge within it. When rotation consistency loss is included for training, where the generator has the same structure as RGBD-GAN [28], the network is unable to capture the correct depth-appearance pair. If rotation consistency loss is removed, the generator is the same with that of StyleGAN2 [19]. Although the FID score on RGB images is lower, the network fails to view the scene from different angles directly and thus lacks 3D knowledge.

4.4   Controllable Image Synthesis

Disentanglement. With the design of the dual-path generator, the latent spaces of the two generators are separate and thus can be sampled independently. This allows for clear disentanglement of the geometry and appearance. Figure 6 shows the cases where the latent codes for depth generation are fixed, and the latent codes are changed for varying appearance. The underlying 3D geometries are the same for all the images within the same row while the styles keep changing. On the contrary, images in Fig. 7 share the same style but the

Fig. 8. Interpolation results regarding both the depth (*i.e.*, the top two rows) and the appearance (*i.e.*, the bottom row). It it noteworthy that interpolation in the latent space is *different* from rotating the viewpoint, as the two depth codes for interpolation are not guaranteed to represent the same geometry. Instead, this figure only verifies the continuity of the latent spaces in our learned model

geometries are various, which is brought by a fixed latent code for G_{rgb} and different latent codes for G_d.

Linearity. To demonstrate that two latent spaces learned by DepthGAN are semantically meaningful, we linearly interpolate between two latent codes from one latent space and fix the latent code from the other latent space. The interpolation results are shown in Fig. 8.

4.5   Discussion

Rotation. Although the choice of rotation axis as the central one relieves the constraint that the camera stays in the same sphere with all scenes located on the center, it brings a large variety of rotation distributions. Thus, during the rotation, the newly generated view may be out of the manifold learned during training, resulting in unsatisfactory images. The access to the prior distribution of the rotation axes from real data may ease the problem. As the current rotation range is $[-15°, 15°]$, we do not take special treatment for occlusion. However, it is an inevitable problem if the angle range is required to be larger, which we leave for future exploration.

Ground-Truth 3D Information. The quality of the generated 3D-aware images highly relies on the performance of the pre-trained depth prediction methods. We notice that for some objects such as light on the ceiling and some windows or paintings on the wall, there is no depth information available (e.g., images in Figs. 6 and 7). This is due to the fact that the pre-trained depth prediction model fails to predict depths for such minute details. Introducing real depth images collected by depth sensors into the training should alleviate this limitation.

5 Conclusion

In this work, we present DepthGAN, which can learn the appearance and the underlying geometry of indoor scenes simultaneously. DepthGAN takes depth as the 3D prior to facilitate the learning of 3D-aware image synthesis. A dual-path generator and a switchable discriminator are carefully designed to make sufficient use of the depth prior. Experimental results demonstrate the superiority of our approach over existing methods from both the image quality and the 3D controllability perspectives.

Acknowledgement. We thank Yinghao Xu and Sida Peng for their fruitful discussions and valuable comments.

References

1. Alhaija, H.A., Mustikovela, S.K., Geiger, A., Rother, C.: Geometric image synthesis. In: Asian Conference on Computer Vision (2018)
2. Ashburner, J., Friston, K.J.: Voxel-based morphometry-the methods. Neuroimage **11**, 805–821 (2000)
3. Brock, A., Donahue, J., Simonyan, K.: Large scale GAN training for high fidelity natural image synthesis. In: International Conference on Learning Representations (2019)
4. Chan, E., Monteiro, M., Kellnhofer, P., Wu, J., Wetzstein, G.: pi-GAN: periodic implicit generative adversarial networks for 3D-aware image synthesis. In: Conference on Computer Vision and Pattern Recognition (2021)
5. Chan, E.R., et al.: Efficient geometry-aware 3D generative adversarial networks. arXiv preprint arXiv:2112.07945 (2021)
6. Chen, X., Cohen-Or, D., Chen, B., Mitra, N.J.: Towards a neural graphics pipeline for controllable image generation. In: Computer Graphics Forum (2021)
7. Cheung, G.K., Kanade, T., Bouguet, J.Y., Holler, M.: A real time system for robust 3D voxel reconstruction of human motions. In: IEEE Conference on Computer Vision and Pattern Recognition (2000)
8. Cordts, M., et al.: The cityscapes dataset for semantic urban scene understanding. In: IEEE Conference on Computer Vision and Pattern Recognition (2016)
9. Deng, Y., Yang, J., Xiang, J., Tong, X.: Gram: Generative radiance manifolds for 3D-aware image generation. In: Conference on Computer Vision and Pattern Recognition (2022)
10. DeVries, T., Bautista, M.A., Srivastava, N., Taylor, G.W., Susskind, J.M.: Unconstrained scene generation with locally conditioned radiance fields. In: International Conference on Computer Vision (2021)
11. Fu, H., Gong, M., Wang, C., Batmanghelich, K., Tao, D.: Deep ordinal regression network for monocular depth estimation. In: IEEE Conference on Computer Vision and Pattern Recognition (2018)
12. Goodfellow, I., et al.: Generative adversarial nets. In: Advances in Neural Information Processing Systems (2014)
13. Gu, J., Liu, L., Wang, P., Theobalt, C.: Stylenerf: A style-based 3D-aware generator for high-resolution image synthesis. In: International Conference on Learning Representations (2022)

14. Härkönen, E., Hertzmann, A., Lehtinen, J., Paris, S.: GANSpace: discovering interpretable GAN controls. In: Advances in Neural Information Processing Systems (2020)
15. Heusel, M., Ramsauer, H., Unterthiner, T., Nessler, B., Hochreiter, S.: GANs trained by a two time-scale update rule converge to a local nash equilibrium. In: Advances in Neural Information Processing Systems (2017)
16. Karras, T., Aila, T., Laine, S., Lehtinen, J.: Progressive growing of GANs for improved quality, stability, and variation. In: International Conference on Learning Representations (2018)
17. Karras, T., et al.: Alias-free generative adversarial networks. In: Advances in Neural Information Processing Systems (2021)
18. Karras, T., Laine, S., Aila, T.: A style-based generator architecture for generative adversarial networks. In: IEEE Conference on Computer Vision and Pattern Recognition (2019)
19. Karras, T., Laine, S., Aittala, M., Hellsten, J., Lehtinen, J., Aila, T.: Analyzing and improving the image quality of StyleGAN. In: IEEE Conference on Computer Vision and Pattern Recognition (2020)
20. Menze, M., Geiger, A.: Object scene flow for autonomous vehicles. In: IEEE Conference on Computer Vision and Pattern Recognition (2015)
21. Michalkiewicz, M., Pontes, J.K., Jack, D., Baktashmotlagh, M., Eriksson, A.: Implicit surface representations as layers in neural networks. In: International Conference on Computer Vision (2019)
22. Mildenhall, B., Srinivasan, P.P., Tancik, M., Barron, J.T., Ramamoorthi, R., Ng, R.: NeRF: representing scenes as neural radiance fields for view synthesis. In: Vedaldi, A., Bischof, H., Brox, T., Frahm, J.-M. (eds.) ECCV 2020. LNCS, vol. 12346, pp. 405–421. Springer, Cham (2020). https://doi.org/10.1007/978-3-030-58452-8_24
23. Silberman, N., Hoiem, D., Kohli, P., Fergus, R.: Indoor segmentation and support inference from RGBD images. In: Fitzgibbon, A., Lazebnik, S., Perona, P., Sato, Y., Schmid, C. (eds.) ECCV 2012. LNCS, vol. 7576, pp. 746–760. Springer, Heidelberg (2012). https://doi.org/10.1007/978-3-642-33715-4_54
24. Nguyen-Phuoc, T., Li, C., Theis, L., Richardt, C., Yang, Y.L.: HoloGAN: Unsupervised learning of 3d representations from natural images. In: International Conference on Learning Representations (2019)
25. Nguyen-Phuoc, T., Richardt, C., Mai, L., Yang, Y.L., Mitra, N.: BlockGAN: Learning 3D object-aware scene representations from unlabelled images. In: Advances in Neural Information Processing Systems (2020)
26. Niemeyer, M., Geiger, A.: Campari: camera-aware decomposed generative neural radiance fields. arXiv preprint arXiv:2103.17269 (2021)
27. Niemeyer, M., Geiger, A.: GIRAFFE: representing scenes as compositional generative neural feature fields. In: Conference on Computer Vision and Pattern Recognition (2021)
28. Noguchi, A., Harada, T.: RGBD-GAN: Unsupervised 3D representation learning from natural image datasets via RGBD image synthesis. In: International Conference on Learning Representations (2020)
29. Ohtake, Y., Belyaev, A., Seidel, H.P.: Ridge-valley lines on meshes via implicit surface fitting. In: ACM SIGGRAPH (2004)
30. Or-El, R., Luo, X., Shan, M., Shechtman, E., Park, J.J., Kemelmacher-Shlizerman, I.: StyleSDF: high-resolution 3D-consistent image and geometry generation. arXiv preprint arXiv:2112.11427 (2021)

31. Ranftl, R., Lasinger, K., Hafner, D., Schindler, K., Koltun, V.: Towards robust monocular depth estimation: mixing datasets for zero-shot cross-dataset transfer. IEEE Trans. Pattern Anal. Mach, Intell (2020)
32. Rusinkiewicz, S., Levoy, M.: Efficient variants of the ICP algorithm. In: International Conference on 3-D digital imaging and modeling (2001)
33. Rusu, R.B., Cousins, S.: 3D is here: point cloud library (PCL). In: IEEE International Conference on Robotics and Automation (2011)
34. Schnabel, R., Wahl, R., Klein, R.: Efficient RANSAC for point-cloud shape detection. In: Computer Graphics Forum (2007)
35. Schwarz, K., Liao, Y., Niemeyer, M., Geiger, A.: GRAF: generative radiance fields for 3D-aware image synthesis. In: Advance Neural Information Processing Systems (2020)
36. Shen, Y., Yang, C., Tang, X., Zhou, B.: InterFaceGAN: interpreting the disentangled face representation learned by GANs. IEEE Trans. Pattern Anal. Mach, Intell (2020)
37. Shen, Y., Zhou, B.: Closed-form factorization of latent semantics in GANs. In: IEEE Conference on Computer Vision and Pattern Recognition (2021)
38. Sitzmann, V., Martel, J.N., Bergman, A.W., Lindell, D.B., Wetzstein, G.: Implicit neural representation with periodic activation functions. In: Advances in Neural Information Processing Systems (2020)
39. Wang, X., Gupta, A.: Generative image modeling using style and structure adversarial networks. In: Leibe, B., Matas, J., Sebe, N., Welling, M. (eds.) ECCV 2016. LNCS, vol. 9908, pp. 318–335. Springer, Cham (2016). https://doi.org/10.1007/978-3-319-46493-0_20
40. Wu, J., Zhang, C., Xue, T., Freeman, W.T., Tenenbaum, J.B.: Learning a probabilistic latent space of object shapes via 3D generative-adversarial modeling. In: Advances in Neural Information Processing Systems (2016)
41. Xu, Y., Peng, S., Yang, C., Shen, Y., Zhou, B.: 3D-aware image synthesis via learning structural and textural representations. In: IEEE Conference on Computer Vision and Pattern Recognition (2022)
42. Yang, C., Shen, Y., Zhou, B.: Semantic hierarchy emerges in deep generative representations for scene synthesis. Int. J. Comput. Vis. **129**(5), 1451–1466 (2021). https://doi.org/10.1007/s11263-020-01429-5
43. Yin, W., Zhang, J., Wang, O., Niklaus, S., Mai, L., Chen, S., Shen, C.: Learning to recover 3D scene shape from a single image. In: IEEE Conference on Computer Vision and Pattern Recognition (2021)
44. Yu, F., Seff, A., Zhang, Y., Song, S., Funkhouser, T., Xiao, J.: LSUN: construction of a large-scale image dataset using deep learning with humans in the loop. arXiv preprint arXiv:1506.03365 (2015)
45. Zhou, Q.Y., Park, J., Koltun, V.: Open3D: a modern library for 3D data processing. arXiv:1801.09847 (2018)
46. Zhu, J., et al.: Low-rank subspaces in GANs. In: Advances in Neural Information Processing Systems (2021)
47. Zhu, J.Y., et al.: Visual object networks: Image generation with disentangled 3D representations. In: Advances in Neural Information Processing Systems (2018)

Deep Portrait Delighting

Joshua Weir[(✉)] [iD], Junhong Zhao[(✉)] [iD], Andrew Chalmers[iD],
and Taehyun Rhee[(✉)] [iD]

Computational Media Innovation Centre, Victoria University of Wellington,
Wellington, New Zealand
{josh.weir,j.zhao,andrew.chalmers,taehyun.rhee}@vuw.ac.nz
https://www.wgtn.ac.nz/cmic

Abstract. We present a deep neural network for removing undesirable
shading features from an unconstrained portrait image, recovering the
underlying texture. Our training scheme incorporates three regularization strategies: masked loss, to emphasize high-frequency shading features; soft-shadow loss, which improves sensitivity to subtle changes in
lighting; and shading-offset estimation, to supervise separation of shading and texture. Our method demonstrates improved delighting quality
and generalization when compared with the state-of-the-art. We further
demonstrate how our delighting method can enhance the performance of
light-sensitive computer vision tasks such as face relighting and semantic
parsing, allowing them to handle extreme lighting conditions.

Keywords: Uniform lighting · Texture recovery · Shadow removal ·
Portrait

1 Introduction

Image delighting is a form of image manipulation that aims to remove unwanted
lighting features from images to recover the underlying texture. Recovering this
underlying texture benefits many light sensitive computer vision tasks such as
face recognition, parsing and relighting. Delighting has seen much research within
these application areas, but they have only addressed the delighting problem
implicitly in their pipeline. None of them have a dedicated delighting solution
independent from their application. For example, the *relighting* application area
has seen rapid progress with deep neural networks able to render convincing non-
lambertian shading effects [19,31,32,41,44,48]. However, the *delighting* phase is
often abstracted. As a consequence, shading features present in the input image
often propagate into the output as distortions. In extreme cases, they can alter
the identity and perceived structure of portraits, affecting face recognition [3,12].

The portrait delighting problem is inherently more difficult due to the
ambiguous combination of lighting and reflectance determining the colour of any

---

Supplementary Information The online version contains supplementary material
available at https://doi.org/10.1007/978-3-031-19787-1_24.

Fig. 1. Given a portrait image (top row), we perform *delighting* (bottom row): removing undesirable lighting characteristics and reconstructing the image under uniform lighting.

pixel. Theoretically, supervised image-to-image translation pipelines can learn this problem given enough labeled image pairs, but acquiring a representative dataset for this task is laborious and expensive. As a result, most researchers are dependant on 3D renderings of scanned human subjects [21,23,36], which consequently leads to poor generalization to real-world images due to oversimplified reflectance and geometry modelling. On the other hand, real human image datasets available to the public suffer from either lighting or subject under-representation. The CMU Multi-PIE dataset [14] for instance, records only 19 lighting conditions for over 300 individuals, while the Extended Yale-B dataset [25] provides 64 lightings for only 38 individuals.

Even with high-quality data [32,41], many challenges still persist, particularly for images exhibiting complex lighting features like reflections and hard shadow boarders; these features have small pixel densities, making common pixel-wise loss functions like ℓ_1 and ℓ_2 ineffective; also, these features are highly embedded in the underlying texture we wish to preserve, making them difficult to remove without losing high-frequency details such as freckles and facial hair. Most recent relighting methods that utilize a delighting phase usually incorporate standard network architectures and loss functions which do not directly address the problems of dataset sparsity or non-lambertian lighting.

We present a fully-supervised portrait delighting method that takes an upper-body portrait lit under an arbitrary illumination, and outputs its reconstruction under uniform white lighting (see Fig. 1).

We localize high-frequency lighting effects in our training data using a guided-filter technique, and incorporate these into our training pipeline using a masked loss function inspired by [19] to emphasize small but visually significant regions of the image.

We estimate a shading offset image, which is the difference between the input image and the ground-truth de-lit image. This facilitates learning of lighting features directly, and greatly improves colour consistency. We also synthesize soft-shaded images from our training data and utilize them in training, applying a small regularization loss to their outputs. This alleviates our dataset bias toward directional lightings, allowing us to remove both hard and soft shadows. We demonstrate how our framework benefits the applications of semantic parsing and relighting.

To summarize, our main contributions are as follows:

- A novel portrait delighting method that can recover the underlying texture of portraits illuminated under a wide range of complex lighting environments.
- Three novel loss functions: *shading-offset* loss, *soft-shadow* loss and *masked* loss that improve our models robustness to unseen lighting environments while preserving image detail.
- Our delighting method can serve as a data normalization tool for improving light-sensitive computer vision tasks such as face relighting and semantic parsing.

2   Related Work

Much work has been made towards removing disrupting light features (*e.g.* dark shadows, specularities) from faces primarily to enhance face recognition systems [6,13], and to increase image quality [5,30,50]. Pioneering works in this field propose optimization methods using Morphable Face models [2,4,7,11,43]. However, relying on parametric models limits their ability to capture non-facial or high-frequency details. Recent deep-learning methods tackle this problem using feature-wise perceptual losses [26] or closed-loop GANs [15] to recover facial details. A drawback of all these methods however is that they either focus exclusively on face regions, perform only one aspect of the delighting process (*e.g.* shadow removal [50], camera-flash removal [5]), or use front-facing illumination as the ground-truth [15,30], ignoring the sharp reflections this causes.

GAN inversion methods [1,10,27,46] enable face editing by projecting images into the latent space of a pre-trained GAN, disentangling lighting, identity, pose and expression attributes such that they can be independently manipulated. These methods can effectively remove sharp shadows and specular reflections, but their reconstructions don't often preserve image content that isn't constrained by editing attributes, particularly high-frequency details such as freckles, and non-face components like clothing.

Intrinsic decomposition methods *delight* by disentangling face [34,36,37] or full-body [21,23] images into geometry, albedo, and lighting via separate convolutional neural networks, where *relighting* can be performed by re-rendering with modified lighting. Estimation errors usually occur, such as when specularities become embedded in the albedo, or when hard shadows are predicted as geometry features. This becomes a bigger problem when relying on synthetic training data with simplified reflectance models, so Sengupta *et al.* [36] proposed a semi-supervised learning framework using reconstruction loss on real images, but conflation between shading, geometry and texture is still present when there's a significant domain-gap between the real and synthetic data.

Other deep learning methods perform portrait relighting directly [19,40,41, 48,51] by co-opting standard encoder-decoder architectures like U-Net [35]. In this manner, both the *delighting* and *relighting* processes are a black box represented by network activations, with supervision applied only to the re-lit image. While this generally leads to more stable and expressive relightings than intrinsic decomposition [41], the neural representation of the delighting task lacks any meaningful form to facilitate supervision. Some researchers attempt to supervise the delighting process by imposing feature-space losses on the network bottleneck [48,51]. While this helps recover global details such as colour and identity, it often preserves local artifacts such as those caused by facial shadows.

The most recent relighting methods utilize an explicit *delighting* phase as a preliminary step [31,32,44], which is the inspiration for our work. Pandey *et al.* [32] use a least-squares GAN [28] with VGG-perceptual loss [20] in their albedo prediction for accurate detail reconstruction, but colour inconsistencies arise when presented with data unseen in training (*e.g.* different clothing patterns). Wang *et al.* [44] supervise the delighting process by predicting the source lighting and a face-parsing map for more robust texture recovery, but their method struggles to remove high-frequency shading artifacts caused by strong directional light.

3   Method

This section overviews how we synthesize training images (Sect. 3.1), and integrate them into our delighting pipeline (Sect. 3.2 to Sect. 3.5).

3.1   Data Processing

We modify the Multi-PIE dataset [14] to synthesize our training data for portrait delighting. We chose this dataset as it is the largest publicly available dataset of real subjects. While other datasets capture more lighting directions [13,38], they lack diversity in terms of pose, expression, and clothing. Figure 2 illustrates our full data processing pipeline.

Each static image in the Multi-PIE database was captured under 19 illuminations; 18 directional flashes, and one non-flash image (room lights only). We first remove the effects the room lights have on our flash images (details are described in the supplementary document), resulting in 18 one-light-at-a-time (OLAT) images [41] (see Fig. 2 (b)), where the target de-lit image (I_{dlt}) is the average over all our OLAT images, with added luminance from our non-flash image. From this, we perform foreground masking and face-parsing. (see Fig. 2 (e)).

We stress that the de-lit image I_{dlt} is different from an albedo image. The difference is that the de-lit image contains subtle light occlusions on non-convex geometric areas such as the ears, nose, and clothing.

To render input images, other learning based methods [32,41,44,48] used Image Based Lighting (IBL). But the mostly front-facing point lights used in

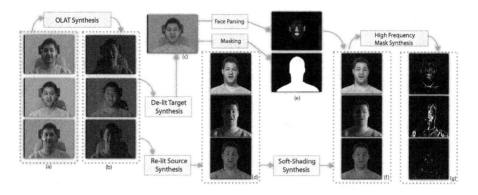

Fig. 2. Our data processing pipeline

Multi-PIE prevent us from adequately sampling 360° environment maps. We instead approximate environment illuminations using a weighted average of two OLAT images, each tinted with different colour temperatures (see Fig. 2 (d)) similar to the VIDIT dataset [18]. This colour assumption in reasonable, as it covers the majority of light found in the real world. High-intensity light images were generated in the same manner by increasing the image brightness, and colour adjusted versions of the de-lit, and room-light-only images were also added to the dataset.

We chose a sample of 140 subjects (70 male, 70 female) under two arbitrary poses, expressions and clothing, producing 240 unique images (220 for training, 20 for testing). We use 1,293 lighting conditions for the training set, and 1,180 for the test set, producing 284,460 images for training, and 25,860 for testing.

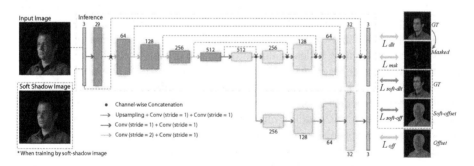

Fig. 3. Overview of our network structure. We predict a shading-offset via a separate decoder during training. All convolutions have a kernel size of 3, and are followed by Instance normalization [42] and PReLU activation [17]. *Tanh* activations are used to produce final outputs.

3.2   Basic Architecture

We train a U-Net [35] based CNN to estimate the ground-truth de-lit image $\mathbf{I}_{dlt}$ from a foreground-masked upper-body portrait $\mathbf{I}_{src}$ by minimizing a perceptual loss term $\mathcal{L}_{perc}$, which based on previous works [20,29,32] has been shown to recover sharp details at multiple scales. This is computed as the ℓ_1 distance between the activations of the VGG-16 network [39] pre-trained on ImageNet [9]. We also apply a small ℓ_1 loss to speed up color convergence:

$$\mathcal{L}_{\mathbf{perc}}(\mathbf{A}, \mathbf{B}) = \sum_{i=1}^{5} \left[\frac{1}{N_i} \| VGG_i(\mathbf{A}) - VGG_i(\mathbf{B}) \|_1 \right] + \frac{0.2}{M} \| \mathbf{A} - \mathbf{B} \|_1, \qquad (1)$$

where $\mathbf{A}$ and $\mathbf{B}$ are images of the same subject with M number of foreground pixels, VGG_i and N_i are the outputs and sizes respectively of the final activations before the ith max-pooling layer in VGG-16. We apply this function to our de-lit output $\mathcal{D}_1(\mathbf{I}_{src})$ as:

$$\mathcal{L}_{\mathbf{dlt}} = \mathcal{L}_{\mathbf{perc}}(\mathbf{I}_{\mathbf{dlt}}, \mathcal{D}_1(\mathbf{I}_{\mathbf{src}})). \qquad (2)$$

This loss alone motivates a direct style transfer from $\mathbf{I}_{src}$ to $\mathbf{I}_{dlt}$ without necessarily learning a physical separation of shading and texture, which consequently leads to increased over-fitting to certain modalities of our training data (*i.e.* lighting distribution, clothing patterns) not fully represented in real-world portraits. Shortcomings of this basic architecture are mitigated via our proposed contributions in Sect. 3.3, 3.4 and 3.5.

3.3   Shading Offset

We add a separate decoder branch $\mathcal{D}_2$ to our network to learn the difference between $\mathbf{I}_{src}$ and $\mathbf{I}_{dlt}$ in the form of a shading-offset image: $\mathbf{I}_{off} = \mathbf{I}_{src} - \mathbf{I}_{dlt}$. These two images differ only in their shading, so learning the difference between them allows our encoder to learn a meaningful separation of shading and texture. Different from some prior works that obtain their results by subtracting the estimated offset image from the input [5,30], we produce our result directly via our delighting decoder $\mathcal{D}_1$ as this makes our model less susceptible to small estimation errors in our shading-offset. Our network pipeline is shown in Fig. 3.

We apply perceptual loss to our offset output $\mathcal{D}_2(\mathbf{I}_{src})$ as:

$$\mathcal{L}_{\mathbf{off}} = \mathcal{L}_{\mathbf{perc}}(\mathbf{I}_{\mathbf{off}}, \mathcal{D}_2(\mathbf{I}_{\mathbf{src}})). \qquad (3)$$

3.4   Soft Shadowed Images

Since the light capture setup of our dataset is too sparse to represent non-directional lighting characteristics such as soft-shadows, we approximate these effects using a guided filter [16] technique to smooth the source image, while preserving edges in the ground-truth image (see Fig. 2 (f)).

$$\mathbf{I_{soft}} = M_{nose} \odot \Omega(\mathbf{I_{src}}, \mathbf{I_{dlt}}, \epsilon)$$
$$+ M_{mouth} \odot \Omega(\mathbf{I_{src}}, \mathbf{I_{dlt}}, \epsilon) \qquad (4)$$
$$+ M_{other} \odot \Omega(\mathbf{I_{src}}, \mathbf{I_{dlt}}, \kappa),$$

where $\Omega(I, R, r)$ is the guided filter function (I is the input image, R is the edge reference, and r is the window radius). To preserve morphological features of the face, we use smaller sized filters for the nose and mouth regions ($\epsilon \leq \kappa$). This is done using the masks M_{nose} and M_{mouth} extracted from face parsing. The resulting image approximates the effect of increasing the area of all lights in the input image, softening sharp shadow boarders and specular highlights.

To regularize the strong influence of $\mathcal{L}_{dlt}$ and $\mathcal{L}_{off}$, we apply a small ℓ_1 loss to the outputs of $\mathbf{I_{soft}}$ as:

$$\mathcal{L}_{\text{soft-dlt}} = \frac{0.6}{M} ||\mathbf{I_{dlt}} - \mathcal{D}_1(\mathbf{I_{soft}})||_1, \qquad (5)$$

$$\mathcal{L}_{\text{soft-off}} = \frac{0.6}{M} ||\mathbf{I_{soft-off}} - \mathcal{D}_2(\mathbf{I_{soft}})||_1, \qquad (6)$$

where $\mathbf{I_{soft-off}} = \mathbf{I_{soft}} - \mathbf{I_{dlt}}$.

3.5   High-Frequency Mask

Similar to [19], We emphasize sharp lighting discontinuities (e.g. shadow boarders, reflections) in our loss function via a weight mask $\mathbf{W}$, which we generate using the gradient difference between $\mathbf{I_{src}}$ and its soft-shadowed version (see Fig. 2 (g)).

$$a = 10 * \max(\Delta \mathbf{I_{src}} - \Delta\Omega(\mathbf{I_{src}}, \mathbf{I_{dlt}}, 15), 0)$$
$$b = median(a) \qquad (7)$$
$$\mathbf{W} = \min(b + gauss(b), 1),$$

where Δ represents the sum of directional gradients along the vertical and horizontal axes. For high-frequency shading features, the gradient of $\mathbf{I_{src}}$ should be higher than its filtered counterpart, which we store in a. Afterwards, we apply a median filter to remove noise, and add a small Gaussian blur to increase its receptive field to neighbouring pixels.

Our method is different from Hou et al. [19], who fit a morphable model to the face, estimate the lighting, and perform ray-casting to produce a shadow mask. Their method is more useful for relighting than delighting, since estimation errors could cause the shadow mask to miss the true shadow boarder. Ours is less vulnerable to estimation errors, and can target areas outside the face.

We compliment the high-level feature activations of $\mathcal{L}_{dlt}$ with this importance mask:

$$\mathcal{L}_{\mathbf{msk}} = \sum_{i=1}^{3} \frac{1}{S_i} ||W_i \odot \left(VGG_i(\mathbf{I_{dlt}}) - VGG_i(\mathcal{D}_1(\mathbf{I_{src}})) \right)||_1, \qquad (8)$$

where W_i is the high-frequency shading mask resized to fit the dimensions of VGG_i, and S_i is the sum of W_i.

Our full pipeline is illustrated in Fig. 3, where the final loss is the sum:

$$\mathcal{L} = \mathcal{L}_{\text{dlt}} + \mathcal{L}_{\text{off}} + \mathcal{L}_{\text{soft-dlt}} + \mathcal{L}_{\text{soft-off}} + \mathcal{L}_{\text{msk}}. \tag{9}$$

4   Results

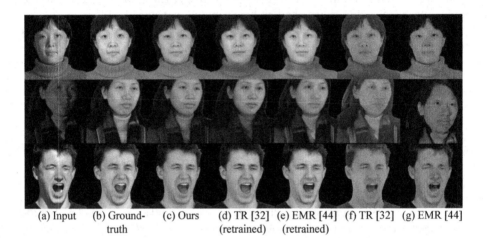

(a) Input   (b) Ground-   (c) Ours   (d) TR [32]   (e) EMR [44]   (f) TR [32]   (g) EMR [44]
                     truth   (retrained)   (retrained)

Fig. 4. Evaluation on our testing dataset. Notice how our method is the only one to remove the scarf shadow in the middle row. A different crop of this subject was used for EMR since their method was trained on only face regions.

4.1   Implementation and Data Setup

We implement our model in PyTorch [33], and train for 4 epochs with a learning rate of 0.0002 using the Adam optimizer [22]. All images are resized to 256×256 resolution, with pixel values normalized to $[-1, 1]$. The average running time of our network for delighting (without shading-offset prediction) is 25 ms on a NVIDIA GTX 1080 GPU.

To prepare the training data in our experiments, we apply random flips along the vertical axis, and random cropping with window sizes chosen uniformly from $[280, 480]$, where 480 covers the entire image. When generating soft-shadow images in Eq. 4, the large filter radius κ is chosen randomly from $[7, 35]$ to increase robustness to different penumbra sizes, while the nose and mouth radius ϵ is fixed at 7.

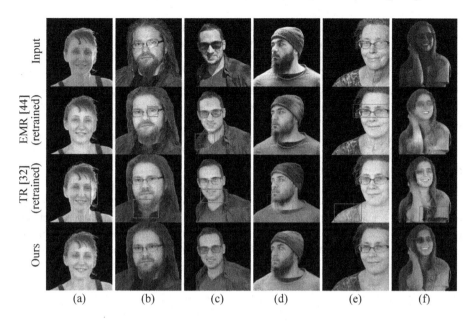

Fig. 5. Evaluations on in-the-wild images. Red boxes emphasize areas in previous works improved by our method. Notice how our method provides the most consistent results in terms of delighting and texture preservation. (Color figure online)

Prior Work. We compare our method with the albedo prediction modules of two state-of-the-art methods: Explicit Multiple Reflectance Channel Modeling (EMR) [44] and Total Relighting (TR) [32], which have demonstrated superior performance over previous related works. TR apply VGG-perceptual loss (see Eq. 1) on the estimated albedo, along with a Least-Squares GAN [28] discriminator to remove high-frequency shading. EMR use only ℓ_1 loss on the albedo, and estimate the source illumination and a face parsing map as auxiliary tasks to improve training stability.

Retrained Models. Besides the pretrained models from prior work, we *retrained* EMR and TR on our dataset, denoted EMR (retrained) and TR (retrained) respectively. We based our implementations of their models on the albedo prediction networks and loss functions described in their papers [32,44]. For EMR (retrained), We use our segmentations in Fig. 2 (e) as the ground-truth for face parsing. and since our pipeline doesn't utilize 360° environment maps, we use our shading-offset decoder $\mathcal{D}_2$ as a substitute for their source lighting estimator. This allows us to use the lighting information found in our shading-offset images to perform this task, just as we do in our method (see Fig. 3).

4.2 Qualitative Evaluations

In Fig. 4, we compare our results qualitatively against these two methods on our testing dataset. From the results, we can see that our method outperforms both

EMR and EMR (retrained) in terms of shadow and specular removal (middle and bottom rows). Our method also recovers large-scale textures, whereas TR and TR (retrained) are prone to incorrect colour estimations as can be seen with the magenta shirt in the top row. Although TR handles most shading effects on the face, it still leaves shadow boarders in some extreme cases (bottom row (f)), while our proposed method can remove them effectively, indicating that our method is generalized to harsh cases.

We show results on in-the-wild images in Fig. 5. Here, we see that our method is able to delight subjects under a wide range of complex illuminations, while also preserving important details such as clothing patterns and facial hair. Our method handles variant conditions more gracefully than other works. For example, TR (retrained) often removes important content details such as beards ((b) and (d)), while EMR (retrained) preserves shadow boarders and sharp reflections ((b) and (c)), and is unstable when images contain large non-face components (f). In (a), we demonstrate another failure case of TR (retrained) where light is coming from behind the subject. Although this lighting condition is missing from our dataset, our model generalizes to this case very well by the contribution of our soft-shadow loss (see Sect. 4.4 for further insight).

4.3   Quantitative Evaluations

Table 1. Results of delighting on our testing dataset. Arrows indicate whether loss is minimized ($\downarrow$) or maximized ($\uparrow$). Our method outperforms prior works on all metrics.

Method	Metric			
	RMSE$\downarrow$	SSIM$\uparrow$	li-SSIM$\uparrow$	LPIPS$\downarrow$
EMR (retrained)	0.047	0.940	0.949	0.047
TR (retrained)	0.056	0.934	0.948	0.048
Proposed	**0.044**	**0.946**	**0.955**	**0.037**

Quantitative evaluations on our testing dataset are shown Table 1, where we compare our delighting performance with TR (retrained) and EMR (retrained) using the following metrics: root mean squared error (RMSE), structural similarity (SSIM) [45] luminance-invariant SSIM (li-SSIM) and learned perceptual image patch similarity (LPIPS) [49] (version 0.1). Our li-SSIM is like traditional SSIM with luminance parameter $\alpha = 0$. This is meant to avoid biases to large-scale colour variations, penalizing contrast and structural errors only. From the results we can see that our method outperforms other works across all metrics. Notably, TR gains a significant improvement on the li-SSIM metric over SSIM, but is still behind our method. These results reflect consistent improvements in most of our qualitative images. More of our results can be found in the supplementary document.

4.4   Ablation Study

Our baseline model consists of our network trained using only the basic loss $\mathcal{L}_{\mathbf{dlt}}$ (see Eq. 2). We add each of our proposed losses onto this model to evaluate their contributions. Quantitative results of all our ablations using our testing dataset are shown in Table 2.

Shading-Offset Loss. We show the benefits provided by our offset decoder $\mathcal{D}_2$ by training an additional model with shading-offset loss $\mathcal{L}_{\mathbf{off}}$ (see Eq. 3) added to our baseline. Qualitative results in Fig. 6 (c) demonstrate more stable texture recovery and light removal when using offset loss. The likely reason is that it could make the learned latent features more discriminative between different colour variations caused by shading and texture. In Table 2 (**Testing Dataset** block), we see that our model with offset loss performs better than our baseline across all metrics, especially RMSE and LPIPS, indicating improved large-scale texture recovery.

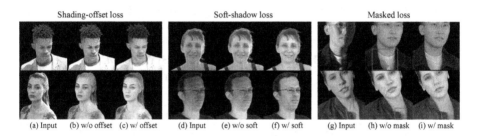

Fig. 6. Ablation study on (a, b, c) shading-offset loss (d, e, f) soft-shadow loss and (g, h, i) masked loss. Red boxes in (h) highlight shadows. (Color figure online)

Soft-Shadow Loss. We add the soft-shadow loss $\mathcal{L}_{\mathbf{soft\text{-}dlt}} + \mathcal{L}_{\mathbf{soft\text{-}off}}$ (see Eq. 5 & 6) to our shading-offset ablation. While our perceptual loss in $\mathcal{L}_{\mathbf{dlt}}$ allows for stable delighting under harsh illumination, it can lead to over-fitting due to the mostly front-facing directional lights from our training data, and can't cope with diffuse illuminations very well. From the results in Fig. 6 (f), we can see that adding our soft-shadow regularization not only improves performance on diffuse illumination conditions (bottom row), but also enables our model to recognize rare edge cases such as in the top row, where we observe that strong light diffractions through the hair and around the left cheek are removed.

To better quantify the benefits of our soft-shadow regularization, we create a new testing dataset (dubbed as **Alt. Lighting**) using the face relighting method of Hou *et al.* [19] to obtain sufficiently variant testing samples. The de-lit images from our original testing dataset were relit under 13 lighting conditions from a wide range of angles not seen in our training dataset. Quantitative results in Table 2 (**Alt. Lighting** block) demonstrates how our soft-shadow loss improves robustness to these conditions, particularly on the RMSE and LPIPS metrics.

Masked Loss. We add our masked loss $\mathcal{L}_{\mathbf{msk}}$ (Eq. 7) to our soft-shadow ablation. As the results in Fig. 6 (i) show, our masked loss improves the removal
of complex shadows that are intertwined with clothing features. It is noteworthy that although the model without $\mathcal{L}_{\mathbf{msk}}$ can handle a wide-range of harsh
illumination conditions situated around the face area, abnormal cast shadows,
particularly around the torso are often preserved. By adding extra importance
to these regions via $\mathcal{L}_{\mathbf{msk}}$, our model reduces the rate at which they get preserved in the result. Quantitative results in Table 2 show that our model without
$\mathcal{L}_{\mathbf{msk}}$ outperforms our proposed method on most metrics, although we can see
the benefit of $\mathcal{L}_{\mathbf{msk}}$ in our testing cases visually. This could mean the weighted
mask creates a bias towards these relatively small regions in the loss function,
at the expense of larger image structures.

Table 2. Ablation results on our loss functions. $\mathcal{L}_{\mathbf{soft}} = \mathcal{L}_{\mathbf{soft\text{-}dlt}} + \mathcal{L}_{\mathbf{soft\text{-}off}}$. We
evaluate against two datasets: Testing Dataset (left block) and Alt. Lighting (right
block), where our testing dataset is the same one used in Table 1.

Method		Metric					
		Testing dataset			Alt. lighting		
		RMSE↓	SSIM↑	LPIPS↓	RMSE↓	SSIM↑	LPIPS↓
(A)	$\mathcal{L}_{\mathbf{dlt}}$	0.056	0.943	0.042	0.040	0.979	0.052
(B)	(A)+$\mathcal{L}_{\mathbf{off}}$	0.046	**0.949**	**0.035**	0.041	0.979	0.052
(C)	(B)+$\mathcal{L}_{\mathbf{soft}}$	0.046	**0.949**	**0.035**	**0.026**	**0.985**	**0.033**
(D)	(C)+$\mathcal{L}_{\mathbf{msk}}$	**0.044**	0.946	0.037	**0.026**	0.984	0.037

4.5   Edge Cases

While the Multi-PIE [14] dataset exhibits much diversity in terms of clothing,
expression and identity, it nonetheless has significant domain biases such as a
95% European/Asian demographic captured under mostly front-facing directional lightings. This can lead to failure cases when presented with dark skin
albedos, and challenging illuminations such as light coming from extreme angles
or shadows cast by foreign objects. In this section, we offer a qualitative evaluation on these under-represented cases.

We present our evaluation on people with dark skin in Fig. 7 (a-c). From the
results, we can see that our method is capable of recovering dark skin textures,
although noticeable lightening is observed.

Dark skin-tone evaluation   Challenging shading evaluation

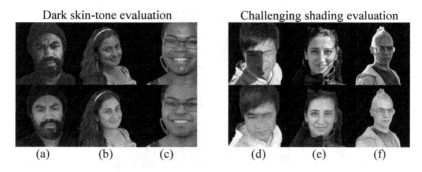

(a)   (b)   (c)   (d)   (e)   (f)

Fig. 7. Edge cases: Our results on images under-represented in our training data. Source images of (d) & (e) are from the evaluation dataset of [50].

In Fig. 7 (d-f), we demonstrate our results on (d, e) irregularly shaped shadows, and (f) lighting from predominantly behind the subject. While no such images were present in our training data, our model nonetheless generalizes to these cases surprisingly well with a few minor artifacts.

5 Applications

(a) Input   (b) Relight 1   (c) Relight 2   (d) Relight 3

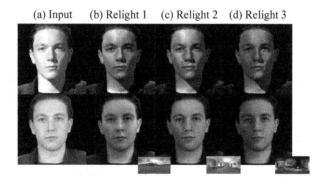

Fig. 8. Relighting. The results of face relighting [19] (top row) before and (bottom row) after our delighting step.

Face Relighting. Figure 8 illustrates the benefits of our delighting as a pre-processing step for other relighting methods. Through our shadow removal and colour normalization, we achieve more natural looking results on HDRI environments. Relighting is performed using the method of Hou *et al.* [19], which we adapt for environment map renderings by taking a weighted sum of 512 unique illuminations (similar to [8]).

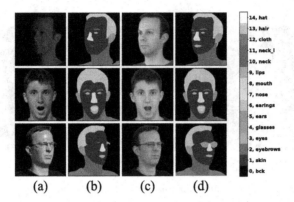

Fig. 9. Face Parsing: Here we show (a) the input image with (b) its semantic parsing output, and (c) our de-lit image with (d) its semantic parsing output.

Face Parsing. Figure 9 illustrates the importance of delighting when performing semantic segmentation on portraits. Our model offers significant performance gains to this task by enhancing the visual clarity of glasses, eyes and mouth regions. Also noteworthy is the neck shadow (middle-row (a)), where without delighting, the neck region outside is classified as clothing (middle-row (b)). The face parser used is a BiSeNet model [47] trained on the CelebAMask-HQ dataset [24].

6   Conclusion

We propose a deep neural network for delighting portrait images under a wide range of illumination conditions. Texture recovery and generalization is improved via estimating shading-offset images, using soft-shadow variants of the input, and our weighted loss function. Each contribution greatly improves delighting performance over the previous work in terms of removing shading features and preserving image content. We tested our model as a useful preprocessing tool for other computer vision tasks.

Failure cases can arise due to biases in the training data as outlined in Sect. 4.5, so future work must consider fairness in terms of physical traits and lighting variations when creating and using datasets. To prevent softening of dark textures (Fig. 7 (f)), future work can also focus on estimating global image statistics, such as ambient light intensity, which will place a lower bound on the darkest regions of the image caused by shading.

Acknowledgment. This work was supported by the Entrepreneurial University Programme from the Tertiary Education Commission, and MBIE Smart Idea Programme by Ministry of Business, Innovation and Employment in New Zealand. We thank all image providers, including Flickr users: "Debarshi Ray", "5of7" and "photographer695", whose photographs were cropped, and processed by our neural network. Sources are provided in the supplementary material.

References

1. Abdal, R., Zhu, P., Mitra, N.J., Wonka, P.: Styleflow: attribute-conditioned exploration of stylegan-generated images using conditional continuous normalizing flows. ACM Trans. Graphics (TOG) **40**(3), 1–21 (2021)
2. Ahmed, A., Farag, A.: A new statistical model combining shape and spherical harmonics illumination for face reconstruction. In: Bebis, G., et al. (eds.) ISVC 2007. LNCS, vol. 4841, pp. 531–541. Springer, Heidelberg (2007). https://doi.org/10.1007/978-3-540-76858-6_52
3. Beveridge, J.R., Bolme, D.S., Draper, B.A., Givens, G.H., Lui, Y.M., Phillips, P.J.: Quantifying how lighting and focus affect face recognition performance. In: 2010 IEEE Computer Society Conference on Computer Vision and Pattern Recognition-Workshops, pp. 74–81. IEEE (2010)
4. Blanz, V., Vetter, T.: A morphable model for the synthesis of 3d faces. In: Proceedings of the 26th Annual Conference on Computer Graphics and Interactive Techniques, pp. 187–194 (1999)
5. Capece, N., Banterle, F., Cignoni, P., Ganovelli, F., Scopigno, R., Erra, U.: Deepflash: turning a flash selfie into a studio portrait. Signal Proces. Image Commun. **77**, 28–39 (2019)
6. Chen, B.C., Chen, C.S., Hsu, W.H.: Face recognition and retrieval using cross-age reference coding with cross-age celebrity dataset. IEEE Trans. Multimedia **17**(6), 804–815 (2015)
7. Chen, X., Wu, H., Jin, X., Zhao, Q.: Face illumination manipulation using a single reference image by adaptive layer decomposition. IEEE Trans. Image Process. **22**(11), 4249–4259 (2013)
8. Debevec, P., Hawkins, T., Tchou, C., Duiker, H.P., Sarokin, W., Sagar, M.: Acquiring the reflectance field of a human face. In: Proceedings of the 27th Annual Conference on Computer Graphics and Interactive Techniques, pp. 145–156 (2000)
9. Deng, J., Dong, W., Socher, R., Li, L.J., Li, K., Fei-Fei, L.: Imagenet: a large-scale hierarchical image database. In: 2009 IEEE Conference on Computer Vision and Pattern Recognition, pp. 248–255. IEEE (2009)
10. Deng, Y., Yang, J., Chen, D., Wen, F., Tong, X.: Disentangled and controllable face image generation via 3d imitative-contrastive learning. In: Proceedings of the IEEE/CVF Conference on Computer Vision and Pattern Recognition, pp. 5154–5163 (2020)
11. Egger, B., et al.: 3d morphable face models-past, present, and future. ACM Trans. Graph. (TOG) **39**(5), 1–38 (2020)
12. Fahmy, G., El-Sherbeeny, A., Mandala, S., Abdel-Mottaleb, M., Ammar, H.: The effect of lighting direction/condition on the performance of face recognition algorithms. In: Biometric Technology for Human Identification III, vol. 6202, p. 62020J. International Society for Optics and Photonics (2006)
13. Georghiades, A.S., Belhumeur, P.N., Kriegman, D.J.: From few to many: Illumination cone models for face recognition under variable lighting and pose. IEEE Trans. Pattern Anal. Mach. Intell. **23**(6), 643–660 (2001)
14. Gross, R., Matthews, I., Cohn, J., Kanade, T., Baker, S.: Multi-pie. Image Vision Comput. **28**(5), 807–813 (2010)

15. Han, X., Yang, H., Xing, G., Liu, Y.: Asymmetric joint gans for normalizing face illumination from a single image. IEEE Trans. Multimedia **22**(6), 1619–1633 (2019)
16. He, K., Sun, J., Tang, X.: Guided image filtering. IEEE Trans. Pattern Anal. Mach. Intell. **35**(6), 1397–1409 (2012)
17. He, K., Zhang, X., Ren, S., Sun, J.: Delving deep into rectifiers: Surpassing human-level performance on imagenet classification. In: Proceedings of the IEEE International Conference on Computer Vision, pp. 1026–1034 (2015)
18. Helou, M.E., Zhou, R., Barthas, J., Süsstrunk, S.: Vidit: virtual image dataset for illumination transfer. arXiv preprint arXiv:2005.05460 (2020)
19. Hou, A., Zhang, Z., Sarkis, M., Bi, N., Tong, Y., Liu, X.: Towards high fidelity face relighting with realistic shadows. In: Proceedings of the IEEE/CVF Conference on Computer Vision and Pattern Recognition, pp. 14719–14728 (2021)
20. Johnson, J., Alahi, A., Fei-Fei, L.: Perceptual losses for real-time style transfer and super-resolution. In: Leibe, B., Matas, J., Sebe, N., Welling, M. (eds.) ECCV 2016. LNCS, vol. 9906, pp. 694–711. Springer, Cham (2016). https://doi.org/10.1007/978-3-319-46475-6_43
21. Kanamori, Y., Endo, Y.: Relighting humans: occlusion-aware inverse rendering for full-body human images. ACM Trans. Graph. (TOG) **37**(6), 1–11 (2018)
22. Kingma, D.P., Ba, J.: Adam: a method for stochastic optimization. arXiv preprint arXiv:1412.6980 (2014)
23. Lagunas, M., et al.: Single-image full-body human relighting (2021)
24. Lee, C.H., Liu, Z., Wu, L., Luo, P.: Maskgan: towards diverse and interactive facial image manipulation. In: IEEE Conference on Computer Vision and Pattern Recognition (CVPR) (2020)
25. Lee, K., Ho, J., Kriegman, D.: Acquiring linear subspaces for face recognition under variable lighting. IEEE Trans. Pattern Anal. Mach. Intelligence **27**(5), 684–698 (2005)
26. Ling, S., Lin, Y., Fu, K., You, D., Cheng, P.: A high-performance face illumination processing method via multi-stage feature maps. Sensors **20**(17), 4869 (2020)
27. Mallikarjun, B., et al.: Photoapp: photorealistic appearance editing of head portraits. ACM Trans. Graph. **40**(4), 1–16 (2021)
28. Mao, X., Li, Q., Xie, H., Lau, R.Y., Wang, Z., Paul Smolley, S.: Least squares generative adversarial networks. In: Proceedings of the IEEE International Conference on Computer Vision, pp. 2794–2802 (2017)
29. Martin-Brualla, R., et al.: Lookingood: enhancing performance capture with real-time neural re-rendering. arXiv preprint arXiv:1811.05029 (2018)
30. Nagano, K., et al.: Deep face normalization. ACM Trans. Graph. (TOG) **38**(6), 1–16 (2019)
31. Nestmeyer, T., Lalonde, J.F., Matthews, I., Lehrmann, A.: Learning physics-guided face relighting under directional light. In: Proceedings of the IEEE/CVF Conference on Computer Vision and Pattern Recognition, pp. 5124–5133 (2020)
32. Pandey, R., et al.: Total relighting: learning to relight portraits for background replacement. ACM Trans. Graph. (TOG) **40**(4), 1–21 (2021)
33. Paszke, A., et al.: Pytorch: an imperative style, high-performance deep learning library. Adv. Neural. Inf. Process. Syst. **32**, 8026–8037 (2019)
34. Qiu, Y., Xiong, Z., Han, K., Wang, Z., Xiong, Z., Han, X.: Learning inverse rendering of faces from real-world videos (2020)
35. Ronneberger, O., Fischer, P., Brox, T.: U-Net: Convolutional networks for biomedical image segmentation. In: Navab, N., Hornegger, J., Wells, W.M., Frangi, A.F. (eds.) MICCAI 2015. LNCS, vol. 9351, pp. 234–241. Springer, Cham (2015). https://doi.org/10.1007/978-3-319-24574-4_28

36. Sengupta, S., Kanazawa, A., Castillo, C.D., Jacobs, D.W.: Sfsnet: learning shape, reflectance and illuminance of facesin the wild'. In: Proceedings of the IEEE Conference on Computer Vision and Pattern Recognition, pp. 6296–6305 (2018)

37. Shu, Z., Yumer, E., Hadap, S., Sunkavalli, K., Shechtman, E., Samaras, D.: Neural face editing with intrinsic image disentangling. In: Proceedings of the IEEE Conference on Computer Vision and Pattern Recognition, pp. 5541–5550 (2017)

38. Sim, T., Baker, S., Bsat, M.: The cmu pose, illumination, and expression (pie) database. In: Proceedings of Fifth IEEE International Conference on Automatic Face Gesture Recognition, pp. 53–58. IEEE (2002)

39. Simonyan, K., Zisserman, A.: Very deep convolutional networks for large-scale image recognition. arXiv preprint arXiv:1409.1556 (2014)

40. Song, G., Cham, T.J., Cai, J., Zheng, J.: Half-body portrait relighting with overcomplete lighting representation. In: Computer Graphics Forum. Wiley Online Library (2021)

41. Sun, T., et al.: Single image portrait relighting. ACM Trans. Graph. **38**(4), 1–79 (2019)

42. Ulyanov, D., Vedaldi, A., Lempitsky, V.: Instance normalization: the missing ingredient for fast stylization. arXiv preprint arXiv:1607.08022 (2016)

43. Wang, Y., et al.: Face relighting from a single image under arbitrary unknown lighting conditions. IEEE Trans. Pattern Anal. Mach. Intell. **31**(11), 1968–1984 (2008)

44. Wang, Z., Yu, X., Lu, M., Wang, Q., Qian, C., Xu, F.: Single image portrait relighting via explicit multiple reflectance channel modeling. ACM Trans. Graph. (TOG) **39**(6), 1–13 (2020)

45. Wang, Z., Bovik, A.C., Sheikh, H.R., Simoncelli, E.P.: Image quality assessment: from error visibility to structural similarity. IEEE Trans. Image Process. **13**(4), 600–612 (2004)

46. Xia, W., Zhang, Y., Yang, Y., Xue, J.H., Zhou, B., Yang, M.H.: Gan inversion: a survey. arXiv preprint arXiv:2101.05278 (2021)

47. Yu, C., Wang, J., Peng, C., Gao, C., Yu, G., Sang, N.: Bisenet: bilateral segmentation network for real-time semantic segmentation (2018)

48. Zhang, L., Zhang, Q., Wu, M., Yu, J., Xu, L.: Neural video portrait relighting in real-time via consistency modeling. arXiv preprint arXiv:2104.00484 (2021)

49. Zhang, R., Isola, P., Efros, A.A., Shechtman, E., Wang, O.: The unreasonable effectiveness of deep features as a perceptual metric. In: Proceedings of the IEEE Conference on Computer Vision and Pattern Recognition, pp. 586–595 (2018)

50. Zhang, X., et al.: Portrait shadow manipulation. ACM Trans. Graph. (TOG) **39**(4), 1–78 (2020)

51. Zhou, H., Hadap, S., Sunkavalli, K., Jacobs, D.W.: Deep single-image portrait relighting. In: Proceedings of the IEEE/CVF International Conference on Computer Vision, pp. 7194–7202 (2019)

Vector Quantized Image-to-Image Translation

Yu-Jie Chen[1,2], Shin-I Cheng[1,2], Wei-Chen Chiu[1,2(✉)] ⓘ, Hung-Yu Tseng[3], and Hsin-Ying Lee[4]

[1] National Chiao Tung University, Hsinchu, Taiwan
`walon@cs.nctu.edu.tw`
[2] MediaTek-NCTU Research Center, Hsinchu, Taiwan
[3] Meta, Menlo Park, USA
[4] Snap Inc., Santa Monica, USA
https://cyj407.github.io/VQ-I2I/

Abstract. Current image-to-image translation methods formulate the task with conditional generation models, leading to learning only the recolorization or regional changes as being constrained by the rich structural information provided by the conditional contexts. In this work, we propose introducing the vector quantization technique into the image-to-image translation framework. The vector quantized content representation can facilitate not only the translation, but also the unconditional distribution shared among different domains. Meanwhile, along with the disentangled style representation, the proposed method further enables the capability of image extension with flexibility in both intra- and inter-domains. Qualitative and quantitative experiments demonstrate that our framework achieves comparable performance to the state-of-the-art image-to-image translation and image extension methods. Compared to methods for individual tasks, the proposed method, as a unified framework, unleashes applications combining image-to-image translation, unconditional generation, and image extension altogether. For example, it provides style variability for image generation and extension, and equips image-to-image translation with further extension capabilities.

Keywords: Image-to-image translation · Vector quantization · Image synthesis · Generative models

1 Introduction

Image-to-image translation (I2I) aims to learn the mapping between different visual domains. Upon being formulated as a conditional generation problem, I2I

Y.-J. Chen and S.-I. Cheng—Equal contribution.

Supplementary Information The online version contains supplementary material available at https://doi.org/10.1007/978-3-031-19787-1_25.

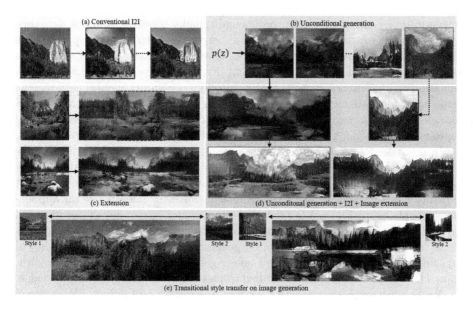

Fig. 1. Applications of Vector Quantized Image-to-Image Translation. Our proposed method enables several applications: (a) conventional image-to-image translation, (b) unconditional image generation, (c) image extension, (d) arbitrary combination of aforementioned operations, *e.g.* translation and extension on unconditionally generated images, and (e) image generation with transitional stylization. Here we use green frame for summer images and blue frame for winter images. (Color figure online)

methods can tackle translation with either paired [13] or unpaired data [34], and perform diverse translations by disentangling the content and style factors of each input domain [12,17,36]. These I2I methods unleash various applications, such as style transfer [11], synthesis from semantic map or layout [2,10,29,32], domain adaptation [6,24], and super-resolution [16].

Most existing I2I methods model the task as a pixel-level conditional generation problem. However, as the conditional contexts are already informative in structure and details, the translation tends to learn simple recolorization or regional transformation without understanding the real target distribution. Is it possible to jointly learn the translation as well as the unconditional distribution to fully exploit the data and make both trainings mutually beneficiary? One intuitive formulation is to define a domain-invariant joint latent distribution, then perform domain-specific maximum likelihood estimation on it. Pixel space is a natural option for the joint latent distribution, yet it struggles to scale due to its computational expensive auto-regressive process. Recently, vector quantization (VQ) technique has shown its effectiveness as an intermediate representation of generative models [7,30]. We thus explore in this work the applicability of adopting vector quantization as the latent representation in the I2I task.

We introduce VQ-I2I, a framework that adopts a vector quantized codebook as an intermediate representation which is able to enable both the image-to-image translation and the unconditional generation of input domains. VQ-I2I consists of a joint domain-invariant content encoder, domain-specific style encoders, and domain-specific decoders. The joint content encoder enforces a shared latent distribution among different domains. The encoded content representation can be further decoded with the style representation obtained from the same input for realizing the self-reconstruction or with that from different inputs for achieving intra- and inter-domain translations. Moreover, with different style representations being given, VQ-I2I is also able to perform diverse translations.

In addition to conventional image-to-image translation, we learn an autoregressive model on the joint quantized content space to unconditionally synthesize the latent content representation. The capability of unconditional content generation with disentangled style representation can unleash several applications: As shown in Fig. 1, VQ-I2I has the multifunctionality for performing I2I, unconditional image synthesis, and image extension. Combining these operations, VQ-I2I can achieve extension on generated samples with the flexibility of stylizing into different domains, and image generation with transitional stylization. These cannot be done by a unified framework to the best of our knowledge.

We conduct extensive quantitative and qualitative evaluations. We measure the realism with the Fréchet inception distance (FID) [9] and subjective study using the AFHQ [4], Yosemite [34], and portrait [17] datasets. On the Cityscapes dataset [5], we use FID metric as well to compare with the I2I methods trained upon paired data. Qualitatively, we demonstrate realistic and diverse I2I translation as well as applications including unconditional generation, image extension, completion, transitional stylization, or combinations over them.

2   Related Work

2.1   Image-to-Image Translation

Image-to-image translation, first addressed in [13], aims at learning the mapping function between the source and the target domain. Following works focus on tackling two major challenges: how to handle unpaired data and how to model diverse translations. Cycle-consistency is adopted to handle unpaired data [20,34], while augmented attribute space is proposed to provide diversity [35]. Following efforts are made to handle both challenges jointly [12,17,18], one-sided translation without cycle-consistency [28], to improve the diversity [21,22], and to better handle geometric transformations [14]. Take a step forward, we propose a framework that can perform not only cross-domain translations, but also enable unconditional generation and image extension using the learned representation.

2.2   Vector Quantized Generative Models

Generative models can be roughly divided into two streams: implicit and explicit density estimation methods. Generative adversarial network, the representative of the implicit method, has been dominant due to its high-fidelity synthesized

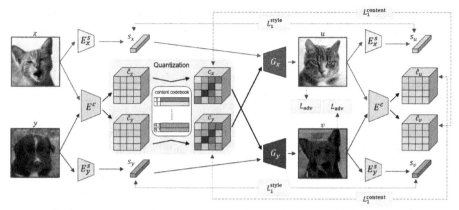

(a) Overall architecture of vector-quantized I2I with disentangled representations.

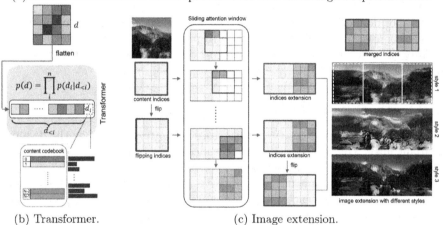

(b) Transformer.   (c) Image extension.

Fig. 2. Method Overview. (a) The proposed framework learns to perform translation with disentangled vector-quantized domain-invariant content and domain-specific style representations. (b) Given the quantized content indices d, we can learn the content distribution in an autoregressive manner using a transformer model. (c) With learned transformer model and the translation model, we can expand an image on both horizontal sides by spatially extending the content map and its flipped one with a sliding attention window. The extended content can be further translated into different styles. (Color figure online)

images, yet suffering from instability in training. On the other hand, explicit methods are more tractable in training but limited to relatively blurred outputs (*e.g.* variational autoencoder (VAE) [15]) or in scaling due to the pixel-level auto-regressive process (*e.g.* PixelRNN [25] and PixelCNN [26]). Recently, vector quantization (VQ) technique has adopted explicit methods to alleviate the scaling issue with quantized latent vectors serving as latent representation [8,27,30,33]. VQGAN then proposes a hybrid framework to first leverage

GAN technique to learn VQ codebook, then adopt transformer [7] to train an auto-regressive model on the learned VQ indices. In this work, we propose adopting VQ technique in the I2I task.

3   Method

As previously motivated, our goal is to leverage the vector quantized codebook, an intermediate representation for 1) image-to-image translation between two visual domains $X \subset \mathbb{R}^{H \times W \times 3}$ and $Y \subset \mathbb{R}^{H \times W \times 3}$ and 2) unconditional generation in each domain. As illustrated in Fig. 2(a), our framework consists of a shared content encoder E^c, a vector quantized content codebook Z, style encoders $\{E_X^s, E_Y^s\}$, generators $\{G_X, G_Y\}$, and discriminators $\{D_X, D_Y\}$. Given an input image, the content encoder E^c extracts the *vector-quantized* domain-invariant representations, while the style encoders E_X^s, E_Y^s compute the domain-specific features for domain X and Y respectively. The generators G_X, G_Y combine the content representation and style feature to produce the image in each domain. Finally, the discriminators D_X, D_Y aim to distinguish between the generated and real images.

3.1   Vector Quantized Content Representation

Our approach leverages the vector quantization strategy to encode the domain-invariant image content information. Specifically, we construct a codebook $Z = \{z_k\}_{k=1}^K$ that consists of learned content codes $z_k \in \mathbb{R}^{n_c}$, where n_c indicates the code dimension. Given a continuous map $\hat{c} \in \mathbb{R}^{h \times w \times n_c}$ extracted by the content encoder E^c, we find for each spatial entry $\hat{c}_{ij} \in \mathbb{R}^{n_c}$ of $\hat{c}$ its closest code in the codebook Z for obtaining the vector quantized content representation c:

$$c = \mathbf{vq}(\hat{c}) := (\arg\min_{z_k \in Z} \|\hat{c}_{ij} - z_k\|) \in \mathbb{R}^{h \times w \times n_c}. \tag{1}$$

Since the quantization operation $\mathbf{vq}$ is not differentiable for gradient back-propagation, we use the straight-through trick [27] that copies the gradient from c to $\hat{c}$. We learn the codebook Z using the self-reconstruction path and the loss function L_{vq} and L_1^{recon}, where

$$L_{vq} = \|sg[\hat{c}] - c\|_2^2 + \|sg[c] - \hat{c}\|_2^2, \tag{2}$$

where $sg[\cdot]$ is the stop-gradient operation. We provide the details of the self-reconstruction path and the loss L_1^{recon} later in Sect. 3.2.

3.2   Diverse Image-to-Image Translation

To enable multi-modal image-to-image translation, our approach learns the disentangled domain-*invariant* content representations and domain-*specific* style features [17,20]. As shown in Fig. 2(a), we use an *shared* encoder E^c to extract the content representation for images of two domains, followed by applying the vector quantization operation $\mathbf{vq}$, and use separate encoders E_X^s, E_Y^s to compute the style features:

$$c_x, s_x = \mathbf{vq}(E^c(x)), E_X^s(x)$$
$$c_y, s_y = \mathbf{vq}(E^c(y)), E_Y^s(y). \tag{3}$$

Since the content space is shared among two domains, we can perform the image-to-image translation by swapping the content representations c_x and c_y. Finally, the generators G_X, G_Y use AdaIN normalization layers [11,12] to combine the swapped content representations and style features to synthesize the translated images $u \in X$ and $v \in Y$:

$$u = G_X(c_y, s_x), \; v = G_Y(c_x, s_y). \tag{4}$$

Image-to-Image Translation Training. We use the discriminators D_X and D_Y to impose the domain adversarial loss L_{adv}. The loss L_{adv} encourages the realism of the translated images u for domain X and v for Y.

Nevertheless, training our model with the domain adversarial loss along cannot guarantee the disentanglement of content and style representations. The content map c may encode the style information, thus the generator ignores the style feature s for synthesizing the translated images. To address this issue, we use the latent style regression loss to enforce the bijection between the style features and the translated images:

$$L_1^{\mathrm{style}} = \|E_X^s(G_X(c_y, s_x)) - s_x\|$$
$$+ \|E_Y^s(G_Y(c_x, s_y)) - s_y\|. \tag{5}$$

We also use the latent content regression loss to facilitate the training:

$$L_1^{\mathrm{content}} = \|E^c(G_X(c_y, s_x)) - c_y\|$$
$$+ \|E^c(G_Y(c_x, s_y)) - c_x\|, \tag{6}$$

where Fig. 2(a) shows the computation flows behind L_1^{style} and L_1^{content}.

Self-reconstruction Training. In addition to image-to-image translation, we also involve a self-reconstruction path (*i.e.* reconstructing an image by using its own content and style representations) during the training stage for two empirical reasons. First, self-reconstruction training is vital for learning a meaningful vector-quantized codebook [27]. Second, it facilitates the overall image-to-image training process. Specifically, we impose the self-reconstruction loss:

$$L_1^{\mathrm{recon}} = \|G_X(c_x, s_x) - x\| + \|G_Y(c_y, s_y) - y\|. \tag{7}$$

As described in Sect. 3.1, we only apply the vector quantization loss L_{vq} (cf. Eq. 2) in the self-reconstruction path. The full objective function of our model (L_D for training discriminators; $L_{E^c, Z, E^s, G}$ for training encoders, codebook, and generators) is then summarized as:

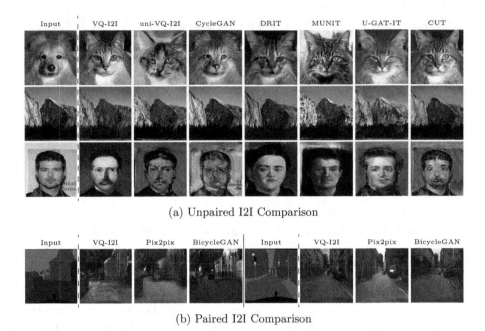

(a) Unpaired I2I Comparison

(b) Paired I2I Comparison

Fig. 3. Qualitative Comparisons with Conventional Image-to-Image Translation Methods. (a) We show the translated results of different methods on three unpaired datasets. From top to bottom rows are dog→cat [4], winter→summer [34], and photo→portrait [17]. (b) Our model is able to handle training with paired data on Cityscapes dataset [5]. For each example set (composed of four columns), the leftmost column shows the semantic segmentation of street scenes, and the other columns show the corresponding generated scenes by various models which are trained on paired data.

$$L_D = L_{\text{adv}},$$
$$L_{E^c, Z, E^s, G} = -\lambda_{\text{adv}} L_{\text{adv}} + \lambda_1^{\text{recon}} L_1^{\text{recon}} + \lambda_{\text{vq}} L_{\text{vq}} \qquad (8)$$
$$+ \lambda_1^{\text{content}} L_1^{\text{content}} + \lambda_1^{\text{style}} L_1^{\text{style}},$$

where λ controls the importance of each loss term. Note that we only optimize the codebook with the vector quantization loss L_{vq} and reconstruction loss L_1^{recon}.

3.3 Unconditional Generation

Vector quantization on the shared content space enables unconditional generation, since we can model the domain-invariant joint (content) distribution using an autoregressive approach [1]. We present the approach in Fig. 2(b). The spatial entries in the content representation c can be represented as a set of indices d in the codebook $Z = \{z_k\}_{k=1}^K$, where $c_{ij} = z_{d_{ij}}$. By ordering the index set d using a particular rule, the content generation task can be formulated as the next-index prediction problem. Specifically, given content indices $d_{<i}$, the goal

Input	Inter-domain	Intra-domain	Input	Completion	+Inter-I2I

Fig. 4. Diverse Image Translation and Completion. (*left*) We demonstrate both inter-domain and intra-domain translations with the query images (leftmost column) combined with various styles on the dog→cat and winter→summer scenarios. (*right*) Given a quarter of an image from AFHQ [4] or Yosemite [34] dataset as the input, we perform image completion AND the inter-domain translation. VQ-I2I is able to not only learn the joint content distribution of both domains, thus achieving reasonable completion, but support the diverse translation via the design of the disentanglement.

is to predict the distribution of next index d_i: $p(d) = \prod_i p(d_i|d_{<i})$. We train a transformer network [7] for this task by maximizing the log-likelihood of the content representation:

$$L_{\text{transformer}} = \mathbb{E}_{x \sim p(x)}[-\log p(d)]. \tag{9}$$

We provide the ordering details (*i.e.* slight difference between training and testing stages as similar to [7]) in the supplementary materials.

During inference, we first generate the complete content representation using the autoregressive next-index prediction process. Then we combine some style features $\{s_x, s_y\}$, and use the generators $\{G_X, G_Y\}$ to synthesize the image for different domains.

3.4 Content Extension

Our autoregressive next-index prediction process enables not only the unconditional content generation, but also *content extension*: extending the content of existing images. We illustrate the process in Fig. 2(c). Specifically, given a vector quantized content representation extracted from an existing image, we use the learned transformer model to spatially extend the content map (red outline). By flipping the content representation horizontally, we can extend the content to the opposite direction using the same process (blue outline). The resultant content map which has been extended (on both horizontal sides) then can be gone through generators together with a style feature to produce the extension.

4   Experiments

We evaluate the proposed framework on image translation, unconditional generation and image extension. We compare VQ-I2I with several representative I2I, image generation and outpainting approaches. We then demonstrate various applications of our framework which seamlessly combine I2I with unconditional image generation, image extension, and transitional stylization. Finally, we conduct the ablation study to understand the efficacy of different design choices.

Datasets. We conduct experiments using both paired and unpaired I2I datasets. For unpaired datasets, we use the Yosemite dataset [34] for the shape-invariant translation, and the AFHQ [4] and portrait [17] datasets for the shape-variant translation task. For paired dataset, we use the Cityscapes dataset [5].

Compared Baselines. For the unpaired I2I setting, we compare our method with CycleGAN [34], DRIT [17], MUNIT [12], and recent CUT [28] and U-GAT-IT [14]. For the paired I2I setting, we make a comparison between our method and Pix2pix [13] as well as BicycleGAN [36]. For unconditional generation, we compare our approach with VQGAN [7]. As for image extension [3,19,31], we consider a representative baseline from Boundless [31]. The training details are provided in the supplement.

Furthermore, to understand the impact of having the latent representation explicitly disentangled, we construct an uni-modal VQ-I2I variant as an additional baseline (denoted as uni-VQ-I2I). Specifically, in such uni-VQ-I2I baseline, we assume that the domain-specific style information is implicitly modeled by the generators $\{G_X, G_Y\}$, thus the domain-specific style features are discarded. Please refer to our supplementary materials for more details.

4.1   Qualitative Evaluation

I2I Translation on Unpaired and Paired Data. The proposed VQ-I2I synthesizes high-quality images on both the shape-invariant (winter-to-summer) and shape-variant (dog-to-cat, photo-to-portrait) datasets, as shown in Fig. 3(a), where it achieves comparable or even better quality in comparison to other representative I2I methods. Moreover, the results of uni-modal VQ-I2I variant (denoted as uni-VQ-I2I), which excludes the disentanglement between content and style information are also provided, where we are able to observe that uni-VQ-I2I encounters the problem of texture inconsistency (*e.g.* there exists different styles in the cat's face on the first row of Fig. 3(a)). The comparison between our VQ-I2I and the uni-VQ-I2I variant reveals that the characteristic of disentanglement enables both content stability and style diversity, where we provide further explorations on uni-VQ-I2I in the supplementary materials. On the other hand, given pairs of semantic segmentation maps and corresponding images as training data, our proposed scheme produces appealing images that correspond to the input segmentation map (cf. Fig. 3(b)). These results validate that our

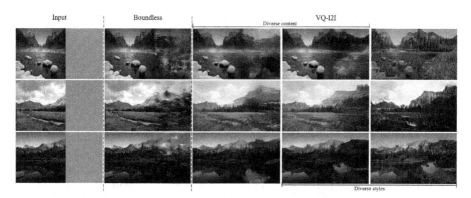

Fig. 5. Qualitative Examples on Image Extension. Example results of image extension on Yosemite [34] datasets, where the comparison with respect to Boundless [31] baseline is also provided. The leftmost column shows the input images for the image extension, where the model takes left portion of size 256×256 for each input image and extends for 192 pixel width toward the right-hand side. VQ-I2I is able to generate smooth and diverse extensions with style variability.

VQ-I2I approach can understand the semantic meaning of labels and synthesize correct instances, such as buildings and vehicles.

Multimodal Translation. Our VQ-I2I framework can also perform style-guided translation that produces diverse (multimodal) I2I results. Since vector-quantized content representation encodes the domain-invariant information while the style features carry the style information, we re-combine the same content with various styles to achieve diverse translations. The results are shown in the left portion of Fig. 4. In addition to the inter-domain I2I, our method can also perform *intra*-domain I2I (as shown in the column labelled as "intra-domain" of Fig. 4, in which we combine the content and style extracted from two images of the same domain), although we do not explicitly involve intra-domain I2I during the training stage.

Diverse Image Extension and Completion. The auto-regressive procedure built upon the content representation of VQ-I2I enables image extension. Specifically, as the content indices on the extended regions are drawn from the conditional distribution predicted by the transformer model, together with the style features being disentangled from content, the resultant extension produced by our VQ-I2I includes the diversity of both content and style (cf. Fig. 5). It is worth noting that the extension results show that VQ-I2I generators would adjust the original image slightly to make the overall appearance of image extension more harmonious. Similar to image extension, our VQ-I2I is able to realize the image completion. We conduct the experiments of image completion on AFHQ [4] and Yosemite [34] dataset and provide some example results in the right portion of Fig. 4, where only a quarter of an image is given as the input. Again, our auto-regressive model and the disentanglement designs are capable of generating diverse content and supporting style variability via combining the translation (*e.g.* inter-domain I2I in the rightmost column of Fig. 4).

4.2   Quantitative Evaluation

We use the Fréchet inception distance (FID) [9] score and natural image quality evaluator (NIQE) [23] to measure the quality of the generated results and compare our proposed method to the existing approaches. Lower FID and NIQE scores indicate better perceptual quality. Moreover, we conduct a user study using the manner of pairwise comparison (*i.e.* our VQ-I2I versus baselines, or the images produced by various methods against the real images).

FID and NIQE. We summarize the FID and NIQE evaluation of unpaired I2I translation in Table 1 and FID measurement of paired I2I translation in Table 3. For unconditional generation, we compute the FID scores on the synthesized image of size 256×256, as shown in Table 2(a). As for image outpainting/extension, we present the quantitative results in Table 2(b), where the FID scores for image extension are computed from the distribution distance between the Yosemite dataset [34] and the rightmost 256×256 pixels of the extended images.

We are able to see that our proposed method performs comparably against the state-of-the-art translation frameworks, generative approach (*i.e.* VQGAN [7]) and the extension baselines (*i.e.* Boundless [31] and Infinity-GAN [19]). Please note that, the main goal of our VQ-I2I is not to achieve superior performance in translation, unconditional generation or extension, instead we aim to facilitate both translation and the unconditional distribution shared among domains in a unified novel framework as well as unleash various interesting applications which other existing works are hard to realize (as described in the next subsection).

User Preference. To better rate the realism of I2I translation and image extension results, we conduct a user study with the manner of pairwise comparison. For I2I translation, each subject (in total $\sim$180 participants) needs to answer the question "Which image is more realistic" given a pair of images (1) sampled from real images and the translated images generated from various I2I baselines or (2) respectively produced by our VQ-I2I and one of the baselines; while for extension, the comparison is conducted between our VQ-I2I and a baseline extension method (*i.e.* Boundless [31]). Figure 6 presents the results of the user study. The performance of VQ-I2I is comparable to those SOTA methods in I2I translation and image extension.

Table 1. Quantitative Comparisons with Unpaired I2I Methods. We measure the FID and NIQE scores across various datasets. VQ-I2I performs comparably to the state-of-the-art methods on unpaired datasets, while enabling applications that cannot be done by these conventional I2I methods.

	FID			NIQE	
	dog→cat	winter→summer	photo→portrait	dog→cat	winter→summer
CycleGAN	76.89	65.71	104.96	**40.65**	53.28
DRIT	35.74	**60.53**	102.52	47.32	**32.76**
MUNIT	33.78	94.78	**94.42**	63.64	35.64
U-GAT-IT	**21.62**	73.89	104.93	59.84	57.44
CUT	22.79	70.41	102.65	48.95	37.29
uni-VQ-I2I	25.65	62.43	99.37	45.41	36.53
VQ-I2I	26.53	63.60	100.29	53.29	35.97

Table 2. Quantitative Comparisons on Applications of VQ-I2I. (a) We evaluate the performance (in terms of FID scores) of unconditional generation on Yosemite [34] dataset via sampling 100 images respectively generated by our VQ-I2I and the VQGAN [7]. (b) Given the input image of size 256×256, we extend it horizontally for 50% and 75% (128 and 192 pixels respectively) toward the right-hand size, where we evaluate the FID scores on the right most portion of size 256×256 of the resultant image (*i.e.* this portion will recover part of the original input image and the extended region). The results show that our model is comparable to the existing extension method while extending for a larger range.

	256×256 generation
VQGAN	127.84
VQ-I2I	127.31

(a) Unconditional Generation.

	Outpaint for 50%	Outpaint for 75%
Boundless [31]	**68.00**	**88.95**
VQ-I2I	77.82	90.05

(b) Image Extension.

4.3 Applications

Unconditional Image Generation and Image Extension. VQ-I2I completes more applications that other existing pixel-level I2I models scarcely achieve, as we adapt the vector quantized representation to the disentangled domain-invariant content space. Combining generated or extended content codes with a replaceable style representation, VQ-I2I can be further utilized in two applications: unconditional image generation and image extension with flexible style modulation in different ways (i.e. the combination between generation/extension and intra- or inter-domain I2I), where we have demonstrate example results in Fig. 4, Fig. 5, and Fig. 1(b)(c)(d). These applications afford to make image synthesis style-oriented, and there are more results provided in the supplementary materials.

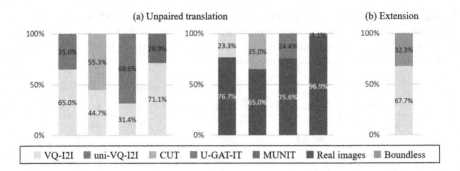

Fig. 6. User Preference Study. We conduct the user study (∼180 participants) to compare VQ-I2I to different existing translation methods in (a), and the Boundless [31] method for the image extension task in (b).

Table 3. Quantitative Comparisons with Paired I2I Methods. We measure FID score for label→cityscapes translation on Cityscapes [5] Dataset.

	FID
	label→cityscapes
Pix2pix [13]	**51.73**
BicycleGAN [36]	93.13
VQ-I2I	74.03

Table 4. Ablation of Varying the Codebook Size and Dimensionality. We measure the FID score for summer→winter translation after 420 epochs of training on each model.

	Codebook size	
	64	512
Dimensionality 64	96.94	96.71
Dimensionality 512	94.38	99.51

Stylized Transitional Generation. In addition to single style modulation on the generated content, we can also perform multi-style transitional transfer via interpolating two styles to produce the style representation. As shown in Fig. 7, We modulate different parts of the content map independently with different proportions by mixing the two styles, and merge all these modulated latents together to generate the transitional stylized output. In detail, for producing smooth and gradually changing effect of stylization, we partition the content map horizontally to 10 equal splits, where some example results are demonstrated in Fig. 1(e) and Fig. 7. More results with different number of splits are provided in the supplementary materials.

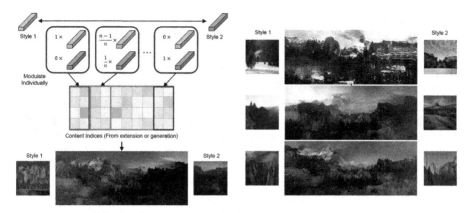

Fig. 7. Advanced Application of our VQ-I2I: Transitional Stylized Image Synthesis. Given two guided styles and the content map (produced from extension or unconditional generation), VQ-I2I is capable of synthesizing images with a smooth and gradually changing stylization effect via blending over two styles.

4.4 Further Investigation

Adding Patchwise Contrastive Loss. As our VQ-I2I framework does not include the cycle consistency as used in CycleGAN [34] or DRIT [17], there could exist a potential concern about being unable to well preserve the geometric information during I2I translation. To address this issue, here we experiment to adopt the patchwise contrastive loss [28], also named as PatchNCE loss, to enhance the content preservation during the training phase. As shown in Fig. 8(a), the performance of using PatchNCE loss is more task-sensitive. Therefore, we consider it as an optional design choice, and use the content/style regression loss as the default design in our framework.

Varying Codebook Size and Dimensionality. To observe the usage of codes in the codebook, we conduct additional ablation on Yosemite dataset [34] by varying the codebook size and the dimensionality of the codebook in VQ-I2I, and the FID scores for summer→winter translation is shown in Table 4. When setting the codebook size as 512 and dimensionality of the codebook as 512, our VQ-I2I model only uses around 35 codes. Besides, when shrinking both codebook size and dimensionality to 64, the codebook utilization grows up to 100%. However, from Table 4, the quantitative differences between different codebook size and dimensionality are imperceptible. Therefore, we suggest that training on Yosemite dataset [34] for a smaller codebook size and dimensionality still maintains its performance and reduces the memory usage of our VQ-I2I model.

FID	VQ-I2I	+PatchNCE
dog→cat	**29.07**	80.72
winter→summer	**65.64**	71.17
photo→portrait	125.37	**114.04**

(a) Ablation study on adopting Patch-NCE Loss in our VQ-I2I model (performance in terms of FID scores).

(b) Qualitative examples of Patchnce loss.

Fig. 8. Ablation of Adding Patchwise Contrastive Loss (PatchNCE Loss). (a) We compute the FID scores with the same input content and style images on AFHQ, Yosemite, and Portrait datasets. The quantitative results reveal that PatchNCE loss makes a strong improvement on photo→portrait translation. (b) Given the input content and style images, the visual results show that PatchNCE loss is beneficial for our VQ-I2I model for preserving the content information on Portrait dataset [17].

5 Conclusion

In this paper, we introduce VQ-I2I, a novel image-to-image translation framework equipped with disentangled and discrete representations. In particular, our method learns a vector-quantized codebook for capturing the domain-invariant content information of input domains, in which such codebook enables the learning of the content distribution via an autoregressive model built upon the transformer network. Upon having comparable quantitative and qualitative performance at image-to-image translation with respect to several baselines, VQ-I2I is especially novel to have multifunctionality integrated into a unified framework, including image-to-image translation, unconditional generation, image extension, transitional stylization, and the combinations of the applications above.

Acknowledgement. This project is supported by MediaTek Inc., MOST (Ministry of Science and Technology, Taiwan) 111-2636-E-A49-003 and 111-2628-E-A49-018-MY4. We are grateful to the National Center for High-performance Computing for computer time and facilities.

References

1. Chen, M., et al.: Generative pretraining from pixels. In: International Conference on Machine Learning (ICML) (2020)
2. Cheng, Y.-C., Lee, H.-Y., Sun, M., Yang, M.-H.: Controllable image synthesis via SegVAE. In: Vedaldi, A., Bischof, H., Brox, T., Frahm, J.-M. (eds.) ECCV 2020. LNCS, vol. 12352, pp. 159–174. Springer, Cham (2020). https://doi.org/10.1007/978-3-030-58571-6_10
3. Cheng, Y.C., Lin, C.H., Lee, H.Y., Ren, J., Tulyakov, S., Yang, M.H.: InOut: diverse image outpainting via GAN inversion. In: IEEE Conference on Computer Vision and Pattern Recognition (CVPR) (2022)

4. Choi, Y., Uh, Y., Yoo, J., Ha, J.W.: StarGAN V2: diverse image synthesis for multiple domains. In: IEEE Conference on Computer Vision and Pattern Recognition (CVPR) (2020)
5. Cordts, M., et al.: The cityscapes dataset for semantic urban scene understanding. In: IEEE Conference on Computer Vision and Pattern Recognition (CVPR) (2016)
6. Deng, W., Zheng, L., Ye, Q., Kang, G., Yang, Y., Jiao, J.: Image-image domain adaptation with preserved self-similarity and domain-dissimilarity for person re-identification. In: IEEE Conference on Computer Vision and Pattern Recognition (CVPR) (2018)
7. Esser, P., Rombach, R., Ommer, B.: Taming transformers for high-resolution image synthesis. In: IEEE Conference on Computer Vision and Pattern Recognition (CVPR) (2021)
8. Han, L., et al.: Show me what and tell me how: video synthesis via multimodal conditioning. In: IEEE Conference on Computer Vision and Pattern Recognition (CVPR) (2022)
9. Heusel, M., Ramsauer, H., Unterthiner, T., Nessler, B., Hochreiter, S.: GANs trained by a two time-scale update rule converge to a local Nash equilibrium. In: Advances in Neural Information Processing Systems (NeurIPS) (2017)
10. Huang, H.-P., Tseng, H.-Y., Lee, H.-Y., Huang, J.-B.: Semantic view synthesis. In: Vedaldi, A., Bischof, H., Brox, T., Frahm, J.-M. (eds.) ECCV 2020. LNCS, vol. 12357, pp. 592–608. Springer, Cham (2020). https://doi.org/10.1007/978-3-030-58610-2_35
11. Huang, X., Belongie, S.: Arbitrary style transfer in real-time with adaptive instance normalization. In: IEEE International Conference on Computer Vision (ICCV) (2017)
12. Huang, X., Liu, M.-Y., Belongie, S., Kautz, J.: Multimodal unsupervised image-to-image translation. In: Ferrari, V., Hebert, M., Sminchisescu, C., Weiss, Y. (eds.) ECCV 2018. LNCS, vol. 11207, pp. 179–196. Springer, Cham (2018). https://doi.org/10.1007/978-3-030-01219-9_11
13. Isola, P., Zhu, J.Y., Zhou, T., Efros, A.A.: Image-to-image translation with conditional adversarial networks. In: IEEE Conference on Computer Vision and Pattern Recognition (CVPR) (2017)
14. Kim, J., Kim, M., Kang, H., Lee, K.: U-GAT-IT: unsupervised generative attentional networks with adaptive layer-instance normalization for image-to-image translation. In: International Conference on Learning Representations (ICLR) (2020)
15. Kingma, D.P., Welling, M.: Auto-encoding variational Bayes. In: International Conference on Learning Representations (ICLR) (2013)
16. Ledig, C., et al.: Photo-realistic single image super-resolution using a generative adversarial network. In: IEEE Conference on Computer Vision and Pattern Recognition (CVPR) (2017)
17. Lee, H.-Y., Tseng, H.-Y., Huang, J.-B., Singh, M., Yang, M.-H.: Diverse image-to-image translation via disentangled representations. In: Ferrari, V., Hebert, M., Sminchisescu, C., Weiss, Y. (eds.) ECCV 2018. LNCS, vol. 11205, pp. 36–52. Springer, Cham (2018). https://doi.org/10.1007/978-3-030-01246-5_3
18. Lee, H.Y., et al.: DRIT++: diverse image-to-image translation via disentangled representations. Int. J. Comput. Vis. (IJCV) **128**, 2402–2417 (2020)
19. Lin, C.H., Lee, H.Y., Cheng, Y.C., Tulyakov, S., Yang, M.H.: InfinityGAN: towards infinite-pixel image synthesis. In: International Conference on Learning Representations (ICLR) (2021)

20. Liu, M.Y., Breuel, T., Kautz, J.: Unsupervised image-to-image translation networks. In: Advances in Neural Information Processing Systems (NeurIPS) (2017)
21. Mao, Q., Lee, H.Y., Tseng, H.Y., Ma, S., Yang, M.H.: Mode seeking generative adversarial networks for diverse image synthesis. In: IEEE Conference on Computer Vision and Pattern Recognition (CVPR) (2019)
22. Mao, Q., Tseng, H.Y., Lee, H.Y., Huang, J.B., Ma, S., Yang, M.H.: Continuous and diverse image-to-image translation via signed attribute vectors. Int. J. Comput. Vis. (IJCV) **130**, 517–549 (2022)
23. Mittal, A., Soundararajan, R., Bovik, A.C.: Making a "completely blind" image quality analyzer. IEEE Signal Process. Lett. **20**, 209–212 (2012)
24. Murez, Z., Kolouri, S., Kriegman, D., Ramamoorthi, R., Kim, K.: Image to image translation for domain adaptation. In: IEEE Conference on Computer Vision and Pattern Recognition (CVPR) (2018)
25. van den Oord, A., Kalchbrenner, N.: Pixel RNN. In: International Conference on Machine Learning (ICML) (2016)
26. van den Oord, A., Kalchbrenner, N., Vinyals, O., Espeholt, L., Graves, A., Kavukcuoglu, K.: Conditional image generation with pixelCNN decoders. In: Advances in Neural Information Processing Systems (NeurIPS) (2016)
27. van den Oord, A., Vinyals, O., Kavukcuoglu, K.: Neural discrete representation learning. In: Advances in Neural Information Processing Systems (NeurIPS) (2017)
28. Park, T., Efros, A.A., Zhang, R., Zhu, J.-Y.: Contrastive learning for unpaired image-to-image translation. In: Vedaldi, A., Bischof, H., Brox, T., Frahm, J.-M. (eds.) ECCV 2020. LNCS, vol. 12354, pp. 319–345. Springer, Cham (2020). https://doi.org/10.1007/978-3-030-58545-7_19
29. Park, T., Liu, M.Y., Wang, T.C., Zhu, J.Y.: Semantic image synthesis with spatially-adaptive normalization. In: IEEE Conference on Computer Vision and Pattern Recognition (CVPR) (2019)
30. Razavi, A., van den Oord, A., Vinyals, O.: Generating diverse high-fidelity images with VQ-VAE-2. In: Advances in Neural Information Processing Systems (NeurIPS) (2019)
31. Teterwak, P., et al.: Boundless: generative adversarial networks for image extension. In: IEEE International Conference on Computer Vision (ICCV) (2019)
32. Tseng, H.-Y., Lee, H.-Y., Jiang, L., Yang, M.-H., Yang, W.: RetrieveGAN: image synthesis via differentiable patch retrieval. In: Vedaldi, A., Bischof, H., Brox, T., Frahm, J.-M. (eds.) ECCV 2020. LNCS, vol. 12353, pp. 242–257. Springer, Cham (2020). https://doi.org/10.1007/978-3-030-58598-3_15
33. Zhang, Z., et al.: UFC-BERT: unifying multi-modal controls for conditional image synthesis (2021)
34. Zhu, J.Y., Park, T., Isola, P., Efros, A.A.: Unpaired image-to-image translation using cycle-consistent adversarial networks. In: IEEE International Conference on Computer Vision (ICCV) (2017)
35. Zhu, J.Y., et al.: Multimodal image-to-image translation by enforcing bi-cycle consistency. In: Advances in Neural Information Processing Systems (NeurIPS) (2017)
36. Zhu, J.Y., et al.: Toward multimodal image-to-image translation. In: Advances in Neural Information Processing Systems (NeurIPS) (2017)

The Surprisingly Straightforward Scene Text Removal Method with Gated Attention and Region of Interest Generation: A Comprehensive Prominent Model Analysis

Hyeonsu Lee$^{(\boxtimes)}$ (ID) and Chankyu Choi (ID)

NAVER Corp, Seongnam-si, South Korea
{hyeon-su.lee,chankyu.choi}@navercorp.com

Abstract. Scene text removal (STR), a task of erasing text from natural scene images, has recently attracted attention as an important component of editing text or concealing private information such as ID, telephone, and license plate numbers. While there are a variety of different methods for STR actively being researched, it is difficult to evaluate superiority because previously proposed methods do not use the same standardized training/evaluation dataset. We use the same standardized training/testing dataset to evaluate the performance of several previous methods after standardized re-implementation. We also introduce a simple yet extremely effective Gated Attention (GA) and Region-of-Interest Generation (RoIG) methodology in this paper. GA uses attention to focus on the text stroke as well as the textures and colors of the surrounding regions to remove text from the input image much more precisely. RoIG is applied to focus on only the region with text instead of the entire image to train the model more efficiently. Experimental results on the benchmark dataset show that our method significantly outperforms existing state-of-the-art methods in almost all metrics with remarkably higher-quality results. Furthermore, because our model does not generate a text stroke mask explicitly, there is no need for additional refinement steps or sub-models, making our model extremely fast with fewer parameters. The dataset and code are available at https://github.com/naver/garnet.

1 Introduction

Scene text removal (STR) is a task of erasing text from natural scene images, which is useful for privacy protection, text editing in images/videos, and Augmented Reality (AR) translation.

Supplementary Information The online version contains supplementary material available at https://doi.org/10.1007/978-3-031-19787-1_26.

A lot of current STR research utilizes the deep learning method. Early studies [22] attempted to erase all text in an image without using a text region mask, but this produced imprecise, dissatisfactory results such as some regions being blurred out. Furthermore, it was impossible to selectively erase letters in a certain region with this method, which drastically limited applications. In order to overcome these limitations, recent studies [17,21] in the field built text removal models after generating text region masks through manual or automatic means. This lead to comparably better quality results and wider applications.

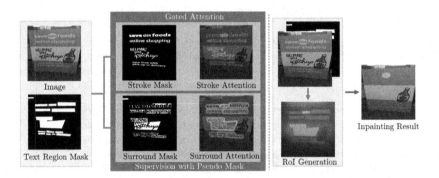

Fig. 1. Visualization of the proposed model's input and output. Visualization of the results of applying attention to the text stroke and applying attention to the text stroke surrounding regions found by the GA. We obtained satisfying STR results using RoIG.

However, these past studies did not use the same standardized training dataset and evaluation dataset, making it impossible to evaluate performance in a fair manner. For instance, some papers trained on different subsets of synthetic data [5]. Furthermore, they do not have the same input image dimensions, which affects speed, accuracy, and model parameters. Thus, it is difficult to select a previously proposed model for the unique requirements of a specific application when there is no standardized comparison of the model size and speed.

In this paper, we perform a fair comparison with both qualitative (text removal quality) and quantitative (model size, inference time) metrics by reimplementing prominent previously proposed models with the same standards, training them with the same training dataset, and evaluating them on the same evaluation dataset. Furthermore, we introduce the *Gated Attention* (GA) and the *Region of Interest Generation* (RoIG). GA is a simple yet extremely effective method that uses attention to focus on the text stroke and its surroundings, while RoIG drastically improves the training efficiency of the STR model.

Figure 1 shows GA found the Text Stroke Region (TSR) and the Text Stroke Surrounding Region (TSSR) after training the pseudo mask and how RoIG generates the results only for text mask regions. More details are described in Sect. 4.

Table 1. A side-by-side comparison of how our proposed model differs from previous works.

Method	Use text box	Selective removal	Stage	Stroke localization	Surround localization	RoI generation
EnsNet [22]	x	x	1	x	x	x
MTRNet [18]	o	o	1	x	x	x
MTRNet++ [17]	o	o	2	o	x	x
EraseNet [9]	x	x	2	x	x	x
Tang et al. [15]	o	o	2	o	x	o(crop)
Ours	o	o	1	o	o	o

Our contributions can be summarized as follows:

1. We re-implemented previously proposed methods with the same input image size, trained them on the same training dataset, and compared their performance with the same evaluation dataset. We also performed comparisons of speed and model size.
2. We proposed a method called GA which not only distinguishes between the background and the text stroke, but also utilizes attention to identify both TSR and TSSR. To the best of our knowledge, there is no other previous study that considered both TSR and TSSR while performing STR.
3. We proposed a method called RoIG, which helps the network focus on only the region with text instead of the entire image to train the model more efficiently.
4. The proposed method generates higher quality results than existing state-of-the-art methods on both synthetic and real datasets. It is also significantly lighter and faster than most other prominent previously proposed popular methods because our model does not have additional refinement steps or sub-models needed to generate a text stroke mask.

2   Related Works

The major trend in scene text removal before the emergence of deep learning was traditional rule-based methods [2,16], which are often hand-crafted and require prior domain knowledge.

Recently, deep learning-based text removal has been proposed by adopting popular GAN-based methods. EnsNet [22] is simple and fast because it doesn't need any auxiliary inputs. However, its results are blurry and of low quality. Furthermore, its practical applications are limited because it is impossible to only erase text in a certain region. MTRNet [18] requires the generation of text box region masks through manual or automatic means. These studies show that text region masks can improve the network's performance but cannot guarantee high-quality results. MTRNet++ [17] and EraseNet [9] proposed a coarse-refinement two-stage network. While the results are of higher quality, the model is much bigger, slower, and too complicated.

Zdenek *et al.* [21] used an auxiliary text detector to retrieve the text box mask, then attempted to erase text through a general inpainting method. However, they were unable to generate results of satisfactory quality because they did not consider qualities specific to text like the text strokes. Tang *et al.* [15] can train their model in a rather efficient manner because they erase text by cropping only the text regions. However, this method ignores all global contexts other than the cropped region and has difficulty precisely cropping the region of curved texts.

Table 2. A comparison of our model's performance with previous STR models on real data (SCUT-EnsText [9]). We compare the results claimed in the original papers of the previous prominent STR models with the results we re-measured in an equal environment. We realize that there are fundamental differences in models and that it would be unfair to evaluate them under potentially disadvantageous circumstances. To make things truly fair, we also pasted the text region (produced by the model using a text box mask) over the original image, then evaluated the results yet again. The values right of "/" reflect experiments conducted after this adjustment was made. Our model produces superior results in all metrics regardless. The best score is highlighted in bold.

| | Method | Train data | Size | Image-Eval | | |
				PSNR	SSIM	AGE
Reported	Scene Text Eraser [12]	Real	256	25.47/-	90.14/-	6.01/-
	EnsNet [22]	Real	512	29.53/-	92.74/-	4.16/-
	MTRNet [18]	Syn(75%)	256	-/-	-/-	-/-
	MTRNet++ [17]	Syn(95%)	256	-/-	-/-	-/-
	EraseNet [9]	Real	512	**32.30**/37.26	**95.42**/96.86	**3.02**/-
Our experiment	EnsNet [22]	Real+Syn	512	31.05/32.99	94.78/95.16	2.67/1.85
	MTRNet [18]	Real+Syn	256	30.61/36.06	89.85/95.72	3.92/1.21
		Real+Syn	512	32.46/36.89	95.86/96.41	3.12/0.97
	MTRNet++ [17]	Real+Syn	256	35.29/37.40	96.31/96.68	1.26/1.09
		Real+Syn	512	34.86/36.50	96.32/96.51	1.48/1.35
	EraseNet [9]	Real+Syn	512	30.54/37.16	96.27/97.53	3.07/1.20
	EraseNet [9] + M	Real+Syn	512	34.29/40.18	97.73/97.98	2.28/0.69
	Ours	Real+Syn	512	-/**41.37**	-/**98.46**	-/**0.64**

We designed a model that can focus on only the text region area while also managing to take the global context of the entire image into consideration. This model does not require an additional refinement process or a sub-model dedicated to text stroke localization, leading to a drastically faster and lighter model.

3   Comprehensive Prominent Model Analysis

In this section, we analyze the difference between previous methods and ours. We also show the results of evaluating previous methods with one standardized dataset.

Table 1 outlines the specific differences between our method and previously proposed methods. First, our proposed method takes a text box mask as the input to our model, leading to results of significantly better quality as well as the option for users to selectively erase only the text that they wish to. Second, our proposed method can localize the TSR and TSSR to erase text in a surgical manner. Finally, the use of RoIG makes our proposed method show off significantly better results than previous methods without even implementing a coarse-refinement 2-stage network.

Table 3. A comparison of our model's performance with previous STR models on real data (SCUT-EnsText [9]). R, P, and F refer to recall, precision and F-score, respectively. The best score is highlighted in bold.

| | Method | Train data | Size | Detection-Eval | | | GPU | |
				P	R	F	Time (ms)	Params
Reported	Scene Text Eraser [12]	Real	256	40.9	5.9	10.2		
	EnsNet [22]	Real	512	68.7	32.8	44.4	–	12.4M
	MTRNet [18]	Syn(75%)	256	–	–	–	–	54.4M
	MTRNet++ [17]	Syn(95%)	256	–	–	–	–	18.7M
	EraseNet [9]	Real	512	53.2	4.6	8.5	–	–
Our experiment	EnsNet [22]	Real+Syn	512	73.1	54.7	62.6	12.0	12.4M
	MTRNet [18]	Real+Syn	256				21.9	50.3M
		Real+Syn	512	69.8	41.1	51.2	51.3	
	MTRNet++ [17]	Real+Syn	256				51.3	
		Real+Syn	512	58.6	20.5	30.4	238.7	
	EraseNet [9]	Real+Syn	512	40.8	6.3	10.9	47.4	17.8M
	EraseNet [9] + M	Real+Syn	512	37.3	6.1	10.3	47.4	17.8M
	Ours	Real+Syn	512	**15.5**	**1.0**	**1.8**	**14.9**	**12.4M**

Table 2, Table 3 and Table 4 show the performance of several previous STR methods on real and synthetic data. Details of the experiment and re-implementation are in Sect. 5 and Sect. 5.3.

After performing an objectively fair comparison, we found that our method generates higher quality results than existing state-of-the-art methods on both

Table 4. A comparison of our model's performance with previous STR models on synthetic data (Oxford [5]). The notation is the same as Table 2 and Table 3.

| Method | Input size | Image-Eval | | | Detection-Eval | | |
		PSNR	SSIM	AGE	P	R	F
Original image	512				71.3	51.5	59.8
EnsNet [22]	512	36.67/39.74	97.71/97.94	1.25/0.77	55.1	14.0	22.3
MTRNet [18]	256	30.96/37.69	90.95/95.83	4.17/1.22			
	512	35.49/40.03	97.10/97.69	2.18/0.80	58.6	13.7	22.2
MTRNet++ [17]	256	37.40/40.26	97.02/97.25	0.86/0.79			
	512	38.31/40.64	97.82/97.94	0.81/0.73	64.0	15.9	25.4
EraseNet [9]	512	34.35/41.73	98.01/98.62	1.81/0.66	30.8	0.6	1.2
EraseNet [9] + M	512	36.37/42.98	98.50/**98.75**	1.72/0.56	31.4	**0.0**	1.4
Ours	512	-/**43.64**	-/98.64	-/**0.55**	18.9	0.1	**0.3**

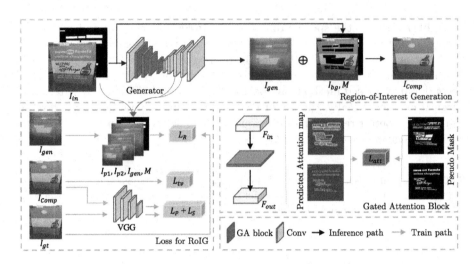

Fig. 2. The overall architecture of the proposed model. The top box, Region-of-Interest Generation, shows that our model generates images focusing only on the text region. The box on the bottom left, Loss for RoIG, how our loss function helps the model focus only on the text region. The L_R only use I_{gen} with M, and L_{tv}, L_p and L_s use I_{comp}, so the outside of text regions in I_{gen} does not participate in loss calculation. On bottom right box, The GA calculates the stroke attention and stroke surrounding attention mask for every layer of the generator's encoder, then automatically calculates the importance of each. Each attention map is supervised by pseudo mask. More detailed information on GA is in Fig. 3.

synthetic and real datasets. It is also significantly lighter and faster than any other method except EnsNet [22].

4   Methodology

4.1   Motivation

Previous STR models [17,18] used a text box region as well as a text stroke region in an attempt to perform precise text removal. However, TSSR was mostly overlooked. After finding inspiration from observing how humans must alternate between paying attention to the text stroke regions and the surrounding regions of the text while manually performing STR, we devised the GA. Meanwhile, because all previous studies performed STR on the entire image, artifacts frequently occurred in non-text regions. Thus, we devised the RoIG, which allows our STR model to only generate a result image from within the text box region instead of wasting resources attempting to perform STR on the full image.

4.2   Model Architecture

Figure 2 shows the architecture of our proposed method. The generator G takes the image and corresponding text box mask as its input and produces a non-text image that is visually plausible. Following GAN-based methods [7,11], the discriminator D takes both the input of the generator and the target images as its input and differentiates between real images and images produced by the generator. The objective functions of the generator and discriminator are as follows:

$$L_{adv} = \mathbb{E}_x \left[log D(x, G(x)) \right] \tag{1}$$

$$L_D = \mathbb{E}_{x,y} \left[log D(x, y) \right] + \mathbb{E}_x \left[(1 - log D(x, G(x))) \right] \tag{2}$$

where x is the input image concatenated with a text box mask and y is the target image.

Generator. The generator has an FCN-ResNet18 backbone and skip connections [13] between the encoder and decoder. The model is composed of five convolution layers paired with five deconvolution layers with a kernel size of 4, stride of 2, and padding of 1. The convolution pathway is composed of two residual blocks [6], which contains the proposed Gated Attention (GA) module.

Discriminator. For training, we use a local-aware discriminator proposed in EnsNet [22], which only penalizes the erased text patches. The discriminator is trained with locality-dependent labels, which indicate text stroke regions in the output tensor. It guides the discriminator to focus on text regions.

4.3 Gated Attention

Localizing both the TSR and TSSR is imperative to perform surgical STR. In order to do this without letting the size of our model blow up, we used spatial attention [20] instead of a separate image segmentation branch [17]. Table 2 shows how our proposed model is significantly faster and smaller than MTR-Net++ [17].

Figure 3 shows the architecture of the GA. The module takes the feature map as its input and generates a TSR and TSSR feature map, then adjusts the proportion of these two feature maps through gate parameters. The process of GA is as follows:

$$F_i' = (MaxPool(F_i^{In}) \oplus AvgPool(F_i^{In}) \oplus M_{box}) \tag{3}$$

$$\begin{aligned} F_i^t &= W_i^t \cdot F_i' \\ F_i^s &= W_i^s \cdot F_i' \end{aligned} \tag{4}$$

$$A_i^{out} = \sigma(\alpha_i F_i^t + \beta_i F_i^s) \tag{5}$$

$$F_i^{out} = F_i^{In} A_i^{out} \tag{6}$$

where i denotes the ith layer in the encoder, F^{In} and F^{out} denote the input and output feature maps, W^t and W^s denote 7×7 convolution filters that extract TSR and TSSR features, F^t and F^s denote extracted feature maps for localized TSR and TSSR, α and β denote gate parameters, and σ and A^{out} denote the Sigmoid activation function and the attention score map.

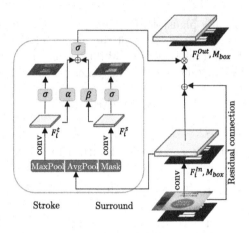

Fig. 3. Architecture of the Gated Attention. The GA calculates the TSR and TSSR attention masks, then automatically calculates the importance of each through gate parameters α, β.

Figure 1 shows the text box mask, TSR mask, and the TSSR mask necessary for this. We generated a pseudo text stroke mask, automatically calculated by taking the pixel value difference between the input image and ground truth image. The TSR masks help train the TSR attention module to distinguish the TSR. The TSSR mask is the intersection region between the exterior of the pseudo text stroke mask and the interior of the text box mask. It makes the TSSR attention module train to focus on the colors and textures of the TSSR. Note that the attention method that we used differs from S. Woo *et al.* [20], which was trained in a weakly-supervised manner. We show that our method is superior in Sect. 5.4. The GA learns the gate parameter on its own, and can thus adjust the respective attention ratios allocated to TSR and TSSR during training. The loss function is designed to be applied only within the text box regions, not to the entire score map. The loss function to train GA is as follows:

$$
\begin{aligned}
L_{att}^t &= \begin{cases} -Gt_i^t log(S_i^t) & \text{if } M_i^{Box} > 0, \\ 0 & \text{otherwise} \end{cases} \\
L_{att}^s &= \begin{cases} -Gt_i^s log(S_i^s) & \text{if } M_i^{Box} > 0, \\ 0 & \text{otherwise} \end{cases}
\end{aligned}
\tag{7}
$$

$$
L_{att} = L_{att}^t + L_{att}^s
\tag{8}
$$

where Gt_i^t and Gt_i^s are the ith pixel of the ground truth mask representing TSR and TSSR. S_i^t and S_i^s are the ith pixel of the TSR and TSSR attention score maps. M_i^{Box} is text box mask. The shape of M, Gt and S is $(H_n, W_n, 1)$. H_n and W_n denote the sizes of feature map in the nth layer.

4.4   Region-of-Interest Generation

Most STR methods attempt to perform in-painting of TSR as well as reconstruction of the entire image. However, our approach can skip the reconstruction of non-masked regions altogether because if the text box region is given as an input to the STR model, there is no need to render the entire image for the output. Therefore, we modified the loss function so that our model's generator only has to focus on the text box region during training. Note that the generator's loss is only calculated with respect to the text box region. Every other region is considered *don't care* and therefore is irrelevant during training. Because all regions other than the text box obtained from the generator's output value are not used, we blurred them in Fig. 3. This makes training significantly easier.

In total, we modified four loss functions for RoIG: RoI Regression, Perceptual, Style, and Total Variation Loss.

RoI Regression Loss. We modified regression loss to only consider text regions. The proposed *Region-of-Interest Regression Loss* is defined as:

$$L_R(M, I_{out}, I_{gt}) = \sum_{i=1}^{n} \lambda_i \odot M_i \parallel I_{out(i)} - I_{gt(i)} \parallel_1 \tag{9}$$

where $I_{out(i)}$ is the output of the *ith* deconvolution pathway. We use the output of the 3rd, 4th, and 5th layers in the deconvolution pathway. M_i and $I_{gt(i)}$ are the box mask and ground truth that was resized to the same scale as $I_{out(i)}$. λ_i is the weight for scale. We set λ_i to 0.6, 0.8, 1.0.

Perceptual Loss. Perceptual Loss [8] is used to make the generated image more realistic. It reduces the difference between the high-level features of the two images extracted using the pre-trained ImageNet. Some works [9,10,22] used a composited output to address the discrepancy between text-erased regions and the background. We use only the text regions in the generated image by pasting it into the input image. We address the discrepancy between the text box and input images by designing the loss function to use two composited outputs generated using the box and stroke mask, respectively. Perceptual Loss is defined as:

$$I_{boxComp} = I_{in}(1 - M_{Box}) + I_{out}M_{Box}$$
$$I_{StrokeComp} = I_{in}(1 - M_{Stroke}) + I_{out}M_{Stroke} \tag{10}$$

$$L_P = \sum_{n=1}^{N-1} \parallel A_n(I_{boxComp}) - A_n(I_{gt}) \parallel_1$$
$$+ \sum_{n=1}^{N-1} \parallel A_n(I_{StrokeComp}) - A_n(I_{gt}) \parallel_1 \tag{11}$$

where I_{in}, I_{out} refer to the input image and the generator's output image, $I_{BoxComp}$, $I_{StrokeComp}$ are generated images composited of box mask M_{box} and stroke mask M_{stroke} respectively. I_{gt} is the ground truth image, and A_n refers to the activation of the nth layer in network. We use the pool1, pool2 and pool3 layers of the VGG-16 [14] pretrained on ImageNet [3].

Style Loss. Style Loss [4] considers the global texture of the entire image and is used to further improve the visual quality of the output. It is calculated using the Gram matrix of feature maps. Like perceptual loss, we use two composited outputs. Style loss is defined as:

$$
L_S = \sum_{n=1}^{N-1} \| \frac{1}{H_n W_n C_n} [(\phi(A_n(I_{boxComp})) - \phi(A_n(gt))] \|_1
$$
$$
+ \sum_{n=1}^{N-1} \| \frac{1}{H_n W_n C_n} [(\phi(A_n(I_{strokeComp})) - \phi(A_n(gt))] \|_1
$$
(12)

where $\phi(x) = x^T x$ is a gram matrix operator and A_n is the activation of the nth layers of VGG-16 [14]. We use same layers as Perceptual Loss.

Total Variation Loss. J. Johnson *et al.* [8] proposed total variation loss for global denoising. As our model generates images using RoIG method, the loss function uses composited images which are generated using box masks. The total variation loss is as follows:

$$
L_t = \sum_{i,j} \| I_{Comp}^{i,j+1} - I_{Comp}^{i,j} \|_1
$$
$$
+ \| I_{Comp}^{i+1,j} - I_{Comp}^{i,j} \|_1
$$
(13)

where i,j are the pixel positions.

Total Loss Function. The total loss function for train generator is as follows:

$$
L_G = \lambda_r L_R + \lambda_p L_P + \lambda_s L_S + \lambda_t L_t
$$
$$
+ \lambda_{adv} L_{adv} + \lambda_{att} L_{Att}
$$
(14)

where λ_r to λ_{att} represent the weight of each loss. We set λ_i to 100, 0.5, 50.0, 25.0, 1, 10.

5   Experiment

5.1   Dataset and Evaluation Metrics

For training and evaluation, we use both synthetic and real datasets.

Synthetic Data. The Oxford Synthetic text dataset [5] is adopted for training and evaluation. The dataset contains around 800,000 images composed of 8,000 text-free images. We randomly selected 95% images for training, selected 10,000 images for testing, and used the rest for validation. Note that the background images in the train set and test set are mutually exclusive.

Real Data. SCUT-EnsText [9] is a real dataset for scene text removal. The dataset, which was manually generated from Chinese and English text images, contains 2,749 train and 813 test images. In this paper, we adopted these images for training and evaluation.

Preprocessing. We need stroke-level segmentation masks to train our model. However, the existing datasets do not provide stroke-level segmentation masks. Therefore, we created it automatically by calculating the pixel value difference between the input image and the ground truth image. To suppress noise, we set a threshold of 25.

We combined synthetic and real datasets. In total, we used 738,113 images for training. The test set was used separately to distinguish between performance on real and synthetic datasets.

Evaluation Metrics. T. Nakamura *et al.* [12] proposed an evaluation method using an auxiliary text detector. An auxiliary text detector obtains detection results on the images with text removed. Then, it evaluates the model performance by calculating Precision, Recall, and F-score values. A lower value means that the texts are better erased. In this paper, we use Detection Eval [19] as an evaluation metric and CRAFT [1] as an auxiliary detector. However, that method only indicates how much text has been erased, not output quality. S. Zhang *et al.* [22] proposed using the evaluation method that is used in image inpainting. They used PSNR, SSIM, MSE, AGE, pEPs, pCEPs to evaluate image quality. The higher the value of PSNR and SSIM, and the lower the value of other metrics, the better the quality of the output image. We use PSNR, SSIM, and AGE for evaluation.

5.2 Implementation Details

We trained our model for 6 epochs with batch size 30 on the combined dataset. The Adam optimizer with β (0.9, 0.999) was used. We set the initial learning rate to 0.0005 and divided it by 5 every 50,000 steps. PyTorch and NVIDIA Tesla M40 GPUs were used in all experiments.

5.3 Re-implementation Details

In this section, we provide details of the re-implementation of previous methods. We re-implemented EnsNet [22] by modifying the code implemented in mxnet and trained the model with the same hyperparameters used in their paper. We re-implemented MTRNet [18] by converting the code implemented in TensorFlow to PyTorch and trained the model with the same batch size and epoch as mentioned in their paper. We used the official implementation of MTRNet++ [17] and EraseNet [9]. We trained them with the same batch size and epoch as mentioned in their respective papers. EraserNet [9] + M refers the model using mask as its input with image.

5.4   Ablation Study

In this section, we validate the effectiveness of our contributions: Gated Attention (GA) and Region-of-Interest Generation (RoIG)

BaseLine. We combined the text region mask with the EnsNet [22] model to make a baseline. This results in a quality improvement of the output image as well as the added functionality of flexibly removing only specific characters at the user's discretion. All of our experiments including the baseline use a 4-channel input by concatenating the 3-channel RGB and a 1-channel mask.

Table 5. Result quality comparison with ablation studies. SA, TSRA, TSSRA, GA, and RoIG refer to Simple Attention [20], Text Stroke Region Attention, Text Stroke Surrounding Region Attention, Gated Attention, and Region of Interest Generation respectively. The notation is the same as Table 2.

Method	Image Eval			Detection Eval		
	PSNR	SSIM	AGE	P	R	F
Original Image				79.8	67.2	73.0
EnsNet [22]	31.05/32.99	94.78/95.16	2.67/1.85	73.1	54.7	62.6
BaseLine (EnsNet + M)	35.65/37.28	96.67/96.78	1.50/0.89	63.3	33.2	43.6
BaseLine + SA	35.73/37.24	96.53/96.62	1.52/0.89	65.5	34.4	45.1
BaseLine + TSRA	36.07/38.06	97.23/97.34	1.46/0.84	54.3	21.1	30.4
BaseLine + TSSRA	36.12/38.45	97.29/97.46	1.51/0.84	39.1	6.0	10.5
BaseLine + GA	36.38/38.82	97.56/97.72	1.46/0.80	30.6	3.9	6.9
BaseLine + RoIG	-/40.82	-/98.19	-/0.66	27.5	3.1	5.6
BaseLine + GA + RoIG	**-/41.37**	**-/98.46**	**-/0.64**	**15.5**	**1.0**	**1.8**

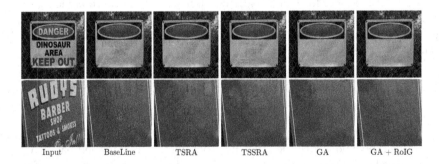

Input   BaseLine   TSRA   TSSRA   GA   GA + RoIG

Fig. 4. Comparison of the output results after ablation. Image from left to right: Input, Baseline, Baseline + TSRA, Baseline + TSSRA, Baseline + GA, and Baseline + GA + RoIG.

Attention. First, we performed the following three experiments to observe the effect of each attention on the STR results: Simple Attention (SA) [20], Text

Stroke Region Attention (TSRA), and Text Stroke Surrounding Region Attention (TSSRA).

Table 5 shows how the application of only SA does not improve the quality of STR. Figure 5's(a) shows how the application of only SA in STR does not lead to localization of the text stroke and the surrounding regions of the text stroke properly. However, both the TSRA and the TSSRA obtained higher PSNR and SSIM results. In particular, the evaluation result of the TSSRA was significantly better than the TSRA in Detection Eval. This shows that TSSR is more important. Figure 4 demonstrates that results produced by the TSSRA reduce more artifacts than the TSRA. TSRA helps locate the TSR because it is a target for inpainting, but is inappropriate when focusing on the TSSR to erase text. On the other hand, TSSRA is appropriate to focus on the TSSR, and the model utilizes this to fill the TSR and generate higher quality output.

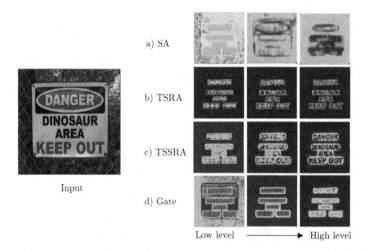

Fig. 5. Visualization of the attention masks for differing encoding layers. The images progressively go from low-level features to high-level features. (a) is a visualization of simple spatial attention [20], (b) and (c) is a visualization of the text stroke and the surrounding attention that the GA generated. (d) is a visualization of how the GA used the gate parameter to aggregate TSA and TSSA. Spatial attention does a poor job finding the text strokes and the surrounding regions if simply applied in STR. In comparison, we can see that the GA pays more attention to the text stroke regions in low-level features while paying more attention to the surrounding regions of the text in high-level features.

Furthermore, We found that having the GA module picking optimal ratios from an ensemble of both TSRA and TSSRA, then aggregating the features as appropriate was super effective. As Table 5 and Fig. 4 clearly demonstrate, the evaluation scores of STR were highest when using the GA while also leaving almost no artifacts behind on the resulting image. In Fig. 5, the GA module puts more emphasis on the surrounding regions of text strokes rather than the text strokes as it approaches higher-level features.

Region of Interest Generation. In order to measure the effects of RoIG, we performed experiments with and without its use. Table 5 demonstrates that RoIG significantly improves the quality of STR in all metrics. Figure 4 shows that models with the application of RoIG left almost no artifacts with the best overall results.

5.5   Comparison with Previous Methods

As we mentioned in Sect. 3, the proposed model maintains a competitive edge with speed and model size while significantly improving the result quality than previous STR models on both real and synthetic data. The first row of Fig. 6 shows that EnsNet [22] and EraseNet [9], both of which do not use an explicit text box region, only partially erase text. The fourth row of the Figure shows that MTRNet [18] and MTRNet++ [17] do not successfully remove all text from complex backgrounds without leaving behind artifacts or partially-erased text. However, our proposed model with GA and RoIG successfully outputs high-quality STR images without residual artifacts from images with small text, curved text, and text on complex backgrounds, even without additional refinement.

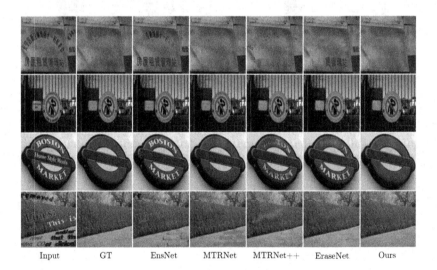

Input   GT   EnsNet   MTRNet   MTRNet++   EraseNet   Ours

Fig. 6. Comparison of output image results with other prominent STR models. Image from left to right: Input, Ground Truth, EnsNet [22], MTRNet [18], MTRNet++ [17], EraseNet [9], and Ours.

6   Conclusion

Although there was a lot of progress in the STR area, it was difficult to establish the superiority of a model from the previously proposed methods because there was no standardized and fair way to evaluate performance. In this paper, we re-implemented prominent previously proposed methods, trained and evaluated on respective standardized datasets, and evaluated their accuracies, model size, and inference time in an objectively fair manner. We proposed a simple yet highly impactful STR method with Gated Attention (GA) and Region-of-Interest Generation (RoIG). GA uses attention on the text strokes and the surrounding region's colors and textures to surgically erase text from images. RoIG makes the generator focus on only the region with text instead of the entire image for more efficient training. Our method significantly outperforms all existing state-of-the-art methods on all benchmark datasets in terms of inference time and output image quality.

Acknowledgements. We wish to thank Osman Tursun for providing codes of MTR-Net and MTRNet++.

References

1. Baek, Y., Lee, B., Han, D., Yun, S., Lee, H.: Character region awareness for text detection. In: Proceedings of the IEEE/CVF Conference on Computer Vision and Pattern Recognition, pp. 9365–9374 (2019)
2. Bertalmio, M., Bertozzi, A.L., Sapiro, G.: Navier-stokes, fluid dynamics, and image and video inpainting. In: Proceedings of the 2001 IEEE Computer Society Conference on Computer Vision and Pattern Recognition, CVPR 2001, vol. 1, p. I. IEEE (2001)
3. Deng, J., Dong, W., Socher, R., Li, L.J., Li, K., Fei-Fei, L.: ImageNet: a large-scale hierarchical image database. In: 2009 IEEE Conference on Computer Vision and Pattern Recognition, pp. 248–255. IEEE (2009)
4. Gatys, L.A., Ecker, A.S., Bethge, M.: A neural algorithm of artistic style. arXiv preprint arXiv:1508.06576 (2015)
5. Gupta, A., Vedaldi, A., Zisserman, A.: Synthetic data for text localisation in natural images. In: Proceedings of the IEEE Conference on Computer Vision and Pattern Recognition, pp. 2315–2324 (2016)
6. He, K., Zhang, X., Ren, S., Sun, J.: Deep residual learning for image recognition. In: Proceedings of the IEEE Conference on Computer Vision and Pattern Recognition, pp. 770–778 (2016)
7. Isola, P., Zhu, J.Y., Zhou, T., Efros, A.A.: Image-to-image translation with conditional adversarial networks. In: Proceedings of the IEEE Conference on Computer Vision and Pattern Recognition, pp. 1125–1134 (2017)
8. Johnson, J., Alahi, A., Fei-Fei, L.: Perceptual losses for real-time style transfer and super-resolution. In: Leibe, B., Matas, J., Sebe, N., Welling, M. (eds.) ECCV 2016. LNCS, vol. 9906, pp. 694–711. Springer, Cham (2016). https://doi.org/10.1007/978-3-319-46475-6_43
9. Liu, C., Liu, Y., Jin, L., Zhang, S., Luo, C., Wang, Y.: EraseNet: end-to-end text removal in the wild. IEEE Trans. Image Process. **29**, 8760–8775 (2020)

10. Liu, G., Reda, F.A., Shih, K.J., Wang, T.-C., Tao, A., Catanzaro, B.: Image inpainting for irregular holes using partial convolutions. In: Ferrari, V., Hebert, M., Sminchisescu, C., Weiss, Y. (eds.) ECCV 2018. LNCS, vol. 11215, pp. 89–105. Springer, Cham (2018). https://doi.org/10.1007/978-3-030-01252-6_6
11. Mirza, M., Osindero, S.: Conditional generative adversarial nets. arXiv preprint arXiv:1411.1784 (2014)
12. Nakamura, T., Zhu, A., Yanai, K., Uchida, S.: Scene text eraser. In: 2017 14th IAPR International Conference on Document Analysis and Recognition (ICDAR), vol. 1, pp. 832–837. IEEE (2017)
13. Ronneberger, O., Fischer, P., Brox, T.: U-Net: convolutional networks for biomedical image segmentation. In: Navab, N., Hornegger, J., Wells, W.M., Frangi, A.F. (eds.) MICCAI 2015. LNCS, vol. 9351, pp. 234–241. Springer, Cham (2015). https://doi.org/10.1007/978-3-319-24574-4_28
14. Simonyan, K., Zisserman, A.: Very deep convolutional networks for large-scale image recognition. arXiv preprint arXiv:1409.1556 (2014)
15. Tang, Z., Miyazaki, T., Sugaya, Y., Omachi, S.: Stroke-based scene text erasing using synthetic data for training. IEEE Trans. Image Process. 30, 9306–9320 (2021)
16. Telea, A.: An image inpainting technique based on the fast marching method. J. Graph. Tools 9(1), 23–34 (2004)
17. Tursun, O., Denman, S., Zeng, R., Sivapalan, S., Sridharan, S., Fookes, C.: MTR-Net++: one-stage mask-based scene text eraser. Comput. Vis. Image Underst. 201, 103066 (2020)
18. Tursun, O., Zeng, R., Denman, S., Sivapalan, S., Sridharan, S., Fookes, C.: MTR-Net: a generic scene text eraser. In: 2019 International Conference on Document Analysis and Recognition (ICDAR), pp. 39–44. IEEE (2019)
19. Wolf, C., Jolion, J.M.: Object count/area graphs for the evaluation of object detection and segmentation algorithms. Int. J. Doc. Anal. Recogn. 8(4), 280–296 (2006)
20. Woo, S., Park, J., Lee, J.-Y., Kweon, I.S.: CBAM: Convolutional block attention module. In: Ferrari, V., Hebert, M., Sminchisescu, C., Weiss, Y. (eds.) ECCV 2018. LNCS, vol. 11211, pp. 3–19. Springer, Cham (2018). https://doi.org/10.1007/978-3-030-01234-2_1
21. Zdenek, J., Nakayama, H.: Erasing scene text with weak supervision. In: Proceedings of the IEEE/CVF Winter Conference on Applications of Computer Vision, pp. 2238–2246 (2020)
22. Zhang, S., Liu, Y., Jin, L., Huang, Y., Lai, S.: Ensnet: Ensconce text in the wild. In: Proceedings of the AAAI Conference on Artificial Intelligence, vol. 33, pp. 801–808 (2019)

Free-Viewpoint RGB-D Human Performance Capture and Rendering

Phong Nguyen-Ha[1(✉)], Nikolaos Sarafianos[2], Christoph Lassner[2], Janne Heikkilä[1], and Tony Tung[2]

[1] Center for Machine Vision and Signal Analysis, University of Oulu, Oulu, Finland
phong.nguyen@oulu.fi
[2] Meta Reality Labs Research, Sausalito, USA
https://www.phongnhhn.info/HVS_Net

Abstract. Capturing and faithfully rendering photorealistic humans from novel views is a fundamental problem for AR/VR applications. While prior work has shown impressive performance capture results in laboratory settings, it is non-trivial to achieve casual free-viewpoint human capture and rendering for unseen identities with high fidelity, especially for facial expressions, hands, and clothes. To tackle these challenges we introduce a novel view synthesis framework that generates realistic renders from unseen views of any human captured from a single-view and sparse RGB-D sensor, similar to a low-cost depth camera, and without actor-specific models. We propose an architecture to create dense feature maps in novel views obtained by sphere-based neural rendering, and create complete renders using a global context inpainting model. Additionally, an enhancer network leverages the overall fidelity, even in occluded areas from the original view, producing crisp renders with fine details. We show that our method generates high-quality novel views of synthetic and real human actors given a single-stream, sparse RGB-D input. It generalizes to unseen identities, and new poses and faithfully reconstructs facial expressions. Our approach outperforms prior view synthesis methods and is robust to different levels of depth sparsity.

1 Introduction

Novel view synthesis of rigid objects or dynamic scenes has been a very active topic of research recently with impressive results across various tasks [42,45,62]. However, synthesizing novel views of humans in motion requires methods to handle dynamic scenes with various deformations which is a challenging task [62, 67]; especially in those regions with fine details such as the face or the clothes [46,

---

P. Nguyen-Ha—This work was conducted during an internship at Meta Reality Labs Research.

Supplementary Information The online version contains supplementary material available at https://doi.org/10.1007/978-3-031-19787-1_27.

S. Avidan et al. (Eds.): ECCV 2022, LNCS 13676, pp. 473–491, 2022.
https://doi.org/10.1007/978-3-031-19787-1_27

50,63,66]. In addition, prior work usually relies on a large amount of cameras [5, 42], expensive capture setups [51], or inference time on the order of several minutes per frame. This work aims to tackle these challenges using a compact, yet effective formulation.

We propose a novel **Human View Synthesis Network (HVS-Net)** that generates high-fidelity rendered images of clothed humans using a commodity RGB-D sensor. The challenging requirements that we impose are: i) generalization to new subjects at test-time as opposed to models trained per subject, ii) the ability to handle dynamic behavior of humans in unseen poses as opposed to animating humans using the same poses seen at training, iii) the ability to handle occlusions (either from objects or self-occlusion), iv) capturing facial expressions and v) the generation of high-fidelity images in a live setup given a single-stream, sparse RGB-D input (similar to a low-cost, off-the-shelf depth camera).

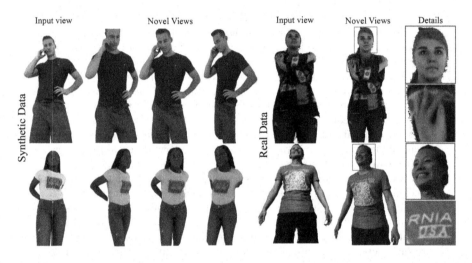

Fig. 1. *Overview.* We present a Human View Synthesis model that predicts novel views of humans from a single-view, sparse RGB-D input. Our method renders high quality novel views of both, synthetic and real humans at 1K resolution without per-subject fine tuning.

HVS-Net takes as input a single, sparse RGB-D image of the upper body of a human and a target camera pose and generates a high-resolution rendering from the target viewpoint (see Fig. 1). The first key differentiating factor of our proposed approach compared to previous approaches is that we utilize depth as an additional input stream. While the input depth is sparse and noisy it still enables us to utilize the information seen in the input view and hence simplifying the synthesis of novel views. To account for the sparseness of the input, we opted for a sphere-based neural renderer that uses a learnable radius to create a denser, warped image compared to simply performing geometry warping from one view to the other. When combined with an encoder-decoder architecture and trained end-to-end, our approach is able to synthesize novel views of unseen individuals and to in-paint areas that are not visible from the main input view. However, we observed that while this approach works well with minimal occlusions it has a hard time generating high-quality renderings when there are severe occlusions, either from the person moving their hands in front of their body or if they're holding various objects. Thus, we propose to utilize a single additional occlusion-free image and warp it to the target novel view by establishing accurate dense correspondences between the two inputs. A compact network can be used for this purpose, which is sufficient to refine the final result and generate the output prediction. We train the entire pipeline end-to-end using photometric losses between the generated and ground-truth pair of images. In addition, we use stereoscopic rendering to encourage view-consistent results between close-by viewpoints. To train HVS-Net, we rely on high-quality synthetic scans of humans that we animated and rendered from various views. A key finding of our work is that it generalizes very well to real data captured by a 3dMD scanner system with a level of detail in the face or the clothes that are not seen in prior works [31, 32, 51]. In summary, the contributions of this work are:

- A robust sphere-based synthesis network that generalizes to multiple identities without per-human optimization.
- A refinement module that enhances the self-occluded regions of the initial estimated novel views. This is accomplished by introducing a novel yet simple approach to establish dense surface correspondences for the clothed human body that addresses key limitations of DensePose which is usually used for this task.
- State-of-the-art results on dynamic humans wearing various clothes, or accessories and with a variety of facial expressions of both, synthetic and real-captured data.

2   Related Work

View synthesis for dynamic scenes, in particular for humans, is a well-established field that provides the basis for this work. Our approach builds on ideas from point-based rendering, warping, and image-based representations.

View Synthesis. For a survey of early image-based rendering methods, we refer to [56, 60]. One of the first methods to work with video in this field is presented in [9] and uses a pre-recorded performance in a multi-view capturing setup to create the free-viewpoint illusion. Zitnick et al. [70] similarly use a multi-view capture setup for viewpoint interpolation. These approaches interpolate between recorded images or videos. Ballan et al. [4] coin the term 'video-based rendering': they use it to interpolate between hand-held camera views of performances. The strong generative capabilities of neural networks enable further extrapolation and relaxation of constraints [18, 22, 28, 41]. Zhou et al. [69] introduce Multi-Plane Images (MPIs) for viewpoint synthesis and use a model to predict them from low-baseline stereo input and [17, 57] improve over the original baseline and additionally work with camera arrays and light fields. Broxton et al. [8] extend the idea to layered, dynamic meshes for immersive video experiences whereas Bansal et al. [5] use free camera viewpoints, but multiple cameras. With even stronger deep neural network priors, [64] performs viewpoint extrapolation from a single view, but for static scenes, whereas [62,67] can work with a single view in dynamic settings with limited motion. Bemana et al. [6] works in static settings but predicts not only the radiance field but also lighting given varying illumination data. Chibane et al. [14] trade instant depth predictions and synthesis for the requirement of multiple images. Alternatively, volumetric representations [38,39] can also being utilized for capturing dynamic scenes. All these works require significant computation time for optimization, multiple views or offline processing for the entire sequence.

3D & 4D Performance Capture. While the aforementioned works are usually scene-agnostic, employing prior knowledge can help in the viewpoint extrapolation task: this has been well explored in the area of 3D & 4D Human Performance Capture. A great overview of the development of the *Virtualized Reality* system developed at CMU in the 90s is presented in [29]. It is one of the first such systems and uses multiple cameras for full 4D capture. Starting from this work, there is a continuous line of work refining and improving over multi-view capture of human performances [1,15,35,70]. Relightables [21] uses again a multi-camera system and adds controlled lighting to the capture set up, so that the resulting reconstructed performances can be replayed in new lighting conditions. The authors of [27] take a different route: they find a way to use bundle adjustment for triangulation of tracked 3D points and obtain results with sub-frame time accuracy. Broxton et al. [8] is one of the latest systems for general-purpose view interpolation and uses a multi-view capture system to create a layered mesh representation of the scene. Many recent works apply neural radiance fields [42,65] to render humans at novel views. Li et al. [36] use a similar multi-view capture system to train a dynamic Neural Radiance Field. Kwon et al. [31] learn generalizable neural radiance fields based on a parametric human body model to perform novel view synthesis. However, this method fails to render high-quality cloth details or facial expressions of the human. Both of these systems use multiple cameras and are unable to transmit performance in real-time. Given multi-view input frames or videos, recent works on rendering animate humans from novel views show impressive results [46,50,51,66]. However such methods can be prohibitively expensive to run ([46] runs at 1 min/frame) and cannot generalize to unseen humans but instead create a dedicated model for each human that they need to render.

Human View Synthesis Using RGB-D. A few methods have been published recently that tackle similar scenarios: LookingGood [40] re-renders novel viewpoints of a captured individual given a single RGB-D input. However, their capture setup produces dense geometry which makes this a comparatively easy task: the target views do not deviate significantly from the input views. A recent approach [48] uses a frontal input view and a large number of calibration images to extrapolate novel views. This method relies on a keypoint estimator to warp the selected calibrated image to the target pose, which leads to unrealistic results for hands, occluded limbs, or for large body shapes.

Point-Based Rendering. We assume a single input RGB-D sensor as a data source for our method. This naturally allows us to work with the depth data in a point-cloud format. To use this for end-to-end optimization, we build on top of ideas from differentiable point cloud rendering. Some of the first methods rendered point clouds by blending discrete samples using local blurring kernels: [25,37,54]. Using the differentiable point cloud rendering together with convolutional neural networks naturally enables the use of latent features and a deferred rendering layer, which has been explored in [33,64]. Recent works on point-based rendering [2,30] use a point renderer implemented in OpenGL, then use a neural network image space to create novel views. Ruckert et al. [55] use purely pixel-sized points and finite differences for optimization. We are directly building on these methods and use the Pulsar renderer [33] in our method together with an additional model to improve the point cloud density.

Warping Representations. To correctly render occluded regions, we warp the respective image regions from an unoccluded posture to the required posture. Debevec et al. [16] is one of the first methods to use "projective texture-mapping" for view synthesis. Chaurasia et al. [11] uses depth synthesis and local warps to improve over image-based rendering. The authors of [19] take view synthesis through warping to its extreme: they solely use warps to create novel views or synthesize gaze. Recent methods [45,53,61] use 3D proxies together with warping and a CNN to generate novel views. All these methods require either creation of an explicit 3D proxy first, or use of image-based rendering. Instead, we use the dynamic per-frame point cloud together with a pre-captured, unoccluded image to warp necessary information into the target view during online processing.

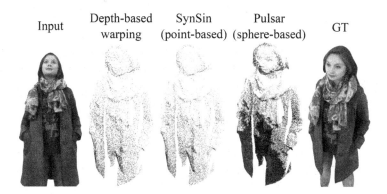

Fig. 2. *Comparison of 3D point cloud transformations.* From a single RGB-D input, we obtain the warped image using: a depth-based warping transformation [34,40], the neural point-based renderer SynSin [64] and the neural sphere-based Pulsar renderer [33]. The novel image warped by Pulsar is significantly denser.

3   HVS-Net Methodology

The goal of our method is to create realistic novel views of a human captured by a single RGB-D sensor (with sparse depth, similar to a low-cost RGB-D camera), as faithful and fast as possible. We assume that the camera parameterization of the view to generate is known. Still, this poses several challenges: 1) the information we are working with is incomplete, since not all regions that are visible from the novel view can be observed by the RGB-D sensor; 2) occlusion adds additional regions with unknown information; 3) even the pixels that are correctly observed by the original sensor are sparse and exhibit holes when viewed from a different angle. We tackle the aforementioned problems using an end-to-end trainable neural network with two components.

First, given an RGB-D image parameterized as its two components RGB I_v and sparse depth D_v taken from the input view v, a sphere-based view synthesis model S produces dense features of the target view and renders the resulting RGB image from the target camera view using a global context inpainting network G (see Sect. 3.1). However, this first network can not fully resolve all occlusions: information from fully occluded regions is missing (e.g., rendering a pattern on a T-shirt that is occluded by a hand). To account for such cases, we optionally extend our model with an enhancer module E (see Sect. 3.2). It uses information from an unoccluded snapshot of the same person, estimates the dense correspondences between the predicted novel view and occlusion-free input view, and then refine the predicted result.

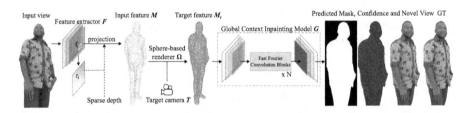

Fig. 3. *Sphere-based view synthesis network architecture.* The feature predictor F learns radius and feature vectors of the sphere set S. We then use the sphere-based differentiable renderer Ω to densify the learned input features M and warp them to the target camera T. The projected features M_t are passed through the global context inpainting module G to generate the foreground mask, confidence map and novel image. Brighter confidence map colors indicate lower confidence.

3.1   Sphere-Based View Synthesis

The goal of this first part of our pipeline is to render a sparse RGB-D view of a human as faithfully as possible from a different perspective. Of the aforementioned artifacts, it can mostly deal with the inherent sparsity of spheres caused due to the depth foreshortening: from a single viewpoint in two neighboring pixels, we only get a signal at their two respective depths—no matter how much they differ. This means that for every two pixels that have a large difference in depth and are seen from the side, large gaps occur. For rendering human subjects, these "gaps" are of limited size, and we can address the problem to a certain extent by using a sphere-based renderer for view synthesis.

Sphere-Based Renderer. Given the depth of every pixel from the original viewpoint as well as the camera parameters, these points can naturally be projected into a novel view. This makes the use of depth-based warping or of a differentiable point- or sphere-renderer a natural choice for the first step in the development of the view synthesis model. The better this renderer can transform the initial information into the novel view, the better; this projection step is automatically correct (except for sensor noise) and not subject to training errors.

In Fig. 2, we compare the density of the warped images from a single sparse RGB-D input using three different methods: depth-based warping [34], point-based rendering [64] and sphere-based rendering [33]. Depth based warping [34] represents the RGD-D input as a set of pixel-sized 3D points and thus, the correctly projected pixels in the novel view are very sensitive to the density of the input view. The widely-used differentiable point-based renderer [64] introduces a global radius-per-point parameter which allows to produce a somewhat denser images. Since it uses the same radius for all points, this comes, however, with a trade-off: if the radius is selected too large, details in dense regions of the input image are lost; if the radius is selected too small, the resulting images get sparser in sparse regions. The recently introduced, sphere-based Pulsar renderer [33] not only provides the option to use a per-sphere radius parameter, but it also provides gradients for these radiuses, which enables us to set them dynamically. As depicted in Fig. 2, this allows us to produce denser images compared to the other methods. Figure 3 shows an overview of the overall architecture of our method. In a first step, we use a shallow set of convolutional layers F to encode the input image I_v to a d-dimensional feature map $M = F(I_v)$. From this feature map, we create a sphere representation that can be rendered using the Pulsar renderer. This means that we have to find position p_i, feature vector f_i, and radius r_i for every sphere $i \in 1, .., N$ when using N spheres (for further details about the rendering step, we refer to [33]). The sphere positions p_i can trivially be inferred from camera parameters, pixel index and depth for each of the pixels. We choose the features f_i as the values of M at the respective pixel position; we infer r_i by passing M to another convolution layer with a sigmoid activation function to bound its range. This leads to an as-dense-as-possible projection of features into the target view, which is the basis for the following steps.

Global Context Inpainiting Model. Next, the projected features are converted to the final image. This remains a challenging problem since several "gaps" in the re-projected feature images M_t cannot be avoided. To address this, we design an efficient encoder-decoder-based inpainting model G to produce the final renders. The encoding bottleneck severely increases the receptive field size of the model, which in turn allows it to correctly fill in more of the missing information. Additionally, we employ a series of Fast Fourier Convolutions (FFC) [13] to take into account the image-wide receptive field. The model is able to hallucinate missing pixels much more accurately compared to regular convolution layers [58].

Photometric Losses. The sphere-based view synthesis network S not only predicts an RGB image I_p of the target view, but also a foreground mask I_m and a confidence map I_c which can be used for compositing and error correction, respectively. We then multiply the predicted foreground mask and confidence map with the predicted novel image: $I_p = I_p * I_m * I_c$. However, an imperfect mask I_m may bias the network towards unimportant areas. Therefore, we predict a confidence mask I_c as a side-product of the G network to dynamically assign less weight to "easy" pixels, whereas "hard" pixels get higher importance [40].

All of the aforementioned model components are trained end-to-end using the photometric loss $\mathcal{L}_{photo}$, which is defined as: $\mathcal{L}_{photo} = \mathcal{L}_i + \mathcal{L}_m$. $\mathcal{L}_i$ is the combination of an ℓ_1, perceptual [12] and hinge GAN [20] loss between the estimated new view I_p and the ground-truth image I_{GT}. $\mathcal{L}_m$ is the binary cross-entropy loss between the predicted and ground-truth foreground mask. We found that this loss encourages the model to predict sharp contours in the novel image. The two losses lead to high-quality reconstruction results for single images. However, we note that stereoscopic rendering of novel views requires matching left and right images for both views. Whereas the above losses lead to *plausible* reconstructions, they do not necessarily lead to sufficiently consistent reconstructions for close-by viewpoints. We found a two-step strategy to address this issue: 1) Instead of predicting a novel image of a single viewpoint, we train the model to predict two nearby novel views. To obtain perfectly consistent depth between both views, we use the warping operator W from [26] to warp the predicted image and the depth from one to the nearby paired viewpoint. 2) In the second step, we define a multi-view consistency loss $\mathcal{L}_c$ as:

$$\mathcal{L}_c = ||I_p^L - W(I_p^R)||_1, \tag{1}$$

where I_p^L and I_p^R are predicted left and right novel views. With this, we define the photometric loss as follows:

$$\mathcal{L}_{photo} = \mathcal{L}_i + 0.5 \times \mathcal{L}_m + 0.5 \times \mathcal{L}_c. \tag{2}$$

3.2   Handling Occlusions

The sphere-based view synthesis network S predicts plausible novel views with high quality. However, if the person is holding an object such as a wallet (c.t. Fig. 4) or if their hands are obstructing large parts of their torso, then the warped transformation will result in missing points in this region (as discussed in Fig 2). This leads to low-fidelity texture estimates for those occluded regions when performing novel view synthesis with a target camera that is not close to the input view. Hence, to further enhance the quality of the novel views, we introduce two additional modules: *i)* an *HD-IUV* predictor D to predict dense correspondences

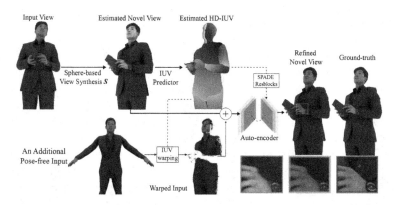

Fig. 4. *IUV-based image refinement.* Using an additional occlusion-free input, we refine the initial estimated novel view by training the Enhancer network E. We infer the dense correspondences of both, predicted novel view and occlusion-free image, using a novel *HD-IUV* module. The occlusion-free image is warped to the target view and then refined by an auto-encoder. The refined novel view shows crisper results on the occluded area compared to the initially estimated render.

between an RGB image (render of a human) and the 3D surface of a human body template, and *ii)* a refinement module R to warp an additional occlusion-free input (e.g. , a selfie in a practical application) to the target camera and enhance the initial estimated novel view to tackling the self-occlusion issue.

HD-IUV Predictor D. We first estimate a representation that maps an RGB image of a human to the 3D surface of a body template [24,43,44,59]. One could use DensePose [43] for this task but the estimated IUV (where I reflects the body part) predictions cover only the naked body instead of the clothed human and are inaccurate as they are trained based on sparse and noisy human annotations. Instead, we build our own IUV predictor and train it on synthetic data for which we can obtain accurate ground-truth correspondences. With pairs of synthetic RGB images and ground-truth dense surface correspondences, we train a UNet-like network that provides dense surface (i.e. IUV) estimates *for each pixel* of the clothed human body. For each pixel p in the foreground image, we predict 3-channeled (RGB) color p' which represents the correspondence (the colors in such a representation are unique which makes subsequent warping easy). Thus, we treat the whole problem as a multi-task classification problem where each task (predictions for the I, U, and V channels) is trained with the following set of losses: a) multi-task classification loss for each of the 3 channels (per-pixel classification label) and b) silhouette loss. In Fig. 5 we show that, unlike DensePose, the proposed HD-IUV module accurately establishes fine level correspondences for the face and hand regions while capturing the whole clothed human and thus making it applicable for such applications. Once this model is pre-trained, we merge it with the rest of the pipeline and continue the training procedure by using the initially estimated novel view I_p ad an input to an encoder-decoder

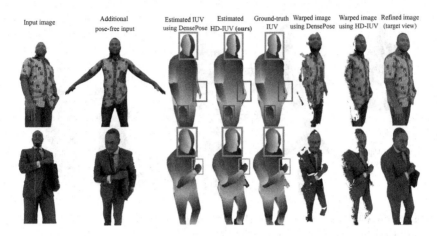

Fig. 5. *Dense correspondence visualization.* Texture warping with DensePose results in inaccurate and distorted images in the target view due to incorrect IUV estimates (enhanced by the fact that it targets the naked body). Our proposed HD-IUV representation covers the human body including clothing, captures facial and hand details with high accuracy, and results in less distorted renderings in the target view. We stack this warped image with the initially estimated target-view synthesized image and provide it as input for the Enhancer network to obtain the final results.

architecture that contains three prediction heads (for the I, U, and V channels). An in-depth discussion on the data generation, network design, and training is provided in the supplementary material.

Warping Representations and View Refinement. The predicted *HD-IUV* in isolation would not be useful for the task of human view synthesis. However, when used along with the occlusion-free RGB input, it allows us to warp all visible pixels to the human in the target camera T and obtain a partial warped image I_w. For real applications this occlusion-free input can be a selfie image— there are no specific requirements to the body pose for the image. In Fig. 5 we compare DensePose results with the proposed HD-IUV module. DensePose clearly produces less accurate and more distorted textures.

In the next step, we stack I_p and I_w and pass the resulting tensor to a refinement module. This module addresses two key details: a) it learns to be robust to artifacts that are originating either from the occluded regions of the initially synthesized novel view as well as texture artifacts that might appear due to the fact that we rely on HD-IUV dense correspondences for warping and b) it is capable of synthesizing crisper results in the occluded regions as it relies on both the initially synthesized image as well as the warped image to the target view based on HD-IUV. The refinement module is trained using the photometric loss $\mathcal{L}_{photo}$ between the refined novel images and ground truths. All details regarding training and image warping, as well as the full network architecture, can be found in the supplementary material.

4   Experiments

Datasets. The proposed approach is trained solely on synthetic data and evaluated quantitatively and qualitatively on both, synthetic and real data. For training, we use the RenderPeople dataset [52], which has been used extensively [3,7,10,23,47,49,59] for human reconstruction and generation tasks. Overall, we use a subset of 1000 watertight meshes of people wearing a variety of garments and in some cases holding objects such as mugs, bags or mobile phones. Whereas this covers a variety of personal appearances and object interaction, all of these meshes are static—the coverage of the pose space is lacking. Hence, we augment the dataset by introducing additional pose variations: we perform non-rigid registration for all meshes, rig them for animation and use a set of pre-defined motions to animate them. With this set of meshes *and* animations, we are able to assemble a set of high-quality ground-truth RGB-D renders as well as their corresponding IUV maps for 25 views per frame using Blender. We use a 90/10 train/test split based on identities to evaluate whether our model can generalize well to unseen individuals.

In addition to the synthetic test set, we also assemble a real-world test dataset consisting of 3dMD 4D scans of people in motion. The 3dMD 4D scanner is a full-body scanner that captures unregistered volumetric point clouds 60 Hz. We use this dataset solely for testing to investigate how well our method handles the domain gap between synthetic and real data. The 3dMD data does not include object interactions, but is generally noisier and has complex facial expressions. To summarize: our training set comprises 950 static scans in their original pose and ~10000 posed scans after animation. Our test set includes 50 static unseen identities along with 1000 animated renders and 3000 frames of two humans captured with a 3dMD full-body scanner.

Novel Viewpoint Range. We assume a scenario with a camera viewpoint at a lower level in front of a person (e.g. , the camera sitting on a desk in front of the user). This is a more challenging scenario than LookingGood [40] or Volumetric Capture [48] use, but also a realistic one: it corresponds to everyday video conference settings. At the same time, the target camera is moving freely in the frontal hemisphere around the person (Pitch & Roll: $[-45^o, 45^o]$, $L_x : [-1.8m, 1.8m]$, $L_y : [1.8m, 2.7m]$, $L_z : [0.1m, 2.7m]$ in a Blender coordinate system). Thus, the viewpoint range is significantly larger per input view than in prior work.

Baselines. In this evaluation, we compare our approach to two novel view synthesis baselines by comparing the performance in generating single, novel-view RGB images. To evaluate the generalization of HVS-Net, we compare it with LookingGood [40]. Since there is no available source code of LookingGood, we reimplemented the method for this comparison and validated in various synthetic and real-world settings that this implementation is qualitatively equivalent to what is reported in the original paper (we include comparison images in the supp.mat.). We followed the stereo set up of LookingGood and use a dense depth map to predict the novel views. Furthermore, we compare HVS-Net with the recently proposed view synthesis method SynSin [64], which estimates monocular depth using a depth predictor. To create fair evaluation conditions, we replace this depth predictor and either provide dense or sparse depth maps as inputs directly. While there are several recently proposed methods in the topic of human-view synthesis; almost all are relying on either proprietary data captured in lab environments [48], multi-view input streams [31,36,51,66] and most importantly none of these works can generalize to new human identities (or for the case of Neural Body [51] not even new poses) at testing time which our proposed HVS-Net can accomplish. Furthermore, inferring new views in a real-time manner is far from solved for most these works. In contrast, our method focuses more on a practical approach of single view synthesis, aiming to generalize to new identities and unseen poses while being fast at inference time. Hence we stick to performing quantitative comparisons against LookingGood [40] and SynSin [64] and we do not compare it with NeRF-based approaches [31,36,51,66] as such comparisons are not applicable.

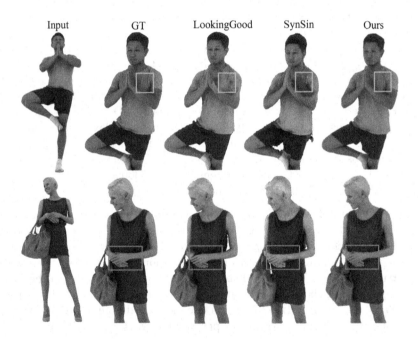

Fig. 6. *Qualitative comparison.* Examples of generated novel views by HVS-Net and state-of-the-art methods on the test set of the RenderPeople [52] dataset. As opposed to all other methods, LookingGood [40] uses dense input depth.

Metrics. We report the PSNR, SSIM, and perceptual similarity (LPIPS) [68] of view synthesis between HVS-Net and other state-of-the-art methods.

4.1 Results

In Table 1 and Fig. 6, we summarize the quantitative and qualitative results for samples from the RenderPeople dataset. We first compare the full model HVS-Net against a variant HVS-Net[†], which utilizes a dense map as an input. We observe no significant differences between the predicted novel views produced by HVS-Net when trained using either sparse or dense depth input. This confirms the effectiveness of the sphere radius predictor: it makes HVS-Net more robust w.r.t. input point cloud density.

Table 1. *Quantitative results on synthetic and real images.* For all datasets, the metrics are averaged across all views. Methods with a † symbol are using dense input depth. Both HVS-Net and HVS-Net[†] achieve the best results compared to other view synthesis methods. We observe a slight drop of performance without using the proposed Enhancer module.

Method	RenderPeople (static)			RenderPeople (animated)			Real 3dMD Data		
	LPIPS↓	SSIM↑	PSNR↑	LPIPS↓	SSIM↑	PSNR↑	LPIPS↓	SSIM↑	PSNR↑
LookingGood[†] [40]	0.24	0.925	25.32	0.25	0.912	24.53	0.29	0.863	25.12
SynSin[†] [64]	0.31	0.851	24.18	0.35	0.937	23.64	0.35	0.937	22.18
SynSin [64]	0.52	0.824	22.45	0.55	0.853	20.86	0.65	0.819	19.92
HVS-Net (w/o Enhancer)	0.18	0.986	28.54	0.19	0.926	26.24	0.20	0.910	26.25
HVS-Net[†]	**0.14**	**0.986**	**28.56**	**0.17**	**0.958**	27.41	**0.20**	**0.918**	**26.47**
HVS-Net	0.15	**0.986**	28.54	**0.17**	0.955	**27.45**	**0.20**	**0.918**	**26.47**

Fig. 7. *Generalization to real-world examples.* Our method generalizes well to real-world 4D data and shows robustness w.r.t to different target poses. These results are produced using HVS-Net, trained solely on synthetic data without further fine-tuning.

In a next step, we evaluate HVS-Net against the current top performing single view human synthesis methods [40,64], which do not require per-subject finetuning. Even though we use dense depth maps as input to LookingGood[†] [40], the method still struggles to produce realistic results if the target pose deviates significantly from the input viewpoint. In the 1st row of Fig. 6, LookingGood† [40] also struggles to recover clean and accurate textures of the occluded regions behind the hands of the person. Although both SynSin [64] and HVS-Net utilize the same sparse depth input, the rendered target images are notably different. SynSin [64] not only performs poorly on the occluded regions but also produces artifacts around the neck of the person, visible in the 2nd row of Fig. 6. In contrast, our method is not only able to render plausible and realistic novel views, but creates them also faithful w.r.t. the input views. Notice that HVS-Net is able to predict fairly accurate hair for both subjects given very little information.

In a last experiment, we test the generalization ability of our method on real-world 4D data, shown in Fig. 7. Being trained only on synthetic data, this requires generalization to novel identity, novel poses, and bridging the domain gap. In the 4D scans, the subjects are able to move freely within the capture volume. We use a fixed, virtual 3D sensor position to create the sparse RGB-D input stream for HVS-Net. The input camera is placed near the feet of the subjects and is facing up. As can be seen in Fig. 1 and Fig. 7, HVS-Net is still able to perform novel view synthesis with high quality. Despite using sparse input depth, our method is able to render realistic textures on the clothes of both subjects. In addition, facial expressions such as opening the mouth or smiling are also well-reconstructed, despite the fact that the static or animated scans used to train our network did not have a variety of facial expressions. The quality of the results obtained in Fig. 7 demonstrates that our approach can render high-fidelity novel views of real humans in motion. We observe that the generated novel views are also temporally consistent across different target view trajectories. For additional results and video examples, we refer to the supplementary material.

Table 2. *Left: Ablation study.* Reconstruction accuracy on the RenderPeople testing set. *Right: Reconstruction accuracy and inference speed* using different levels of input depth sparsity.

Method Variant	LPIPS↓	SSIM↑	PSNR↑	Input depth (%)	Run-time↑ (fps)	LPIPS↓	SSIM↑	PSNR↑
No Sphere Repres.	0.22	0.934	26.15	5	**25**	0.17	0.985	28.27
No Global Context	0.21	0.954	26.82	10	22	0.15	0.986	28.54
No Enhancer	0.18	0.967	27.92	25	21	**0.14**	0.986	28.55
HVS-Net (full)	**0.15**	**0.986**	**28.54**	100	20	0.14	0.986	**28.56**

4.2 Ablation Studies and Discussion

Model Design. Table 2 (left) and Fig. 8 summarize the quantitative and qualitative performance for different model variants on the test set of the Render-People dataset [52]. HVS-Net without the sphere-based representation does not produce plausible target views (see, for example, the rendered face, which is blurry compared to the full model). This is due to the high level of sparsity of the input depth, which leads to a harder inpainting problem for the neural network that addresses this task. Replacing the Fast Fourier Convolution residual blocks of the global context inpainting model with regular convolution layers leads to a drop in render quality in the occluded region (red box). Using the proposed model architecture, but without the enhancer (5th column of Fig. 8) leads to a loss of detail in texture. In contrast, the full proposed model using the Enhancer network renders the logo accurately. Note that this logo is completely occluded by the human's hands so it is non-trivial to render the logo using a single input image.

Sparse Depth Robustness. In Fig. 9, we show novel view synthesis results using different levels of sparsity of the input depth maps. We first randomly sample several versions of the sparse input depth and HVS-Net to process them. Our method is able to maintain the quality of view synthesis despite strong reductions in point cloud density. This highlights the importance of the proposed sphere-based rendering component and the enhancer module. As can be seen in Table 2 (right), we observe a slight drop of performance when using 5% or 10% of the input maps. To balance between visual quality and rendering speed, we suggest that using 25% of the input depth data is sufficient to achieve similar results compared to using the full data.

Inference Speed. For AR/VR applications, a prime target for a method like the one proposed, runtime performance is critical. At test time, HVS-Net generates 1024×1024 images at 21FPS using a single NVIDIA V100 GPU. This speed can be further increased with more efficient data loaders and an optimized implementation that uses the NVIDIA TensorRT engine. Finally, different depth sparsity levels do not significantly affect the average runtime of HVS-Net, which is a plus compared to prior work.

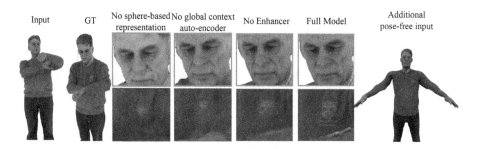

Fig. 8. *Qualitative ablation study.* Comparison of the ground-truth with predicted novel views by several variants of the proposed HVS-Net.

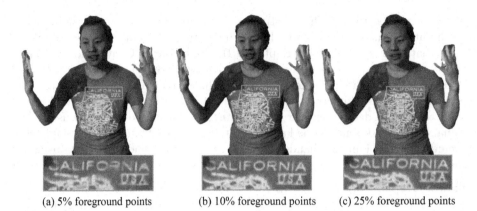

(a) 5% foreground points   (b) 10% foreground points   (c) 25% foreground points

Fig. 9. *HVS-Net sparsity robustness.* We randomly sample (a) 5%, (b) 10% and (c) 25% of dense depth points as input depth map and use it as an input for HVS-Net to predict novel views. The text in the T-shirt is reconstructed at high-fidelity with 25% of the depth points utilized.

5   Conclusion

We presented HVS-Net, a method that performs novel view synthesis of humans in motion given a single, sparse RGB-D source. HVS-Net uses a sphere-based view synthesis model that produces dense features of the target view; these are then utilized along with an autoencoder to complete the missing details of the target viewpoints. To account for heavily occluded regions, we propose an enhancer module that uses an additional unoccluded view of the human to provide additional information and produce high-quality results based on an novel IUV mapping. Our approach generates high-fidelity renders at new views of unseen humans in various new poses and can faithfully capture and render facial expressions that were not present in training. This is especially remarkable, since we train HVS-Net only on synthetic data; yet it achieves high-quality results across synthetic and real-world examples.

Acknowledgements. The authors would like to thank Albert Para Pozzo, Sam Johnson and Ronald Mallet for the initial discussions related to the project.

References

1. de Aguiar, E., Stoll, C., Theobalt, C., Ahmed, N., Seidel, H.P., Thrun, S.: Performance capture from sparse multi-view video. In: TOG (2008)
2. Aliev, K.-A., Sevastopolsky, A., Kolos, M., Ulyanov, D., Lempitsky, V.: Neural point-based graphics. In: Vedaldi, A., Bischof, H., Brox, T., Frahm, J.-M. (eds.) ECCV 2020. LNCS, vol. 12367, pp. 696–712. Springer, Cham (2020). https://doi.org/10.1007/978-3-030-58542-6_42
3. Alldieck, T., Pons-Moll, G., Theobalt, C., Magnor, M.: Tex2shape: detailed full human body geometry from a single image. In: ICCV (2019)

4. Ballan, L., Brostow, G.J., Puwein, J., Pollefeys, M.: Unstructured video-based rendering: interactive exploration of casually captured videos. In: SIGGRAPH (2010)
5. Bansal, A., Vo, M., Sheikh, Y., Ramanan, D., Narasimhan, S.: 4D visualization of dynamic events from unconstrained multi-view videos. In: CVPR (2020)
6. Bemana, M., Myszkowski, K., Seidel, H.P., Ritschel, T.: X-fields: implicit neural view-, light- and time-image interpolation. In: SIGGRAPH Asia (2020)
7. Bhatnagar, B.L., Tiwari, G., Theobalt, C., Pons-Moll, G.: Multi-garment net: learning to dress 3D people from images. In: ICCV (2019)
8. Broxton, M., et al.: Immersive light field video with a layered mesh representation. TOG **39**, 861–8615 (2020)
9. Carranza, J., Theobalt, C., Magnor, M.A., Seidel, H.P.: Free-viewpoint video of human actors. TOG **22**, 569–577 (2003)
10. Chaudhuri, B., Sarafianos, N., Shapiro, L., Tung, T.: Semi-supervised synthesis of high-resolution editable textures for 3D humans. In: CVPR (2021)
11. Chaurasia, G., Duchene, S., Sorkine-Hornung, O., Drettakis, G.: Depth synthesis and local warps for plausible image-based navigation. TOG (2013)
12. Chen, Q., Koltun, V.: Photographic image synthesis with cascaded refinement networks. In: ICCV (2017)
13. Chi, L., Jiang, B., Mu, Y.: Fast fourier convolution. In: NeurIPS (2020)
14. Chibane, J., Bansal, A., Lazova, V., Pons-Moll, G.: Stereo radiance fields (SRF): learning view synthesis from sparse views of novel scenes. In: CVPR (2021)
15. Collet, A., et al.: High-quality streamable free-viewpoint video. TOG **34**, 1–13 (2015)
16. Debevec, P., Yu, Y., Borshukov, G.: Efficient view-dependent image-based rendering with projective texture-mapping. In: Eurographics Rendering Workshop (1998)
17. Flynn, J., et al.: DeepView: view synthesis with learned gradient descent. In: CVPR (2019)
18. Flynn, J., Neulander, I., Philbin, J., Snavely, N.: Deep stereo: learning to predict new views from the world's imagery. In: CVPR (2016)
19. Ganin, Y., Kononenko, D., Sungatullina, D., Lempitsky, V.: DeepWarp: photorealistic image resynthesis for gaze manipulation. In: Leibe, B., Matas, J., Sebe, N., Welling, M. (eds.) ECCV 2016. LNCS, vol. 9906, pp. 311–326. Springer, Cham (2016). https://doi.org/10.1007/978-3-319-46475-6_20
20. Goodfellow, I., et al.: Generative adversarial nets. In: NeurIPS (2014)
21. Guo, K., et al.: The relightables: volumetric performance capture of humans with realistic relighting. TOG **38**, 1–19 (2019)
22. Huang, Z., et al.: Deep volumetric video from very sparse multi-view performance capture. In: Ferrari, V., Hebert, M., Sminchisescu, C., Weiss, Y. (eds.) ECCV 2018. LNCS, vol. 11220, pp. 351–369. Springer, Cham (2018). https://doi.org/10.1007/978-3-030-01270-0_21
23. Huang, Z., Xu, Y., Lassner, C., Li, H., Tung, T.: ARCH: animatable reconstruction of clothed humans. In: CVPR (2020)
24. Ianina, A., Sarafianos, N., Xu, Y., Rocco, I., Tung, T.: BodyMap: learning full-body dense correspondence map. In: CVPR (2022)
25. Insafutdinov, E., Dosovitskiy, A.: Unsupervised learning of shape and pose with differentiable point clouds. In: NeurIPS (2018)
26. Jaderberg, M., Simonyan, K., Zisserman, A., Kavukcuoglu, K.: Spatial transformer networks. In: NeurIPS (2015)
27. Joo, H., et al.: Panoptic studio: a massively multiview system for social motion capture. In: ICCV (2015)

28. Kalantari, N.K., Wang, T.C., Ramamoorthi, R.: Learning-based view synthesis for light field cameras. TOG **35**, 1–10 (2016)
29. Kanade, T., Rander, P., Narayanan, P.: Virtualized reality: constructing virtual worlds from real scenes. IEEE MultiMedia **4**, 34–47 (1997)
30. Kopanas, G., Philip, J., Leimkühler, T., Drettakis, G.: Point-based neural rendering with per-view optimization. In: Computer Graphics Forum (2021)
31. Kwon, Y., Kim, D., Ceylan, D., Fuchs, H.: Neural human performer: learning generalizable radiance fields for human performance rendering. In: NeurIPS (2021)
32. Kwon, Y., et al.: Rotationally-temporally consistent novel view synthesis of human performance video. In: Vedaldi, A., Bischof, H., Brox, T., Frahm, J.-M. (eds.) ECCV 2020. LNCS, vol. 12349, pp. 387–402. Springer, Cham (2020). https://doi. org/10.1007/978-3-030-58548-8_23
33. Lassner, C., Zollhofer, M.: Pulsar: efficient sphere-based neural rendering. In: CVPR (2021)
34. Le, H.A., Mensink, T., Das, P., Gevers, T.: Novel view synthesis from a single image via point cloud transformation. In: BMVC (2020)
35. Li, H., et al.: Temporally coherent completion of dynamic shapes. TOG **31**, 1–11 (2012)
36. Li, T., et al.: Neural 3D video synthesis. In: CVPR (2021)
37. Lin, C.H., Kong, C., Lucey, S.: Learning efficient point cloud generation for dense 3D object reconstruction. In: AAAI (2018)
38. Lombardi, S., Simon, T., Saragih, J., Schwartz, G., Lehrmann, A., Sheikh, Y.: Neural volumes: learning dynamic renderable volumes from images. TOG (2019)
39. Lombardi, S., Simon, T., Schwartz, G., Zollhoefer, M., Sheikh, Y., Saragih, J.: Mixture of volumetric primitives for efficient neural rendering. TOG (2021)
40. Martin-Brualla, R., et al.: Lookingood: enhancing performance capture with real-time neural re-rendering. TOG (2018)
41. Meshry, M., et al.: Neural rerendering in the wild. In: CVPR (2019)
42. Mildenhall, B., Srinivasan, P.P., Tancik, M., Barron, J.T., Ramamoorthi, R., Ng, R.: NeRF: representing scenes as neural radiance fields for view synthesis. In: Vedaldi, A., Bischof, H., Brox, T., Frahm, J.-M. (eds.) ECCV 2020. LNCS, vol. 12346, pp. 405–421. Springer, Cham (2020). https://doi.org/10.1007/978-3-030-58452-8_24
43. Neverova, N., Alp Güler, R., Kokkinos, I.: Dense pose transfer. In: Ferrari, V., Hebert, M., Sminchisescu, C., Weiss, Y. (eds.) ECCV 2018. LNCS, vol. 11207, pp. 128–143. Springer, Cham (2018). https://doi.org/10.1007/978-3-030-01219-9_8
44. Neverova, N., Novotny, D., Khalidov, V., Szafraniec, M., Labatut, P., Vedaldi, A.: Continuous surface embeddings. In: NeurIPS (2020)
45. Nguyen, P., Karnewar, A., Huynh, L., Rahtu, E., Matas, J., Heikkila, J.: RGBD-net: predicting color and depth images for novel views synthesis. In: 3DV (2021)
46. Noguchi, A., Sun, X., Lin, S., Harada, T.: Neural articulated radiance field. In: ICCV (2021)
47. Palafox, P., Sarafianos, N., Tung, T., Dai, A.: SPAMs: structured implicit parametric models. In: CVPR (2022)
48. Pandey, R., et al.: Volumetric capture of humans with a single RGBD camera via semi-parametric learning. In: CVPR (2019)
49. Patel, P., Huang, C.H.P., Tesch, J., Hoffmann, D.T., Tripathi, S., Black, M.J.: AGORA: avatars in geography optimized for regression analysis. In: CVPR (2021)
50. Peng, S., et al.: Animatable neural radiance fields for modeling dynamic human bodies. In: ICCV (2021)

51. Peng, S., et al.: Neural body: implicit neural representations with structured latent codes for novel view synthesis of dynamic humans. In: CVPR (2021)
52. RenderPeople: http://renderpeople.com
53. Riegler, G., Koltun, V.: Free view synthesis. In: Vedaldi, A., Bischof, H., Brox, T., Frahm, J.-M. (eds.) ECCV 2020. LNCS, vol. 12364, pp. 623–640. Springer, Cham (2020). https://doi.org/10.1007/978-3-030-58529-7_37
54. Roveri, R., Rahmann, L., Oztireli, C., Gross, M.: A network architecture for point cloud classification via automatic depth images generation. In: CVPR (2018)
55. Rückert, D., Franke, L., Stamminger, M.: Adop: Approximate differentiable one-pixel point rendering. arXiv preprint arXiv:2110.06635 (2021)
56. Shum, H., Kang, S.B.: Review of image-based rendering techniques. In: Visual Communications and Image Processing (2000)
57. Srinivasan, P.P., Tucker, R., Barron, J.T., Ramamoorthi, R., Ng, R., Snavely, N.: Pushing the boundaries of view extrapolation with multiplane images. In: CVPR (2019)
58. Suvorov, R., et al.: Resolution-robust large mask inpainting with Fourier convolutions. In: WACV (2022)
59. Tan, F., et al.: Humangps: geodesic preserving feature for dense human correspondences. In: CVPR (2021)
60. Tewari, A., et al.: State of the art on neural rendering. In: Computer Graphics Forum (2020)
61. Thies, J., Zollhöfer, M., Theobalt, C., Stamminger, M., Nießner, M.: IGNOR: Image-guided Neural Object Rendering. In: ICLR (2020)
62. Tretschk, E., Tewari, A., Golyanik, V., Zollhöfer, M., Lassner, C., Theobalt, C.: Non-rigid neural radiance fields: reconstruction and novel view synthesis of a dynamic scene from monocular video. In: ICCV (2021)
63. Wang, T., Sarafianos, N., Yang, M.H., Tung, T.: Animatable neural radiance fields from monocular RGB-D. arXiv preprint arXiv:2204.01218 (2022)
64. Wiles, O., Gkioxari, G., Szeliski, R., Johnson, J.: Synsin: end-to-end view synthesis from a single image. In: CVPR (2020)
65. Xie, Y., et al.: Neural fields in visual computing and beyond (2021)
66. Xu, H., Alldieck, T., Sminchisescu, C.: H-nerf: neural radiance fields for rendering and temporal reconstruction of humans in motion. In: NeurIPS (2021)
67. Yoon, J.S., Kim, K., Gallo, O., Park, H.S., Kautz, J.: Novel view synthesis of dynamic scenes with globally coherent depths from a monocular camera. In: CVPR (2020)
68. Zhang, R., Isola, P., Efros, A.A., Shechtman, E., Wang, O.: The unreasonable effectiveness of deep features as a perceptual metric. In: CVPR (2018)
69. Zhou, T., Tucker, R., Flynn, J., Fyffe, G., Snavely, N.: Stereo magnification: Learning view synthesis using multiplane images. TOG (2018)
70. Zitnick, C., Kang, S.B., Uyttendaele, M., Winder, S., Szeliski, R.: High-quality video view interpolation using a layered representation. TOG **23**, 600–608 (2004)

Multiview Regenerative Morphing with Dual Flows

Chih-Jung Tsai[1], Cheng Sun[1,2], and Hwann-Tzong Chen[1,3(✉)] (iD)

[1] National Tsing Hua University, Hsinchu, Taiwan
htchen@cs.nthu.edu.tw
[2] ASUS AICS Department, Taipei, Taiwan
[3] Aeolus Robotics, Taipei, Taiwan

Abstract. This paper aims to address a new task of image morphing under a multiview setting, which takes two sets of multiview images as the input and generates intermediate renderings that not only exhibit smooth transitions between the two input sets but also ensure visual consistency across different views at any transition state. To achieve this goal, we propose a novel approach called Multiview Regenerative Morphing that formulates the morphing process as an optimization to solve for rigid transformation and optimal-transport interpolation. Given the multiview input images of the source and target scenes, we first learn a volumetric representation that models the geometry and appearance for each scene to enable the rendering of novel views. Then, the morphing between the two scenes is obtained by solving optimal transport between the two volumetric representations in Wasserstein metrics. Our approach does not rely on user-specified correspondences or 2D/3D input meshes, and we do not assume any predefined categories of the source and target scenes. The proposed view-consistent interpolation scheme directly works on multiview images to yield a novel and visually plausible effect of multiview free-form morphing. Code: https://github.com/jimtsai23/MorphFlow

1 Introduction

Image morphing is an appealing visual effect that transforms one image into another with coherent intermediate results showing smooth transitions. It has wide applications in visualization, special visual effect, and virtual reality. Conventional morphing methods consist of three steps: *i*) acquire user-specified landmark-based or dense correspondences, *ii*) warp each image into an intermediate layout based on the aligned correspondences, and *iii*) blend the two aligned images with the respective weights. The need of user-specified correspondences is unfavorable and sometimes even impossible when image contents are very dissimilar. On the other hand, as observed in [47], simple warping functions may cause unnatural appearances when characterizing complex deformations. Later

---

Supplementary Information The online version contains supplementary material available at https://doi.org/10.1007/978-3-031-19787-1_28.

S. Avidan et al. (Eds.): ECCV 2022, LNCS 13676, pp. 492–509, 2022.
https://doi.org/10.1007/978-3-031-19787-1_28

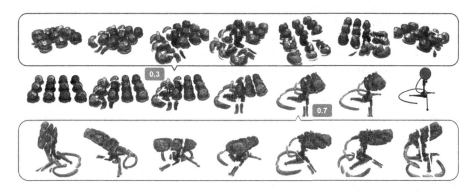

Fig. 1. Multiview regenerative morphing. The middle row shows an example of *Multiview regenerative morphing* from the source (left) to the target (right). The top row shows multiview rendering when the blending weight is 0.3. The bottom row shows multiview rendering with blending weight 0.7.

approaches like *regenerative morphing* [48], and more recently the GAN-based methods such as [29,30], are able to relax the requirement of explicitly specified correspondences and produce intriguing effects of image-to-image warping and interpolation.

In this work, we build upon the success of prior techniques and aim to solve a more challenging task of image morphing to render multiview morphs between two structurally unaligned and visually unrelated scenes, as shown in Fig. 1. More precisely, the new task we seek to address can be described as follows: Consider two sets of images, each taken from a scene of arbitrary categories under various viewpoints. The goal of our task is to build a model that has the capability of producing multiview morphs, such that, *i*) at any chosen viewpoint, it can generate a sequence of transitions as in standard image morphing, while *ii*) at any given transition moment, it can present multiview renderings of the intermediate morphing scene. To achieve this goal, we propose to learn a model comprising volumetric scene representations for rendering morphs. Each scene is represented by a 4D volume, where each voxel contains RGB and alpha (opacity) values. The representation is learned in a coarse-to-fine manner. First, we use a coarse voxel grid to locate the probable occupancy of the scene, and then we use a finer voxel grid to optimize for the details. Instead of using implicit functions or any kind of neural network, we adopt an explicit differentiable volume rendering scheme to reduce computation time.

Suppose that we have derived the aforementioned representations from the multiview images of the source and target scenes. Now, to proceed with the morphing task, we need to fuse the two representations into an intermediate one that can be used to render novel views of the morphs for any given transition moment. Without relying on predefined correspondences, we adopt an optimal transport mechanism that models intermediate transitional representations as interpolations of the original two representations. The interpolations resemble a mixture of scaffolds of the two scenes for querying and blending the volumetric representations. As free-form regenerative morphing may result in shattered structures during transitions, we regularize the morphing process by enforcing a rigid transformation to avoid generating fragmented morphs. With the interpolated representation and rigid transformation, our method can render a sequence of coherent morphs of the two category-independent input scenes. The rendered morphs can be displayed from different viewpoints at different transition moments.

We summarize the main ideas of this work as follows:

1. This paper presents an optimization-based method that tackles a new task of multiview regenerative morphing as illustrated in Fig. 1. The proposed method takes multiview images as the input; no 2D or 3D meshes are needed.
2. Our approach does not assume the categories of and the affinities between the source and target images, nor does it require any predefined correspondences between them.
3. Our approach adopts the mechanism of optimal transport to get an interpolated volume for rendering transitional multiview morphs. We also include a rigid transformation in the morphing process to favor 'structure-preserving' morphs when possible.
4. Our method is efficient in learning and rendering. It can learn a morphing renderer from scratch (directly from the input images) in 30 min. For morphing and rendering, the learned renderer can generate one novel-view morph per second.

2 Related Work

Image Morphing. Image morphing aims at transforming a source image to a target image smoothly and with natural-looking in-between results. Traditional approaches [33,56] use image warping and color interpolation with predefined dense correspondence. In particular, [2] enforces the transition to be as rigid as possible, and [47] considers camera viewpoint to prevent distortion. Patch-based methods [10,48] are later proposed to synthesize in-between images using source and target patches under temporal coherence constraints. Recently, Generative Adversarial Networks (GAN) has shown impressive image generation results by learning a projection from latent space to image space. Image morphing can then be achieved by simple linear blending in the GAN latent space [1,19,44,58]. Simon and Aberdam [50] further propose to solve the Wasserstein barycenter problem constrained on GAN latent space to achieve smooth transitions and natural-looking in-between results. Optimal transport has also been used to produce morphing between simple 2D geometries [4–6,51]. However, the morphs lack textures as in nature images. Our method is different from the above morphing methods in that we generate 3D representations and render view-consistent morphs in novel views. Perhaps the most similar to us are the multilevel free-form deformation morphing techniques described in [56]. While they still depend on 3D primitives and human-labeled correspondence, our morphing technique is fully automatic and unsupervised.

Volume Renderer from Multiview Images. Reconstructing a volumetric scene representation that supports novel-view synthesis from a set of images is a long-standing task with steady progress [15,17]. NeRF [41] has recently revolutionized this task by incorporating the coordinate-based multilayer perceptrons (MLP) to represent each spatial point's color and volume density implicitly. The MLP model is trained to minimize the photometric loss on the observed views with differentiable volume rendering. Many follow-up works of NeRF are proposed to achieve better qualities on background [63], surface [43], multi-resolution [3], imperfect input poses [27,36,40], fewer input views [7,53,62], and dynamic scene [20,35,39,59]. Despite the high quality and flexibility, NeRF still has a disadvantage of its lengthy training and rendering run-time. To improve rendering speed, many methods are proposed to convert the trained implicit MLP representations to explicit voxel-grid or hybrid representations [21,23,55,61]. The improvement on training run-time relies on cross-scene pre-training [7,53,62] or external depth [11,38]. Our morphing algorithm is agnostic to the underlying scene representation reconstruction techniques. For ease of use, our representation explicitly models the scene with voxels, similar to DirectVoxGO [52]. We reconstruct the source and target scene representations from their respective image sets, and use the learned volumetric representations for morphing.

Shape Interpolation. Given two (or more) shapes, shape interpolation aims at generating their in-between shape specifying by a composition percentage, which enables a smooth deformation from one shape to the other. Traditional methods try to recover the shape space manifold [24,25,31,54], and then the shape interpolation becomes a geodesic-path searching problem on the manifold. As shape manifold recovering is challenging, some recent approaches [12,13] directly find the deformation field from source to target shapes but with isometric (zero-divergence, constant volume) assumption. NeuroMorph [14] uses neural networks to predict the correspondences and the deformation field, which works well even for non-isometric pairs. Generative neural networks have recently achieved good results in 3D by building shape latent spaces and using deep decoders to map latent codes to voxels [57], signed distance fields [28,45], point clouds [9,34,49], or meshes [8]; shape interpolation is then achieved by linearly blending the latent codes of the two shapes. These shape-interpolation techniques typically take 3D as input (*e.g.*, mesh), and GAN-based methods further require large datasets, while our method only needs two sets of images that capture the source and the target shapes. Janati *et al.* and Solomon *et al.* have considered interpolations between shapes as Wasserstein barycenters [26,51]. However, their computations do not include morph in appearances. To facilitate learning on more complicated shapes, we take a different strategy and morph the scenes with the Wasserstein flow [18]. We further regularize the flow with rigidity constraints, and use these local and global dual flows to achieve multiview regenerative morphing.

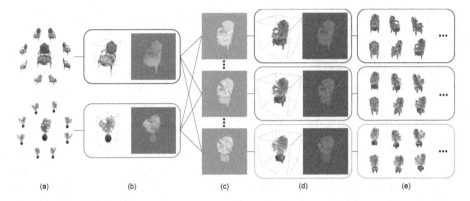

Fig. 2. An overview of our method. (a) The input multiview images of the source and target scenes. (b) The volumetric representation of each scene comprises color and opacity information. (c) A sequence of the morphed point sets with different blending weights. They are interpolations between source and target scene in Wasserstein metrics. (d) The volumetric representations generated by morphing. (e) Examples of multiview morphing rendered at arbitrary viewing directions. It can be seen that the rendering results are view-consistent—at any moment, the intermediate morph can be viewed as an actual scene and does not exhibit any conflicts across views. (Color figure online)

3 Overview

Consider the images $\mathcal{I}^{\mathcal{S}}$ and $\mathcal{I}^{\mathcal{T}}$ collected from the source and target scenes with camera poses $\zeta^{\mathcal{S}}$ and $\zeta^{\mathcal{T}}$. Our method can generate morphed images between $\mathcal{I}^{\mathcal{S}}$ and $\mathcal{I}^{\mathcal{T}}$ given arbitrary weights t and viewing angles θ, ϕ. The method has two phases. In the first phase, we establish volumetric representations for each scene with the purpose of generating view-consistent morphs from unaligned images. The representation comprises opacity and color. In the second phase, we use optimal transport to generate morphs between the derived volumetric representations from the first phase. The morphing process is controlled by rigid transformation (RT) flow and optimal transport (OT) flow. RT flow preserves the impression of the source scene during morphing, preventing shattered generation. OT flow deforms the source scene into the target without the need of correspondences. By applying the two flows, we obtain a morphed representation and render view-consistent morphs in any views. Figure 2 illustrates the pipeline of our method.

Formally, we use a volume $\mathcal{V}$ to model a scene by mapping a 3D position $\mathbf{x} = (x, y, z)$ to its corresponding opacity α and color $\mathbf{c} = (r, g, b)$ as

$$\mathcal{V} : \mathbb{R}^3 \rightarrow \mathbb{R}^4, \quad \mathcal{V}(\mathbf{x}) = (\mathcal{V}_\alpha(\mathbf{x}), \mathcal{V}_\mathbf{c}(\mathbf{x})) , \tag{1}$$

where $\mathcal{V}_\alpha(\mathbf{x})$ retrieves the opacity α and $\mathcal{V}_\mathbf{c}(\mathbf{x})$ yields the color $\mathbf{c}$. Based on Eq. (1) we build volumes $\mathcal{V}^{\mathcal{S}}$ and $\mathcal{V}^{\mathcal{T}}$ for representing the source and target scenes. The opacity can be used to filter out negligible voxels. The balance between granularity and efficiency is controlled by a threshold δ_α, *i.e.*, voxel v_i in $\mathcal{V}_\alpha$ is collected if $\alpha_i > \delta_\alpha$. The collection of points and opacity values serves as a shape representation, which can be expressed as a weighted point set $\mathcal{P} = \{(\omega_i, \mathbf{x}_i)\}_{i=1}^N$, where $\omega_i = \alpha_i / \sum_j^N \alpha_j$, associating each point with a weight derived from the opacity by normalization. In this way, we create a source shape $\mathcal{S}$ from the source volume $\mathcal{V}^{\mathcal{S}}$ and a target shape $\mathcal{T}$ from the target volume $\mathcal{V}^{\mathcal{T}}$, where both are in the form of a weighted point set as described above. While we transform the opacity volume into the source shape $\mathcal{V}^{\mathcal{S}}$, for each point i in the source set, its color $\mathbf{c}_i$ can be gathered from the color volume $\mathcal{V}_c^{\mathcal{S}}$. The point colors are preserved for blending the final appearances.

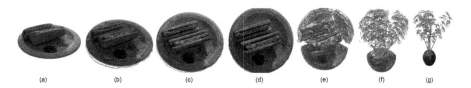

(a)   (b)   (c)   (d)   (e)   (f)   (g)

Fig. 3. To visualize the different effects of the rigid transformation (RT) flow $\mathbf{f}$ and the optimal transport (OT) flow $\mathbf{g}$, we deliberately apply them one after the other. (a–d): we apply only $\mathbf{f}$, which aligns the source's pose with the target's without changing the shape. (d–g): we apply only $\mathbf{g}$, which deforms from aligned source shape to target shape. In our method, the two flows are jointly applied during the entire morphing process.

To morph between the source shape $\mathcal{S}$ and the target shape $\mathcal{T}$, we develop a two-step algorithm using the *2–Wasserstein distance*. In the first step, we seek a rigid transformation $\hat{\Psi}$ that minimizes the distance between the source and target shapes in the Wasserstein space. We solve for $\hat{\Psi}$ using gradient descent on unbiased Sinkhorn divergence SD. A more detailed algorithm is described in Sect. 5.1. The second step finds an interpolation between the rigidly transformed source shape $\hat{\Psi}(\mathcal{S})$ and the target shape $\mathcal{T}$. We obtain the interpolation via a gradient-guided step on SD [18,22], which is interpreted as a displacement vector from the source shape to the target shape. As a result, two flows are created, namely, the RT flow $\mathbf{f}$ and the OT flow $\mathbf{g}$, by comparing the transformation $\hat{\Psi}(\mathcal{S})$ to the source shape and the target shape to the transformed source. A morphed shape can then be generated by

$$\mathcal{M}_t \leftarrow \mathcal{S} + \mathbf{f}(t) + \mathbf{g}(t), \tag{2}$$

where the flow parameter $t \in [0, 1]$ controls the progression of $\mathbf{f}$ and $\mathbf{g}$ at each transition moment t to create the expected morphs. Figure 3 visualizes the effects of the two flows.

We use the morphed shape $\mathcal{M}_t$ to query color volume of target $\mathcal{V}_c^{\mathcal{T}}$ and generate blended colors $\mathcal{V}_c^{\mathcal{M}_t}$ along with colors of source points. The morphed shape is voxelized to form $\mathcal{V}^{\mathcal{M}_t} = (\mathcal{V}_\alpha^{\mathcal{M}_t}, \mathcal{V}_c^{\mathcal{M}_t})$ so that we can render view-consistent morphing images using $\mathcal{V}^{\mathcal{M}_t}$ under arbitrary viewing directions.

In short, the proposed Multiview Regenerative Morphing provides an on-the-fly morphing renderer trained on two different scenes without any correspondences. The main idea is to extend image morphing from single-view to multiview and generate view-consistent multiview morphs. We introduce an efficient learning strategy in Sect. 4 to derive volume representations from multiview images. We use optimal transport in 2–Wasserstein space to merge the volumes of the two scenes. The algorithm for computing the morphs is detailed in Sect. 5.

4   Volume Renderer

We lift the image morphing problem from single-view to multiview by learning volumetric representations for the source and target scenes. The representation contains shape and appearance, and is learned from multiview images and their camera poses. Below, we briefly introduce how to reconstruct such a scene representation from the calibrated input images and our design choices.

To obtain volumes as in Eq. (1), we adopt differentiable volume rendering to optimize the opacity and color volumes for each scene. Given the image poses $\zeta^{\mathcal{S}}$ and $\zeta^{\mathcal{T}}$, we assume a pinhole camera and generate rays emitted from the camera center, based on each pixel's position. During rendering, points are sampled along a ray and queried with the scene representation to produce a series of colors and volume densities. The densities are converted into alpha values via $\alpha_i = 1 - \exp(-\sigma_i \delta_i)$ for the follow-up alpha compositing to accumulate the point queries into a single ray color:

$$\hat{C}(\mathbf{r}) = \sum_{i=1}^{N} T_i \alpha_i \mathbf{c}_i , \quad T_i = \prod_{j=1}^{i-1} (1 - \alpha_j) , \qquad (3)$$

where $\mathbf{r}$ is the camera ray on which the N discrete points are sampled, T_i is the accumulated transmittance from ray emission to the current sample i, and δ_i is the distance between adjacent samples. The scene representation is optimized by minimizing the photometric mean squared error

$$\mathcal{L} = \frac{1}{K} \sum_{m=1}^{K} \left\| \hat{C}(\mathbf{r}_m) - C(\mathbf{r}_m) \right\|_2^2 , \qquad (4)$$

where K is the mini-batch size, $\hat{C}$ is the rendered color, and C is the observed pixel color.

There are two common volumetric representations: i) voxel grids, which explicitly parameterize the 3D scene as grid values, and ii) multilayer perceptrons (MLP), which implicitly learn the mapping via MLP weights. We opt to use the explicit voxel grid to model the scene for faster convergence and for the convenience of latter usage in morphing. During training, we learn volumes of opacity $\mathcal{V}_\alpha(\mathbf{x})$ and color $\mathcal{V}_\mathbf{c}(\mathbf{x})$ for each scene, while each sample on the rays is trilinearly interpolated with neighboring voxels. We note, however, that the trained implicit representations [21,61] can easily be turned into volumetric representations and used with our morphing algorithm.

5   Wasserstein Morphing Flow

With learned volumetric representations $\mathcal{V}^\mathcal{S}$ and $\mathcal{V}^\mathcal{T}$, we develop a differentiable morphing algorithm that welds two volumes into a morphed volume for rendering. Since the source and target scenes are not constrained to be in one category and may be very dissimilar, optimal transport is used to deform the source scene into the target. Specifically, we use Sinkhorn divergence (SD), a regularized optimal transport objective that minimizes 2–Wassertein distance between two point sets. The source and target shapes $\mathcal{S}$ and $\mathcal{T}$ are created as weighted point sets collected from the volumes $\mathcal{V}^\mathcal{S}$ and $\mathcal{V}^\mathcal{T}$. Note that our method assumes the two weighted point sets to be positive discrete measures such that they can be compared in Wasserstein metrics. Therefore, the weights of the point sets have been normalized to make them discrete probability distributions, *i.e.*, the weights $\{\omega_i^\mathcal{S}\}_{i=1}^{N^\mathcal{S}}$ and $\{\omega_j^\mathcal{T}\}_{j=1}^{N^\mathcal{T}}$ of the source and target shapes satisfy $\sum_i^{N^\mathcal{S}} \omega_i^\mathcal{S} = 1$ and $\sum_j^{N^\mathcal{T}} \omega_j^\mathcal{T} = 1$. More precisely, such a discrete measure can be expressed as a sum of weighted Dirac mass, and we thus have $\mathcal{S} = \sum_{i=1}^{N^\mathcal{S}} \omega_i^\mathcal{S} \Delta_{\mathbf{x}_i^\mathcal{S}}$ and $\mathcal{T} = \sum_{j=1}^{N^\mathcal{T}} \omega_j^\mathcal{T} \Delta_{\mathbf{x}_j^\mathcal{T}}$, where Δ is the Dirac delta function that can be thought of as an indicator of occupancy at a given point $\mathbf{x}$.

Our aim now is to obtain a 3D morphing renderer, where at the core is a morphed volumetric representation $\mathcal{V}^{\mathcal{M}_t}$. We view the morphing process as solving an optimal transport problem for moving mass from a source distribution to a target distribution. To generate smooth transitions, we design a flow-based morphing scheme, which comprises the rigid transformation (RT) flow and optimal transport (OT) flow. The RT flow pushes the source scene toward the target scene by applying rotation and translation. As the rigid transformation is global, it can preserve the original appearance. On the other hand, the OT flow provides smooth local deformations but may change the topology of the shape. In what follows, we first describe how to compute the RT flow for globally registering the two shapes, and then we detail the algorithm of OT flow, as well as the complete scheme of dual-flow based morphing.

5.1   Rigid Transformation Flow

With the two shapes $\mathcal{S}$ and $\mathcal{T}$ expressed as two discrete measures that are derived from the source and target volumes, we estimate the rigid transformation $\Psi \in$ SE(3) by minimizing the Sinkhorn divergence SD [18] between the transformed source measure and the target measure:

$$\hat{\Psi} = \arg\min_{\Psi} \mathrm{SD}\left(\Psi(\mathcal{S}), \mathcal{T}\right), \tag{5}$$

where the transformed measure $\Psi(\mathcal{S})$ is defined in the form of weighted sum of Dirac mass by

$$\Psi(\mathcal{S}) = \sum_{i=1}^{N^\mathcal{S}} \omega_i^\mathcal{S} \Delta_{\Psi(\mathbf{x}_i^\mathcal{S})}. \tag{6}$$

The rigid transformation Ψ comprises rotation $\mathbf{R}$ and translation $\mathbf{z}$. The initial states $\mathbf{R}^{(0)} = \mathbf{I}_3$ and $\mathbf{z}^{(0)} = \mathbf{0}$ are updated by the gradients $\nabla_{\mathbf{R}}\mathrm{SD}(\Psi(\mathcal{S}), \mathcal{T})$ and $\nabla_{\mathbf{z}}\mathrm{SD}(\Psi(\mathcal{S}), \mathcal{T})$. Note that directly applying the gradient update to $\mathbf{R}$ may lead to an unconstrained projection matrix, and therefore we replace the gradient-updated matrix with identity singular values via singular value decomposition (SVD). Specifically, we compute

$$(\mathbf{R} - \nabla_{\mathbf{R}}\mathrm{SD}(\Psi(\mathcal{S}), \mathcal{T})) = \mathbf{U}\boldsymbol{\Sigma}\mathbf{V}^{\mathsf{T}}, \tag{7}$$

and use $\mathbf{U}\mathbf{V}^{\mathsf{T}}$ as a surrogate for the new rotation matrix. Finally, we can solve Eq. (5) for the estimated transformation $\hat{\Psi}$, and the transformation flow $\mathbf{f}$ parameterized by the time step $t \in [0, 1]$ is then given by

$$\mathbf{f}(t) = t \cdot (\hat{\Psi}(\mathcal{S}) - \mathcal{S}), \tag{8}$$

which is used in Eq. (2) to provide the progression on the source shape $\mathcal{S}$. For brevity, the definition of Sinkhorn divergence SD and the derivation of the gradients $\nabla_{\mathbf{R}}\mathrm{SD}$ and $\nabla_{\mathbf{z}}\mathrm{SD}$ are omitted here. More details can be found in the supplementary material.

5.2   Optimal Transport Flow

The rigid transformation makes the two measures $\hat{\Psi}(\mathcal{S})$ and $\mathcal{T}$ distribute in similar loci in $\mathbb{R}^3$. We can further find a smooth deformation between them using optimal transport. In our method, the interpolation is achieved by adding the gradient that minimizes the Sinkhorn divergence between the transformed shape and the target. Given a time step $t \in [0, 1]$ as a blending weight, the optimal transport flow can be written as

$$\mathbf{g}(t) = -t \cdot \nabla_{\mathbf{x}}\mathrm{SD}(\hat{\Psi}(\mathcal{S}), \mathcal{T}), \tag{9}$$

which facilitates on-the-fly rendering with a varying time step t. Instead of applying the two flows $\mathbf{f}$ and $\mathbf{g}$ sequentially, we advance the two flows simultaneously on $t \in [0, 1]$ as shown in Eq. (2), and get the morphed measure $\mathcal{M}_t$ as the morphed shape. The disentanglement of the two flows in Wasserstein metrics enables the morphing to evolve as rigidly as possible even under inevitable topology changes.

   We voxelize the morphed measure $\mathcal{M}_t$ into $\mathcal{V}_\alpha^{\mathcal{M}_t}$ by collecting the points with histograms. Each point allocates its weight to the bin according to its position. The histogram is then transformed to a volume. The discretization has to be implemented with care to prevent aliasing. Here we first spread each point in $\mathcal{M}_t$ to its eight neighbors on the grid. The weight also splits into eight according to the distances between points. The histogram gathers weights of all points and generates $\mathcal{V}_\alpha^{\mathcal{M}_t}$. $\mathcal{M}_{t=1}$ is used to query the color volume $\mathcal{V}_c^{\mathcal{T}}$ for the corresponding target colors. The color of each morphed point is blended between the source color and target color. The morphed color volume $\mathcal{V}_c^{\mathcal{M}_t}$ is generated using a histogram method similar to $\mathcal{V}_\alpha^{\mathcal{M}_t}$.

6   Results

We evaluate our method on real and synthetic datasets, including **Synthetic–NeRF** [42], **Synthetic–NSVF** [37], **Tanks&Temples** [32] and **BlendedMVS** [60]. Each datasets contains scenes with surrounded imaging and their camera poses.

6.1   Multiview Regenerative Morphing

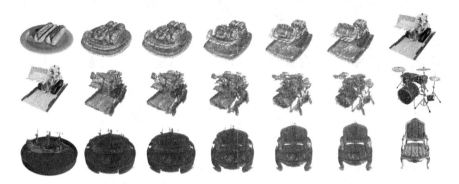

Fig. 4. Morphing of scenes in Synthetic–NeRF. Each row shows a smooth transition from the source (left) to the target (right).

Each scene may contain a number of objects, where the morphing between scenes needs to divide or merge objects smoothly. Figure 1 shows morphed images from materials with different colors and reflections into a single microphone. The top and bottom rows show the different views of the morph frozen at $t = 0.3$ and $t = 0.7$. The rendering results are view-consistent, *i.e.*, at the frozen moment, the intermediate morph can be viewed as an actual coherent scene and exhibits no conflicts across different views.

Fig. 5. Morphing of scenes in Synthetic–NSVF. Each row shows a smooth transition from the source (left) to the target (right).

Figure 4 shows morphing with scenes in Synthetic-NeRF. We demonstrate smooth transitions between three different sets of source and target scenes. Especially, in the middle row a lego truck morphs into a drum set, with very complex detail. In Fig. 5, we show morphing results in Synthetic-NSVF. Due to the limitation of space, we provide more results of multiview rendering in videos in the supplementary material.

Fig. 6. Morphing of scenes in BlendedMVS. All the rows exhibit the same transition, while each row shows renderings under some view.

Figure 6 demonstrates the morphing between 'Statue' and 'Character'. Both scenes are from BlendedMVS. The three rows represent the same transition rendered in different viewpoints. Two people in the Statue gradually get close to each other and merge into a single person. Also the scene becomes colorful, from metallic texture into custom and makeup. Since the training images have abundant specular lighting, resulting in noises and heavy shadows, we therefore use thresholding and color distribution manipulation on the morphed volumes to compensate the flaw.

Fig. 7. Morphing of scenes in Tanks&Temples. All the rows exhibit the same transition, but in different views.

Figure 7 shows transitions between 'Caterpillar' and 'Truck'. Both scenes are from Tanks&Temples. The two shapes are reconstructed with collected images,

Fig. 8. Morphing between real and synthetic scenes. We demonstrate view-consistent morphing by rendering in three different views.

under varying lighting conditions. The direct result has the floating noise around the shapes. We use simple thresholding to remove the noise in the space of the volume. Each row shows renderings of the transition under some view. The morphs are view-consistent in each column. Each column relates to some blending weight. In addition to morphing real scenes, we demonstrate morphing between real and synthetic scenes. Figure 8 shows the morphs from a real truck to synthetic 'Spaceship'. Likewise, Fig. 9 interpolates between 'Caterpillar' and 'Character' from BlendedMVS. The real to synthetic scene morphing is achieved under our normal setting.

Fig. 9. Morphing between real and synthetic scenes. The morphs are rendered with three viewpoints, one in each row.

Limitations on Real Scenes. Our dual-flow multiview morphing is a model-free method and therefore not restricted to specific data domains. Figure 7 shows the morphing between two real scenes, where our method generates reasonable transitions between the two very dissimilar scenes. However, due to the lighting changes across views, the rendered morphs contain more noise in comparison with synthetic data. The quality of multiview morphing relies on the reconstructed volumetric representations of the original scenes. Those reconstruction artifacts tend to remain during the entire morphing process. Similar artifacts are observed in Fig. 6, where the reflection of the surface affects the reconstruction.

6.2   Comparisons with Other Approaches

We validate that our morph generation is geometry-aware in contrast to other correspondence-free morphing algorithms. We compare our method with a 3D-based method, Debiased Sinkhorn [26], and a 2D-based method, Deep Image Analogy [16]. To compare with the 2D-based method, we use the optimized volume of each scene to generate pose-aligned images as their input. NeRF's eight scenes are used for evaluation. We randomly sample camera poses from the upper hemisphere for different scenes and render the morphs with varying transition weights. For each transition, we use COLMAP to solve *structure from motion*. As a result, 83.3% of the morphing images generated by our method are successfully registered by COLMAP, which means our method can mostly render 3D consistent morphs. On the other hand, Debiased Sinkhorn [26] generates blur images, resulting in poor reconstruction: Only 16.7% of its morphing images can be successfully registered by COLMAP. Deep Image Analogy [16] generates visually pleasing morphs, but it is not robust to view changes, as mentioned in [47]: not surprisingly, no consistent 3D structure can be reconstructed.

6.3   Ablation Study

We evaluate different aspects of our method, especially the effect of the RT flow and the OT flow. We compare the rendering of direct optimal-transport morphing with our rigid-transformation-enabled OT morphing. We also demonstrate the effect when only one of the RT flow or the OT flow is applied.

Rigid Transformation Flow: As mentioned in [2,14,46], achieving as-rigid-as-possible transformation is an appealing property for morphing. The property preserves the original shape during transition and thus helps to produce plausible intermediate results. Unlike previous methods that generate meshes for deformation, here we formulate the rigid transformation as a flow in Wasserstein space. Figure 10 shows comparisons of transitions with or without the rigid transformation flow. The first row shows renderings using our full model with both the RT and OT flows. The second row shows renderings without the RT flow; only the OT flow is used. It can be seen that morphing without the RT flow directly moves each point toward the target and results in shattered rendering. Such an effect is unsatisfactory as the edge of the plate falls into pieces. In contrast, our method gradually transforms the plate so that the hotdog and the ficus have their poses aligned. During transformation, the OT flow simultaneously performs deformation in local regions so the texture can better resemble the target. We also compare to the baseline, where no flows but simple blending is applied, as shown in the third row. Morphing with cross-dissolve leads to ghost effects.

Optimal Transport Flow: Our flow composition method generates smooth morphing by jointly performing rigid transformation and OT deformation. Here we examine the effect when only one of the two flows is used. This can be manipulated by the blending weight t in each flow. As previously shown in Fig. 3, we use only the RT flow in the first half of morphing, and the OT flow in the

Fig. 10. Comparisons between three morphing method. First row shows our dual-flow morphing. Second row uses only OT flow. Third row use no flow, but simple blending.

second half. The source scene is on the left, and the subsequent three images are only affected by the RT flow. We can see that the flow aligns the poses of the source and target shapes. When the two scenes are fully aligned, the OT flow deforms the hotdog into the ficus, as shown in the fifth and sixth images.

7   Conclusion

This paper introduces Multiview Regenerative Morphing—a new method that integrates volume rendering and optimal transport to address a new task of multiview image morphing. Our method can produce interesting morphing effects that have not yet been demonstrated by previous image-based morphing methods. From the multiview images of two category-agnostic scenes without predefined correspondences, our method learns volumetric representations to render free-form morphs that can be visualized from arbitrary perspectives at any transition moment. We decouple the morphing process into two flows in Wasserstein metrics: one governs the rigid transformation and the other models the correspondences and deformations. The two flows estimated via optimization then jointly provide as-rigid-as-possible transformation under required topological and morphological changes between the two shapes. Our method is fast in training; it takes less than half an hour to learn the morphing renderer from both scenes, which otherwise might need 30x longer time if learned by typical neural rendering methods. The learned morphing renderer can readily generate on-the-fly multiview morphs showcasing new visual effects.

Acknowledgements. This work was supported in part by the MOST grants 110-2634-F-007-027 and 111-2221-E-001-011-MY2 of Taiwan. We are grateful to National Center for High-performance Computing for providing computational resources and facilities.

References

1. Abdal, R., Qin, Y., Wonka, P.: Image2stylegan: how to embed images into the stylegan latent space? In: 2019 IEEE/CVF International Conference on Computer Vision, ICCV 2019, Seoul, Korea (South), October 27–November 2, 2019, pp. 4431–4440. IEEE (2019)
2. Alexa, M., Cohen-Or, D., Levin, D.: As-rigid-as-possible shape interpolation. In: Proceedings of the 27th Annual Conference on Computer Graphics and Interactive Techniques, pp. 157–164 (2000)
3. Barron, J.T., Mildenhall, B., Tancik, M., Hedman, P., Martin-Brualla, R., Srinivasan, P.P.: Mip-nerf: a multiscale representation for anti-aliasing neural radiance fields. In: ICCV (2021)
4. Benamou, J.D., Carlier, G., Cuturi, M., Nenna, L., Peyré, G.: Iterative bregman projections for regularized transportation problems. SIAM J. Sci. Comput. **37**(2), A1111–A1138 (2015)
5. Bonneel, N., Peyré, G., Cuturi, M.: Wasserstein barycentric coordinates: histogram regression using optimal transport. ACM Trans. Graph. **35**(4), 1–71 (2016)
6. Bonneel, N., Van De Panne, M., Paris, S., Heidrich, W.: Displacement interpolation using lagrangian mass transport. In: Proceedings of the 2011 SIGGRAPH Asia Conference, pp. 1–12 (2011)
7. Chen, A., et al.: Mvsnerf: fast generalizable radiance field reconstruction from multi-view stereo. In: ICCV (2021)
8. Cheng, S., Bronstein, M.M., Zhou, Y., Kotsia, I., Pantic, M., Zafeiriou, S.: Meshgan: non-linear 3d morphable models of faces. arxiv CS.CV 1903.10384 (2019)
9. Cosmo, L., Norelli, A., Halimi, O., Kimmel, R., Rodolà, E.: LIMP: learning latent shape representations with metric preservation priors. In: Vedaldi, A., Bischof, H., Brox, T., Frahm, J.-M. (eds.) ECCV 2020. LNCS, vol. 12348, pp. 19–35. Springer, Cham (2020). https://doi.org/10.1007/978-3-030-58580-8_2
10. Darabi, S., Shechtman, E., Barnes, C., Goldman, D.B., Sen, P.: Image melding: combining inconsistent images using patch-based synthesis. ACM Trans. Graph. (TOG) **31**(4), 1–10 (2012)
11. Deng, K., Liu, A., Zhu, J., Ramanan, D.: Depth-supervised nerf: Fewer views and faster training for free. arxiv CS.CV 2107.02791 (2021)
12. Eisenberger, M., Cremers, D.: Hamiltonian dynamics for real-world shape interpolation. In: Vedaldi, A., Bischof, H., Brox, T., Frahm, J.-M. (eds.) ECCV 2020. LNCS, vol. 12349, pp. 179–196. Springer, Cham (2020). https://doi.org/10.1007/978-3-030-58548-8_11
13. Eisenberger, M., Lähner, Z., Cremers, D.: Divergence-free shape correspondence by deformation. Comput. Graph. Forum **38**(5), 1–12 (2019)
14. Eisenberger, M., et al.: Neuromorph: unsupervised shape interpolation and correspondence in one go. In: Proceedings of the IEEE/CVF Conference on Computer Vision and Pattern Recognition, pp. 7473–7483 (2021)
15. De Bonet, J.S., Viola, P.: Poxels: probabilistic voxelized volume reconstruction. In: ICCV (1999)
16. Liao, J., Yao, Y., Yuan, L., Hua, G., Kang, S.B.: Visual attribute transfer through deep image analogy. arXiv:1705.01088 (2017)
17. Szeliski, R., Golland, P.: Stereo Matching with Transparency and Matting. Int. J. Comput. Vis. **32**, 45–61 (1999). https://doi.org/10.1023/A:1008192912624

18. Feydy, J., Séjourné, T., Vialard, F.X., Amari, S.I., Trouvé, A., Peyré, G.: Interpolating between optimal transport and mmd using sinkhorn divergences. In: The 22nd International Conference on Artificial Intelligence and Statistics, pp. 2681–2690. PMLR (2019)
19. Fish, N., Zhang, R., Perry, L., Cohen-Or, D., Shechtman, E., Barnes, C.: Image morphing with perceptual constraints and STN alignment. Comput. Graph. Forum **39**(6), 303–313 (2020)
20. Gao, C., Saraf, A., Kopf, J., Huang, J.: Dynamic view synthesis from dynamic monocular video. In: ICCV (2021)
21. Garbin, S.J., Kowalski, M., Johnson, M., Shotton, J., Valentin, J.P.C.: Fastnerf: high-fidelity neural rendering at 200fps. arxiv CS.CV 2103.10380 (2021)
22. Genevay, A., Peyré, G., Cuturi, M.: Learning generative models with sinkhorn divergences. In: International Conference on Artificial Intelligence and Statistics, pp. 1608–1617. PMLR (2018)
23. Hedman, P., Srinivasan, P.P., Mildenhall, B., Barron, J.T., Debevec, P.E.: Baking neural radiance fields for real-time view synthesis. In: ICCV (2021)
24. Heeren, B., Rumpf, M., Schröder, P., Wardetzky, M., Wirth, B.: Splines in the space of shells. Comput. Graph. Forum **35**(5), 111–120 (2016)
25. Heeren, B., Rumpf, M., Wardetzky, M., Wirth, B.: Time-discrete geodesics in the space of shells. Comput. Graph. Forum **31**(5), 1755–1764 (2012)
26. Janati, H., Cuturi, M., Gramfort, A.: Debiased sinkhorn barycenters. In: International Conference on Machine Learning, pp. 4692–4701. PMLR (2020)
27. Jeong, Y., Ahn, S., Choy, C., Anandkumar, A., Cho, M., Park, J.: Self-calibrating neural radiance fields. In: ICCV (2021)
28. Jiang, C.M., Marcus, P.: Hierarchical detail enhancing mesh-based shape generation with 3d generative adversarial network. arxiv CS.CV 1709.07581 (2017)
29. Karras, T., Laine, S., Aila, T.: A style-based generator architecture for generative adversarial networks. In: IEEE Conference on Computer Vision and Pattern Recognition, CVPR 2019, Long Beach, CA, USA, 16–20 June 2019, pp. 4401–4410. Computer Vision Foundation/IEEE (2019)
30. Karras, T., Laine, S., Aittala, M., Hellsten, J., Lehtinen, J., Aila, T.: Analyzing and improving the image quality of stylegan. In: 2020 IEEE/CVF Conference on Computer Vision and Pattern Recognition, CVPR 2020, Seattle, WA, USA, 13–19 June 2020, pp. 8107–8116. Computer Vision Foundation/IEEE (2020)
31. Kilian, M., Mitra, N.J., Pottmann, H.: Geometric modeling in shape space. ACM Trans. Graph. **26**(3), 64 (2007)
32. Knapitsch, A., Park, J., Zhou, Q.Y., Koltun, V.: Tanks and temples: Benchmarking large-scale scene reconstruction. ACM Trans. Graph. (ToG) **36**(4), 1–13 (2017)
33. Lerios, A., Garfinkle, C.D., Levoy, M.: Feature-based volume metamorphosis. In: Proceedings of the 22nd Annual Conference on Computer Graphics and Interactive Techniques, pp. 449–456 (1995)
34. Li, C., Zaheer, M., Zhang, Y., Póczos, B., Salakhutdinov, R.: Point cloud GAN. In: Deep Generative Models for Highly Structured Data, ICLR 2019 Workshop, New Orleans, Louisiana, United States, 6 May 2019, OpenReview.net (2019)
35. Li, Z., Niklaus, S., Snavely, N., Wang, O.: Neural scene flow fields for space-time view synthesis of dynamic scenes. In: CVPR (2021)
36. Lin, C., Ma, W., Torralba, A., Lucey, S.: BARF: bundle-adjusting neural radiance fields. In: ICCV (2021)
37. Liu, L., Gu, J., Lin, K.Z., Chua, T., Theobalt, C.: Neural sparse voxel fields. In: NeurIPS (2020)

38. Liu, Y., et al.: Neural rays for occlusion-aware image-based rendering. arxiv CS.CV 2107.13421 (2021)
39. Martin-Brualla, R., Radwan, N., Sajjadi, M.S.M., Barron, J.T., Dosovitskiy, A., Duckworth, D.: Nerf in the wild: neural radiance fields for unconstrained photo collections. In: CVPR (2021)
40. Meng, Q., et al.: Gnerf: Gan-based neural radiance field without posed camera. In: ICCV (2021)
41. Mildenhall, B., Srinivasan, P.P., Tancik, M., Barron, J.T., Ramamoorthi, R., Ng, R.: Nerf: representing scenes as neural radiance fields for view synthesis. In: ECCV (2020)
42. Mildenhall, B., Srinivasan, P.P., Tancik, M., Barron, J.T., Ramamoorthi, R., Ng, R.: NeRF: representing scenes as neural radiance fields for view synthesis. In: The European Conference on Computer Vision (ECCV) (2020)
43. Oechsle, M., Peng, S., Geiger, A.: UNISURF: unifying neural implicit surfaces and radiance fields for multi-view reconstruction. arxiv CS.CV 2104.10078 (2021)
44. Pan, X., Zhan, X., Dai, B., Lin, D., Loy, C.C., Luo, P.: Exploiting deep generative prior for versatile image restoration and manipulation. In: Vedaldi, A., Bischof, H., Brox, T., Frahm, J.-M. (eds.) ECCV 2020. LNCS, vol. 12347, pp. 262–277. Springer, Cham (2020). https://doi.org/10.1007/978-3-030-58536-5_16
45. Park, J.J., Florence, P., Straub, J., Newcombe, R.A., Lovegrove, S.: Deepsdf: learning continuous signed distance functions for shape representation. In: IEEE Conference on Computer Vision and Pattern Recognition, CVPR 2019, Long Beach, CA, USA, 16–20 June 2019, pp. 165–174. Computer Vision Foundation/IEEE (2019)
46. Schaefer, S., McPhail, T., Warren, J.: Image deformation using moving least squares. In: ACM SIGGRAPH 2006 Papers, pp. 533–540 (2006)
47. Seitz, S.M., Dyer, C.R.: View morphing. In: Proceedings of the 23rd Annual Conference on Computer Graphics and Interactive Techniques, pp. 21–30 (1996)
48. Shechtman, E., Rav-Acha, A., Irani, M., Seitz, S.: Regenerative morphing. In: 2010 IEEE Computer Society Conference on Computer Vision and Pattern Recognition, pp. 615–622. IEEE (2010)
49. Shu, D.W., Park, S.W., Kwon, J.: 3d point cloud generative adversarial network based on tree structured graph convolutions. In: 2019 IEEE/CVF International Conference on Computer Vision, ICCV 2019, Seoul, Korea (South), October 27–November 2, 2019, pp. 3858–3867. IEEE (2019)
50. Simon, D., Aberdam, A.: Barycenters of natural images - constrained wasserstein barycenters for image morphing. In: 2020 IEEE/CVF Conference on Computer Vision and Pattern Recognition, CVPR 2020, Seattle, WA, USA, 13–19 June 2020, pp. 7907–7916. Computer Vision Foundation/IEEE (2020)
51. Solomon, J., et al.: Convolutional wasserstein distances: efficient optimal transportation on geometric domains. ACM Trans. Graph. (TOG) 34(4), 1–11 (2015)
52. Sun, C., Sun, M., Chen, H.T.: Direct voxel grid optimization: super-fast convergence for radiance fields reconstruction. arXiv preprint arXiv:2111.11215 (2021)
53. Wang, Q., et al.: Ibrnet: learning multi-view image-based rendering. In: CVPR (2021)
54. Wirth, B., Bar, L., Rumpf, M., Sapiro, G.: A continuum mechanical approach to geodesics in shape space. Int. J. Comput. Vis. 93(3), 293–318 (2011)
55. Wizadwongsa, S., Phongthawee, P., Yenphraphai, J., Suwajanakorn, S.: Nex: real-time view synthesis with neural basis expansion. In: CVPR (2021)
56. Wolberg, G.: Image morphing: a survey. Visual Comput. 14(8), 360–372 (1998)

57. Wu, J., Zhang, C., Xue, T., Freeman, B., Tenenbaum, J.: Learning a probabilistic latent space of object shapes via 3d generative-adversarial modeling. In: Advances in Neural Information Processing Systems 29: Annual Conference on Neural Information Processing Systems 2016, 5–10 December 2016, Barcelona, Spain, pp. 82–90 (2016)
58. Wu, Z., Nitzan, Y., Shechtman, E., Lischinski, D.: Stylealign: analysis and applications of aligned stylegan models. arxiv CS.CV 2110.11323 (2021)
59. Xian, W., Huang, J., Kopf, J., Kim, C.: Space-time neural irradiance fields for free-viewpoint video. In: CVPR (2021)
60. Yao, Y., et al.: Blendedmvs: a large-scale dataset for generalized multi-view stereo networks. In: Proceedings of the IEEE/CVF Conference on Computer Vision and Pattern Recognition, pp. 1790–1799 (2020)
61. Yu, A., Li, R., Tancik, M., Li, H., Ng, R., Kanazawa, A.: Plenoctrees for real-time rendering of neural radiance fields. In: ICCV (2021)
62. Yu, A., Ye, V., Tancik, M., Kanazawa, A.: pixelNeRF: neural radiance fields from one or few images. https://arxiv.org/abs/2012.02190 (2020)
63. Zhang, K., Riegler, G., Snavely, N., Koltun, V.: Nerf++: analyzing and improving neural radiance fields. arxiv CS.CV 2010.07492 (2020)

Hallucinating Pose-Compatible Scenes

Tim Brooks[(✉)] and Alexei A. Efros

UC Berkeley, Berkeley, USA
tim@timothybrooks.com

Abstract. What does human pose tell us about a scene? We propose a
task to answer this question: given human pose as input, hallucinate a
compatible scene. Subtle cues captured by human pose—action semantics, environment affordances, object interactions—provide surprising
insight into which scenes are compatible. We present a large-scale generative adversarial network for pose-conditioned scene generation. We significantly scale the size and complexity of training data, curating a massive meta-dataset containing over 19 million frames of humans in everyday environments. We double the capacity of our model with respect to
StyleGAN2 to handle such complex data, and design a pose conditioning mechanism that drives our model to learn the nuanced relationship
between pose and scene. We leverage our trained model for various applications: hallucinating pose-compatible scene(s) with or without humans,
visualizing incompatible scenes and poses, placing a person from one
generated image into another scene, and animating pose. Our model
produces diverse samples and outperforms pose-conditioned StyleGAN2
and Pix2Pix/Pix2PixHD baselines in terms of accurate human placement
(percent of correct keypoints) and quality (Fréchet inception distance).

1 Introduction

Human pose can reveal a lot about a scene. For example, mime artists[1] invoke
vivid scenes in a viewer's mind through pose and movement alone, despite performing on a bare stage. The viewer is able to imagine the invisible objects
and scene elements because of the strong relationship between human poses and
scenes learned through a lifetime of daily observations.

Psychologists have long been interested in understanding this symbiotic relationship between human and scene [6,24]. J.J. Gibson proposed the notion of
affordances [24], which can be described as "opportunities for interactions" fur-
nished by the environment. In computer vision, affordances have been used to
provide a functional description of the scene. Given an image, a number of
approaches try to predict likely human poses these scenes afford [15,23,29,30].

---

[1] For those unfamiliar with mime artists, here is a wonderful example performance:
https://youtu.be/FPMBV3rd_hI.

Supplementary Information The online version contains supplementary material
available at https://doi.org/10.1007/978-3-031-19787-1_29.

S. Avidan et al. (Eds.): ECCV 2022, LNCS 13676, pp. 510–528, 2022.
https://doi.org/10.1007/978-3-031-19787-1_29

This work, on the other hand, considers the opposite problem: given a human pose as input, the goal is to hallucinate scene(s) that are compatible with that pose. Consider Fig. 1. A push-up pose (top) places severe constraints on the space of compatible scenes: they must not only be semantically compatible (e.g., gym, exercise room), but also have compatible spatial affordances (enough floor space or appropriate equipment). Objects in the scene can afford interaction with the human (e.g., squishing down an exercise ball). Other poses might not appear as constraining, but even a simple standing pose (bottom)—head looking down, hands reaching in, legs occluded—is actually a strong indicator of a cooking scene, and signals that an object (e.g., countertop) must be occluding the legs.

Input poses   Sample output scenes

Fig. 1. Given a human pose as input, the goal of this paper is to hallucinate scene(s) that are compatible with that pose. Our model can generate isolated scenes as well as scenes containing humans.

Rather than explicitly model scene affordances and contextual compatibility, we employ a modern large-scale generative model (based on a souped-up Style-GAN2 [44] architecture) to *discover* these relationships implicitly, from data. While GANs have performed well at capturing disentangled visual models in specialized scenarios (e.g., faces, churches, categories from ImageNet [16]), they have not been demonstrated *in situ*, on complex, real-world data across varying environments.

We curate a massive meta-dataset of humans interacting with everyday environments, containing over 19 million frames. The complexity and scale of data is much higher than common GAN datasets, such as FFHQ [43] (70,000 face images) and ImageNet [16] (1.3M object images). With an appropriate pose conditioning mechanism, increased model capacity, and removal of style mixing, we are able to successfully train a pose-conditioned GAN on this highly complex data. Our model and meta-dataset mark substantial progress leveraging GANs in real-world settings containing humans and diverse environments. Through numerous visual experiments, we demonstrate our model's emergent ability to capture affordances and contextual relationships between poses and scenes.

See our webpage[2] for our supplemental video and code release.

2   Related Work

Scene and Object Affordances. Affordances [24] describe the possible uses of a given object or environment. A significant body of work learns scene affordances, such as where a person can stand or sit, from observing data of humans [14,15, 21,23,28,29,38,49,74]. Overlapping areas of work focus on human interactions with objects [10,25,47,77,82] or synthesize human pose conditioned on an input scene [8,48,71]. We propose the reverse task of hallucinating a scene conditioned on pose.

Pose-Conditioned Human Synthesis. There are a plethora of methods that take a source image (or video) of a human plus a new pose and generate an image of the human in the new pose [1,4,13,50,53,67,72]. Although we too condition on pose, our goals are almost entirely opposite: we aim to generate novel scenes compatible with a given pose, whereas the above methods reuse the scene from the source image/video and only focus on reposing within that provided scene.

GANs for Image Synthesis. Introduced by Goodfellow *et al.* [27] a generative adversarial network (GAN) is an implicit generative model that learns to synthesize data samples by optimizing a minimax objective. The generator is tasked with fooling a discriminator, and the discriminator is tasked with differentiating real and generated samples. Modern GANs are capable of producing high quality images [7,41,43,44]. Image translation [36,72] utilizes conditional GANs [58] to translate from one domain to another. While our task is pose-conditional scene generation, we leverage benefits of modern unconditional GANs [44].

Visual Disentanglement. Disentanglement methods attempt to separate out independent controllable attributes of images. This can be achieved with unsupervised methods [34,37,43,63], or an auxiliary signal [26,51]. Components of image samples can be added, removed and composed using pretrained GANs [5,12]. Recent work has applied similar strategies to image translation models to compose style and content from different images [62]. The most related to us is the work of Ma *et al.* [54], who synthesize images of people, while independently controlling foreground, background, and pose. However, the focus is on generating humans in very tightly cropped images with simple backgrounds, rather than generating scenes with appropriate affordances. Many disentanglement methods assume all images or image attributes can be combined with all others [26,34,37,43,51,54,62,63]. In this work, we seek disentangled representations of pose, human appearance and scene, yet it is essential our model understand which scenes can or cannot be composed with which poses.

---

[2] https://www.timothybrooks.com/tech/hallucinating-scenes.

Contextual Relationships. Many works leverage contextual relationships among objects and scenes [6] to improve vision models such as object recognition and semantic segmentation [18,61,64,70]. Divvala *et al.* [18] explicitly enumerate (Table 1) a taxonomy of possible contextual information. In this paper we are specifically interested in contextual relationships between humans and their environments, and aim to recover them implicitly, from data (Fig. 2).

(a) Humans 3.6M [35]   (b) Kinetics [45]   (c) Humans in Context (ours)

Fig. 2. Dataset comparison. (a) The largest human-centric video dataset with ground truth poses uses a fixed background, missing scene interactions. (b) Action recognition datasets include scenes, but contain videos without people or of close-up content. (c) Our dataset is a massive curation of humans in scenes.

3   Humans in Context Meta-dataset

To study the rich relationship between scenes and human poses requires large-scale data of people interacting with many different environments. Internet videos are a natural source, containing vast data of daily human activities. Unfortunately, large-scale action recognition datasets [17,45,60] include substantial content without humans, as well as close-up footage not of scenes. Most existing human-centric datasets are insufficiently small [3,68], narrow in scene type [22,80], or captured on a fixed background [35].

We therefore curate a meta-dataset of 229,595 video clips, each containing a single person in a scene, sourced from 10 existing human and action recognition video datasets [3,17,19,22,40,45,60,68,76,80], and supplemented with pseudo-ground truth pose obtained using OpenPose [9,11]. Video offers a massive source of real-world data, and ensures all poses of human activity are represented, rather than only poses photographers choose to capture in still images.

Videos are extensively filtered for quality, ensuring satisfactory framerate, bitrate and resolution. 1,509,032 videos (75% of source videos) pass quality filtering. Frames are then filtered with pretrained Keypoint R-CNN [31,75] person detection and OpenPose [9,11] keypoint prediction models. The final dataset only includes clips of at least 30 frames where Keypoint R-CNN detects a single person and OpenPose predicts sufficient keypoints. This results in 19,503,700 frames (7.8% of high quality frames), with each clip averaging 85 frames long.

While we train on images, we split data into partitions based on video clips, reserving 12,800 clips for testing and the remaining 216,795 for training. See the supplement for dataset details.

4   Pose-compatible Scene GAN

We design a conditional GAN [27,58] to produce scenes compatible with human pose. Our network architectures are based on StyleGAN2 [44] and are depicted in Fig. 3. Generating high quality pose-compatible scenes arises from simple yet important modifications: dual pose conditioning, removal of style mixing, and large-scale training. Our model can produce isolated scene images without any human by zeroing out keypoint heatmaps when generating images.

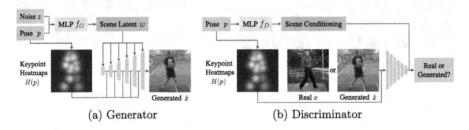

(a) Generator   (b) Discriminator

Fig. 3. Network architectures. Our networks are based on StyleGAN2, with simple modifications to ensure accurately placed humans and compatible scenes. In particular, the conditional generator and discriminator networks utilize pose p via two mechanisms: keypoint heatmaps and pose latent conditioning. Keypoint heatmaps correctly positions a human, and pose latent conditioning drives latent codes w to generate compatible scenes. Multiple plausible scenes can be produced for the same input pose by sampling different noise vectors z.

4.1   Dual Pose Conditioning

The conditional generator G and discriminator D both utilize input pose via two mechanisms: keypoint heatmap conditioning, which specifies spatial placement of a human subject, and pose latent conditioning, which infers compatible scenes. To succeed at our task, humans must be positioned correctly and generated scenes must be compatible. Dual pose conditioning drives strong performance in both respects, and outperforms conditioning on either alone in our ablation experiment (Table 3). Furthermore, dual pose conditioning disentangles control of scene and human pose. We leverage these separate controls for numerous applications: generating scenes without humans, visualizing incompatible scenes and poses, placing a person in a new scene, and animating pose.

Keypoint Heatmaps. Let pose $p = (p_1, ..., p_K)$ denote 2D locations of the $K = 18$ human keypoints detected by OpenPose [9], and let $v = (v_1, ..., v_K)$ indicate visibility of each keypoint. Following the works of [1,4,67], our keypoint heatmaps $H(p)$ consist of radial basis function kernels centered at each keypoint. For heatmap $k \in \{1, ..., K\}$, the intensity at location q is given by Eq. 1. We concatenate heatmaps at each scale of the generator, and at the input of the discriminator. We set $\sigma^2 = \max(0.5, 0.005R^2)$ where R is the spatial resolution of the heatmaps. After training, we generate images of scenes without humans by simply zeroing out all keypoint heatmaps.

$$H_{k,q}(p) = \begin{cases} \exp\left(-\frac{||q-p_k||^2}{2\sigma^2}\right) & \text{if } v_k = 1 \\ 0 & \text{otherwise} \end{cases} \quad (1)$$

Pose Latent Conditioning. To generate compatible scenes, we condition scene latent codes on the input pose. Akin to intermediate latents in StyleGAN2 [44], the scene latent code w controls generation by modulating convolutional weights. To condition the latent code, pose locations and visibility are flattened and mapped to a 512-dimensional input via a learned linear projection. A noise sample $z \sim \mathcal{Z}$ is concatenated with the input vector and passed through a multi-layer perceptron (MLP) f_G to produce a scene latent code $w \in \mathcal{W}$. Multiple plausible scenes can be generated by sampling different noise vectors z for the same pose. The discriminator learns a separate linear projection and MLP f_D.

4.2 Removal of Style Mixing

Style mixing regularization [43,62] encourages disentanglement by randomly mixing intermediate latent codes during training. The technique assumes image attributes at each layer are compatible with all other image attributes (e.g. any face could have any color hair). This assumption is not true when composing scenes and humans, which we visually demonstrate through the incompatible scenes and poses in Fig. 6. This motivates removing style mixing regularization during training, which improves results in our ablation experiments (Table 2).

4.3 Large-scale GAN Training

Typical datasets used with StyleGAN2 (e.g. faces, bedrooms, churches [52,78]) are relatively homogeneous. Increasing model capacity is a natural extension given the diversity and complexity of scene images in our dataset. We find that increasing the channel width of convolutional layers by 2× significantly improves our model (see ablation in Table 2). Following prior work in scaling GANs [7], we also increase minibatch size (from 40 to 120). Concurrent work [59,66] also explores scaling StyleGAN, and proposes strategies such as self-filtering the training dataset [59], progressive growing and leveraging pre-trained classifiers [66].

4.4 Model Details

We train all models at 128×128 resolution with non-saturating logistic loss [27], path length [44] and R_1 [57] regularization, and exponential moving average of generator parameters [41]. We remove spatial noise maps from StyleGAN2 for all models, employ differentiable augmentation of both real and generated images [42,81], and train the discriminator with an additional fake example containing real images with mismatched labels [65]. See the supplement for details.

5   Experiments

Our model hallucinates diverse, high quality images of scenes compatible with input pose. We generate scenes in isolation as well as scenes containing humans, and analyze our model through several visual experiments. Generating scene images is challenging due to the high complexity of data, and our model

Fig. 4. Success cases. Our model learns complex scene-pose relationships. For each input pose, we show many hallucinated scenes, with and without a human. Diverse outputs include a person paddling a kayak (B), lifting a barbell in their hand (G), cleaning the toilet (K), and playing the drums (M). Our model produces multiple plausible scenes for the same pose, providing insight into scenes with related affordances: in the same pose, a person may climb in an indoor gym or on a snowy ledge (F); a person can ride a horse, ride a bicycle, or ride a tractor (L). Please see the appendix for multiple pages of random results.

outperforms Pix2Pix/Pix2PixHD [36,73] and pose-conditioned StyleGAN2 [44] baselines in terms of image quality and accurate human placement. We present characteristic success and failure results in Fig. 4 and Fig. 5. See the supplement for more results, including multiple pages of random uncurated samples.

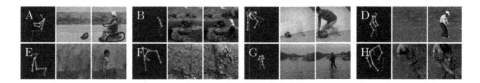

Fig. 5. Failure cases. Causes for failure include: partially generating objects, such as a bike (A); poor overall image quality (B); missing limbs without proper occluders (C); difficulty placing objects, such as a golf club, in a person's hands (D); difficulty hallucinating an object on which to sit (E); overly repetitive textures (F); infeasible scenes, such as walking on water (G); and leaving behind a partial human when hallucinating the scene in isolation (H).

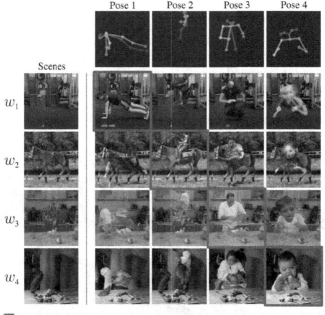

Fig. 6. A central theme of our paper is that scenes must be compatible with human poses to produce realistic images—here we visualize what happens when scenes and poses are *not* compatible. Correctly paired images are shown in blue on the diagonal—a person doing a pushup in a gym, riding a horse, cooking in a kitchen, and a baby leaning on a table. These exemplify interesting relationships between human pose and scene learned by our model. Other images mix scene latent codes with keypoint heatmaps from the wrong pose, often producing unrealistic images. Generating pose-compatible scenes is essential to avoid these incorrect pairings.

5.1 Not All Scenes and Poses Are Compatible

It is essential that we model which scenes are compatible with which poses. A person cannot do a push-up in the middle of a horse, ride atop a kitchen countertop, or be occluded by thin air. These scenarios sound obviously false, yet could occur if the scene and human pose are incompatible. We visualize images generated with correctly and incorrectly paired scenes and poses in Fig. 6.

These examples of incompatible scenes and poses highlight an important difference between our scene data and other datasets commonly used for GAN training, such as cropped faces in the CelebA [52] and FFHQ [43] datasets. Any face can be given glasses, longer or shorter hair, or a darker or lighter skin tone and still remain a feasible image. This enables global disentanglement of attributes, and applications like style mixing, which combines different intermediate latent codes of any two samples (see Fig. 3 of the original StyleGAN paper [43] for a wonderful example). The assumption of compatibility between all attribute pairs no longer holds for data of scenes with humans, which motivates conditioning scene latent codes on pose. Relatedly, we find that removing style mixing from training significantly improves performance (Table 2).

5.2 Scene Occlusion Reasoning

Portions of a human pose may be occluded by foreground objects, such as pieces of furniture. Provided a partially visible human pose, our model hallucinates scenes with foreground objects to occlude portions of the pose not visible. Figure 7a shows an example full-body pose and output scenes. When the legs are not visible in the input pose in 7b, our model produces scenes with occluders blocking the legs, demonstrating its emergent ability to reason about occlusions.

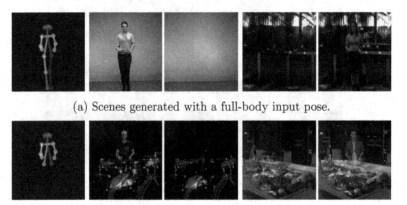

(a) Scenes generated with a full-body input pose.

(b) Scenes generated from the same input pose with legs not visible.

Fig. 7. (a) A full-body input pose and corresponding scenes. (b) When the legs from an otherwise identical pose are hidden, our model hallucinates scenes with foreground objects, such as a drum kit or table, to occlude the missing legs.

Fig. 8. Given an input pose (top left), our method can compose human appearances (top row) and scenes (left column) from different generated images.

Fig. 9. Provided an input pose sequence (top), we infer scenes based on the first pose, then generate animations (middle/bottom) by keeping the scene latent fixed and passing keypoint heatmaps for each subsequent pose.

5.3 Human Appearance and Scene Disentanglement

Section 5.1 demonstrates why complete separation of pose and scene is undesirable. We can, however, disentangle human appearance from scene when both are conditioned on the same pose, as shown in Fig. 8. To achieve this, we optimize a latent code to compose two samples. We minimize perceptual loss [39,79] between person-only crops of the composition and first sample, and scene-only images of the composition and second sample. See the supplement for details.

5.4 Animating Pose

After training, our model is capable of animating pose in a stationary scene. In Fig. 9 we demonstrate a sequence of images generated by fixing the scene and animating the human pose. The scene is inferred from only the first pose, and is limited to small human motion and stationary backgrounds.

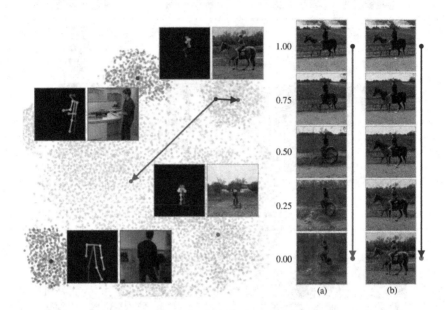

Fig. 10. We contrast truncation via (a) interpolation toward the mean of random latents, and (b) interpolation toward the mean of conditional latent clusters. The left plot shows a t-SNE [55] visualization of latent codes. Gray points are 10,000 random latents. Colored sets of points are each 1000 latent samples conditioned on the same pose. The formation of clusters signifies that different scene latents conditioned on the same pose are close to each other in the intermediate latent space. The dark gray point in the center is the mean of all random latents, and dark colored points are the means for each pose. Beside each cluster is the input pose and image generated using the corresponding mean cluster latent. Conditional truncation (b) works significantly better than unconditional (a).

5.5 Scene Clustering and Truncation

Regions of low density in the data distribution are particularly challenging to model. Quality can be improved (at the loss of some diversity) by sampling from a shrunk distribution [2,7,20,33,46,56]. StyleGAN [43] interpolates intermediate latents w toward the mean $\bar{w} = \mathbb{E}_{z \sim \mathcal{Z}}[w]$ to shrink the sampling distribution, which improves generation quality for models trained on data such as faces. However, on our more complex data, interpolating toward the mean scene latent produces a gray scene rather than improving quality, as shown in Fig. 10a.

In visualizing a t-SNE [55] plot of scene latents in Fig. 10, we observe that latents sampled from different noise vectors z yet conditioned on the same pose p form clusters. We apply conditional truncation by interpolating a latent w toward the conditional mean $\bar{w}_p = \mathbb{E}_{z \sim \mathcal{Z}}[w|p]$, shifting the sample toward the cluster center. Shown in part (b) of Fig. 10, conditional truncation works significantly better for our model. We apply conditional truncation $w' = \bar{w}_p + \psi(w - \bar{w}_p)$ of $\psi = 0.75$ to generated images throughout the paper. Concurrent work [59] proposes a similar method for applying truncation toward the centers of perceptual clusters.

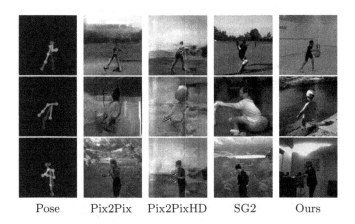

Pose   Pix2Pix   Pix2PixHD   SG2   Ours

Fig. 11. Baseline comparisons. Pix2Pix/Pix2PixHD struggle to produce realistic images; pose-conditioned StyleGAN2 (SG2) often generates humans in the wrong pose; our model generates realistic scenes with humans in the correct pose.

5.6   Baseline Comparisons

Please see Fig. 11 for visual comparisons with baseline methods. Pix2Pix and Pix2PixHD were designed for image translation tasks with stronger conditioning, such as segmentation masks. These methods struggle to produce reasonable images on our more challenging task and dataset. StyleGAN2 (SG2) with latent pose conditioning provides a stronger baseline, but still has notable issues with image quality and often places humans in the incorrect pose. These observations are corroborated by metric performance in two respects: how accurately human subjects are positioned, and how realistic generated scenes look.

To succeed at our task, a model must both put a human in the correct pose and generate a compatible scene. Table 1 compares our model with Pix2Pix, Pix2PixHD and StyleGAN2 baselines on these metrics, demonstrating that our model achieves superior performance. Note that StyleGAN2 [44] is primarily an unconditional GAN. The public code release and follow-up work [42] support class-conditional generation. We refer to the version of our model with only pose latent conditioning as StyleGAN2, since it is the most straightforward extension of StyleGAN2 for our task.

Accurate Human Positioning. PCKh [3] measures the percent of correct pose keypoints (within a radius relative to the head size), where a higher percent is better. We use OpenPose [9] to extract poses from generated images for comparison with input poses. PCKh is computed on a held out test set, ensuring accurate placement of new poses not seen during training.

Realistic Scene Images. FID—Fréchet inception distance [32]—measures realism by comparing distributions of Inception network [69] features between the training dataset and generated images. Lower FID scores are better and correlate with higher quality, more realistic images.

Table 1. Baseline metric comparisons. We report PCKh (higher is better) as a measure of how accurately humans are positioned, and FID (lower is better) as a measure of how realistic generated scenes look. Our model outperforms Pix2Pix, Pix2PixHD and pose-conditioned StyleGAN2 baselines on both metrics. While the poor performance of baselines may appear surprising, note that our task is much more challenging than standard conditional generation tasks: the dataset is diverse and complex, and conditioning on pose requires the network to infer scene contents and layout.

	PCKh ↑	FID ↓
Pix2Pix	48.4	71.2
Pix2PixHD	73.8	149.7
StyleGAN2 (with pose latent conditioning)	32.4	16.6
Ours	84.2	5.9

Table 2. StyleGAN2 ablation. We enumerate modifications relative to a pose-conditioned StyleGAN2 baseline. In particular, removing style mixing, conditioning on keypoint heatmaps, augmenting discriminator inputs and passing a fake mismatched example to the discriminator, and increasing scale all contribute to our final model.

	PCKh ↑	FID ↓
StyleGAN2 (with pose latent conditioning)	32.4	16.6
– Style mixing	36.4	11.6
+ Keypoint heatmaps	79.8	12.2
+ Augmentation, mismatch	80.7	12.1
+ Large scale (Ours)	84.2	5.9

Table 3. Pose conditioning ablation. We contrast three options for pose conditioning: only conditioning the latent on pose, only conditioning on keypoint heatmaps, and dual conditioning of both latents and heatmaps. We conduct this ablation on the smaller version of our model. We find that keypoint heatmap conditioning is crucial for accurately placing a human (PCKh), whereas latent conditioning improves the quality of scene generation (FID). We condition with both mechanisms in our final model, which has the best metric trade-off, and enables separating control of human position and scene generation after training.

Conditioning method	PCKh ↑	FID ↓
Latent only	36.4	11.6
Heatmap only	79.7	15.1
Both	79.8	12.2

5.7 Ablations

We present two ablation experiments. Table 2 enumerates changes relative to a pose-conditioned StyleGAN2 baseline, demonstrating improvements gained by our simple yet important modifications. Table 3 compares three options for pose conditioning: latents only, keypoint heatmaps only, and dual conditioning of both. Keypoint heatmaps are necessary to accurately position a human in the scene, which is shown by a substantially higher PCKh. Latent conditioning improves quality, which is shown by a lower FID score. We condition with both mechanisms—in addition to offering the best trade-off in metric performance, dual conditioning enables applications of disentanglement, such as generating scenes without humans or visualizing incompatible scenes and poses.

6 Discussion

Limitations. Our dataset and model only consider images with a single human subject. Dataset curation is limited by the performance of Keypoint R-CNN [31, 75] and OpenPose [9,11] when filtering videos for humans. Training depends on OpenPose to correctly predict poses. Our model does not consider human movement when inferring scenes.

Societal Impact. There is some risk of this or future generative models being used to create fake and misleading content. Our model also inherits any demographic bias present in the existing datasets used to source our training data.

Conclusion. In this paper, we present a new task: provided a human pose as input, hallucinate the possible scene(s) which are compatible with that input pose. Strong relationships between humans, objects and environments dictate which scenes afford a given pose. Many prior works study human affordances from the angle of predicting possible poses given an input scene—we study the other side of the same coin, and hallucinate scenes that afford an input pose.

We demonstrate the emergent ability of our model to capture affordance relationships between scenes and poses. This work marks a significant step toward using GANs to represent complex real-world environments. We hope it will motivate the broader research community to leverage modern generative approaches for scene understanding and modeling.

Acknowledgements. We thank William Peebles, Ilija Radosavovic, Matthew Tancik, Allan Jabri, Dave Epstein, Lucy Chai, Toru Lin, Shiry Ginosar, Angjoo Kanazawa, Vickie Ye, Karttikeya Mangalam, and Taesung Park for insightful discussion and feedback. Tim Brooks is supported by the National Science Foundation Graduate Research Fellowship under Grant No. 2020306087. Additional support for this project is provided by DARPA MCS, and SAP.

References

1. Aberman, K., Shi, M., Liao, J., Lischinski, D., Chen, B., Cohen-Or, D.: Deep video-based performance cloning. In: Computer Graphics Forum. vol. 38, pp. 219–233. Wiley Online Library (2019)
2. Ackley, D.H., Hinton, G.E., Sejnowski, T.J.: A learning algorithm for boltzmann machines. Cogn. Sci. **9**(1), 147–169 (1985)
3. Andriluka, M., Pishchulin, L., Gehler, P., Schiele, B.: 2d human pose estimation: new benchmark and state of the art analysis. In: Proceedings of the IEEE Conference on computer Vision and Pattern Recognition, pp. 3686–3693 (2014)
4. Balakrishnan, G., Zhao, A., Dalca, A.V., Durand, F., Guttag, J.: Synthesizing images of humans in unseen poses. In: Proceedings of the IEEE Conference on Computer Vision and Pattern Recognition, pp. 8340–8348 (2018)
5. Bau, D., et al.: Gan dissection: visualizing and understanding generative adversarial networks. In: Proceedings of the International Conference on Learning Representations (ICLR) (2019)
6. Biederman, I.: On the semantics of a glance at a scene. In: Perceptual Organization (1981)
7. Brock, A., Donahue, J., Simonyan, K.: Large scale GAN training for high fidelity natural image synthesis. In: International Conference on Learning Representations (2019). https://openreview.net/forum?id=B1xsqj09Fm
8. Cao, Z., Gao, H., Mangalam, K., Cai, Q., Vo, M., Malik, J.: Long-term human motion prediction with scene context. In: ECCV (2020)
9. Cao, Z., Hidalgo, G., Simon, T., Wei, S.E., Sheikh, Y.: Openpose: realtime multi-person 2d pose estimation using part affinity fields. IEEE Trans. Pattern Anal. Mach. Intell. **43**(1), 172–186 (2019)
10. Cao, Z., Radosavovic, I., Kanazawa, A., Malik, J.: Reconstructing hand-object interactions in the wild. arXiv e-prints pp. arXiv-2012 (2020)
11. Cao, Z., Simon, T., Wei, S.E., Sheikh, Y.: Realtime multi-person 2d pose estimation using part affinity fields. In: Proceedings of the IEEE Conference on Computer Vision and Pattern Recognition, pp. 7291–7299 (2017)
12. Chai, L., Wulff, J., Isola, P.: Using latent space regression to analyze and leverage compositionality in GANs. In: International Conference on Learning Representations (2021)

13. Chan, C., Ginosar, S., Zhou, T., Efros, A.A.: Everybody dance now. In: Proceedings of the IEEE/CVF International Conference on Computer Vision, pp. 5933–5942 (2019)
14. Chuang, C.Y., Li, J., Torralba, A., Fidler, S.: Learning to act properly: predicting and explaining affordances from images. In: Proceedings of the IEEE Conference on Computer Vision and Pattern Recognition, pp. 975–983 (2018)
15. Delaitre, V., Fouhey, D., Laptev, I., Sivic, J., Gupta, A., Efros, A.: Scene semantics from long-term observation of people. In: Proceedings of 12th European Conference on Computer Vision (2012)
16. Deng, J., Dong, W., Socher, R., Li, L.J., Li, K., Fei-Fei, L.: ImageNet: a large-scale hierarchical image database. In: CVPR09 (2009)
17. Diba, A., et al.: Large scale holistic video understanding. In: Vedaldi, A., Bischof, H., Brox, T., Frahm, J.-M. (eds.) ECCV 2020. LNCS, vol. 12350, pp. 593–610. Springer, Cham (2020). https://doi.org/10.1007/978-3-030-58558-7_35
18. Divvala, S.K., Hoiem, D., Hays, J.H., Efros, A.A., Hebert, M.: An empirical study of context in object detection. In: 2009 IEEE Conference on computer vision and Pattern Recognition, pp. 1271–1278. IEEE (2009)
19. Epstein, D., Chen, B., Vondrick, C.: Oops! predicting unintentional action in video. In: Proceedings of the IEEE/CVF Conference on Computer Vision and Pattern Recognition, pp. 919–929 (2020)
20. Fan, A., Lewis, M., Dauphin, Y.: Hierarchical neural story generation. arXiv preprint arXiv:1805.04833 (2018)
21. Fouhey, D.F., Delaitre, V., Gupta, A., Efros, A.A., Laptev, I., Sivic, J.: People watching: human actions as a cue for single view geometry. In: Fitzgibbon, A., Lazebnik, S., Perona, P., Sato, Y., Schmid, C. (eds.) ECCV 2012. LNCS, vol. 7576, pp. 732–745. Springer, Heidelberg (2012). https://doi.org/10.1007/978-3-642-33715-4_53
22. Fouhey, D.F., Kuo, W.C., Efros, A.A., Malik, J.: From lifestyle vlogs to everyday interactions. In: Proceedings of the IEEE Conference on Computer Vision and Pattern Recognition, pp. 4991–5000 (2018)
23. Fouhey, D.F., Wang, X., Gupta, A.: In defense of the direct perception of affordances (2015)
24. Gibson, J.J.: The Ecological Approach to Visual Perception. Houghton Mifflin, Boston (1979)
25. Gkioxari, G., Girshick, R., Dollár, P., He, K.: Detecting and recognizing human-object interactions. In: Proceedings of the IEEE Conference on Computer Vision and Pattern Recognition, pp. 8359–8367 (2018)
26. Goetschalckx, L., Andonian, A., Oliva, A., Isola, P.: Ganalyze: toward visual definitions of cognitive image properties. In: Proceedings of the IEEE/CVF International Conference on Computer Vision, pp. 5744–5753 (2019)
27. Goodfellow, I.J., et al.: Generative adversarial networks. arXiv preprint arXiv:1406.2661 (2014)
28. Grabner, H., Gall, J., Van Gool, L.: What makes a chair a chair? In: Proceedings of IEEE Computer Society Conference on Computer Vision and Pattern Recognition, pp. 1529–1536, June 2011. https://doi.org/10.1109/CVPR.2011.5995327
29. Gupta, A., Satkin, S., Efros, A.A., Hebert, M.: From 3d scene geometry to human workspace. In: Computer Vision and Pattern Recognition (CVPR) (2011)
30. Hassan, M., Ghosh, P., Tesch, J., Tzionas, D., Black, M.J.: Populating 3D scenes by learning human-scene interaction. In: Proceedings IEEE/CVF Conf. on Computer Vision and Pattern Recognition (CVPR), June 2021

31. He, K., Gkioxari, G., Dollár, P., Girshick, R.: Mask r-cnn. In: Proceedings of the IEEE International Conference on Computer Vision, pp. 2961–2969 (2017)
32. Heusel, M., Ramsauer, H., Unterthiner, T., Nessler, B., Hochreiter, S.: Gans trained by a two time-scale update rule converge to a local nash equilibrium. In: Proceedings of the 31st International Conference on Neural Information Processing Systems, NIPS 2017, pp. 6629–6640 (2017)
33. Holtzman, A., Buys, J., Du, L., Forbes, M., Choi, Y.: The curious case of neural text degeneration. arXiv preprint arXiv:1904.09751 (2019)
34. Härkönen, E., Hertzmann, A., Lehtinen, J., Paris, S.: Ganspace: discovering interpretable GAN controls. In: Proceedings of NeurIPS (2020)
35. Ionescu, C., Papava, D., Olaru, V., Sminchisescu, C.: Human3. 6m: large scale datasets and predictive methods for 3d human sensing in natural environments. IEEE Trans. Pattern Anal. Mach. Intell. **36**(7), 1325–1339 (2013)
36. Isola, P., Zhu, J.Y., Zhou, T., Efros, A.A.: Image-to-image translation with conditional adversarial networks. In: Computer Vision and Pattern Recognition (CVPR), 2017 IEEE Conference on (2017)
37. Jahanian, A., Chai, L., Isola, P.: On the "steerability" of generative adversarial networks. In: International Conference on Learning Representations (2020)
38. Jiang, Y., Koppula, H., Saxena, A.: Hallucinated humans as the hidden context for labeling 3d scenes. In: Proceedings of the IEEE Conference on Computer Vision and Pattern Recognition (CVPR), June 2013
39. Johnson, J., Alahi, A., Fei-Fei, L.: Perceptual losses for real-time style transfer and super-resolution. In: Leibe, B., Matas, J., Sebe, N., Welling, M. (eds.) ECCV 2016. LNCS, vol. 9906, pp. 694–711. Springer, Cham (2016). https://doi.org/10.1007/978-3-319-46475-6_43
40. Kanazawa, A., Zhang, J.Y., Felsen, P., Malik, J.: Learning 3d human dynamics from video. In: Proceedings of the IEEE/CVF Conference on Computer Vision and Pattern Recognition, pp. 5614–5623 (2019)
41. Karras, T., Aila, T., Laine, S., Lehtinen, J.: Progressive growing of GANs for improved quality, stability, and variation. In: International Conference on Learning Representations (2018), https://openreview.net/forum?id=Hk99zCeAb
42. Karras, T., Aittala, M., Hellsten, J., Laine, S., Lehtinen, J., Aila, T.: Training generative adversarial networks with limited data. arXiv preprint arXiv:2006.06676 (2020)
43. Karras, T., Laine, S., Aila, T.: A style-based generator architecture for generative adversarial networks. In: Proceedings of the IEEE/CVF Conference on Computer Vision and Pattern Recognition, pp. 4401–4410 (2019)
44. Karras, T., Laine, S., Aittala, M., Hellsten, J., Lehtinen, J., Aila, T.: Analyzing and improving the image quality of stylegan. In: Proceedings of the IEEE/CVF Conference on Computer Vision and Pattern Recognition, pp. 8110–8119 (2020)
45. Kay, W., et al.: The kinetics human action video dataset. arXiv preprint arXiv:1705.06950 (2017)
46. Kingma, D.P., Dhariwal, P.: Glow: generative flow with invertible 1×1 convolutions. In: NeurIPS, pp. 10236–10245 (2018). http://papers.nips.cc/paper/8224-glow-generative-flow-with-invertible-1x1-convolutions
47. Koppula, H.S., Gupta, R., Saxena, A.: Learning human activities and object affordances from RGB-d videos. Int. J. Robot. Res. **32**(8), 951–970 (2013)
48. Lee, J., Chai, J., Reitsma, P.S., Hodgins, J.K., Pollard, N.S.: Interactive control of avatars animated with human motion data. In: Proceedings of the 29th Annual Conference on Computer Graphics and Interactive Techniques, pp. 491–500 (2002)

49. Li, X., Liu, S., Kim, K., Wang, X., Yang, M.H., Kautz, J.: Putting humans in a scene: learning affordance in 3d indoor environments. In: CVPR (2019)
50. Li, Y., Huang, C., Loy, C.C.: Dense intrinsic appearance flow for human pose transfer. In: Proceedings of the IEEE/CVF Conference on Computer Vision and Pattern Recognition (CVPR), June 2019
51. Li, Y., Singh, K.K., Ojha, U., Lee, Y.J.: Mixnmatch: multifactor disentanglement and encoding for conditional image generation. In: CVPR (2020)
52. Liu, Z., Luo, P., Wang, X., Tang, X.: Deep learning face attributes in the wild. In: Proceedings of International Conference on Computer Vision (ICCV), December 2015
53. Ma, L., Jia, X., Sun, Q., Schiele, B., Tuytelaars, T., Van Gool, L.: Pose guided person image generation. In: Advances in Neural Information Processing Systems, pp. 405–415 (2017)
54. Ma, L., Sun, Q., Georgoulis, S., Van Gool, L., Schiele, B., Fritz, M.: Disentangled person image generation. In: The IEEE International Conference on Computer Vision and Pattern Recognition (CVPR), June 2018
55. Van der Maaten, L., Hinton, G.: Visualizing data using t-SNE. J. Mach. Learn. Res. **9**(11), 2579–2605 (2008)
56. Marchesi, M.: Megapixel size image creation using generative adversarial networks (2017)
57. Mescheder, L., Nowozin, S., Geiger, A.: Which training methods for GANs do actually converge? In: International Conference on Machine Learning (ICML) (2018)
58. Mirza, M., Osindero, S.: Conditional generative adversarial nets. arXiv preprint arXiv:1411.1784 (2014)
59. Mokady, R., et al.: Self-distilled stylegan: towards generation from internet photos. arXiv preprint arXiv:2202.12211 (2022)
60. Monfort, M., et al.: Moments in time dataset: one million videos for event understanding. IEEE Trans. Pattern Anal. Mach. Intell. **42**(2), 502–508 (2019)
61. Mottaghi, R., et al.: The role of context for object detection and semantic segmentation in the wild. In: Proceedings of the IEEE Conference on Computer Vision and Pattern Recognition, pp. 891–898 (2014)
62. Park, T., Zhu, J.Y., Wang, O., Lu, J., Shechtman, E., Efros, A.A., Zhang, R.: Swapping autoencoder for deep image manipulation. In: Advances in Neural Information Processing Systems (2020)
63. Peebles, W., Peebles, J., Zhu, J.Y., Efros, A.A., Torralba, A.: The hessian penalty: a weak prior for unsupervised disentanglement. In: Proceedings of European Conference on Computer Vision (ECCV) (2020)
64. Rabinovich, A., Vedaldi, A., Galleguillos, C., Wiewiora, E., Belongie, S.: Objects in context. In: 2007 IEEE 11th International Conference on Computer Vision, pp. 1–8. IEEE (2007)
65. Reed, S., Akata, Z., Yan, X., Logeswaran, L., Schiele, B., Lee, H.: Generative adversarial text to image synthesis. In: International Conference on Machine Learning, pp. 1060–1069. PMLR (2016)
66. Sauer, A., Schwarz, K., Geiger, A.: Stylegan-xl: scaling stylegan to large diverse datasets. arXiv preprint arXiv:2202.00273 (2022)
67. Siarohin, A., Sangineto, E., Lathuiliere, S., Sebe, N.: Deformable GANs for pose-based human image generation. In: Proceedings of the IEEE Conference on Computer Vision and Pattern Recognition, pp. 3408–3416 (2018)

68. Sigurdsson, G.A., Varol, G., Wang, X., Farhadi, A., Laptev, I., Gupta, A.: Hollywood in homes: crowdsourcing data collection for activity understanding. In: Leibe, B., Matas, J., Sebe, N., Welling, M. (eds.) ECCV 2016. LNCS, vol. 9905, pp. 510–526. Springer, Cham (2016). https://doi.org/10.1007/978-3-319-46448-0_31

69. Szegedy, C., et al.: Going deeper with convolutions. In: Proceedings of the IEEE Conference on Computer Vision and Pattern Recognition, pp. 1–9 (2015)

70. Torralba, A., Murphy, K.P., Freeman, W.T., Rubin, M.A.: Context-based vision system for place and object recognition. In: Computer Vision, IEEE International Conference on. vol. 2, pp. 273–273. IEEE Computer Society (2003)

71. Wang, J., Xu, H., Xu, J., Liu, S., Wang, X.: Synthesizing long-term 3d human motion and interaction in 3d scenes (2020)

72. Wang, T.C., et al.: Video-to-video synthesis. In: Advances in Neural Information Processing Systems (NeurIPS) (2018)

73. Wang, T.C., Liu, M.Y., Zhu, J.Y., Tao, A., Kautz, J., Catanzaro, B.: High-resolution image synthesis and semantic manipulation with conditional GANs. In: Proceedings of the IEEE Conference on Computer Vision and Pattern Recognition (2018)

74. Wang, X., Girdhar, R., Gupta, A.: Binge watching: scaling affordance learning from sitcoms. In: CVPR (2017)

75. Wu, Y., Kirillov, A., Massa, F., Lo, W.Y., Girshick, R.: Detectron2. https://github.com/facebookresearch/detectron2 (2019)

76. Xu, N., et al.: Youtube-vos: sequence-to-sequence video object segmentation. In: Proceedings of the European Conference on Computer Vision (ECCV), pp. 585–601 (2018)

77. Yao, B., Fei-Fei, L.: Modeling mutual context of object and human pose in human-object interaction activities. In: 2010 IEEE Computer Society Conference on Computer Vision and Pattern Recognition, pp. 17–24 (2010)

78. Yu, F., Zhang, Y., Song, S., Seff, A., Xiao, J.: Lsun: construction of a large-scale image dataset using deep learning with humans in the loop. CoRR abs/1506.03365 (2015). http://dblp.uni-trier.de/db/journals/corr/corr1506.html#YuZSSX15

79. Zhang, R., Isola, P., Efros, A.A., Shechtman, E., Wang, O.: The unreasonable effectiveness of deep features as a perceptual metric. In: CVPR (2018)

80. Zhang, W., Zhu, M., Derpanis, K.G.: From actemes to action: a strongly-supervised representation for detailed action understanding. In: Proceedings of the IEEE International Conference on Computer Vision, pp. 2248–2255 (2013)

81. Zhao, S., Liu, Z., Lin, J., Zhu, J.Y., Han, S.: Differentiable augmentation for data-efficient GAN training. arXiv preprint arXiv:2006.10738 (2020)

82. Zhu, Y., Fathi, A., Fei-Fei, L.: Reasoning about object affordances in a knowledge base representation. In: Fleet, D., Pajdla, T., Schiele, B., Tuytelaars, T. (eds.) ECCV 2014. LNCS, vol. 8690, pp. 408–424. Springer, Cham (2014). https://doi.org/10.1007/978-3-319-10605-2_27

Motion and Appearance Adaptation
for Cross-domain Motion Transfer

Borun Xu[1], Biao Wang[2], Jinhong Deng[1], Jiale Tao[1], Tiezheng Ge[2], Yuning Jiang[2],
Wen Li[1(✉)], and Lixin Duan[1]

[1] University of Electronic Science and Technology of China, Chengdu, China
xbr_2017@std.uestc.edu.cn, liwenbnu@gmail.com
[2] Alibaba Group, Hangzhou, China
{eric.wb,tiezheng.gtz,mengzhu.jyn}@alibaba-inc.com

Abstract. Motion transfer aims to transfer the motion of a driving video to a source image. When there are considerable differences between object in the driving video and that in the source image, traditional single domain motion transfer approaches often produce notable artifacts; for example, the synthesized image may fail to preserve the human shape of the source image (*cf*. Fig. 1 (a)). To address this issue, in this work, we propose a Motion and Appearance Adaptation (MAA) approach for cross-domain motion transfer, in which we regularize the object in the synthesized image to capture the motion of the object in the driving frame, while still preserving the shape and appearance of the object in the source image. On one hand, considering the object shapes of the synthesized image and the driving frame might be different, we design a shape-invariant motion adaptation module that enforces the consistency of the angles of object parts in two images to capture the motion information. On the other hand, we introduce a structure-guided appearance consistency module designed to regularize the similarity between the corresponding patches of the synthesized image and the source image without affecting the learned motion in the synthesized image. Our proposed MAA model can be trained in an end-to-end manner with a cyclic reconstruction loss, and ultimately produces a satisfactory motion transfer result (*cf*. Fig. 1 (b)). We conduct extensive experiments on human dancing dataset Mixamo-Video to Fashion-Video and human face dataset Vox-Celeb to Cufs; on both of these, our MAA model outperforms existing methods both quantitatively and qualitatively.

1 Introduction

Given a source image and a driving video of the same object, motion transfer (*a.k.a.* image animation) aims to generate a synthesized video that mimics the motion of the driving video while preserving the appearance of the source image. It recently received increasing attention, due to its potential applications in real-world scenarios, such as face swapping [31,38,39], dance transferring [5], *etc.*

Many works in this field focus on the single-domain motion transfer [30–32], where the driving video and source image come from the same domain. However, in real applications, there are often requirements to transfer motion among different domains.

S. Avidan et al. (Eds.): ECCV 2022, LNCS 13676, pp. 529–545, 2022.
https://doi.org/10.1007/978-3-031-19787-1_30

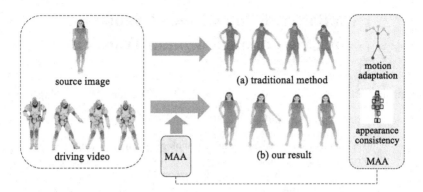

Fig. 1. Motion transfer results. (a) is generated by traditional motion transfer model trained on source domain videos only and (b) is generated by our proposed MAA model

For example, as shown in Fig 1, the e-commerce companies might be interested in animating a fashion model to attract consumers by learning the robot dance from a Mixamo character. However, due to the differences in shape and cloth between the Mixamo character and the fashion model, traditional single-domain motion transfer approaches often produce notable artifacts in the synthesized image (*e.g.*, failing to preserve the human shape of the fashion model (*cf*. Fig. 1 (a))).

To this end, in the present work, we study the cross-domain motion transfer problem and propose a novel Motion and Appearance Adaptation (MAA) approach to address this issue. Specifically, traditional motion transfer methods usually take two arbitrary frames of the same video as source image and driving frame for learning motion with a reconstruction loss, because the two frames share the same appearance and shape. However, such training mode cannot be directly applied to the cross-domain motion transfer because no ground-truth is available. In our proposed MAA approach, we build a cyclic reconstruction pipeline inspired by CycleGAN [42] and cross-identity [19]. In particular, given a source image and a driving frame obtained from different domains, we first obtain a synthesized image using a basic motion transfer (MT) model, *e.g.*, the model in [31] or [32]. We next arbitrarily take another frame from the driving video as the source image and the synthesized image as a driving frame, and input them into the basic MT model to produce the second synthesized image. Because the second synthesized image should mimic both the motion and appearance of original driving frame, a cyclic reconstruction loss can be applied for training. In this way, we obtain a motion transfer model for cross-domain motion transfer.

Moreover, since the source image and driving frame are drawn from different domains, while the topology of the object structure (*e.g.*, the skeleton) is similar, the configurations of the object structure (*e.g.*, the human body shape) often deviate. When doing motion transfer, we should be aware of such difference and keep the object shape of synthesized image be similar to the source image while unaffected by the driving frame. For this purpose, we design a shape-invariant motion adaptation module and a structure-guided appearance consistency module to regularize the basic motion transfer model.

Specifically, in the shape-invariant motion adaptation module, we design an angle consistency loss to enforce the angles of the corresponding object parts in the synthesized image to be similar to those of the driving frame, such that the motion of this frame can be mimicked well without changing the object shape. In the structure-guided appearance consistency module, we extract image patches from the synthesized image and the source images based on the object structure and enforce the corresponding patches to be similar; this ensures that the appearance of the synthesized image and the source image are consistent, even though the motions of the two images are different.

The entire process can be trained in an end-to-end manner, and finally our MAA model can effectively perform motion transfer across domains while also properly preserving the shape and appearance of the object (*cf*. Fig. 1 (b)). We validate our proposed approach on two pairs of datasets: the human body datasets Mixamo-Video to Fashion-Video [40] and the human face datasets Vox-Celeb [28] to Cufs [37]. Extensive experimental results demonstrate the effectiveness of our proposed approach. Our source code will be released soon.

2 Related Work

Motion Transfer: Current motion transfer approaches can be categorized into two types: model-based and model-free approaches. The model-based approaches mainly focus on human body pose transfer [3,26,27], which utilize a pre-trained pose estimator or key point detector to extract the pose of driving image as a guidance information. And a number of researchers followed such setting [16,20,22,23,29,43]. Moreover, a series of works apply this model-based pattern on human facial expression transfer [4,7,13]. Like body pose transfer, they also employ a pre-trained facial landmark detector to model the facial expression.

The model-free approaches [19,30–32,34] does not rely on pre-trained third-party models, and extend the model-based method to arbitrary objects. Aliaksandr *et al.* [30] proposed a model-free motion transfer model Monky-Net that can apply motion transfer on arbitrary objects with an unsupervised key point detector trained by reconstruction loss [18]. Aliaksandr *et al.* [31] further improved Monkey-Net to FOMM to solve the large motion problem. The unsupervised key point detector is also utilized in FOMM, with local affine transformations being added for motion modeling. A generator module is utilized to generate final result with the warped source image feature. Subin *et al.* [19] proposed pose attention mechanism with an unsupervised key point detector to model motion. Recently, Aliaksandr *et al.* [32] improved FOMM with an advanced motion model and background motion model to MRAA. Although promising results are achieved for the single domain motion transfer, these methods might suffer from performance degradation when the source image and driving video come from different domains, where a considerable appearance difference often exists. Recently, Wang *et al.* [36] used encoder based motion transfer approach which can be applied to the cross-domain scenario, and better results are achieved compared with the single domain motion transfer Monkey-Net model. In contrast, our proposed MAA approach is a general framework, and can integrate traditional motion transfer model like FOMM and MRAA to produce excellent results for large motion.

Domain Adaptation: Many works have been proposed to handle the scenario where the training and test data comes from different domains for different computer vision tasks , *e.g.*, classification [8], semantic segmentation [10,11,24,25], object detection [6,9], pose estimation [21,41], *etc.* A majority of works were developed to learn domain-invariant features using the domain adversarial learning [12,35]. Cross-domain motion transfer is more complicated, since we need to capture motion from the the driving video while preserving the appearance from the source domain. Nevertheless, the strategies proposed in traditional domain adaptation works might be useful to help motion transfer. For example, we apply the cyclic training pipeline inspired by CycleGAN [42], and build our patch-based appearance consistency module based on Patch-GAN [17].

3   Methodology

In this section, we present our Motion and Appearance Adaptation approach for cross-domain motion transfer. Formally, let us denote a driving video as $V_d = \{I_d^i|_{i=1}^T\}$, where each I_d^i is a driving frame, while a source image is denoted as I_s; thus, the task of motion transfer is to synthesize a new video $\hat{V}_d = \{\hat{I}_d^i|_{i=1}^T\}$, where each $\hat{I}_d$ adequately captures the object motion in the corresponding driving frame I_d^i while also preserving the object appearance of the source image I_s.

The appearance of an object roughly consists of two aspects, *shape* and *texture*. The shape largely refers to its geometric property (*e.g.*, length, slimness, etc.), while the texture usually means how the object looks like regardless of its shape (*e.g.*, dresses with different colors). Traditional motion transfer methods generally assume that the driving frame and the source image are derived from the same domain, where they implicitly suppose the object shapes are similar. Consequently, when the source image is derived from a new domain with different object shapes, these methods often fail to preserve the shape of the object in the source image.

In this work, we study the cross-domain motion transfer problem, in which the source image and driving frame are from different domains. In other words, there might be considerable differences in appearance between them in terms of both shape and texture. An example is given in Fig. 1, where both the clothes and body shapes of the fashion model and the Mixamo character exhibit notable differences.

In what follows, we first present an overview of the pipeline of our proposed MAA approach in Sect. 3.1, after which we present the shape-invariant motion adaptation (SIMA) module and structure-guided appearance consistency (SGAC) module in Sect. 3.2 and Sect. 3.3 respectively; these effectively learning the motion and appearance from the driving frame and source image, respectively.

3.1   Overview

We design a cyclic training pipeline for cross-domain motion transfer, as shown in the right-hand part of Fig. 2. The pipeline consists of a basic motion transfer model, our proposed shape-invariant motion adaptation module and structure-guided appearance consistency module, and a cyclic loss.

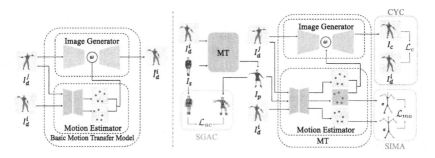

Fig. 2. The pipeline of our proposed method. The left-hand side is the architecture of the traditional single domain motion transfer model FOMM [31], which is used as a basic motion transfer model in our approach. Moreover, the right-hand side is the framework of our proposed MAA method where we design a cyclic reconstruction loss (CYC), a shape-invariant motion adaptation (SIMA) and a structure-guided appearance consistency (SGAC) module

Basic Motion Transfer Model: The basic motion transfer (MT) model follows the traditional motion transfer model [31,32]. We here illustrate the basic MT model by taking FOMM [31] as an example, and other models like [32] can be similarly integrated into our pipeline.

As shown in the left-hand part of Fig. 2, traditional motion transfer methods typically employ a reconstruction training mode for learning and synthesizing motion. During the training phase, they select two arbitrary frames from the driving video as the source image and driving frame, which are used as input of the MT model. For each image, the motion keypoints and their local affine transformation are extracted using a motion estimator, where the motion keypoints can be conceptualized as the centroids of moving object parts. The dense motion flow from the source image to the driving frame can therefore be estimated using their motion keypoints and affine transformations. In the next step, the dense motion flow is used to warp the feature map of the source image, and produce the synthesized image $\hat{I}_d^i$ using the image generator. A perceptual loss is used as the reconstruction loss after the image generator to ensure that the synthesized image $\hat{I}_d^i$ fully reconstructs the driving frame I_d^i as in [31]:

$$\mathcal{L}_r = \sum_{l=k}^{K} |F_l(\hat{I}_d^i) - F_l(I_d^i)| \tag{1}$$

where $F_l(\cdot)$ is feature map output by the l-th layer of a pre-trained VGG-19 network [33].

Researchers have proposed different method [32] to improve the motion estimator in order to more precisely extract motion information, yet the motion representation (*i.e.*, motion keypoints and affine transformations) remains similar. In the interests of simplicity, we depict only the motion keypoints in Fig. 2, which are related to our MAA approach. Readers can refer to [31] for further details.

Cyclic Training Pipeline: In cross-domain motion transfer, the source image and driving video are obtained from different domains. So it is undesirable to pick a frame in the driving video as a source image and apply the reconstruction loss after the image generator, as the model will inevitably be overfitted to the driving video, which will lead to artifacts in the synthesized image.

To address this issue, we build a cyclic reconstruction framework inspired by the CycleGAN [42] and cross-identity [19]. As shown in the right-hand side of Fig. 2, we employ two basic MT models that share the same parameters. Given a source image I_s and a driving frame I_d^i, we first obtain a synthesized image I_p using the basic MT model. Since there is no ground-truth for the synthesized image, the reconstruction loss cannot be used for I_p.

We then take the synthesized image I_p as a driving frame, along with an arbitrary frame I_d^j as the source image, and these are input again into the basic MT model to produce another synthesized image I_c. Intuitively, I_c should mimic the motion of I_p, as well as I_d^i, since we expect I_p to mimic the motion of I_d^i. At the same time, I_c should also preserve the appearance of I_d^j, as well as I_d^i, which is derived from the same driving video as I_d^j. This allows us to employ I_d^i and the cyclically generated I_c to create a reconstruction loss for training. More specifically, we employ the perceptual loss similarly as in Eq. (1):

$$\mathcal{L}_c = \sum_{l=k}^{K} |F_l(I_c) - F_l(I_d^i)| \tag{2}$$

While the cyclic reconstruction loss enables us to train the motion transfer model in the cross-domain setting, this is only a weak supervision that cannot fully guarantee a satisfactory result. We therefore further introduce the shape-invariant motion adaptation module and patch-based appearance model to regularize the motion transfer process, which will be explained in more detail below.

3.2   Shape-invariant Motion Adaptation

Due to the significant appearance difference between the source image I_s and the driving frame I_d, the generated synthesized image I_p often fails to adequately capture the object motion in the driving frame I_d. We therefore propose to directly regularize the object pose in I_p with that in I_d based on the extracted motion keypoints.

However, due to the diversity of the object shapes in I_p and I_d, it is undesirable to directly regularize the consistency of their keypoint positions. We therefore propose to discover the intrinsic topology of the object, then regularize the included angles between adjacent object parts of two objects.

Structure Topology Discovery: To discover the intrinsic object topology, for each driving video, we employ a pre-trained basic MT model to extract the motion keypoints of all frames in the video. Because the motion keypoints roughly describe the objects' moving body parts, two keypoints can be considered to be adjacent if their distance does not change substantially between different frames.

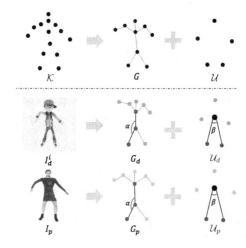

Fig. 3. Illustration of our shape-invariant motion adaptation module. The top row show the structure topology, and the bottom two rows represents the motion adaptation stage using structured and unstructured keypoints

Formally, given a driving frame I_d, we denote its motion keypoints as $\mathcal{K}_d = \{\mathbf{k}_d^i|_{i=1}^K\}$, where K is the number of motion keypoints. For each pair of keypoints $\mathbf{k}_d^i$ and $\mathbf{k}_d^j$, we calculate their ℓ_2 distance $d_{i,j} = \ell_2(\mathbf{k}_d^i, \mathbf{k}_d^j)$, where $i \neq j$. The average distance across all frames of all driving videos can then be computed as $\bar{d}_{i,j} = \frac{1}{T}\sum_{t=1}^T d_{i,j}^{(t)}$, where $d_{i,j}^{(t)}$ is the distance in the t-th frame, while T is the total number of video frames. Finally, we calculate the total distance diversity of $\mathbf{k}_d^i$ and $\mathbf{k}_d^j$ as follows:

$$v_{i,j} = \sum_{t=1}^T |d_{i,j}^{(t)} - \bar{d}_{i,j}| \tag{3}$$

Intuitively, the distance diversity describes the stability of the connection between two keypoints. The smaller the distance diversity $v_{i,j}$, the higher the likelihood that the two keypoints will be adjacent. We then use the distance diversities to construct a structure topology graph G, where the nodes are keypoints, and the edges are defined according to the distance diversities. Specifically, we define the edge value as follows:

$$e_{i,j} = \begin{cases} \frac{(v_{i,j}-\eta)^2}{\eta^2}, & v_{i,j} < \eta, \\ 0, & \text{otherwise} \end{cases} \tag{4}$$

where η is a threshold, and we filter out the edges with high distance diversities, as these imply that the two keypoints are unlikely to be adjacent. Note that the edge value $e_{i,j}$ is within the range of $[0, 1]$. It can be seen as a measurement of the strength of the connection between two keypoints. We will demonstrate that it can also be used as a weight when we regularize the keypoints between driving frame and synthesized image.

Moreover, it is possible that not all keypoints are connected in a single graph; we select the largest graph as our structure topology graph G. We refer to the keypoints in

G as the structured keypoints and the others as unstructured keypoints. For improved convenience of presentation, we denote the set of structured keypoints as $\mathcal{S}$ and their edges as $\mathcal{E}$, the structure topology graph can be presented as $G = \{\mathcal{S}, \mathcal{E}\}$. For unstructured keypoints, we retain only the keypoints and discard their edges, since their connectivities are weak, and denote the set of unstructured keypoints as $\mathcal{U}$. We present an illustration of the structure topology discovery in the top row of Fig. 3.

Regularizing Structured Keypoints: Given a driving frame I_d and the synthesized I_p, we extract their keypoints $\mathcal{K}_d = \{k_d\}$ and $\mathcal{K}_p = \{k_p\}$ using the basic MT model. To regularize the keypoints in the driving frame I_d and the synthesized I_p, we instantiate the structure topology G using the extracted keypoints $\mathcal{K}_d$ and $\mathcal{K}_p$, respectively. Taking the driving frame as an example, the instantiated graph is presented as $G_d = \{\mathcal{S}_d, \mathcal{E}_d\}$; here, $\mathcal{S}_d$ is the set of structured keypoints in I_d, while $\mathcal{E}_d$ is the set of corresponding edges which are calculated based on the Euclidean distances between keypoints. The instantiated graph of the synthesized image $G_p = \{\mathcal{S}_d, \mathcal{E}_d\}$ can be similarly defined. We illustrate the instantiated graphs in the top of Fig. 3.

When examining the structured keypoints, we can observe that considerable differences exist in terms of object shape; this validates our analysis that it is not preferable to directly regularize the keypoint positions. However, the pose can be portrayed as the included angle of each triplet of the connected keypoints in the structure graph.

Specifically, taking the driving frame as an example, let us define a triplet of connected keypoints as $t_d = \{k_d^i, k_d^j, k_d^k\}$, where both k_d^j and k_d^k are connected to k_d^i. We further denote the set of all keypoint triplets in G_d as $\mathcal{T}_d = \{t_d^n|_{n=1}^N\}$, where N is the total number of triplets. Similarly, we define the set of triplets for the synthesized image as $\mathcal{T}_p = \{t_p^n|_{n=1}^N\}$.

For each triplet t_d^n (*resp.*, t_p^n), we calculate its included angle and denote it by $\alpha(t_d^n)$ (*resp.*, $\alpha(t_p^n)$). We then regularize the consistency of the corresponding included angles for structured keypoints in the driving frame and the synthesized image as follows:

$$\mathcal{L}_{rs} = \frac{1}{N} \sum_{n=1}^N \gamma_n |\alpha(t_d^n) - \alpha(t_p^n)| \tag{5}$$

where γ_n is the weight for the n-th triplet. We calculate γ_n using the edge values in the topology graph G. Specifically, given any triplet $t = \{k^i, k^j, k^k\}$ in the topology graph, the weight is computed as $\gamma = e_{i,j}e_{i,k}$. As the edge represents the strength of the connections between two keypoints, it is reasonable to employ the multiplication of the two edges that formed the included angle as the weight for regularization.

Regularizing Unstructured Keypoints: Similarly, given a driving frame I_d and the synthesized I_p, we identify their unstructured keypoints $\mathcal{U}_d$ and $\mathcal{U}_p$, respectively. Since these unstructured keypoints are disjoint, we constrain them by encoding their included angles with the object centroid. Taking the driving frame as an example, for each pair of keypoints (k_d^i, k_d^j) in $\mathcal{U}_d$, we construct a triplet $\hat{t}_d = (k_d^i, k_d^c, k_d^j)$ in which k_d^c is the object centroid, and further denote the included angle as $\beta(\hat{t}_d)$. We similarly define the corresponding included angle for the synthesized image as $\beta(\hat{t}_p)$. We then regularize the consistency of the corresponding included angles for the structured keypoints in the driving frame and the synthesized image as follows:

$$\mathcal{L}_{ru} = \frac{1}{\hat{N}} \sum_{n=1}^{\hat{N}} |\beta(\hat{\mathbf{t}}_d^n) - \beta(\hat{\mathbf{t}}_p^n)| \tag{6}$$

where $\hat{N}$ is the number of constructed triplets using structured keypoints in each image.

Combining the loss of structured and unstructured keypoints, the total loss of our shape-invariant motion adaptation loss can be written as follows:

$$L_{ma} = L_{rs} + L_{ru} \tag{7}$$

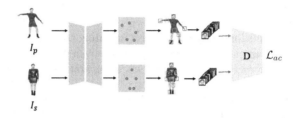

Fig. 4. Illustration of our structure-guided appearance consistency module

3.3 Structure-Guided Appearance Consistency Module

We now consider how the appearance of the synthesized image I_p might be enforced to be similar to that of the source image I_s. Note that the object poses in I_p and I_s are different, as we have enforced I_p to mimic the pose of the driving frame. We therefore propose structure-guided appearance consistency module to regularize the appearance consistency of object parts in I_p and I_c to avoid impacting the learned object pose of I_p

In particular, we use the predicted motion keypoints to extract image patches of fixed size from both images. After collecting the patches from I_p (*resp.*, I_s), a discriminator $\mathcal{D}$ is then introduced to enforce the appearance consistency between the corresponding patches by means of an adversarial training strategy, as shown in Fig. 4. The aim of the discriminator is to determine whether the input patches are from I_p or I_s by minimizing a cross-entropy loss, while the generation model $\mathcal{B}$ (*i.e.*, the basic MT model) aims at generating pseudo-images $\mathcal{B}(I_s)$, which are difficult to distinguish from the source image I_s by maximizing the cross-entropy loss. Formally, we express the loss of our patch-based appearance consistency module as follows:

$$L_{ac} = \log \mathcal{D}(V(I_s)) + \log(1 - \mathcal{D}(V(\mathcal{B}(I_s)))) \tag{8}$$

where $V(\cdot)$ represents the patch extraction operation.

3.4 Summary

We combine all losses together to train our proposed MAA model in an end-to-end manner. The overall objective function can be written as follows,

$$\mathcal{L} = \mathcal{L}_r + \mathcal{L}_c + \lambda_{ma}\mathcal{L}_{ma} - \lambda_{ac}\mathcal{L}_{ac} \tag{9}$$

where λ_{ma} and λ_{ac} are tradeoff parameters used to balance the losses. Due to the existence of the discriminator, we optimize the overall loss in an adversarial training manner, *i.e.*, $\min_{\mathcal{B}} \max_{\mathcal{D}} \mathcal{L}$. Detailed training loop is presented in Supplementary materials.

4 Experiment

4.1 Datasets

We conduct experiments for two types of object including human body and human face. For the human body animation, we transfer motion from Mixamo-Video to Fashion-Video, and for human face animation we transfer motion from Vox-Celeb to CUHK Face Sketch (Cufs).

Mixamo-Video Dataset is a synthetic human dancing video dataset newly constructed by ourselves. We collect 15 characters of 3D human body models and 46 dancing sequences from the mixamo [1] website, then render dancing videos for these characters and dancing sequences, leading to $15 \times 46 = 690$ videos in total with resolution of 256×256. We split ten of the characters as training set and the rest as test set, *i.e.* 460 and 230 videos, respectively. Details of the dataset are presented in supplementary materials, and we will release the dataset soon.

Fashion-Video Dataset is a video dataset for showing clothes. It contains 500 training videos and 100 testing videos with size of 256×256. Although it is a video dataset, We use it as an image dataset by selecting one frame per video randomly in training stage.

Vox-Celeb is a video dataset of human talking. It consists of $12,331$ training videos and 444 testing videos resized to 256×256.

CUHK Face Sketch (Cufs) is an image dataset of human face sketches. The dataset contains 305 images where training set and test set have 250 and 45 images *resp.*. Each image is a sketch drawn by an artist based on a photo taken in a frontal pose with a natural expression. We also resize those images into the size of 256×256.

Table 1. Quantity results comparison of our method with source only FOMM model and MRAA model. The lower FID and AED values are the better

	Mixamo ⟶ Fashion		Vox ⟶ Cufs	
	FID ↓	AED ↓	FID ↓	AED ↓
MRAA	177.3	0.376	127.1	0.764
FOMM	175.9	0.359	112.5	0.693
Ours (MRAA)	72.1	0.289	86.5	0.627
Ours (FOMM)	**61.7**	**0.274**	**50.1**	**0.573**

4.2 Quantitative Results

Metrics: As the ground-truth video are not available in cross-domain motion transfer, to quantitatively assess the synthesized videos, we employ two metrics for evaluate generative models as follows

- **Fréchet Inception distance (FID)** [15] This score indicates the overall quality of generated frames, it compares the feature statistics of generated frames and real images, then calculates the distance between them.
- **Average Euclidean Distance (AED)** [31] Considering the generated images share the same identity with source images, we utilize AED to evaluate the identity similarity between them. It also computes the feature distance between two input images. Specifically, a pre-trained person re-identification network [14] and a pre-trained facial identification network [2] are used to extract identity feature representations for human body and human face dataset, respectively.

Results: As aforementioned, unsupervised motion transfer models like FOMM [31] or MRAA [32] can be integrated into our MAA framework as the basic motion transfer model. We conduct experiments by respectively using FOMM and MRAA as our basic motion transfer model, and take the original FOMM and MRAA as the corresponding baseline for comparison. For both methods, the baseline models are trained on the driving video dataset without considering the cross-domain issue. Note that the newly proposed modules in our MAA model are only used in training stage, and the model in the test stage share the same architecture with the baseline FOMM or MRAA model.

We report the results for Mixamo-Video $\rightarrow$ Fashion-Video and Vox $\rightarrow$ Cufs in Table 1. Comparing with the FOMM and MRAA model, our proposed MAA approach achieves a much better performance. In particular, compared with FOMM, we achieve a FID of 61.7 and an AED of 0.274 for Mixamo-Video $\rightarrow$ Fashion-Video, and 50.1 *vs.* 112.5 and 0.573 *vs.* 0.693 for Vox $\rightarrow$ Cufs, respectively. Compared with MRAA, we achieve a FID of 72.1 and an AED of 0.289 for Mixamo-Video $\rightarrow$ Fashion-Video, and 86.5 *vs.* 127.1 and 0.627 *vs.* 0.764 for Vox $\rightarrow$ Cufs, respectively. Note that, for both FID and AED metrics, smaller value is better. The large improvement indicates that the cross-domain motion transfer is challenging for the traditional FOMM and MRAA method, while our MAA model works well on the cross-domain scenario. We observe that MRAA performs worse than FOMM in the cross-domain motion transfer task, although the previous work shows MRAA usually performs better than FOMM in the traditional single-domain motion transfer [32]. This possibly dues to that the PCA based motion estimation in the MRAA method is non-parametric and less flexible for cross-domain motion transfer.

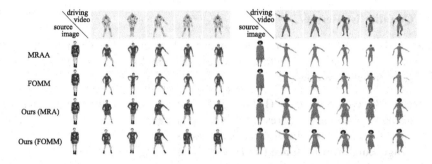

Fig. 5. Visualization results of FOMM, MRAA and ours method on human body datasets

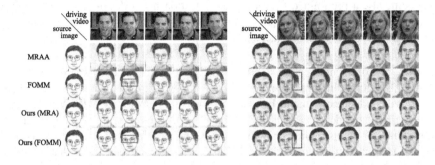

Fig. 6. Visualization results of FOMM, MRAA and ours method on human face datasets

4.3   Qualitative Results

We visualize the generated results to gain an intuitive assessment of FOMM, MRAA and our MAA models for cross-domain human body and human face animation in Fig. 5 and Fig. 6, respectively. In each figure, two pairs of results are visualized in the left and right parts, respectively. Driving frames extracted from the test video are displayed on the top row, while the source images are showed at the most left column of each part.

For human body animation, as shown in Fig. 5, the results generated by the FOMM and MRAA model obviously suffer from domain shift problem. Although the motion of driving video is roughly captured, the human body shape of source image is rarely preserved, and notable artifacts can be observed in almost all frames of the synthesized video. In contrast, our MAA model is able to capture the motion of the driving frames while properly preserving the appearance of the source image.

For human face animation, as shown in Fig. 6, the FOMM and MRAA model could generate results with a rough motion of driving frames and a similar facial appearance with source image. However, the quality of synthesized image are not satisfactory where artifacts are obvious to observe. For example, artifacts on glasses and heads can be observed for FOMM results as highlighted in the red bounding boxes. These differences in qualitative results clearly demonstrate the effectiveness of our proposed MAA model for cross-domain motion transfer.

4.4 Ablation Study

To study the impact of our proposed modules, we further conduct ablation study on both human body and human face datasets. The FOMM is used as the basic motion transfer model. The quantitative results are shown in Table 2, where w/o 'CYC', 'w/o SIMA' and 'w/o SGAC' means removing the cyclic training pipeline, shape-invariant motion adaptation and structure-guided appearance consistency of FOMM model, respectively.

Table 2. Ablation results comparison of FOMM and our ablated models.

	Mixamo ⟶ Fashion		Vox ⟶ Cufs	
	FID ↓	AED ↓	FID ↓	AED ↓
FOMM	175.9	0.359	112.5	0.693
w/o CYC	136.9	0.354	74.1	0.633
w/o SIMA	80.2	0.303	60.7	0.622
w/o SGAC	67.7	0.284	55.2	0.603
Ours (FOMM)	**61.7**	**0.274**	**50.1**	**0.573**

For both human body and human face animation, as shown in Table 2, we observe considerable performance drops on both AED and FID for w/o CYC, which again confirms the importance to explicitly consider the cross-domain issue when performing motion transfer across domains. Similar observations can be obtained on human face dataset, which is also confirmed in the qualitative results in Fig. 7. Other ablation settings w/o SIMA and w/o SGAC also degrade the performance considerably, which validates the necessity of using the two modules for generating satisfactory synthesized video in cross-domain motion transfer.

To show the effect of each module intuitively, we further visualize the synthesized results in Fig. 7. We observe that the result of w/o CYC has richer details than that of FOMM model. For example, the face and the clothes are clearer. However, compared with our final MAA result, it still drops important motion and appearance information. Moreover, we observe the result of w/o SIMA are able to preserve relative rich appearance information, however, the pose of driving frames are not transferred properly without the help of motion consistency module. For example, artifacts can be observed for the poses of the arms and heads as highlighted in the blue rectangles. And, on the third row, w/o SGAC performs well in pose transferring but fails to preserve source image appearance without SGAC, especially for the details of human face as highlighted in red rectangles. These observations confirm the effectiveness of the modules proposed in our MAA approach.

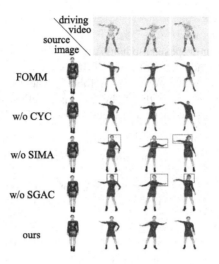

Fig. 7. Visualized ablation study results on the human body datasets

4.5 User Study

To further evaluate our model, we additionally conduct a user study. In particular, we randomly select 50 pairs of source domain driving videos and target domain source images for both human body animation and human face animation, and generate result videos in an ablation setting. For each dataset, we compare results of our final MAA model with those of FOMM and three ablation methods, respectively. The comparison are evaluated by 25 users according to three aspects, motion, appearance (denoted as app. in Table 3) and overall, respectively.

The user preferences are shown in Table 3. We observe that all scores are above 0.5, which means our results are preferred by the majority of users for all aspects in all settings. For motion aspect, fewer users prefer w/o SIMA than other settings when compared with MAA model on both datasets (0.748 vs. 0.717 and 0.679 for human body, and 0.571 vs. 0.704 for human face), which indicates SIMA improves the motion of generated results. For appearance aspect, fewer user prefer w/o SGAC than other ablation settings when compared with MAA model in human body dataset (0.715 vs. 0.711 and 0.699), which indicates SGAC contributes to appearance of generated results.

Table 3. User study results. We compare the Ours (FOMM) model to every ablation model, and the values represent the user preferences to Ours (FOMM) model

	Mixamo ⟶ Fashion			Vox ⟶ Cufs		
	Motion	Appearance	Overall	Motion	Appearance	Overall
FOMM	0.888	0.983	0.978	0.845	0.792	0.875
w/o CYC	0.717	0.699	0.732	0.571	0.615	0.626
w/o SIMA	0.748	0.711	0.702	0.704	0.655	0.675
W/o SGAC	0.679	0.715	0.725	0.593	0.617	0.575

5   Conclusion

In this paper, we propose a Motion and Appearance Adaptation (MAA) approach for cross-domain motion transfer. In MAA, we design a shape-invariant motion adaptation module to enforce the consistency of the angles of object parts in two images to capture the motion information. Meanwhile, we introduce a structure-guided appearance consistency module to regularize the similarity between the patches of the synthesized image and the source image. The experimental results demonstrates the effectiveness of our proposed method.

Acknowledgement. This work is supported by the Major Project for New Generation of AI under Grant No. 2018AAA0100400, the National Natural Science Foundation of China (Grant No. 62176047), Sichuan Science and Technology Program (No. 2021YFS0374, 2022YFS0600), Beijing Natural Science Foundation (Z190023), and Alibaba Group through Alibaba Innovation Research Program. This work is also partially supported by the Science and Technology on Electronic Information Control Laboratory.

References

1. Adobe's mixamo website. https://www.mixamo.com Accessed 3 Nov 2021
2. Amos, B., Ludwiczuk, B., Satyanarayanan, M., et al.: Openface: a general-purpose face recognition library with mobile applications. CMU School Comput. Sci. **6**(2), 20 (2016)
3. Balakrishnan, G., Zhao, A., Dalca, A.V., Durand, F., Guttag, J.: Synthesizing images of humans in unseen poses. In: Proceedings of the IEEE Conference on Computer Vision and Pattern Recognition, pp. 8340–8348 (2018)
4. Burkov, E., Pasechnik, I., Grigorev, A., Lempitsky, V.: Neural head reenactment with latent pose descriptors. In: Proceedings of the IEEE/CVF Conference on Computer Vision and Pattern Recognition, pp. 13786–13795 (2020)
5. Chan, C., Ginosar, S., Zhou, T., Efros, A.A.: Everybody dance now. In: Proceedings of the IEEE/CVF International Conference on Computer Vision, pp. 5933–5942 (2019)
6. Chen, Y., Li, W., Sakaridis, C., Dai, D., Van Gool, L.: Domain adaptive faster R-CNN for object detection in the wild. In: Proceedings of the IEEE Conference on Computer Vision and Pattern Recognition, pp. 3339–3348 (2018)
7. Chen, Z., Wang, C., Yuan, B., Tao, D.: Puppeteergan: arbitrary portrait animation with semantic-aware appearance transformation. In: Proceedings of the IEEE/CVF Conference on Computer Vision and Pattern Recognition, pp. 13518–13527 (2020)

8. Chu, T., Liu, Y., Deng, J., Li, W., Duan, L.: Denoised maximum classifier discrepancy for source-free unsupervised domain adaptation. In: Thirty-Sixth AAAI Conference on Artificial Intelligence (AAAI-22) (2022)

9. Deng, J., Li, W., Chen, Y., Duan, L.: Unbiased mean teacher for cross-domain object detection. In: Proceedings of the IEEE/CVF Conference on Computer Vision and Pattern Recognition, pp. 4091–4101 (2021)

10. Dong, J., Cong, Y., Sun, G., Fang, Z., Ding, Z.: Where and how to transfer: Knowledge aggregation-induced transferability perception for unsupervised domain adaptation. IEEE Trans. Pattern Anal. Mach. Intell. 1–1 (2021)

11. Dong, J., Cong, Y., Sun, G., Zhong, B., Xu, X.: What can be transferred: unsupervised domain adaptation for endoscopic lesions segmentation. In: IEEE/CVF Conference on Computer Vision and Pattern Recognition (CVPR), pp. 4022–4031, June 2020

12. Ganin, Y., Lempitsky, V.: Unsupervised domain adaptation by backpropagation. In: International Conference on Machine Learning, pp. 1180–1189. PMLR (2015)

13. Gu, K., Zhou, Y., Huang, T.: FLNET: landmark driven fetching and learning network for faithful talking facial animation synthesis. In: Proceedings of the AAAI Conference on Artificial Intelligence, vol. 34, pp. 10861–10868 (2020)

14. Hermans, A., Beyer, L., Leibe, B.: In defense of the triplet loss for person re-identification. arXiv preprint arXiv:1703.07737 (2017)

15. Heusel, M., Ramsauer, H., Unterthiner, T., Nessler, B., Hochreiter, S.: Gans trained by a two time-scale update rule converge to a local Nash equilibrium. In: Advances in Neural Information Processing Systems, vol. 30 (2017)

16. Huang, Z., Han, X., Xu, J., Zhang, T.: Few-shot human motion transfer by personalized geometry and texture modeling. In: Proceedings of the IEEE/CVF Conference on Computer Vision and Pattern Recognition, pp. 2297–2306 (2021)

17. Isola, P., Zhu, J.Y., Zhou, T., Efros, A.A.: Image-to-image translation with conditional adversarial networks. In: Proceedings of the IEEE Conference on Computer Vision and Pattern Recognition, pp. 1125–1134 (2017)

18. Jakab, T., Gupta, A., Bilen, H., Vedaldi, A.: Unsupervised learning of object landmarks through conditional image generation. In: Proceedings of the 32nd International Conference on Neural Information Processing Systems, pp. 4020–4031 (2018)

19. Jeon, S., Nam, S., Oh, S.W., Kim, S.J.: Cross-identity motion transfer for arbitrary objects through pose-attentive video reassembling. In: Vedaldi, A., Bischof, H., Brox, T., Frahm, J.-M. (eds.) ECCV 2020. LNCS, vol. 12369, pp. 292–308. Springer, Cham (2020). https://doi.org/10.1007/978-3-030-58586-0_18

20. Kappel, M., et al.: High-fidelity neural human motion transfer from monocular video. In: Proceedings of the IEEE/CVF Conference on Computer Vision and Pattern Recognition, pp. 1541–1550 (2021)

21. Li, C., Lee, G.H.: From synthetic to real: Unsupervised domain adaptation for animal pose estimation. In: Proceedings of the IEEE/CVF Conference on Computer Vision and Pattern Recognition, pp. 1482–1491 (2021)

22. Li, Y., Huang, C., Loy, C.C.: Dense intrinsic appearance flow for human pose transfer. In: Proceedings of the IEEE/CVF Conference on Computer Vision and Pattern Recognition, pp. 3693–3702 (2019)

23. Liu, W., Piao, Z., Min, J., Luo, W., Ma, L., Gao, S.: Liquid warping GAN: a unified framework for human motion imitation, appearance transfer and novel view synthesis. In: Proceedings of the IEEE/CVF International Conference on Computer Vision, pp. 5904–5913 (2019)

24. Liu, Y., Deng, J., Gao, X., Li, W., Duan, L.: Bapa-net: boundary adaptation and prototype alignment for cross-domain semantic segmentation. In: The IEEE International Conference on Computer Vision (ICCV), October 2021

25. Liu, Y., Deng, J., Tao, J., Chu, T., Duan, L., Li, W.: Undoing the damage of label shift for cross-domain semantic segmentation. In: The IEEE Conference on Computer Vision and Pattern Recognition(CVPR) (2022)
26. Ma, L., Jia, X., Sun, Q., Schiele, B., Tuytelaars, T., Van Gool, L.: Pose guided person image generation. In: Advances in Neural Information Processing Systems, vol. 30 (2017)
27. Ma, L., Sun, Q., Georgoulis, S., Van Gool, L., Schiele, B., Fritz, M.: Disentangled person image generation. In: Proceedings of the IEEE Conference on Computer Vision and Pattern Recognition, pp. 99–108 (2018)
28. Nagrani, A., Chung, J.S., Zisserman, A.: Voxceleb: a large-scale speaker identification dataset. arXiv preprint arXiv:1706.08612 (2017)
29. Ren, Y., Yu, X., Chen, J., Li, T.H., Li, G.: Deep image spatial transformation for person image generation. In: Proceedings of the IEEE/CVF Conference on Computer Vision and Pattern Recognition, pp. 7690–7699 (2020)
30. Siarohin, A., Lathuilière, S., Tulyakov, S., Ricci, E., Sebe, N.: Animating arbitrary objects via deep motion transfer. In: Proceedings of the IEEE/CVF Conference on Computer Vision and Pattern Recognition, pp. 2377–2386 (2019)
31. Siarohin, A., Lathuilière, S., Tulyakov, S., Ricci, E., Sebe, N.: First order motion model for image animation. In: Advances in Neural Information Processing Systems, vol. 32, pp. 7137–7147 (2019)
32. Siarohin, A., Woodford, O.J., Ren, J., Chai, M., Tulyakov, S.: Motion representations for articulated animation. In: Proceedings of the IEEE/CVF Conference on Computer Vision and Pattern Recognition, pp. 13653–13662 (2021)
33. Simonyan, K., Zisserman, A.: Very deep convolutional networks for large-scale image recognition. arXiv preprint arXiv:1409.1556 (2014)
34. Tao, J., Wang, B., Xu, B., Ge, T., Jiang, Y., Li, W., Duan, L.: Structure-aware motion transfer with deformable anchor model. In: Proceedings of the IEEE/CVF Conference on Computer Vision and Pattern Recognition, pp. 3637–3646 (2022)
35. Tzeng, E., Hoffman, J., Saenko, K., Darrell, T.: Adversarial discriminative domain adaptation. In: Proceedings of the IEEE Conference on Computer Vision and Pattern Recognition, pp. 7167–7176 (2017)
36. Wang, C., Xu, C., Tao, D.: Self-supervised pose adaptation for cross-domain image animation. IEEE Transa Artif. Intell. 1(1), 34–46 (2020)
37. Wang, X., Tang, X.: Face photo-sketch synthesis and recognition. IEEE Trans. Pattern Anal. Mach. Intell. 31(11), 1955–1967 (2008)
38. Wiles, O., Koepke, A.S., Zisserman, A.: X2Face: a network for controlling face generation using images, audio, and pose codes. In: Ferrari, V., Hebert, M., Sminchisescu, C., Weiss, Y. (eds.) ECCV 2018. LNCS, vol. 11217, pp. 690–706. Springer, Cham (2018). https://doi.org/10.1007/978-3-030-01261-8_41
39. Xu, B., et al.: Move as you like: image animation in e-commerce scenario. In: Proceedings of the 29th ACM International Conference on Multimedia, pp. 2759–2761 (2021)
40. Zablotskaia, P., Siarohin, A., Zhao, B., Sigal, L.: DWNET: dense warp-based network for pose-guided human video generation. arXiv preprint arXiv:1910.09139 (2019)
41. Zhang, S., Zhao, W., Guan, Z., Peng, X., Peng, J.: Keypoint-graph-driven learning framework for object pose estimation. In: Proceedings of the IEEE/CVF Conference on Computer Vision and Pattern Recognition, pp. 1065–1073 (2021)
42. Zhu, J.Y., Park, T., Isola, P., Efros, A.A.: Unpaired image-to-image translation using cycle-consistent adversarial networks. In: Proceedings of the IEEE International Conference on Computer Vision, pp. 2223–2232 (2017)
43. Zhu, Z., Huang, T., Shi, B., Yu, M., Wang, B., Bai, X.: Progressive pose attention transfer for person image generation. In: Proceedings of the IEEE/CVF Conference on Computer Vision and Pattern Recognition, pp. 2347–2356 (2019)

Layered Controllable Video Generation

Jiahui Huang[1,2(✉)] [iD], Yuhe Jin[1] [iD], Kwang Moo Yi[1] [iD], and Leonid Sigal[1,2,3,4]

[1] University of British Columbia, Endowment Lands, Canada
{gabrie20,yuhejin,kmyi,lsigal}@cs.ubc.ca
[2] Vector Institute for AI, Toronto, Canada
[3] CIFAR AI Chair, Toronto, Canada
[4] NSERC CRC Chair, Toronto, Canada

Abstract. We introduce layered controllable video generation, where we, without any supervision, decompose the initial frame of a video into foreground and background layers, with which the user can control the video generation process by simply manipulating the foreground mask. The key challenges are the unsupervised foreground-background separation, which is ambiguous, and ability to anticipate user manipulations with access to only raw video sequences. We address these challenges by proposing a two-stage learning procedure. In the first stage, with the rich set of losses and dynamic foreground size prior, we learn how to separate the frame into foreground and background layers and, conditioned on these layers, how to generate the next frame using VQ-VAE generator. In the second stage, we fine-tune this network to anticipate edits to the mask, by fitting (parameterized) control to the mask from future frame. We demonstrate the effectiveness of this learning and the more granular control mechanism, while illustrating state-of-the-art performance on two benchmark datasets. Our project website/code can be found at gabriel-huang.github.io/layered_controllable_video_generation.

1 Introduction

Advances in deep generative models have led to impressive results in image and video synthesis. Typical forms of such models, including Variational Autoencoder (VAE) [32], Generative Adversarial (GAN) [19] and recurrent (RNN) [43] formulations, can produce complex and highly realistic content. However, synthesis of realistic images/videos, without the ability to control the depicted content in them, has limited practical utility. This has led to a variety of conditional generative tasks and formulations.

In the image domain, both coarse- (*e.g.*, sentence [70]) and fine-level (*e.g.*, layout [72] and instance attribute [18]) control signals have been explored. The progress on the video side, on the other hand, has generally been more modest, in part due to an added challenge of synthesizing temporally coherent content.

Supplementary Information The online version contains supplementary material available at https://doi.org/10.1007/978-3-031-19787-1_31.

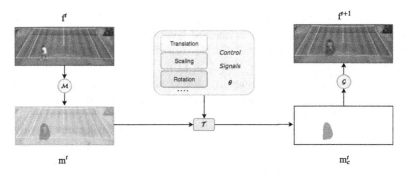

Fig. 1. Layered Controllable Video Generation. Illustration of the proposed task, where a frame at time t is first decomposed into a foreground/background layers, using the learned mask network $\mathcal{M}$, and then the user is allowed to modify this mask with control signals θ (*e.g.*, by shifting it by Δ_t) to control the next generated frame realized by generator $\mathcal{G}$. The foreground source and target mask are illustrated in red and blue. (Color figure online)

Future frame prediction [12,16,17,34,38,55–57,64] conditions the future generated frames on one (or a couple) seed frame(s). But this provides very limited control as the object(s), or person(s), depicted in the conditioned frame can move in a multitude of ways, particularly as predictions are made longer into the future. To address this, a number of methods condition future frames on the action [22,52] and object [40] label. Still, they only provide very coarse global video-level control (Fig. 1).

More recent approaches focus on the ability to control the video content on-the-fly at the frame-level. Typically, these methods are formulated as conditional auto-regressive (or recurrent) models that generate one frame at a time, conditioned, for example, on the discrete action label [30] or keypoint-based human pose specification [54,58,68] (*e.g.*, obtained from a target video source [69]). Such methods, however, require dense per-frame annotation of actions or poses at training time, which are costly to obtain and make it challenging to employ such approaches in realistic environments. The task of *playable video generation* [39], has been introduced to address these limitations. In playable video generation, the discrete action space is *discovered* in an unsupervised manner and can then be used as a conditioning signal in an auto-regressive probabilistic generative model. While obtaining impressive results, with no intermediate supervision and allowing frame-level control over the generative process, [39] is inherently limited to a single subject and a small set of discrete action controls.

Thus, in this work, we aim to allow richer and more granular control over the generated video content, while similarly requiring no supervision of any kind – *i.e.*, having only raw videos as training input, as in [39]. To do so, we make an observation (inspired by early works in vision [28,33,59]) that a video can be effectively decomposed into foreground / background layers. The background layer corresponds to static (or slowly changing) parts of the scene. The foreground layer, on the other hand, evolves as the dominant objects move. Importantly, the foreground mask itself contains position, size and shape information necessary to both characterize and render the moving objects appropriately. Hence, this foreground mask can in itself be leveraged as a useful level of abstraction that allows both intuitive and simple control over the generative process.

With the above intuition, we therefore propose an approach that automatically learns to segment video frames into foreground and background, and at the same time generates future frames conditioned on the segmentation and the past frame, iteratively. To allow on-the-fly control of video generation, we expose the foreground mask to the user for manipulation – *e.g.*, translations of the mask leads to corresponding in-plane motion of the object, resizing it would lead to the depth away/towards motion, and changes in the shape of the mask can control the object pose.

From the technical perspective, the challenge of formulating such a method is two fold. First, unsupervised foreground/background layer segmentation is highly ambiguous, particularly if the background is not assumed to be static and the foreground motions can be small. Second, user input needs to be anticipated to ensure model learns how to *react* to changes in the mask, without explicit access to such information. To this end, we propose a two-stage learning procedure. In the first stage, the network learns how to perform foreground/background separation and, conditioned on this layered representation, future frame prediction. Specifically, we introduce a set of sophisticated losses and a dynamic prior to learn how to predict a foreground mask and leverage VQ-VAE [44] to predict foreground and background latent content which is then fused and decoded to the next frame. In the second stage, we simulate user input and fine-tune the generative model such that this user input can be appropriately handled.

Contributions: Our contributions are multi-fold.

– From raw video data, our model learns to generate foreground/background separation masks in an unsupervised manner. We then leverage the foreground layer as a flexible (parametric) user control mechanism for the generative process. This provides both richer and more intuitive control compared to action vectors [39] or sparse trajectories [21].
– To effectively train our model we introduce two-stage training: the first stage tasked with learning how to separate layers and perform future frame prediction; the second, to adopting and anticipating user control.
– To prevent over-/under-segmentation, we regularize layer separation with sparsity loss and dynamic mask size prior.
– Finally, we validate our approach on multiple datasets and show that we are able to generate state-of-the-art results and, at the same time, allow higher level of control over the generated content without any supervision.

2   Related Works

Video Generation. Early video generation techniques proposed to generate a video as a whole. Most of these convert a noise vector, sampled from simple distribution or a prior, to a video, using a GAN [2,55,61] or VAE [6,12] formulation. More recent architectures leveraged transformer-based formulations [41,46,64,67] that have generally resulted in higher quality video outputs. As an alternative to 3D (transposed) convolution techniques, that generate all frames at once, recurrent auto-regressive variants [29] have also been explored. While these early works largely focused on the video quality and resolution, more recently, the focus has shifted to conditioned or controlled video generation.

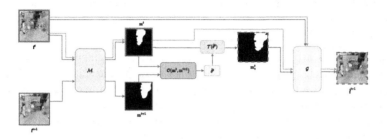

Fig. 2. Illustration of the Proposed Two-stage Training. The flow of Stage I is represented in orange and Stage II in green. $\mathcal{M}$ denotes the *mask network*, which estimates the foreground/background mask, and $\mathcal{G}$ denotes the *generator network* that takes the current frame and a mask to produce the next frame. $\mathcal{O}(\cdot)$ is the optimization procedure described in Eq. (12). $\mathcal{T}(\cdot)$ is a differentiable function that transforms a mask to a target shape using *user control signal* θ^t. (Color figure online)

Video Generation with Global Control. Future frame prediction considers the task of generating a video conditioned on a few starting seed frames. Early approaches to future prediction employed deterministic models [16,38,56,65] that failed to model uncertainty in the future induced by variability of unfolding events. To overcome this limitation, later methods, based on VAE [6,12], GAN [35,36], and probabilistic formulations [66], attempted to introduce real-world stochasticity into the generative process. Action label conditioning, in combination with seed frames or not, where a video sequence is generated conditioned on an input action label [31,62] is also popular; some such approaches leverage disentangled factored representations (*e.g.*, of subject identity and action [22]). Other types of global conditioning signals include action-object tuples [40]. However, these methods, collectively, require action annotations for training, and, more importantly, do not allow control at the frame-level.

Video Generation with Frame-Level Control. More granular control, at the frame-level, has also been explored. Pose-guided generative models first generate a sequence sparse [54,58,68] or dense [69] keypoint human poses, either predictively [54,58] or from a source video [69], and then use these for conditional generation of respective video frames. However, these methods are only applicable to videos of human subjects and require either pose annotations or a pre-trained pose detector. Alternatively, individual frames can be conditioned on action labels [11,30,42]. However, these methods require dense frame-level annotations, which are only available in limited environments such as video games. Closest to our work, Menapace *et al.* proposed Playable Video Generation [39], which, in an unsupervised manner, discovers semantically consistent actions meanwhile generating the next frame conditioned on the past frames and an input action, thus providing user control to the generation process. However, their method is limited to a small set of discrete action controls and explicitly assumes a single moving object. In contrast, our method allows richer and more granular control, and can be used to generate, and control, videos with multiple objects.

Video Generation with Pixel-Level Control. It is worth mentioning that some prior works attempted to use dense semantic segmentation [45,60] and sparse trajectories [21] to control video generation. While such approaches allow granular control at a pixel-level, these representations are incredibly difficult for a user to produce or modify. We also use a form of (foreground) segmentation for control, however, it is unsupervised, class-agnostic, and can be easily controlled either parametrically or non-parametrically.

Unsupervised Object Segmentation. Layered representations have long history in computer vision [28,33,59], and are supported by neuroscience evidence [63]. Traditional techniques for this rely on feature clustering [1,5,24] and statistical background modeling and subtraction [9,14,50]. Such techniques work best for videos where the background is (mostly) static, lighting fixed and the foreground is fast moving; we refer readers to [20] for an extensive analysis and discussion. More recent techniques have focused on generative formulations for the task. In particular, Bielski *et al.* proposed the PerturbGAN [8], their model generates the foreground and background layers separately, and uses a perturbation strategy to enforce the generation of semantically meaningful masks. Related, MarioNette [49] learns to decompose scenes into a background and a learned dictionary of sprites. Other approaches focus on separation of videos into natural layers (*e.g.*, to factorize secondary effects such as shadows and reflections [4]) and to control which layer to attend to [3]. Similar to [55], we decompose frames and separately model foreground/background content that is then composed/fused together to produce video. However, unlike [55] and others, we allow the user to have frame-level control over the foreground mask using both intuitive parametric and nonparametric controls.

3   Method

Our fundamental goal is frame-level controllable video generation. We address this by proposing a model that first segments the input image into foreground / background layers using a mask and then allows user to control the generative process by applying (parameterized) modifications to this mask. While it is possible to train such a model directly, we find that it is difficult in such a case to learn to disentangle foreground/background mask prediction and the controlled generation process (see our ablation study in Sect. 4.2 for details).

We address this by separating the training into two stages (see Fig. 2). Similar to other vision domains (*e.g.*, recognition [7,10], multi-modal learning [37,51]), where it was shown that pre-training was helpful for a number of downstream tasks, we introduce a related pre-training (Stage I) task and formulation, which ultimately helps in our final controllable video generation (Stage II).

Intuitively, **Stage I** pre-training, which we describe in Sect. 3.1, learns how to preform foreground/background separation, using $\mathcal{M}(\cdot)$, and, conditioned on this layered representation, is optimized for future frame prediction. This stage does not consider controlability within the generation process. Instead, it is designed to predict the most *likely* future frame(s) given a single input frame. Note that the task is effectively one of forecasting, but without attempting to model full distribution over potential futures, rather just a single most likely video trajectory exemplified by the observed video itself.

Given the pre-trained foreground/background mask predictor $\mathcal{M}(\cdot)$ and generator $\mathcal{G}(\cdot)$, learned in Stage I, in **Stage II** (see Sect. 3.2) we effectively *fine-tune* this model to take into account user control. Notably, Stage I and Stage II share foreground/background separation, but in Stage I the generator $\mathcal{G}(\cdot)$ learns the expected *dynamics* as part of its generative process. The main goal of Stage II then is to fine-tune $\mathcal{G}(\cdot)$ in such a way as to allow direct conditioning of said dynamics based on user controlled edits to the mask.

3.1 Stage I: Pre-training for Mask-Based Generation

We first train our method with the focus of generating high-quality foreground-background segmentation masks, **without** introducing controllability into the generation process. Doing so requires two main objectives: (1) high-quality estimation of the current frame's segmentation mask $\mathbf{m}^t$; and (2) pretraining a frame generator for mask-based future frame prediction and generation. Writing the two objectives as loss terms $\mathcal{L}_{\text{mask}}$ and $\mathcal{L}_{\text{img}}$, respectively, the total loss for the first stage training of our method $\mathcal{L}_{\text{total}}$ can be written as:

$$\mathcal{L}_{\text{total}} = \mathcal{L}_{\text{mask}} + \mathcal{L}_{\text{img}}. \tag{1}$$

We detail each loss term in the following.

Regularizing the Mask. To train our method to generate proper masks without any supervision, we regularize the mask with three losses: (1) $\mathcal{L}_{\text{bg}}$ – the contents of the background should not change; (2) $\mathcal{L}_{\text{fg}}$ – there should be as little amount of foreground as possible since classifying all pixels as foreground provides a trivial solution for $\mathcal{L}_{\text{bg}}$; and (3) $\mathcal{L}_{\text{bin}}$ – the masking should be binary for effective separation. We therefore write the mask regularization loss $\mathcal{L}_{\text{mask}}$ as

$$\mathcal{L}_{\text{mask}} = \lambda_{\text{bg}}\mathcal{L}_{\text{bg}} + \lambda_{\text{fg}}\mathcal{L}_{\text{fg}} + \lambda_{\text{bin}}\mathcal{L}_{\text{bin}}, \tag{2}$$

where λ_{bg}, λ_{fg}, and λ_{bin} are the hyperparameters controlling how much each loss term affects the mask regularization.

— $\mathcal{L}_{\text{bg}}$. We aim to ensure that the mask correctly identifies the background, *i.e.*, non-moving parts of the scene. Hence, we simply define it as the amount of change in the masked out (background) region between two consecutive frames. We write

$$\mathcal{L}_{\text{bg}} = \left\| (1 - \mathbf{m}^t) \odot \mathbf{f}^t - (1 - \mathbf{m}^t) \odot \mathbf{f}^{t+1} \right\|_1, \tag{3}$$

where $\odot$ denotes the elementwise multiplication. Note that we define this loss using the ℓ_1 norm, as changes in the scene are not strictly restricted to the mask – *e.g.* shadows of moving objects can occur in the background, or other scenic changes, such as a global illumination change can happen – and the ℓ_1 norm leaves room for the method to incorporate these changes if necessary.

— $\mathcal{L}_{\text{fg}}$. As mentioned earlier, $\mathcal{L}_{\text{bg}}$ alone, leaves room for a trivial solution—assigning $\mathbf{m}^t{=}1$ results in $\mathcal{L}_{\text{bg}}{=}0$ regardless of the values of $\mathbf{f}^t$ and $\mathbf{f}^{t+1}$. This could be avoided by enforcing an additional loss term that penalizes having too many foreground pixels, but a naïve regularization is not sufficient, as the amount of the actual foreground pixels may drastically change from frame to frame – *e.g.*, robot arm moving close to the camera vs. further away.

We thus propose to regularize based on the amount of evident change between $\mathbf{f}^t$ and $\mathbf{f}^{t+1}$, approximated using simple background subtraction. Specifically, for a pixel index (i, j), if we denote whether the pixel changed between the two consecutive frames $\mathbf{f}^t$ and $\mathbf{f}^{t+1}$ as

$$\boldsymbol{\mu}_{ij}^t = \begin{cases} 1, & \text{if } \left\| \mathbf{f}_{ij}^{t+1} - \mathbf{f}_{ij}^t \right\|_1 > \tau \\ 0, & \text{otherwise} \end{cases}, \tag{4}$$

where τ is a threshold for controlling the sensitivity, we can use the average of $\boldsymbol{\mu}_{ij}^t$ as a rough estimate for how much of the pixels should be foreground, dynamically for each consecutive frame. We thus write

$$\mathcal{L}_{\text{fg}} = \max \left\{ 0, \left\| \mathbb{E}_{ij} \left[\mathbf{m}_{ij}^t \right] - \mathbb{E}_{ij} \left[\boldsymbol{\mu}_{ij}^t \right] \right\|_1 \right\}, \tag{5}$$

where $\mathbf{m}_{ij}^t$ is the generated mask value at pixel index (i, j) at frame t, and $\boldsymbol{\mu}_{ij}^t$ is the variation between adjacent frames exceeds the threshold. We note that the balance between the hyperparameter settings related to $\mathcal{L}_{\text{fg}}$ and $\mathcal{L}_{\text{bg}}$ is important since they govern how the loss behaves—*e.g.* a wider mask that covers all potential changes in the scene, or a tight mask that only focuses on the actual changing locations. We empirically set the ratio between λ_{bg} and λ_{fg} to be 100:1, and we gradually decrease this ratio in training for faster convergence.

— $\mathcal{L}_{\text{bin}}$. Finally, there is one last loophole for the network to cheat its way through the two mask regularizors – by producing intermediate values in the range $[0, 1]$. In fact, we found in our early experiments that *soft masks* allows the deep network to encode information of the image in the mask alone, and any modification on the mask will result in degenerated frames (see *Supp. Mat.* Section D). Therefore, we encourage the mask to be binary [8]:

$$\mathcal{L}_{\text{bin}} = \min\{\mathbf{m}^t, (1 - \mathbf{m}^t)\}. \tag{6}$$

At test time we clip these pseudo-binary masks to 0/1 with a 0.5 threshold.

Learning to Predict the Next Frame. To pretrain the generator for next frame prediction, we follow the VQGAN [15] framework. The choice of VQGAN is motivated by two factors: (1) its ability to preserve spatial information in the latent space, allowing effective masking that we need, and (2) its illustrated high-quality image generation performance. We note that VQGAN itself, or its use in this context, is *not* a contribution we claim. We therefore write

$$\mathcal{L}_{\text{img}} = \lambda_{\text{VQ}}\mathcal{L}_{\text{VQ}} + \lambda_{\text{GAN}}\mathcal{L}_{\text{GAN}} + \lambda_{\text{percept}}\mathcal{L}_{\text{percept}}, \tag{7}$$

where λ_{VQ}, λ_{GAN}, λ_{percept} are hyperparameters controlling influence of terms.
— $\mathcal{L}_{\text{VQ}}$. Following the original VQ-VAE [44] formulation we write

$$\mathcal{L}_{\text{VQ}} = \left\| \mathbf{f}^{t+1} - \hat{\mathbf{f}}^{t+1} \right\|_1 + \left\| \text{sg}\left[\mathcal{E}(\mathbf{f}^t) \right] - \mathbf{z}_{\mathbf{q}} \right\|_2^2 + \left\| \text{sg}\left[\mathbf{z}_{\mathbf{q}} \right] - \mathcal{E}(\mathbf{f}^t) \right\|_2^2, \tag{8}$$

where $\mathbf{f}^{t+1}$ and $\hat{\mathbf{f}}^{t+1}$ are true and estimated next frame, sg[·] denotes the stop gradient operation, and $\mathbf{z}_{\mathbf{q}}$ is the quantized latent variable of the VQ-VAE. Note that we use the ℓ_1 norm, instead of the ℓ_2, for the reconstruction part of the loss, as we emprically found it to be more stable in training.
— $\mathcal{L}_{\text{GAN}}$. We train a discriminator $\mathcal{C}$ with the architecture from [25] and aim to improve the generation quality. We therefore write

$$\mathcal{L}_{\text{GAN}} = \log\left(\mathcal{C}\left(\mathbf{f}^{t+1} \right) \right) + \log\left(1 - \mathcal{C}\left(\hat{\mathbf{f}}^{t+1} \right) \right). \tag{9}$$

For the hyperparameter for this loss λ_{GAN}, we follow [15] and apply a dynamic weighing strategy, which stabilizes training.
— $\mathcal{L}_{\text{percept}}$. We use a pretrained VGG-16 network [48] to extract deep features and compute the perceptual loss. Denoting the deep feature extraction process as $\mathcal{V}$ we write

$$\mathcal{L}_{\text{percept}} = \left\| \mathcal{V}(\hat{\mathbf{f}}^{t+1}) - \mathcal{V}(\mathbf{f}^{t+1}) \right\|_2. \tag{10}$$

3.2 Stage II: Fine-tuning for Controllability

While the model trained in Sect. 3.1 is a generative model conditioned on the latent mask $\mathbf{m}^t$, it cannot be immediately used with any arbitrary mask – the generator $\mathcal{G}$ would expect a mask that aligns perfectly with $\mathbf{f}^t$, whereas our user edited mask $\mathbf{m}_c^t$ will not. In other words, we need a way to simulate user input, in terms of mask modifications, and incorporate it into the training. Therefore, we now discuss the second stage of our training setup, where we shift our focus to imbue our method with controllability.

We turn our attention to the assumption that the changes between the masks of two consecutive frames $\mathbf{m}^t$ and $\mathbf{m}^{t+1}$ can be approximated by a differentiable transformation function $\mathcal{T}(\cdot)$:

$$\mathbf{m}^{t+1} \approx \mathcal{T}\left(\mathbf{m}^t, \boldsymbol{\theta}^t \right). \tag{11}$$

In the crudest form, $\mathcal{T}(\cdot)$ can simply be shifting the mask $\mathbf{m}^t$ by Δx and Δy in horizontal and vertical directions, respectively, or for example, be an affine transformation. Both can be implemented differentiably as a parametric coordinate transformation on $\mathbf{m}^t$ [26,27]. We utilize this differentiabilty to find the "ground-truth" control signal $\hat{\boldsymbol{\theta}}^t$ (a.k.a., the *pseudo user control*) using the following optimization procedure:

$$\hat{\boldsymbol{\theta}}^t \equiv \arg\min_{\theta} \left\| \mathbf{m}^{t+1} - \mathcal{T}\left(\mathbf{m}^t, \boldsymbol{\theta}^t\right) \right\|. \tag{12}$$

Now, we can apply this control signal $\hat{\boldsymbol{\theta}}^t$ to the current mask $\mathbf{m}^t$ to obtain the *pseudo user-edited mask* using $\mathcal{T}(\cdot)$, and finetune the network which results in now *controllable* video generation:

$$\tilde{\mathbf{f}}^{t+1} = \mathcal{G}\left(\mathbf{f}^t, \left\lfloor \mathcal{T}\left(\mathbf{m}^t, \hat{\boldsymbol{\theta}}^t\right)\right\rfloor_{0.5}\right). \tag{13}$$

where $\mathcal{G}$ denotes the frame generator, and $\lfloor\cdot\rfloor_{0.5}$ denoting the binarization operation. We then use $\tilde{\mathbf{f}}^{t+1}$ in our loss functions to fine-tune.

One noteworthy aspect of this second stage training is that, because we binarize the mask, no gradient flows through to $\mathcal{M}$. this leads to $\mathcal{L}_{\text{bg}}$, $\mathcal{L}_{\text{fg}}$, and $\mathcal{L}_{\text{bin}}$ not affecting training. While the latter two can be dropped since they are purely on how the mask network $\mathcal{M}$ behaves, completely dropping $\mathcal{L}_{\text{bg}}$ now has the danger of the generated image ignoring the mask. Hence, we replace $\mathbf{f}^{t+1}$ in Eq. (3) with $\tilde{\mathbf{f}}^{t+1}$, so that the generated image still obeys the mask conditioning. Hence, for the second stage training, instead of $\mathcal{L}_{\text{bg}}$, we utilize $\mathcal{L}'_{\text{bg}}$ where

$$\mathcal{L}'_{\text{bg}} = \left\|(1 - \mathbf{m}^t) \odot \mathbf{f}^t - (1 - \mathbf{m}^t) \odot \tilde{\mathbf{f}}^{t+1}\right\|_1, \tag{14}$$

which now enforces our fine-tuned generator $\mathcal{G}$ to still obey the provided mask.

3.3   Framework

As illustrated in Fig. 2, our method is composed of mainly two components: the mask network and the generator network. The *mask network* $\mathcal{M}$ takes as input an image frame $\mathbf{f}^t \in \mathbb{R}^{3 \times W \times H}$ at time t and outputs a mask:

$$\mathbf{m}^t = \mathcal{M}\left(\mathbf{f}^t\right), \tag{15}$$

where $\mathbf{m}^t \in \mathbb{R}^{W \times H}$, segmenting the foreground layer and the background layer. Then, our *generator network* $\mathcal{G}$ takes the mask and the frame as input to generate the next frame:

$$\hat{\mathbf{f}}^{t+1} = \mathcal{G}\left(\mathbf{f}^t, \mathbf{m}^t\right). \tag{16}$$

As shown in Fig. 3, the generator $\mathcal{G}$ can be written as a composite function of an encoder $\mathcal{E}$, a decoder $\mathcal{D}$, and a learnable discrete code book $\mathcal{Z} = \{\mathbf{z}_k\}_{k=1}^K \subset \mathbb{R}^{n_z}$, where n_z is the dimension of each code, we encode the foreground layer and the background layer separately in our pipeline, and then merge them in the latent space before feeding them into the decoder:

$$\mathbf{z}_{fg}^t = \mathcal{E}\left(\mathbf{f}^t \odot \mathbf{m}^t\right)$$
$$\mathbf{z}_{bg}^t = \mathcal{E}\left(\mathbf{f}^t \odot (1 - \mathbf{m}^t)\right) \quad , \qquad (17)$$
$$\hat{\mathbf{f}}^{t+1} = \mathcal{D}\left(\mathsf{q}\left(\mathbf{z}_{fg}^t\right) + \mathsf{q}\left(\mathbf{z}_{bg}^t\right)\right)$$

where $\odot$ is the element-wise product and q is the element-wise quantization function defined as:

$$\mathsf{q}(\mathbf{z})_{ij} := \arg\min_{\mathbf{z}_k \in \mathcal{Z}} \|\mathbf{z}_{ij} - \mathbf{z}_k\| \,, \qquad (18)$$

where i and j are the row and column indices. $\mathcal{G}$ and $\mathcal{M}$ are then used to generate the video via auto-regression.

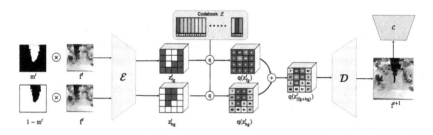

Fig. 3. Frame Generator $\mathcal{G}$. We employ VQGAN framework for the generator $\mathcal{G}$, comprising of an encoder $\mathcal{E}$, decoder $\mathcal{D}$, a learnable discrete codebook $\mathcal{Z}$ and a discriminator $\mathcal{C}$. We encode the foreground and the background layers separately and then merge them in the latent space before feeding them to the decoder $\mathcal{D}$ which generates the next frame in the sequence.

With the above pipeline, during testing time, we enable on-the-fly user control by modifying the current frame mask $\mathbf{m}^t$ to create $\mathbf{m}_c^t$ and using it in place of $\mathbf{m}^t$ in Eq. (16). Mathematically, we write

$$\mathbf{m}_c^t = \mathcal{T}\left(\mathbf{m}^t, \boldsymbol{\theta}^t\right), \qquad (19)$$

where $\mathcal{T}(\cdot)$ is the mask controlling operation described in Sect. 3.2. Other forms of control are also possible, including more granular non-parametric manipulation of the mask (see section "Action Mimicking" on our project website).

4   Experiments

While our method is *not* limited to a "single-agent" assumption, *i.e.*, single dominant moving agent in the scene, previous work, and notably [39] which is the closest and the most competitive baseline, are. Hence, for fair comparison, we adopt the single-agent setup for the majority of our experiments. We train/test on the following datasets:

BAIR Robot Pushing Dataset [13]. This dataset contains 44K video clips (256×256 resolution) of a single robot arm agent pushing toys on a flat surface.

Tennis Dataset [39]. This dataset contains 900 clips extracted from two full tennis matches on YouTube. These clips are cropped such that only the half of the court is visible. The resolution of each frame is (96×256).

4.1 Results

Evaluation Protocol. We compare our model against other conditional generative methods, focusing on quality of reconstructed sequences and controlability. We evaluate our model under three control protocols: two parametric (position, affine) and one non-parametric (direct non-differentiable control over mask):

- Ours /w `position` control: We first use our trained mask network $\mathcal{M}$ to extract masks from the ground truth test sequences, then we use Eq. (19) to approximate those masks with control for generation, where θ is restricted to `positional` parameters, *i.e.*, x and y translation.
- Ours /w `affine` control: Similar to above, here our θ employs full `affine` transformation parameters, *i.e.*, translation, rotation, scaling and shearing.
- Ours /w `non-param` control: We use masks predicted from ground truth test sequences themselves to condition our generation. These masks can change at a pixel-level, hence constituting non-parametric control.

For testing, we generate video sequences conditioned on the first frame f_0 and the user input $\Theta = \{\theta_u^t\}_{t=1}^T$ in all cases.

Metrics. To quantitatively evaluate our results, we consider standard metrics:

- *Learned Perceptual Image Patch Similarity (LPIPS)* [71]: LPIPS measures the perceptual distance between generated and ground truth frames.
- *Fréchet Inception Distance (FID)* [23]: FID calculates the Fréchet distance between multivariate Gaussians fitted to the feature space of the Inception-v3 network of generated and ground truth frames.
- *Fréchet Video Distance (FVD)* [53]: FVD extends FID to the video domain. In addition to the quality of each frame, FVD also evaluates the temporal coherence between generated and ground truth sequences.
- *Average Detection Distance (ADD)* [39]: ADD first uses Faster-RCNN [47] to detect the target object in both generated and ground truth frames, then calculates the Euclidean distance between the bound box centers.
- *Missing Detection Rate (MDR)* [39]: MDR reports percentage of unsuccessful detections in generated vs. successful detections in ground truth sequences.
- *Rooted Mean Square Error of Displacement (RMSED)* [39]: RMSED, which we define, reports the RMSE of the displacement of ground truth locations vs. generated locations. See Fig. 4 for more details.

LIPIPS, FID and FVD measure the quality of generated videos. ADD and MDR measure how the action label conditions the generated video, and RMSED measures the precision of control.

Baselines. CADDY [39] is the only unsupervised video generation method that allows frame-level user conditioning, thus we use it as our main baseline. We also include results of other frame-level conditioned methods: MoCoGAN [52], SAVP [36], and their high-resolution adaptations MoCoGAN+ and SAVP+ from [39].

Table 1. Results on the *BAIR* Dataset

Method	LPIPS ↓	FID ↓	FVD ↓	RMSED ↓
MoCoGAN [52]	0.466	198	1380	–
MoCoGAN+ (from [39])	0.201	66.1	849	0.211
SAVP [36]	0.433	220	1720	–
SAVP+ (from [39])	**0.154**	**27.2**	303	0.109
CADDY [39]	0.202	35.9	423	0.132
Ours /w **position** control	0.202	28.5	333	0.059
Ours /w **affine** control	0.201	30.1	**292**	0.035
Ours /w **non-param** control	0.176	29.3	293	**0.021**

Table 2. Results on the *Tennis* Dataset

Method	LPIPS ↓	FID ↓	FVD ↓	ADD ↓	MDR ↓
MoCoGAN [52]	0.266	132	3400	28.5	20.2
MoCoGAN+ (from [39])	0.166	56.8	1410	48.2	27.0
SAVP [36]	0.245	156	3270	10.7	19.7
SAVP+ (from [39])	0.104	25.2	223	13.4	19.2
CADDY [39]	0.102	13.7	239	8.85	1.01
Ours /w **position** control	0.122	10.1	215	4.30	**0.300**
Ours /w **affine** control	0.115	11.2	207	3.40	0.317
Ours /w **non-param** control	**0.100**	**8.68**	**204**	**1.76**	0.306

Quantitative Results. We report the results on the *BAIR* dataset in Table 1. We highlight that in terms of RMSED score, our method achieved the highest precision of control (more than ×5 improvement compared to other baselines). In terms of generation quality, with similar level of abstraction of ground truth information (ours: 6 continuous `affine` control parameters, CADDY: 7 discrete action labels), our model outperformed CADDY on all three evaluated metrics by a large margin, demonstrating that our model is of better generation quality. With `non-param` control, our generation quality is comparable to the SAVP+.

Table 2 shows the results on the *Tennis* dataset. In terms of generation quality (LPIPS, FID, FVD), all our adaptations outperformed all other comparing methods. Specifically, in terms of FID score, our model is up to 37% better than the closest baseline ([39]). Further, in terms of control precision, our method achieves the lowest error on ADD and MDR (improvement of 80% & 70% respectively), indicating our method is able to generate consistent players with accurate control. One can see that simple positional control works much better here compared to the *BAIR* dataset. This can be attributed to largely in plane motion of the subject.

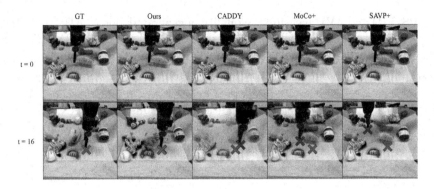

Fig. 4. Qualitative Results On the BAIR Dataset. We labelled the robot arm positions for half of the testing sequences (128 videos out of 256) from both generated videos and ground truth videos. As illustrated on the figure: GT location is marked green and generated locations are marked red. We calculate RMSE of the displacement between GT locations and generated locations to arrive at the RMSED score. (Color figure online)

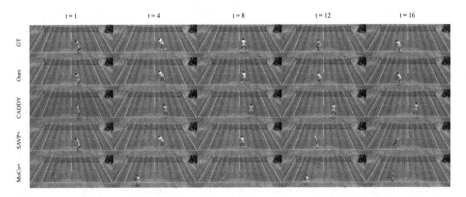

Fig. 5. Qualitative Results On the Tennis dataset.

Qualitative Results. In Fig. 4 and Fig. 5, we show generated sequences on the *BAIR* and *Tennis* dataset (we used our model with `affine` control for both cases shown). In terms of image quality, our method is superior to competitors. In terms of control accuracy, unlike other competing methods, our method is able to precisely place the robot arm and the tennis player in the correct position.

In Fig. 6, we show the results of our model reacting to different user control signals. On the *Tennis* dataset, our method not only moves the player in the correct direction, it's also able to generate plausible motions of the player itself. On the *BAIR* dataset, our model is able to "hullcinate" what's missing in the original frame and generate frames with respect to the control signal, *i.e.*, in the "down, 35 pixels" example, our model successfully generates the upper part of the robot arm, not available on the input frame.

Fig. 6. Effectiveness of Control. Illustrated is how our model precisely reacts to different controlling signals starting from the same initial frame. We illustrate position parameters; however, other affine control parameters are also possible (*e.g.*, scale, rotation and shear).

In Fig. 7, we show that our method is capable of generating and controlling videos with multiple moving objects by simply overlaying two individually controlled mask sequences together, *i.e.*, producing 2 and 3 players in this example. As far as we know, our method is the only video generation method that allows frame-level control of multiple objects acting in the same scene. We provide more visual results in the *Supp. Mat.*

Fig. 7. Control of Multiple Agents. Our method is able to generate videos with multiple moving objects that can be controlled individually by their respective masks.

4.2 Ablation Study

Mask Losses. Here we explore impact of our key design choices have on the quality of generated results and the foreground mask. We show quantitative and qualitative results in Fig. 8. The background loss $\mathcal{L}_{bg}$ enforces the network to generate meaningful masks, without it, the mask network fails to generate a reasonable mask (all zeros). The foreground constraint $\mathcal{L}_{fg}$ shrinks the mask as much as possible. Without this term the network learns a travail solution, where $\mathcal{L}_{bg}$ in Eq. (3) becomes 0 – labeling everything as the foreground (all ones). When computing the foreground loss $\mathcal{L}_{fg}$, we introduce a dynamic mask size prior. We ablate this choice by instead using a fixed global prior of 0.15 as in [8]. Visuals show that if we do not use dynamic prior, the network tends to generate masks with a fixed size, which leads to hollow masks for samples with larger foreground. To prevent information leaking from soft masks, we binarize the masks with thresholding the mask value, without the binary loss $\mathcal{L}_{bin}$, some pixels on the mask fails to pass the threshold and leave some defects on the binarized mask. Overall, the ablations show that all our design choices are important.

One-Stage Training vs. Two-Stage Training. Breaking our training procedure into two stages is a crucial design for the performance of our method. As described in Eq. (12), a well-trained mask generator is a prerequisite for finding the *pseudo user control* $\hat{\theta}$, which we use to introduce controllability to our model. Nevertheless, we still experimented with training the model with one single shot (training Stage II directly by replacing $\mathbf{f}^t$ with $\mathbf{f}^{t+1}$). This leads to vastly poorer performance during test time (Fig. 8, "single-stage training").

Method	LPIPS ↓	FID ↓	FVD ↓
w/o $\mathcal{L}_{fg}$	0.333	60.1	816
w/o $\mathcal{L}_{bg}$	0.306	97.1	796
w/o $\mathcal{L}_{bin}$	0.222	59.2	398
w/o mask prior	0.208	55.0	**279**
single-stage training	0.608	302.3	6614
full	**0.176**	**29.3**	293

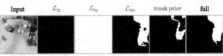

Fig. 8. Ablation of Design Choices. We ablate various loss terms, the use of dynamic mask prior and the two-stage training design. Thumbnails below illustrate the effect these components have on the estimated mask itself; full model producing the most coherent mask.

5 Conclusions

We have introduced layered controllable video generation, an unsupervised method that decomposes frames into foreground and background, with which the user can control the generative process at a frame-level by altering the foreground mask. Our core contributions are the framework itself, and the two-stage training strategy that allows our model to learn to both separate and control on its own. We show that various degrees of control can be implemented with our method, from parametric (position, affine) to complete non-parametric control with the mask. Our results on *BAIR* and *Tennis* datasets show that our method outperforms the state-of-the-art in both quality and control.

Acknowledgements. This work was funded, in part, by the Vector Institute for AI, Canada CIFAR AI Chair, NSERC CRC and an NSERC Discovery and Discovery Accelerator Grants. Resources used in preparing this research were provided, in part, by the Province of Ontario, the Government of Canada through CIFAR, and companies sponsoring the Vector Institute www.vectorinstitute.ai/#partners. Additional hardware support was provided by John R. Evans Leaders Fund CFI grant and Compute Canada under the Resource Allocation Competition awards of the two PIs. We would also like to thank Willi Menapace for his help in answering our questions and helping with fair comparisons to [39].

References

1. Achanta, R., Shaji, A., Smith, K., Lucchi, A., Fua, P.V., Süsstrunk, S.: Slic superpixels compared to state-of-the-art superpixel methods **34**, 2274–2282 (2012)
2. Acharya, D., Huang, Z., Paudel, D.P., Gool, L.V.: Towards high resolution video generation with progressive growing of sliced wasserstein gans. ArXiv preprint (2018)

3. Alayrac, J.B., Carreira, J., Arandjelovic, R., Zisserman, A.: Controllable attention for structured layered video decomposition. In: International Conference on Computing Vision (2019)
4. Alayrac, J.B., Carreira, J., Zisserman, A.: The visual centrifuge: model-free layered video representations. In: IEEE Conference on Computing Vision Pattern Recognition (2019)
5. Alexe, B., Deselaers, T., Ferrari, V.: ClassCut for unsupervised class segmentation. In: Daniilidis, K., Maragos, P., Paragios, N. (eds.) ECCV 2010. LNCS, vol. 6315, pp. 380–393. Springer, Heidelberg (2010). https://doi.org/10.1007/978-3-642-15555-0_28
6. Babaeizadeh, M., Finn, C., Erhan, D., Campbell, R.H., Levine, S.: Stochastic variational video prediction (2018)
7. Bao, H., Dong, L., Wei, F.: Beit: Bert pre-training of image transformers. arXiv preprint arXiv:2106.08254 (2021)
8. Bielski, A., Favaro, P.: Emergence of object segmentation in perturbed generative models. In: Advances in Neural Information Processing Systems (2019)
9. Bouwmans, T., Porikli, F., Höferlin, B., Vacavant, A.: Background Modeling and Foreground Detection for Video Surveillance. CRC Press (2014)
10. Chen, M., et al.: Generative pretraining from pixels. In: International Conference on Machine Learning (ICML), pp. 1691–1703 (2020)
11. Chiappa, S., Racanière, S., Wierstra, D., Mohamed, S.: Recurrent environment simulators. In: International Conference on Learning Representation (2017)
12. Denton, E.L., Fergus, R.: Stochastic video generation with a learned prior. In: International Conference on Machine Learning (2018)
13. Ebert, F., Finn, C., Lee, A.X., Levine, S.: Self-supervised visual planning with temporal skip connections. In: Conference on Robot Learning (2017)
14. Elgammal, A., Harwood, D., Davis, L.: Non-parametric model for background subtraction. In: Vernon, D. (ed.) ECCV 2000. LNCS, vol. 1843, pp. 751–767. Springer, Heidelberg (2000). https://doi.org/10.1007/3-540-45053-X_48
15. Esser, P., Rombach, R., Ommer, B.: Taming transformers for high-resolution image synthesis. In: Conference on Computer Vision and Pattern Recognition (2021)
16. Finn, C., Goodfellow, I., Levine, S.: Unsupervised learning for physical interaction through video prediction. In: Advances in Neural Information Processing Systems (2016)
17. Franceschi, J.Y., Delasalles, E., Chen, M., Lamprier, S., Gallinari, P.: Stochastic latent residual video prediction. In: International Conference on Machine Learning (2020)
18. Frolov, S., Sharma, A., Hees, J., Federico Raue, T.K., Dengel, A.: Attrlostgan: attribute controlled image synthesis from reconfigurable layout and style. In: German Conference on Pattern Recognition (2021)
19. Goodfellow, I., et al.: Generative adversarial nets. In: Advances in Neural Information Processing Systems (2014)
20. Goyette, N., Jodoin, P.M., Porikli, F., Konrad, J., Ishwar, P.: changedetection.net: a new change detection benchmark dataset. In: IEEE Conference on Computer Vision and Pattern Recognition (2012)
21. Hao, Z., Huang, X., Belongie, S.: Controllable video generation with sparse trajectories. In: IEEE Conference on Computer Vision and Pattern Recognition (2018)
22. He, J., Lehrmann, A., Marino, J., Mori, G., Sigal, L.: Probabilistic video generation using holistic attribute control. In: Ferrari, V., Hebert, M., Sminchisescu, C., Weiss, Y. (eds.) ECCV 2018. LNCS, vol. 11209, pp. 466–483. Springer, Cham (2018). https://doi.org/10.1007/978-3-030-01228-1_28

23. Heusel, M., Ramsauer, H., Unterthiner, T., Nessler, B., Hochreiter, S.: GANs trained by a two time-scale update rule converge to a local Nash equilibrium. In: Advances in Neural Information Processing Systems (2017)
24. Hochbaum, D.S., Singh, V.: An efficient algorithm for co-segmentation. In: ICCV, pp. 269–276 (2009)
25. Isola, P., Zhu, J.Y., Zhou, T., Efros, A.A.: Image-to-image translation with conditional adversarial networks (2017)
26. Jaderberg, M., Simonyan, K., Zisserman, A., Kavukcuoglu, K.: Spatial transformer networks. In: Advances in Neural Information Processing Systems (2015)
27. Jiang, W., Sun, W., Tagliasacchi, A., Trulls, E., Yi, K.M.: Linearized multisampling for differentiable image transformation. In: IEEE Conference on Computer Vision and Pattern Recognition (2019)
28. Jojic, N., Frey, B.J.: Learning flexible sprites in video layers. In: IEEE Conference on Computer Vision and Pattern Recognition (2001)
29. Kalchbrenner, N., et al.: Video pixel networks. In: Precup, D., Teh, Y.W. (eds.) Proceedings of the 34th International Conference on Machine Learning. Proceedings of Machine Learning Research, vol. 70, pp. 1771–1779. PMLR, 06–11 August 2017
30. Kim, S.W., Zhou, Y., Philion, J., Torralba, A., Fidler, S.: Learning to simulate dynamic environments with Gamegan. In: IEEE Conference on Computer Vision and Pattern Recognition (2020)
31. Kim, Y., Nam, S., Cho, I., Kim, S.J.: Unsupervised keypoint learning for guiding class-conditional video prediction. In: Advances in Neural Information Processing Systems (2019)
32. Kingma, D.P., Welling, M.: Auto-encoding variational Bayes. In: International Conference on Learning Representations (2013)
33. Kumar, M.P., Torr, P.H., Zisserman, A.: Learning layered motion segmentations of video (2008)
34. Kumar, M., Dumitru Erhan, M.B., Finn, C., Levine, S., Dinh, L., Kingma, D.: Videoflow: a conditional flow-based model for stochastic video generation. In: International Conference on Learning Representations (2020)
35. Kwon, Y.H., Park, M.G.: Predicting future frames using retrospective cycle GAN (2019)
36. Lee, A.X., Zhang, R., Ebert, F., Abbeel, P., Finn, C., Levine, S.: Stochastic adversarial video prediction. ArXiv preprint (2018)
37. Lu, J., Batra, D., Parikh, D., Lee, S.: VilBERT: pretraining task-agnostic visiolinguistic representations for vision-and-language tasks. In: Conference on Neural Information Processing Systems (NeurIPS) (2019)
38. Mathieu, M., Couprie, C., LeCun, Y.: Deep multi-scale video prediction beyond mean square error. In: International Conference on Learning Representations (2016)
39. Menapace, W., Lathuilière, S., Tulyakov, S., Siarohin, A., Ricci, E.: Playable video generation. In: IEEE Conference on Computer Vision and Pattern Recognition (2021)
40. Nawhal, M., Zhai, M., Lehrmann, A., Sigal, L., Mori, G.: Generating videos of zero-shot compositions of actions and objects. In: Vedaldi, A., Bischof, H., Brox, T., Frahm, J.-M. (eds.) ECCV 2020. LNCS, vol. 12357, pp. 382–401. Springer, Cham (2020). https://doi.org/10.1007/978-3-030-58610-2_23
41. Neimark, D., Bar, O., Zohar, M., Asselmann, D.: Video transformer network. ArXiv preprint (2021)

42. Oh, J., Guo, X., Lee, H., Lewis, R.L., Singh, S.: Action-conditional video prediction using deep networks in Atari games. In: Advances in Neural Information Processing Systems (2015)
43. van den Oord, A., Kalchbrenner, N., Kavukcuoglu, K.: Pixel recurrent neural networks. In: International Conference on Learning Representations (2016)
44. van den Oord, A., Vinyals, O., Kavukcuoglu, K.: Neural discrete representation learning. In: Advances in Neural Information Processing Systems (2017)
45. Pan, J., Wang, C., Jia, X., Shao, J., Sheng, L., Yan, J., Wang, X.: Video generation from single semantic label map. In: IEEE Conference on Computer Vision and Pattern Recognition, pp. 3728–3737, June 2019
46. Rakhimov, R., Volkhonskiy, D., Artemov, A., Zorin, D., Burnaev, E.: Latent video transformer. In: International Joint Conference on Computer Vision, Imaging and Computer Graphics Theory and Applications (2021)
47. Ren, S., He, K., Girshick, R.B., Sun, J.: Faster R-CNN: towards real-time object detection with region proposal networks. IEEE Trans. Pattern Anal. Mach. Intell. **39**, 1137–1149 (2015)
48. Simonyan, K., Zisserman, A.: Very deep convolutional networks for large-scale image recognition. CoRR abs/1409.1556 (2015)
49. Smirnov, D., Gharbi, M., Fisher, M., Guizilini, V., Efros, A.A., Solomon, J.: Marionette: self-supervised sprite learning. In: Advances in Neural Information Processing Systems (2021)
50. Stauffer, C., Grimson, W.: Adaptive background mixture models for real-time tracking. In: IEEE Conference on Computer Vision and Pattern Recognition (1999)
51. Su, W., Zhu, X., Cao, Y., Li, B., Lu, L., Wei, F., Dai, J.: VL-BERT: pre-training of generic visual-linguistic representations. In: International Conference on Learning Representations (ICLR) (2020)
52. Tulyakov, S., Liu, M.Y., Yang, X., Kautz, J.: MoCoGAN: decomposing motion and content for video generation. In: IEEE Conference on Computer Vision and Pattern Recognition (2018)
53. Unterthiner, T., van Steenkiste, S., Kurach, K., Marinier, R., Michalski, M., Gelly, S.: Towards accurate generative models of video: a new metric & challenges. ArXiv preprint (2018)
54. Villegas, R., Yang, J., Zou, Y., Sohn, S., Lin, X., Lee, H.: Learning to generate long-term future via hierarchical prediction. In: International Conference on Machine Learning (2017)
55. Vondrick, C., Pirsiavash, H., Torralba, A.: Generating videos with scene dynamics. In: Advances in Neural Information Processing Systems (2016)
56. Vondrick, C., Pirsiavash, H., Torralba, A.: Anticipating the future by watching unlabeled video. In: International Conference on Learning Representations (2017)
57. Vondrick, C., Torralba, A.: Generating the future with adversarial transformers. In: IEEE Conference on Computer Vision and Pattern Recognition (2017)
58. Walker, J., Marino, K., Gupta, A., Hebert, M.: The pose knows: video forecasting by generating pose futures. In: International Conference on Computer Vision (2017)
59. Wang, J.Y., Adelson, E.H.: Representing Moving Images with Layers (1994)
60. Wang, T.C., et al.: Video-to-video synthesis. In: Advances in Neural Information Processing Systems (2019)
61. Wang, Y., Bilinski, P.T., Brémond, F., Dantcheva, A.: G3an: disentangling appearance and motion for video generation (2020)

62. Wang, Y., Bilinski, P.T., Brémond, F., Dantcheva, A.: Imaginator: conditional spatio-temporal GAN for video generation. In: IEEE Winter Conference on Application Computer Vision (2020)
63. Webster, M.A.: Color vision: appearance is a many-layered thing. In: Current Biology (2009)
64. Weissenborn, D., Tackstrom, O., Uszkoreit, J.: Scaling autoregressive video models. In: International Conference on Learning Representations (2020)
65. Wichers, N., Villegas, R., Erhan, D., Lee, H.: Hierarchical long-term video prediction without supervision. In: ICML (2018)
66. Xue, T., Wu, J., Bouman, K.L., Freeman, W.T.: Visual dynamics: stochastic future generation via layered cross convolutional networks. IEEE Trans. Pattern Anal. Mach. Intell. **41**, 2236–2250 (2019)
67. Yan, W., Zhang, Y., Abbeel, P., Srinivas, A.: VideoGPT: video generation using vq-vae and transformers. ArXiv preprint (2021)
68. Yang, C., Wang, Z., Zhu, X., Huang, C., Shi, J., Lin, D.: Pose guided human video generation. In: Ferrari, V., Hebert, M., Sminchisescu, C., Weiss, Y. (eds.) ECCV 2018. LNCS, vol. 11214, pp. 204–219. Springer, Cham (2018). https://doi.org/10.1007/978-3-030-01249-6_13
69. Zablotskaia, P., Siarohin, A., Zhao, B., Sigal, L.: Dwnet: dense warp-based network for pose-guided human video generation. In: British Machine Vision Conference (2019)
70. Zhang, H., et al.: StackGAN: text to photo-realistic image synthesis with stacked generative adversarial networks. In: International Conference on Machine Learning (2017)
71. Zhang, R., Isola, P., Efros, A.A., Shechtman, E., Wang, O.: The unreasonable effectiveness of deep features as a perceptual metric. In: IEEE Conference on Computer Vision and Pattern Recognition (2018)
72. Zhao, B., Meng, L., Yin, W., Sigal, L.: Image generation from layout. In: IEEE Conference on Computer Vision and Pattern Recognition (2019)

Custom Structure Preservation in Face Aging

Guillermo Gomez-Trenado[1]($\boxtimes$)(iD), Stéphane Lathuilière[2](iD), Pablo Mesejo[1], and Óscar Cordón[1]

[1] DaSCI Research Institute, DECSAI, University of Granada, Granada, Spain
{guillermogomez,pmesejo,ocordon}@ugr.es
[2] LTCI, Télécom-Paris, Intitute Polytechnique de Paris, Palaiseau, France
stephane.lathuiliere@telecom-paris.fr

Abstract. In this work, we propose a novel architecture for face age editing that can produce structural modifications while maintaining relevant details present in the original image. We disentangle the style and content of the input image and propose a new decoder network that adopts a style-based strategy to combine the style and content representations of the input image while conditioning the output on the target age. We go beyond existing aging methods allowing users to adjust the degree of structure preservation in the input image during inference. To this purpose, we introduce a masking mechanism, the CUstom Structure Preservation module, that distinguishes relevant regions in the input image from those that should be discarded. CUSP requires no additional supervision. Finally, our quantitative and qualitative analysis which include a user study, show that our method outperforms prior art and demonstrates the effectiveness of our strategy regarding image editing and adjustable structure preservation.

Keywords: Face aging · Image editing · Style-base architecture

1 Introduction

Face age editing [7,17,39], or aging, consists in automatically modifying an input face image to alter the age of the depicted person while preserving identity. Over the last few years, this problem has attracted a growing interest because of its numerous applications. In particular, it is used in the movie production industry to edit actors' faces or in forensic facial approximation to reconstruct the faces of missing people. The advances in deep learning methods unlock the development of fully automatic edition algorithms that avoid hours of makeup and post-production retouching.

Recent deep learning approaches adopt an encoder-decoder architecture [3,8,23, 26,39,40,43,45]. The image is encoded in a latent space that can be modified depending on the target age and fed to a decoder that generates the output image. The overall network is usually trained using a combination of losses that assess image quality, identity preservation, and age matching. However, despite the success of all these approaches,

Supplementary Information The online version contains supplementary material available at https://doi.org/10.1007/978-3-031-19787-1_32.

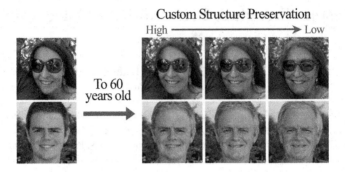

Fig. 1. The user can choose the degree of structure preservation at inference time. Facial morphology transformations are more profound as we move to the right (lower structure preservation).

face editing remains challenging, and current methods usually fail when faced with sizeable differences between the age of the person displayed in the input image and the target age. Indeed, most approaches [3,8,39,40,43,45] only superficially modify the skin's texture while the face's shape is kept unchanged. These approaches fail with significant age gaps since face shape can change significantly during a lifetime. Few methods try to go beyond some limited age gaps, but they either consider only a tightly cropped face region [17,39] or require specific pre-processing involving an image segmentation step [26].

This work proposes a novel framework that allows profound structural changes in facial transformations. This framework achieves a realistic image transformation with age gaps that imply changes in head shape or hair growth. In addition, we argue that the face editing task is an ill-posed problem because every person gets older in a different and non-deterministic way: some people drastically change, while others are easily recognizable in old photographs. In this sense, we propose a methodology that allows the user to adjust, at inference time, the degree of structure preservation. Thus, the user can provide an image and obtain different transformations where the structure (*i.e.*, face shape or hair growth) is preserved at different levels. Figure 1 shows some qualitative results obtained with our method. Furthermore, the user can choose different degrees of structure preservation: with high preservation, the model only changes the texture, while with lower preservation, the shape of the face is also modified.

The contributions of this paper can be summarized as follows:

– We propose a novel architecture for face age editing that can produce structural modifications in the input image while maintaining relevant details present in the original image. We take advantage of recent advances in image-to-image (I2I) translation [10,20] and unconditional image generation [12] to design our architecture. We disentangle the style and content of the input image, and we propose a new decoder network that adopts a style-based strategy to combine the style and content representations of the input image while conditioning the output on the target age.
– We go beyond existing aging methods allowing the user to adjust the degree of structure preservation in the input image at inference time. To this aim, we introduce a

masking mechanism, through a so-called CUstom Structure Preservation (CUSP) module, that identifies the relevant regions in the input image that should be preserved and those where details are irrelevant to the task. Importantly, our mechanism for adjustable structural preservation does not require additional training supervision.

- Experimentally, we show that our method outperforms existing approaches in three publicly available high-resolution datasets and demonstrate the effectiveness of our mechanism for adjusting structure preservation.[1]

2 Related Work

Most recent approaches for **face aging** adopt a similar strategy based on an encoder-decoder architecture [3,8,23,26,39,40,43,45]. In these methods, the input image is projected onto a latent space where content is manipulated before decoding the output image. Some methods [2,40] add an identity term to the total loss to better ensure the preservation of the identity during the translation process. These methods principally differ in the choice of the network architecture and the manner the latent representation is manipulated. For instance, Wang *et al.* [39] introduce a recurrent neural network to iteratively alter the image, while in [43], the latent image representation is modified using a simple affine transformation. Re-AgingGAN [23] employs an age modulator that outputs transformations that are applied then to the decoder, and Or-El *et al.* [26] adopt a multi-domain translation formulation, showing that segmentation information can be leveraged to improve aging. In our work, we adopt an encoder-decoder framework similar to [8,43]. However, our approach goes beyond existing methods that generate a single image for a given image-target age pair. Indeed, we offer the user the possibility to adjust the degree of structure preservation during translation, and, in this way, we can output a set of plausible resulting facial images.

Our method also leverages recent advances from the **I2I translation** research area. I2I translation consists in learning a mapping between two visual domains. In the pioneering work of Isola *et al.* [11], an encoder-decoder network is trained using a dataset composed of image pairs from the two domains. Later, many works addressed I2I translation in an unpaired setting, assuming two independent sets of images of each domain [6,21,46]. These works, of which cycleGAN [46] is a paradigmatic example, mainly focus on introducing regularization mechanisms when training the I2I translation models. Another research direction is designing more advanced architectures to improve image quality or obtain several possible outputs for a given input [10,20,47]. Disentangling style and content information has led to both higher image quality and diversity [10,28]. We adopt a similar strategy in order to allow custom structure preservation. Thanks to this strategy, our CUPS module can act on the spatial information passing through the content branch while preserving style information.

Style-based architectures recently attracted much attention for the problem of unconditional image generation. In particular, StyleGAN2 [12] is now used in many face manipulation tasks [30,42]. In the case of face aging, [2] uses a pretrained Style-GAN2 model [12] equipped with a pSp encoder [30], and an age classifier [32] to tailor

---

[1] Code and pretrained models are available at https://github.com/guillermogotre/CUSP.

an age editing model with unlabeled data. In StyleGAN2, a network maps a Gaussian latent space onto style vectors; these vectors are later combined via a convolutional network to produce the output image. Finally, the synthesis network aggregates the style vectors through modulation operations. We take inspiration from the StyleGAN2 generator to design a novel decoder that combines the input style and the target age with the content representation via weight demodulation.

Regarding the more general **image editing** problem, our method shares similarities with several approaches employing masking mechanisms or attention maps to preserve relevant parts in the input image [1,18,29,38]. For instance, mask consistency is employed in [18] to improve multi-domain translations. As in our approach, masks are estimated using the guided backpropagation (GB) algorithm [36]. In the case of facial images, a mask is employed in GANimation [29] to different regions that should be preserved and those that should be modified to change the facial expression. In GANimation, masks are predicted by the main network, while we employ an auxiliary network and GB [36] to obtain the mask.

3   Proposed Method

In this work, we address the face age editing problem. Therefore, our goal is to train a network able to transform an input image $\mathbf{X}$, such that the person depicted looks like being of the target age a_t. At training time, we assume that we have at our disposal a dataset composed of I face images of resolution $H \times W$, such that $\mathbf{X}_i \in \mathbb{R}^{H \times W \times 3}, i = 1, ..., I$ with their corresponding age label $a_i \in \{1, ..N\}$. Note that the age labels are automatically obtained using a pre-trained age classifier. Similar to previous approaches [43,45], we employ the DEX classifier [33].

One of the main difficulties lies in modifying the relevant details in the input image while preserving non-age-related regions. To this aim, we introduce a style-based architecture detailed in Sect.3.1. In contrast to previous works, the CUSP module allows the user to indicate the desired level of structure preservation through two parameters: $\sigma_m > 0$ and $\sigma_g > 0$. These parameters act locally and globally, respectively, as detailed later. The proposed CUSP module is described in Sect. 3.2. Finally, we present the whole training procedure in Sect. 3.3.

3.1   Style-based Encoder-Decoder

As illustrated in Fig. 2, our architecture employs five different networks: (1) A style encoder E_s extracts a style representation s_i of the input image $\mathbf{X}_i$. E_s discards any spatial information via global-average-pooling at the last layer. The use of a style encoder allows global information to be used at any location in the decoder. (2) A content encoder E_c outputs a tensor $\mathbf{c}$ describing the content of the input image. Contrary to E_s, the content encoder preserves spatial and local information. In our case, the use of separated style and content encoders is justified by the fact that our CUSP module should not affect the image style s_i but only the structure of the image. (3) An 8-layers fully connected network, E_a, embeds the target age a_t: $\tilde{a}_t = E_a(a_t)$. (4) An image

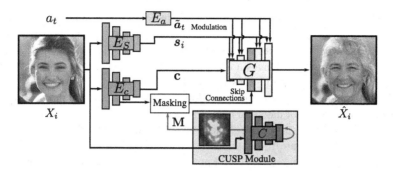

Fig. 2. Illustration of the proposed approach. A style encoder E_s extracts a style representation of the input image $\mathbf{X}_i$. A content encoder E_c encodes spatial information. Target age a_t is embedded using a multi-layer perceptron E_a. Our generator G outputs the image $\hat{\mathbf{X}}_i$ by combining the input style and content representations conditioned on the target age. Our CUSP module predicts a blurring mask $\mathbf{M}$ applied to the skip connections to allow the user to choose a CUstom level of structure preservation.

generator G estimates the output image $\hat{\mathbf{X}}_i$ by combining the style and content representations with the target age embedding $\tilde{a}_t$. (5) Finally, our CUSP module allows the user to choose the level of structure preservation. This module predicts a mask $\mathbf{M}$ used to act on the skip connections between the content encoder and the decoder. More precisely, we blur the regions indicated by the mask $\mathbf{M}$ to propagate only the non-age-related structural information to the decoder.

Our image generator G is designed to combine the outputs of the style and content encoders with the target age embedding. Its architecture is inspired by StyleGAN2 [16], which achieves state-of-the-art performance in unconditional image generation. However, we provide several modifications to tailor the architecture to the aging task. G comprises a sequence of elementary blocks (see *Supplementary Materials* for illustration). Differently from StyleGAN2, each block takes three inputs: the former block output feature map, the style encoding s_i, and the class embedding $\tilde{a}_t$. Each block is composed of two sub-blocks. In the first one, we use the style vector s_t to modulate the convolution operations as in [16]. In the second one, the age embedding is used for modulation. Up-sampling is applied to the input of the first sub-block. Similarly to [16], random noise is summed to the feature maps between each sub-block, while scaling and bias parameters (*i.e.*, w and b) are learned for each sub-block.

Note that all blocks are combined following the *input skips* architecture of StyleGAN2, where a layer named *tRGB* is introduced. Such layer predicts intermediate images at every resolution scaled and added to generate the final image. tRGB is also conditioned on the age embedding. A single skip connection is introduced before the last block, contrarily to U-Net [31] that includes them in every layer.

3.2 CUSP Module

Skip connections (SC) [31] are efficient tools to provide high-frequency information to the decoder allowing accurate reconstruction [11]. High frequencies carry accurate spatial information that favors pixel-to-pixel alignment between inputs and outputs, as, for instance, needed in segmentation. However, previous works [35] show they are not suited for tasks where the input and output images are not pixel-to-pixel aligned. For example, input and output images are aligned when the age gap is small in the aging task. However, this assumption does not hold in every image region with significant gaps. This misalignment is particularly predominant in areas other than the background since facial morphology or hairstyle may change.

Therefore, we propose to control the amount of structural information that flows through the SC. This control is obtained by blurring the feature maps going through them. Nevertheless, every region should not be treated in the same way. For instance, depending on the task, the user may prefer to preserve the background while blurring the foreground to loosen conditioning on the input image in this region. Therefore, we propose a specific mechanism to identify relevant image regions for the translation.

Mask Estimation. We employ an additional classification network C, pretrained to recognize the age of the person depicted on an image. We use the DEX classifier [33] again. Since DEX is pretrained on 224×224, the input image is rescaled to this resolution. Then, we apply the GB algorithm [36] to obtain a tensor $\mathbf{B} \in \mathbb{R}^{224 \times 224 \times 3}$, where locations with higher norm correspond to regions predominantly used by the classifier. In other words, $\mathbf{B}$ pinpoints relevant regions for the age classification task. GB points out the key areas to recognize the age and should, therefore, be modified by the aging network. Importantly, GB is usually used to visualize the regions that influence one specific network output (*i.e.*, one specific class) [36]. In our case, we apply GB to the sum of the classification layer before softmax normalization to obtain class-independent masks. We select GB [36] over other approaches [25, 34, 37] since it is a fast, simple, and strongly supported method for visualization.

We need to transform $\mathbf{B}$ to obtain a mask $\mathbf{M} \in [0, 1]^{224 \times 224}$. We proceed in several steps. First, we average $\mathbf{B}$ over the RGB channels, take the absolute value, and apply Gaussian blur to get smoother maps. In this way, we obtain a tensor $\tilde{\mathbf{B}} \in \mathbb{R}_{>0}^{224 \times 224}$ that indicates relevant regions. To obtain values in $[0, 1]$, we need to normalize $\tilde{\mathbf{B}}$. Our preliminary experiments showed that after normalizing by twice the variance σ of $\tilde{\mathbf{B}}$ (over the locations), relevant areas for the aging task are close to 1 or above. We apply clipping to bring all those important regions to 1. Formally the mask values are computed as follows:

$$\mathbf{M} = \min\left(\frac{\tilde{\mathbf{B}}}{2 \times \sigma}, 1\right) \tag{1}$$

where *min* denotes the element-wise minimum. Next, we detail how this mask is employed in our encoder-decoder architecture.

Skip Connection Blurring. Assuming a feature map $\mathbf{F}_c \in \mathbb{R}^{H' \times W' \times C}$ provided by the content encoder E_c, we resize $\mathbf{M}$ to the dimension of $\mathbf{F}_c$ obtaining a mask $\tilde{\mathbf{M}} \in [0, 1]^{H' \times W'}$. We then blur $\mathbf{F}_c$ using two different Gaussian kernels with variance $\sigma_m > 0$ and $\sigma_g > 0$. The variance σ_m is applied in the region indicated by $\mathbf{M}$, while σ_g is used over the whole feature map. The motivation for this choice is that the user can choose to alter structure preservation locally, globally, or both. At training time, σ_m and σ_g are sampled randomly to force the generator G to perform well for any blur parameter. At test time, both values might be provided by the user. Formally, the blurred feature map is computed as follows:

$$\tilde{\mathbf{F}}_c = \tilde{\mathbf{M}} \circ (\mathbf{F}_c * \mathbf{k}_m) + (1 - \tilde{\mathbf{M}}) \circ (\mathbf{F}_c * \mathbf{k}_g) \tag{2}$$

where $*$ denotes the convolution operation, $\circ$ is the Hadamard product, and $\mathbf{k}_m$ and $\mathbf{k}_g$ are the Gaussian kernels of variances σ_m and σ_g.

3.3 Overall Training Procedure

Training facial age editing models is particularly challenging since paired images are unavailable. Therefore, similarly to [23,26,43], our training strategy is either focused on reconstruction (when the target age matches the input age) or I2I translation (when the target age is different). Also, similar to [23,26,43], training is performed using a set of complementary losses described below.

Reconstruction Loss ($\mathcal{L}_r$). When the target age a_t is equal to the image age a_i, we expect to reconstruct the input image. We, therefore, adopt an L1 reconstruction loss:

$$\mathcal{L}_r = \|T(\mathbf{X}_i, a_i) - \mathbf{X}_i\|_1 \tag{3}$$

where T denotes the whole aging network, which output is the scaled addition of every *tRGB* block.

Age Fidelity Losses ($\mathcal{L}_D, \mathcal{L}_C$). Following [5], we use a conditional discriminator D to asses that generated images correspond to the target age a_t. More precisely, we employ the discriminator architecture of StyleGAN2 equipped with a multiclass prediction head, together with the training loss $\mathcal{L}_D$ defined in [24].

We employ a loss $\mathcal{L}_C$ that assesses age matching using the same pretrained classifier C used in the CUSP module to complement the adversarial loss. Furthermore, $\mathcal{L}_C$ is implemented using the Mean-Variance loss [27], a classification loss tailored for age estimation.

Cycle-Consistency Loss ($\mathcal{L}_{cy}$). Following [46], we adopt a cycle consistency $\mathcal{L}_{cy}$ to force the network to preserve details that are not specific to the age (*e.g.,* , background or face identity). $\mathcal{L}_{cy}$ is given by:

$$\mathcal{L}_{cy} = \|\mathbf{X}_i - T(T(\mathbf{X}_i, a_t), a_i)\|_1 \tag{4}$$

Full Objective. Finally, the total cost function can be written

$$\min_{M} \max_{D} \lambda_r \mathcal{L}_r + \lambda_C \mathcal{L}_C + \lambda_D \mathcal{L}_D + \lambda_{cy} \mathcal{L}_{cy} \tag{5}$$

where $\lambda_r, \lambda_C, \lambda_D,$ and λ_{cy} are constant weights.

4   Experiments

4.1   Evaluation Protocol and Implementation

Every paper employs different metrics, datasets, and tasks in the aging literature. Therefore, we include a large set of metrics, datasets, and tasks in our experiments to allow
comparison with most existing methods.

Datasets. In this paper, we employ three widely-used, publicly available high-
resolution datasets for face aging and analysis:

- *FFHQ-RR*: Initially proposed in [43], this aging dataset based on FFHQ [15] comprises of 48K images depicting people from 20 to 69 years old. Because of this
  Restricted age Range, we refer to this dataset as *FFHQ-RR*. Images are downsampled to 224 × 224.
- *FFHQ-LS*: Tis aging dataset, introduced in [26], is composed of the 70K images
  from FFHQ [15], manually labeled in 10 age clusters that try to capture both geometric and appearance changes throughout a person's life: 0–2, 3–6, 7–9, 10–14,
  15–19, 20–29, 30–39, 40–49, 50–69 and 70+ years old. Consequently, this dataset is
  referred to as *FFHQ-LS* because of its *LifeSpan* age range. The resolution of these
  images is 256 × 256 pixels.
- *CelebA-HQ* [13,22]: It consists of 30K images at 1024 × 1024 resolution, which we
  downsample to 224 × 224 pixels. The only age-related label in the dataset is *young*,
  which can be either true or false.

The use of *FFHQ-RR* and *FFHQ-LS* may seem redundant since they are both based
on the FFHQ dataset, but we perform distinct experiments on both datasets to allow
comparison with existing state-of-the-art methods (which report results on at least one
of them).

Tasks. We employ two tasks to evaluate the performance:

- *Young → Old*: as in [43], we sample 1000 images belonging to the "young" category
  and translate them to a target age of 60. This task is only performed on CelebA-HQ.
- *Age group comparison*: similarly to [23], we consider different age groups: (20–29),
  (30–39), (40–49), and (50–69) on *FFHQ-RR* and additionally (0–2), (3–6), (7,9),
  (15,19) on *FFHQ-LS*. We again sample the first 1000 test images and translate every
  one of them into the central age of each of the four different age groups (25, 35, 45,
  and 55, respectively).

Metrics. We choose metrics to evaluate the two main aspects of the aging task. Firstly,
the translated/generated images must preserve the content of the input image in terms
of identity, facial expression, and background. Secondly, the age translation might be
accurate. In particular, we adopt the following metrics:

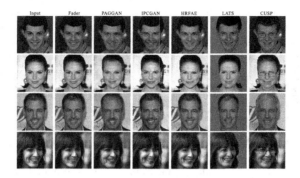

Fig. 3. Comparison with state-of-the-art on CelebA-HQ for the *Young → Old* task employing a target age of 60 years old.

- *LPIPS* [44] measures the perceptual similarity when the target age coincides with the input image age.
- *Age Mean Absolute Error (MAE)*. We employ a pretrained and independent age estimation network to compare the predicted age with the target age given an input image. As we already use the DEX pretrained classifier [33] at training time, we utilize Face++ API[2]. Experiments show that DEX is more biased towards younger age predictions than Face++. Therefore, reporting the MAE to the input target age a_t would be biased. To compensate for this DEX-Face++ misalignment, we estimate the age of the original images with Face++ and compute the mean for each group. We then report the distance between the mean group predicted age and the transformed image predicted age.
- *Kernel-Inception Distance* [4] *(KID)* assesses that the generated images are similar to real ones for similar ages. While FID [9] is adopted in [23], we adopted KID as it is better suited for smaller datasets. We report the KID between original and generated images within the same age groups.
- *Gender*, *Smile*, and *Face expression* preservation and *Blurriness*: Face++ provides these metrics to evaluate input image preservation and quality. *Gender*, *Smile*, and *Face expression* preservation are reported in percentages as in [43].

Implementation Details. We use the same training settings as StyleGAN2-ADA [14] with $\lambda_r = 10$, $\lambda_C = 0.06$, $\lambda_D = 1$, $\lambda_{cy} = 10$. The optimizer used is Adam with $lr = 0.0025$ and $\beta_1 = 0$, $\beta_2 = 0.99$. FFHQ-RR and CelebA-HQ models are trained for 65 epochs with a batch size of 18. FFHQ-LS is trained for 140 epochs with a batch size of 16. All experiments are run on a single Nvidia A100 GPU.

4.2   Comparison with State-of-the-Art

From our literature review (Sect. 2), we identify HRFAE [43] and LATS [26] as the two main competing methods. Indeed, Re-aging GAN [23] cannot be included in the

---

[2] Face++ Face detection API: https://www.faceplusplus.com/ (last visited on September 25, 2022).

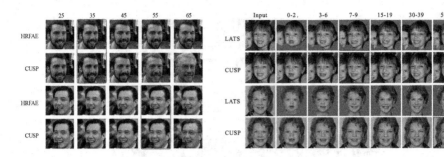

Fig. 4. Qualitative comparison with HRFAE. The images corresponding to the input ages are highlighted with red frames. (Color figure online)

Fig. 5. LATS comparison for different age targets. The images corresponding to the input ages are highlighted with red frames. (Color figure online)

Table 1. User study on four different aspects of image aging comparing CUSP.

	Age accuracy			Identity preservation			Overall quality			Natural progression
	20–29	50–69	Added	20–29	50–69	Added	20–29	50–69	Added	–
CUSP	**60.2**	**72.9**	**66.6**	**50.8**	**63.7**	**57.3**	**55.8**	**67.7**	**61.8**	**60.6**
HRFAE [43]	17.5	15.6	16.6	24.4	24.0	24.2	21.7	20.6	21.1	24.9
LATS [26]	22.3	11.5	16.9	24.8	12.3	18.5	22.5	11.7	17.1	14.5

comparison since neither the code nor the age classifier used for evaluation is publicly available. Since HRFAE and LATS report experiments on different datasets and follow different protocols, we perform experiments using the two tasks previously described. First, we follow HRFAE [43], which employs the *Young → Old* task on *CelebA-HQ*. In this case, the performance of FaderNet [19], PAG-GAN [41], IPC-GAN [40], and HRFAE (on 1024×1024 resolution images) is reported in [43] and is included in our experimental comparison. Second, we employ the *age group comparison* task to allow better comparison with LATS [26] on the most challenging *FFHQ-LS* dataset. Indeed, since no automatic quantitative evaluation is reported on the *FFHQ-LS* in [26], we chose the *age group comparison* task that provides richer analysis than the *Young → Old* task.

User Study. We conducted a study on 80 different users comparing CUPSP with HRFAE [43] and LATS [26]) on the young-to-old and old-to-young tasks on FFHQ-RR. Similarly to [26], we asked about user preferences regarding identity preservation, target age accuracy, realism, the naturalness of the age transition, and overall preference.

As seen in Table 1, CUSP outperforms HRFAE [43] and LATS [26] in every single category by a large margin (CUSP was selected globally in 62% of cases, compared to 22% and 17%, respectively). Furthermore, CUSP's results depict people of the target age with greater accuracy while maintaining the source image identity. On top of that, it outputs higher quality images, and the progression seems more natural and realistic.

Qualitative Comparison. In Fig. 3, we show a qualitative comparison with the state-of-the-art evaluated on the *celebA-HQ* dataset, where we transform the input image to

Table 2. Quantitative comparison on *CelebA-HQ* for the *Young* → *Old* task employing a target age of 60. CUSP HP (High preservation) is run with $\sigma_m = \sigma_g = 1.8$.

Method	Predicted age	Blur	Gender	Smiling	Neutral	Happy
Real images	*68.23 ± 6.54*	2.40	–	–	–	–
FaderNet	44.34 ± 11.40	9.15	97.60	95.20	90.60	92.40
PAGGAN	49.07 ± 11.22	3.68	95.10	93.10	90.20	91.70
IPCGAN	49.72 ± 10.95	9.73	96.70	93.60	89.50	91.10
HRFAE	54.77 ± 8.40	**2.15**	97.10	**96.30**	**91.30**	**92.70**
HRFAE-224	51.87 ± 9.59	5.49	**97.30**	95.50	88.30	92.50
LATS	55.33 ± 9.33	4.77	96.55	92.70	83.77	88.64
CUSP HP	**67.76 ± 5.38**	2.53	93.20	88.70	79.80	84.60

Table 3. Quantitative comparison with LATS on the FFHQ-LS dataset for the *age group comparison* task. CUSP CP (Custom preservation) and LP (Low preservation) are run with $(\sigma_m, \sigma_g) = (8, 4.5)$ and $(\sigma_m, \sigma_g) = (8, 8)$ respectively.

	Age MAE							Gender preservation (%)						
	0–2	3–6	7–9	15–19	30–39	50–69	Mean	0–2	3–6	7–9	15–19	30–39	50–69	Mean
LATS	7.68	8.91	6.59	5.19	**8.23**	**5.73**	7.05	72.2	70.6	74.2	**93.7**	**93.9**	**93.9**	**83.1**
CUSP CP	6.89	**8.26**	7.67	6.70	10.67	10.86	8.51	**74.5**	69.3	**78.1**	88.3	92.1	85.9	81.4
CUSP LP	**6.49**	9.29	**5.59**	**4.99**	8.36	5.74	**6.74**	69.0	**76.0**	**78.1**	87.4	86.1	80.1	79.4

the age of 60 years old. First, we observe that Fader, PAGGAN, and IPCGAN generate images with important artifacts. On the contrary, HRFAE, LATS, and our approach generate consistent images with only minor artifacts. However, only CUSP produces images that correspond to the correct target age. Other methods generate images where people look younger than expected since they are unable to make suitable structural changes. Furthermore, LATS operates only in the foreground, requiring a previous masking procedure; for this reason, in Fig. 3, the outputs related to LATS display a constant gray background. In addition, CUSP can preserve identity and non-age-related details.

We also perform a qualitative comparison with the two main competitors: HRFAE on *FFHQ-RR* in Fig. 4 and with LATS on *FFHQ-LS* in Fig. 5. We show that CUSP achieves more profound facial structure modifications (*e.g.*, thin face shapes that grow wider and wrinkled skin) and hair color transformation. The age progression is smooth. Close ages produce almost identical pictures, but global age progression seems realistic and natural. Regarding LATS (Fig. 5), we see that we obtain similar performance while our method has four major advantages: (1) it operates directly on the entire image and deals with backgrounds and clothing; (2) it does not require an externally trained image segmentation network; (3) CUSP employs a single network while LATS uses a separate network for each gender; and (4) it offers user control as shown in our ablation study (see Sect. 4.3).

Quantitative Comparison. In Table 2, we report a quantitative comparison evaluated on the *CelebA-HQ* dataset employing the *Young* → *Old* task. Every model has been trained on *FFHQ-RR*. Regarding HRFAE, we report the performance obtained with models trained and tested at 224×224 and 1024×1024 resolutions (referred to as HRFAE-224 and HRFAE, respectively). We used the available code for LATS to train a model on this dataset. We also report (first row) the mean age predicted by the Face++ classifier when feeding the images of the age class 60 according to the DEX classifier used at training time. We observe an 8.23-year discrepancy. In other words, to generate images that look similar to those labeled as 60 at training time, we need to predict images that the Face++ classifier will perceive on average as 68.23 years old. These experiments confirm that CUSP outperforms other methods, being the only method that substantially modifies the image to adjust the person's target age.

In addition, CUSP ranks second in terms of Blur, quantifying the good quality of our images. For instance, the performance of HRFAE-224 worsens the predicted age with respect to its 1024×1024 counterpart and deteriorates noticeably in the Blur metric, suggesting a severe drop in the generated image quality. Interestingly, the more profound and realistic transformations yielded by CUSP and LATS imply slightly worse scores according to the preservation metrics. Indeed, preservation metrics suffer from the increased ability to make structural changes to pictures. However, this drop in quantitative fidelity is not manifested in the user study or qualitative results (Figs. 5 and 4). Two hypotheses can explain this discrepancy between qualitative and quantitative results. First, several biases can impact the results (*e.g.*, sports clothing is replaced for formal clothes at higher ages, and glasses appear in older targets as well). In addition, there may also be some expression-related biases in different age groups. Second, the CUSP module more frequently targets the image's mouth and eye areas. Those areas are the most related to facial expression detection, and their blurring might negatively affect facial expression preservation.

We report in Table 3 a comparison with LATS, both trained and evaluated on the *FFHQ-LS* dataset. The results support the qualitative analysis performed regarding Figs. 4 and 5. Our proposed method is on par with LATS performance concerning the aging task and achieves those results while preserving numerous image details. CUSP with low preservation even outperforms LATS in terms of Mean Age-MAE. We also notice that our approach obtains similar performance in terms of gender preservation while employing a single model and not using gender annotations as in [26].

4.3   Ablation Study

Architecture Ablation. We consider four variants of our approach where we ablate the skip connections and the style encoder. In (i), the style encoder is not used; an *Average Pooling layer* replaces E_s on top of the output from E_c. *(ii)* employs a style encoder but no skip connections, while *(iii)* employs skip connections in every layer. Finally, *(iv)* follows the proposed architecture employing skip connections in the second-to-last layer only. In order to make an unbiased evaluation of the architecture and not the masking operation performed by CUSP, we report the performance of CUSP with high preservation $(\sigma_m, \sigma_g) = (7.1, 0.0)$, as *(ii)* applies no masking.

Table 4. Ablation study: impact of the skip connections (*SC.*) and the style encoder.

	LPIPS	Age MAE	Mean KID
(i) No style encoder	0.84	6.21	0.0163
(ii) No *SC.*	1.70	**6.17**	0.0109
(iii) SC. at every layer	1.85	6.34	0.0175
(iv) Full	**0.78**	6.29	**0.0089**

Table 5. Ablation study: impact of the masking strategy used in CUSP.

	LPIPS	Age MAE	Mean KID
Top-class GB	1.25	6.19	0.0145
Class-indep. (Ours)	**0.78**	**6.29**	**0.0089**

Results shown in Table 4 suggest that a separate style encoder, as in our *Full* model *(iv)*, yields better reconstruction (lower LPIPS) and similar aging performance (Age MAE and Mean KID) than using a single encoder for both content and style as in *(i)*. Regarding skip connections, not using them leads to an important reconstruction error (see high LPIPS) since the network cannot reconstruct the image details. However, skip connections in every layer also results in low reconstruction performance. We hypothesize that the model faces optimization issues. More specifically, adding skip connections on every layer dramatically increases the decoder's complexity (approximately doubling its number of parameters), making the network slower and harder to train.

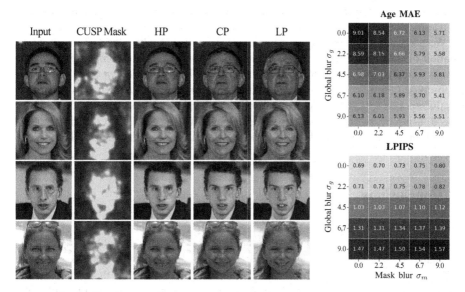

Fig. 6. Impact of the kernel value: images obtained with High, Low, and Custom structure preservation (LP, HP, and CP). HP:$(\sigma_m, \sigma_g) = (0, 0)$; CP:$(\sigma_m, \sigma_g) = (9, 0)$; HP:$(\sigma_m, \sigma_g) = (9, 9)$. The second column shows the mask estimated by CUSP.

Fig. 7. CUSP parameters and impact on Age MAE (left) and LPIPS $\times$ 10 (right).

CUSP Module Analysis. In Fig. 6, we qualitatively evaluate the impact of the kernel values used in CUSP. We compare images obtained with Low, Custom, and High structure preservation (referred to as LP, CP, and HP), where we use kernel values ranging from $\sigma = 0$ to $\sigma = 9$. We also display the mask **M** estimated by the CUSP module. We observe that when the user provides low kernel values (*i.e.*, higher preservation), the shape of the face is kept, while with higher kernel values, the network has the freedom to change its shape. The impact is clearly visible on the neck and chin of the women in the second and last row.

The visualization of the mask shows that our approach identifies those regions that change with age (chin, mouth, and forehead). We also quantitatively measure the impact of each kernel parameter. In Fig. 7, we report the Age MAE and LPIPS while changing the local and global blur parameters. By increasing the local blur, we can see that CUSP achieves a significantly lower age error while keeping a small reconstruction error. On the contrary, using global blur to improve the age performance (*i.e.*, reduce the age MAE) implies a substantial increase in the LPIPS metric, reflecting some loss of details. Overall, these experiments demonstrate the conflicting nature of aging and reconstruction performances. These observations further justify our motivation to offer the user the possibility of controlling this trade-off, thereby demonstrating the value of CUSP and its masking strategy. The ability to modify both σ_m and σ_g with different values allows us to achieve the same age-accurate transformation results while minimizing the reconstruction performance drop.

We complete this analysis with an ablation study regarding the GB-based computation of the CUSP masks. More precisely, two strategies are compared: in *Top-1 class*, we apply GB on the most-activated class, while in *class-independent*, we adopt the proposed strategy of taking the sum of the classification layer before softmax. Results reported in Tab. 5 demonstrate that the class-independent strategy performs best. Indeed, using every class output from the age classifier might benefit the masking, as every age-related feature is relevant for the translation, not only those involving its current age.

5   Conclusions

We present a novel architecture for face age editing that can produce structural facial modifications while preserving relevant details in the original image. Our proposal has two main contributions. First, we propose a style-based strategy to combine the style and content representations of the input image while conditioning the output on the target age. Second, we present a Custom Structure Preservation (CUSP) module that allows users to adjust the degree of structure preservation in the input image at inference time. We validate our approach by comparing six state-of-the-art solutions and employing three datasets. Our results suggest that our method generates more natural-looking, age-accurate transformed images and allows more profound facial changes while adequately preserving identity and modifying only age-related aspects. An extensive user study further confirmed this analysis. We plan to extend CUSP to other image editing tasks in future works.

References

1. Ak, K.E., Lim, J.H., Tham, J.Y., Kassim, A.A.: Attribute manipulation generative adversarial networks for fashion images. In: IEEE/CVF ICCV (2019)
2. Alaluf, Y., Patashnik, O., Cohen-Or, D.: Only a matter of style: age transformation using a style-based regression model. ACM Trans. Graph. **40**(4), 1–12 (2021)
3. Antipov, G., Baccouche, M., Dugelay, J.L.: Face aging with conditional generative adversarial networks. In: IEEE ICIP (2017)
4. Bińkowski, M., Sutherland, D.J., Arbel, M., Gretton, A.: Demystifying mmd gans. arXiv preprint arXiv:1801.01401 (2018)
5. Choi, Y., Choi, M., Kim, M., Ha, J.W., Kim, S., Choo, J.: Stargan: unified generative adversarial networks for multi-domain image-to-image translation. In: IEEE/CVF CVPR (2018)
6. Fu, H., Gong, M., Wang, C., Batmanghelich, K., Zhang, K., Tao, D.: Geometry-consistent generative adversarial networks for one-sided unsupervised domain mapping. In: IEEE/CVF CVPR (2019)
7. Fu, Y., Guo, G., Huang, T.S.: Age synthesis and estimation via faces: a survey. IEEE T-PAMI **32**(11), 1955–1976 (2010)
8. He, Z., Kan, M., Shan, S., Chen, X.: S2gan: share aging factors across ages and share aging trends among individuals. In: IEEE/CVF ICCV (2019)
9. Heusel, M., Ramsauer, H., Unterthiner, T., Nessler, B., Hochreiter, S.: Gans trained by a two time-scale update rule converge to a local nash equilibrium. In: Proceedings of the 31st International Conference on Neural Information Processing Systems, pp. 6629–6640 (2017)
10. Huang, X., Liu, M.Y., Belongie, S., Kautz, J.: Multimodal unsupervised image-to-image translation. In: IEEE/CVF ECCV (2018)
11. Isola, P., Zhu, J.Y., Zhou, T., Efros, A.A.: Image-to-image translation with conditional adversarial networks. In: IEEE/CVF CVPR (2017)
12. Karras, T., Laine, S., Aittala, M., Hellsten, J., Lehtinen, J., Aila, T.: Analyzing and improving the image quality of stylegan. In: IEEE Conference on Computer Vision and Pattern Recognition (2020)
13. Karras, T., Aila, T., Laine, S., Lehtinen, J.: Progressive growing of gans for improved quality, stability, and variation. In: ICLR (2017)
14. Karras, T., Aittala, M., Hellsten, J., Laine, S., Lehtinen, J., Aila, T.: Training generative adversarial networks with limited data. arXiv preprint arXiv:2006.06676 (2020)
15. Karras, T., Laine, S., Aila, T.: A style-based generator architecture for generative adversarial networks. In: IEEE/CVF CVPR (2019)
16. Karras, T., Laine, S., Aittala, M., Hellsten, J., Lehtinen, J., Aila, T.: Analyzing and improving the image quality of stylegan. In: IEEE/CVF CVPR (2020)
17. Kemelmacher-Shlizerman, I., Suwajanakorn, S., Seitz, S.M.: Illumination-aware age progression. In: IEEE/CVF CVPR (2014)
18. Kim, D., Khan, M.A., Choo, J.: Not just compete, but collaborate: local image-to-image translation via cooperative mask prediction. In: IEEE/CVF CVPR (2021)
19. Lample, G., Zeghidour, N., Usunier, N., Bordes, A., Denoyer, L., et al.: Fader networks: manipulating images by sliding attributes. In: Neurips (2017)
20. Lee, H.Y., Tseng, H.Y., Huang, J.B., Singh, M., Yang, M.H.: Diverse image-to-image translation via disentangled representations. In: IEEE/CVF ECCV (2018)
21. Liu, M.Y., Breuel, T., Kautz, J.: Unsupervised image-to-image translation networks. In: Neurips (2017)
22. Liu, Z., Luo, P., Wang, X., Tang, X.: Deep learning face attributes in the wild. In: IEEE/CVF ICCV (2015)
23. Makhmudkhujaev, F., Hong, S., Park, I.K.: Re-aging gan: toward personalized face age transformation. In: IEEE/CVF ICCV (2021)

24. Miyato, T., Koyama, M.: cGANs with projection discriminator. arXiv preprint arXiv:1802.05637 (2018)
25. Muhammad, M.B., Yeasin, M.: Eigen-cam: class activation map using principal components. In: 2020 International Joint Conference on Neural Networks (IJCNN), pp. 1–7. IEEE (2020)
26. Or-El, R., Sengupta, S., Fried, O., Shechtman, E., Kemelmacher-Shlizerman, I.: Lifespan age transformation synthesis. In: IEEE/CVF ECCV (2020)
27. Pan, H., Han, H., Shan, S., Chen, X.: Mean-variance loss for deep age estimation from a face. In: IEEE/CVF CVPR (2018)
28. Park, T., et al.: Swapping autoencoder for deep image manipulation. Adv. Neural Inf. Process. Syst. **33**, 7198–7211 (2020)
29. Pumarola, A., Agudo, A., Martinez, A.M., Sanfeliu, A., Moreno-Noguer, F.: Ganimation: anatomically-aware facial animation from a single image. In: IEEE/CVF ECCV (2018)
30. Richardson, E., et al.: Encoding in style: a StyleGAN encoder for image-to-image translation. In: IEEE Conference on Computer Vision and Pattern Recognition (2021)
31. Ronneberger, O., Fischer, P., Brox, T.: U-net: convolutional networks for biomedical image segmentation. In: Navab, N., Hornegger, J., Wells, W.M., Frangi, A.F. (eds.) MICCAI 2015. LNCS, vol. 9351, pp. 234–241. Springer, Cham (2015). https://doi.org/10.1007/978-3-319-24574-4_28
32. Rothe, R., Timofte, R., Gool, L.V.: Deep expectation of real and apparent age from a single image without facial landmarks. Int. J. Comput. Vis. **126**(2–4), 144–157 (2018)
33. Rothe, R., Timofte, R., Van Gool, L.: Dex: deep expectation of apparent age from a single image. In: IEEE/CVF ICCV-W (2015)
34. Selvaraju, R.R., Cogswell, M., Das, A., Vedantam, R., Parikh, D., Batra, D.: Grad-cam: visual explanations from deep networks via gradient-based localization. In: IEEE/CVF ICCV (2017)
35. Siarohin, A., Sangineto, E., Lathuiliere, S., Sebe, N.: Deformable gans for pose-based human image generation. In: IEEE/CVF CVPR (2018)
36. Springenberg, J.T., Dosovitskiy, A., Brox, T., Riedmiller, M.A.: Striving for simplicity: the all convolutional net. In: 3rd International Conference on Learning Representations, ICLR 2015, San Diego, CA, USA, 7–9 May 2015, Workshop Track Proceedings (2015)
37. Srinivas, S., Fleuret, F.: Full-gradient representation for neural network visualization. Adv. Neural Inf. Process. Syst. **32** (2019)
38. Tang, H., Xu, D., Sebe, N., Yan, Y.: Attention-guided generative adversarial networks for unsupervised image-to-image translation. In: IJCNN (2019)
39. Wang, W., et al.: Recurrent face aging. In: IEEE/CVF CVPR (2016)
40. Wang, Z., Tang, X., Luo, W., Gao, S.: Face aging with identity-preserved conditional generative adversarial networks. In: IEEE/CVF CVPR (2018)
41. Yang, H., Huang, D., Wang, Y., Jain, A.K.: Learning face age progression: a pyramid architecture of gans. In: IEEE/CVF CVPR (2018)
42. Yao, X., Newson, A., Gousseau, Y., Hellier, P.: A latent transformer for disentangled face editing in images and videos. In: IEEE/CVF ICCV (2021)
43. Yao, X., Puy, G., Newson, A., Gousseau, Y., Hellier, P.: High resolution face age editing. In: IEEE ICPR (2021)
44. Zhang, R., Isola, P., Efros, A.A., Shechtman, E., Wang, O.: The unreasonable effectiveness of deep features as a perceptual metric. In: CVPR (2018)
45. Zhang, Z., Song, Y., Qi, H.: Age progression/regression by conditional adversarial autoencoder. In: IEEE/CVF CVPR (2017)
46. Zhu, J.Y., Park, T., Isola, P., Efros, A.A.: Unpaired image-to-image translation using cycle-consistent adversarial networks. In: IEEE/CVF ICCV (2017)
47. Zhu, J.Y., et al.: Multimodal image-to-image translation by enforcing bi-cycle consistency. In: Neurips (2017)

Spatio-Temporal Deformable Attention Network for Video Deblurring

Huicong Zhang[1] , Haozhe Xie[2] , and Hongxun Yao[1(✉)]

[1] Harbin Institute of Technology, Harbin, China
huicongzhang@outlook.com,
h.yao@hit.edu.cn
[2] Tencent AI Lab, Shenzhen, China
haozhexie@tencent.com
https://vilab.hit.edu.cn/projects/stdan/

Abstract. The key success factor of the video deblurring methods is to compensate for the blurry pixels of the mid-frame with the sharp pixels of the adjacent video frames. Therefore, mainstream methods align the adjacent frames based on the estimated optical flows and fuse the alignment frames for restoration. However, these methods sometimes generate unsatisfactory results because they rarely consider the blur levels of pixels, which may introduce blurry pixels from video frames. Actually, not all the pixels in the video frames are sharp and beneficial for deblurring. To address this problem, we propose the spatio-temporal deformable attention network (STDANet) for video delurring, which extracts the information of sharp pixels by considering the pixel-wise blur levels of the video frames. Specifically, STDANet is an encoder-decoder network combined with the motion estimator and spatio-temporal deformable attention (STDA) module, where motion estimator predicts coarse optical flows that are used as base offsets to find the corresponding sharp pixels in STDA module. Experimental results indicate that the proposed STDANet performs favorably against state-of-the-art methods on the GoPro, DVD, and BSD datasets.

Keywords: Video deblurring · Pixel-wise blur levels · Spatio-temporal deformable attention

1 Introduction

In the past few years, hand-held image capturing devices, such as smartphones and action cameras, have been pervasive in our daily life. The camera shake and high-speed movement in dynamic scenes often generate undesirable blur in the video. The blurry video significantly reduces the visual quality and degrades performance in many subsequent vision tasks, including tracking [9,21], video stabilization [20], and SLAM [17]. Therefore, it is extremely attractive to develop an effective method to deblur videos for above mentioned human perception and high-level vision tasks.

© The Author(s), under exclusive license to Springer Nature Switzerland AG 2022
S. Avidan et al. (Eds.): ECCV 2022, LNCS 13676, pp. 581–596, 2022.
https://doi.org/10.1007/978-3-031-19787-1_33

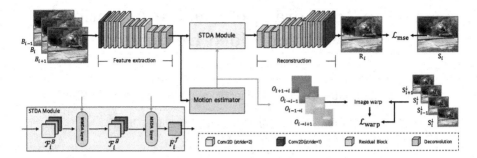

Fig. 1. The overview of STDANet, which takes three adjacent frames as input and restores the sharp mid-frame. Note that $\mathbf{S}^{\downarrow}_{i-1}$, $\mathbf{S}^{\downarrow}_i$, and $\mathbf{S}^{\downarrow}_{i+1}$ are the corresponding downsampled ground truth sharp frames of $\mathbf{S}_{i-1}$, $\mathbf{S}_i$, and $\mathbf{S}_{i+1}$, respectively.

Unlike image deblurring, video deblurring methods exploit additional information in the temporal domain. The key success factor of the video deblurring methods is to compensate for the blurry pixels of the mid-frame with the sharp pixels of the adjacent video frames. Traditional video deblurring methods [1,3,12,38] often model motion blur by optical flow. Then those methods jointly estimate the optical flow and latent frames under the constraints by some hand-crafted priors.

Early deep learning methods [13,24,30,35] directly concatenate the multi-frames features to restore the mid-frame based on the CNN. However, those methods do not take full advantage of the information of the video frames because they explicitly considering the alignment of video frames. The recent mainstream deep learning methods [19,25] align the video frames by optical flows and directly generate the sharp frames by fusing aligned frames. However, they are less effective for the frames whose pixels contain large displacements because they may introduce blurry pixels that are not beneficial for blurring. EDVR [35] computes the pixel-wise similarity in multiple frames and restores the pixels in the mid-frame with high-similarity pixels in the video frames. However, the pixels of high similarity in the adjacent frames are also blurry for the blurry pixels in the mid-frame, which are not beneficial for deblurring.

To solve these issues, we propose spatio-temporal deformable attention network (STDANet), which extracts the information of sharp pixels by considering the pixel-wise blur levels of the video frames. Specifically, STDANet is based on an encoder-decoder network combined with motion estimator and spatio-temporal deformable attention (STDA) module. First, the encoder extracts the multi-features from multiple input frames. Then, the motion estimator predicts coarse optical flows between consecutive video frames given the multi-features generated by the encoder. After that, the estimated optical flows and the extracted features are fed to STDA module to generate the fused features by aggregating the information of the sharp pixels from the extracted multi-features. Different from recent methods [19,25], where the optical flows are used to align the adjacent frames, the optical flows are used as base offsets in the

STDA module, which reduces the degradation of deblurring results caused by inaccurate optical flows. Finally, the decoder restores the sharp mid-frame based on the fused features.

The main contributions are summarized as follows:

- We propose a spatio-temporal deformable attention (STDA) module which aggregates the information of sharp pixels in the input consecutive video frames and eliminates the effects of blurry pixels introduced from input consecutive video frames.
- We present a spatio-temporal deformable attention network (STDANet) equipped with motion estimator and the proposed STDA module, where motion estimator predicts coarse optical flows and provides base offsets to find sharp pixels in adjacent frames.
- We quantitatively and qualitatively evaluate STDANet on the DVD, GoPro, and BSD datasets. The experimental results indicate that STDANet performs favorably against state-of-the-art methods with comparable computational complexity.

2   Related Work

2.1   Single-Image Deblurring

The traditional single image deblurring methods [15,18,22,28,33] assume a uniform blur kernel and design various natural image priors to compensate for the ill-posed blur removal process. However, these methods do not have the ability to handle the non-uniform blur. To solve the non-uniform blur problem, one group of methods [6–8,36,40] extends the degree of freedom of the blur model from uniform to non-uniform in a limited way compared to the dense matrix. Another group of methods [2,10–12] introduces additional segmentations into blur models or adopt motion estimation-based deblurs.

With the development of deep learning, many CNN-based methods are proposed to solve dynamic scene deblurring. Gong [5] adopt a fully-convolutional deep neural network (FCN) to directly estimate the motion flow from the blurry image and restore the unblurred image from the estimated motion flow. Sun [32] use CNN to estimate the motion blur field. With the emergence of large datasets for single image deblurring, several works [16,23,26,34,41] use CNN to directly generate clear images from blurry images in an end-to-end manner. Nah [23] use a multi-scale method for single image deblurring. However, the parameters between each scale are not shared, which leads to a huge amount of parameters. To solve this problem, SRN [34] introduces a deblur network with skip connections where the parameters are shared in each scale. DeblurGAN-v2 [16] uses an end-to-end generative adversarial network (GAN) for single image motion deblurring and introduces the Feature Pyramid Network into single image deblurring. DMPHN [41] introduces the hierarchical multi-patch (MP) model for deblurring and improves deblur performance. MT-RNN [26] uses an RNN with recursive feature maps for progressive deblurring over iterations.

2.2   Multi-image Deblurring

Several methods utilize multiple images to solve dynamic scene deblurring from videos. The traditional methods [1,3,12,38] jointly estimate the optical flow and blur kernel to restored frames with the some hand-crafted priors. However, the proposed priors usually lead to complex energy functions which are difficult to solve. In addition, Su [30] align the consecutive frames and then the Convolutional Neural Networks are used to restored images. Kim [13] propose a recurrent neural network to fuse the concatenation of the multi-frames features. Wieschollek [37] develop a recurrent network to recurrently use the features from the previous frame in multiple scales. Wang [35] achieve better alignment performance base on deformable convolution. Zhou [44] use the dynamic filters to align the consecutive frames. Pan [25] introduce a temporal sharpness prior to improve the ability of the deblur network. Zhang [42] develop a adversarial loss and spatial-temporal 3D convolutions to improve latent frame restoration. Recently, ARVo [19] uses self-attention to capture the pixel correlation of the consecutive frames. However, those methods rarely consider the different blur levels of each frame, which make they do not take full advantage of the sharpness pixel information in the video frames.

3   The Proposed Method

The proposed STDANet aims to restore the sharp mid-frame $\mathbf{R}_i$ given three consecutive blurry frames $\mathcal{B}_i = \{\mathbf{B}_k\}_{k=i-1}^{i+1}$. As shown in Fig. 1, it contains four components: the feature extraction network, the motion estimator, the STDA module, and the reconstruction network, where the feature extraction network and the reconstruction network follow the encoder-decoder architecture. First, the feature extraction network generates the extracted features $\mathcal{F}_i^b = \{\mathbf{F}_k^b\}_{k=i-1}^{i+1}$ for $\mathcal{B}_i$. Then, the motion estimator predicts the optical flows $\mathcal{O}_i = \{\mathbf{O}_{k\to k+1}|k = i-1, i\} \cup \{\mathbf{O}_{k+1\to k}|k = i-1, i\}$ between the two adjacent frames $\mathbf{B}_k$ and $\mathbf{B}_{k+1}$. Next, the STDA module takes $\mathcal{F}_i^b$, $\mathcal{O}_i$ as input and generates the fused features $\mathbf{F}_i^f$ by aggregating the features of low-blur-level pixels in the consecutive frames. Finally, the reconstruction network restores the sharp frame $\mathbf{R}_i$ for $\mathbf{B}_i$. Except STDANet, we also propose STDANet-Stack, which uses a cascaded strategy [25] to stack STDANet and takes five adjacent blurry frames $\{\mathbf{B}_k\}_{k=i-2}^{i+2}$ as input.

3.1   Motion Estimator

Previous video deblurring methods [19,25] that use optical flows to align two adjacent frames to the mid-frame, which requires accurate optical flows generated by heavyweight neural networks such as PWC-Net [31]. In contrast, optical flows are used as the base offsets in the STDA module, which are more robust to the errors in estimated optical flows. Therefore, we propose the motion estimator that predicts coarse optical flows between two adjacent frames with much

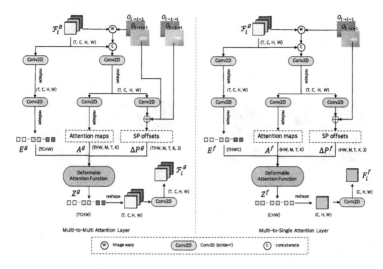

Fig. 2. The detailed network structure of the MMA and MSA layers. Note that "SP Offsets" denotes "the offsets of sampling points".

smaller computational complexity. To accelerate the computational complexity, the motion estimator generates the optical flows that are of $1/4$ sizes the input images. Consequently, the motion estimator is $1/70$ the size of PWC-Net. Compared to existing methods for optical flow estimation [4,31,39], the motion estimator does not use any time-consuming layers such as correlation layer [4], cost volume layer [31,39].

Specifically, the motion estimator consists of stacked four convolutional layers with kernel sizes of 3 and strides of 1. Given the three adjacent image features $\mathcal{F}_i^b$, the motion estimator generates four optical flows $\mathcal{O}_i = \{\mathbf{O}_{k \rightarrow k+1} | k = i - 1, i\} \cup \{\mathbf{O}_{k+1 \rightarrow k} | k = i - 1, i\}$, where $\mathbf{O}_{m \rightarrow n}$ represents the optical flow from the m-th frame to the n-th frame.

3.2 Spatio-temporal Deformable Attention Module

To extract the information of sharp pixels from consecutive video frames, we propose spatio-temporal deformable attention (STDA) module. As shown in Fig. 2, there are two layers in the STDA module that aggregates features in a coarse-to-fine manner, named Multi-to-Multi attention (MMA) layer and Multi-to-Single attention (MSA) layer. Figure 3 gives an illustration how the MMA and MSA layers extract image features of sharp pixels.

Multi-to-Multi attention Layer. The multi-to-multi attention layer takes the image features of three consecutive frames $\mathcal{F}_i^b$ as input and generates the coarse aggregated image features $\mathcal{F}_i^g = \left\{ \mathbf{F}_k^g | \mathbf{F}_k^g \in \mathbb{R}^{C \times H \times W} \right\}_{k=i-1}^{i+1}$, where C, H, and W represent the number of channels, height, and width of the image features, respectively.

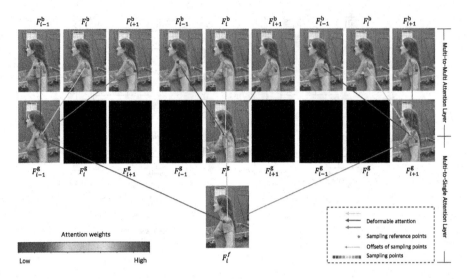

Fig. 3. The illustration of MMA and MSA layers. The colors of the sampling points denotes the corresponding attention weights, where higher attention weights indicate that the sampling points are sharper. First, the MMA layer extracts the information of sharp pixels from multi-features $\mathcal{F}_i^b$ and generates the features of adjacent frames $\mathcal{F}_i^g = \{\mathbf{F}_k^g|\mathbf{F}_k^g\}_{k=i-1}^{i+1}$. Second, the MSA layer generates the fused features $\mathbf{F}_i^f$ by aggregating the information of sharp pixels from $\mathcal{F}_i^g$. (Color figure online)

Step 1. The image features $\mathcal{F}_i^b = \{\mathbf{F}_k^b|\mathbf{F}_k^b \in \mathbb{R}^{C \times H \times W}\}_{k=i-1}^{i+1}$ of adjacent frames are aligned to the mid-frame with the estimated optical flows $\mathcal{O}_i$ and produces $\mathcal{F}_i^w = \{\mathbf{F}_k^w|\mathbf{F}_k^w \in \mathbb{R}^{C \times H \times W}\}$, where $\mathbf{F}_k^w$ is

$$\mathbf{F}_k^w = \text{Warp}(\mathbf{F}_k^b, \mathbf{O}_{i \to k}), k = i - 1, i + 1 \tag{1}$$

where "Warp" denotes the backward warp with operation.

Step 2. The concatenated features $\mathbf{F}_i^c \in \mathbb{R}^{(2T-1) \times C \times H \times W}$ are generated by concatenating the aligned features $\mathcal{F}_i^w$ and image features $\mathcal{F}_i^b$, where T denotes the number of frames in the sliding window.

Step 3. Given $\mathbf{F}_i^c$ and $\mathcal{F}_i^b$ as input, the attention maps $\mathbf{A}^g \in \mathbb{R}^{Q \times M \times T \times K}$, the offsets of sampling points $\Delta \mathbf{P}^g \in \mathbb{R}^{Q \times M \times T \times K \times 2}$, and the flatten features $\mathbf{E}^g \in \mathbb{R}^{THWC}$ are generated, where $Q = THW$. M, T, and K represent the number of attention heads, the number of sampling points, and the number of frames, respectively. The attention maps $\mathbf{A}^g$ are used to measure the sharpness of the pixels, which are normalized by $\sum_{t=1}^{T} \sum_{k=1}^{K} \mathbf{A}_{mtqk}^g = 1$. The offsets of sampling points $\Delta \mathbf{P}^g$ and estimated optical flows $\mathcal{O}_i$ are used to find the corresponding pixels in the adjacent frames, where $\mathcal{O}_i$ provides the base offsets. $\Delta \mathbf{P}^g$, $\mathbf{A}^g$, and $\mathcal{O}_i$ are generated as following

$$\Delta \mathbf{P}^g = \mathcal{C}^o_{\text{MMA}}(\mathbf{F}^c_i)$$
$$\mathbf{A}^g = \mathcal{C}^m_{\text{MMA}}(\mathbf{F}^c_i)$$
$$\mathbf{E}^g = \text{Concat}(\mathcal{C}^l_{\text{MMA}}(\mathbf{F}^b_{i-1}), \mathcal{C}^l_{\text{MMA}}(\mathbf{F}^b_i), \mathcal{C}^l_{\text{MMA}}(\mathbf{F}^b_{i+1})) \tag{2}$$

where "Concat" denotes the concatenation operation. $\mathcal{C}^m_{\text{MMA}}, \mathcal{C}^o_{\text{MMA}}, \mathcal{C}^l_{\text{MMA}}$ represent different convolution layers. The attention map $\mathbf{A}^g$, offsets of sampling points $\Delta \mathbf{P}^g$ and flatten features $\mathbf{E}^g$ are fed to the deformable attention function $\mathcal{D}$ [45] and produces the fused features $\mathbf{Z}^g \in \mathbb{R}^{TCHW}$.

$$\mathbf{Z}^g = \mathcal{D}(\mathbf{A}^g, \phi(\Delta \mathbf{P}^g, \mathcal{O}_i), \mathbf{E}^g), \tag{3}$$

where ϕ represents the operation that adds the estimated optical flows to the offsets of sampling points $\Delta \mathbf{P}^g$. In 3, the optical flows is used as based offsets, which reduces the degradation of deblurring results caused by inaccurate optical flows.

Step 4. $\mathbf{Z}^g$ is reshaped and splitted into $\left\{ \mathbf{F}^h_k | \mathbf{F}^h_k \in \mathbb{R}^{C \times H \times W} \right\}^{i+1}_{k=i-1}$. The final fused features $\mathcal{F}^g_i = \left\{ \mathbf{F}^g_k | \mathbf{F}^g_k \in \mathbb{R}^{C \times H \times W} \right\}^{i+1}_{k=i-1}$ are generated as following

$$\mathbf{F}^g_k = \mathcal{C}^g_{\text{MMA}}(\mathbf{F}^h_k) \tag{4}$$

where $\mathcal{C}^g_{\text{MMA}}$ denotes a convolutional layer.

Multi-to-Single Attention Layer. The multi-to-single attention layer takes the coarse aggregated image features $\mathcal{F}^g_i$ as input and generates the fused features $\mathbf{F}^f_i$ for the mid-frame. Similar to the MMA layer, the MSA layer aggregates information of sharp pixels from the adjacent frames. However, in the MSA layer, the aggregated features are only propagated to the mid-frame. Therefore, in the MSA layer, the fused features $\mathbf{Z}^f \in \mathbb{R}^{CHW}$ is generated as following

$$\mathbf{Z}^f = \mathcal{D}(\mathbf{A}^f, \phi(\Delta \mathbf{P}^f, \{\mathbf{O}_{k \to i} | k = i - 1, i + 1\}), \mathbf{E}^f) \tag{5}$$

where $\mathbf{A}^f \in \mathbb{R}^{HW \times M \times T \times K}$, $\Delta \mathbf{P}^f \in \mathbb{R}^{HW \times M \times T \times K \times 2}$, and $\mathbf{E}^f \in \mathbb{R}^{TCHW}$ are the attention maps, the offsets of sampling points, and flatten features obtained as in the MMA layer. The fused features $\mathbf{F}^f_i$ is obtained as following

$$\mathbf{F}^f_i = \mathcal{C}^f_{\text{MSA}}(\mathbf{F}^n_i) \tag{6}$$

where $\mathcal{C}^f_{\text{MSA}}$ denotes a convolutional layer. $\mathbf{F}^n_i \in \mathbb{R}^{C \times H \times W}$ is reshaped from $\mathbf{Z}^f$.

3.3 Feature Extraction and Reconstruction Networks

Feature Extraction Network. The feature extraction network generates image features $\mathcal{F}^b_i$ from blurry images $\mathcal{B}_i$. It consists of three convolutional blocks, two of which have a convolution layer with the stride of 2 followed by three residual blocks with LeakyReLU as the activation function. The first convolutional block has a convolution layer with the stride of 1 followed by three residual blocks with LeakyReLU as the activation function.

Reconstruction Network. The reconstruction network is used to restore the sharp mid-frame $\mathbf{R}_i$ by taking the fused features from STDA module as input. It consists of three convolutional blocks, two of which have one deconvolutional layer with the stride of 2 and three residual blocks with LeakyReLU as the activation function. The last convolutional block has one convolutional layer with the stride of 1 and three residual blocks with LeakyReLU as the activation function.

3.4   Cascaded Progressive Deblurring

Inspired by TSP [25], we propose STDANet-Stack by stacking STDANet in a cascaded manner [25]. It takes five adjacent blurry video frames $\{\mathbf{B}_k\}_{k=i-2}^{i+2}$ as input and restores the sharp mid-frame $\mathbf{R}_i$.

Specifically, STDANet-Stack restores $\mathbf{R}_i$ in two steps. First, it produces $\hat{\mathbf{R}}_{i-1}$ by taking $\{\mathbf{B}_k\}_{k=i-2}^{i}$ as input. Similarly, $\hat{\mathbf{R}}_i$ and $\hat{\mathbf{R}}_{i+1}$ are restored by taking $\{\mathbf{B}_k\}_{k=i-1}^{i+1}$ and $\{\mathbf{B}_k\}_{k=i}^{i+2}$ as inputs, respectively. Next, $\mathbf{R}_i$ is generated by taking $\left\{\hat{\mathbf{R}}_k\right\}_{k=i-1}^{i+1}$ as input.

3.5   Loss Functions

We employ two loss functions to train STDANet and STDFANet-Stack.
MSE Loss represents the distance between the restored frame R and its corresponding ground truth sharp frame S, which is formulated as

$$\mathcal{L}_{mse} =\| \mathbf{R} - \mathbf{S} \|^2 \tag{7}$$

Warp Loss is introduced to train the motion estimator in an unsupervised manner, which is computed as

$$\mathcal{L}_{warp} =\| \mathbf{S}_i^\downarrow - \text{Warp}(\mathbf{S}_j^\downarrow, \mathbf{O}_{i\rightarrow j}) \|^2 \tag{8}$$

where $\mathbf{S}_i^\downarrow$ and $\mathbf{S}_j^\downarrow$ are the two downsampled frames. $\mathbf{O}_{i\rightarrow j}$ represents the estimated optical flow from $\mathbf{S}_j^\downarrow$ and $\mathbf{S}_i^\downarrow$. "Warp" denotes the backward warp operation.
Total Loss are defined as

$$\mathcal{L}_{total} = \mathcal{L}_{mse} + \gamma\mathcal{L}_{warp} \tag{9}$$

where γ controls the weights of the two loss functions.

Table 1. The quantitative results on the DVD dataset. Note that "Ours*" denotes STDANet-Stack.

Method	SRN	IFI-RNN-L	STFAN	EDVR	TSP	PVDNet	ARVo	Ours	Ours*
PSNR	30.53	31.67	31.24	31.82	32.13	32.31	32.80	32.63	**33.05**
SSIM	0.8940	0.9160	0.9340	0.9160	0.9270	0.9260	0.9352	0.9300	**0.9374**

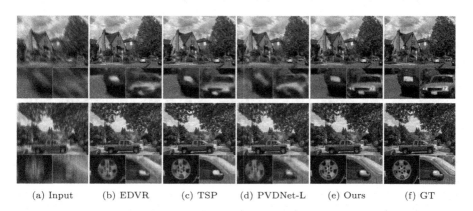

(a) Input　　(b) EDVR　　(c) TSP　　(d) PVDNet-L　　(e) Ours　　(f) GT

Fig. 4. The qualitative results on the DVD dataset. Note that "GT" stands for ground truth.

4   Experiments

4.1   Datasets

DVD. The DVD dataset [30] contains 71 videos (6,708 blurry-sharp pairs), splitting into 61 training videos (5,708 pairs) and 10 testing videos (1,000 pairs).

GoPro. The GoPro dataset [23] contains 3,214 pairs of blurry images and sharp images at 1280×720 resolution, where 2,103 and 1,111 pairs of blurry images and sharp images are used for training and testing, respectively.

BSD. The BSD dataset [43] is a real-world video deblur dataset, which contains three sub-datasets with different sharp exposure time - blurry exposure time.

4.2   Evaluation Metrics

For fair comparisons, STDANet-Stack use same cascaded progressive structure like TSP [25] and ARVo [19]. In the experiments, we use both peak signal-to-noise ratio (PSNR) and structural similarity (SSIM) as quantitative evaluation metrics for testing set. Moreover, GMACs (Giga multiply-add operations per second) is used to evaluate the computational complexity.

4.3   Implementation Details

To achieve better trade-off between video deblurring quality and computational efficiency, the M, K, T are set as 4, 12 and 3, respectively. γ is set to 0.05.

Table 2. The quantitative results on the GoPro dataset. Note that "Ours*" denotes STDANet-Stack.

Method	SRN	IFI-RNN-L	STFAN	EDVR	TSP	PVDNet	PVDNet-L	Ours	Ours*
PSNR	30.61	31.05	28.59	31.54	31.67	31.52	31.98	**32.29**	**32.62**
SSIM	0.9080	0.9110	0.8608	0.9260	0.9279	0.9210	0.9280	**0.9313**	**0.9375**
GMACs	1175	1,425	504	2739	6450	1004	1755	1677	6000

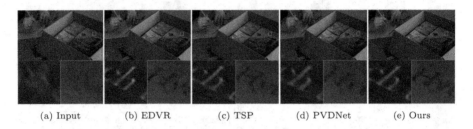

   (a) Input   (b) EDVR   (c) TSP   (d) PVDNet   (e) Ours

Fig. 5. The qualitative results of real blur images from the DVD dataset. There are no corresponding ground truth for the real blur images.

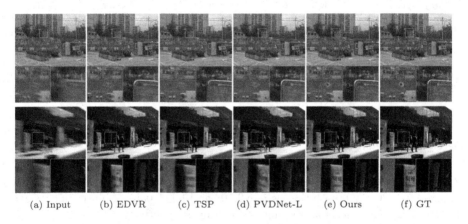

  (a) Input   (b) EDVR   (c) TSP   (d) PVDNet-L   (e) Ours   (f) GT

Fig. 6. The qualitative results on the GoPro dataset. Note that "GT" stands for ground truth.

The network is implemented with PyTorch [27][1]. The network is trained with a batch size of 8 on four NVIDIA Geforce RTX 2080 Ti GPUs. The initial learning rate is set to 10^{-4}. The network is optimized using Adam optimizer [14] with $\beta_1 = 0.9$ and $\beta_2 = 0.999$. We randomly crop the input images into patches with resolutions of 256×256, along with random flipping or rotation during training.

---

[1] The source code is available at https://github.com/huicongzhang/STDAN.

Table 3. The quantitative results on the BSD dataset. Note that "Ours*" denotes STDANet-Stack.

Method	1 ms–8 ms		2 ms–16 ms		3 ms–24 ms	
	PSNR	SSIM	PSNR	SSIM	PSNR	SSIM
IFIRNN	33.00	0.9330	31.53	0.9190	30.89	0.9170
ESTRNN	33.36	0.9370	31.95	0.9250	31.39	0.9260
EDVR	32.79	0.9264	31.99	0.9129	31.53	0.9192
TSP	33.62	0.9419	32.19	0.9285	31.68	0.9266
PVDNet-L	33.93	0.9392	32.46	0.9290	31.87	0.9293
Ours	**34.21**	**0.9446**	**33.13**	**0.9388**	**32.65**	**0.9409**
Ours*	**34.32**	**0.9456**	**33.27**	**0.9420**	**32.83**	**0.9443**

Table 4. The quantitative results on the GoPro dataset in terms of PSNR and SSIM when replacing MMA and MSA layers with the concatenation operation.

MMA			✓	✓
MSA		✓		✓
PSNR	30.12	31.15	31.18	**32.29**
SSIM	0.8950	0.9146	0.9152	**0.9313**

4.4 Experimental Results

The DVD Dataset. To evaluate the performance of the proposed method, we compare it with the state-of-the-art methods. Table 1 shows the quantitative results on the DVD dataset [30], where IFI-RNN-L [29] is larger IFI-RNN [24]. The proposed method outperforms the state-of-the-art methods in term of PSNR and SSIM. Compared to the best state-of-the-art method ARVo, the proposed STDANet-Stack improves the PSNR and SSIM by 0.25dB and 0.0022, respectively. Figure 4 shows several examples in the testing set, which indicates that existing state-of-the-art methods are less effective when the inputs contain heavy blur. We further compare the proposed method with state-of-the-art methods on the real blur images from the DVD dataset. Figure 5 shows that the proposed method generates sharper images with more visual details, which demonstrates the superiority of removing the unknown real blur in dynamic scenes robustly.

The GoPro Dataset. We compare STDANet to the state-of-the-art video deblurring methods on the GoPro dataset [23]. As show in Table 2, the proposed STDANet and STDANet-Stack perform favorably against the state-of-the-art methods in terms of PSNR and SSIM. Compared to the PVDNet-L [29], STDANet achieves higher PSNR and SSIM with lower computational complexity. STDANet-Stack achieves 0.95dB higher PSNR than TSP [25] with lower computational complexity, where the STDANet-Stack use the same cascaded progressive structure as TSP [25]. As shown in Fig. 6, the proposed method restores better image details and structures.

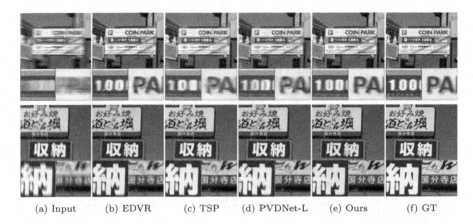

(a) Input   (b) EDVR   (c) TSP   (d) PVDNet-L   (e) Ours   (f) GT

Fig. 7. The qualitative results on the 2 ms–16 ms subset of the BSD dataset. Note that "GT" stands for ground truth.

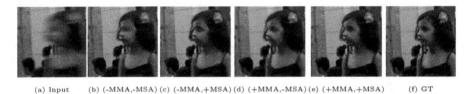

(a) Input   (b) (-MMA,-MSA) (c) (-MMA,+MSA) (d) (+MMA,-MSA) (e) (+MMA,+MSA)   (f) GT

Fig. 8. The qualitative results when replacing MMA and MSA layers with the concatenation operation. Note that "+" and "-" denote "with" and "without", respectively. "GT" stands for ground truth.

The BSD Dataset. We compared the our method to the state-of-the-art methods on BSD dataset [43]. For a fair comparison, the EDVR [35], TSP [25], and PVDNet-L [29] are trained with their open-sourced implementations. In Table 3, our method achieves the best results on all the three subsets in terms of PSNR and SSIM. The qualitative results are shown in Fig. 7, which indicate that our method restores much sharper images.

5   Analysis and Discussions

5.1   Effectiveness of the STDA Module

MMA and MSA Layers. The STDA Module contains two main components: the MMA and MSA layers, which aggregates information of sharp pixels from adjacent frames. To validate the effectiveness of the STDA module, the MMA and MSA layers are replaced with the concatenate operation. In the concatenate operation, the information from all pixels are introduced from adjacent frames. Table 4 shows the qualitative comparison when the MMA or MSA layer

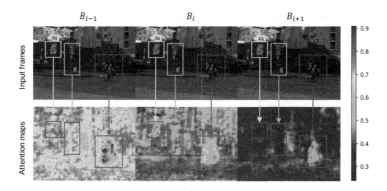

Fig. 9. The visualization of the attention maps in the MSA layer. Sharper pixels have larger attention weights. (**zoom in for best view**).

Table 5. The quantitative results on the GoPro dataset in terms of PSNR, SSIM, and GMACs with different numbers of sampling points.

#Sampling Points	$K = 1$	$K = 8$	$K = 12$	$K = 16$
PSNR	31.64	32.12	32.29	**32.32**
SSIM	0.9183	0.9288	0.9313	**0.9319**
GMACs	**1520**	1620	1677	1735

is removed. Specially, when both MMA or MSA layers are removed, the estimated optical flows are used to align the features from adjacent frames. The experimental results shows that the networks perform worse without the help of the information of sharp pixels extracted by the MMA and MSA layers. Figure 8 shows the qualitative comparison on the GoPro dataset. The network is less effective to restore sharp details when both MMA and MSA layers are removed. Figure 9 gives the visualization of the attention maps in the MSA layer, which shows that sharper pixels have larger attention weights. For example, the man riding a bicycle (highlighted with a red bounding box) is blurry in B_{i-1}, and thus the corresponding regions are of low weights in the attention maps. In contrast, B_i have larger weights for this region. To conclude, the proposed STDA module effectively aggregates the information of sharp pixels from adjacent frames.

Sampling Points. To investigate the effect of numbers of sampling points in the STDA module, we compare the performance with different numbers of sampling points. As shown in Table 5, larger number of sampling points leads to better restoration results but also heavier computational cost. Specially, the STDA Module degenerates to the temporal attention when $K = 1$, which causes severe degeneration in restoration results. The PSNR only increases 0.03 dB when increasing the number of sampling points from 12 to 16. Therefore, we set $K = 12$ due to the trade-off between the computational cost and restoration performance.

Table 6. The quantitative results on the GoPro dataset in terms of PSNR, SSIM, and GMACs with different numbers of attention heads.

#Attention Heads	$M = 1$	$M = 4$	$M = 8$
PSNR	32.13	32.29	**32.34**
SSIM	0.9294	0.9313	**0.9322**
GMACs	**1548**	1677	1849

Table 7. The quantitative results on the GoPro dataset in terms of PSNR, SSIM, and GMACs with different optical flow estimators.

Estimator	None	PWC-Net	Motion Estimator
PSNR	31.58	**32.36**	32.29
SSIM	0.9176	**0.9326**	0.9313
GMACs	**1632**	2352	1677

Attention Heads. The number of attention heads is one of the important hyperparameter in the deformable attention function. We also compare the effect with different numbers of attention heads in Table 6. As the number of attention heads increases, the PSNR, SSIM, and GMACs increase. Considering the trade-off between the computational complexity and restoration performance, we choose the number of attention heads $M = 4$.

5.2   Effectiveness of the Motion Estimator

To evaluate the effectiveness of the motion estimator, we compare the video deblur results with different optical flow estimators. As shown in Table 7, removing the optical flow estimator causes considerable degeneration. Although STDANet with PWC-Net [31] archives the best results, it also leads to high computational cost. STDANet with the proposed motion estimator archives the best trade-off between the deblur results and computational complexity.

6   Conclusions

In this paper, we propose STDANet for video deblurring. The main motivation of this work is that not all the pixels in the video frames are sharp and beneficial for deblurring. Therefore, the proposed STDANet extracts the information of sharp pixels by considering the pixel-wise blur levels of the video frames. Different from mainstream video debulr methods that requires accurate optical flows to align two adjacent frames to the mid-frame, the coarse optical flows are estimated by a lightweight motion estimator and are used as the base offsets to find the corresponding sharp pixels in the adjacent frames. Experimental results indicate that the proposed STDANet performs favorably against state-of-the-art methods on the GoPro, DVD, and BSD datasets.

Acknowledgement. This work is supported by the National Key R&D Program of China (No. 2021ZD0110901).

References

1. Bar, L., Berkels, B., Rumpf, M., Sapiro, G.: A variational framework for simultaneous motion estimation and restoration of motion-blurred video. In: ICCV (2007)
2. Couzinie-Devy, F., Sun, J., Alahari, K., Ponce, J.: Learning to estimate and remove non-uniform image blur. In: CVPR (2013)
3. Dai, S., Wu, Y.: Motion from blur. In: CVPR (2008)
4. Dosovitskiy, A., et al.: Flownet: learning optical flow with convolutional networks. In: ICCV (2015)
5. Gong, D., et al.: From motion blur to motion flow: a deep learning solution for removing heterogeneous motion blur. In: CVPR (2017)
6. Gupta, A., Joshi, N., Zitnick, C.L., Cohen, M.F., Curless, B.: Single image deblurring using motion density functions. In: ECCV (2010)
7. Harmeling, S., Hirsch, M., Schölkopf, B.: Space-variant single-image blind deconvolution for removing camera shake. In: NIPS (2010)
8. Hirsch, M., Schuler, C.J., Harmeling, S., Schölkopf, B.: Fast removal of non-uniform camera shake. In: ICCV (2011)
9. Jin, H., Favaro, P., Cipolla, R.: Visual tracking in the presence of motion blur. In: CVPR (2005)
10. Kim, T.H., Ahn, B., Lee, K.M.: Dynamic scene deblurring. In: ICCV (2013)
11. Kim, T.H., Lee, K.M.: Segmentation-free dynamic scene deblurring. In: CVPR (2014)
12. Kim, T.H., Lee, K.M.: Generalized video deblurring for dynamic scenes. In: CVPR (2015)
13. Kim, T.H., Lee, K.M., Schölkopf, B., Hirsch, M.: Online video deblurring via dynamic temporal blending network. In: ICCV (2017)
14. Kingma, D.P., Ba, J.: Adam: a method for stochastic optimization. In: ICLR (2015)
15. Krishnan, D., Tay, T., Fergus, R.: Blind deconvolution using a normalized sparsity measure. In: CVPR (2011)
16. Kupyn, O., Martyniuk, T., Wu, J., Wang, Z.: DeblurGAN-v2: deblurring (orders-of-magnitude) faster and better. In: ICCV (2019)
17. Lee, H.S., Kwon, J., Lee, K.M.: Simultaneous localization, mapping and deblurring. In: ICCV (2011)
18. Levin, A., Weiss, Y., Durand, F., Freeman, W.T.: Efficient marginal likelihood optimization in blind deconvolution. In: CVPR (2011)
19. Li, D., et al.: Arvo: learning all-range volumetric correspondence for video deblurring. In: CVPR (2021)
20. Matsushita, Y., Ofek, E., Ge, W., Tang, X., Shum, H.: Full-frame video stabilization with motion inpainting. TPAMI **28**(7), 1150–1163 (2006)
21. Mei, C., Reid, I.D.: Modeling and generating complex motion blur for real-time tracking. In: CVPR (2008)
22. Michaeli, T., Irani, M.: Blind deblurring using internal patch recurrence. In: ECCV (2014)
23. Nah, S., Kim, T.H., Lee, K.M.: Deep multi-scale convolutional neural network for dynamic scene deblurring. In: CVPR (2017)
24. Nah, S., Son, S., Lee, K.M.: Recurrent neural networks with intra-frame iterations for video deblurring. In: CVPR (2019)

25. Pan, J., Bai, H., Tang, J.: Cascaded deep video deblurring using temporal sharpness prior. In: CVPR (2020)
26. Park, D., Kang, D.U., Kim, J., Chun, S.Y.: Multi-temporal recurrent neural networks for progressive non-uniform single image deblurring with incremental temporal training. In: ECCV (2020)
27. Paszke, A., et al.: Pytorch: an imperative style, high-performance deep learning library. In: NeurIPS (2019)
28. Ren, W., Cao, X., Pan, J., Guo, X., Zuo, W., Yang, M.: Image deblurring via enhanced low-rank prior. In: TIP (2016)
29. Son, H., Lee, J., Lee, J., Cho, S., Lee, S.: Recurrent video deblurring with blur-invariant motion estimation and pixel volumes. ACM Trans. Graph **40**(5), 1–18 (2021)
30. Su, S., Delbracio, M., Wang, J., Sapiro, G., Heidrich, W., Wang, O.: Deep video deblurring for hand-held cameras. In: CVPR (2017)
31. Sun, D., Yang, X., Liu, M., Kautz, J.: Pwc-net: Cnns for optical flow using pyramid, warping, and cost volume. In: CVPR (2018)
32. Sun, J., Cao, W., Xu, Z., Ponce, J.: Learning a convolutional neural network for non-uniform motion blur removal. In: CVPR (2015)
33. Sun, L., Cho, S., Wang, J., Hays, J.: Edge-based blur kernel estimation using patch priors. In: ICCP (2013)
34. Tao, X., Gao, H., Shen, X., Wang, J., Jia, J.: Scale-recurrent network for deep image deblurring. In: CVPR (2018)
35. Wang, X., Chan, K.C.K., Yu, K., Dong, C., Loy, C.C.: EDVR: video restoration with enhanced deformable convolutional networks. In: CVPR Workshops (2019)
36. Whyte, O., Sivic, J., Zisserman, A., Ponce, J.: Non-uniform deblurring for shaken images. In: CVPR (2010)
37. Wieschollek, P., Hirsch, M., Schölkopf, B., Lensch, H.P.A.: Learning blind motion deblurring. In: ICCV (2017)
38. Wulff, J., Black, M.J.: Modeling blurred video with layers. In: Fleet, D.J., Pajdla, T., Schiele, B., Tuytelaars, T. (eds.) ECCV (2014)
39. Xu, J., Ranftl, R., Koltun, V.: Accurate optical flow via direct cost volume processing. In: CVPR (2017)
40. Xu, L., Zheng, S., Jia, J.: Unnatural L0 sparse representation for natural image deblurring. In: CVPR (2013)
41. Zhang, H., Dai, Y., Li, H., Koniusz, P.: Deep stacked hierarchical multi-patch network for image deblurring. In: CVPR (2019)
42. Zhang, K., Luo, W., Zhong, Y., Ma, L., Liu, W., Li, H.: Adversarial spatio-temporal learning for video deblurring. In: TIP (2019)
43. Zhong, Z., Gao, Y., Zheng, Y., Zheng, B.: Efficient spatio-temporal recurrent neural network for video deblurring. In: ECCV (2020)
44. Zhou, S., Zhang, J., Pan, J., Zuo, W., Xie, H., Ren, J.S.J.: Spatio-temporal filter adaptive network for video deblurring. In: ICCV (2019)
45. Zhu, X., Su, W., Lu, L., Li, B., Wang, X., Dai, J.: Deformable DETR: deformable transformers for end-to-end object detection. In: ICLR (2021)

NeuMesh: Learning Disentangled Neural Mesh-Based Implicit Field for Geometry and Texture Editing

Bangbang Yang[1], Chong Bao[1], Junyi Zeng[1], Hujun Bao[1], Yinda Zhang[2(✉)], Zhaopeng Cui[1(✉)], and Guofeng Zhang[1(✉)]

[1] State Key Lab of CAD&CG, Zhejiang University, Hangzhou, China
zhangguofeng@zju.edu.cn
[2] Google, Mountain View, USA

Abstract. Very recently neural implicit rendering techniques have been rapidly evolved and shown great advantages in novel view synthesis and 3D scene reconstruction. However, existing neural rendering methods for editing purposes offer limited functionality, *e.g.*, rigid transformation, or not applicable for fine-grained editing for general objects from daily lives. In this paper, we present a novel mesh-based representation by encoding the neural implicit field with disentangled geometry and texture codes on mesh vertices, which facilitates a set of editing functionalities, including mesh-guided geometry editing, designated texture editing with texture swapping, filling and painting operations. To this end, we develop several techniques including learnable sign indicators to magnify spatial distinguishability of mesh-based representation, distillation and fine-tuning mechanism to make a steady convergence, and the spatial-aware optimization strategy to realize precise texture editing. Extensive experiments and editing examples on both real and synthetic data demonstrate the superiority of our method on representation quality and editing ability. Code is available on the project webpage: https://zju3dv.github.io/neumesh/.

Keywords: Neural rendering · Mesh-based representation · Scene editing · View synthesis · 3D deep learning

1 Introduction

Neural implicit field has achieved great success in 3D reconstruction and free-viewpoint rendering, and becomes a promising solution to take the place of traditional 3D shape and texture representation, *e.g.*, point cloud or textured mesh, due to its phenomenal rendering quality. However, for 3D modeling and

---

B. Yang and C. Bao—Contributed equally to this work.

Supplementary Information The online version contains supplementary material available at https://doi.org/10.1007/978-3-031-19787-1_34.

S. Avidan et al. (Eds.): ECCV 2022, LNCS 13676, pp. 597–614, 2022.
https://doi.org/10.1007/978-3-031-19787-1_34

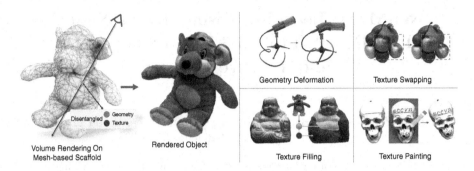

Fig. 1. NeuMesh. We present a novel representation for volumetric neural rendering, which encodes the neural implicit field with disentangled geometry and texture features on a mesh scaffold. With the locally separated latent codes, our representation enables a series of editing functionalities, including mesh-guided geometry deformation, designatable texture swapping, filling and painting.

CG creation, artists still prefer to use mesh-based workflow across daily works. For instance, in modern 3D CG software (*e.g.*, Blender, Maya and 3ds Max), polygon mesh-based representations can be precisely controlled and edited, *i.e.*, texturing with UV-map and changing shapes by altering vertices and faces, with all the previewed modification accurately reflected in the final rendering product. Despite great progress made to improve the flexibility of the neural implicit field, including handling dynamic scenes [30,31], becoming scene agnostic [41,52], fast rendering [24,39], and scalability improvement [47,56], the support of neural implicit field towards editing is still limited, *e.g.*, on a very specific semantic category [5,13,20] or purely rigid transformation [10,60,65]. One plausible reason is that particular network encoding structures (*e.g.*, coordinate-based MLP, voxels or scattered point cloud) are not compatible with fine-grained scene editing such as non-rigid geometry deforming and texture editing for a local region of interest, and thus cannot satisfy the broad demands of artistic creation.

In this paper, we propose a novel neural implicit representation, NeuMesh, to facilitate editing in both 3D modeling and texturing. Our representation bares the following properties to seamlessly integrate with existing common workflow for 3D editing: **1)** The neural representation encodes scene with a series of vertex-bounded codes on a mesh scaffold and MLP-based decoders, instead of a pure MLP, point clouds or voxels, and can be deformed together with the mesh. During the volume rendering, the implicit field is decoded via interpolation of these codes. By doing so, any modification to the mesh geometry or local codes would precisely reflect the rendering output. **2)** The geometry and appearance representations are disentangled, *i.e.*, encoded in two separate latent codes, such that texture can be transferred across geometry by replacing the appearance code from one another. As shown in Fig. 1, our representation supports non-rigid object deformation with a handful approach (*e.g.*, deforming with a mesh proxy), and provides various fashions of texture editing, including texture swapping of irregular mesh segments, texture filling at a specific area with pattern from a pre-captured object, and a user-friendly texture painting which reflects the philosophy of 'what you get is what you see'.

However, learning and deploying such representation for rendering and editing is non-trivial. **1)** Unlike voxel-based representation [17], naïve trilinear code interpolation is not sufficient to measure spatial variation since we dedicate to encoding the implicit field with a set of 'single layer' codes on mesh vertices, and the inner/outer queries along the direction perpendicular to the surface lacks spatial distinguishability (*i.e.*, failing to determine positive or negative direction when crossing through a mesh face). A possible workaround is to complement the network input with signed distance to the mesh surface [18], which, however, is not always available, especially on non-watertight/ill-defined geometries. To tackle this challenge, we propose to maintain a set of learnable sign indicators for mesh vertices. Then, for each query point along the ray, we compute a signed distance from nearby vertices by weighting the projected distances on the indicators. In this way, our representation is completely agnostic to arbitrary mesh typologies (*e.g.*, non-watertight or non-manifold meshes). During the training process, these sign indicators are continuously adjusted to best fit the optimization objective. **2)** Although such vertex-bounded and geometry-texture disentangled representation merits good flexibility on editing purpose, it does not preserve spatial continuity as MLP-based methods [23,51,61] and thus easily suffers from unstable training. To mitigate this problem, we employ a distillation and fine-tuning training scheme, which leverages a pre-trained implicit field to guide the optimization of our representation. In this way, we transfer a baked MLP-based implicit model into NeuMesh, the first neural rendering model that naturally inherits editable capability from the flexible mesh-based workflow. **3)** To fulfill the demand for flexible and user-friendly texture editing operations (*e.g.*, propagating 2D image painting to the 3D field), a naïve approach is fine-tuning with a single image. However, this might let the network overfit to a specific view and the rendered images from other views degrade (*e.g.*, introducing noticeable artifacts as shown in Fig. 8 (b)). In order to solve this challenge, we propose a spatial-aware optimization strategy that is naturally derived from our representation, in which we select the affected texture codes with several probing rays from painted pixels to the mesh surface, and only fine-tuning these codes during the optimization. Therefore, we can precisely transfer the painting to the desired region while maintaining other parts unchanged.

The contributions of our paper can be summarized as follows. 1) We present a novel mesh-based neural implicit representation which aims to break the barrier between volumetric neural rendering and mesh-based 3D modeling and texturing workflow, and delivers a set of editing functionalities, including mesh-guided geometry editing, designated texture editing with texture swapping, filling and painting operations. To make the representation locally editable both on geometry and texture, we design to encode the implicit field into mesh vertices, where each vertex possesses disentangled geometry and texture features of its local space. 2) We analyze the technical challenges and develop several techniques to enhance the spatial distinguishability with learnable sign indicators, ensure a steady training with distillation and fine-tuning mechanism, and improve the texture editing precision with spatial-aware optimization strategy. **3)** Extensive experiments and impressive editing examples on both real and synthetic datasets demonstrate that our method achieves photo-realistic rendering quality, and is flexible and powerful at geometry and texture editing of the neural implicit field.

2   Related Works

Mesh-based Representation and Rendering. In computer vision and graphics, polygon mesh has been widely used in 3D scene modeling and rendering [1,11,19]. Traditional methods utilize multi-view geometry and numerical theories to reconstruct surface meshes of a captured scene [15,40,50,58]. Recently, more attention has been paid to neural network based scene reconstruction [25,45] and texture learning [7,48,55]. However, existing mesh-based rendering pipelines usually require UV-mapping to build correspondences between meshes vertices and texture maps, which limits the applicability from representing scenes with complex topology and delicate structure. Another line of methods uses MVS based mesh as a geometry proxy for image feature aggregation [37,38], but requires nearby source images to be warped back to the mesh surface and is not feasible for high-level editing operations. Instead of storing textures in a flat 2D map or warping-based view synthesis, our method directly encodes appearance information on 3D vertices, and is more flexible in representing complex objects whose UV-maps are difficult to be unwrapped.

Neural Rendering. Given a set of image captures, neural rendering methods [4,62] aim to render photo-realistic images of novel views. NeRF [23] takes advantages of volume rendering to boost rendering quality, which inspires a lot of works, including surface reconstruction [28,51,61], human modeling [18,32], pose estimation [63], scene understanding [59] and relighting [2,44,66,67], etc.. To further increase network capacity and reduce computation, many works propose to decompose scene into local representations, such as multiple tiny networks [35], point clouds [29] and voxels [17,39]. Although these works explicitly encode scenes in a 3D spatial structure, they are not designed to be easily manipulated as polygon meshes, thus not capable of high-level applications like geometry and texture editing.

Neural Scene Editing. Scene editing is a popular topic in computer vision and photography. Early methods mainly focus on editing a single static view by inserting [14], compositing [33], moving [16,42] objects or changing lighting [22] for an existing photograph. With the development of neural rendering, many works start to edit scenes with movable [10,60,65] and deformable [64] objects, changeable colors, shapes [20,57] and textures [55]. However, existing methods are either limited to object-level rigid transformation [10,60,65], not generalize to out-of-distribution categories [3,5,20,27,46,53], restricts its representation to simple shapes [55] or orthographic projection [36], or does not support fine-grained texture editing [20,26,44,62,64,66]. By contrast, we pick up triangle mesh as a scaffold to encode the scene, since the mesh can be edited conveniently and intuitively in mature industry software, and the region of interest on the mesh can be precisely selected by vertices. Built upon this, our method delivers the capability of non-rigid geometry editing and fine-grained texture editing.

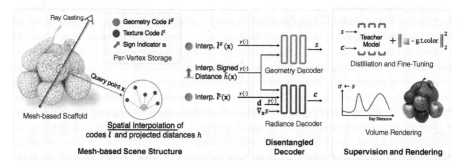

Fig. 2. Overview. We encode neural implicit field on a mesh-based scaffold, where each vertex possesses a geometry and texture code l^g, l^t, and a sign indicator $\mathbf{n}$ for computing projected distance h. For a query point x along a casted camera ray, we retrieve interpolated codes and signed distances from the nearby mesh vertices, and forward to the geometry/radiance decoder to obtain SDF value s and color $\mathbf{c}$.

3   Method

We introduce NeuMesh, a novel scene representation that encodes neural implicit field at a mesh-based scaffold. As demonstrated in Fig. 2, instead of learning the entire scene as a whole in a coordinate-based network, we leverage 3D mesh structure by decomposing the scene into a set of local-vertex-bounded implicit fields (Sect. 3.1), where each vertex stores geometry and texture information of its neighboring local space. Motivated by previous works [23,35,51], we adopt the volume rendering technique to render pixels, and employ a distillation and fine-tuning training scheme to encode the neural implicit field into the mesh surface (Sect. 3.2). During the rendering stage, we retrieve interpolated codes and learnable signed distances (*i.e.*, projected distances to the mesh vertices, which complements spatial distinguishability) from the mesh, and use two separated MLPs to decode geometry (*i.e.*, SDF values) and radiance color. In this way, the scene representation is locally aligned to the mesh, and the geometry and color are encoded in two separated latent spaces, which naturally derives the approaches of mesh-guided geometry deforming and designatable texture editing (Sect. 3.3).

3.1   Neural Mesh-based Implicit Field

Mesh-based Representation. As illustrated in Fig. 2, we use a mesh-based scaffold to model the neural implicit field. First, we reconstruct the target object using out-of-box NeuS [51] and marching cubes [21], which yields a triangle mesh with about 50K~150K vertices. Then, for each vertex $\mathbf{v}$ on the mesh, we store a set of learnable parameters, including a geometry code l^g, a texture code l^t and a sign indicator $\mathbf{n}$ ($\mathbf{n}$ helps to identify relative position, and will be introduced later). In a typically volume rendering process that sample points $\mathbf{x}$ along the ray, we first find K nearest vertices $\{\mathbf{v}_k | k = 1, 2, ..., K\}$ for each point $\mathbf{x}$, and perform spatial interpolation to obtain the interpolated codes $\tilde{l}^g(\mathbf{x})$, $\tilde{l}^t(\mathbf{x})$ and signed distances $\tilde{h}(\mathbf{x})$. Specifically, we adopt inverse distance weighting based interpolation [34], as:

$$\tilde{l}(\mathbf{x}) = \frac{\sum_{k=1}^{K} w_k l_k}{\sum_{k=1}^{K} w_k}, \quad w_k = \frac{1}{||\mathbf{v}_k - \mathbf{x}||}. \tag{1}$$

Then, we forward all these variables to geometry decoder F_G and radiance decoder F_R to obtain the SDF value $s = F_G\left(\tilde{l}^g, \tilde{h}\right)$ and color $\mathbf{c} = F_R\left(\tilde{l}^t, \tilde{h}, \mathbf{d}, \nabla_\mathbf{x} s\right)$ at point $\mathbf{x}$, where $\mathbf{d}$ is the viewing direction, $\nabla_\mathbf{x} s$ is the gradient of the SDF w.r.t query position. Different from the previous methods [51,61,62], we replace the global coordinate $\mathbf{x}$ with locally retrieved codes $\tilde{l}^g, \tilde{l}^t$ and sign distances $\tilde{h}$, where $\tilde{h}$ complements spatial distinguishability without hurting the locality of the representation. Note that we also apply positional encoding $\gamma(\cdot)$ [23] to the interpolated codes, distance and direction before feeding them into the MLP, but we omit it in the equations for brevity. Following the formulation of NeuS [51] and quadrature rules [23], we render the pixel $\hat{C}(\boldsymbol{r})$ with points $\{\mathbf{x}_i | i = 1, ..., N\}$ along the ray $\boldsymbol{r}$ as:

$$\hat{C}(\boldsymbol{r}) = \sum_{i=1}^{N} T_i \alpha_i \mathbf{c}_i, \quad T_i = \prod_{j=1}^{i-1}(1 - \alpha_j), \quad \alpha_j = \max\left(\frac{\Phi_s(s_i) - \Phi_s(s_{i+1})}{\Phi_s(s_i)}, 0\right), \tag{2}$$

where T is accumulated transmittance, Φ_s is the cumulative distribution of logistic distribution, and α is opacity derived from adjacent SDF.

Learnable Sign Indicator for Interpolated Signed Distance. To complement the spatial distinguishability of the network query along the direction perpendicular to the surface (*i.e.*, inside or outside the mesh), we introduce a *learnable* sign indicator $\mathbf{n}_k$ for each vertex $\mathbf{v}_k$ that aids at computing interpolated signed distances for spatial query points. Indeed, the sign indicator plays a similar role as vertex normal (*i.e.*, initialized with vertex normal), but is continuously adjusted during the training process to best fit the target loss. The computation of interpolated signed distance $\tilde{h}(\mathbf{x})$ is defined as:

$$\tilde{h}(\mathbf{x}) = \frac{\sum_{k=1}^{K} w_k h_k}{\sum_{k=1}^{K} w_k}, \quad h_k = \mathbf{p}_k \cdot \frac{\omega^n \mathbf{n}_k + \omega_k^p \mathbf{p}_k}{\omega^n + \omega_k^p}, \quad \mathbf{p}_k = \mathbf{x} - \mathbf{v}_k, \tag{3}$$

where w_k is inverse distance weighting as defined in Eq. (1), ω^n and ω_k^p controls the influence between sign indicator and point-to-vertex vector $\mathbf{p}_k$, and we empirically set $\omega^n = 0.1, \omega_k^p = ||\mathbf{p}_k||$. Intuitively, when the sample points are far from the surface, $\tilde{h}(\mathbf{x})$ is numerically close to the point-to-surface distance; otherwise, when the sample points are getting close to the surface, $\tilde{h}(\mathbf{x})$ would be gradually perturbed by learnable sign indicators.

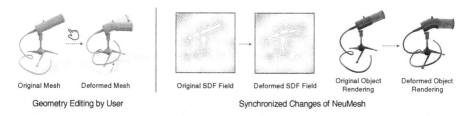

Original Mesh   Deformed Mesh   Original SDF Field   Deformed SDF Field   Original Object   Deformed Object
                                                                                             Rendering   Rendering

Geometry Editing by User   Synchronized Changes of NeuMesh

Fig. 3. Mesh-guided Geometry editing. By simply deforming the corresponding mesh, the change will synchronously take effect on the implicit field, and the rendered object will also be deformed accordingly.

3.2   Optimizing Mesh-based Implicit Field

Distillation and Fine-Tuning. We observe that training NeuMesh from scratch leads to artifacts and converges to sub-optimal results (see Fig. 8 and Table 3). Inspired by Reiser *et al.* [35], we apply a distillation and fine-tuning training scheme, *i.e.*, we supervise NeuMesh simultaneously with the output from a coordinate-based teacher model (*e.g.*, NeuS), and also the images. For a batched training rays $r \in R$, we defined the distillation loss L_d and photometric fine-tuning loss L_f as:

$$\mathcal{L}_{\mathrm{d}} = \sum_{r \in R} \sum_{i \in N} ||s_i - s_i^t|| + ||c_i - c_i^t||, \quad \mathcal{L}_{\mathrm{f}} = \sum_{r \in R} ||\hat{C}(r) - C(r)||_2^2 \qquad (4)$$

where s_i^t and c_i^t are the SDF value and color from the teacher model, and $C(r)$ is the ground-truth pixel color from images. By leveraging distillation and fine-tuning, we smoothly transfer a pure MLP-based neural implicit model into a flexible and editable mesh-based representation, and the final model even produces better appearance details, as shown in our experiment (Sect. 4.2).

Regularization. As introduced in Sect. 3.1, we dynamically adjust a set of per-vertex sign indicators during the training process. To ensure a smooth convergence, we empirically apply a regularization to the sign indicator by slightly encouraging them being close to pre-computed vertex normal n^t, as: $\mathcal{L}_{\mathrm{rs}} = \sum_k ||n_k - n_k^t||_2^2$. Besides, as suggested by Gropp *et al.* [8], we add an Eikonal loss to regularize the norm of the spatial gradients to 1, as: $\mathcal{L}_{\mathrm{re}} = \sum_k || \, ||\nabla_{\mathbf{x}_k} s_k|| - 1||_2^2$.
The final loss is then defined as:

$$\mathcal{L}_{\mathrm{total}} = \lambda_{\mathrm{d}} \mathcal{L}_{\mathrm{d}} + \lambda_{\mathrm{f}} \mathcal{L}_{\mathrm{f}} + \lambda_{\mathrm{rs}} \mathcal{L}_{\mathrm{rs}} + \lambda_{\mathrm{re}} \mathcal{L}_{\mathrm{re}}, \qquad (5)$$

where we set $\lambda_{\mathrm{d}} = 1.0$, $\lambda_{\mathrm{f}} = 1.0$, $\lambda_{\mathrm{rs}} = 0.01$ and $\lambda_{\mathrm{re}} = 0.01$.

3.3   Mesh-guided Geometry Editing

In NeuMesh, since the neural implicit field has been tightly aligned to the mesh surface, any manipulation on mesh vertices would directly take effect on the field and the volume rendering results. Therefore, to perform geometry editing with

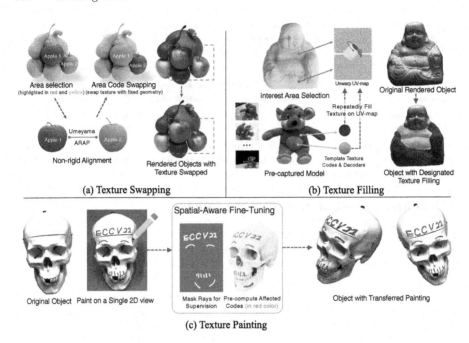

Fig. 4. Designatable Texture editing. By exchanging texture codes (and decoders), our representation delivers various texture editing pipelines on neural implicit field. (Icon credit: Flaticon [6]).

a NeuMesh-based scene, users are only required to edit the corresponding mesh, which can be easily accomplished by interactively moving a few vertices with out-of-box mesh deforming methods (*e.g.*, as-rigid-as-possible, or ARAP [43]), or 3D modeling software like Blender. We show an example of the geometry editing in Fig. 3, where we first deform the microphone by bending its head and lifting the wire on the corresponding mesh. Then, to maintain the local consistency of signed distance (Sect. 3.1), for each transformed vertex, we also compute a relative rotation of the surface normal and compensate the rotation according to its sign indicator (Sect. 3.1). Without any fine-tuning, the microphone's implicit field has been deformed in the meantime, and we can easily render the deformed view (see Fig. 3). Please refer to the supplementary materials for more details.

3.4 Designatable Texture Editing

Until then, texture editing on the neural implicit model is still an open problem. Previous methods tend to replace the entire materials by swapping the appearance branch [44,62,66], changing a uniformed color [20], or learning an editable UV mapping for simple and plump shapes [55]. However, in real texturing of 3D modeling software, artists are used to working with a mesh-based workflow, which allows them to select a partial region of an object and modify it with arbitrary colors and material properties. We propose to mimic such pipelines by introducing a designatable texture editing, where the selection of mesh vertices can be used to precisely guide the texture editing on the region of interest. The core step of our texture editing is that we update the latent texture code l^t ('material properties') and the binding decoder F_R ('rendering palette') at the selected region. As shown in Fig. 4, we deliver three ways of texture editing:

1) Texture swapping by exchanging textures between two objects through 3D geometry (*e.g.*, swapping textures of two apples in Fig. 4). Users are first asked to mark out the source and target object on the mesh, which can be done by mature 3D model software, or point-based instance segmentation [54]. Then, given a putative point matches with interactive annotation [68], we perform non-rigid 3D alignment to the source and target object with Umeyama [49] and ARAP [43]. Finally, we transfer texture codes by assigning each target vertex with code interpolated from nearby source vertices.

2) Texture filling by filling a targeting object area with repeated textures from a pre-captured model (*e.g.*, assigning part of Buddha with two furry materials from a teddy bear as shown in Fig. 4). In real applications, artists might want to try out some materials from a daily captured scene or pre-built material library, or want to fill some areas (*e.g.*, floor and walls) with uniformed materials. Therefore, we build a compatible workflow for the standard texturing operation, where we first construct a UV map for the user-interest areas, and then repeatedly fill the mapped vertices (*e.g.*, chest and cloth of Buddha) with template textures from a pre-captured model (*e.g.*, gray and brown hairs from the teddy bear).

3) Texture painting from a single 2D view to the 3D field. Users paint an arbitrary pattern or put some text on a captured image, and we can transfer these paintings into the 3D neural implicit field and freely preview in rendered novel views (*e.g.*, painting ECCV logo on a skull in Fig. 4). Compared to Neu-Tex [55] that might be difficult to edit on the desired position due to distorted UV-mapping, our method delivers a more natural editing way, *i.e.*, what you draw and see is what you get. However, it is not trivial to precisely control the painting transferring with only one image, since the overfitting of a single image might lead to appearance drifting at unconstrained views, which inevitably introduces artifacts in rendered novel views. To tackle this issue, we propose a *spatial-aware optimization mechanism*. Specifically, we first shoot rays through the painted pixels to obtain the surface points and find the affected texture codes of nearby vertices around the points. During the fine-tuning stage, we optimize by minimizing photometric loss of rendered pixels and painted pixels, and only backward gradient of these codes while detaching the others. Besides, to improve the training efficiency and the painting consistency across views, we restrict training rays inside a slightly dilated paint mask, and also augment with random viewing directions at the input to the radiance decoder.

Please refer to the supplementary materials for more details.

4 Experiments

4.1 Datasets

We evaluate our method on the real captured DTU [12] dataset and NeRF 360° Synthetic dataset. For the DTU dataset, we follow the setting of IDR [62] by

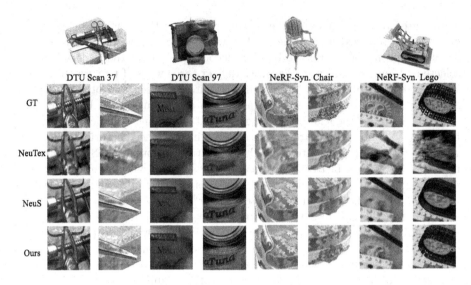

Fig. 5. We show rendering examples of NeuTex [55], NeuS [51] and our method on the DTU dataset and the NeRF 360° synthetic dataset.

Table 1. We compare rendering quality with NeuS [51] and NeuTex [55] on the DTU dataset and the NeRF 360° synthetic dataset.

Methods	DTU			NeRF 360° synthetic		
	PSNR ↑	SSIM ↑	LPIPS ↓	PSNR ↑	SSIM ↑	LPIPS ↓
NeuTex [55]	26.080	0.893	0.196	25.718	0.914	0.109
NeuS [51]	26.352	0.909	0.176	30.588	**0.960**	0.058
Ours	**28.289**	**0.921**	**0.117**	**30.945**	0.951	**0.043**

using 15 scenes with images of 1600 × 1200 resolution and foreground masks for experiments. To facilitate the metric evaluation for both rendering and mesh quality, we randomly select 10% images as test split and use the remaining images for training. For NeRF 360° Synthetic dataset, we follow the official split and choose 4 representative scenes for evaluation, including thin structures (Mic), complex shapes (Lego) and rich textures (Chair and Hotdog).

4.2   Comparison of Rendering and Mesh Quality

We first compare the rendering and mesh reconstruction quality of our representation with the baseline method NeuS [51] and the SOTA texture-editable implicit neural rendering method NeuTex [55]. Following previous works [23,51,62], we use PSNR, SSIM and LPIPS to measure the rendering quality, and use Chamfer distance to measure the reconstructed mesh quality. Please note that for mesh quality comparison, we use a subset (training split) of images, while NeuS takes all images for training in their paper, so the result

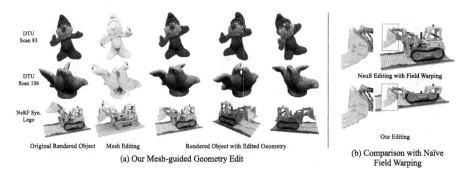

DTU Scan 83

DTU Scan 106

NeRF Syn. Lego

NeuS Editing with Field Warping

Our Editing

Original Rendered Object   Mesh Editing   Rendered Object with Edited Geometry

(a) Our Mesh-guided Geometry Edit

(b) Comparison with Naïve Field Warping

Fig. 6. We show examples of mesh-guided geometry editing in (a) and also compare with naïve field warping solution in (b).

Table 2. We compare mesh quality (Chamfer distance) with NeuS [51] on the DTU dataset. Note that we use training split of images instead of full images, so the result of NeuS [51] is different from the original paper.

Method	DTU Scan ID															
	24	37	40	55	63	65	69	83	97	105	106	110	114	118	122	Avg.
NeuTex [55]	2.078	5.038	3.477	1.039	3.744	2.078	3.201	2.163	5.104	1.828	1.951	4.319	1.177	3.100	1.921	2.815
NeuS* [51]	1.544	**1.224**	1.065	**0.665**	1.286	**0.825**	0.904	1.350	1.320	0.855	0.987	1.328	**0.487**	**0.636**	**0.678**	1.010
Ours	**1.112**	1.262	**0.988**	0.674	**1.224**	0.835	**0.878**	**1.232**	**1.304**	**0.741**	**0.963**	**1.239**	0.558	0.645	0.739	**0.960**

is slightly different. As demonstrated in Fig. 5 and Table 1, our method is comparable or even better than NeuS and NeuTex on rendering quality. To achieve texture editing, NeuTex attempts to encode all the textures in a single continuous UV space, which works for plump objects (*e.g.*, plush toys or Buddah as shown in its paper) but struggles to reconstruct objects with complex shapes (*e.g.*, scissors in DTU Scan 37 and gears in NeRF-Synthetic Lego as shown in Fig. 5). We consider that because NeuTex tries to memorize all textures in the single continuous UV-map by using a simplified Atlas-Net [9] (*i.e.*, one atlas), which limits its representation of complex shapes. For NeuS, as it pursues better mesh reconstruction quality than novel view synthesis, the details of rendered images are slightly blurred and smoothed. By contrast, our representation not only delivers the capability of geometry and texture editing, but also shows clear appearance detail (see Fig. 5) and maintains mesh quality on par with NeuS (see Table 2).

4.3 Experiment on Geometry Editing

We now show the result of mesh-guided geometry editing in Fig. 6 (a), where we simply deform meshes with Blender, and the rendered objects are deformed simultaneously. Since an implicit field can be trivially deformed with non-rigid warping, we also compare our editing with a naïve field warping solution that is directly applied to NeuS, which bends the query points from the deformed space to the original space by computing interpolated warping with the offsets from 3 nearest vertices of the extracted mesh. As shown in Fig. 6 (b), the object

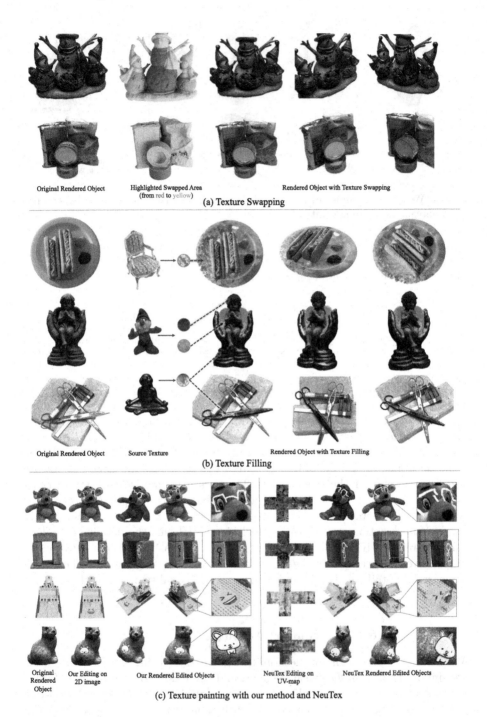

Fig. 7. We show texture editing examples on the DTU dataset and the NeRF 360° synthetic dataset.

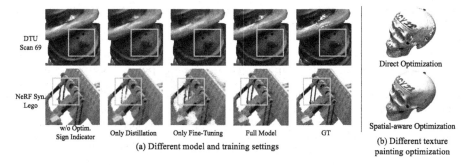

Fig. 8. We show the rendering results with different settings in (a) and also show the effectiveness of spatial-aware optimization for texture painting in (b).

boundary of the field warping results is much jaggier than ours, which proves the necessity of our mesh-based representation on this task.

4.4   Experiment on Texture Editing

To the best of our knowledge, prior to our work, only NeuTex [55] supports texture editing of the neural implicit field by painting on 2D UV texture. However, due to the distorted UV mapping, we find NeuTex hard to perform all the editing operations like ours, so we only compare it on the texture painting task.

Texture Swapping. We present 2 examples of texture swapping in Fig. 7 (a), where the textures of the snowman's body and the packaging of cans have been seamlessly swapped, and even the details (*e.g.*, texts on the cans) have been clearly transferred into the target object, while the geometry is kept unchanged. This demonstrates that our representation successfully disentangles geometry and texture in two spaces, and the disentangled texture representation can be seamlessly integrated into new shapes.

Texture Filling. We show 3 examples of texture filling in Fig. 7 (b), in which the targeting areas are repeatedly filled with template texture code and decoder from previously captured source models. It is worth noting that even though the source template only covers a small area of texture codes, we can still observe view-dependent effects (*e.g.*, golden metal naturally exhibits specular reflections at different views).

Texture Painting. In Fig. 7 (c), we exhibit 4 examples of texture painting, and also conduct similar editing with NeuTex [55] by painting on the unwrapped UV-map. For NeuTex, as the learned texture mapping is somehow irregular and distorted (*e.g.*, the head of the teddy bear is separated in UV-map), we find it hard to paint at the desired location. In the second row (painting on bricks), we have to adjust the painting position back and forth to get a reasonable editing result. Besides, due to the mapping issue explained in Sect. 4.2, NeuTex cannot picture a clear result when editing on complex shapes (Lego in the second row). On the contrary, our method offers a user-friendly editing pipeline by directly

Table 3. We perform ablation studies on the model design and training strategy with DTU Scan 69 and NeRF 360° synthetic lego.

Config.	DTU 69			NeRF 360° Synthetic Lego		
	PSNR ↑	SSIM ↑	LPIPS ↓	PSNR ↑	SSIM ↑	LPIPS ↓
w/o Optim. Sign indicator	26.633	0.940	0.119	27.631	0.922	0.053
Only distillation	26.599	0.936	0.144	25.606	0.901	0.081
Only fine-tuning	26.258	0.926	0.132	23.583	0.876	0.135
Full model	**27.254**	**0.946**	**0.113**	**27.881**	**0.926**	**0.046**

painting on 2D images and then transferring the painting into the 3D implicit field.

4.5   Ablation Studies

Learnable Sign Indicator. We first inspect the effectiveness of the proposed learnable sign indicator in each vertex. Specifically, we set sign indicators as constant vertex normal without adjusting during the training process and evaluate the model both qualitatively and quantitatively. As demonstrated in Fig. 8 (a) and Table 3 (first row), online adjusting sign indicators consistently improves the image quality. By the way, we notice that the PSNR improvement on real data (DTU Scan 69) is more significant than the synthetic one (Lego). We consider that the mesh quality (and vertex normal) of real data is worse than the synthetic data due to sensor noises, which degrades the rendering quality, while the learnable sign indicator helps to mitigate this issue.

Distillation and Fine-Tuning Training Scheme. We then study the necessity of distillation and fine-tuning training scheme by ablating one of them during model training. As shown in Fig. 8 (a) and Table 3 (second and third row), by enabling distillation only, the rendered image is blurry than the full model ones. When using fine-tuning without distillation, the rendering result ends up with noticeable artifacts. These results suggest that both distillation and fine-tuning are indispensable when training our mesh-based representation.

Spatial-Aware Optimization in Texture Painting. We also evaluate the proposed spatial-aware optimization in the texture painting task and visualize the comparison in Fig. 8 (b). It is clear that when naïvely optimizing painting with a single image, the model will overfit to the specific viewpoint, and the change to the texture codes might break the appearance consistency, which results in visual artifacts when rendering the implicit field from the side view. By introducing a spatial-aware optimization mechanism, we successfully avoid such artifacts and obtain the modified field while maintaining other parts untouched.

5   Conclusion

We have proposed a novel mesh-based neural representation, which supports high-fidelity volume rendering, and flexible geometry and texture editing. Specifically, we encode the neural implicit field into a mesh scaffold, where each mesh vertex possesses learnable geometry and texture code for its neighboring local space. One limitation of our method is that we do not model fine-grained lighting effects such as shadowing and specular reflection of a certain lighting environment, which can be improved by introducing material and lighting estimation in future works. Besides, due to the reliance on mesh scaffold, we cannot represent objects that fails during reconstruction (*e.g.*, smoke or liquid).

Acknowledgment. This work was partially supported by NSF of China (No. 61932003, No. 62102356).

References

1. Akenine-Moller, T., Haines, E., Hoffman, N.: Real-Time Rendering. AK Peters/CRC Press, Boca Raton (2019)
2. Boss, M., Braun, R., Jampani, V., Barron, J.T., Liu, C., Lensch, H.: NeRD: neural reflectance decomposition from image collections. In: Proceedings of the IEEE/CVF International Conference on Computer Vision, pp. 12684–12694 (2021)
3. Chen, A., et al.: Mvsnerf: fast generalizable radiance field reconstruction from multi-view stereo. In: Proceedings of the IEEE/CVF International Conference on Computer Vision, pp. 14124–14133 (2021)
4. Dellaert, F., Yen-Chen, L.: Neural volume rendering: NeRF and beyond. arXiv preprint arXiv:2101.05204 (2020)
5. Deng, Y., Yang, J., Tong, X.: deformed implicit field: modeling 3d shapes with learned dense correspondence. In: Proceedings of the IEEE/CVF Conference on Computer Vision and Pattern Recognition, pp. 10286–10296 (2021)
6. Freepik: Flaticon. https://www.flaticon.com/ (2022). Accessed 19 July 2022
7. Gao, D., Chen, G., Dong, Y., Peers, P., Xu, K., Tong, X.: Deferred neural lighting: free-viewpoint relighting from unstructured photographs. ACM Trans. Graph. (TOG) **39**(6), 1–15 (2020)
8. Gropp, A., Yariv, L., Haim, N., Atzmon, M., Lipman, Y.: Implicit geometric regularization for learning shapes. In: Proceedings of Machine Learning and Systems 2020, pp. 3569–3579 (2020)
9. Groueix, T., Fisher, M., Kim, V.G., Russell, B.C., Aubry, M.: A papier-mâché approach to learning 3d surface generation. In: Proceedings of the IEEE Conference on Computer Vision and Pattern Recognition, pp. 216–224 (2018)
10. Guo, M., Fathi, A., Wu, J., Funkhouser, T.: Object-centric neural scene rendering. arXiv preprint arXiv:2012.08503 (2020)
11. Izadi, S., et al.: Kinectfusion: real-time 3D reconstruction and interaction using a moving depth camera. In: Proceedings of the 24th Annual ACM Symposium on User Interface Software and Technology, pp. 559–568 (2011)
12. Jensen, R., Dahl, A., Vogiatzis, G., Tola, E., Aanæs, H.: Large scale multi-view stereopsis evaluation. In: 2014 IEEE Conference on Computer Vision and Pattern Recognition, pp. 406–413. IEEE (2014)
13. Kania, K., Yi, K.M., Kowalski, M., Trzciński, T., Tagliasacchi, A.: CoNeRF: controllable neural radiance fields. In: Proceedings of the IEEE/CVF Conference on Computer Vision and Pattern Recognition, pp. 18623–18632 (2022)
14. Karsch, K., Hedau, V., Forsyth, D.A., Hoiem, D.: Rendering synthetic objects into legacy photographs. ACM Trans. Graph. **30**(6), 157 (2011)

15. Kazhdan, M.M., Bolitho, M., Hoppe, H.: Poisson surface reconstruction. In: Proceedings of Eurographics Symposium on Geometry Processing, pp. 61–70 (2006)

16. Kholgade, N., Simon, T., Efros, A., Sheikh, Y.: 3D object manipulation in a single photograph using stock 3d models. ACM Trans. Graph. (TOG) **33**(4), 1–12 (2014)

17. Liu, L., Gu, J., Zaw Lin, K., Chua, T.S., Theobalt, C.: Neural sparse voxel fields. Adv. Neural Inf. Process. Syst. **33**, 15651–15663 (2020)

18. Liu, L., Habermann, M., Rudnev, V., Sarkar, K., Gu, J., Theobalt, C.: Neural actor: neural free-view synthesis of human actors with pose control. ACM Trans. Graph. (TOG) **40**(6), 1–16 (2021)

19. Liu, S., Li, T., Chen, W., Li, H.: Soft rasterizer: a differentiable renderer for image-based 3d reasoning. In: The IEEE International Conference on Computer Vision (ICCV), October 2019

20. Liu, S., Zhang, X., Zhang, Z., Zhang, R., Zhu, J.Y., Russell, B.: Editing conditional radiance fields. In: Proceedings of the IEEE/CVF International Conference on Computer Vision, pp. 5773–5783 (2021)

21. Lorensen, W.E., Cline, H.E.: Marching cubes: a high resolution 3d surface construction algorithm. ACM SIGGRAPH Comput. Graph. **21**(4), 163–169 (1987)

22. Luo, J., Huang, Z., Li, Y., Zhou, X., Zhang, G., Bao, H.: NIID-net: adapting surface normal knowledge for intrinsic image decomposition in indoor scenes. IEEE Trans. Visual. Comput. Graph. **26**(12), 3434–3445 (2020)

23. Mildenhall, B., Srinivasan, P.P., Tancik, M., Barron, J.T., Ramamoorthi, R., Ng, R.: NeRF: representing scenes as neural radiance fields for view synthesis. In: Vedaldi, A., Bischof, H., Brox, T., Frahm, J.-M. (eds.) ECCV 2020. LNCS, vol. 12346, pp. 405–421. Springer, Cham (2020). https://doi.org/10.1007/978-3-030-58452-8_24

24. Müller, T., Evans, A., Schied, C., Keller, A.: Instant neural graphics primitives with a multiresolution hash encoding. ACM Trans. Graph. **41**(4), 102:1–102:15 (2022)

25. Murez, Z., van As, T., Bartolozzi, J., Sinha, A., Badrinarayanan, V., Rabinovich, A.: Atlas: end-to-end 3d scene reconstruction from posed images. In: Vedaldi, A., Bischof, H., Brox, T., Frahm, J.-M. (eds.) ECCV 2020. LNCS, vol. 12352, pp. 414–431. Springer, Cham (2020). https://doi.org/10.1007/978-3-030-58571-6_25

26. Niemeyer, M., Geiger, A.: Giraffe: representing scenes as compositional generative neural feature fields. In: Proceedings of the IEEE/CVF Conference on Computer Vision and Pattern Recognition, pp. 11453–11464 (2021)

27. Oechsle, M., Mescheder, L., Niemeyer, M., Strauss, T., Geiger, A.: Texture fields: learning texture representations in function space. In: Proceedings of the IEEE/CVF International Conference on Computer Vision, pp. 4531–4540 (2019)

28. Oechsle, M., Peng, S., Geiger, A.: UNISURF: unifying neural implicit surfaces and radiance fields for multi-view reconstruction. In: International Conference on Computer Vision (ICCV) (2021)

29. Ost, J., Laradji, I., Newell, A., Bahat, Y., Heide, F.: Neural point light fields. arXiv preprint arXiv:2112.01473 (2021)

30. Park, K., et al.: Nerfies: deformable neural radiance fields. In: Proceedings of the IEEE/CVF International Conference on Computer Vision, pp. 5865–5874 (2021)

31. Park, K., et al.: HyperNeRF: a higher-dimensional representation for topologically varying neural radiance fields. ACM Trans. Graph. **40**(6) (2021)

32. Peng, S., et al.: Neural body: implicit neural representations with structured latent codes for novel view synthesis of dynamic humans. In: Proceedings of the IEEE/CVF Conference on Computer Vision and Pattern Recognition, pp. 9054–9063 (2021)

33. Pérez, P., Gangnet, M., Blake, A.: Poisson Image Editing. ACM Trans. Graph. **22**(3), 313–318 (2003)
34. Qi, C.R., Su, H., Mo, K., Guibas, L.J.: PointNet: deep learning on point sets for 3d classification and segmentation. In: Proceedings of the IEEE conference on Computer Vision and Pattern Recognition, pp. 652–660 (2017)
35. Reiser, C., Peng, S., Liao, Y., Geiger, A.: KiloNeRF: speeding up neural radiance fields with thousands of tiny MLPs. In: Proceedings of the IEEE/CVF International Conference on Computer Vision, pp. 14335–14345 (2021)
36. Rematas, K., Ferrari, V.: Neural voxel renderer: learning an accurate and controllable rendering tool. In: Proceedings of the IEEE/CVF Conference on Computer Vision and Pattern Recognition, pp. 5417–5427 (2020)
37. Riegler, G., Koltun, V.: Free view synthesis. In: Vedaldi, A., Bischof, H., Brox, T., Frahm, J.-M. (eds.) ECCV 2020. LNCS, vol. 12364, pp. 623–640. Springer, Cham (2020). https://doi.org/10.1007/978-3-030-58529-7_37
38. Riegler, G., Koltun, V.: Stable view synthesis. In: Proceedings of the IEEE/CVF Conference on Computer Vision and Pattern Recognition, pp. 12216–12225 (2021)
39. Fridovich-Keil, S., Yu, A., Tancik, M., Chen, Q., Recht, B., Kanazawa, A.: Plenoxels: radiance fields without neural networks. In: CVPR (2022)
40. Schönberger, J.L., Frahm, J.: Structure-from-motion revisited. In: Proceedings of IEEE Conference on Computer Vision and Pattern Recognition, pp. 4104–4113. IEEE Computer Society (2016)
41. Schwarz, K., Liao, Y., Niemeyer, M., Geiger, A.: GRAF: generative radiance fields for 3d-aware image synthesis. Adv. Neural Inf. Process. Syst. **33**, 20154–20166 (2020)
42. Shetty, R.R., Fritz, M., Schiele, B.: Adversarial scene editing: automatic object removal from weak supervision. Adv. Neural Inf. Process. Syst. **31** (2018)
43. Sorkine, O., Alexa, M.: As-rigid-as-possible surface modeling. In: Symposium on Geometry Processing, vol. 4, pp. 109–116 (2007)
44. Srinivasan, P.P., Deng, B., Zhang, X., Tancik, M., Mildenhall, B., Barron, J.T.: NeRV: neural reflectance and visibility fields for relighting and view synthesis. In: Proceedings of the IEEE/CVF Conference on Computer Vision and Pattern Recognition, pp. 7495–7504 (2021)
45. Sun, J., Xie, Y., Chen, L., Zhou, X., Bao, H.: NeuralRecon: real-time coherent 3d reconstruction from monocular video. In: Proceedings of the IEEE/CVF Conference on Computer Vision and Pattern Recognition, pp. 15598–15607 (2021)
46. Sun, J., et al.: FENeRF: face editing in neural radiance fields. In: Proceedings of the IEEE/CVF Conference on Computer Vision and Pattern Recognition, pp. 7672–7682 (2022)
47. Tancik, M., et al.: Block-NeRF: scalable large scene neural view synthesis. In: Proceedings of the IEEE/CVF Conference on Computer Vision and Pattern Recognition, pp. 8248–8258 (2022)
48. Thies, J., Zollhöfer, M., Nießner, M.: Deferred neural rendering: image synthesis using neural textures. ACM Trans. Graph. (TOG) **38**(4), 1–12 (2019)
49. Umeyama, S.: Least-squares estimation of transformation parameters between two point patterns. IEEE Trans. Pattern Anal. Mach. Intell. **13**(04), 376–380 (1991)
50. Waechter, M., Moehrle, N., Goesele, M.: Let there be color! large-scale texturing of 3d reconstructions. In: Fleet, D., Pajdla, T., Schiele, B., Tuytelaars, T. (eds.) ECCV 2014. LNCS, vol. 8693, pp. 836–850. Springer, Cham (2014). https://doi.org/10.1007/978-3-319-10602-1_54

51. Wang, P., Liu, L., Liu, Y., Theobalt, C., Komura, T., Wang, W.: NeuS: learning neural implicit surfaces by volume rendering for multi-view reconstruction. In: NeurIPS (2021)
52. Wang, Q., et al.: IBRNet: learning multi-view image-based rendering. In: Proceedings of the IEEE/CVF Conference on Computer Vision and Pattern Recognition, pp. 4690–4699 (2021)
53. Wang, T.Y., Su, H., Huang, Q., Huang, J., Guibas, L.J., Mitra, N.J.: Unsupervised texture transfer from images to model collections. ACM Trans. Graph. **35**(6), 1–177 (2016)
54. Wang, W., Yu, R., Huang, Q., Neumann, U.: SGPN: similarity group proposal network for 3d point cloud instance segmentation. In: Proceedings of the IEEE Conference on Computer Vision and Pattern Recognition, pp. 2569–2578 (2018)
55. Xiang, F., Xu, Z., Hasan, M., Hold-Geoffroy, Y., Sunkavalli, K., Su, H.: NeuTex: neural texture mapping for volumetric neural rendering. In: Proceedings of the IEEE/CVF Conference on Computer Vision and Pattern Recognition, pp. 7119–7128 (2021)
56. Xiangli, Y., et al.: CityNeRF: Building NeRF at city scale. arXiv preprint arXiv:2112.05504 (2021)
57. Xie, C., Park, K., Martin-Brualla, R., Brown, M.: FiG-NeRF: figure-ground neural radiance fields for 3d object category modelling. In: 2021 International Conference on 3D Vision (3DV), pp. 962–971. IEEE (2021)
58. Xu, Q., Tao, W.: Multi-scale geometric consistency guided multi-view stereo. In: Proceedings of IEEE Conference on Computer Vision and Pattern Recognition, pp. 5483–5492 (2019)
59. Yang, B., et al.: Neural rendering in a room: amodal 3d understanding and free-viewpoint rendering for the closed scene composed of pre-captured objects. ACM Trans. Graph. **41**(4), 101:1–101:10 (2022)
60. Yang, B., et al.: Learning object-compositional neural radiance field for editable scene rendering. In: Proceedings of the IEEE/CVF International Conference on Computer Vision, pp. 13779–13788 (2021)
61. Yariv, L., Gu, J., Kasten, Y., Lipman, Y.: Volume rendering of neural implicit surfaces (2021)
62. Yariv, L., et al.: Multiview neural surface reconstruction by disentangling geometry and appearance. Adv. Neural Inf. Process. Syst. **33**, 2492–2502 (2020)
63. Yen-Chen, L., Florence, P., Barron, J.T., Rodriguez, A., Isola, P., Lin, T.Y.: iNeRF: inverting neural radiance fields for pose estimation. In: 2021 IEEE/RSJ International Conference on Intelligent Robots and Systems (IROS), pp. 1323–1330. IEEE (2021)
64. Yuan, Y.J., Sun, Y.T., Lai, Y.K., Ma, Y., Jia, R., Gao, L.: NeRF-editing: geometry editing of neural radiance fields. In: Proceedings of the IEEE/CVF Conference on Computer Vision and Pattern Recognition (CVPR), pp. 18353–18364, June 2022
65. Zhang, J., et al.: Editable free-viewpoint video using a layered neural representation. ACM Trans. Graph. (TOG) **40**(4), 1–18 (2021)
66. Zhang, X., Srinivasan, P.P., Deng, B., Debevec, P., Freeman, W.T., Barron, J.T.: NeRFactor: neural factorization of shape and reflectance under an unknown illumination. ACM Trans. Graph. (TOG) **40**(6), 1–18 (2021)
67. Zhao, B., et al.: Factorized and controllable neural re-rendering of outdoor scene for photo extrapolation. In: Proceedings of the 30th ACM International Conference on Multimedia (2022)
68. Zhou, Q.Y., Park, J., Koltun, V.: Open3D: a modern library for 3d data processing. arXiv preprint arXiv:1801.09847 (2018)

NeRF for Outdoor Scene Relighting

Viktor Rudnev[1,2]([✉]) [ID], Mohamed Elgharib[1], William Smith[3] [ID], Lingjie Liu[1], Vladislav Golyanik[1], and Christian Theobalt[1] [ID]

[1] MPI for Informatics, SIC, Saarbrücken, Germany
vrudnev@mpi-inf.mpg.de
[2] Saarland University, SIC, Saarbrücken, Germany
[3] University of York, Saarbrücken, Germany

Abstract. Photorealistic editing of outdoor scenes from photographs requires a profound understanding of the image formation process and an accurate estimation of the scene geometry, reflectance and illumination. A delicate manipulation of the lighting can then be performed while keeping the scene albedo and geometry unaltered. We present NeRF-OSR, *i.e.*, the first approach for outdoor scene relighting based on neural radiance fields. In contrast to the prior art, our method allows simultaneous editing of illumination and camera viewpoint using only a collection of outdoor photos shot in uncontrolled settings. Moreover, it enables direct control over the scene illumination, as defined through a spherical harmonics model. For evaluation, we collect a new benchmark dataset of several outdoor sites photographed from multiple viewpoints and at different times. For each time, a 360° environment map is captured together with a colour-calibration chequerboard to allow accurate numerical evaluations on real data against ground truth. Comparisons against SoTA show that NeRF-OSR enables controllable lighting and viewpoint editing at higher quality and with realistic self-shadowing reproduction. (see the project web page https://4dqv.mpi-inf.mpg.de/NeRF-OSR/)

1 Introduction

Controllable lighting editing of real scenes from photographs is a long-standing and challenging problem, with several applications in virtual and augmented reality [8,20,21,27,46]. It requires explicit modelling of the image formation process and an accurate estimation of the material properties and scene illumination. Such scene decomposition enables manipulating the lighting in isolation while maintaining the integrity of the remaining scene components (*e.g.*, albedo and geometry.) While several methods for controllable lighting editing exist, some solutions are dedicated to a specific class of objects such as human faces [18,34] and human bodies [8,21]. Other solutions are designed for processing either

---

Supplementary Information The online version contains supplementary material available at https://doi.org/10.1007/978-3-031-19787-1_35.

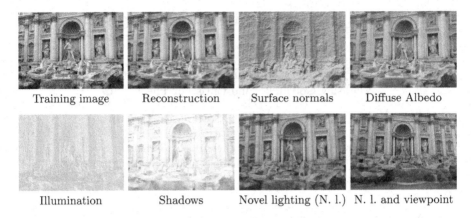

| Training image | Reconstruction | Surface normals | Diffuse Albedo |

| Illumination | Shadows | Novel lighting (N. l.) | N. l. and viewpoint |

Fig. 1. NeRF-OSR is the first neural radiance fields approach for outdoor scene relighting. We learn a neural representation of the scene geometry, diffuse albedo and illumination-dependent shadows from a set of images capturing the same site from different viewpoints and at different times. The learnt intrinsics enable simultaneous editing of both the scene's lighting and viewpoint.

indoor [7,20,30,33,45,49] or outdoor [1,5,27,44,46] scenes. Due to the very different nature of indoor and outdoor data, methods for relighting them were largely treated separately in the literature. In this work, we focus on outdoor scene relighting. Unlike existing methods [1,5,27,44,46], our approach is the first to simultaneously edit both scene illumination and camera viewpoint.

The recently proposed Neural Radiance Fields (NeRF) [23] is a powerful neural 3D scene representation capable of self-supervised training from 2D images recorded by a calibrated monocular camera [26,39,40,48]. At test time, NeRF can produce photorealistic novel scene views. While there were a few attempts to extend NeRF for lighting editing [3,19,33,35,49], existing approaches are either designed for a specific object class [35], require known or single illumination condition for training [33,49] or they do not model important outdoor illumination effects such as cast shadows [3]. Most existing NeRF-based relighting methods [3,33,35,49] are not designed for outdoor scenes captured in uncontrolled settings. An exception to this is, at first sight, NeRF in the Wild (NeRF-W) [19] trained from uncontrolled images, factoring per-image appearance into an embedding space. However, NeRF-W and more recent follow-ups [4,37] do not perform intrinsic image decomposition and thus semantically meaningful parametric control of lighting, shadows or even albedo is not possible.

This paper addresses the shortcomings of existing methods and presents NeRF-OSR, *i.e.,* the first approach based on neural radiance fields that can change both illumination and camera viewpoint of outdoor scenes photographed in uncontrolled settings, in a high-quality and semantically meaningful way; see Fig. 1. Our approach models the image formation process, disentangling the input image into its intrinsic components and scene illumination. It also contains a dedicated network for learning shadows, whose realistic reproduction is crucial for high-quality outdoor scene relighting. NeRF-OSR is trained in a self-supervised manner on multiple images of a site photographed from different viewpoints and under different illuminations. We evaluate our method qualitatively and quantitatively on a variety of outdoor scenes and show that it outperforms state of the art. Aspects of the novelty of our work include:

- NeRF-OSR, *i.e.*, the first method using neural radiance fields for outdoor scene relighting supporting simultaneous and semantically meaningful editing of scene illumination and camera viewpoint. Our model has explicit control over the scene intrinsics, including local shading, shadows and even albedo.
- Our method learns a neural scene representation that decomposes the scene into spatial occupancy, illumination, shadowing and diffuse albedo reflectance. It is trained in a self-supervised manner from outdoor data captured from various viewpoints and at different illuminations.
- A new and biggest in literature benchmark dataset for outdoor scene relighting. It includes eight buildings photographed from 3240 viewpoints and at 110 different times. In addition, it is the first one that includes colour-calibrated 360° environment maps, which allows accurate numerical evaluations.

2 Related Work

Scene Relighting. There are several methods for outdoor illumination editing [1,5,10–12,27,36,44,46,47]. Some of them focus on integrating objects into images in an illumination-consistent manner [10,44], while others process the full scene [1,5,11,12,27,36,46,47]. Duchene *et al.* [5] estimate scene reflectance, shading and visibility from multiple views shot at fixed lighting. They produce novel relighting effects such as moving cast shadows. Barron *et al.* [1] formulate inverse rendering through statistical inference. Given a single RGB image of an object, their method estimates the most likely shape normals, reflectance, shading and illumination that can reproduce the examined image. They assume piecewise smooth and low-entropy reflectance images and isotropic (with frequent bends) surfaces. Philip *et al.* [27] guide relighting via a proxy geometry estimated from multi-view images. Their neural network translates image-space buffers of the examined scene into the desired relighting. The buffers include shadow masks (estimated from the extracted geometry), normal maps and illumination components. Philip *et al.*'s method is trained with high-quality synthetic data. However, it is primarily designed to edit only the illumination of the input and not the camera viewpoint. Furthermore, their illumination model is limited to sun lighting and can not handle other cases such as cloudy skies.

Yu and Smith [47] estimate the albedo, normals and lighting of an outdoor scene from just a single image; lighting is modelled through spherical harmonics (SH) with a statistical model as a prior. Relighting is then achieved by editing the reconstructed illumination (using a low-frequency model). Yu *et al.* [46] train a method for scene relighting given a single image in a self-supervised manner on a large corpus of uncontrolled outdoor images. A neural renderer takes the original albedo and geometry, the target shading and the target shadowing, and relights the scene; a dedicated network predicts the target shadows. Next, residuals of the inverse rendering are also supplied to the neural renderer as input to better capture scene details. Impressive results are shown visually and validated numerically on a new benchmark dataset. In contrast to our approach, neither Yu *et al.* [46] nor Yu and Smith [47] can edit the camera viewpoint.

Recently, there were efforts in developing relighting methods using NeRF backbone [3,33,35,49]. Most of these methods operate in a setting different from ours, *i.e.*, they either require input images with a single illumination condition [49], assume a known illumination during training [33] or are designed for a specific class of objects such as faces [35]. The closest to our technique is NeRD by Boss *et al.* [3], in the sense it can operate on images of the same scene shot under different illuminations. Here, the spatially varying BRDF of the examined scene is estimated through the help of physically-based rendering. To allow fast rendering at arbitrary viewpoints and illumination, the learnt reflectance volume is converted into a relightable texture mesh. Unlike our NeRF-OSR, NeRD does not explicitly model shadows, which are crucial for high-quality outdoor scene relighting. Furthermore, it requires the examined object to be at a similar distance from all views—an assumption that can not be easily satisfied for outdoor photographs captured in an uncontrolled setup.

Style-Based Editing. Scene relighting techniques are distantly related to style-based category of appearance editing methods [4,15,17,19,22,24,31,37]. Unlike relighting methods, the latter do not have a physical understanding of the scene illumination and seek to edit the overall appearance at once. Hence, they lack explicit parametric control over the local shading and shadows. In contrast, our NeRF-OSR performs scene intrinsic decomposition and seeks to edit illumination in isolation from albedo and geometry. It also directly models illumination-based shadows, which is crucial for high-quality outdoor relighting. Next, our intrisic decomposition allow editing applications that are not possible by style-based methods by any means (*e.g.*, inserting objects by editing the albedo channel separately and then relighting the entire composited new scene).

3   Method

NeRF-OSR takes as input multiple RGB images of a single scene, shot at different timings and from different viewpoints. It then renders the examined scene from an arbitrary viewpoint and under various illuminations. Our method estimates the scene intrinsics explicitly and has direct access to the scene illumination. It also includes a dedicated component for predicting shadows, *i.e.*, an essential feature of outdoor scene illumination.

An overview of NeRF-OSR is shown in Fig. 2. At its heart is a neural radiance fields (NeRF), *i.e.*, a neural implicit scene representation for volumetric rendering. Our method is trained in a self-supervised manner on outdoor data captured in uncontrolled settings and can render photorealistic views. Next, we describe in Sect. 3.1 the NeRF model [23] without view-dependent effects, which we build upon. We then discuss our illumination model and how it is adapted

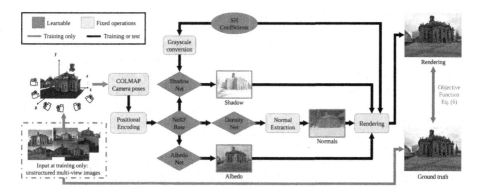

Fig. 2. Our NeRF-OSR uses outdoor images of a site photographed in an uncontrolled setting (dashed green) to recover a relightable implicit scene model. It learns the scene intrinsics and illumination as expressed by the SH coefficients. Here, a dedicated neural component learns shadows. During the test, NeRF-OSR can synthesise novel images at arbitrary camera viewpoints and scene illumination; the user directly supplies the desired camera pose and the scene illumination, either from an environment map or via SH coefficients. (Color figure online)

in a volumetric-based representation in Sect. 3.2–3.3. The objective function is presented in Sect. 3.4, followed by a discussion of the training details (Sect. 3.5).

3.1   Neural Radiance Fields (NeRF)

For each point $\mathbf{x}$ in 3D space, NeRF [23] defines its density $\sigma(\mathbf{x})$ and colour $\mathbf{c}(\mathbf{x})$. To render an image, a ray is cast from the camera origin $\mathbf{o}$, in a direction $\mathbf{d}$ corresponding to each of the output pixels. N_{depth} points $\{\mathbf{x}_i\}_{i=1}^{N_{\text{depth}}}$ are sampled along each ray, where $\mathbf{x}_i = \mathbf{o} + t_i\mathbf{d}$ and $\{t_i\}_{i=1}^{N_{\text{depth}}}$ are the corresponding ray depths. The final colour in the image space $\mathbf{C}(\mathbf{o}, \mathbf{d})$ is obtained by integrating the density and colour along the ray $(\mathbf{o}, \mathbf{d})$ as follows:

$$\mathbf{C}(\mathbf{o}, \mathbf{d}) = \mathbf{C}\left(\{\mathbf{x}_i\}_{i=1}^{N_{\text{depth}}}\right) = \sum_{i=1}^{N_{\text{depth}}} T(t_i)\alpha(\sigma(\mathbf{x}_i)\delta_i)\mathbf{c}(\mathbf{x}_i), \tag{1}$$

where $T(t_i) = \exp\left(-\sum_{j=1}^{N_{\text{depth}}-1} \sigma(\mathbf{x}_j)\delta_j\right)$, $\delta_i = t_{i+1}-t_i$, and $\alpha(y) = 1-\exp(-y)$. The depths $\{t_i\}_{i=1}^{N_{\text{depth}}}$ are selected using stratified sampling from the uniform distribution, spanning the depths along $(\mathbf{o}, \mathbf{d})$ starting from the near and ending at the far camera plane. Both density $\sigma(\mathbf{x})$ and colour $\mathbf{c}(\mathbf{x})$ are modelled using MLPs, and the final rendering is trained in a self-supervised manner using the observed ground-truth per-pixel colours.

To better capture small details, NeRF uses *hierarchical volume sampling* for $\{t_i\}_{i=1}^{N_{\text{depth}}}$, *i.e.*, instead of performing a single rendering pass, points are first sampled in stratified manner. The densities at these points are then used for importance sampling in the final pass. The final model is thus learnt by supervising the rendered pixel colours of both passes with the ground-truth colours.

3.2 Spherical Harmonics NeRF

While (1) allows for high-quality free viewpoint synthesis, $\mathbf{c}(\mathbf{x})$ are defined only through an MLP that does not encode the lighting. In other words, such formulation learns a Lambertian model of the scene under a fixed lighting. The more generalised model with view direction dependencies [23] learns a slice of the apparent BRDF at a fixed illumination. Nonetheless, this learnt representation still does not have a semantic meaning of the underlying scene intrinsics and has no direct control over the lighting.

To allow relighting, we introduce an explicit 2nd-order Spherical Harmonics (SH) illumination model [2] and redefine the rendering Eq. (1) as follows:

$$\mathbf{C}\left(\{\mathbf{x}_i\}_{i=1}^{N_{\text{depth}}}, \mathbf{L}\right) = \mathbf{A}\left(\{\mathbf{x}_i\}_{i=1}^{N_{\text{depth}}}\right) \odot \mathbf{Lb}\left(\mathbf{N}\left(\{\mathbf{x}_i\}_{i=1}^{N_{\text{depth}}}\right)\right), \qquad (2)$$

where $\odot$ denotes elementwise multiplication. $\mathbf{A}(\mathbf{x}) \in \mathbb{R}^3$ is the accumulated albedo colour, generated in the similar way as in (1), *i.e.*, by integrating the output of an albedo MLP. $\mathbf{L} \in \mathbb{R}^{9 \times 3}$ is the per-image learnable SH coefficients, and $\mathbf{b}(\mathbf{n}) \in \mathbb{R}^9$ is the SH basis. $\mathbf{N}(\mathbf{x})$ is the surface normal computed from the accumulated ray density. It is defined as

$$\mathbf{N}\left(\{\mathbf{x}_i\}_{i=1}^{N_{\text{depth}}}\right) = \frac{\hat{\mathbf{N}}\left(\{\mathbf{x}_i\}_{i=1}^{N_{\text{depth}}}\right)}{\left\|\hat{\mathbf{N}}\left(\{\mathbf{x}_i\}_{i=1}^{N_{\text{depth}}}\right)\right\|^2}, \qquad (3)$$

where $\hat{\mathbf{N}}\left(\{\mathbf{x}_i\}_{i=1}^{N_{\text{depth}}}\right) = \sum_{i=1}^{N_{\text{depth}}} \left(\frac{\partial}{\partial \mathbf{x}_i}\sigma(\mathbf{x}_i)\right) \odot T(t_i)\alpha(\sigma(\mathbf{x}_i)\delta_i). \qquad (4)$

To extract $\mathbf{N}$, we first differentiate the density of points on the ray with respect to the original x-, y-, z-components of the ray samples, accumulate them over all N_{depth} samples on the ray with weights $T(t_i)\alpha(\sigma(\mathbf{x}_i)\delta_i)$, and normalise the resulting vector to a unit sphere. Note that in (2), we render in screen space using screen space albedo and normals accumulated from the neural volume. The accumulation makes the albedo and surface normal estimates less noisy and aids convergence. It also means we only make a single shading calculation rather than the alternative of one per sample point and accumulating shaded colours.

All terms of (2) are learnable except for the SH basis $\mathbf{b}(\cdot)$ and the normal extraction operator $\mathbf{N}(\cdot)$, which are based on fixed, explicit models. The proposed lighting model integration allows for explicit relighting by varying $\mathbf{L}$. While it accounts for Lambertian effects, it lacks direct shadow generation, which is crucial for modelling and subsequent relighting of outdoor scenes.

3.3 Shadow Generation Network

To allow for explicit shadow control during relighting, we introduce a dedicated shadow model $S\left(\{\mathbf{x}_i\}_{i=1}^{N_{\text{depth}}}, \mathbf{L}\right)$ and extend the rendering Eq. (2):

$$\mathbf{C}\left(\{\mathbf{x}_i\}_{i=1}^{N_{\text{depth}}}, \mathbf{L}\right) = S\left(\{\mathbf{x}_i\}_{i=1}^{N_{\text{depth}}}, \mathbf{L}\right)\mathbf{A}\left(\{\mathbf{x}_i\}_{i=1}^{N_{\text{depth}}}\right) \odot \mathbf{Lb}\left(\mathbf{N}\left(\{\mathbf{x}_i\}_{i=1}^{N_{\text{depth}}}\right)\right). \qquad (5)$$

The shadow model is defined with a scalar computed by an MLP $s(\mathbf{x}, \mathbf{L}) \in [0, 1]$. The final shadow value is computed by accumulating along the ray into $S\left(\{\mathbf{x}_i\}_{i=1}^{N_{\text{depth}}}, \mathbf{L}\right) \in [0, 1]$, in the same way as in (1). Figure 2 shows the high-level diagram of the proposed NeRF-OSR. Note that the shadow prediction network takes as input the SH coefficients in their grey-scale version, *i.e.*, $\mathbf{L} \in \mathbb{R}^{1 \times 9}$ and not $\mathbb{R}^{3 \times 9}$. This is motivated by the fact that shadows depend only on the spatial light distribution. Unlike traditional ray-tracing approaches as the one used in [27,33], our shadow estimator operates much more efficiently, through just same single forward pass as albedo and geometry. We argue that it is a strength, as it makes the method much more computationally scalable, while still allowing us to relight using completely new illumination conditions.

3.4 Objective Function

We optimise the following loss function:

$$\mathcal{L}(\mathbf{C}, \mathbf{C}^{(\text{GT})}, S) = \text{MSE}(\mathbf{C}, \mathbf{C}^{(\text{GT})}) + \lambda \text{MSE}(S, 1), \tag{6}$$

where $\text{MSE}(\cdot, \cdot)$ is the mean squared error. The first term is a reconstruction loss defined on the estimated colour $\mathbf{C}$ and the corresponding ground truth $\mathbf{C}^{(\text{GT})}$. The second term regularises shadows. The shadow network S absorbs all greyscale lighting effects that cannot be explained by SH. To limit it to learning only shadows, we select the largest value of the regularisation strength λ that does not degrade the PSNR of the reconstructed images. Experimentation shows that removing the regulariser usually leads to S learning all the illumination components, except for the chromaticity—thus making the SH lighting useless.

3.5 Training and Implementation Details

Our self-supervised model is trained on RGB images of an outdoor scene photographed from various viewpoints and under different illumination. We next describe several strategies for training our method and their importance.

Frequency Annealing. We noticed empirically that training the model as-is leads to noisy normal maps. Above some threshold on the number of the positional encoding (PE) frequencies, the initially generated noise (at the start of the training) becomes very hard to manipulate; it hardly converges to the correct geometry. Hence, we alleviate this by using the annealing scheme slightly modified from Deformable NeRF [26], *i.e.*, we add an annealing coefficient $\beta_k(n)$ to each of the PE components $\gamma_k(\mathbf{x})$: $\gamma_k'(\mathbf{x}) = \gamma_k(\mathbf{x})\beta_k(n)$, where $\beta_k(n) = \frac{1}{2}(1 - \cos(\pi \text{clamp}(\alpha - i + N_{\text{fmin}}, 0, 1)))$, $\alpha(n) = (N_{\text{fmax}} - N_{\text{fmin}})\frac{n}{N_{\text{anneal}}}$, n is the current training iteration, N_{fmax} is the total number of used PE frequencies (the proposed model uses 12), N_{fmin} is the number of used PE frequences at the start (we use 8), N_{anneal} is tuned empirically to $3 \cdot 10^4$ for all sequences. This training strategy enables significantly improved geometry predictions.

Fig. 3. Sample views from the new benchmark dataset for outdoor scene relighting. The dataset has 3240 views captured in 110 different recording sessions.

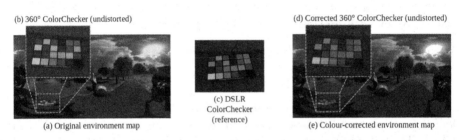

Fig. 4. For each recording session, we capture a colour chequerboard with the DSLR and 360° cameras. We colour-correct the 360° maps to match the DSLR.

Ray Direction Jitter. To improve the generalisability of NeRF-OSR, we apply a sub-pixel jitter to the ray direction. Here, instead of shooting in the pixel centres, a jitter ψ is used as follows: $x_i = \mathbf{o} + t_i(\mathbf{d} + \psi)$. We sample ψ uniformly, such that the resulting ray still confines to the boundaries of its designated pixel.

Shadow Network Input Jitter. Since the shadows are generated in a learning-based fashion instead of using direct geometric approaches, there remains the possibility of overfitting to the training lightings. To mitigate this effect, we add a slight normal noise ε to the environment coefficients as input of the shadow generation network:

$$S'\left(\{\mathbf{x}_i\}_{i=1}^{N_{\text{depth}}}, \mathbf{L}\right) = S\left(\{\mathbf{x}_i\}_{i=1}^{N_{\text{depth}}}, \mathbf{L} + \varepsilon\right), \tag{7}$$

where $\varepsilon \sim \mathcal{N}(0, 0.025I)$. (7) can be interpreted as a locality condition, *i.e.*, in similar lighting conditions, shadows should not be too different. This allows the model to learn smoother transitions between different lightings.

Implementation. We use NeRF++ [48] with the background network disabled as the code base and work within the unit sphere bounds of the foreground network. For training and evaluation, we use two Nvidia Quadro RTX 8000 GPUs. We train the model for $5 \cdot 10^5$ iterations using a batch size of 2^{10} rays, which takes ≈ 2 days.

4   A New Benchmark for Outdoor Scene Relighting

Several datasets for outdoor sites exist [9,14,15,32,46]. Most of them [9,14, 15,32] were collected with the task of 3D scene reconstruction in mind and not relighting. Hence, they mostly contain publicly available photos collected in uncontrolled settings. Furthermore, they do not provide environment maps, which are important for evaluating relighting techniques numerically on real data against ground truth. Examples of such datasets are the PhotoTourism [9,32] and the MegaDepth [14]. The MegaDepth dataset consists of multi-view images of several sites that were initially a part of the Landmarks10k dataset [13]. Here, the depth signal is extracted using COLMAP [29] and the multi-view stereo (MVS) approach [28]. While MegaDepth was originally released as a benchmark for single-view depth extraction, it was used by Yu *et al.* [46] (one of the most recent relighting works). However, it can only evaluate methods qualitatively.

To allow for numerical evaluation on real data against ground truth, Yu *et al.* [46] recorded *one site* from different viewpoints and at different times of the day using a DSLR camera, along with the environment maps. Unfortunately, this benchmark is limited in two ways: First, it contains a single site. Second, the captured environment maps were not colour-corrected with respect to the DSLR camera of the main recordings. *Hence, numerical results obtained with this dataset would always differ from the ground truth by an unknown, possibly nonlinear, colour transformation.* Therefore, any error metric must first compute an optimal transformation (Yu *et al.* [46] used a per-colour channel linear scaling). This makes it hard to separate the behaviour of the examined relighting methods from the corrective behaviour of this normalisation.

Hence, we present a new benchmark for outdoor scene relighting. Our dataset is the first of its kind in terms of size and the ability to perform accurate numerical evaluations on real data against ground truth. It is much larger than Yu *et al.* [46], containing eight sites captured from various viewpoints using a DSLR camera (3240 viewpoints captured in 110 different recording sessions). Multiple recording sessions were performed for each site, at different times of the day; all sessions cover different weathers, including sunny and cloudy days. We also capture a 360° shot of the environment map for each session. Unlike Yu *et al.* [46], we explicitly account for the colour calibration between the environment maps and the DSLR camera of the main recordings. To this end—for every session in the test set—we also simultaneously capture the "GretagMacbeth ColorChecker" colour calibration chart with the DSLR and the 360° cameras. We then apply the second-order method of Finlayson *et al.* [6] to colour-correct the environment maps by calibrating their ColorChecker values to the ColorChecker values of the corresponding DSLR image. Finally, we manually align the environment maps to the world coordinates using COLMAP [29] reconstructions of each site.

Figure 3 shows samples from the various sites from our dataset and the corresponding environment maps. See Fig. 4 for the colour-corrected environment maps. Note that we target scenes with minimal specular effects (such as brick or wood buildings). All data was captured in exposure brackets of five photos ranging from –3 to +3 EV for the DSLR photos and from –2 to +2 EV for the environment maps. We used the darkest capture for the 360° environment maps so that the sun is least overexposed. For the ColorChecker calibration with DSLR, we use images that are dark enough so that the white cells of the chequerboard are not overexposed. The DSLR image resolution is 5184×3456 pixel, while the resolution of the environment maps is 5660×2830 pixel.

5   Results

We evaluate the performance of NeRF-OSR on various real-world sites. We examine three sites from our newly proposed dataset and the Trevi Fountain from the PhotoTourism dataset [19]. These scenes include a variety of features. This includes large and small scale details as the sculptures (Site 1, Trevi), structural details such as trees and umbrellas (Site 3), a piecewise-smooth surfaces casting a lot of shadows on itself (Site 2), water (Trevi) and surrounding buildings (in all). Furthermore, Trevi Fountain shows performance on data collected completely from the internet through crowdsourcing. Note that only qualitative evaluation on Trevi Fountain is possible due to the absence of environment maps. We also evaluate the various design choices of our method in an ablative study.

Among existing scene relighting methods (see Sect. 2), we primarily compare against Yu et al. [46] and Philip et al.[27] as they handle a similar type of input data like ours; outdoor scenes photographed in uncontrolled settings and have a direct semantic understanding of the scene illumination. We note however, despite this, both Yu et al. [46] and Philip et al.[27] are designed to edit only the illumination of the input image, while our method can edit both the illumination and the viewpoint. This makes both these methods [27,46] not direct competitors to our method, but still the most related in literature. NeRV [33] can not be applied to our data as their setup is fundamentally different from ours. It requires a training scene to be illuminated by known lighting while our technique uses data shot in unknown lightings. We also do not compare quantitatively against NeRF-W [19] or other style-based methods as they do not perform intrinsic decomposition, don't have a physical understanding of the scene illumination and can not edit lighting according to an environment map. In contrast, the intrinsic decomposition of NeRF-OSR enables applications that are inaccessible for style-based methods. For instance, we show how we can edit the albedo of an examined scene and relighting the entire resulting composited scene (Fig. 7-middle). We also show how our method can achieve real-time rendering with conventional computer graphics methods using the extracted mesh and albedo (Fig. 7-left).

We note that Boss et al. [3] (NeRD) requires the examined object to be at a similar distance from all views—an assumption that is fundamentally violated for outdoor data captured in uncontrolled setup and in our data. As confirmed by the authors of NeRD, this makes the reconstruction nearly impossible for our data. Furthermore, attempting to run NeRD on our data by the paper authors resulted in ray distance variation that is very large, requiring a large number of samples per ray. Thus it was computationally infeasible to process our data with

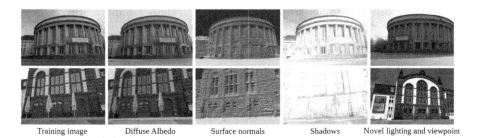

| Training image | Diffuse Albedo | Surface normals | Shadows | Novel lighting and viewpoint |

Fig. 5. Our NeRF-OSR renders photorealistic novel views and simultaneously edits lighting. It also estimates the underlying scene semantics including a dedicated shadows component. Moreover, it can also synthesise hard shadows.

NeRD. NeRF-OSR is the first method that can simultaneously edit the viewpoint and lighting of outdoor sites using neural radiance fields. It also extracts the underlying scene intrinsics and has a dedicated illumination-based shadow component It produces photorealistic results and significantly outperforms state of the art. It is also not limited by synthesising soft shadows only and can synthesise novel hard shadows as well (see Figs. 1 and 5).

Data Pre-Processing. Since NeRF-OSR does not aim to synthesise dynamic objects and discards them (*e.g.,* cars, people and bikes) from the training stage. Although we attempted to reduce their presence during our recordings, the uncontrolled nature of the data makes eliminating them during capture impossible. We, therefore, use the segmentation method of Tao *et al.* [38] to obtain high-quality masks of such objects. Furthermore, even though NeRF-OSR can synthesise the sky and vegetation (*e.g.,* trees), it is not possible to evaluate their predictions due to their highly varying appearance, especially when recordings sessions span different weather seasons. Hence, we also estimate the masks of these regions and exclude them from our evaluation. For Sites 1–3, we keep five recording sessions for testing and use the rest for training. The resulting training/test splits are: 160/95 views for Site 1, 301/96 views for Site 2 and 258/96 views for Site 3.

Relighting with Ground-Truth Environments. We quantitatively evaluate the parametric lighting control of our method and show that it can reproduce novel lighting using lighting coefficients extracted from environment maps. From each recording session of our dataset, we select one photo from the test set as the source. With Site 1, this gives five source images in total. We render all five images at the observed viewpoints and illumination directly. However, for Philip *et al.* [27] and Yu *et al.* [46], only the illumination of a given image can be edited. Hence, for each source image, we relight it using the illumination of the four other source images of the same site. We then cross-project the output to the camera viewpoint from which the target illumination was extracted. This is done by utilising the COLMAP reconstructions. We use segmentation masks to evaluate performance on regions where consistent predictions can be made and

Table 1. Quantitative evaluation of the relighting capabilities of different techniques. We report the metrics for Sites 1, 2 and 3 from our dataset. Our technique significantly outperforms related methods [27, 46]. "d/s" and "u/s" are shorthands for "downscaled" and "upscaled", respectively. Bottom left: ablation study of our various design choices. Our full model achieves the best result.

Method	PSNR ↑	MSE ↓	MAE ↓	SSIM ↑	Method	PSNR ↑	MSE ↓	MAE ↓	SSIM ↑
Site 1					Site 2				
Yu *et al.* [46]	18.71	0.014	0.088	0.4	Yu *et al.* [46]	15.43	0.031	0.136	0.363
Philip *et al.* [27] (d/s)	17.37	0.019	0.105	0.429	Philip *et al.* [27] (d/s)	11.85	0.07	0.21	0.184
Ours (d/s)	**19.86**	**0.011**	**0.08**	**0.626**	Ours (d/s)	**15.83**	**0.026**	**0.128**	**0.556**
Yu *et al.* [46] (u/s)	17.87	0.017	0.097	0.378	Yu *et al.* [46] (u/s)	15.28	0.032	0.138	0.385
Philip *et al.* [27]	16.63	0.023	0.113	0.367	Philip *et al.* [27]	12.34	0.065	0.2	0.272
Ours	**18.72**	**0.014**	**0.09**	**0.468**	Ours	**15.43**	**0.029**	**0.133**	**0.517**
No shadows	17.82	0.017	0.101	0.418	Site 3				
No annealing	17.16	0.02	0.108	0.324	Yu *et al.* [46]	15.84	0.028	0.123	0.392
No ray jitter	18.43	0.015	0.093	0.433	Philip *et al.* [27] (d/s)	12.85	0.054	0.169	0.164
No shadow jitter	18.28	0.016	0.095	0.413	Ours (d/s)	**17.38**	**0.021**	**0.106**	**0.576**
No shadow regulariser	17.62	0.018	0.105	0.373	Yu *et al.* [46] (u/s)	15.17	0.033	0.133	0.376
					Philip *et al.* [27]	12.28	0.062	0.179	0.319
					Ours	**16.65**	**0.024**	**0.114**	**0.501**

Fig. 6. Relighting using ground-truth environment map. Since Philip *et al.* [27] and Yu *et al.* [46] can not edit the camera viewpoint—unlike NeRF-OSR—we cross-project their result on the ground-truth view. Our approach captures the illumination significantly better than related methods. See Table 1 for the corresponding numerical evaluations.

cross-projected for other methods. This is usually the main building. Here, we compute several metrics, including MSE, MAE and SSIM. For SSIM we report the average over the segmentation mask, using scikit-image [41] implementation. Here, we use an SSIM metric with a window size of 5 and the segmentation mask eroded by the same window size to remove the impact of the pixels outside of the mask on the metric value.

Table 1 reports the results of this experiment (the averages over all evaluated images). Figure 6 shows several ground-truth images and views rendered by the compared methods. NeRF-OSR outperforms related techniques quantitatively and qualitatively. While our method and Philip *et al.*generate results at 1280 × 844 pixel, Yu *et al.* can only generate results at 303 × 200 pixel. Hence in Table 1 we also compare against Yu *et al.* in a setting where we downscale the output of our method to Yu *et al.*'s default resolution (see d/s in Table 1). Despite this, our

Fig. 7. (Left) Real-time interactive rendering of the extracted model in VR (screen capture is overlayed over the display for clearer image). Here, the sunlight illuminates the left side of the building and casts clear hard shadows on the right side. And an example of editing the scene albedo (middle) and shadows (right), independently of illumination and other intrinsics.

method still outperforms Yu *et al.*which shows our more superior performance is not due to differences in the output resolution. We note that comparing against style-based methods like NeRF-W here is not feasible as they do not have a semantic representation of light and thus can not edit the light according to an environment map.

Ablation Study. We evaluate the design choices of our method through an ablation study. We follow the same evaluation procedure as for the relighting comparison and report results as an average taken over all output images. For our approach, that are five images of Site 1. For Philip *et al.* [27], that are 20 images in total. Table 1-(bottom) reports the PSNR, MSE, MAE and SSIM of various tested settings. Results show that the best performance is obtained by using the full version of NeRF-OSR. We note that since all metrics in Table 1 are computed only over masked regions, they are expected to be of a higher performance if the entire image was evaluated, while filling the unmasked regions with black.

Real-time Interactive Rendering in VR. In contrast to style-based methods such as [19], our rendering is an explicit function of geometry, albedo, shadow, and the lighting conditions (see Eq. 5). Our model provides direct access to albedo and geometry. The lighting and shadows can be generated from the geometry using multiple, potentially non-differentiable, lighting models at render-time. Hence, if we can extract geometry and albedo at sufficient resolution, we can use them without the slow NeRF ray-marching at little to no loss of quality, compared to the original neural model.

We extract the geometry and albedo from the learned model of Site 1 as a mesh using Marching Cubes [16] at resolution 1000^3 voxels. Then we use them in our interactive VR renderer implemented with C++, OpenGL and SteamVR. The lighting model consists of the sun and a simple geometry-based shadow map [43]: $\mathbf{C}_{\text{interactive}} = \mathbf{C}_{\text{ambient}} + s \odot \mathbf{C}_{\text{sun}} \max\{0, \mathbf{D}_{\text{sun}}^T \mathbf{N}\}$, where $\mathbf{C}_{\text{ambient}}$ is the ambient colour, s is 0 when the rendered point is occluded and 1 if not, according to the shadow map, $\mathbf{C}_{\text{sun}}$ is the colour of the sun, $\mathbf{D}_{\text{sun}}$ is the direction of the sun and $\mathbf{N}$ is the normal of the mesh. The user can interactively move in the scene and control the sun direction with their controllers. The demo runs in real-time on a desktop computer with an Intel i7-4770 CPU, an Nvidia GeForce GTX 970 (4 GB VRAM) GPU and an Oculus Rift S HMD. The system RAM usage of the application is below 3 GB. We provide an extensive demo in the supplementary video and show an extract in Fig. 7-(left).

Albedo and Shadow Editing. Another application of our intrinsic decomposition is to edit the scene albedo, without affecting the illumination or shadows. Such application is not possible by style-based methods by any means, *e.g.*, NeRF-W [19], as they do not perform image decomposition. In Fig. 7-(middle), we replace the announcement poster in Site 3 with an ECCV 2022 poster. Note how the replaced poster looks natural with the rest of the scene. In Fig. 7-(right), we edit the shadow strength post-render. Please find extended video results of this experiment in our supplementary video, where we also show relighting results with the composited announcement poster.

6   Discussion and Conclusion

We have shown that the second-order SH lighting model is capable of producing plausible relightings. While sunlit environments can contain shadows not well represented by a second-order SH, we believe our learned shadow component compensates for this (see Figs. 7, right, for examples of novel hard shadows). Nevertheless, the SH illumination model can still be restricted in terms of high-frequency illumination, specularities and spatially varying illumination. Capturing such effects would enable reconstruction of view-dependent effects and more challenging scenes, including nighttime conditions.

Despite our method outperforms related approaches numerically and visually, some blur could exist. We believe this is due to some inaccuracies in geometry estimation. More specifically, we learn a disentangled representation of the image intrinsics, allowing many novel applications (Fig. 7). This, however, requires precise geometry, as even tiny bumps in the learned geometry can lead to significant change in normals and, hence, errors in the computed illumination. Hence the model can smooth some parts of the geometry in favour of having more accurate lighting. This leads to overall better relighting results, compared to other methods as shown in Table 1. Nevertheless, future work can further improve results by examining more sophisticated geometry models (*e.g.*, a hybrid volume density or implicit surface representation [25,42]). Our method needs only a set of in-the-wild photos taken from different times and views. To this end, we have evaluated our approach on our newly collected dataset and on the "Trevi" scene from [9,32]. This scene was collected completely from the Internet and is widely used in literature [15,19,22]. Recall that ground-truth environment maps are only used for evaluation (as in Sect. 5) and not required for our method to work. While the datasets we examined show practical use-cases of our method, future work could investigate using as few as a single illumination condition during test. Finally, incorporating more priors of the outdoor scenes could be an interesting future research direction.

Concluding Remarks. We presented the first method for simultaneous novel view and novel lighting generation of outdoor scenes captured from uncontrolled settings. We have shown that posed images with varying illumination are sufficient to train a neural representation of scene intrinsics and estimate per-image illumination. Our method outperforms related techniques subjectively and quantitatively on several sequences, including the newly collected benchmark dataset with ground-truth environment maps.

Acknowledgements. We thank Christen Millerdurai for the help with the dataset recording. This work was supported by the ERC Consolidator Grant 4DRepLy (770784).

References

1. Barron, J.T., Malik, J.: Shape, illumination, and reflectance from shading. IEEE Trans. Pattern Anal. Mach. Intell. (TPAMI) **37**(8), 1670–1687 (2015)
2. Basri, R., Jacobs, D.W.: Lambertian reflectance and linear subspaces. IEEE TPAMI **25**(2), 218–233 (2003)
3. Boss, M., Braun, R., Jampani, V., Barron, J.T., Liu, C., Lensch, H.P.: Nerd: neural reflectance decomposition from image collections. In: International Conference on Computer Vision (ICCV) (2021)
4. Chen, X., et al.: Hallucinated neural radiance fields in the wild. In: Proceedings of the IEEE/CVF Conference on Computer Vision and Pattern Recognition, pp. 12943–12952 (2022)
5. Duchêne, S., et al.: Multiview intrinsic images of outdoors scenes with an application to relighting. ACM Trans. Graph. **34**(5) (2015)
6. Finlayson, G.D., Mackiewicz, M., Hurlbert, A.: Color correction using root-polynomial regression. IEEE Trans. Image Process. **24**(5), 1460–1470 (2015)
7. Garon, M., Sunkavalli, K., Hadap, S., Carr, N., Lalonde, J.F.: Fast spatially-varying indoor lighting estimation. In: Computer Vision and Pattern Recognition (CVPR) (2019)
8. Guo, K., et al.: The relightables: volumetric performance capture of humans with realistic relighting. ACM Trans. Graph. **38**(6), 1–19 (2019)
9. Jin, Y., et al.: Image matching across wide baselines: from paper to practice. Int. J. Comput. Vis. 1–31 (2020). https://doi.org/10.1007/s11263-020-01385-0
10. Karsch, K., Hedau, V., Forsyth, D., Hoiem, D.: Rendering synthetic objects into legacy photographs. ACM Trans. Graph. **30**(6), 1–12 (2011)
11. Laffont, P.Y., Bousseau, A., Paris, S., Durand, F., Drettakis, G.: Coherent intrinsic images from photo collections. ACM Trans. Graph. **31**(6) (2012)
12. Lalonde, J.F., Efros, A.A., Narasimhan, S.G.: Webcam clip art: appearance and illuminant transfer from time-lapse sequences. ACM Trans. Graph. **28**(5), 1–10 (2009)
13. Li, Y., Snavely, N., Huttenlocher, D., Fua, P.: Worldwide pose estimation using 3d point clouds. In: European Conference on Computer Vision (ECCV), pp. 15–29 (2012)
14. Li, Z., Snavely, N.: Megadepth: learning single-view depth prediction from internet photos. In: Computer Vision and Pattern Recognition (CVPR) (2018)
15. Li, Z., Xian, W., Davis, A., Snavely, N.: Crowdsampling the plenoptic function. In: European Conference on Computer Vision (ECCV) (2020)

16. Lorensen, W.E., Cline, H.E.: Marching cubes: a high resolution 3d surface construction algorithm. ACM siggraph Comput. Graph. **21**(4), 163–169 (1987)
17. Luan, F., Paris, S., Shechtman, E., Bala, K.: Deep photo style transfer. In: Computer Vision and Pattern Recognition (CVPR), pp. 6997–7005 (2017)
18. Mallikarjun, B.R., et al.: Photoapp: photorealistic appearance editing of head portraits. ACM Trans. Graph. **40**(4), 44 (2021)
19. Martin-Brualla, R., Radwan, N., Sajjadi, M.S.M., Barron, J.T., Dosovitskiy, A., Duckworth, D.: NeRF in the wild: neural radiance fields for unconstrained photo collections. In: Computer Vision and Pattern Recognition (CVPR) (2021)
20. Meka, A., et al.: Lime: live intrinsic material estimation. In: Computer Vision and Pattern Recognition (CVPR) (2018)
21. Meka, A., et al.: Deep relightable textures - volumetric performance capture with neural rendering. In: ACM Trans. Graph. (Proceedings SIGGRAPH Asia), **39**(6), 1–21 (2020)
22. Meshry, M., Goldman, D.B., Khamis, S., Hoppe, H., Pandey, R., Snavely, N., Martin-Brualla, R.: Neural re-rendering in the wild. In: Computer Vision and Pattern Recognition (CVPR), pp. 6871–6880 (2019)
23. Mildenhall, B., Srinivasan, P.P., Tancik, M., Barron, J.T., Ramamoorthi, R., Ng, R.: Nerf: representing scenes as neural radiance fields for view synthesis. In: European Conference on Computer Vision (ECCV) (2020)
24. Nam, S., Ma, C., Chai, M., Brendel, W., Xu, N., Kim, S.: End-to-end time-lapse video synthesis from a single outdoor image. In: Computer Vision and Pattern Recognition (CVPR), pp. 1409–1418 (2019)
25. Oechsle, M., Peng, S., Geiger, A.: Unisurf: unifying neural implicit surfaces and radiance fields for multi-view reconstruction. In: International Conference on Computer Vision (ICCV) (2021)
26. Park, K., et al.: Nerfies: deformable neural radiance fields. In: International Conference on Computer Vision (ICCV) (2021)
27. Philip, J., Gharbi, M., Zhou, T., Efros, A.A., Drettakis, G.: Multi-view relighting using a geometry-aware network. ACM Trans. Graph. **38**(4), 1–78 (2019)
28. Schönberger, J.L., Zheng, E., Frahm, J.M., Pollefeys, M.: Pixelwise view selection for unstructured multi-view stereo. In: European Conference on Computer Vision (ECCV)m pp. 501–518 (2016)
29. Schönberger, J.L., Frahm, J.M.: Structure-from-motion revisited. In: Computer Vision and Pattern Recognition (CVPR), pp. 4104–4113 (2016)
30. Sengupta, S., Gu, J., Kim, K., Liu, G., Jacobs, D.W., Kautz, J.: Neural inverse rendering of an indoor scene from a single image. In: International Conference on Computer Vision (ICCV) (2019)
31. Shih, Y., Paris, S., Durand, F., Freeman, W.T.: Data-driven hallucination of different times of day from a single outdoor photo. ACM Trans. Graph. **32**(6), 1–11 (2013)
32. Snavely, N., Seitz, S.M., Szeliski, R.: Photo tourism: Exploring photo collections in 3d. ACM Trans. Graph. **25**(3), 835–846 (2006)
33. Srinivasan, P.P., Deng, B., Zhang, X., Tancik, M., Mildenhall, B., Barron, J.T.: Nerv: neural reflectance and visibility fields for relighting and view synthesis. In: Computer Vision and Pattern Recognition (CVPR) (2021)
34. Sun, T., et al.: Single image portrait relighting. ACM Trans. Graph. **38**(4), 1–79 (2019)
35. Sun, T., Lin, K.E., Bi, S., Xu, Z., Ramamoorthi, R.: Nelf: neural light-transport field for portrait view synthesis and relighting. In: Eurographics Symposium on Rendering (2021)

36. Sunkavalli, K., Matusik, W., Pfister, H., Rusinkiewicz, S.: Factored time-lapse video. ACM Trans. Graph. **26**(3), 101-es (2007)
37. Tancik, M., et al.: Block-NeRF: scalable large scene neural view synthesis. In: Proceedings of the IEEE/CVF Conference on Computer Vision and Pattern Recognition, pp. 8248–8258 (2022)
38. Tao, A., Sapra, K., Catanzaro, B.: Hierarchical multi-scale attention for semantic segmentation. arXiv preprint arXiv:2005.10821 (2020)
39. Tewari, A., et al.: Advances in neural rendering. arXiv e-prints (2021)
40. Tretschk, E., Tewari, A., Golyanik, V., Zollhöfer, M., Lassner, C., Theobalt, C.: Non-rigid neural radiance fields: Reconstruction and novel view synthesis of a dynamic scene from monocular video. In: International Conference on Computer Vision (ICCV) (2021)
41. Van der Walt, S., et al.: scikit-image: image processing in python. PeerJ **2**, e453 (2014)
42. Wang, P., Liu, L., Liu, Y., Theobalt, C., Komura, T., Wang, W.: Neus: learning neural implicit surfaces by volume rendering for multi-view reconstruction. In: Neural Information Processing Systems (NeurIPS) (2021)
43. Williams, L.: Casting curved shadows on curved surfaces. In: Proceedings of the 5th Annual Conference on Computer Graphics and Interactive Techniques, pp. 270–274 (1978)
44. Xing, G., Zhou, X., Peng, Q., Liu, Y., Qin, X.: Lighting simulation of augmented outdoor scene based on a legacy photograph. In: Computer Graphics Forum (2013)
45. Xu, Z., Sunkavalli, K., Hadap, S., Ramamoorthi, R.: Deep image-based relighting from optimal sparse samples. ACM Trans. Graph. **37**(4), 126 (2018)
46. Yu, Y., Meka, A., Elgharib, M., Seidel, H.P., Theobalt, C., Smith, W.: Self-supervised outdoor scene relighting. In: European Conference on Computer Vision (ECCV) (2020)
47. Yu, Y., Smith, W.A.: Inverserendernet: learning single image inverse rendering. In: Computer Vision and Pattern Recognition (CVPR) (2019)
48. Zhang, K., Riegler, G., Snavely, N., Koltun, V.: Nerf++: analyzing and improving neural radiance fields. arXiv:2010.07492 (2020)
49. Zhang, X., Srinivasan, P.P., Deng, B., Debevec, P., Freeman, W.T., Barron, J.T.: Nerfactor: neural factorization of shape and reflectance under an unknown illumination. ACM Trans. Graph **40**(6), 1–18 (2021)

CoGS: Controllable Generation and Search from Sketch and Style

Cusuh Ham[1(✉)], Gemma Canet Tarrés[2], Tu Bui[2], James Hays[1], Zhe Lin[3], and John Collomosse[2,3]

[1] Georgia Institute of Technology, Atlanta, USA
{cusuh,hays}@gatech.edu
[2] University of Surrey, Guildford, UK
{g.canettarres,t.v.bui,j.collomosse}@surrey.ac.uk
[3] Adobe Inc., San Jose, USA
zlin@adobe.com

Abstract. We present CoGS, a novel method for the style-conditioned, sketch-driven synthesis of images. CoGS enables exploration of diverse appearance possibilities for a given sketched object, enabling decoupled control over the structure and the appearance of the output. Coarse-grained control over object structure and appearance are enabled via an input sketch and an exemplar "style" conditioning image to a transformer-based sketch and style encoder to generate a discrete codebook representation. We map the codebook representation into a metric space, enabling fine-grained control over selection and interpolation between multiple synthesis options before generating the image via a vector quantized GAN (VQGAN) decoder. Our framework thereby unifies search and synthesis tasks, in that a sketch and style pair may be used to run an initial synthesis which may be refined via combination with similar results in a search corpus to produce an image more closely matching the user's intent. We show that our model, trained on the 125 object classes of our newly created Pseudosketches dataset, is capable of producing a diverse gamut of semantic content and appearance styles.

Keywords: Image generation · Sketch · Style · Generative search

1 Introduction

Generative artwork is a transforming creative practice, driven by advances in deep networks that enable diverse and realistic image synthesis [9,10,28,40,54,59]. Sketches offer an intuitive modality to both control visual synthesis, and to search visual content [4,44,50]. However sketches are inherently ambiguous, offering incomplete descriptions of users' intent [11,12]. While a sketch may

---

C. Ham and G. C. Tarres—Equal contribution.

Supplementary Information The online version contains supplementary material available at https://doi.org/10.1007/978-3-031-19787-1_36.

Fig. 1. Images synthesized using CoGS. On the left we demonstrate the ability to control the style of the output for a given sketch, and on the right we demonstrate the ability to control the structure via multiple sketches for a given style image.

communicate a rough approximation of structure or semantic layout, it offers limited fine-grained control over appearance or texture. This presents a barrier to the practical use of sketches as a tool for controlled search and synthesis. Thus, sketches are limited to serendipitous "content discovery" use cases, rather than as a tool to obtain an image with content and appearance matching a users' target intent.

Inspired by the recently proposed DALL-E architecture [40] for diverse, text-driven image synthesis, we introduce CoGS, a novel method for diverse, sketch-driven image synthesis with fine-grained control over structure and appearance (Fig. 1). We make the following technical contributions:

1. Style-conditioned sketch-based synthesis using Vector Quantized GANs (VQGANs) [16]. We synthesize images via a VQGAN decoder, driven by discrete codebook representations generated via a transformer-based sketch and style encoder. We show decoupled coarse-grained control over the structure and appearance of the synthesized image using a provided sketch and exemplar "style" image. This enables users to explore multiple appearance possibilities for a given sketched object, and to do so for a diverse set of object classes.
2. Unified embedding for search and synthesis. We propose a variational auto-encoder (VAE) [36] module that maps codebook representations into a metric space. The embedding thus enables fine-tuning the appearance of the synthesized image either by exploring the local space, or by interpolating between existing similar images within a search corpus.
3. Paired sketch-image dataset. To enable training and evaluation of our model, we create a novel paired dataset of images, hand-drawn sketches, and "pseudosketches" derived from the images via an automated process and graded via crowdsourcing. Specifically, the dataset comprises of 113,370 images and corresponding pseudosketches. A subset of 9,580 images map to the existing Sketchy Database [46], providing 5–10 free-hand sketches for each image.

2   Related Work

Sketch-based image generation was first explored through non-parametric patch-based approaches initially developed for texture synthesis [2,14,23,57], and extended to image synthesis via visual analogy [24] and interactive montage [7]. These methods recall and blend textures sampled from a training set, guided by labeled sketches or semantic maps [3]. With the advent of deep learning, image translation networks, such as CycleGAN [65] and pix2pix [33,56], were exploited to map sketches directly to photos for a single class [20,48,58]. These methods were later adapted for high-resolution portrait synthesis (e.g., via convolutional inversion [21] and semantic priors [60]). SketchyGAN [9] and ContextualGAN [37] proposed multi-class extensions for creating low-resolution outputs of individual objects, building upon the success of conditional generative adversarial networks (cGANs) [38]. Zhu *et al.* proposed a manifold exploration technique to improve controllability of cGAN synthesis [64]. Gao *et al.* showed that these approaches were insufficient for compositing scenes using objects of differing classes, proposing to fuse scene graph representations with cGANs for direct sketch-to-scene image synthesis [17]. Scene compositions were also produced by mapping sketches to semantic segmentation maps [43] using SPADE [39].

Conditional image generation has been explored for multiple input modalities. Text embeddings [41,42] have been incorporated into both generator and discriminator of cGANs for keyword-driven synthesis, and extended to natural language phrases. A two-stage generator guided via attention from input features was proposed in [54]. Images have been used for conditioning upsampling [18] and inpainting [22], while sketches mostly focus on image colorization [31]. The use of bounding box layouts [52,53,62] and scene graphs [1,34] has also been explored. Recently, transformers have been explored for image translation and completion tasks, and combined with the discrete codebook representations of VQGAN [16]. DALL-E [40] exploited VQGAN to learn a direct mapping from natural language to image for diverse image synthesis. Our work builds upon a similar concept, exchanging text for sketches and further conditioning the codebook generation on an appearance (style). In this sense, our approach is aligned to recent instance-conditioned (IC) synthesis. IC-GAN [6] performed retrieval using descriptors derived from the source image, using an average of spatial semantic maps derived from the top results. Our method optionally leverages retrieval to interpolate between images similar to the synthesized output to fine-tune appearance. Recently, PoE-GAN [28] has also explored multiple guidance cues (sketch, semantic map, text) to synthesize images. Our approach differs both in our transformer architecture and in our two-stage control, offering coarse-grained conditioning on synthesis using a sketch and style image and using a continuous embedding for fine-grained refinement.

3   Methodology

CoGS accepts a labelled sketch (a raster, and an associated semantic class label), and a style image as inputs. Given these three conditioning signals, our goal is to synthesize an image of the specified class that combines the general structure of the sketch with the colors and textures of the style image, providing users

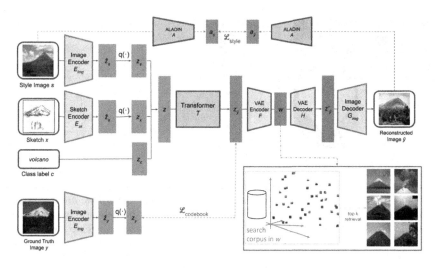

Fig. 2. Illustration of the full CoGS pipeline. We use VQGANs (shown in green) to learn two codebooks, one for sketches and one for images, which are combined with a tokenized representation of a class label. Then, we use a transformer with an auxiliary style loss (shown in red) to learn to composite the inputs into a predicted codebook, which can be subsequently decoded by the image VQGAN decoder to synthesize an image. The input structure and style images offer coarse-grained control over the synthesis. Finally, we use a VAE (shown in orange) to map the codebook to and from a latent space that enables fine-grained refinement of the output image via retrieval or interpolation of results from a search corpus. (Color figure online)

with decoupled control over both the content and style of the output image. An optional additional step can be used to further refine the output.

A major challenge for this task is understanding the correspondence between the input sketch and style image. That is, the network must understand how to appropriately map textures from the style image to the corresponding regions of the sketch. We propose to feed an additional input of a class label to resolve correspondence ambiguities, and to clarify the object of interest. CoGS consists of three components (Fig. 2): 1) VQGANs for encoding the sketch and style inputs; 2) a transformer network for sketch- and style-conditioned image synthesis; and 3) a VAE to further refine output through retrieval and interpolation.

3.1   Paired Pseudosketches Dataset

While several datasets of human-drawn sketches exist, they are often too small, span only a single or small number of categories [27,51], contain simplistic stock images [20,27,51], or lack paired image correspondences [15,35]. For our work, we require a large-scale dataset of sketch-image pairs across diverse categories in order to learn to synthesize images that capture the structural characteristics of the input sketch. The most relevant existing dataset is the Sketchy Database [47], which contains over 75K sketches of 12.5K images across 125 categories.

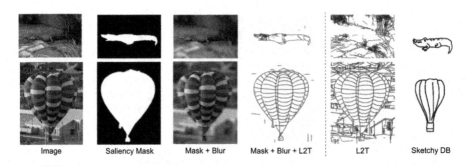

Image   Saliency Mask   Mask + Blur   Mask + Blur + L2T   L2T   Sketchy DB

Fig. 3. Stages in the creation of the Pseudosketches dataset. Given an input image, we generate its saliency mask with [29], blur the non-salient regions, and extract the edgemap of the masked image using L2T [32]. Without the saliency mask blurring, we would retain background details not present in hand-drawn sketches. (Color figure online)

The limited number of images in Sketchy DB makes it difficult for a network to learn to synthesize a diverse set of photorealistic outputs. Thus, we propose to create a new dataset of sketch-like edgemaps, or "pseudosketches", extracted from a larger pool of images to create a large set of explicit sketch-image pairs. We take the 12.5K images from Sketchy DB and query additional images from similar ImageNet categories (based on WordNet distances) to run through our automated extraction pipeline shown in Fig. 3.

For each image, we generate a saliency mask using an example-based open-set panoptic segmentation network [29], and blur the non-salient regions with a large Gaussian kern. Then, we run a line drawing extractor [32] to create pseudosketches of the masked images. Blurring is an essential step in making the pseudosketches more similar in nature to hand-drawn sketches. Without blurring, we retain extraneous background details as shown in Fig. 3. Finally, we present the input image and corresponding pseudosketch side-by-side to Amazon Mechanical Turk (AMT) workers, and ask them to rate how representative the pseudosketch is of the image on a scale of 1 (lowest) to 5 (highest), given the semantic class of the image. After filtering out any pseudosketches with a score below 3, we have 113,370 pseudosketch-image pairs across the original 125 categories of Sketchy DB.

3.2   Learning Effective Encodings for Input Modalities

Given a sketch x, style image s, and class label c as inputs, we need to encode and combine each modality into a single input to the network that synthesizes a composite image. Inspired by recent success with visual sequential modeling [13,40], we want to represent the inputs as a sequence of tokens. We use two vector quantized generative adversarial networks (VQGANs), one trained on pseudosketches and one trained on images, to encode the sketch and style inputs, respectively, into codebook representations. The class label is encoded as a single token representing the index of the class.

Each VQGAN consists of an encoder E, decoder G, and discriminator D. The goal of E and G is to learn the best way to represent an image $x \in \mathbb{R}^{H \times W \times 3}$ as a collection of codebook entries $z_x \in \mathbb{R}^{h \times w \times n_z}$, where n_z is the dimensionality of codes. The discriminator D is trained to ensure that the learned codebook $\mathcal{Z} = \{z_k\}_{k=1}^{K} \subset \mathbb{R}^{n_z}$ is both as rich and compressed as possible, encouraging high quality generative outputs through adversarial training with G.

A quantization step $q(\cdot)$ converts the continuous output of the encoder $\hat{z}_x = E(x) \in \mathbb{R}^{h \times w \times n_z}$ into its discretized form $z_x = q(\hat{z}_x)$ by considering the closest codebook entry z_k to each spatial code of $\hat{z}_x$. Then, G can be trained so that we obtain a reconstruction $\hat{x} = G(z_x)$ as close as possible to the original image x. An additional patch-based discriminator D and perceptual loss are applied to $\hat{x}$ to further ensure that the codebook entries capture details that contribute to improving the overall quality and realism of the image.

3.3 Synthesizing Images with Transformers

In order to train the CoGS transformer, we must select an image that represents the style connecting each sketch to its corresponding image. We use a pre-trained ALADIN [45] style encoder A to pre-compute the style embedding $a_y = A(y)$ for all images y in the Pseudosketches dataset. Then we compute the pairwise Euclidean distance between all style embeddings $\{a_y^c\}$ belonging to the same class c to find the nearest neighbor s for each y. By limiting the nearest neighbor to come from within the same class, s is guaranteed to be semantically and stylistically similar to y but not necessarily share the same fine-grained details of the object (e.g., pose, orientation, scale). Thus, the network must learn a more complex relationship between the sketch and style inputs to apply the textures of the style image to the corresponding regions of the sketch.

After determining an appropriate style image for each sketch-image pair, we now have inputs of a sketch x, style image s, and class label c whose combined conditioning signals should produce a target image y. We freeze the weights of the VQGANs described in Sect. 3.2, and use the encoders E_{sk} and E_{img} to encode the sketch and style images into $\hat{z}_x$ and $\hat{z}_s$, respectively. The quantization step $q(\cdot)$ discretizes the representations into sequences of indices from the codebook $z_x = q(\hat{z}_x)$ and $z_s = q(\hat{z}_s)$ by replacing each code by its index in their respective codebooks, $\mathcal{Z}_{sk}$ and $\mathcal{Z}_{img}$, before concatenating them with the tokenized class encoding z_c into a single input z to a transformer T.

Given a concatenated input z for some input (x, s, c), the transformer T aims to learn $z_y = E_{img}(y)$, which is the codebook representation of the image y corresponding to the input pseudosketch x. The likelihood of the target sequence z_y is modeled autoregressively as

$$p(z_y | x, s, c) = \prod_i p(z_y^i | z_y^{<i}, x, s, c). \tag{1}$$

Thus, the codebook loss is defined as maximizing the log-likelihood:

$$\mathcal{L}_{\text{codebook}} = \mathbb{E}_{y \sim p(y)}[-\log p(z_y | x, s, c)]. \tag{2}$$

We also introduce an auxiliary style loss $\mathcal{L}_{style}$ to further reinforce the style constraint on the generated image. We decode the predicted codebook $z_{\hat{y}} = T(\{z_x, z_s, z_c\})$ using the image VQGAN decoder G_{img} into image $\hat{y} = G_{img}(z_{\hat{y}})$, and compute its ALADIN style embedding $a_{\hat{y}} = A(\hat{y})$ as well as the ALADIN embedding of the input style image $a_s = A(s)$. Then, the style loss is defined as:

$$\mathcal{L}_{\text{style}} = \text{MSE}(a_s, a_{\hat{y}}), \tag{3}$$

where $\text{MSE}(i, j)$ is the mean squared error between i and j. Thus, the total transformer loss is defined as the sum of the two losses weighted by λ_T:

$$\mathcal{L}_{\text{transformer}} = \mathcal{L}_{\text{codebook}} + \lambda_T \mathcal{L}_{\text{style}}. \tag{4}$$

3.4 Refining Outputs with VAEs

The decoded image from the transformer's output $\hat{y} = G_{img}(z_{\hat{y}})$ may not always lead to the exact result in mind, whether a consequence of the difficulty of the task or the user wanting to make additional modifications. Therefore, we provide an optional step to further refine the generated image via retrieval or synthesis.

Freezing the VQGANs and transformer from the earlier stages of the pipeline, we train a variational autoencoder (VAE) [36] with encoder F and decoder H to map the encoding $z_{\hat{y}}$ to a more compact representation $w = F(z_{\hat{y}}) \in \mathbb{R}^d$. The VAE is trained on synthesis and search to ensure w, corresponding to the latent space $W \subset \mathbb{R}^d$, constitutes a unified embedding for both tasks.

The VAE encoder output w is a distribution represented by two vectors: the mean $w_\mu \in \mathbb{R}^d$, and standard deviation $w_\sigma \in \mathbb{R}^d$. The traditional evidence lower bound (ELBO) objective of VAEs, defined in Eq. 5, encourages a smooth latent space W that can both reconstruct an input image and synthesize novel outputs:

$$\mathcal{L}_{\text{ELBO}} = \mathcal{L}_{\text{KL}} - \mathcal{L}_{\text{rec}} = \min \mathbb{E}_p[\log p(w|z_{\hat{y}}^q) - \log g(w)] - \mathbb{E}_q \log g(z_{\hat{y}}^q|w). \tag{5}$$

While $\mathcal{L}_{ELBO}$ enables synthesis, we must also enforce metric properties on W to enable it for retrieval. We leverage self-supervised contrastive learning [8,26,30,55] to map embeddings of structurally similar images close together in latent space and embeddings of dissimilar images far apart:

$$\mathcal{L}_{\text{contrastive}} = -\sum_{i \in I} \log \frac{\exp(w_i \cdot w_{j(i)}/\tau)}{\sum_{a \in I \setminus i} \exp(w_i \cdot w_a/\tau)}. \tag{6}$$

For an anchor i in a batch of $2N$ elements, a positive $j(i)$ is an image generated by the CoGS transformer using the same sketch and different style images as i, while the remaining $2(N-1)$ elements are negatives generated using different sketches and completely different style images. Thus, the VAE loss is defined as the sum of the two losses weighted by λ_V:

$$\mathcal{L}_{\text{VAE}} = \mathcal{L}_{\text{ELBO}} + \lambda_V \mathcal{L}_{\text{contrastive}}. \tag{7}$$

4 Experiments

We describe our experimental setup, evaluation protocols, and comparisons to baseline methods. We evaluate the quality and ability of the transformer component of CoGS to enable both style and structure controllability in Sect. 4.2–4.6, and investigate the VAE refinement step in Sect. 4.7.

4.1 Experimental Setup

Network Architectures and Training. We use an ImageNet pre-trained VQGAN from [16] for E_{img} and G_{img}, and train another VQGAN on the Pseudosketches dataset for E_{sk}. Both VQGANs are trained with the same settings: the encoder E transforms $x \in \mathbb{R}^{256 \times 256 \times 3}$ to codebooks $z_x \in \mathbb{R}^{16 \times 16 \times 256}$, and the decoder G reconstructs z_x into $\hat{x} \in \mathbb{R}^{256 \times 256 \times 3}$. For the transformer, we use 16 layers with 16 attention heads, and a vocabulary size $|\mathcal{Z}| = 1024$. The dimensionality of the sketch and style codebooks is $n_{\hat{z}_x} = n_{\hat{z}_s} = 256$ and class token $n_{z_c} = 1$, so the concatenated sequences of quantized inputs are of length $n_z = 513$. We randomly partition the Pseudosketches dataset into 102,024 training and 11,346 validation examples, and use a weighting term $\lambda_T = 1$ in Eq. 4.

The VAE encoder F compresses input codebooks $z_y \in \mathbb{R}^{16 \times 16 \times 256}$ into $w \in \mathbb{R}^{1024}$, and the decoder H reconstructs the input back to $z'_y \in \mathbb{R}^{16 \times 16 \times 256}$. With the three components of CoGS being trained sequentially, the VAE is trained using a two-stage approach where training is bootstrapped by only using the contrastive loss $\mathcal{L}_{contrastive}$ for training the encoder F. When the loss converges, both F and H are further trained using the dual loss in Eq. 7 with $\lambda_V = 10^6$.

Evaluation Metrics. Two popular metrics to assess the quality of generative networks are the Fréchet Inception Distance (FID) [25] and Learned Perceptual Image Patch Similarity (LPIPS) [61]. FID computes the distance between the distribution of synthesized validation images and the distribution of the ground truth images corresponding to the input sketches. Thus, the lower the FID, the more similar the generated images are to the distribution of real images.

LPIPS measures the distance between image patches, and is used to quantify a generative network's ability to produce diverse outputs for a set of inputs. We subsample the top 10% validation pseudosketches (as determined by AMT scores during data collection) and their respective style images and class labels, sample 5 output images per input, and average the LPIPS on all unique pairs of outputs.

We further evaluate CoGS using style and structure distance metrics (described in Sect. 4.4), and AMT to crowd-source quality assessments of CoGS and the baselines. For the latter we source both: 1) user preference for our approach versus baseline approaches, and 2) a subjective evaluation methodology akin to Gao et al. [17] that scores how "realistic" each image is, and how "faithful" each synthesized image is to its conditional inputs and ground truth image.

4.2   Dataset Partitioning

To approximate the difficulty for the CoGS transformer to synthesize the various categories of the Pseudosketches dataset, we compute the FID on the validation set for each of the 125 classes individually, and group them into 3 sets: the "simple" partition contains classes with the 41 lowest FIDs, "medium" with the 42 middle FIDs, "complex" with the 42 highest FIDs. We then combine the images from all the classes within each partition and recompute the overall FID score of each partition, shown in Table 1. We note that there is a correlation between the number of images in a class and its FID score, i.e., classes with higher number of images tend to produce lower FIDs.

4.3   Comparison to Baseline Methods

We briefly describe the baseline methods used for evaluations. Other methods such as PoE-GAN [28] and Scribbler [48] align with our work, but lack public code or models for comparison.

Neural Style Transfer (NST) [19] takes as input a content image x_c and style image x_s, and synthesizes an image $\hat{x}$ where the style of x_s is "transferred" onto x_c by optimizing it to match texture (Gram matrix) statistics from x_s using mid-late layer activations of VGG-16 [49]. We use the pseudosketch as the input.

SketchyGAN [9] is a sketch-conditioned GAN trained on Sketchy DB and edgemaps of Flickr images. We substitute Flickr edgemaps for the Pseudosketches dataset to train SketchyGAN as the data and model are not public.

iSketchNFill [20] is a method for inpainting partial sketches and synthesizing images from them. We train the generative component on the Pseudosketches dataset, though the wide class diversity proved challenging for this network.

Instance-Conditioned GAN (IC-GAN) [6] is a conditional GAN that leverages instances of images or text to condition the generated output, guided by pre-computed descriptors from the source images and their top retrieval results. We use the Pseudosketches dataset to train two variants that still compute the descriptors on photorealistic images: 1) add the corresponding pseudosketch's descriptors as an additional conditioning instance, and 2) replace the main photorealistic input with its corresponding pseudosketch.

CoCosNet v2. [63] is a patch-based method that uses examplar images from different domains, such as edge maps or semantic maps, to translate into high-quality photorealistic images.

Quantitative evaluations using FID and LPIPS are provided in Table 1, and qualitative evaluations based on AMT experiments in Table 2. Our method produces the best FID and comparable LPIPS, indicating good overall quality of images while avoiding mode collapse by producing diverse outputs. The slightly

Fig. 4. Representative examples generated by the CoGS transformer stage for the simple, medium, and complex partitions.

higher LPIPS score from SketchyGAN is likely due to the output being constrained only by the sketch input, whereas the solution space of CoGS is reduced by imposing additional style and class conditions.

We ran three crowd-sourced AMT evaluations: 1) *Realism*: We present output images (e.g. Fig. 5) to participants, and ask them to rate the realism on a scale 1 ("very dissatisfied") to 5 ("very satisfied") using 3 participants per image. We consider only responses with consensus, where the maximum and minimum ratings deviate by at most 1 point; 2) *Fidelity*: We similarly measure the fidelity, or "faithfulness", of the output to the ground truth image. Although scores are generally low (2–3 on the scale), they are similar to scores reported in other papers following this methodology (e.g., [17,43]); and 3) *Preference*: We ask 3 participants per image to indicate their preferred method in terms of overall fidelity to the ground truth image (gt), fidelity of structure to the sketch (sk) and to the style (st). Responses with majority consensus (2 out of 3 agree) are included. CoGS outperforms the baselines in the majority of the AMT studies, and produces the highest ratings overall. NST achieves higher ratings on faithfulness and SketchyGAN on structure only on the complex set.

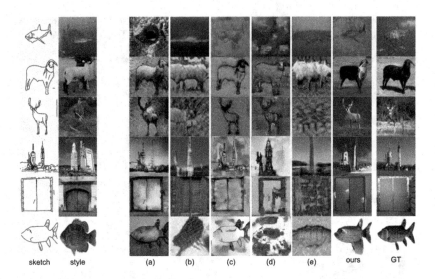

sketch   style   (a)   (b)   (c)   (d)   (e)   ours   GT

Fig. 5. Representative outputs from: a) SketchyGAN [9], b) IC-GAN [6] (sketch), c) NST [19], d) CoCosNetv2 [63], e) IC-GAN [6] (sketch, style image), and CoGS. Note that neither (a) or (b) use the input style image. We emphasize that the synthesized image may not necessarily aim to look exactly like the "ground truth" image due the stylistic differences between the style and ground truth images (e.g., second row style image contains a white sheep, but the ground truth image is a black sheep).

Table 1. Quantitative comparison of FID and LPIPS scores. The symbol next to each method denotes the input(s): $\circ = \{x\}$, $\diamond = \{x, c\}$, $\bullet = \{x, s\}$, and $\star = \{x, s, c\}$, where x is a sketch, s is a style image, and c is the class label.

Method	Simple		Medium		Complex		Overall	
	FID↓	LPIPS↑	FID↓	LPIPS↑	FID↓	LPIPS↑	FID↓	LPIPS↑
SketchyGAN ∘	210.919	**0.534**	256.167	**0.555**	330.649	0.550	213.762	**0.541**
IC-GAN ⋄	103.106	0.183	161.924	0.209	293.037	0.185	109.336	0.190
iSketchNFill ⋄	497.693	1e−8	522.380	9e−9	548.525	1e−8	506.790	1e−8
NST •	118.839	0.308	167.651	0.302	262.893	0.311	114.707	0.307
CoCosNetv2 •	153.371	0.399	204.983	0.404	279.311	0.403	160.705	0.401
IC-GAN ⋆	131.750	0.180	176.123	0.216	293.086	0.189	130.235	0.190
CoGS (ours) ⋆	**43.896**	0.500	**95.539**	0.547	**201.230**	**0.616**	**50.630**	0.521

4.4   Style and Structure Controllability

While the realism of the generated images is important, the main goal of the CoGS transformer is to provide users decoupled control over both the style and structure of the generated image across a diverse set of categories.

Style Controllability. For measuring the stylistic similarity of the generated image to the input style image, we use $d_{style}(s, \hat{y}) = d(a_s, a_{\hat{y}})$, where a_s and $a_{\hat{y}}$

Table 2. Qualitative evaluations within each partition and across the overall validation set based on AMT experiments. The considered evaluation metrics are: fidelity to ground truth (F), realism of the generated image (R), preference overall given the ground truth image (pref. (gt)), preference based on structure, given the input sketch (pref.(sk)), preference based on style, given the style image (pref.(st)). Values for pref. (gt)/(sk)/(st) correspond to percentages of workers' selection as the best option, while F and R correspond to workers' rankings of outputs from 1 (worst) to 5 (best).

Method	Simple					Medium					Complex				
	F	R	pref. (gt)	pref. (sk)	pref. (st)	F	R	pref. (gt)	pref. (sk)	pref. (st)	F	R	pref. (gt)	pref. (sk)	pref. (st)
SketchyGAN ∘	2.68	2.67	23.37	36.06	7.04	2.52	2.87	31.72	38.32	10.84	2.64	2.83	36.21	**49.89**	16.19
IC-GAN ∘	2.49	2.52	5.66	0.56	22.26	2.27	2.63	6.04	0.82	21.76	2.36	2.62	3.60	0.67	16.97
iSketchNFill ∘	1.96	1.95	0.00	0.00	0.00	1.87	2.13	0.18	0.74	0.25	1.65	1.99	0.00	0.00	0.26
NST •	2.69	2.67	10.53	10.73	21.13	2.52	2.85	8.35	7.71	21.76	**2.68**	2.82	10.07	9.89	23.39
IC-GAN ⋆	2.34	2.39	4.32	0.84	15.60	2.14	2.46	4.21	0.67	11.60	2.26	2.51	5.28	0.9	10.80
CoGS (ours) ⋆	**2.94**	**3.04**	**55.93**	**51.80**	**33.96**	**2.67**	**3.05**	**49.44**	**52.40**	**33.78**	2.58	**2.92**	44.60	38.65	**32.39**

Table 3. Evaluations for style and structure controllability using distance metrics (ALADIN and Chamfer distances, respectively) and AMT human evaluations where participants are asked to score the fidelity of the output to the input style and sketch on a scale of 1 (low) to 5 (high).

Partition	Style↓	AMT Style↑	Structure↓	AMT Structure↑
Simple	1.085	2.655	2.627	3.117
Medium	1.136	2.374	2.004	2.805
Complex	1.104	2.450	1.519	2.393
Overall	1.100	2.501	2.387	2.776

are the style encodings from ALADIN [45] of the style and synthesized images, respectively, and $d(i, j)$ is the Euclidean distance between i and j. In Table 3 we show that the mean style distance of each partition are all within a small epsilon of each other, which agrees with AMT style similarity evaluations, where workers are asked to rate how well the generated image matches the style image on a scale of 1 (worst) to 5 (best). Three participants rate each image, and responses with majority consensus, i.e., min/max scores within 1 point, are included.

We demonstrate style control in Fig. 6. We algorithmically select the 5 style images for each sketch by taking the top $N = 100$ nearest neighbors within the same class in ALADIN style space to the ground truth image corresponding to the input sketch, clustering the N style vectors into $k = 5$ clusters using k-means, and finding the image closest to each of the k centroids. We show that the output images capture the variations in the style image.

Structure Controllability. To quantitatively measure the structural fidelity of the generated image to the input sketch, we compute the Chamfer distance $d_{structure}(x, e_{\hat{y}}) = \frac{1}{n} \sum_{i \in x} v_i$, where x is the input sketch, $e_{\hat{y}}$ is the edgemap extracted from the generated image $\hat{y}$ using the Canny edge detector [5], and v_i is the distance transform value of $e_{\hat{y}}$. We only sum over the black pixel coordinates i of x to measure the structural coherence of just the target object. In Table 3

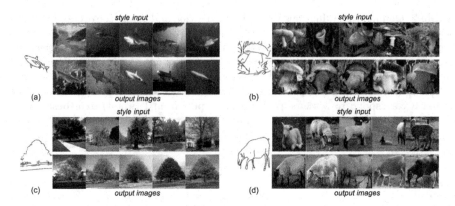

Fig. 6. Style controllability. For each subfigure (a–d), we generate outputs using a single pseudosketch and 5 different style images.

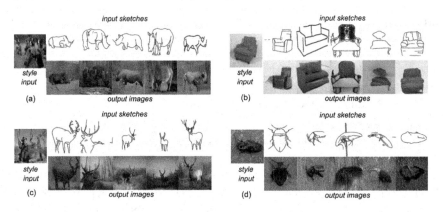

Fig. 7. Structure controllability. For each subfigure (a–d), we generate outputs using 5 pseudosketches for a single style image.

we report the mean structure distance and AMT evaluations, where workers are asked to rate how well the generated image matches the input sketch contours on a scale of 1 (worst) to 5 (best) with consensus.

We qualitatively demonstrate the ability to control the structure of the output by sampling the top 10% of pseudosketches for each class (as determined by the AMT score during data collection), and randomly sample one style image from within the same class. We visualize the results in Fig. 7 to show that the synthesized image is guided by the contours of the input sketch.

4.5 Generalization to Hand-Drawn Sketches

Pseudosketches are similarly minimalistic to human-drawn sketches, but their contours have direct pixel correspondences to their ground truth images, whereas human-drawn sketches may come from users of varying skill levels and are more

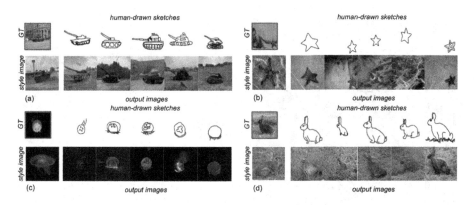

Fig. 8. Generalization. For each subfigure (a–d), we generate outputs using 5 sketches from Sketchy DB [46] corresponding to a ground truth image (framed in grey) with a given style image to demonstrate generalization to hand-drawn sketches.

abstract. Although CoGS is only trained on pseudosketches, we show that it is able to generalize to some of the higher quality human-drawn sketches from Sketchy DB (see Fig. 8). Because there is a strong correlation in the number of examples used to compute the FID, we first randomly sample a subset of Sketchy DB similar in size to the Pseudosketches validation set. Our method achieves an FID of 81.820 on Sketchy DB, compared to 50.630 on Pseudosketches, highlighting a gap in the quality of synthesis results from using hand-drawn sketches compared to pseudosketches due to the domain gap and abstraction.

4.6   Transformer Ablation Study

We investigate the impact of various components of the CoGS transformer using FID, LPIPS, style distance, and structure distance. We show in Table 4 that both the class label c and style loss $\mathcal{L}_{style}$ are important in generating a diverse set of realistic outputs that are consistent with the stylistic and structural conditions.

Table 4. Ablation study on the inputs and losses for our proposed method.

Inputs	Losses	FID↓	LPIPS↑	Style↓	Structure↓
sketch, style	codebook	58.202	0.519	1.117	3.063
sketch, style	codebook, style	58.327	0.514	1.129	2.823
sketch, style, label	codebook	52.992	0.518	1.120	3.043
sketch, style, label	codebook, style	50.630	0.521	1.100	2.387

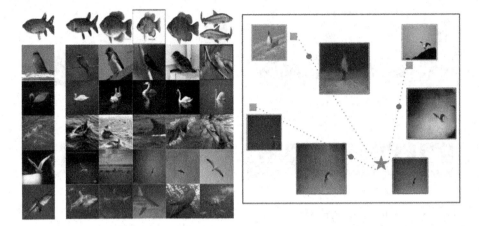

Fig. 9. (a) Top 5 images retrieved from the validation set using generated images as queries. (b) Depiction of the latent space exploration for synthesizing new results. Blue ⋆: query image, green ■: retrieval results, red ●: interpolation results.(Color figure online)

4.7   Fine-Grained Control via Image Retrieval and Interpolation

We first evaluate the latent space at offering users similar images, given the transformer output and its class label. We use a VAE trained on the query image class and retrieve the closest validation images from the same class. We show in Fig. 9(a) how, independent of the style and quality of the query image, the retrieved images have the most similar structure. This behavior is validated by AMT evaluations, where 3 participants were asked to rate each retrieved image of a synthesized query image as relevant or not. The majority vote was used to calculate precision@k of 0.533, 0.535, 0.530, 0.520, 0.425 at $k = \{1, 5, 10, 15, 20\}$, respectively, showing how workers agreed on the relevance of up to around the top 15 images, which demonstrates local coherency. Next, we consider the top 3 retrieved images and synthesize new ones by interpolating between each of retrieved images and the query image. We sample 50 images for each interpolation pair, and filter them using an FID threshold of 120 to ensure the quality of the results. We depict the interpolation scheme in Fig. 9(b), showing a few of the many variations the user can create by mixing attributes from two images.

5   Conclusion

We introduce CoGS, a method for image synthesis across a diverse set of categories that provides control over the style and structure of the output image. In order to learn effective controllable synthesis, we collect a large-scale dataset of "pseudosketch"-image correspondences using an automated pipeline. Our approach produces images with higher fidelity to the given style and structure constraints, while also producing diverse and more realistic images within those conditions. We also learn a unified embedding for search and synthesis, which enables further refinement of the generated images via retrieval or interpolation.

Limitations. To broaden the application of our method, we would like to use inputs of hand-drawn sketches from a wide range of skill levels. As shown in Sect. 4.5, we are able to generalize to examples of hand-drawn sketches with better artistry, but there is still a gap in the FID computed on sketches across skill levels. While this is understandable due to domain gap from the pseudosketches training data, there is room for improvement in resolving abstraction of structure present in hand-drawn sketches.

References

1. Ashual, O., Wolf, L.: Specifying object attributes and relations in interactive scene generation. In: Proceedings of the CVPR (2019)
2. Barnes, C., Shechtman, E., Finkelstein, A., Goldman, D.B.: PatchMatch: a randomized correspondence algorithm for structural image editing. ACM Trans. Graph. **28**(3), 24 (2009)
3. Barnes, C., Zhang, F.-L.: A survey of the state-of-the-art in patch-based synthesis. Comput. Visual Media **3**(1), 3–20 (2016). https://doi.org/10.1007/s41095-016-0064-2
4. Bui, T., Ribeiro, L., Collomosse, J., Ponti, M.: Sketching out the details: Sketch-based image retrieval using convolutional neural networks with multi-stage regression. Comput. Graph. **71**, 77–87 (2018)
5. Canny, J.: A computational approach to edge detection. IEEE Trans. Pattern Anal. Mach. Intell. **6**, 679–698 (1986)
6. Casanova, A., Careil, M., Verbeek, J., Drozdzal, M., Romero-Soriano, A.: Instance-conditioned gan. arXiv preprint arXiv:2109.05070 (2021)
7. Chen, T., Cheng, M.M., Tan, P., Shamir, A., Hu, S.M.: Sketch2Photo: Internet image montage. Proc ACM SIGGRAPH **28**(5), 124 (2009)
8. Chen, T., Kornblith, S., Norouzi, M., Hinton, G.: A simple framework for contrastive learning of visual representations. In: III, H.D., Singh, A. (eds.) Proceedings of the 37th International Conference on Machine Learning. Proceedings of Machine Learning Research, vol. 119, pp. 1597–1607. PMLR, 13–18 July 2020. https://proceedings.mlr.press/v119/chen20j.html
9. Chen, W., Hays, J.: SketchyGAN: towards diverse and realistic sketch to image synthesis. In: The IEEE Conference on Computer Vision and Pattern Recognition (CVPR), June 2018
10. Chen, X., Duan, Y., Houthooft, R., Schulman, J., Sutskever, I., Abbeel, P.: Info-GAN: interpretable representation learning by information maximizing generative adversarial nets. In: 30th Conference on Neural Information Processing Systems (NIPS 2016), Barcelona, Spain, June 2016
11. Collomosse, J., Bui, T., Wilber, M., Fang, C., Jin, H.: Sketching with style: Visual search with sketches and aesthetic context. In: Proceedings of the ICCV (2017)
12. Collomosse, J.P., McNeill, G., Watts, L.: Free-hand sketch grouping for video retrieval. In: Proceedings of the ICPR (2008)
13. Dosovitskiy, A., et al.: An image is worth 16x16 words: transformers for image recognition at scale. arXiv preprint arXiv:2010.11929 (2020)

14. Efros, A., Freeman, W.: Image quilting for texture synthesis and transfer. In: Proceedings of the SIGGRAPH (2001)
15. Eitz, M., Hays, J., Alexa, M.: How do humans sketch objects? ACM Trans. Graph. (Proc. SIGGRAPH) **31**(4), 44:1–44:10 (2012)
16. Esser, P., Rombach, R., Ommer, B.: Taming transformers for high-resolution image synthesis (2020)
17. Gao, C., Liu, Q., Xu, Q., Wang, L., Liu, J., Zou, C.: SketchyCOCO: image generation from freehand scene sketches. In: Proceedings of the IEEE/CVF Conference on Computer Vision and Pattern Recognition (CVPR), June 2020
18. Gao, H., Chen, Z., Huang, B., Chen, J., Li, Z.: Image super-resolution based on conditional generative adversarial network. IET Image Proc. **14**(13), 3006–3013 (2020)
19. Gatys, L.A., Ecker, A.S., Bethge, M.: Image style transfer using convolutional neural networks. In: 2016 IEEE Conference on Computer Vision and Pattern Recognition (CVPR), pp. 2414–2423 (2016). https://doi.org/10.1109/CVPR.2016.265
20. Ghosh, A., et al.: Interactive sketch & fill: multiclass sketch-to-image translation. In: Proceedings of the IEEE International Conference on Computer Vision (2019)
21. Gucluturk, Y., Guclu, U., van Lier, R., van Gerven, M.A.: Convolutional sketch inversion. In: Proceedings of the ECCV Workshop on Vision and Art (VISART) (2016)
22. Guo, X., Yang, H., Huang, D.: Image inpainting via conditional texture and structure dual generation. In: Conference: 2021 IEEE/CVF International Conference on Computer Vision (ICCV) (2021)
23. Hays, J., Efros, A.A.: Scene completion using millions of photographs. ACM Trans. Graph. **26**(3), 4 (2007)
24. Hertzmann, A., Jacobs, C.E., Oliver, N., Curless, B., Salesin, D.H.: Image analogies. In: Proceedings of the ACM SIGGRAPH. pp. 327–340 (2001)
25. Heusel, M., Ramsauer, H., Unterthiner, T., Nessler, B., Hochreiter, S.: GANs trained by a two time-scale update rule converge to a local Nash equilibrium. Adv. Neural Inf. Process. Syst. **30** (2017)
26. Hjelm, R.D., et al.: Learning deep representations by mutual information estimation and maximization. In: International Conference on Learning Representations (2019). https://openreview.net/forum?id=Bklr3j0cKX
27. Hospedales, T., Song, Y.Z.: Sketch me that shoe. In: 2016 IEEE Conference on Computer Vision and Pattern Recognition (CVPR), January 2016
28. Huang, X., Mallya, A., Wang, T.C., Liu, M.Y.: Multimodal conditional image synthesis with product-of-experts GANs (2021)
29. Hwang, J., Oh, S.W., Lee, J., Han, B.: Exemplar-based open-set panoptic segmentation network. CoRR abs/2105.08336 (2021). https://arxiv.org/abs/2105.08336
30. Hénaff, O.J., Razavi, A., Doersch, C., Eslami, S.M.A., Oord, A.v.d.: Data-efficient image recognition with contrastive predictive coding (2019). https://arxiv.org/abs/1905.09272, cite arxiv:1905.09272
31. Iizuka, S., Simo-Serra, E., Ishikawa, H.: Let there be color!: Joint end-to-end learning of global and local image priors for automatic image colorization with simultaneous classification. ACM Trans. Graph. (Proc. of SIGGRAPH 2016) **35**(6) (2016)
32. Inoue, N., Ito, D., Xu, N., Yang, J., Price, B., Yamasaki, T.: Learning to trace: expressive line drawing generation from photographs. Comput. Graph. Forum **38**(7), 69–80 (2019)

33. Isola, P., Zhu, J., Zhou, T., Efros, A.A.: Image-to-image translation with conditional adversarial networks. In: 2017 IEEE Conference on Computer Vision and Pattern Recognition (CVPR), pp. 5967–5976 (2017). https://doi.org/10.1109/CVPR.2017.632

34. Johnson, J., Gupta, A., Fei-Fei, L.: Image synthesis from reconfigurable layout and style. In: Proceedings of the CVPR (2018)

35. Jongejan, J., Rowley, H., Kawashima, T., Kim, J., Fox-Gieg, N.: The quick, draw! A.I. experiment (2016). https://quickdraw.withgoogle.com/

36. Kingma, D.P., Welling, M.: Auto-encoding variational bayes. ArXiv e-prints, December 2013

37. Lu, Y., Wu, S., Tai, Y.W., Tang, C.K.: Image generation from sketch constraint using contextual GAN. In: The European Conference on Computer Vision (ECCV), September 2018

38. Mirza, M., Osindero, S.: Conditional generative adversarial nets. arXiv preprint arXiv:1411.1784 (2014)

39. Park, T., Liu, M.Y., Wang, T.C., Zhu, J.Y.: Semantic image synthesis with spatially-adaptive normalization. In: Proceedings of the IEEE Conference on Computer Vision and Pattern Recognition (2019)

40. Ramesh, A., et al.: Zero-shot text-to-image generation. arXiv preprint arXiv:2102.12092 (2021)

41. Reed, S., Akata, Z., Mohan, S., Tenka, S., Schiele, B., Lee, H.: Learning what and where to draw. In: Advances in Neural Information Processing Systems (NIPS) (2016)

42. Reed, S., Akata, Z., Yan, X., Logeswaran, L., Schiele, B., Lee, H.: Generative adversarial text-to-image synthesis. In: Proceedings ICML (2016)

43. Ribeiro, L., Bui, T., Collomosse, J., Ponti, M.: Scene designer: a unified model for scene search and synthesis from sketch. In: Proceedings of CVPRW on Sketch and Human Expressivity (SHE) (2021)

44. Ribeiro, L.S.F., Bui, T., Collomosse, J., Ponti, M.: Sketchformer: transformer-based representation for sketched structure. In: Proceedings of CVPR (2020)

45. Ruta, D., et al.: Aladin: all layer adaptive instance normalization for fine-grained style similarity. In: 2021 IEEE/CVF International Conference on Computer Vision (ICCV), pp. 11906–11915 (2021)

46. Sangkloy, P., Burnell, N., Ham, C., Hays, J.: The sketchy database: learning to retrieve badly drawn bunnies. ACM Trans. Graph. **35**(4), 119 (2016)

47. Sangkloy, P., Burnell, N., Ham, C., Hays, J.: The sketchy database: Learning to retrieve badly drawn bunnies. ACM Trans. Graph. **35**(4) (2016). https://doi.org/10.1145/2897824.2925954, https://doi.org/10.1145/2897824.2925954

48. Sangkloy, P., Lu, J., Fang, C., Yu, F., Hays, J.: Scribbler: controlling deep image synthesis with sketch and color. In: Proceedings of the IEEE Conference on Computer Vision and Pattern Recognition, pp. 5400–5409 (2017)

49. Simonyan, K., Zisserman, A.: Very deep convolutional networks for large-scale image recognition. arXiv preprint arXiv:1409.1556 (2014)

50. Song, J., Song, Y.Z., Xiang, T., Hospedales, T., Ruan, X.: Deep multi-task attribute-driven ranking for fine-grained sketch-based image retrieval. In: British Machine Vision Conference (2016)

51. Song, J., Yu, Q., Song, Y.Z., Xiang, T., Hospedales, T.M.: Deep spatial-semantic attention for fine-grained sketch-based image retrieval. In: Proceedings of the IEEE International Conference on Computer Vision (ICCV), October 2017

52. Sun, W., Wu, T.: Image synthesis from reconfigurable layout and style. In: Proceedings of CVPR (2019)

53. Sylvain, T., Zhang, P., Bengio, Y., Hjelm, D., Sharma, S.: Object-centric image generation from layouts. arXiv preprint arXiv:2003.07449 (2020)
54. Tang, H., Liu, H., Xu, D., Torr, P., Sebe, N.: Attentiongan: unpaired image-to-image translation using attention-guided generative adversarial networks. arXiv preprint arXiv:1911.11897 (2019)
55. Tian, Y., Krishnan, D., Isola, P.: Contrastive multiview coding. CoRR abs/1906.05849 (2019). https://arxiv.org/abs/1906.05849
56. Wang, T.C., Liu, M.Y., Zhu, J.Y., Tao, A., Kautz, J., Catanzaro, B.: High-resolution image synthesis and semantic manipulation with conditional GANs. In: Proceedings of the IEEE Conference on Computer Vision and Pattern Recognition (2018)
57. Wexler, Y., Shechtman, E., Irani, M.: Space-time video completion. In: Proceedings of the 2004 IEEE Computer Society Conference on Computer Vision and Pattern Recognition, 2004. CVPR 2004. vol. 1, pp. I-I. IEEE (2004)
58. Xian, W., et al.: TextureGAN: controlling deep image synthesis with texture patches. arXiv preprint arXiv:1706.02823 (2017)
59. Xue, Y., Guo, Y.-C., Zhang, H., Xu, T., Zhang, S.-H., Huang, X.: Deep image synthesis from intuitive user input: a review and perspectives. Comput. Visual Media 8(1), 3–31 (2021). https://doi.org/10.1007/s41095-021-0234-8
60. Yang, Y., Hossain, M.Z., Gedeon, T., Rahman, S.: S2FGAN: semantically aware interactive sketch-to-face translation. arXiv preprint arXiv:2011.14785 (2020)
61. Zhang, R., Isola, P., Efros, A.A., Shechtman, E., Wang, O.: The unreasonable effectiveness of deep features as a perceptual metric. In: Proceedings of the IEEE conference on computer vision and pattern recognition, pp. 586–595 (2018)
62. Zhao, B., Meng, L., Yin, W., Sigal, L.: Image generation from layout. In: Proceedings of CVPR (2019)
63. Zhou, X., et al.: Full-resolution correspondence learning for image translation. CoRR abs/2012.02047 (2020). https://arxiv.org/abs/2012.02047
64. Zhu, J.Y., Krahenbuhl, P., Shechtman, E., Efros, A.A.: Generative visual manipulation on the natural image manifold. In: Proceedings of ECCV (2016)
65. Zhu, J.Y., Park, T., Isola, P., Efros, A.A.: Unpaired image-to-image translation using cycle-consistent adversarial networks. arXiv preprint arXiv:1703.10593 (2017)

HairNet: Hairstyle Transfer with Pose Changes

Peihao Zhu[1]([✉]), Rameen Abdal[1], John Femiani[2], and Peter Wonka[1]

[1] KAUST, Thuwal, Saudi Arabia
`peihao.zhu@kaust.edu.sa`
[2] Miami University, Oxford, OH 45056, USA

Abstract. We propose a novel algorithm for automatic hairstyle transfer, specifically targeting complicated inputs that do not match in pose. The input to our algorithm are two images, one for the hairstyle and one for the identity (face). We do not require any additional inputs such as segmentation masks. Our algorithm consists of multiple steps and we contribute three novel components. The first contribution is the idea to include baldification into hairstyle editing pipelines to simplify inpainting of background and face regions covered by hair. The second contribution is a novel embedding algorithm that can handle both pose changes and semantic image blending. The third contribution is the *hairnet* architecture that semantically blends the hairstyle and identity images, performing multiple tasks jointly, such as baldification of the identity image, transformation estimation between the two images, warping, and hairstyle copying. Our results show a clear improvement over current state of the art methods in both quantitative and qualitative results. Code and data will be released.

Keywords: Hairstyle transfer · GANs · StyleGAN · Deep learning · Image editing

1 Introduction

Choosing a new hairstyle is an important decision, and for many applications that range from marketing to social media and entertainment, there is a need to 'try on' different hairstyles. For this problem we are given a reference image I_{hair} and an identity image I_{ident} which includes a face and background. The goal is to generate a new image I_{mix} that is as similar as possible to I_{ident} with the hair of I_{hair}, but to keep the generated image plausible as a realistic portrait image. The problem is challenging because hair is complex in its interaction with the environment – it is reflective and translucent, and the lighting of the hair including scattering and shadows must be consistent with the I_{ident}. It may pass

Supplementary Information The online version contains supplementary material available at https://doi.org/10.1007/978-3-031-19787-1_37.

Hair

Identity

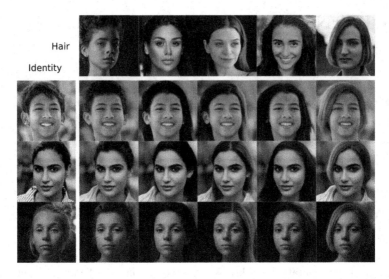

Fig. 1. Given an image of a hairstyle (top row) and a face (left column) which may be in a *different* pose than the hair, we seamlessly transfer the hair onto the face image. Unlike previous approaches, details are preserved even when hair and face images have different poses and head-shapes.

in front of or behind the face, ears, and clothing . It may be backlit, may have sub-pixel strands, and it reflects light anisotropically (Fig. 1).

The current state of the art is Barbershop [44], which uses StyleGAN to invert I_{hair} and I_{ident}, aligns both images to a target segmentation mask, and then copies the activations of an early style-block that correspond to the 'hair' into I_{ident}. However, Barbershop has two main limitations that we try to address. First, the pipeline is inherently limited to work for two input images I_{hair} and I_{ident} that have similar pose. Second, it requires a target segmentation mask to determine the shape of the hair region in relation to the face – but the mask is produced by a naïve heuristic approach that may result in impossible hair shapes, e.g. when complex inpainting of the mask would be needed to deal with translation, scale, occlusion, or disocclusion of the face in I_{ident}.

To overcome these limitations we propose three novel concepts in our work. First, we introduce baldification as a pipeline stage in hairstyle editing. Baldificaton can address disocclusion issues in I_{ident} and provides a dedicated stage to consistently inpaint occluded background and face regions.

Second, we propose a novel embedding algorithm for hairstyle transfer combining two recent embedding algorithms. StyleGAN has biases towards certain hairstyles and so GAN-inversion tends to lose many of the unique characteristics of a hairstyle. Barbershop [44] addresses this by optimizing the activations of an early style-block (called an F-code) of the StyleGAN generator. This embedding is compatible with SOTA semantic image blending algorithms, such as combining hair and face images, but it cannot work for pose changes. On the other hand, generator fine tuning proposed by PTI [31] is a high quality embedding method that is great for pose changes, but it doesn't work with SOTA image blending algorithms. The reason is that PTI generates a separate generator for each input image, but a single generator is required for image blending. Our solution is to combine F code embedding with generator fine tuning [31] to create the first high quality embedding method that allows for both pose changes as well as image blending.

Third, we propose an image blending architecture, called Hairnet, that is a learned substitute for the heuristic segmentation map editing step in Barbershop. In particular, hairnet is capable to learn how to make difficult semantic decisions that are required when merging a hairstyle image with an identity image. For example, should long hair pass infront or behind the shoulders or ears. Hairnet is trained to jointly perform multiple steps required to blend the two input images, such as baldification, estimating translation and scale between two images, warping, and hairstyle copying. The most interesting aspect about hairnet is that it can be trained in an unsupervised fashion, since there are no ground truth images telling us how to transfer hair from one image to another.

In summary, we make the following contributions: 1) We extend the current SOTA method barbershop to handle input images with different pose. 2) We introduce the concept of baldification to tackle disocclusion problems of the background and face regions. 3) We combine two SOTA GAN embedding algorithms to enable pose changes and image blending. 4) We introduce an image blending architecture hairnet that can be trained in an unsupervised manner.

2   Prior Art

State-of-the-Art GANs. Generative adversarial Networks (GANs) [15,29] have seen recent improvements to the loss functions, architecture as well as availability of high quality datasets [22]. Karras together with a team of changing co-authors developed the current state of the art GAN called StyleGAN [19–23]. These GANs are trained on quality high quality datasets like FFHQ [22], AFHQ [8] and LSUN objects [41]. Apart from photo-realistic image synthesis, StyleGAN learns a rich latent space which has been used to perform various downstream tasks such as image editing [1,4,27,33]. Moreover, the architecture of StyleGAN is now used to model other tasks like unsupervised dense correspondences [28] and 3D GANs which are able to generate high-resolution multi-view-consistent images along with approximate 3D geometry [7,13,26]. In the context of this work, we build upon a StyleGAN based hair editing framework, BarberShop [44], to improve the generalization capabilities and the speed of the framework. Based on the issues section in the official released repository of Style-GAN3 [21], there is no solid evidence as of now that it is better than StyleGAN2 in real image projection and editing quality, hence, we use StyleGAN2 [23] to build and compare our framework. This is also important for a comparison, as none of the competing methods uses StyleGAN3.

GAN Latent Space Projection and GAN-Based Editing. In order to extract meaningful information from a GAN, there are two important components: projecting existing images into a GAN latent space and extracting latent directions to edit an images. First, to enable image editing, image embedding/projection is used as a technique to project real images into the GAN's latent space. In the StyleGAN domain, Image2StyleGAN [1] uses the extended $W+$ latent-space to project a real image into the StyleGAN latent space using optimization. Other methods like II2S [46] and PIE [36] improve the reconstruction-editing quality trade-off. Other works like IDInvert [43], pSp [30], e4e [37], and Restyle [5] use encoders and identity preserving loss functions to maintain the semantic meaning of the embedding. PTI [31] and HyperStyle [6] modify the generator weights to better map out-of-distribution images. Secondly, image editing frameworks extract important semantic directions in a pretrained GAN. Related to StyleGAN, the editing frameworks [4,16,33,35] analyze the linear and non-linear nature of the underlying W and $W+$ spaces. We use Style-Flow to perform two editing operations critical to our HairNet framework i.e. "Bald" and "Pose" edits. StyleSpace [39] proposes to edit images in StyleSpace S. Another important area of StyleGAN based editing is CLIP based image editing [2,14,27] and domain transfer [10,45].

Hair Editing Using GANs. Using StyleGAN, there are broadly two types of hair editing frameworks. The first category uses hair segmentation information to edit the hairstyles, e.g. [32,34,44]. We call such methods as semantic region based methods. Recent works use unsupervised analysis on the feature maps [3,11] to identify semantic regions. Editing in Style [11] uses k-means clustering. Based on the Editing in Style method, two relevant works, StyleFusion [18] and Retrieve in Style [9] modify the hair styles using StyleSpace [39] editing and interpolation. Apart from the segmentation and semantic region based methods, some other methods like StyleCLIP [27] and HairCLIP [38] use the CLIP model to modify the hairstyle of a person based on the text prompts. Our method falls in the realm of segmentation based methods. In this work we quantitatively and qualitatively compare our method with the segmentation and semantic region based methods in Sect. 4.3.

3 Method

3.1 Background

We build on the approach of Barbershop [44], and we briefly summarize the key points here but encourage the reader to consult the source for details. The main idea of Barbershop is to use a new latent space for images that allows for spatial control of image features. The new FS-code starts with the frequently used $W+$ latent representation of an image [1], but replaces the first $7 \cdot 512$ elements of the embedding vector by the activations of style-block 7 of the StyleGAN generator. The new latent code consists of a $32 \times 32 \times 512$ tensor of activations F and a vector S with the remaining elements of the $W+$ embedding. The new F code has more degrees of freedom than the $W+$ latent-code, and an optimization process is used to find values of F that are similar to the original $W+$ based activations, but that also improve the reconstruction of the input image. That optimization process is described in [44] and we omit it here for brevity, as it is not critical for understanding our contribution.

The tensor F can be understood as a 32×32 coarse spatial representation of an image, and so copying and pasting regions of the F-code allows coarse details (called *structure*) of the image to be transferred. This allows the Barbershop to preserve the shape of medium-to-large features, such as the shape of a hair region or the structure of large curls or braids. The style code S used by barbershop is global, and so an optimization process was described to find a single S that best captures the appearance (color and texture) of the hair in I_{hair} in regions of hair while preserving the appearance of I_{ident} in other parts of the image.

A key challenge in transferring the hairstyle from one image to another is that the pose and head-geometry of the subject in I_{hair} may be different from I_{ident}. In order to address this issue, Barbershop aligned each image to a common target-segmentation mask. The target mask was constructed heuristically, and so occasionally implausible target masks were created. In addition, the alignment mainly works for smaller translations, but cannot compensate for larger pose differences (Fig. 2).

3.2 Overview

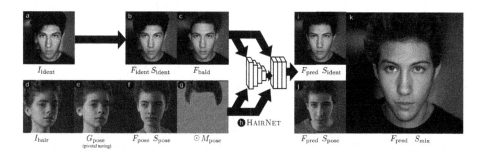

Fig. 2. Overview of hair blending pipeline. Input identity image (a) is embedded into FS-space (b) and then a latent-edit is used to baldify the image and get its F-code (c). Another *hair* input image (d) cannot be embedded the same way, so pivotal tuning is used to make a new generator (e) and then after a pose-edit the generator is used to get FS codes (f) for the hair. The hair is masked (g) and then (c) and (g) are inputs to a new HAIRNET network (h) that predicts a blended F code. The identity image's global style information (i) and the hair's style (j) are combined to produce final output (k).

We borrow the idea of FS-codes from Barbershop, but we propose to use an alternative approach to deal with changes in pose and head shape. Our approach eliminates the need to generate a target segmentation mask, resulting in a more robust approach to hairstyle transfer. Rather than a target mask, we first *remove hair* from I_{ident} to make a new image I_{bald} using a latent-edit, eliminating the need to handle disocclusion after this step. The baldification method is described in Sect. 3.3.

Furthermore, prior approaches to hairstyle transfer struggle when I_{hair} and I_{ident} are from very different viewpoints or head-poses. We use a latent-edit to change the pose of I_{hair} to match the pose of I_{ident}. We adapt the pivotal tuning approach [31] to fine-tune a StyleGAN generator in order to create a detailed pose-edited hair image, I_{pose}. Our pose-editing approach is described in Sect. 3.4.

Next, we train a new network called HAIRNET to blend the F-codes. Let F_{pose} and S_{pose} denote the FS-code of the pose-edited hair image I_{pose}, and similarly let F_{bald} and S_{bald} denote the FS-code of I_{bald}. Then

$$F_{\text{mix}} = \text{HAIRNET}(F_{\text{pose}}, F_{\text{bald}}). \tag{1}$$

The purpose of HAIRNET is to use unsupervised machine learning rather than heuristics to learn how to copy hair from one image into another. A key element of our approach is a process for training HAIRNET in an unsupervised way, which we discuss in detail in Sect. 3.5.

Proceeding with the overview of our method, once F_{mix} is determined, the corresponding style-code S_{mix} must also be determined by mixing elements of S_{ident} and S_{hair}. This is important to preserve the color and texture of the hair, as described in Sect. 3.6.

Finally, the image I_{mix} is found by applying the StyleGAN generator with the activations style-block 7 set to F_{mix} and using S_{mix} as for the remaining style blocks.

3.3   Baldification

A key element of our approach to hair transfer is removal of the existing hair in I_{ident}, which we call 'baldification'. To do so, we first find a latent code W_{ident} in $W+$ space by applying GAN inversion to I_{ident}. In addition, we find an F-code F_{ident} for the identity image to capture a more detailed and spatially-aligned representation. Then we use the StyleFlow [4] method to generate a latent code W_{bald}. StyleFlow in $W+$ space may cause details other than head-hair to change – for example, facial hair may be removed, and the expression or facial features may slightly change. In addition, $W+$ space does not have the capacity to reconstruct all images as well as FS space does, so after the edit we use W_{bald} to find an initial F-code $F_{\text{bald}}^{\text{init}}$. Then we use an automatic segmentation method, BiseNET [40], to label pixels. A binary hair mask for the hair-region is formed and then down sampled bicubically to a shape of 32×32 pixels (so that each pixel is a real-valued number between zero and one) to form M_{hair}. Then

$$F_{\text{bald}} = (1 - M_{\text{hair}})F_{\text{ident}} + M_{\text{hair}}F_{\text{bald}}^{\text{init}}. \tag{2}$$

where the expression is evaluated for each pixel. Note that the latent edit using StyleFlow only modifies the early layers of $W+$ as described in [4, Section 6.2.3], and so the S-code of the baldified image is simply the latter elements of W_{bald}.

3.4 Pose-Editing

A major failure-mode of hair-transfer is when the hair and face images are from different poses, however automatically changing the pose of the hair image while also preserving its structure is extremely challenging. Latent editing approaches, such as StyleFlow [4], are capable of changing the pose by modifying a $W+$ latent code, and therefore have a limited capacity to reconstruct details in the edited image. For baldification, Eq. (2) works only because F_{ident} and F_{bald} are spatially aligned, however this will not be the case if a latent-edit is used to change the hair pose.

We address this by using pivotal tuning (PTI) [31], which refines a generator G to create a specialized generator with slightly different weights, which does a better job at reconstructing details of a specific image. We adapt this approach to create a generator G_{pose} that is refined with frozen weights for all the StyleGAN blocks that *follow* the F-code, but the first $m = 7$ blocks are free to adapt to better-reconstruct I_{hair}. Because the FS code uses only the activations of style block m, the two generators produce identical images for the same FS-codes. We use StyleFlow to generate a latent-code W_{pose} from W_{hair} and then use the activations of layer m of $G_{\mathrm{pose}}(W_{\mathrm{pose}})$ as F_{pose}

3.5 Training HairNet

Overview In order to train HairNet we first describe an unsupervised way to generate inputs and the desired output of the hair-transfer network. We use images from the FFHQ [22] dataset as a source of training images for hair transfer. In order to do hair-transfer, we consider the FFHQ images to be an ideal result for I_{mix}, and we generate I_{bald} and I_{pose}[1] from it using latent edits and augmentation (described in Sect. 3.5). From these we train HairNet to minimize a reconstruction loss (Sect. 3.5), so that the image generated using the FS-code predicted by HairNet is perceptually similar to I_{mix}. Finally, we describe the network architecture of HairNet.

Hair Image Generation. The process of generating I_{bald} was described in Sect. 3.3. In order to train HairNet we need to do an inverse problem of hair transfer – we need to generate an image that preserves the same hair as I_{mix} but that is different elsewhere. We rely on a semantic segmentation of I_{hair} to determine a binary hair mask that is zero in regions that are not hair and one elsewhere, then downsample the mask bicubically to match the spatial dimensions (32×32) of the F-code and multiply it by the downsampled mask to produce an image with meaningful information only in the hair region.

---

[1] In this case the $I_{\mathrm{pose}} = I_{\mathrm{hair}}$ as the pose is perfectly aligned by construction.

Even after masking out the F-code while preserving the hair, the results are still different from the inputs one expects to provide HAIRNET for inference because the hair is still perfectly aligned to with the head-shape of I_{bald} (and the desired output in I_{mix}). In order to ensure that the network does not simply learn to copy the hair, we apply the following augmentations to I_{hair}: a) We apply a random translation to I_{hair}, drawn from a truncated normal distribution with $\sigma = 0.2$. b) We apply a random log-normally-distributed scale to I_{hair}, drawn from the same distribution. The augmentation applies a transform with parameters,

$$T_{hair} = (t_x, t_y, s_x, s_y) \tag{3}$$

where t_x and t_y are translations and s_x and s_y are the logarithms of the scale applied to the hair image. Note that we represent the transformation using the *log scale* so that all parameters are normally distributed with zero-mean.

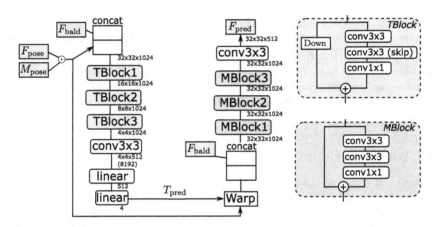

Fig. 3. The HAIRNET architecture. The input F_{pose} is multiplied by a hair-segmentation mask M_{pose} and concatenated channelwise with F_{bald}. The result is passed through the same set of residual blocks used by StyleGAN2 discriminator and a final convolution reduces the features to 512 channels, then two fully-connected layers predict T_{pred}. The F_{pose} tensor is warped and concatenated with F_{bald} before passing through a similar number of residual blocks. We call the StyleGAN2 blocks with strided convolution *TBlocks*, and the residual blocks without downsampling or strided convolution are called *MBlocks*.

Network Architecture. The architecture of HAIRNET is inspired by the Style-GAN2 discriminator network. The inputs to the network are the F-codes F_{bald} and F_{pose} along with a segmentation mask, M_{pose}, for the hair in I_{bald} computed using BiseNet [40]. The network first predicts a spatial transformation, T_{pred}, which is then used to warp the masked F_{pose} tensor so that the features corresponding to hair are positioned properly relative to F_{bald}. The spatial transformer network portion of the architecture uses residual blocks that are identical to the ones used in the StyleGAN2 discriminator, indicated in Fig. 3 as TBlocks. After three residual TBlocks, a final convolution is used to reduce channels to 512, before fully-connected layers predict the transformation (T_{pred}). Then a spatial transformation is applied to the masked activations of F_{pose}, warping it so that the features corresponding to hair are in new spatial locations (ideally aligned to F_{bald}).

Rather than simply copy/pasting the F-code values as was done in Barbershop [44], we use another residual 'blending' network to predict F_{pred}. The aim is to allow the network to learn when hair should cover the face or be occluded by it. The residual blocks of this network do not downsample their inputs, and they are labeled as MBlocks in Fig. 3. A final convolution applied to the output of the last MBlock reduces the channels from 1024 down to 512, producing F_{pred} as output.

Loss Function. The goal of the HAIRNET network is to generate an F-code,

$$F_{\mathrm{pred}} = \mathrm{HAIRNET}(F_{\mathrm{pose}}, F_{\mathrm{bald}}), \tag{4}$$

so that the generated image

$$I_{\mathrm{pred}} = G(F_{\mathrm{pred}}, S_{\mathrm{ident}}), \tag{5}$$

where the function $G(\cdot)$ is the StyleGAN image generator, is perceptually as similar as possible to I_{mix}. Our training process augments the hair image by applying a translation scale transformation, T_{hair}, to the hair image, and architecture predicts the same spatial transformation, T_{pred} in order to align hair. We use the L_2 loss between the two transformation parameters in order to encourage the network to learn the correct transformation. We also use L_{PIPS} [42] as a perceptual similarity metric. A secondary goal is to minimize the L_2-error between I_{pred} and I_{mix}, and finally, we also want to keep the L_2 error between the latent code F_{pred} and F_{mix} small. We minimize the following loss function:

$$L_{\mathrm{rec}} = L_2(T_{\mathrm{hair}}, T_{\mathrm{pred}}) + \lambda_1 L_{\mathrm{PIPS}}(I_{\mathrm{mix}}, I_{\mathrm{pred}}) \\ + \lambda_2 L_2(F_{\mathrm{mix}}, F_{\mathrm{pred}}) + \lambda_3 L_2(I_{\mathrm{mix}}, I_{\mathrm{pred}}) \tag{6}$$

The contribution of different loss terms are evaluated empirically in supplemental materials.

Preventing Overfitting. The F-codes contain significantly more information than a $W+$ code for an image, and it is possible for a system such as the one we described to learn how to infer the missing information about an image's original hair from a 'baldified' input image. This means that our training process if capable of ignoring F_{hair} completely in order to generate F_{mix} using baldified images if I_{mix} is always the same as I_{ident}. We address this by randomly using I_{bald} as I_{mix}. With probability p we replace F_{hair} with zeros, and we use I_{bald} and F_{bald} as I_{mix} and F_{mix} when evaluating the losses. We find $p = 0.5$ produces reasonable results.

3.6   Styling

RetreiveInStyle (RiS) [9] and EncodeInStyle (EiS) [30] use a fast method to interpolate latent codes by first selecting a set of elements of a latent code in style-space. Style-space latent-codes modulate the channels of each stylegan block. RiS and EiS threshold total activatiosn *within* a masked region to determine which latent code elements are relevant to that region. In order to edit the region-of-interest, the latent code elements that are not relevant are frozen and the others are free to change.

One important caveat is that the layer relevance approach uses elements in *style space* rather then $W+$ space. We build on this approach and we also use style-space. This slightly changes our interpretation of the S-code in FS-space as the style space elements are the result of an affine transform applied to the $W+$ vector.

The methods of RiS and EiS use an effective heuristic to change the relevant elements of a latent code - however we find that their approach, while fast, would cause unexpected changes to the color of the hair. Instead, wo use the same optimization criteria as Barbershop (the masked LPIPS function, L_{mask}) to solve for the relevant portions of the code. This loss function is the same function used by Barbershop [44, Section 3.5] for style mixing, with one modification. Let that R be a mask so that $R_i = 1$ if the corresponding element S_i of the style code is relevant, and $R_i = 0$ otherwise. We use the mask to change only the relevant parts of the stylecode using optimization. Assume that, for some vector C_{mix}, we have $S_{mix} = (1 - R)S + RC_{mix}$, and

$$C_{mix} = \arg\min_C L_{mask}((1 - R)S + RC), \tag{7}$$

where L_{mask} is calculated using I_{pose} and I_{bald} using M_{pose} as the segmentation mask.

4   Evaluation

4.1   Metrics

Quantitative evaluation of the success of an image editing task has always been a challenging task. A successful edit does two things; it correctly preserves some attributes while changing others (e.g. the face vs the hair), and it also produces a high quality image as output (e.g. free of artifacts). In our case, a successful edit reconstructs the *face* from one image (I_{ident}) and it correctly constructs the hair from another image (I_{hair}) after a pose change.

We use the FID [17] of the generated images and the II2S dataset as a quantitative measure of the quality of generated images. The FID is a standard metric for evaluating GANs and if the generated I_{mix} images are not similarly distributed to the II2S dataset then they will have high FID scores, and that may indicate low quality results. However, the FID is a poor approximation to human perception of the quality of the input. Several other attempts have been made to quantitatively evaluate the quality of a generated image. The Naturalness Image Quality Estimator (NIQE) [25] measures overall image quality (not specific to face). Precision, Recall, and Realism were introduced in [24]. Precision and recall check whether edited images are on the same manifold as a ground-truth set of image, Realism is simply the distance of an image from the manifold of training images. These methods check the overall quality of an image, but do not capture whether the hair and face are preserved. ArcFace [12] measures the edit's ability to preserve the face of the identity image after the edits. However, we are unaware of any automatic quantitative way to evaluate whether the hairstyle was correctly transferred with a pose that matches the face image. For this problem, we must rely on a user-study and human perception.

4.2   Ablation Study

Many parts of our method are necessary to get any meaningful result, for example we cannot evaluate the effect of using pivotal tuning in isolation because without it a pose edit that produces an F-code that is different from the $W+$ embedding is not even possible. The main contributions we can ablate are the HAIRNET and our optimization method for mixing the S-codes, which is presented in Fig. 4.

Qualitative results of ablating HAIRNET are shown in Fig. 4(top), which highlights the importance of human perception in evaluating images, as we expect most readers would agree that the 'w/o HAIRNET' row of Fig. 4 is significantly worse than the last row, which uses HAIRNET. However, the quantitative metrics are nearly identical; FID without hairnet is 56.4 (vs 55.6 with HAIRNET). The NIQE is 11.82 (vs 11.85), the Precision is 95.5% (vs 97%), Recall is 57% (vs 60%) and Realism is 1.21 (vs 1.26%). Quantitatively, we show that HAIRNET only slightly changes each metric, even though the visual results are significant. We evaluate ablating the style mixing step qualitatively in Fig. 4(bottom), which shows that style mixing with L_{mask} better preserves the colors of the hair and face.

4.3   Comparison

We compare our results against several recent state-of-the-art hair editing methods; Barbershop [44], Retrieve in Style [9], Style Fusion [18], and MichiGAN [34]. Quantitative results are shown in Table 2, however as mentioned previously these metrics do not capture human perception of whether the edit was successful. For example methods that simply copy the face achieve high ArcFace scores but produce images with very undesirably cut&paste artifacts. The same is true for other quantitative metrics, however, our method is within the top-2 for most quality metrics.

Hair

Identity

w/o HairNet

w HairNet

Hair

Identity

w/o L_{mask}

w L_{mask}

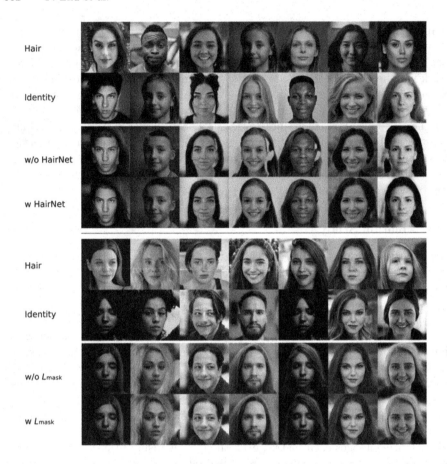

Fig. 4. Qualitative ablation studies: (top) results with, and without using the novel HAIRNET component proposed in our process, demonstrating both inpainting and handling of occlusions; (bottom) the effect of style-mixing using L_{mask} to produce face and hair colors that are more similar to the corresponding source images.

For a more reliable picture of our performance, we conducted a user study with Amazon Mechanical Turk. For each competing method, 764 image pairs (theirs vs ours) or (ours vs theirs) were shown to human subjects and they were asked which image better reconstructed the hair, the face, and which had the highest overall quality. These quantitative results are shown in Table 1 and it is clear that for the editing tasks of preserving hair and face our method dominates. With respect to overall image quality, we are nearly a tie with StyleFusion, however StyleFusion generates images restricted to a latent space that only has high quality images at the cost of reconstructing hair and face accurately. Many additional qualitative results are included in the supplemental materials.

Table 1. User-study results comparing to Barbershop [44], Retreive In Style [9], Michi-GAN [34], and StyleFusion [18]. Our method outperforms all others for reconstruction tasks. We only lose to StyleFusion (by less than 1%) for the overall quality question, however this is expected because images in the restricted StyleGAN latent space can be more generic and attractive than images with good reconstruction, which many users will perceive as high quality. However, we do significantly better at hair reconstruction than StyleFusion.

Method	Face Rec.		Hair Rec.		Overall Qual.	
	Theirs	Ours	Theirs	Ours	Theirs	Ours
Barbershop	32%	**68%**	30%	**70%**	24%	**76%**
RetrieveInStyle	43%	**57%**	43%	**57%**	42%	**58%**
MichiGAN	8%	**92%**	4%	**96%**	2%	**98%**
StyleFusion	49%	**51%**	46%	**54%**	**51%**	49%

Table 2. Quantitative evaluation of different methods using the following metrics: NIQE [25], ArcFace [12], FID [17], precision [24], recall [24], and realism [24]. The best result is bold and the second best is underlined.

Method	NIQE↓	ArcFace↑	FID↓	Precision↑	Recall↑	Realism↑
Barbershop	12.52	<u>0.78</u>	**47.34**	0.93	**0.83**	**1.30**
RetrieveInStyle	12.18	0.58	60.74	0.96	0.31	1.17
MichiGAN	**11.65**	**0.88**	84.66	0.58	<u>0.72</u>	1.09
StyleFusion	12.12	0.56	68.46	**0.98**	0.19	1.19
Ours	<u>11.85</u>	0.66	<u>55.60</u>	<u>0.97</u>	0.60	<u>1.26</u>

Qualitative Results. In addition to the quantitative results and the user study, several examples of our results compared with competing methods are shown in Fig. 5. We observe that the results visually agree with the user study, and both StyleFusion and our approach produce high quality results. However, our approach does a better job at preserving salient qualities of the hair and face.

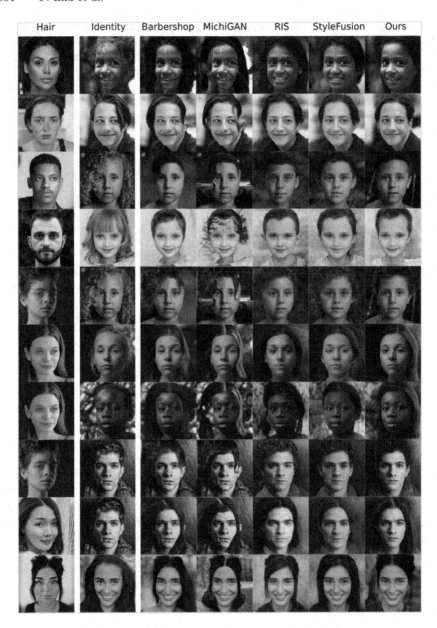

Fig. 5. Main comparison between competing methods. Notice that our method and StyleFusion both produce very realistic results, but our method preserves the appearance of the hair with high fidelity to the source image.

5   Conclusion

We propose a novel algorithm for automatic hairstyle transfer, specifically targeting complicated inputs that do not match in pose. We introduced three main technical contributions to tackle this challenge. First, we introduce the concept of baldification to hairstyle editing pipelines. Second, we propose a novel embedding algorithm that combines the advantages of two recent state-of-the-art methods. Third, we propose the hairnet architecture that automatically combines two images at inference time and can be trained in an unsupervised fashion. In future work, we would like to extend our framework to recent 3D GANs such as EG3D or GRAM.

References

1. Abdal, R., Qin, Y., Wonka, P.: Image2stylegan: how to embed images into the style-GAN latent space? In: Proceedings of the IEEE/CVF International Conference on Computer Vision, pp. 4432–4441. IEEE, Seoul, Korea (2019)

2. Abdal, R., Zhu, P., Femiani, J., Mitra, N.J., Wonka, P.: Clip2stylegan: Unsupervised extraction of StyleGAN edit directions. CoRR abs/2112.05219 (2021). https://arxiv.org/abs/2112.05219

3. Abdal, R., Zhu, P., Mitra, N.J., Wonka, P.: Labels4Free: unsupervised segmentation using StyleGAN. In: Proceedings of the IEEE/CVF International Conference on Computer Vision (ICCV), pp. 13970–13979, October 2021

4. Abdal, R., Zhu, P., Mitra, N.J., Wonka, P.: StyleFlow: attribute-conditioned exploration of StyleGAN-generated images using conditional continuous normalizing flows. ACM Trans. Graph. **40**(3) (2021). https://doi.org/10.1145/3447648, https://doi.org/10.1145/3447648

5. Alaluf, Y., Patashnik, O., Cohen-Or, D.: Restyle: a residual-based StyleGAN encoder via iterative refinement. In: Proceedings of the IEEE/CVF International Conference on Computer Vision (ICCV), October 2021

6. Alaluf, Y., Tov, O., Mokady, R., Gal, R., Bermano, A.H.: HyperStyle: StyleGAN inversion with hypernetworks for real image editing. CoRR abs/2111.15666 (2021). https://arxiv.org/abs/2111.15666

7. Chan, E.R., et al.: Efficient geometry-aware 3d generative adversarial networks. In: EEE Conference on Computer Vision and Pattern Recognition (CVPR) (2021)

8. Choi, Y., Uh, Y., Yoo, J., Ha, J.W.: StarGAN v2: diverse image synthesis for multiple domains. In: Proceedings of the IEEE Conference on Computer Vision and Pattern Recognition (2020)

9. Chong, M.J., Chu, W.S., Kumar, A., Forsyth, D.: Retrieve in style: unsupervised facial feature transfer and retrieval. In: In International Conference on Computer Vision (2021)

10. Chong, M.J., Forsyth, D.A.: JojoGAN: one shot face stylization. CoRR abs/2112.11641 (2021). https://arxiv.org/abs/2112.11641

11. Collins, E., Bala, R., Price, B., Süsstrunk, S.: Editing in style: uncovering the local semantics of GANs. In: 2020 IEEE/CVF Conference on Computer Vision and Pattern Recognition (CVPR) (2020)

12. Deng, J., Guo, J., Niannan, X., Zafeiriou, S.: ArcFace: additive angular margin loss for deep face recognition. In: CVPR (2019)

13. Deng, Y., Yang, J., Xiang, J., Tong, X.: GRAM: generative radiance manifolds for 3D-aware image generation (2021)
14. Gal, R., Patashnik, O., Maron, H., Chechik, G., Cohen-Or, D.: StyleGAN-nada: clip-guided domain adaptation of image generators. arXiv preprint arXiv:2108.00946 (2021)
15. Goodfellow, I.J., et al.: Generative adversarial networks. In: Advances in Neural Information Processing Systems 27 (NIPS 2014) (2014)
16. Härkönen, E., Hertzmann, A., Lehtinen, J., Paris, S.: Ganspace: Discovering interpretable GAN controls. arXiv preprint arXiv:2004.02546 (2020)
17. Heusel, M., Ramsauer, H., Unterthiner, T., Nessler, B., Hochreiter, S.: GANs trained by a two time-scale update rule converge to a local Nash equilibrium. In: Proceedings of the 31st International Conference on Neural Information Processing Systems. NIPS 2017, pp. 6629–6640, Curran Associates Inc., Red Hook, NY, USA (2017)
18. Kafri, O., Patashnik, O., Alaluf, Y., Cohen-Or, D.: StyleFusion: a generative model for disentangling spatial segments. ACM Trans. Graph. (2021)
19. Karras, T., Aila, T., Laine, S., Lehtinen, J.: Progressive growing of GANs for improved quality, stability, and variation. In: ICLR (2017)
20. Karras, T., Aittala, M., Hellsten, J., Laine, S., Lehtinen, J., Aila, T.: Training generative adversarial networks with limited data. In: Proceedings of NeurIPS (2020)
21. Karras, T., et al.: Alias-free generative adversarial networks (2021)
22. Karras, T., Laine, S., Aila, T.: A style-based generator architecture for generative adversarial networks. IEEE Trans. Pattern Anal. Mach. Intell. **43**(12), 4217–4228 (2021)
23. Karras, T., Laine, S., Aittala, M., Hellsten, J., Lehtinen, J., Aila, T.: Analyzing and improving the image quality of StyleGAN. In: Proceedings of CVPR (2020)
24. Kynkäänniemi, T., Karras, T., Laine, S., Lehtinen, J., Aila, T.: Improved precision and recall metric for assessing generative models. CoRR abs/1904.06991 (2019)
25. Mittal, A., Soundararajan, R., Bovik, A.C.: Making a "completely blind" image quality analyzer. IEEE Signal Process. Lett. **20**(3), 209–212 (2013). https://doi.org/10.1109/LSP.2012.2227726
26. Or-El, R., Luo, X., Shan, M., Shechtman, E., Park, J.J., Kemelmacher-Shlizerman, I.: StyleSDF: high-resolution 3D-consistent image and geometry generation (2021)
27. Patashnik, O., Wu, Z., Shechtman, E., Cohen-Or, D., Lischinski, D.: StyleCLIP: Text-driven manipulation of StyleGAN imagery. In: 2021 IEEE/CVF International Conference on Computer Vision (ICCV) (2021)
28. Peebles, W., Zhu, J.Y., Zhang, R., Torralba, A., Efros, A., Shechtman, E.: GAN-supervised dense visual alignment (2021)
29. Radford, A., Metz, L., Chintala, S.: RDCGAN: Unsupervised representation learning with deep convolutional generative adversarial networks. In: 2018 9th Conference on Artificial Intelligence and Robotics and 2nd Asia-Pacific International Symposium (2015)
30. Richardson, E., et al.: Encoding in style: a StyleGAN encoder for image-to-image translation. arXiv preprint arXiv:2008.00951 (2020)
31. Roich, D., Mokady, R., Bermano, A.H., Cohen-Or, D.: Pivotal tuning for latent-based editing of real images. arXiv preprint arXiv:2106.05744 (2021)
32. Saha, R., Duke, B., Shkurti, F., Taylor, G.W., Aarabi, P.: LOHO: latent optimization of hairstyles via orthogonalization. In: 2021 IEEE/CVF Conference on Computer Vision and Pattern Recognition (CVPR) (2021)

33. Shen, Y., Yang, C., Tang, X., Zhou, B.: InterFaceGAN: Interpreting the disentangled face representation learned by GANs. IEEE Trans. Pattern Anal. Mach. Intell. (2020)
34. Tan, Z., et al.: Michigan. ACM Trans. Graph. **39**(4) (2020). https://doi.org/10.1145/3386569.3392488, https://doi.org/10.1145/3386569.3392488
35. Tewari, A., et al.: StyleRig: rigging styleGAN for 3D control over portrait images. In: Proceedings of the IEEE/CVF Conference on Computer Vision and Pattern Recognition, pp. 6142–6151 (2020)
36. Tewari, A., et al.: PIE: portrait image embedding for semantic control. ACM Trans. Graph. **39** (2020). https://doi.org/10.1145/3414685.3417803
37. Tov, O., Alaluf, Y., Nitzan, Y., Patashnik, O., Cohen-Or, D.: Designing an encoder for styleGAN image manipulation. arXiv preprint arXiv:2102.02766 (2021)
38. Wei, T., et al.: HairCLIP: design your hair by text and reference image. In: CVPR (2021)
39. Wu, Z., Lischinski, D., Shechtman, E.: StyleSpace analysis: disentangled controls for StyleGAN image generation. arXiv preprint arXiv:2011.12799 (2020)
40. Yu, C., Wang, J., Peng, C., Gao, C., Yu, G., Sang, N.: BiSeNet: bilateral segmentation network for real-time semantic segmentation. In: Ferrari, V., Hebert, M., Sminchisescu, C., Weiss, Y. (eds.) ECCV 2018. LNCS, vol. 11217, pp. 334–349. Springer, Cham (2018). https://doi.org/10.1007/978-3-030-01261-8_20
41. Yu, F., Zhang, Y., Song, S., Seff, A., Xiao, J.: LSUN: construction of a large-scale image dataset using deep learning with humans in the loop. arXiv preprint arXiv:1506.03365 (2015)
42. Zhang, R., Isola, P., Efros, A.A., Shechtman, E., Wang, O.: The unreasonable effectiveness of deep features as a perceptual metric. In: CVPR (2018)
43. Zhu, J., Shen, Y., Zhao, D., Zhou, B., In-domain GAN inversion for real image editing: In-domain GAN inversion for real image editing. In: Vedaldi, A., Bischof, H., Brox, T., Frahm, J.-M. (eds.) ECCV 2020. LNCS, vol. 12362, pp. 592–608. Springer, Cham (2020). https://doi.org/10.1007/978-3-030-58520-4_35
44. Zhu, P., Abdal, R., Femiani, J., Wonka, P.: Barbershop: GAN-based image compositing using segmentation masks. ACM Trans. Graph. **40**(6) (2021). https://doi.org/10.1145/3478513.3480537, https://doi.org/10.1145/3478513.3480537
45. Zhu, P., Abdal, R., Femiani, J., Wonka, P.: Mind the gap: domain gap control for single shot domain adaptation for generative adversarial networks. In: International Conference on Learning Representations (2022). https://openreview.net/forum?id=vqGi8Kp0wM
46. Zhu, P., Abdal, R., Qin, Y., Femiani, J., Wonka, P.: Improved StyleGAN embedding: where are the good latents? (2020)

Unbiased Multi-modality Guidance
for Image Inpainting

Yongsheng Yu[1,3], Dawei Du[2], Libo Zhang[1,3,4(✉)], and Tiejian Luo[3]

[1] Institute of Software Chinese Academy of Sciences, Beijing, China
libo@iscas.ac.cn
[2] Kitware, New York, USA
[3] University of Chinese Academy of Sciences, Beijing, China
yuyongsheng19@mails.ucas.ac.cn, tjluo@ucas.ac.cn
[4] Nanjing Institute of Software Technology, Beijing, China

Abstract. Image inpainting is an ill-posed problem to recover missing or damaged image content based on incomplete images with masks. Previous works usually predict the auxiliary structures (*e.g.*, edges, segmentation and contours) to help fill visually realistic patches in a multi-stage fashion. However, imprecise auxiliary priors may yield biased inpainted results. Besides, it is time-consuming for some methods to be implemented by multiple stages of complex neural networks. To solve this issue, we develop an end-to-end multi-modality guided transformer network, including one inpainting branch and two auxiliary branches for semantic segmentation and edge textures. Within each transformer block, the proposed multi-scale spatial-aware attention module can learn the multi-modal structural features efficiently via auxiliary denormalization. Different from previous methods relying on direct guidance from biased priors, our method enriches semantically consistent context in an image based on discriminative interplay information from multiple modalities. Comprehensive experiments on several challenging image inpainting datasets show that our method achieves state-of-the-art performance to deal with various regular/irregular masks efficiently. The code is available at https://github.com/yeates/MMT.

Keywords: Biased prior · Multi-modality guidance · Auxiliary denormalization · Image inpainting

1 Introduction

Image inpainting aims to repair missing or damaged image content based on known information of an image. It has been applied on many real-world scenarios, such as image editing [1,3], unwanted object removal [8,30], and old photo restoration [31].

Supplementary Information The online version contains supplementary material available at https://doi.org/10.1007/978-3-031-19787-1_38.

Following the assumption that corrupted images have adequate knowledge for inpainting [19,42], modern image inpainting methods [19,20,25,27,39] employ an encoder-decoder architecture. Concretely, they focus on various contextual attention mechanisms to learn the known visible content and fill the missing region. However, this assumption does not hold if the image is damaged by larger masks. It is difficult to provide sufficient semantically consistent information for realistic image inpainting based on known area in a RGB image.

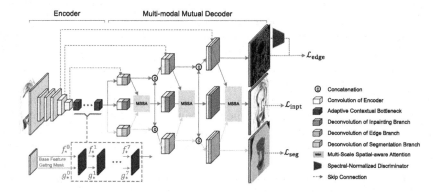

Fig. 1. Architecture of our Multi-modality guided Transformer that couples various modalities including RGB image, semantic segmentation, and edge textures.

Therefore, recent approaches [4,21,25,32,39] have made great efforts to introduce auxiliary priors, such as *edges*, *segmentation*, and *contours*, to facilitate improving image inpainting performance. However, they still suffer from the biased prior issue by using predicted auxiliary structures to guide image inpainting intermediately. Without ground-truth in testing phase, such direct guidance is inevitably biased, resulting in more deviations and errors for image inpainting. On the other hand, previous works [23,32] are usually divided into multiple stages of neural networks under the U-Net architecture. If each stage contains a complex subnetwork, it is time-consuming for potential real-world inpainting applications. This problem becomes more prominent when extending to video inpainting. For example, Liu *et al.* [23] tackle the image inpainting problem by a two-stage process, *i.e.*, two individual U-Nets for rough inpainting and refinement inpainting, yielding the running speed of only 1.37 FPS.

To solve the above issues, we propose a new multi-modality guided transformer network for image inpainting. As shown in Fig. 1, it follows the U-Net style [28] encoder-decoder architecture. In the encoder, we first develop the adaptive contextual bottlenecks for better context reasoning. To adapt to the current image content and missing region, the gating mask is updated to weight different dilated convolutions to enhance base features. Then, the multi-modal mutual decoder is proposed to decode the enhanced features into three modalities, *i.e.*, RGB image, and corresponding semantic segmentation and edge textures. It consists of one image inpainting branch and two auxiliary branches for semantic segmentation and edge textures. Unlike existing approaches based on direct guidance from predicted auxiliary structures, we focus on jointly learning the unbiased discriminative interplay information among the three branches. Specifically, the proposed multi-scale spatial-aware attention mechanism integrates multi-modal feature maps via auxiliary denormalization to reduce duplicated and noisy content for image inpainting. Supervised by ground-truth RGB images, semantic segmentation and edge maps, the whole network is trained in an end-to-end fashion efficiently. Note that segmentation and edge annotations can be provided by the off-the-shelf algorithms [6, 25].

As shown in Fig. 2, previous image inpainting methods fail to restore correct faces and buildings based on either biased edge [25] or segmentation [32] prior. On the contrary, our method still achieves robust results even though the glasses are not repaired in edge prior (see Ours* in the 1st row of Fig. 2) or the roof shape is not predicted correctly in segmentation prior (see Ours* in the 2nd row of Fig. 2). It demonstrates that our method can extract discriminative unbiased context information to guide image inpainting. To verify the effectiveness of our method, the experiment is conducted on three datasets including CelebA-HQ [17, 18], OST [35] and CityScapes [7]. The results show our method achieves the state-of-the-art image inpainting performance. For example, our method obtains the best FID score on the CelebA-HQ dataset with both regular and irregular masks, yeilding ~ 2 gain over the second best performer CTSDG [12]. By using segmentation results from DeepLabv3+ [6], our method still performs well on those datasets without segmentation annotation (e.g., Places2 [50]).

Contributions. 1. We propose an end-to-end multi-modality guided transformer to learn interplay information from multiple modalities including RGB image, edge textures and semantic segmentation. 2. We develop the multi-scale spatial-aware attention mechanism with auxiliary denormalization to capture compact and discriminative multi-modal features to guide unbiased image inpainting. 3. Comprehensive results on several datasets demonstrate the effectiveness of our unbiased multi-modality guidance, especially for irregular masks.

2   Related Work

Image Inpainting. Mainstream image inpainting methods employ the encoder-decoder architecture based on the U-Net [28]. For example, Pathak et al. [27] introduces an adversarial network [11] to help train the U-Net and mitigate the blurring caused by the pixel-level averaging property of a reconstruction loss. After that, Contextual Attention (CA) [42] is a two-stage coarse-to-fine model to weight known region as the reference of mission region. Using partial conv [22], Recurrent Feature Reasoning (RFR) [19] applies multiple iterations at the bottleneck of the encoder from outside to inside for large corrupt areas. Different from partial conv [22] with a heuristic mask update step to standard convolution, Gated Conv (GC) [43] improves this mask update process with a learnable convolution layer.

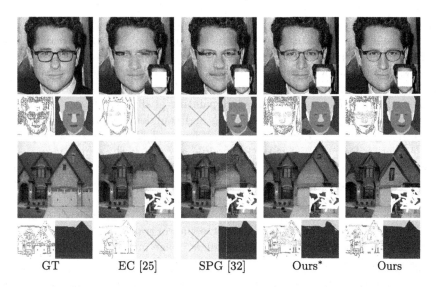

| GT | EC [25] | SPG [32] | Ours* | Ours |

Fig. 2. Influence of biased prior guidance. ✗ means no edge prior for SPG [32] and segmentation prior for EC [25]. Ours* denotes the variant of our multi-modality guided image inpainting method with inaccurate edge and segmentation priors by reducing the loss weights of two auxiliary branches by 30 times.

To better exploit context between missing and uncorrupted regions, GLILC [16] first introduces multiple residual modules [13] of dilation convolution [41] as the bottleneck in the encoder. However, it may bring the "gridding" problem [5,34] due to only sampling non-zero positions. That is, a single constant dilation rate results in either sparse convolution kernels (large hole rate) or difficulty crossing over large masks (small hole rate). To this end, Wang et al. [36] develop a generative multi-column network for image inpainting. Recently, Zeng et al. [46] propose the AOT blocks to aggregate contextual transformations from various receptive fields, which capture both informative distant image contexts and rich patterns of interest. Different from above methods, we introduce a new adaptive contextual bottleneck in the encoder, where the dynamic gating updating weights different pathways of dilated convolutions based on various masks.

Image Inpainting with Auxiliary Structures. Due to the ill-posed nature of reconstructing missing regions, additional structural priors (e.g., edges, segmentation, and contours) are used to facilitate image inpainting models for more realistic results. Edge Connect (EC) [25] relies on the corrupted canny edge image to deliver finer inpainting results. Cao and Fu [4] introduce an extra encoder to infer precise wireframe sketches to bypass the pool coherence of canny edge. According to the style and spatial consistency of semantic segmentation, Segmentation Prediction and Guidance network (SPG) [32] is a two-stage based segmentation and RGB image inpainting model, where DeepLabv3+ [6] is used to estimate the segmentation of corrupted image. Another work [39] is a new three-stage based model to locate and fill foreground object and its contour by disentangling the inter-object intersection.

However, the above multi-stage methods are usually time-consuming. For better efficiency, the Semantic Guidance and Evaluation (SGE) network [20] couples with segmentation and image inpainting at different layers of decoder, where the segmentation after completing and confidence scoring guides image inpainting by semantic normalization [26]. Liao *et al.* [21] propose the Semantic-wise Attention Propagation (SWAP) module to capture the semantic relevance between segmentation and image textures in non-local operation. Recently, Yang *et al.* [40] predict explicit edge embedding with an attention mechanism to facilitate image inpainting by the multi-task learning strategy. It worth mentioning that most aforementioned works use estimated auxiliary structures as the direct guidance of image inpainting. On the contrary, we develop the multi-head spatial-aware attention module to guide image inpainting based on jointly learned discriminative features from unbiased auxiliary priors.

Transformers in Image Inpainting. Inspired by Vision Transformer [10], recent methods [9,44] decode the long-range dependencies between input features for better image inpainting. Deng *et al.* [9] learn relations between the corrupted and uncorrupted regions and exploit their respective internal closeness. Yu *et al.* [44] introduce the bidirectional autoregressive transformer that enables bidirectionally modeling of contextual information of missing regions. In contrast, our method propose a new multi-modality guided transformer to capture interplay information across three modalities.

3   Multi-modality Guided Transformer

The original image $\mathbf{I}$ is degraded as a corrupted image $\mathbf{I}_m = \mathbf{I} \odot (1 - \mathbf{M})$, where the pixel values in the missing region $\mathbf{M}$ equal to 0 are defined as invisible pixels. Our goal is to produce semantically reasonable and visually realistic reconstructed images $\mathbf{I}_{\mathrm{pred}}$ with the input of the corrupted image $\mathbf{I}_m$. Similar to previous works [16,19,27,43], we retain the U-Net style encoder-decoder architecture. As illustrated in Fig. 1, the multi-modality guided transformer contains an encoder with adaptive contextual bottlenecks, and a multi-modal mutual encoder with multi-scale spatial-aware attention, described in detail as follows.

3.1   Encoder with Adaptive Contextual Bottlenecks

For better context reasoning, the multi-stream structure is used in the encoder to weight dilated convolutions and encode the current image content and missing region. Unlike simply stacking parameters in previous ASPP [5] and AOT [46], we develop a stack of Adaptive Contextual Bottlenecks (ACB) to adapt to the specific mask shape size and image context by dynamic gating. As shown in Fig. 3, the ACB module consists of four parallel pathways of convolutional layers with different dilation rate and one gating mask to weight dilated convolutions.

In this way, the encoder can enlarge the perceptual field of convolutions and find the most plausible pathway according to the current missing region.

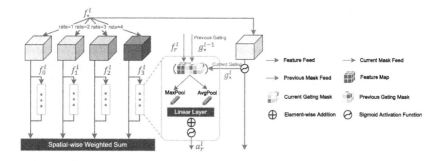

Fig. 3. Structure of adaptive contextual bottlenecks in the encoder.

Given the corrupt image $\mathbf{I}_m$, the base features f_*^0 and gating g_*^0 are initialized by the last layer (gated conv) of encoder. Then f_*^l and g_*^l at each layer is updated by the ACB block. The gating mask g_*^l is used to estimate the probability of missing region based on the feature map at the l-th layer ($l = 1, \cdots, L$), $i.e.$, $g_*^l = \mathrm{gconv}(f_*^l)$, where gconv denotes the gated conv operation [43]. In terms of each pathway with dilation rate r, we compute the dilated feature maps f_r^l based on f_*^l and corresponding weight a_r^l. Similar to [38], the spatial-wise weight a_r^l is calculated based on both average and max pooling of concatenation of dilated feature maps f_r^l and gating masks g_*^l, g_*^{l-1}, $i.e.$, $a_r^l = \sigma(\mathrm{fc}(\mathrm{avg}(g_r^l)) + \mathrm{fc}(\mathrm{max}(g_r^l)))$, where σ is the sigmoid function, and avg and max are the average and maximal pooling respectively. fc denotes the fully-connected layer, and the gating mask for each pathway is calculated as $g_r^l = \mathrm{conv}([f_r^l; g_*^l; g_*^{l-1}])$. Finally, the feature map at the $(l + 1)$-th ACB layer is updated by the spatial-wise weighted summation of f_r^l as

$$f_*^{l+1} = \sum_{r \in R} \frac{\exp(a_r^l)}{\sum_{r \in R} \exp(a_r^l)} \cdot f_r^l + f_*^l, \tag{1}$$

where R denotes the set of different dilation rates. The fractional term denotes element-wise product between dilated feature map f_r^l and attention vector a_r^l, weighting dilation block based on mask and image context. For simplicity, we omit the subscript l in the following sections.

3.2 Multi-modal Mutual Decoder

Given enhanced features f_*, the decoder use stacks of transformer blocks to learn the structural multi-modal information jointly. It consists of three branches, $i.e.$, one *inpainting branch* to recover the damaged image, and two *auxiliary branches* with additional segmentation and edge priors.

As shown in Fig. 1, within each transformer block, we first calculate the attention among feature maps from three branches by the proposed Multi-Scale Spatial-aware Attention (MSSA). Then, the enhanced features are split to combine the previous feature maps in each branch for attention calculation at next stage. Note that the skip connections between the encoder and decoder are used to prevent network degradation. After three stages, we predict the inpainted image $\mathbf{I}_{\text{pred}}$, edge and segmentation maps. Thus we leverage the structural features from auxiliary branches to enforce the model focus on discriminative interplay features for more realistic image inpainting.

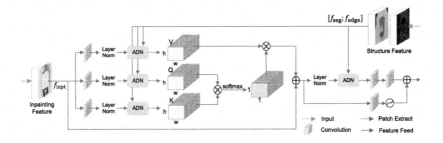

Fig. 4. Illustration of multi-scale spatial-aware attention.

To learn mutual features from different modalities, it is intuitive to simply concatenate or add the feature maps in three branches. Nevertheless, such strategies may introduce duplicated and noisy content for image inpainting. To effectively integrate compact features from *auxiliary branches*, we introduce a new Multi-Scale Spatial-aware Attention (MSSA) mechanism as follows.

Multi-scale Spatial-Aware Attention. Based on the encoded feature maps f_*, we use $f_{\text{inpt}}, f_{\text{edge}}, f_{\text{seg}}$ to denote the input feature maps for the inpainting branch, edge branch, and segmentation branch, respectively. As illustrated in Fig. 4, we combine the feature maps from three branches by the following Auxiliary DeNormalization (ADN):

$$\text{ADN}(f_{\text{inpt}}|[f_{\text{edge}}; f_{\text{seg}}]) = \gamma \odot \text{LN}(f_{\text{inpt}}) + \beta, \tag{2}$$

where $[;]$ denotes the matrix concatenation along channel dimension, and $\odot$ the element-wise multiplication. LN denotes layer normalization [2]. γ and β are the affine transformation parameters learned by two convolutional layers based on $[f_{\text{edge}}; f_{\text{seg}}]$ (see the top-right corner of Fig. 4). In this way, the multi-modal features are merged based on context from auxiliary structures that varies with respect to different spatial location.

Then, the merged features are embedded into query Q, key K and value V. Similar to [45], the embedded feature map is spatially split into N patches, *i.e.*, $P_i \in \mathbb{R}^{h \times w \times c}(i = 1, \ldots, N)$, where h, w, c denote the height, width and channel of patches respectively. The normalized self-attention $\alpha_{i,j}$ between patches i and

j can be calculated as $\alpha_{i,j} = \text{softmax}(\frac{Q_i \cdot K_j^T}{\sqrt{h \cdot w \cdot c}})$, $i, j \in 1, \ldots, N$. Note that we can perform multi-head self-attention like [10]. Thus the feature map of each patch is updated in a non-local form, $i.e.$, $\hat{P}_i = \sum_{j=1}^{N} \alpha_{i,j} V_j$.

Comparison Between Existing Denormalization Methods. Our ADN is related to two previous denormalization methods including AdaIN [15] and SPADE [26]. As shown in Fig. 5, we compare the networks of three denormalization methods. However, they are different in two aspects:

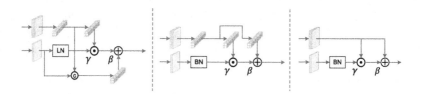

Fig. 5. (a) Our Auxiliary DeNormalization (ADN). (b) SPatially-Adaptive DEnormalization (SPADE) [26]. (c) Adaptive Instance Normalization (AaIN) [15]. LN, BN and IN denote layer, batch and instance normalizations respectively.

- AdaIN [15] and SPADE [26] learn the affine transformation parameters $\{\gamma, \beta\}$ based on the predicted auxiliary structures. Without ground-truth in testing phase, the predicted auxiliary structures are inevitably biased and result in inferior performance. In contrast, our ADN is based on the multi-modal features from two auxiliary branches.
- AdaIN [15] leverages the image's mean and variance instead of learnable affine parameters. SPADE [26] learns the spatial style of features by two convolutions after Batch Normalization. However, we combine features from both inpainting and auxiliary branches to learn the affine parameters.

Gated Feed-Forward. Finally, we piece all feature maps $\hat{P}_i$ together and reshape them with the original scale of input inpainting features f_{inpt}. Following the gated feed-forward layer, we can output the final feature maps for inpainted image prediction. Similar to gated conv [43], the gated feed-forward layer can ease the color discrepancy problem by detecting potentially corrupted and uncorrupted regions.

3.3 Optimization

To train our network, the overall loss consists of three terms, $i.e.$,

$$\mathcal{L} = \mathcal{L}_{\text{inpt}} + \lambda_{\text{edge}}\mathcal{L}_{\text{edge}} + \lambda_{\text{seg}}\mathcal{L}_{\text{seg}}, \tag{3}$$

where $\mathcal{L}_{\text{inpt}}$, $\mathcal{L}_{\text{edge}}$ and $\mathcal{L}_{\text{seg}}$ denote the loss terms for inpainting branch, edge branch and segmentation branch respectively. λ_{edge} and λ_{seg} are the balancing factors. The inpainting loss $\mathcal{L}_{\text{inpt}}$ follows the work in [22]. Similar to [25], we use both binary cross-entropy and adversarial loss functions to train the edge branch, *i.e.*,

$$\mathcal{L}_{\text{edge}} = w_1 \mathcal{L}_{\text{BCE}} + \mathcal{L}_{\text{adv}}, \tag{4}$$

where w_1 is the balancing weight. $\mathcal{L}_{\text{BCE}} = \frac{1}{N} \sum_{i=1}^{N} -[\mathbf{C}_{\text{gt}}^i \log \mathbf{C}_{\text{pred}}^i + (1 - \mathbf{C}_{\text{gt}}^i) \log(1 - \mathbf{C}_{\text{pred}}^i)]$ predicts the edge structure, and $\mathcal{L}_{\text{adv}} = -\mathbb{E}\left[\mathbf{D}\left(\mathbf{C}_{\text{pred}}\right)\right]$ justifies if the predicted edge is fake or real. $\mathbf{C}_{\text{pred}}$ is the probability map between 0 and 1 for the reconstructed edge while $\mathbf{C}_{\text{gt}}$ is the ground-truth edge based on the canny operator [25]. $\mathbf{D}$ denotes the spectral normalization discriminator [24] that is composed of five convolutional layers. For the segmentation branch, we use the cross-entropy loss denoted by $\mathcal{L}_{\text{seg}} = \frac{1}{N} \sum_{i=1}^{N} -\mathbf{S}_{\text{gt}}^i \log \mathbf{S}_{\text{pred}}^i$, where $\mathbf{S}_{\text{gt}}^i$ and $\mathbf{S}_{\text{pred}}^i$ denote the ground-truth category and predicted probability for pixel i.

Table 1. Quantitative comparison with the state-of-the-art approaches on CelebA-HQ. Easy, medium, and hard irregular masks denote the mask with coverage ratio of $10\% \sim 20\%$, $30\% \sim 40\%$, and $50\% \sim 60\%$, respectively. ↑ higher is better, and ↓ lower is better. Best and second best results are **highlighted** and underlined.

Mask type		Irregular			Regular
		Easy	Medium	Hard	
PSNR↑	GC [43]	29.30	25.72	23.77	25.75
	RFR [19]	29.22	26.12	24.31	24.85
	CMGAN [48]	29.06	25.79	23.90	24.33
	ICT [33]	28.07	24.56	22.70	24.51
	CTSDG [12]	<u>29.59</u>	<u>26.59</u>	<u>24.69</u>	<u>26.56</u>
	Ours	**29.94**	**26.88**	**25.12**	**26.70**
SSIM↑	GC [43]	0.96	0.93	0.90	0.90
	RFR [19]	0.96	<u>0.94</u>	0.91	0.87
	CMGAN [48]	**0.97**	<u>0.94</u>	0.91	0.87
	ICT [33]	0.96	0.92	0.89	0.87
	CTSDG [12]	**0.97**	<u>0.94</u>	<u>0.92</u>	<u>0.91</u>
	Ours	**0.97**	**0.95**	**0.93**	**0.92**
FID↓	GC [43]	15.00	18.41	21.28	22.45
	RFR [19]	7.37	10.74	13.45	14.35
	CMGAN [48]	6.80	11.85	14.12	12.91
	ICT [33]	<u>6.54</u>	11.80	15.93	<u>11.90</u>
	CTSDG [12]	7.80	<u>10.14</u>	<u>13.30</u>	14.52
	Ours	**6.47**	**9.32**	**11.61**	**11.40**

4 Experiment

We compare our method with state-of-the-arts on three large-scale datasets. An extensive ablation study is conducted to investigate the important designs in our model. All experiments are conducted on two 24 G TITAN RTX GPUs.

Datasets. CelebA-HQ dataset [17,18] is a large-scale face image dataset with $30K$ HD face images, where each image has a semantic segmentation mask corresponding to 19 facial categories. Outdoor dataset (OST) [35] includes 9,900 training images and 300 testing images for 8 semantic categories, which are obtained from the outdoor scene photography collection. Cityscapes dataset [7] contains 5,000 street view images belonging to 20 categories. We expand the number of training images in this dataset, *i.e.*, 2,975 images from the training set and 1,525 images from the test set are used for training, and 500 images from the validation set are used for testing. In addition, the Places2 dataset [50] contains 10 million images covering more than 400 different types of scenes. We generate both regular and irregular masks to verify the ability of image inpainting methods. For regular masks, we draw a 128×128 centered square mask for CelebA-HQ and OST, and a 96×96 centered square mask for Cityscape. For irregular masks, we settle masks from [19] for CelebA-HQ and masks from [22] for Cityscape and OST.

Evaluation Metrics. Similar to the previous works [20,46], we use three metrics as follows. Peak Signal to Noise Ratio (PSNR) is an objective evaluation metric to assess the quality of generate images. Structural Similarity Index (SSIM) [37] uses the mean as an estimate of luminance, standard deviation as an estimate of contrast, and covariance as a measure of structural similarity to compare the difference between the generated and original images. Frechet Inception Distance (FID) [14] evaluates the accuracy and diversity of generated images. Notably, the Inception network [29] is used to extract the image features when calculating the FID score, and then calculate its mean and covariance matrix to estimate the distance between the ground-truth and generated data distribution. According to [47], deep metrics like FID are close to human perception.

| Masked | GC | CMGAN | ICT | CTSDG | Ours | GT |

Fig. 6. Qualitative results of existing methods on CelebA-HQ.

Masked   Predicted Edge   Predicted Seg.   Predicted RGB   Masked   Predicted Edge   Predicted Seg.   Predicted RGB

Fig. 7. Qualitative results of our method on Cityscape (1 to 4 columns) and OST (5 to 8 columns).

4.1   Implementation Details

Our model is supervised by auxiliary structures including edge textures and semantic segmentation. With regard to edge structure, we employ the canny detection method [25] to generate edges of images. Besides, the CelebA-HQ, CityScapes and OST datasets all contain hand-crafted semantic segmentation, hence we can easily adopt these official labels for the segmentation part. More details of implementation are shown in the supplementary.

Table 2. Quantitative comparison with previous auxiliary prior guided approaches on OST and Cityscapes datasets.

Method	Auxiliary prior	OST						CityScapes					
		Regular			Irregular			Regular			Irregular		
		PSNR↑	SSIM↑	FID↓	PSNR↑	SSIM↑	FID↓	PSNR↑	SSIM↑	FID↓	PSNR↑	SSIM↑	FID↓
EC [25]	edge	19.32	0.76	41.25	19.12	0.74	42.27	21.71	0.76	19.87	17.63	0.72	39.04
SPG [32]	seg	18.04	0.70	45.31	17.85	0.74	50.03	20.14	0.71	23.21	16.41	0.67	43.63
SGINet [1]	seg.	–	–	–	–	–	–	25.74	0.87	23.02	18.53	0.77	57.53
SGE [20]	seg.	20.53	**0.81**	40.67	19.46	0.76	39.14	23.41	0.85	18.67	17.78	0.74	41.45
SWAP [21]	edge, seg.	21.18	**0.81**	**38.15**	20.31	0.80	36.74	23.89	0.84	18.14	17.86	0.76	38.18
Ours w/o seg.	edge	20.91	0.76	41.85	21.48	0.80	39.00	25.10	0.86	19.33	19.17	<u>0.78</u>	37.50
Ours w/o edge	seg	<u>21.80</u>	<u>0.77</u>	40.96	<u>22.58</u>	<u>0.81</u>	<u>36.03</u>	<u>25.95</u>	<u>0.87</u>	<u>17.85</u>	**20.49**	0.79	<u>34.79</u>
Ours	edge, seg.	**21.84**	<u>0.77</u>	<u>40.15</u>	**23.15**	**0.82**	**35.77**	**26.13**	**0.88**	**17.52**	<u>20.43</u>	0.79	**33.45**

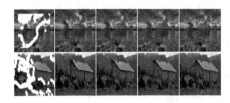

Fig. 8. Visual comparisons on Places2. From left to right: input, GC [43], EC [25], our method, and ground truth.

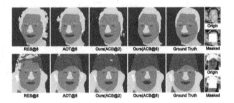

RES@8   AOT@8   Ours(ACB@2)   Ours(ACB@8)   Ground Truth   Masked
Origin

RES@8   AOT@8   Ours(ACB@2)   Ours(ACB@8)   Ground Truth   Masked
Origin

Fig. 9. Segmentation results with different bottlenecks on CelebA-HQ dataset with 128 × 128 regular center masks.

4.2   Result Analysis

We compare our model with several state-of-the-art methods including GC [43], RFR [19], CMGAN [48], ICT [33], CTSDG [12], SPG [32], SGINet [1], SGE [20], and SWAP [21]. A quantitative comparison is carried out on three datasets in terms of both regular and irregular masks with different coverage ratios. Full comparison results [23,25,46,49] we put in the appendix.

From Table 1, our method achieves the best or comparable performance among state-of-the-art image inpainting approaches that may not adopt auxiliary priors. Our method produces much better FID score than others for both regular and irregular masks, indicating that our inpainted results are more realistic. In Table 2, we compare several auxiliary prior guided inpainting approaches [20,21,25,32]. For a fair comparison with the methods relying on only one auxiliary structure, we construct two variants, denoted by "Ours w/o seg." and "Ours w/o edge". Compared with existing methods, our method achieves considerable gain respective to PSNR and FID especially on irregular masks. This is because our method focuses on the interplay representation from three modalities rather than directly guiding the image inpainting branch by predicted auxiliary structures (see Table 4).

In addition, we provide some visual examples on the CelebA-HQ dataset in Fig. 6. It can be seen that our method can generate more semantically consistent results compared with other approaches. More learned auxiliary priors of our method from CityScapes and OST datasets are visualized in Fig. 7.

Table 3. Contribution of two auxiliary branches in our method.

Edge branch	Segmentation branch	PSNR↑	SSIM↑	FID↓
✗	✗	25.88	0.90	12.36
✗	✓	26.47	0.91	11.42
✓	✗	26.19	0.90	11.95
✓	✓	**26.70**	**0.92**	**11.40**

Table 4. Comparison with different attention mechanisms.

Variant	Biased prior	Attention mechanism	PSNR↑	SSIM↑	FID↓
MMT-1	✓	Concat	26.17	0.89	20.01
MMT-2	✓	AdaIN [15]	26.17	0.89	21.71
MMT-3	✓	SPADE [26]	26.29	0.90	14.60
MMT-4	✓	MSSA+ADN	26.24	0.91	12.59
MMT-5	✗	MSSA+add	26.37	0.91	12.64
MMT-6	✗	MSSA+conv	26.50	0.91	11.90
MMT-7	✗	MSSA+AdaIN [15]	26.36	0.91	12.81
MMT-8	✗	MSSA+SPADE [26]	26.42	0.91	12.17
MMT-9	✗	MSSA+ADN	**26.70**	**0.92**	**11.40**

Additional Results on Places2. Similar to SGE [20] and SWAP [21], we also conduct additional experiment on the Places2 dataset [50] for a comprehensive evaluation. Since there is no ground-truth segmentation, we use the segmentation results by DeepLabv3+ [6] to supervise the segmentation branch in our model. As shown in Fig. 8, the visual results show that our method still generate realistic inpainted images without ground-truth segmentation labels.

4.3   Ablation Study

To verify the effectiveness of the proposed modules in our network, the ablation experiments are carried out on the CelebA-HQ dataset.

Contribution of Auxiliary Branches. In Table 3, we construct three variants to verify the contribution of two auxiliary branches in our method. By learning from two auxiliary modalities, our method considerably outperforms the non-auxiliary variant w.r.t PSNR, SSIM, and FID. In addition, semantic segmentation contributes slightly more to image inpainting than edge textures. In summary, our Multi-Modal Mutual Decoder enriches semantic content on the inpainting branch by cross-attending segmentation and edge structures.

Biased Prior Guidance. Different from previous works [20,21,25,32,39] relying on biased prior guidance from predicted auxiliary structures, we jointly learn the interplay information of multi-modal features across the three branches and guide image inpainting based on ADN. To demonstrate its effectiveness, we construct four variants that are directly guided by predicted auxiliary structures. In practice, we first add one convolutional layer at different stages to predict the auxiliary structures (Fig. 1), and then combine multi-modal features (Fig. 4).

In Table 4, MMT-1 denotes concatenating predicted structures with feature maps in the inpainting branch. MMT-2, MMT-3 and MMT-4 denote that we use AdaIN [15], SPADE [26], and MSSA with ADN to calculate the affine transformation parameters (γ, β) based on predicted structures, respectively. Compared with our method without biased prior guidance (*i.e.*, MMT-9), the FID score is significantly reduced based on predicted auxiliary structures. The results support our statement that predicted structures may introduce additional noises in image inpainting intermediately without ground-truth.

Effectiveness of Multi-scale Spatial-Aware Attention. To verify the effectiveness of Multi-Scale Spatial-aware Attention (MSSA), we construct four baseline feature fusion strategies from MMT-5 to MMT-8 in Table 4. MMT-5 means that we directly perform element-wise summation on features from three branches, while MMT-6 means that we splice the features from three branches together and then fuse them by two convolutional layers.

From Table 4, our MSSA performs the best in terms of three metrics. Compared with simple addition or convolution, our MSSA can provide reliable cross-attention among multiple modalities to guide high-quality reconstructed images. We also replace ADN by AdaIN [15] and SPADE [26] in MSSA for MMT-7 and MMT-8 respectively. The results show that our ADN performs better than previous normalization methods, demonstrating its effectiveness.

Effectiveness of Adaptive Contextual Bottlenecks. In Table 5, we compare our Adaptive Contextual Bottlenecks (ACB) with the vanilla ResNet block [13] and the recently proposed AOT [46]. ACB@L ($L = 2, 4, 6, 8$) denotes L layers of ACB modules; RES@8 and AOT@8 denote 8 ResNet blocks [13] or 8 AOT blocks [46] respectively. † means quadrupling the channels of feature maps in ResNet blocks or copying base feature maps for different pathways in AOT blocks. The results show that the performance of ACB is improved along with the number of blocks is increased from 2 to 8. Using 8 ResNet or AOT blocks achieves similarly as that using 4 ACB blocks. It is worth mentioning that ResNet and AOT blocks have less number of channels of feature maps in each pathway. For a fair comparison, we construct two variants †RES@8 and †AOT@8 with the same channels as our ACB blocks. However, more channels in feature maps do not help improve the performance by using ResNet or AOT blocks. We speculate that the gating updating scheme in our ACB can reduce the influence of redundant noisy context with more channels of feature maps.

Besides, the mean of category-wise intersection-over-union (mIoU) [6] is another metric to validate the influence of bottleneck modules on segmentation inpainting. Our ACB module ($L \geq 4$) still outperforms other two blocks by more than 2%. The segmentation results in Fig. 9 also show that our ACB module generates more accurate segmentation performance. If the number of bottlenecks are increased, some isolated errors in segmentation can be removed (see the 3rd and 4th columns in Fig. 9).

Table 5. Comparison between different bottlenecks.

Bottleneck	PSNR↑	SSIM↑	FID↓	mIoU%↑
RES@8	26.48	0.91	12.54	61.93
†RES@8	26.23	0.91	13.26	60.11
AOT@8	26.51	0.91	11.61	63.68
†AOT@8	26.29	0.91	14.17	62.28
ACB@2	26.48	0.91	12.18	63.54
ACB@4	26.60	0.91	12.24	65.84
ACB@6	26.61	0.91	12.09	66.16
ACB@8	**26.70**	**0.92**	**11.40**	**67.13**

Table 6. Efficiency of image inpainting networks.

Method	Params (M)	MACs (G)	Speed (FPS)
SPG [32]	119.64	58.68	2.03
EC [25]	27.06	122.67	**67.21**
CTSDG [12]	52.15	**17.67**	36.99
RFR [19]	31.22	206.12	15.56
CSA [23]	132.11	55.23	1.37
RES@8 [13]	**22.76**	96.10	40.82
AOT@8 [46]	27.48	100.93	30.96
Ours (ACB@2)	**22.76**	96.10	40.88
Ours (ACB@8)	51.09	125.11	29.49

Efficiency Comparison. From Table 6, we compare the number of parameters, computational complexity (MACs), and the running speed (FPS) of existing methods. Two-stage based SPG [32] and CSA [23], composed of complex sub-networks at each stage, run much more slowly than end-to-end methods. In contrast, EC [25] consists of two simple sub-networks for edge prediction and image inpainting, resulting in fast running speed but inferior performance. RFR [19] is an end-to-end model but predicts the inpainted results by the decoding heads recurrently. In terms of bottlenecks in the encoder, our ACB@2 achieves similar performance as AOT@8 with faster speed. By using 8 blocks, our method is still efficient with state-of-the-art performance among end-to-end methods.

Limitation Discussion. Although our model generates promising results in most cases, it fails to recognize and recover unseen semantic knowledge, hence produces strange artifacts in complex scenes with large masks. Note that this weakness also affects other methods. It indicates that image inpainting model requires not only generative but also recognition capability. For example, our method can synthesize the human silhouette but lacks precise semantic details.

5 Conclusion

In this paper, we propose an end-to-end Multi-modality Guided Transformer for image impainting, which enriches coupled spatial features from shared multi-modal representations (*i.e.*, RGB image, semantic segmentation and edge textures). The proposed Multi-Scale Spatial-aware Attention can integrate compact discriminative features from multiple modalities via Auxiliary DeNormalization. Meanwhile, we introduce the Adaptive Contextual Bottlenecks in the encoder to enhance context reasoning for more semantically consistent inpainted results for the missing region. To the best of our knowledge, our scientific value lies in first analyzing the biased prior problem in image inpainting.

Acknowledgements and Declaration of Conflicting Interests. This work was supported by the Key Research Program of Frontier Sciences, CAS, Grant No. ZDBS-LY-JSC038. Libo Zhang was supported Youth Innovation Promotion Association, CAS (2020111). Dr. Du and his employer received no financial support for the research, authorship, and/or publication of this article.

References

1. Ardino, P., Liu, Y., Ricci, E., Lepri, B., Nadai, M.D.: Semantic-guided inpainting network for complex urban scenes manipulation. In: ICPR, pp. 9280–9287 (2020)
2. Ba, L.J., Kiros, J.R., Hinton, G.E.: Layer normalization. CoRR abs/1607.06450 (2016)
3. Barnes, C., Shechtman, E., Finkelstein, A., Goldman, D.B.: Patchmatch: a randomized correspondence algorithm for structural image editing. TOG **28**, 24 (2009)
4. Cao, C., Fu, Y.: Learning a sketch tensor space for image inpainting of man-made scenes. In: ICCV (2021)
5. Chen, L., Papandreou, G., Schroff, F., Adam, H.: Rethinking atrous convolution for semantic image segmentation. CoRR abs/1706.05587 (2017)
6. Chen, L.-C., Zhu, Y., Papandreou, G., Schroff, F., Adam, H.: Encoder-decoder with atrous separable convolution for semantic image segmentation. In: Ferrari, V., Hebert, M., Sminchisescu, C., Weiss, Y. (eds.) ECCV 2018. LNCS, vol. 11211, pp. 833–851. Springer, Cham (2018). https://doi.org/10.1007/978-3-030-01234-2_49
7. Cordts, M., et al.: The cityscapes dataset for semantic urban scene understanding. In: CVPR, pp. 3213–3223 (2016)

8. Criminisi, A., Pérez, P., Toyama, K.: Object removal by exemplar-based inpainting. In: CVPR, pp. 721–728 (2003)
9. Deng, Y., Hui, S., Zhou, S., Meng, D., Wang, J.: Learning contextual transformer network for image inpainting. In: MM. pp. 2529–2538 (2021)
10. Dosovitskiy, A., et al.: An image is worth 16x16 words: transformers for image recognition at scale. In: ICLR (2021)
11. Goodfellow, I.J., et al.: Generative adversarial nets. In: NeurIPS, pp. 2672–2680 (2014)
12. Guo, X., Yang, H., Huang, D.: Image inpainting via conditional texture and structure dual generation. In: ICCV, pp. 14114–14123 (2021)
13. He, K., Zhang, X., Ren, S., Sun, J.: Deep residual learning for image recognition. In: CVPR, pp. 770–778 (2016)
14. Heusel, M., Ramsauer, H., Unterthiner, T., Nessler, B., Hochreiter, S.: GANs trained by a two time-scale update rule converge to a local Nash equilibrium. In: NeurIPS, pp. 6626–6637 (2017)
15. Huang, X., Belongie, S.J.: Arbitrary style transfer in real-time with adaptive instance normalization. In: ICCV, pp. 1510–1519 (2017)
16. Iizuka, S., Simo-Serra, E., Ishikawa, H.: Globally and locally consistent image completion. TOG **36**, 107:1-107:14 (2017)
17. Karras, T., Aila, T., Laine, S., Lehtinen, J.: Progressive growing of GANs for improved quality, stability, and variation. In: ICLR (2018)
18. Lee, C., Liu, Z., Wu, L., Luo, P.: MaskGAN: Towards diverse and interactive facial image manipulation. In: CVPR, pp. 5548–5557 (2020)
19. Li, J., Wang, N., Zhang, L., Du, B., Tao, D.: Recurrent feature reasoning for image inpainting. In: CVPR, pp. 7757–7765 (2020)
20. Liao, L., Xiao, J., Wang, Z., Lin, C.-W., Satoh, S.: Guidance and evaluation: semantic-aware image inpainting for mixed scenes. In: Vedaldi, A., Bischof, H., Brox, T., Frahm, J.-M. (eds.) ECCV 2020. LNCS, vol. 12372, pp. 683–700. Springer, Cham (2020). https://doi.org/10.1007/978-3-030-58583-9_41
21. Liao, L., Xiao, J., Wang, Z., Lin, C., Satoh, S.: Image inpainting guided by coherence priors of semantics and textures. In: CVPR, pp. 6539–6548 (2021)
22. Liu, G., Reda, F.A., Shih, K.J., Wang, T.-C., Tao, A., Catanzaro, B.: Image inpainting for irregular holes using partial convolutions. In: Ferrari, V., Hebert, M., Sminchisescu, C., Weiss, Y. (eds.) ECCV 2018. LNCS, vol. 11215, pp. 89–105. Springer, Cham (2018). https://doi.org/10.1007/978-3-030-01252-6_6
23. Liu, H., Jiang, B., Xiao, Y., Yang, C.: Coherent semantic attention for image inpainting. In: ICCV, pp. 4169–4178 (2019)
24. Miyato, T., Kataoka, T., Koyama, M., Yoshida, Y.: Spectral normalization for generative adversarial networks. In: ICLR (2018)
25. Nazeri, K., Ng, E., Joseph, T., Qureshi, F.Z., Ebrahimi, M.: Edgeconnect: structure guided image inpainting using edge prediction. In: ICCVW, pp. 3265–3274 (2019)
26. Park, T., Liu, M., Wang, T., Zhu, J.: Semantic image synthesis with spatially-adaptive normalization. In: CVPR, pp. 2337–2346 (2019)
27. Pathak, D., Krähenbühl, P., Donahue, J., Darrell, T., Efros, A.A.: Context encoders: feature learning by inpainting. In: CVPR, pp. 2536–2544 (2016)
28. Ronneberger, O., Fischer, P., Brox, T.: U-Net: convolutional networks for biomedical image segmentation. In: Navab, N., Hornegger, J., Wells, W.M., Frangi, A.F. (eds.) MICCAI 2015. LNCS, vol. 9351, pp. 234–241. Springer, Cham (2015). https://doi.org/10.1007/978-3-319-24574-4_28
29. Salimans, T., Goodfellow, I.J., Zaremba, W., Cheung, V., Radford, A., Chen, X.: Improved techniques for training GANs. In: NeurIPS, pp. 2226–2234 (2016)

30. Shetty, R., Fritz, M., Schiele, B.: Adversarial scene editing: automatic object removal from weak supervision. In: NeurIPS, pp. 7717–7727 (2018)
31. Song, L., Cao, J., Song, L., Hu, Y., He, R.: Geometry-aware face completion and editing. In: AAAI, pp. 2506–2513 (2019)
32. Song, Y., Yang, C., Shen, Y., Wang, P., Huang, Q., Kuo, C.J.: SPG-Net: segmentation prediction and guidance network for image inpainting. In: BMVC, p. 97 (2018)
33. Wan, Z., Zhang, J., Chen, D., Liao, J.: High-fidelity pluralistic image completion with transformers. In: ICCV, pp. 4672–4681 (2021)
34. Wang, P., et al.: Understanding convolution for semantic segmentation. In: WACV, pp. 1451–1460 (2018)
35. Wang, X., Yu, K., Dong, C., Loy, C.C.: Recovering realistic texture in image super-resolution by deep spatial feature transform. In: CVPR, pp. 606–615 (2018)
36. Wang, Y., Tao, X., Qi, X., Shen, X., Jia, J.: Image inpainting via generative multi-column convolutional neural networks. In: NeurIPS, pp. 329–338 (2018)
37. Wang, Z., Bovik, A.C., Sheikh, H.R., Simoncelli, E.P.: Image quality assessment: from error visibility to structural similarity. TIP **13**, 600–612 (2004)
38. Woo, S., Park, J., Lee, J.-Y., Kweon, I.S.: CBAM: convolutional block attention module. In: Ferrari, V., Hebert, M., Sminchisescu, C., Weiss, Y. (eds.) ECCV 2018. LNCS, vol. 11211, pp. 3–19. Springer, Cham (2018). https://doi.org/10.1007/978-3-030-01234-2_1
39. Xiong, W., et al.: Foreground-aware image inpainting. In: CVPR, pp. 5840–5848 (2019)
40. Yang, J., Qi, Z., Shi, Y.: Learning to incorporate structure knowledge for image inpainting. In: AAAI, pp. 12605–12612 (2020)
41. Yu, F., Koltun, V., Funkhouser, T.A.: Dilated residual networks. In: CVPR, pp. 636–644 (2017)
42. Yu, J., Lin, Z., Yang, J., Shen, X., Lu, X., Huang, T.S.: Generative image inpainting with contextual attention. In: CVPR, pp. 5505–5514 (2018)
43. Yu, J., Lin, Z., Yang, J., Shen, X., Lu, X., Huang, T.S.: Free-form image inpainting with gated convolution. In: ICCV, pp. 4470–4479 (2019)
44. Yu, Y., et al.: Diverse image inpainting with bidirectional and autoregressive transformers. In: MM, pp. 69–78 (2021)
45. Zeng, Y., Fu, J., Chao, H.: Learning joint spatial-temporal transformations for video inpainting. In: Vedaldi, A., Bischof, H., Brox, T., Frahm, J.-M. (eds.) ECCV 2020. LNCS, vol. 12361, pp. 528–543. Springer, Cham (2020). https://doi.org/10.1007/978-3-030-58517-4_31
46. Zeng, Y., Fu, J., Chao, H., Guo, B.: Aggregated contextual transformations for high-resolution image inpainting. CoRR abs/2104.01431 (2021)
47. Zhang, R., Isola, P., Efros, A.A., Shechtman, E., Wang, O.: The unreasonable effectiveness of deep features as a perceptual metric. In: CVPR, pp. 586–595 (2018)
48. Zhao, S., et al.: Large scale image completion via co-modulated generative adversarial networks. In: ICLR (2021)
49. Zheng, C., Cham, T., Cai, J.: Pluralistic image completion. In: CVPR, pp. 1438–1447 (2019)
50. Zhou, B., Lapedriza, À., Khosla, A., Oliva, A., Torralba, A.: Places: a 10 million image database for scene recognition. TPAMI **40**, 1452–1464 (2018)

Intelli-Paint: Towards Developing More Human-Intelligible Painting Agents

Jaskirat Singh[1,2](✉) , Cameron Smith[2] , Jose Echevarria[2] ,
and Liang Zheng[1]

[1] Australian National University, Canberra, Australia
{jaskirat.singh,liang.zheng}@anu.edu.au
[2] Adobe Research, San Jose, USA
{casmith,echevarr}@adobe.com

Abstract. Stroke based rendering methods have recently become a popular solution for the generation of stylized paintings. However, the current research in this direction is focused mainly on the improvement of final canvas quality, and thus often fails to consider the intelligibility of the generated painting sequences to actual human users. In this work, we motivate the need to learn more human-intelligible painting sequences in order to facilitate the use of autonomous painting systems in a more interactive context (*e.g.* as a painting assistant tool for human users or for robotic painting applications). To this end, we propose a novel painting approach which learns to generate output canvases while exhibiting a painting style which is more relatable to human users. The proposed painting pipeline Intelli-Paint consists of **1)** a progressive layering strategy which allows the agent to first paint a natural background scene before adding in each of the foreground objects in a progressive fashion. **2)** We also introduce a novel sequential brushstroke guidance strategy which helps the painting agent to shift its attention between different image regions in a semantic-aware manner. **3)** Finally, we propose a brushstroke regularization strategy which allows for ∼60–80% reduction in the total number of required brushstrokes without any perceivable differences in the quality of generated canvases. Through both quantitative and qualitative results, we show that the resulting agents not only show enhanced efficiency in output canvas generation but also exhibit a more natural-looking painting style which would better assist human users express their ideas through digital artwork.

1   Introduction

Paintings form a key medium through which humans express their ideas and emotions. Nevertheless, the creation of finer-quality art is often quite challenging and requires a considerable amount of time on part of the human painter.

Supplementary Information The online version contains supplementary material available at https://doi.org/10.1007/978-3-031-19787-1_39.

Fig. 1. Developing a more human-relatable painting style. (Left) Painting sequence visualization which demonstrates that our method exhibits higher resemblance with the human painting style as opposed to previous state of the art. (Right) This resemblance is achieved through **1)** a progressive layering strategy which allows for a more human-like evolution of the canvas, **2)** a sequential attention mechanism which focuses on different image regions in a coarse-to-fine fashion and **3)** a brushstroke regularization formulation which allows our method to obtain detailed results while using significantly fewer brushstrokes ($\sim$1/20 as compared to Paint Transformer [28] in above).

One way to address this problem is to develop autonomous painting agents which can assist human painters to better express their ideas in a quick and concise fashion. To this end, there is a growing research interest [8, 9, 11, 14, 18, 23, 28–30, 35, 44, 48, 49] in teaching machines *"how to paint"*, in a manner similar to a human painter. For instance, Huang *et al.* [18] use deep reinforcement learning to learn an unsupervised brushstroke decomposition for the creation of non-photorealistic imagery. Zou *et al.*. [49] use gradient descent to optimize over the brushstroke parameters for the entire painting trajectory. Similarly, Liu *et al.*. [28] propose a novel Paint Transformer which formulates the "learning to paint" problem as a feed-forward set prediction problem. Despite their efficacy, existing works often lack semantic understanding of image contents and are invariably reliant on a progressive grid-based division strategy, wherein the painting agent divides the overall image into successively finer grids, and then proceeds to paint each in parallel. This leads to hierarchically bottom-up painting sequences which are quite mechanical and thus not applicable for human users.

In this paper, we propose a novel painting pipeline (*intelli-paint*), which tries to address the need for more human-intelligible painting sequences, by mimicking some commonly found traits of the human painting process. This is achieved in three main ways. **First,** we propose a progressive layering strategy which, much like a human, allows the painting agent to successively draw a given scene in multiple layers. That is, instead of starting to paint the entire scene at once, our method learns to first paint a realistic background scene representation before adding in each of the foreground objects in a progressive layerwise fashion.

Second, the human painting process is often characterized by a localized spatial attention span. For instance, a potential artist would focus on different local image areas while painting distinct parts of the final canvas [47]. This is in sharp contrast with previous works, which either focus on the entire image or several predefined grid blocks [28, 49]. To better mimic the human style, we introduce a sequential brushstroke guidance approach which allows the painting agent to shift its attention between different image areas through a self-learned sequence of localized attention windows. The spatial dimensions and position of the localized attention window are progressively adjusted during the painting process so as to paint a given scene in a coarse-to-fine fashion.

Third, we note that prior works often use a fixed brushstroke budget irrespective of the complexity of the target image. This not only leads to wasteful (and overlapping) brushstroke patterns (refer Fig. 4) but also imparts an artificial painting style to the final agent. To this end, we propose an inference-time brushstroke regularization formulation which removes brushstroke redundancies by regularizing the total number of brushstrokes required for painting a given canvas. Our experiments reveal that this not only leads to a ~60–80% enhancement in the brushstroke decomposition efficiency but also results in more natural looking painting sequences which are easily intelligible by a human painter.

To summarize, this paper makes the following contributions.

- We introduce a progressive layering approach, which much like a human, allows the painting agent to draw a given scene in multiple successive layers.
- We propose a sequential brushstroke guidance strategy which enables the painting agent to focus on different image regions through a learned sequence of coarse-to-fine localized attention windows.
- Finally, we introduce an inference time brushstroke regularization procedure which results in a ~60–80% enhancement in the brushstroke decomposition efficiency and leads to more natural painting sequences. which are better intelligible by a human user.

2   Related Work

Classical Stroke Based Rendering. The problem of teaching machines "how to paint" has been extensively studied in the context of stroke-based rendering (SBR), which focuses on the recreation of non-photorealistic imagery through appropriate positioning and selection of discrete elements such as paint strokes or stipples [15–17, 26, 38, 40, 46]. Classical works for incorporating semantic knowledge into the painting process have also been explored. [2, 5–7] use image saliency to generate a coarse to fine painting sequence in which sailent details (e.g. edges) are preserved in increasing amounts. In contrast, our work uses image saliency to learn a progressive layering strategy in which the agent learns to paint a natural background scene (refer Fig. 2) before adding in the foreground objects in a progressive fashion. [34, 39] use local attention over a heuristically determined window for computation of stroke parameters. Our work differs as it provides an unsupervised approach for learning the optimal movement of this attention window for painting in a coarse-to-fine manner (refer Sect. 4.1).

Supervised Painting Methods. More recent solutions [13,14] adopt the use of recurrent neural networks for computing optimal brushstroke decomposition. However, these methods require access to dense human brushstroke annotations, which limits their applicability to most real world problems. In another work, Zhao *et al.* [47] use a conditional variational autoencoder framework for synthesising time-lapse videos depicting the recreation of a given target image. However, this requires access to painting time-lapse videos from real artists for training. Furthermore, the time-lapse outputs are generated at very low-resolution as compared to the high-resolution sequences generated using our approach.

Unsupervised Painting Methods. In recent years, there has been an increased focus on learning an unsupervised brushstroke decomposition without requiring access to dense human brushstroke annotations. For instance, recent works [11,18,19,29,35,44] use deep reinforcement learning and an adversarial training approach for learning an efficient brushstroke decomposition. Optimization based methods [49] directly search for the optimal brushstroke parameters by performing gradient descent over a novel optimal-transport-based loss function. In another recent work, Liu *et al.* [28] propose a Paint Transformer which formulates the painting problem as a feed-forward stroke set prediction problem.

While the above works show high proficiency in painting high-quality output canvases, the generation of the same invariably depends on a progressive grid-based division strategy. In this setting, the agent divides the overall image into successively finer grids, and then proceeds to paint each of them in parallel. Experimental analysis reveals that this not only reduces the efficiency of the final agent, but also leads to mechanical (grid-based) painting sequences which are not directly applicable to actual human users.

3 Need for More Human-Intelligible Painting Agents

Interactive Applications. The need for stroke-based-rendering methods (as opposed to pixel-based methods [12,21]) is often motivated from the need to mimic the human artistic creation process [3,28,49], which can then be used for development of painting assistant and teaching tools [18,28] for human users. The development of more human-intelligible painting sequences is thus important as it will allow for the use of autonomous painting methods in an interactive context.

Robotic Painting Tasks. Robotic applications for expression of AI creativity are being increasingly explored [22,31,42]. Our contribution is significant in this direction, as our method not only learns a painting sequence which is more interpretable to actual human users, but more importantly it provides an *efficient painting plan* which would allow a robotic agent to paint a vivid scene using significantly less number of brushstrokes as compared to previous works.

Approximating the Manifold of Human Paintings. While sketch based methods for photorealistic image generation [4,29,43,45] have been extensively

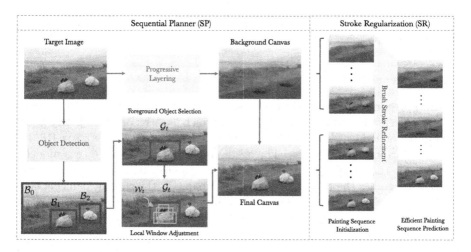

Fig. 2. Method overview. Given a target image I, the *Intelli-Paint* agent first learns to paint a realistic background scene on the canvas. Once the background scene has been painted, the agent then proceeds to progressively add each of the foreground objects using a sequential brushstroke guidance procedure. To do this, the painting agent first uses the convex combination formulation from Eq. 6 to select the foreground object it would like to paint (indicated by object window $\mathcal{G}_t$). The features within each object region are then painted in a coarse-to-fine fashion through a sequence of localized attention windows $\mathcal{W}_t$. Finally, the brushstroke sequence is fed into a stroke-regularization procedure which removes brushstroke redundancies (and overlaps) to output the most efficient painting sequence for each test image.

studied, the use of partially drawn human paintings for image synthesis remains unexplored due to lack of large-scale collection of human (or human-like) painting trajectories. In an exciting concurrent work, Singh *et al.* [36] use the improved human-likeliness of *our* painting sequences in order to perform photorealistic image synthesis and editing from rudimentary user paintings and brushstrokes.

4 Our Method

The *intelli-paint* framework (Fig. 2) is based on a two-stage hybrid optimization strategy which consists of two modules: *sequential-planner* (SP) and *stroke-regularizer* (SR). In the first stage, the *sequential-planner* (SP) learns to predict a coarse but more human-like initialization for the brushstroke sequence $\mathbf{s}_{init}$. The coarse brushstroke sequence initializations are then fed into a gradient descent based *stroke regularization* (SR) procedure, which removes redundant brushstroke patterns and refines the original brushstroke parameters to output the most efficient stroke decomposition $\mathbf{s}_{pred}$ for each test image. This two-stage process can be mathematically formulated as,

$$\mathbf{s}_{init} = SP\left(C_{init}, I_{target}\right) \quad \rightarrow \quad \mathbf{s}_{pred} = SR\left(\mathbf{s}_{init}, I_{target}\right), \tag{1}$$

where C_{init} signifies the blank canvas initialization and I is the target image. In the following sections, we discuss each of the above modules in full detail.

4.1   Sequential Planner

Reinforcement Learning Formulation. The *sequential-planner* (SP) is modelled as a deep reinforcement learning agent which learns a painting policy π predicting vectorized brushstroke parameters $\mathbf{a}_t$ (modeled as a Bézier curve [18,35]) from the current agent state s_t. The agent state s_t at any timestep t is modeled as the tuple $(C_t, I, t, \mathcal{G}_t, \mathcal{W}_t, \mathcal{S}_I, l)$, where C_t is the canvas state, I is the target image, $\mathcal{S}_I$ signifies the target-image saliency map, l is the current painting layer (refer Section 4.1.2), and $(\mathcal{G}_t, \mathcal{W}_t)$ represent the coarse and fine local attention windows for the painting agent respectively (refer Section 4.1.3).

The canvas state C_t is updated using a differentiable neural renderer module, which rasterizes the predicted brushstroke parameters $\mathbf{a}_t$ to output a brushstroke alpha map $S_\alpha(\mathbf{a}_t)$ and its colored rendering $S_{color}(\mathbf{a}_t)$. The canvas updates at each timestep t are then computed as follows,

$$C_{t+1} = C_t \odot (1 - S_\alpha(\mathbf{a}_t)) + S_{color}(\mathbf{a}_t). \qquad (2)$$

We next discuss further details regarding the above formulation which allows our painting agent to generate output canvases while exhibiting some commonly found traits (*e.g.* layering, sequential attention) of the human painting process.

Progressive Layering: The human painting process is often progressive and multi-layered [33,47]. That is, instead of painting everything on the canvas at once, humans often first paint a basic background layer before progressively adding each of the foreground objects on top of it (refer Fig. 1). However, such a strategy is hard to learn using previous works which directly minimize the pixel wise distance the generated canvas C_t and the target image I.

To this end, we propose a progressive layering strategy, which much like a human artist, allows the painted canvas to evolve in multiple successive layers. The objective of the painting agent in the first layer, is to paint a realistic background scene by trying to only focus on the non-salient (background) image areas. In doing so, the salient image regions are painted so as to maximize the efficiency of painting the background contents (*e.g.* salient region corresponding to a bird sitting on a tree would be painted while focusing on tree leaves and branches, as in Fig. 4). Once the background layer is drawn, the painting agent in the successive layer then proceeds to add different foreground objects in a decreasing order of saliency. An illustration of a two layer painting process is shown in Fig. 2. The painting agent first draws a realistic background scene (by focusing only on background image contents like ground, grass, *etc.*), before adding in the foreground objects (sheep) in the second layer.

In order to achieve this layering process, we first divide the overall painting episode into multiple layers as follows,

$$C_{out} = \sum_{l=0}^{L-1} \sum_{t=1}^{T/L} C_t^l \odot (1 - S_\alpha(\mathbf{a}_t^l)) + S_{color}(\mathbf{a}_t^l), \tag{3}$$

where $L = 2$ is the number of layers[1], T is the episode length, $C_0^{l=0}$ signifies an empty canvas, and $C_0^{l=1}$ is initialized as the canvas output $C_{T/L}^{l=0}$ from last layer. Given canvas state C_t, input image I and foreground saliency map S_I, the layerwise painting style can then achieved be achieved by optimizing the following layered reward objective for each layer l,

$$r_t^{layer}(l) = D(I \odot M_I(l), C_{t+1} \odot M_I(l)) - D(I \odot M_I(l), C_t \odot M_I(l)), \tag{4}$$

where $D(I, C_t)$ is the joint conditional Wasserstein GAN [1] discriminator score for image I and canvas C_t [18], and the layered-mask $M_I(l)$ is defined as,

$$M_I(l) = 1 - S_I \odot (1 - l). \tag{5}$$

Sequential Brushstroke Guidance: Human painters often exhibit a localized spatial attention span while focusing on distinct image areas [47]. This is in stark contrast with previous works which either compute stroke decomposition globally over the entire canvas or over a set of predefined grid regions [18,28,49]. To this end, we propose a sequential brushstroke guidance strategy, which allows the reinforcement learning agent to shift its attention between different image regions through a sequence of coarse-to-fine attention windows $\{W_0, W_1 \ldots W_T\}$. In particular, the computation of the localized attention window W_t at any timestep t during the painting process is done in the following broad steps,

– *Foreground object selection:* The RL agent first selects the in-focus foreground object by predicting coordinates of a coarse global attention window G_t. Given an input image I with N foreground objects, we model $G_t = x_t^G, y_t^G, w_t^G, h_t^G$ as a convex combination of each of in-image object bounding box detections $B_i \in \mathbb{R}^4, i \in [1, N]$.

$$G_t = \sum_{i=0}^{N} \alpha_i^t B_i, \quad s.t. \quad \forall t \quad \sum_i \alpha_i^t = 1, \quad \alpha_i^t \geqslant 0. \tag{6}$$

where $\boldsymbol{\alpha}^t = \{\alpha_0^t, \ldots \alpha_N^t\} \in \mathbb{R}^{N+1}$ are the spatial attention parameters predicted by the RL agent at timestep t. B_0 represents an attention window over the entire canvas and is used to switch focus to background image areas.

---

[1] For simplicity, we primarily use $L = 2$ in the main paper. Further details on extending progressive layering to $L > 2$ are provided in Appendix A.2.

- *Local attention window selection*: Within each object window $\mathcal{G}_t$, the agent further learns to sequentially shift its focus on different in-object features through a sequence of coarse-to-fine local attention windows $\mathcal{W}_t$. In particular, given the coarse object window coordinates $\mathcal{G}_t = x_t^{\mathcal{G}}, y_t^{\mathcal{G}}, w_t^{\mathcal{G}}, h_t^{\mathcal{G}}$, the coordinates $\mathcal{W}_t = x_t^{\mathcal{L}}, y_t^{\mathcal{L}}, w_t^{\mathcal{L}}, h_t^{\mathcal{L}}$ for the finer localized attention windows are computed in a Markovian fashion as,

$$x_{t+1}^{\mathcal{L}} = x_{t+1}^{\mathcal{G}} + (x_t^{\mathcal{L}} + \Delta x_t)\, w_{t+1}^{\mathcal{G}}, \tag{7}$$

$$y_{t+1}^{\mathcal{L}} = y_{t+1}^{\mathcal{G}} + (y_t^{\mathcal{L}} + \Delta y_t)\, h_{t+1}^{\mathcal{G}}, \tag{8}$$

$$w_{t+1}^{\mathcal{L}} = (max(1 - \tilde{t}, w_{min}) + \Delta w_t)\, w_{t+1}^{\mathcal{G}}, \tag{9}$$

$$h_{t+1}^{\mathcal{L}} = (max(1 - \tilde{t}, h_{min}) + \Delta h_t)\, h_{t+1}^{\mathcal{G}}, \tag{10}$$

where $\tilde{t} \in [0,1]$ is the normalized episode timestep, (w_{min}, h_{min}) are the minimum attention window dimensions and $(\Delta \mathcal{W}_t = \Delta x_t, \Delta y_t, \Delta w_t, \Delta h_t) \in \mathbb{R}^4$ are successive Markovian [10] updates predicted by the RL agent. The above Markovian update formulation helps ensure spatial closeness of two consecutive local attention windows (Eq. 7, 8), while facilitating a coarse-to-fine adjustment of the spatial attention window dimensions (Eq. 9, 10).
- *Brushstroke parameter adjustment*: Finally, the coordinates of attention window $\mathcal{W}_t$ are used to modify the predicted brushstroke parameters $\mathbf{a}_t^l$ (modeled as Bézier curve), so as to constrain the painting agent to only draw within the local attention window. This procedure can be expressed as,

$$\mathbf{a}_t^l \leftarrow ParamAdjustment(\mathbf{a}_t^l, \mathcal{W}_t). \tag{11}$$

Please refer Appendix C for detailed implementation notes and instructions.

Human-Consistency Penalties: Human artists inherently try to focus on spatially close image areas and try to avoid unnecessary spatial oscillations when painting a given image [47]. In this regard, while the Markovian adjustment procedure introduced in Sect. 4.1.3 ensures the spatial closeness of two consecutive local attention windows $\mathcal{W}_t$, unnecessary movements may still arise due to oscillations between different coarse attention windows $\mathcal{G}_t$. To prevent learning such stroke decompositions we introduce the following spatial penalty,

$$r_t^{spatial} = -\|\mathcal{G}_{t+1} - \mathcal{G}_t\|_F, \tag{12}$$

where $\|.\|_F$ represents the Frobenius norm.

Similarly, human painting sequences are also characterized by the use of same (or similar) color patterns at consecutive timesteps [47]. Thus, in order to mimic this behaviour we propose the following color transition penalty r_t^{color},

$$r_t^{color} = -\|(R, G, B)_{t+1} - (R, G, B)_t\|_F, \tag{13}$$

where $(R, G, B)_t$ represents the brushstroke color prediction at timestep t.

4.2   Brushstroke Regularization

Existing works on autonomous painting systems are often limited to using (an almost) fixed brush stroke budget irrespective of the complexity of the target image. Experiments reveal that this not only reduces the efficiency of the generated painting sequence but also results in redundant (overlapping) brushstroke patterns (Fig. 4) which impart an unnatural painting style to the final agent.

To address this, we propose an inference-time brushstroke regularization strategy which refines and removes redundancies from the initial brushstroke sequence predictions $\mathbf{s}_{init}$ to output the most efficient stroke decomposition $\mathbf{s}_{pred}$ for each test image. To do this, we first associate each brushstroke with an importance vector $\beta_t^l \in [0, 1]$ by modifying the stroke rendering process as,

$$C_{out} = \sum_{l=0}^{L-1} \sum_{t=1}^{T/L} C_t^l \odot (1 - \beta_t^l \, S_\alpha(\mathbf{a}_t^l)) + \beta_t^l \, S_{color}(\mathbf{a}_t^l),$$

where $\beta_t^l = sign(x_t^l)$ and $x_t^l \sim \mathcal{N}(0, 10^{-3})$ is randomly initialized from a normal distribution. We then use gradient descent to optimize the following loss function over both brushstroke parameters $\mathbf{a}_t^l$ and importance vectors β_t^l (through x_t^l)

$$\mathcal{L}_{total}(\mathbf{a}_t^l, x_t^l) = \mathcal{L}_2(I, C_{out}) + \gamma \sum_{l=0}^{L-1} \sum_{t=1}^{T/L} \|\beta_t^l\|_1, \qquad (14)$$

where the backpropagation gradients $\partial \beta_t^l / \partial x_t^l$ are computed as $\sigma(x_t^l)(1 - \sigma(x_t^l))$, $\sigma(.)$ is the sigmoid function and γ balances the weightage between brushstroke refinement and the need to use as few brushstrokes as possible.

5   Implementation Details

Neural Renderer. In this paper, we primarily adopt the *PixelShuffleNet* architecture from Huang *et al.* [18] while designing the neural differentiable renderer. While our approach is not limited to a particular rendering mechanism, we find that as opposed to the opaque brushstroke models used in [28,49], the use of a more naturally blending brushstroke representation from [18], allows our method to mimic the human painting style in a more closer fashion.

Layered Training. The use of progressive layering module requires conditionally training the painting agent policy at each layer while initializing the canvas state with the output from the last layer. In order to save computation time during training, we train the successive layer policies in consecutive batches while using the canvas output from the last layer. Furthermore, we only use $L = 2$ layers at the training time. At inference time, the trained progressive layering policy can then be applied for $L > 2$ layers by appropriately modifying the target image saliency maps. Please refer Appendix A.2 for further details.

Saliency and Bounding Box Predictions. A key component of the Intelli-Paint pipeline is the sequential brushstroke guidance strategy which relies on

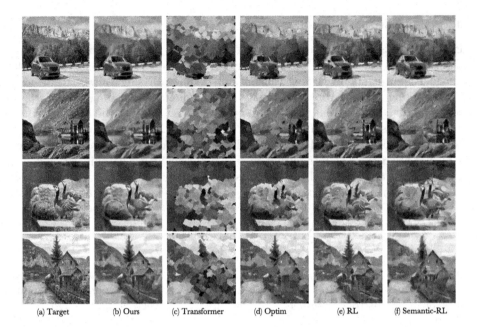

Fig. 3. Qualitative method comparison w.r.t painting efficiency. Comparing final canvas outputs while using ∼ 300 brushstrokes for (b) Ours, (c) Paint Transformer [28], (d) Optim [49], (e) RL [18] and (f) Semantic-RL [35]. We observe that our approach results in more accurate depiction of the fine-grain features in the target image while using a low brushstroke count. Please zoom in for better comparison.

the computation of object saliency and bounding box predictions. In this work, we use a pretrained U-2-Net model [32] model in order to compute foreground saliency predictions. The bounding box predictions are then computed as the union over bounding box outputs from pretrained Yolo-v5 [20] and the overall bounding box for the saliency prediction output.

Overall Training. The RL-based *sequential-planner* (SP) agent is trained using the model-based DDPG algorithm [18] with the following overall reward function for each layer l,

$$r_t^{overall}(l) = r_t^{layer}(l) + \mu\, r_t^{gbp} + \eta\, r_t^{spatial} + \lambda\, r_t^{color}, \tag{15}$$

where r_t^{gbp} is the guided-backpropagation based reward from [35]. The final RL agent is trained for a total of 5M iterations with a batch size of 128.

6   Comparison with State of the Art

In this section, we provide extensive qualitative and quantitative results comparing our method with recent state-of-the-art neural painting methods [18,28,35,49]. First, in Sect. 6.1, we demonstrate the improved painting efficiency of our method in generating detailed paintings when using limited number of brushstrokes. Second, we show that our method leads to painting sequences with increased resemblance with the human painting style (refer Sect. 6.2). Finally, in Sect. 6.3, we provide a discussion of some limitations of our approach in order to aid a more holistic understanding of the proposed method and future directions.

6.1 Painting Efficiency

As discussed in Sect. 3, the ability to learn an *efficient painting plan*, in order to paint detailed output canvases using as few brushstrokes as possible, is essential for most interactive and robotic painting applications [22,31,42]. In this section, we compare the painting efficiency of our approach with previous works while painting under a limited brushstroke budget.

Qualitative Comparison. Figure 3 shows a qualitative comparison between the generated canvases using a low budget of 300 brushstrokes per canvas. Note that due to grid-wise formulation for Paint Transformer [28] and Optim [49], the corresponding results are reported after ∼360 and 330 brushstrokes respectively. We observe that our method results in more accurate depictions of target image (*e.g.* fine-grain features for car, hut, and birds in row 1–3 from Fig. 3) when using a limited number of brushstrokes. In contrast, previous methods often lack an intelligent mechanism for efficient brushstroke distribution across the canvas which leads to poor performance when using a limited brushstroke budget. Surprisingly, we also find that Paint Transformer [28] performs worse than previous methods like Optim [49] when using a small number of brushstrokes.

Quantitative Comparison. Table 1 shows quantitative results on the quality of the finally generated canvases while using ∼ 300 brushstrokes per canvas. The final results are reported in terms of both pixel wise l_2 distance $\mathcal{L}_{pixel}$ and perceptual similarity loss $\mathcal{L}_{pcpt}$ [21] between the final canvas and the target image. The quantitative values show that our method helps in significantly lowering the distance metrics between the painted canvas and the target image as compared to previous works. In particular, we note that for the CUB-Birds dataset [41], our approach leads to a reduction of 30.1%, 25.6% 24.9% and 38.2% in the $\mathcal{L}_{pixel}$ distance metric as compared to RL [18], Semantic-RL [35], Optim [49] and Paint Transformer [27], respectively.

6.2 Resemblance with Human Painting Style

Qualitative Comparison. We demonstrate the practical applicability of our method to actual human users by qualitatively comparing the painting sequences generated by our method with those drawn by actual human artists (refer Fig. 4). We observe that our method bears high resemblance with the human painting style in terms of both layerwise painting evolution and localized attention. In contrast, previous state-of-the-art methods often try to directly minimize the pixel-wise distance between painted canvas and the target image, thereby leading to intermediate canvas states which are less intelligible for a human user.

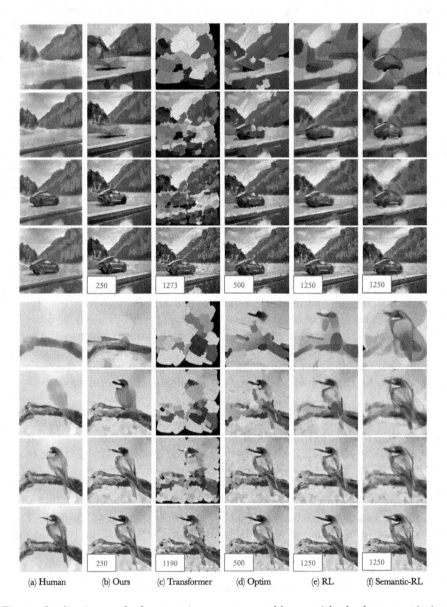

Fig. 4. Qualitative method comparison w.r.t resemblance with the human painting style. We compare different methods (b–f). All painting sequences are generated using a different brushstroke count (indicated in the boxes), so as to ensure similar pixel-wise reconstruction loss with the target image. The corresponding frames for each sequence are computed after ∼ 10%, 40%, 60% and 100% of the overall painting episode. We observe that our method offers higher resemblance with the human painting style (shown in column-a) as compared to previous works.

Table 1. Quantitative Evaluations. (Left) Method comparison w.r.t painting efficiency using a limited brushstroke budget. (Right) User-study results, showing % of painting samples for which human users prefer intelli-paint sequences over previous works.

Method	Stanford cars [24]		CUB-Birds [41]		Intelli-Paint preference	
	$\mathcal{L}_{pixel}$	$\mathcal{L}_{pcpt}$	$\mathcal{L}_{pixel}$	$\mathcal{L}_{pcpt}$	Study A	Study B
RL [18]	78.06	0.54	72.93	0.56	87.17%	83.11%
Semantic [35]	79.98	0.55	68.46	0.55	84.46%	69.09%
Optim [49]	76.52	0.54	67.90	0.53	76.95%	75.41%
Transformer [28]	87.78	0.57	82.43	0.56	91.11%	86.50%
Ours	**56.92**	**0.44**	**50.94**	**0.45**	N/A	N/A

For instance, consider the first example from Fig. 4. Much like a human painter, our method first paints a realistic background representation (consisting of the sky, mountains, river and the ground) before drawing in the foreground car in a coarse-to-fine fashion. This results in a more human-like evolution of the painted canvas which can be easily relatable to actual human artists. In contrast, methods like Paint Transformer [28], Optim [49] and RL [18] directly make brushstrokes based on low-level image features (*e.g.* red brushstrokes for the car in row-1 and head of the bird in row-5). This leads to more bottom-up painting sequences which are different from the human style. Meanwhile, Semantic-RL [35] tries to paint both foreground and background regions in parallel, thereby lacking the semantic painting evolution exhibited by human users.

Quantitative Comparison. We also report quantitative results demonstrating the human-like resemblance of our approach as compared to previous works. To this end, we devise a human user study wherein each human participant is shown a series of paired painting sequences comparing our method with previous works. For each pair, the subject is then asked to select the painting sequence which best resembles the human painting style. The user study is performed in two different variations: **1)** User-Study A, where subjects are provided with a human painting sequence to act as reference in their decision-masking, and **2)** User-Study B, where participants are only shown a pair of artificial painting sequences (ours vs competing method) and are thus asked to make the decision based on their own subjective understanding of the human painting style. User-Study A was conducted across 10 different full-length painting sequences procured from real human artists, while User-Study B uses a set of randomly chosen 100 painting sequences from the CelebA [25] and CUB-Birds [41] datasets. A total of 50 unique Amazon Mechanical Turk subjects were used for both studies.

Results for both user-studies are shown in Table 1. User-Study A reveals that human subjects consider our painting sequences to be closer to those of a *particular human artist* (used as reference). However, as noted in Sect. 6.3, since each person has its own subjective understanding of what a human-like painting style constitutes, it does not answer the broader question on the relatability of these painting sequences from the context of a generic human user. User-study B tries to address this question. While we observe that the corresponding preference scores are lower than User-study A, it provides evidence that our approach is considered more relatable by a majority of human subjects.

6.3   Discussion and Limitations

In this section, we provide a discussion of some limitations of our method in order to facilitate a more holistic understanding of our approach.

Limited Variation in Painting Style. We note that our method only mimics some commonly found traits (progressive layering, coarse to fine localized attention) of the human painting process, and, thus does not claim to be calibrated to the fine-grain variations in the painting styles of each human artist. Nevertheless, as demonstrated in Table 1, we find that our painting style is considered more relatable (over previous works) by majority of human users.

Human-Like vs Human-Intelligible. We note that while our work provides a step towards improving the *human-intelligibility* of the painting sequences over previous works, it does not claim to be *truly human-like*. Several factors *e.g.* limited variation in painting style (discussed above), use of primitive brush-strokes (Bézier curves) *etc.*. contribute to this limitation. This leaves much room for improvement in the development of truly human-like painting agents, which could motivate future work in this area (*e.g.* using advanced stroke representation [37]).

Reliance on Pretrained Image Saliency Models. Our method relies on the computation of image saliency masks for allowing a human-like evolution of the painted canvas. Thus limitations of the pretrained U2-Net [32] model become our limitations. Nevertheless, we note that failure to detect a particular salient object would simply lead to painting the corresponding region in the background layer, and thus does not affect the quality of the final canvas.

Training Requirements. In order to learn a human-relatable style, our method requires self-supervised training on a dataset of *real* images. This is in contrast with Paint transformer [28] which performs self-training on an *artificial* dataset, and Optim [49] which does not require any training. That said, once trained we find that our method is able to generalize across a range of domains at inference time. For instance, we note that all results in Fig. 3, 4 were generated using an Intelli-paint model trained only on the CUB-Birds [41] dataset.

7   Conclusion

In this paper, we emphasize that the practical merits of an autonomous painting system should be evaluated not only by the quality of generated canvas but also by the interpretability of the corresponding painting sequence by actual human artists. To this end, we propose a novel *Intelli-Paint* pipeline, which uses progressive layering to allow for a more human-like evolution of the painted canvas. The painting agent focuses on different image areas through a sequence of coarse-to-fine localized attention windows and is able to paint detailed scenes while using a limited number of brushstrokes. Experiments reveal that in comparison with previous state-of-the-art methods, our approach not only shows improved painting efficiency but also exhibits a painting style which is much more relatable to actual human users. We hope our work opens new avenues for the further development of interactive and robotic painting applications in the real world.

References

1. Arjovsky, M., Chintala, S., Bottou, L.: Wasserstein generative adversarial networks. In: International Conference on Machine Learning, pp. 214–223. PMLR (2017)
2. Bangham, J.A., Gibson, S.E., Harvey, R.W.: The art of scale-space. In: BMVC. pp. 1–10. Citeseer (2003)
3. Chen, L.C., Papandreou, G., Schroff, F., Adam, H.: Rethinking atrous convolution for semantic image segmentation. arXiv preprint arXiv:1706.05587 (2017)
4. Chen, W., Hays, J.: SketchyGAN: towards diverse and realistic sketch to image synthesis. In: Proceedings of the IEEE Conference on Computer Vision and Pattern Recognition, pp. 9416–9425 (2018)
5. Collomosse, J.P., Hall, P.M.: Genetic paint: a search for salient paintings. In: Rothlauf, F., et al. (eds.) EvoWorkshops 2005. LNCS, vol. 3449, pp. 437–447. Springer, Heidelberg (2005). https://doi.org/10.1007/978-3-540-32003-6_44
6. Collomosse, J.P., Hall, P.M.: Salience-adaptive painterly rendering using genetic search. Int. J. Artif. Intell. Tools **15**(04), 551–575 (2006)
7. Collomosse, J., Hall, P.: Painterly rendering using image salience. In: Proceedings 20th Eurographics UK Conference, pp. 122–128. IEEE (2002)
8. Frans, K., Cheng, C.Y.: Unsupervised image to sequence translation with canvas-drawer networks. arXiv preprint arXiv:1809.08340 (2018)
9. Frans, K., Soros, L., Witkowski, O.: Clipdraw: exploring text-to-drawing synthesis through language-image encoders. arXiv preprint arXiv:2106.14843 (2021)
10. Gagniuc, P.A.: Markov Chains: From Theory to Implementation and Experimentation. John Wiley & Sons, Hoboken (2017)
11. Ganin, Y., Kulkarni, T., Babuschkin, I., Eslami, S., Vinyals, O.: Synthesizing programs for images using reinforced adversarial learning. arXiv preprint arXiv:1804.01118 (2018)
12. Gatys, L.A., Ecker, A.S., Bethge, M.: Image style transfer using convolutional neural networks. In: Proceedings of the IEEE Conference on Computer Vision and Pattern Recognition, pp. 2414–2423 (2016)
13. Graves, A.: Generating sequences with recurrent neural networks. arXiv preprint arXiv:1308.0850 (2013)
14. Ha, D., Eck, D.: A neural representation of sketch drawings. arXiv preprint arXiv:1704.03477 (2017)
15. Haeberli, P.: Paint by numbers: abstract image representations. In: Proceedings of the 17th Annual Conference on Computer Graphics and Interactive Techniques, pp. 207–214 (1990)

16. Hertzmann, A.: Painterly rendering with curved brush strokes of multiple sizes. In: Proceedings of the 25th Annual Conference on Computer Graphics and Interactive Techniques, pp. 453–460 (1998)
17. Hertzmann, A., et al.: A survey of stroke-based rendering. IEEE Comput. Graph. Appl. **23**, 70–81. Institute of Electrical and Electronics Engineers (2003)
18. Huang, Z., Heng, W., Zhou, S.: Learning to paint with model-based deep reinforcement learning. In: Proceedings of the IEEE International Conference on Computer Vision, pp. 8709–8718 (2019)
19. Jia, B., Brandt, J., Mech, R., Kim, B., Manocha, D.: LPaintB: learning to paint from self-supervision arXiv preprint arXiv:1906.06841 (2019)
20. Jocher, G., et al.: ultralytics/yolov5: v6.0 - YOLOv5n 'Nano' models, Roboflow integration, TensorFlow export, OpenCV DNN support, October 2021. https://doi.org/10.5281/zenodo.5563715
21. Johnson, J., Alahi, A., Fei-Fei, L.: Perceptual losses for real-time Style transfer and super-resolution. In: Leibe, B., Matas, J., Sebe, N., Welling, M. (eds.) ECCV 2016. LNCS, vol. 9906, pp. 694–711. Springer, Cham (2016). https://doi.org/10.1007/978-3-319-46475-6_43
22. Kite-Powell, J.: This AI robot will paint a canvas at sxsw 2021 (March 2021), https://www.forbes.com/sites/jenniferhicks/2021/03/10/this-ai-robot-will-paint-a-canvas-at-sxsw-2021/?sh=5b1f0d1ab449
23. Kotovenko, D., Wright, M., Heimbrecht, A., Ommer, B.: Rethinking style transfer: from pixels to parameterized brushstrokes. arXiv preprint arXiv:2103.17185 (2021)
24. Krause, J., Stark, M., Deng, J., Fei-Fei, L.: 3D object representations for fine-grained categorization. In: 4th International IEEE Workshop on 3D Representation and Recognition (3dRR-13). Sydney, Australia (2013)
25. Lee, C.H., Liu, Z., Wu, L., Luo, P.: MaskGAN: towards diverse and interactive facial image manipulation. In: IEEE Conference on Computer Vision and Pattern Recognition (CVPR) (2020)
26. Litwinowicz, P.: Processing images and video for an impressionist effect. In: Proceedings of the 24th Annual Conference on Computer Graphics and Interactive Techniques, pp. 407–414 (1997)
27. Liu, B., Gould, S., Koller, D.: Single image depth estimation from predicted semantic labels. In: 2010 IEEE Computer Society Conference on Computer Vision and Pattern Recognition, pp. 1253–1260. IEEE (2010)
28. Liu, S., et al.: Paint transformer: feed forward neural painting with stroke prediction. In: Proceedings of the IEEE/CVF International Conference on Computer Vision, pp. 6598–6607 (2021)
29. Mellor, J.F., et al.: Unsupervised doodling and painting with improved spiral. arXiv preprint arXiv:1910.01007 (2019)
30. Nakano, R.: Neural painters: a learned differentiable constraint for generating brushstroke paintings. arXiv preprint arXiv:1904.08410 (2019)
31. Nemire, B.: Ai painting robot, May 2017. https://developer.nvidia.com/blog/ai-painting-robot/
32. Qin, X., Zhang, Z., Huang, C., Dehghan, M., Zaiane, O., Jagersand, M.: U2-Net: going deeper with nested u-structure for salient object detection. Pattern Recogn. **106**, p. 107404 (2020)
33. Reyner, N.: How to paint with layers - in acrylic & oil, December 2017. https://nancyreyner.com/2017/12/25/what-is-layering-for-painting/
34. Shiraishi, M., Yamaguchi, Y.: An algorithm for automatic painterly rendering based on local source image approximation. In: Proceedings of the 1st International Symposium on Non-photorealistic Animation and Rendering, pp. 53–58 (2000)

35. Singh, J., Zheng, L.: Combining semantic guidance and deep reinforcement learning for generating human level paintings. In: Proceedings of the IEEE/CVF International Conference on Computer Vision (2021)
36. Singh, J., Zheng, L., Smith, C., Echevarria, J.: Paint2pix: interactive painting based progressive image synthesis and editing. In: European Conference on Computer Vision. Springer, Cham (2022)
37. Sochorová, Š, Jamriška, O.: Practical pigment mixing for digital painting. ACM Trans. Graph. **40**(6), 1–11 (2021)
38. Teece, D.: 3D painting for non-photorealistic rendering. In: ACM SIGGRAPH 98 Conference Abstracts and Applications, p. 248 (1998)
39. Treavett, S., Chen, M.: Statistical techniques for the automated synthesis of non-photorealistic images. In: Proceedings of 15th Eurographics UK Conference, pp. 201–210 (1997)
40. Turk, G., Banks, D.: Image-guided streamline placement. In: Proceedings of the 23rd Annual Conference on Computer Graphics and Interactive Techniques, pp. 453–460 (1996)
41. Wah, C., Branson, S., Welinder, P., Perona, P., Belongie, S.: The Caltech-UCSD Birds-200-2011 Dataset. Tech. Rep. CNS-TR-2011-001, California Institute of Technology (2011)
42. Wikipedia contributors: Ai-da (robot) – Wikipedia, the free encyclopedia (2022). https://en.wikipedia.org/w/index.php?title=Ai-Da_(robot)&oldid=1070639724. Accessed 7 Mar 2022
43. Xiang, X., Liu, D., Yang, X., Zhu, Y., Shen, X., Allebach, J.P.: Adversarial open domain adaptation for sketch-to-photo synthesis. In: Proceedings of the IEEE/CVF Winter Conference on Applications of Computer Vision, pp. 1434–1444 (2022)
44. Xie, N., Hachiya, H., Sugiyama, M.: Artist agent: a reinforcement learning approach to automatic stroke generation in oriental ink painting. IEICE Trans. Inf. Syst. **96**(5), 1134–1144 (2013)
45. Yang, S., Wang, Z., Liu, J., Guo, Z.: Controllable sketch-to-image translation for robust face synthesis. IEEE Trans. Image Process. **30**, 8797–8810 (2021)
46. Zeng, K., Zhao, M., Xiong, C., Zhu, S.C.: From image parsing to painterly rendering. ACM Trans. Graph. **29**(1), 1–2 (2009)
47. Zhao, A., Balakrishnan, G., Lewis, K.M., Durand, F., Guttag, J.V., Dalca, A.V.: Painting many pasts: Synthesizing time lapse videos of paintings. In: Proceedings of the IEEE/CVF Conference on Computer Vision and Pattern Recognition, pp. 8435–8445 (2020)
48. Zheng, N., Jiang, Y., Huang, D.: Strokenet: A neural painting environment. In: International Conference on Learning Representations (2018)
49. Zou, Z., Shi, T., Qiu, S., Yuan, Y., Shi, Z.: Stylized neural painting. In: Proceedings of the IEEE/CVF Conference on Computer Vision and Pattern Recognition, pp. 15689–15698 (2021)

Motion Transformer for Unsupervised Image Animation

Jiale Tao[1], Biao Wang[2], Tiezheng Ge[2], Yuning Jiang[2], Wen Li[1(✉)],
and Lixin Duan[1]

[1] School of Computer Science and Engineering and Shenzhen Institute for Advanced
Study, University of Electronic Science and Technology of China, Chengdu, China
liwenbnu@gmail.com
[2] Alibaba Group, Hangzhou, China
{eric.wb,tiezheng.gtz,mengzhu.jyn}@alibaba-inc.com

Abstract. Image animation aims to animate a source image by using
motion learned from a driving video. Current state-of-the-art methods
typically use convolutional neural networks (CNNs) to predict motion
information, such as motion keypoints and corresponding local transformations. However, these CNN based methods do not explicitly model
the interactions between motions; as a result, the important underlying motion relationship may be neglected, which can potentially lead to
noticeable artifacts being produced in the generated animation video.
To this end, we propose a new method, the motion transformer, which
is the first attempt to build a motion estimator based on a vision transformer. More specifically, we introduce two types of tokens in our proposed method: i) image tokens formed from patch features and corresponding position encoding; and ii) motion tokens encoded with motion
information. Both types of tokens are sent into vision transformers to
promote underlying interactions between them through multi-head self
attention blocks. By adopting this process, the motion information can
be better learned to boost the model performance. The final embedded
motion tokens are then used to predict the corresponding motion keypoints and local transformations. Extensive experiments on benchmark
datasets show that our proposed method achieves promising results to
the state-of-the-art baselines. Our source code will be public available.

1 Introduction

Image animation (also known as motion transfer) is a technique that aims to
animate a source image based on the motion information extracted from a given
driving video, such that the generated video can mimic the motion in the driving video while simultaneously retaining the appearance of the target object

---

J. Tao—Work done during an intership at Alibaba Group.

Supplementary Information The online version contains supplementary material
available at https://doi.org/10.1007/978-3-031-19787-1_40.

in the source image. This approach enables people to quickly create innovative content without the need to start from scratch, which can save large amounts time. Motion transfer has gained significant attention from the computer vision community in recent years [5,11,21,27,34,35,37], owing to its wide range of practical applications across entertainment and education such as virtual try-on [2,7], video conferencing [44], e-commerce advertising [49], and so on.

Existing image animation works can be roughly divided into two categories, namely supervised methods and unsupervised methods. In more detail, supervised methods typically focus on the animation of a specific object type (e.g., human body, human face, etc.) and utilize a third-party model to extract structural representations, which might take the forms of 2D keypoints [23,24], 3D meshes [57], 3D optical flow [21] and so on. This type of method has advantages in modeling accurate object structures, but is limited by the object-specific approach to image animation. On the other hand, unsupervised methods [34,35,37] aim to avoid the requirement for object-specific predefined structure representations. These approaches usually learn intermediate motion representations (e.g., keypoints and affine matrices) between two images by warping one image to reconstruct another. Currently proposed unsupervised methods generally comprise of two modules: a motion estimator and an image generator. In these methods, the image generators tend to be quite similar, while the motion estimators are always the research focus and are proven to be quite crucial for animation performance. For example, Siarohin et al. [34] utilize an unsupervised keypoint detector to estimate sparse motions. These authors later boost the performance of their method by adding a head in order to better predict the affine matrix [35]. Moreover, the method proposed in [37] further improves the motion learning process, by entangling a keypoint and the corresponding affine matrix into a single heat-map estimation.

In order to animate arbitrary objects, we follow the unsupervised setting and focus primarily on motion estimation in this work. It is worth noting that all CNN-based methods discussed above fail to consider the interactions between motions, which may prevent these methods from learning robust motion estimators. We believe the robustness of motion estimators can be boosted by the global information of motions. Accordingly, in this work, we make the first attempt to model the global motion information by employing vision transformers in unsupervised image animation. More specifically, we explicitly model the motion (i.e., the keypoint and corresponding affine matrix) as a query token in the transformer, which we refer to as motion tokens and treat as learnable parameters. We further introduce image tokens, which are obtained by projecting the flattened image patch features to the same dimension as the motion tokens. These motion tokens, conditioned on image tokens, are then decoded to final keypoints and affine matrices through several transformer layers. Intuitively, the motion transformer compensates for the lack of prior structural representations by naturally introducing global motion information to assist with part motion learning; this procedure is efficiently implemented through the self-attention mechanism. We can summarize the advantages of the motion transformer in two aspects with reference to different objects: i) For objects with relatively non-rigid motions (such as human body), it learns the set of local motions in a more stable fashion; ii) for objects with relatively rigid motions (such as faces), it exhibits a strong ability to learn global motion patterns simultaneously for all motion tokens.

We conduct extensive experiments on four benchmark datasets, which contain various kinds of objects such as talking heads, human bodies, animals, etc. The superior performance of our proposed method relative to existing baselines clearly demonstrates that global motion information can help to improve the robustness of motion estimators, as well as showing the success of our proposed motion transformer in capturing the global motion information.

2 Related Work

Image Animation: Supervised methods [5,11,19,23,24,26,27,29,30,36,55,56, 60] focus on the animation of a specific object type. Among these, the human body [1,6,13,14,22,29,31,33,52,54] and human face [3,11,12,16,28,41,46–48, 51] are the most popular animation objects. Methods of this kind rely on object-specific landmark detectors, 3D models or other forms of supervision, which are usually pre-trained on a large amount of labeled data. On one hand, the advantage of these methods is that based on the pre-obtained structure representations, it is easier to further learn the warping flow between two images. On the other hand, these methods are also hampered by an obvious limitation, as they are only suitable for a specific object type.

Unsupervised methods [34,35,37,40] have been recently proposed to address the above limitation. These approaches typically leverage a large amount of easy-to-obtain unlabeled web videos and design image reconstruction losses to learn intermediate motion representations (e.g., keypoints and affine matrices). Benefiting from the unsupervised scenario, methods of this kind can be applied to animate a wide range of objects, including human bodies, human faces, animals, etc. It is worth noting that no predefined structural representations of objects are available for training in those methods. Specifically, Monkey-Net [34] proposes to learn intermediate part keypoints as sparse motions by means of the downstream image reconstruction task. Subsequently, FOMM [35] improves on this approach by simultaneously regressing the local affine matrices along with the keypoints of object parts. Moreover, MRAA [37] further improves FOMM by combining the learning of part keypoints and local affine matrices into a single heat-map estimation process. Among these approaches, however, Monkey-Net is limited by the coarsely defined motion model, FOMM suffers from regressing stable affine matrices, while MRAA struggles in modeling relatively rigid motions (e.g., human face) and fails to consider cooperative part motions.

It is worth noting that all above unsupervised methods focus on the motion estimation process in image animation. Our method also lies in this research scope, with a newly proposed motion transformer as the motion estimator. Similar to FOMM, our method also regresses the affine matrices, while it avoids the instability problem by adopting a global-assisted approach; moreover, benefiting from this approach, our method is also better able to handle rigid motions and cooperative local motions when compared with MRAA.

Vision Transformer: Transformers [42] have achieved great success in natural language processing (NLP) community. Recently, they have also achieved promising results in computer vision tasks. Among them, DETR [4] and ViT [9] are the pioneering methods, and have been followed by a series of other vision transformer methods [18,20,43,50,53,58,59]. DETR [4] was the first to introduce transformers to the object detection. It follows the encoder-decoder architecture of traditional transformers in NLP, where object queries are introduced as learnable parameters. ViT [9] proposes a transformer encoder architecture for image classification, which directly splits the image into patches and introduces a learnable classification token to aid in performing the task. Recently, several methods have proposed variant forms of transformers for landmark detection [18,45,50]. Our method is motivated by these recent works, in that we regress the object keypoints as well as affine matrices for image animation.

3 Methodology

In the context of unsupervised image animation, we are given a source image S and a driving video $D = \{Z_i\}$, where Z_i is the i-th video frame. Unless otherwise noted, in the remainder of this work we will use Z to represent a video frame for the sake of simplicity.

3.1 The General Framework for Image Animation

Existing unsupervised image animation methods [34,35,37] generally perform image animation in a frame-by-frame manner. Given the source image and each frame of the driving video, the image animation model outputs a synthesized image that mimics the pose of the object in the driving frame while also preserving the appearance of the object in the source image.

Unsupervised models typically comprise two stages: motion estimation and image generation. The motion estimation stage produces the relative motion (often in the form of optical flow) between each driving video frame and the source image, while the image generation stage warps the source image based on the relative motion in order to generate the synthesized image.

To obtain the relative motion, the motion information of the source image or a driving video frame is first separately predicted and then ensembled to calculate the dense motion flow. More specifically, the motion information of a single image is disentangled as a set of transformations of object parts, each of which is represented by a keypoint and its affine transformation from an latent reference image. A motion estimator is designed to predict the keypoints and the corresponding affine transformations for the input image.

Motion Estimation: The first stage of the general image animation framework involves estimating relative motions between the source image and driving frame. This stage plays a critical role in the process, as the estimation accuracy largely determines the overall quality of the generated video. Existing works [34,35,37] generally follow CNN-style models, where the transformation of each object part

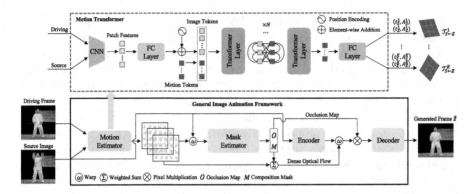

Fig. 1. Overview of the general image animation framework and our proposed motion transformer. Unlike the existing CNN based works [35,37], our motion transformer introduces image tokens and motion tokens, which encode visual and motion information respectively. And those tokens are further sent into multiple transformer layers to mine the underlying interactions between them, the self attention and cross attention are denoted by the straight and curved lines. By using a linear head, the output motion tokens are finally regressed to keypoints and their corresponding affine matrices.

is derived by learning a mapping from the learned feature maps. We contend that current CNN-based methods may not adequately capture the global motion information, as they do not consider interactions between part motions.

Assume there are K parts in an object for either the source image or each driving frame. In the motion estimation stage, the goal is to learn a motion transformation (t^k, A^k) for the k-th part, where $t^k \in \mathbb{R}^{2 \times 1}$ denotes a keypoint (*i.e.*, the centroid of the transformation), $A^k \in \mathbb{R}^{2 \times 2}$ represents the corresponding affine transformation matrix, and $k = 1, ..., K$.

Moreover, since the affine transformation A^k should be applied only to a certain neighboring area (also known as a mask) of the object part rather than the entire image, we constrain the effect of A^k by further learning the corresponding mask M^k. In the literature [35,37], the mask estimator is usually designed as a CNN-based encoder-decoder architecture. Specifically, it takes the warped source image as input and generates the masks $\{M^k|_{k=1}^K\}$'s of K object parts. Furthermore, an additional occlusion map can also be learned to guide the image generator in inpainting the occluded regions. We refer the readers to [35,37] for further details.

Motion Representation: After learning (t_S^k, A_S^k) and (t_Z^k, A_Z^k), we can obtain the following motion flow from the driving frame Z to the source image S for the k-th object part based on the first-order motion model [35,37], as follows:

$$\mathcal{T}_{S \leftarrow Z}^k(c) = t_S^k + A_S^k (A_Z^k)^{-1}(c - t_Z^k), \tag{1}$$

where c denotes any image coordinate in the driving frame.

We next obtain the dense motion flow $\mathcal{T}_{S \leftarrow Z}(c)$ by combining $\mathcal{T}_{S \leftarrow Z}^k(c)$ with masks $M^k(c)$ as linear weights:

$$\mathcal{T}_{S \leftarrow Z}(c) = \sum_{k=1}^{K} M^k(c) \cdot \mathcal{T}_{S \leftarrow Z}^k(c), \tag{2}$$

where $M^k(c)$ is the mask of the k-th object part at the coordinate c and $\sum_{k=1}^{K} M^k(c) = 1$. Moreover, the dense motion flow $\mathcal{T}_{S \leftarrow Z}(c)$ represents that the pixel value at coordinate c of the generated image $\tilde{Z}$ is warped and obtained based on the pixel value at coordinate $\mathcal{T}_{S \leftarrow Z}(c)$ of the source image.

Image Generation: Given the source image S and the dense motion flow $\mathcal{T}_{S \leftarrow Z}(c)$, a Unet-based encoder-decoder generator is introduced to generate the synthesized image $\tilde{Z}$. Specifically, the source image S is passed through the encoder to obtain feature maps, after which it is then warped according to the dense motion flow $\mathcal{T}_{S \leftarrow Z}(c)$. Finally, the decoder learns the synthesized image $\tilde{Z}$ based on the warped feature map.

3.2   Motion Transformer

Since existing CNN-based methods do not explicitly model the interactions between motions, the underlying motion relationship is not fully exploited and cannot be properly captured. We argue that this underlying relationship is critical to the process and helps reduce artifacts in generated animation videos. For instance, when people smile, the movement of their mouths and eyes occurs simultaneously, meaning that they are highly correlated. Considering this limitation of existing CNN-based models, we aim to seek out a better way of modeling the motion interactions.

To address the above issue, we propose to take advantage of the recently proposed vision transformer. We accordingly name our method the *motion transformer*. In a vision transformer layer, raw data are processed to form tokens, which act as the layer input. The underlying relationship among those tokens can be effectively mined through the attention mechanism. As a result of adopting this approach, meaningful embeddings can be learned for those tokens. Our motion transformer employs multiple vision transformer layers.

In our proposed motion transformer for image animation, we explicitly model the motions of object parts as input query tokens (*motion tokens*) to the transformer. We further obtain *image tokens* by projecting image patch features through a fully connected layer and subsequently embedding them with position encoding. By feeding those two types of tokens together into the transformer, the motion tokens are able to utilize the global context information of the entire image through attention with image tokens, which aids in better capturing the interaction between object part motions. Moreover, a linear head is designed in the last transformer layer to directly regress the keypoints and affine matrices of the motions. The entire process is illustrated in Fig. 1.

Tokens: Two types of tokens are introduced in our motion transformer. We first introduce a set of motion tokens, inspired by the recent vision transformers [4]. Each motion token is expected to encode the motion information of an object part (*i.e.*, a keypoint and its affine transformation). These motion tokens are considered as learnable embeddings in our method; we denote them as $\{P_0^k|_{k=1}^K\}$, where $P_0^k \in \mathbb{R}^d$ represents the k-th object part and d is the embedding dimension.

The second type of tokens is the image token. Rather than directly using raw images, we first extract low-level image features by utilizing a CNN model. Subsequently we flatten the patch image features, with each patch projected to dimension d. To maintain the position information of the patch image features, we add the projected features by the absolute position encoding. We refer to the result features as image tokens, denoted as $I_0^n \in \mathbb{R}^d, n = 1, ..., N$.

Multiple Vision Transformers: When an object moves, different object parts are not completely independent. Rather, they often correlate with each other, which, however, was not discussed in existing motion transfer methods [35,37]. To model the relation among tokens, we utilize the natural advantages of vision transformer for building attention. In particular, a motion token is updated via all motion tokens and image tokens, and correspondingly we build two types of attention for the motion tokens. i) self attention for mining the underlying relationship between motion tokens; ii) cross attention for decoding motion tokens to the final keypoints and affine matrices.

Formally, let us denote by P_{l-1}^i a motion token that to be input to the l-th transformer layer, in which P_{l-1}^i is linearly projected to the query, key and value features $Q_{P_{l-1}^i}, K_{P_{l-1}^i}, V_{P_{l-1}^i}$; and similarly for image tokens we have $Q_{I_{l-1}^i}, K_{I_{l-1}^i}, V_{I_{l-1}^i}$. For ease of illustration, we temporally drop the subscript and define the multi-head self attention (MSA) as follow:

$$head^j = \text{softmax}\left(\frac{QW_Q^j\left(KW_K^j\right)^T}{\sqrt{d}}\right)VW_V^j, \tag{3}$$

$$\text{MSA}(Q,K,V) = [head^1, ..., head^h]W_O, \tag{4}$$

W_O, W_Q^j, W_K^j, W_V^j are learnable parameters, and h represents the total number of heads in each transformer layer. In practice, Q is the query from a token, while K, V are from another token. With this definition, the motion token is updated by self attention and cross attention as follows:

$$P_l^i = \sum_j \text{MSA}(Q_{P_{l-1}^i}, K_{P_{l-1}^j}, V_{P_{l-1}^j}) + \sum_j \text{MSA}(Q_{P_{l-1}^i}, K_{I_{l-1}^j}, V_{I_{l-1}^j}), \tag{5}$$

$$P_l^i = \text{FFN}\left(\text{LN}\left(P_l^i\right) + P_l^i\right), \tag{6}$$

where FFN and LN denote the feed forward network and layer normalization respectively. On one hand, the left term of Eq. (5) (*i.e.*, the self attention) indicates that each motion token tends to query all other motion tokens, in this way the underlying relationship between motion tokens could be effectively captured; on the other hand, the motion token can be gradually embedded with motion information through querying the image tokens, as formulated in the right term of Eq. (5) (*i.e.*, the cross attention).

The two types of attention process above enable the efficient interactions between tokens. In implementation, different from recent works [4,20] in which the two types of attention process are separately conducted, we found it more efficient to unify the self attention and cross attention in a single transformer architecture. To do this, we directly concatenate the image tokens and motion tokens as the initial input tokens $F_0 = [P_0^1; ...; P_0^K; I_0^1; ...; I_0^N] \in \mathbb{R}^{(N+K) \times d}$, and use the single self attention process to update the concatenated tokens:

$$F_l^i = \sum_j \mathrm{MSA}(Q_{F_{l-1}^i}, K_{F_{l-1}^j}, V_{F_{l-1}^j}), \tag{7}$$

$$F_l^i = \mathrm{FFN}\left(\mathrm{LN}\left(F_l^i\right) + F_l^i\right). \tag{8}$$

Note that in this procedure, image tokens are also updated with attention to all tokens, while to our observation, this dose not affect the performance, and the unified self attention is more efficient in end-to-end training.

At the final transformer layer, we take out the motion tokens P_L^k from the output token feature F_L, and employ a linear head to directly regress the affine matrix and translation vector. Let $W_h \in \mathbb{R}^{d \times 6}$ denote the parameters of the linear head; the decoded part affine matrix and translation vector of each motion token are then computed as $[A^k, t^k] = P_L^k W_h$.

3.3   Training

We consider the following losses to formulate the objective function of our method. The whole training process is conducted in an end-to-end fashion.

Perceptual Loss: Following FOMM and MRAA, we adopt the multi-resolution perceptual loss [15] defined with a pre-trained VGG-19 [38] network. Given the driving frame Z with resolution index i, the generated image $\tilde{Z}$, and the feature extractor ϕ with layer index l, the perceptual loss can be written as follows:

$$\mathcal{L}_{per} = \sum_i \sum_l \left\| \phi_l(Z_i) - \phi_l(\tilde{Z}_i) \right\|_1. \tag{9}$$

Equivariance Loss: Following FOMM and MRAA, the equivariance loss is adopted here. Given a random geometric transformation $\mathbf{T}$ and a driving image Z, this loss can be written as follows:

$$\mathcal{L}_{equi} = \sum_k \left\| \mathbf{T}(t_Z^k) - t_{\mathbf{T}(Z)}^k \right\|_1. \tag{10}$$

Background Losses: To handle those situations in which the background is not static, we follow the MRAA in utilizing a background predictor network to predict the background motion flow. To facilitate the separation of the background and foreground motion learning, inspired by [10], we adopt the following losses:

$$\mathcal{L}_{mask} = \left\| M^0 - 1 \right\|_1 + \sum_{k \neq 0} \left\| M^k - 0 \right\|_1, \tag{11}$$

and

$$\mathcal{L}_{con} = \sum_{k \neq 0} \sum_c M^k(c) \cdot \left(c - u^k\right)^2 / \operatorname{sum}\left(M^k\right), \tag{12}$$

where $u^k = \sum_c M^k(c) \cdot c / \operatorname{sum}\left(M^k\right)$. Intuitively, the mask loss $\mathcal{L}_{mask}$ is used to constrain the portion of foreground and background area in an image, and the foreground motion mask is considered to be more focused under the constraint of the concentration loss $\mathcal{L}_{con}$.

Overall Loss: We combine all of the above losses to formulate the overall object function of our proposed vision transformer, as follows:

$$\mathcal{L} = \mathcal{L}_{per} + \mathcal{L}_{equi} + \lambda(\mathcal{L}_{mask} + \mathcal{L}_{con}). \tag{13}$$

It should be noted here we adopt only the above background losses $\mathcal{L}_{mask}$ and $\mathcal{L}_{con}$ for the TaichiHD dataset with $\lambda = 0.1$ to handle the dynamic background change in TaichiHD videos in the experiments. As videos from other datasets typically have a static background, we omit $\mathcal{L}_{mask}$ and $\mathcal{L}_{con}$ from the overall loss for those datasets.

4   Experiments

Datasets: The following benchmark datasets are used in our experiments:

- VoxCeleb [25]: A talking head dataset consisting of 20047 videos. All videos are cropped and resized to 256×256.
- TaiChiHD [35]: This dataset contains 3120 videos. All videos are cropped and resized to 256×256.
- TED-talks [37]: This is a talking show dataset containing 1255 videos. All videos are cropped and resized to 384×384.
- MGIF [34]: This dataset contains 1000 cartoon animal videos, all of which are resized to 256×256, following [35].

Evaluation Metrics: We follow [35,37] in evaluating the video reconstruction quality, where videos are reconstructed with appearance representations by using their first frame and motion representations w.r.t. all frames. Four commonly used evaluation metrics are listed below.

- L1 distance: The average L1 distance between the generated and ground-truth video frames.
- Average keypoint distance (AKD): The average distance of detected keypoints between the generated and ground-truth video frames. This metric is designed for evaluating the pose quality of generated videos.
- Missing keypoint rate (MKR): The percentage of keypoints that are not detected in the generated video frames but do exist in the ground-truth.
- Average Euclidean distance (AED): The average Euclidean distance between generated and ground-truth video frames, as in the feature space. This metric evaluates the identity information of the generated video frames.

Implementation Details: For image generation, we adopt Unet [32] to construct the mask predictor and generative encoder-decoder. Skip connections are added in the encoder-decoder architecture similar to [34,37]. For motion estimation, we adopt the first three stages of the HRNet-W32 encoder [39] pretrained on ImageNet [8] as the CNN backbone in our model. After the CNN encoder has been applied, image features are down-sampled by a scale factor of 4. We utilize a 12-layer standard transformer encoder architecture. Moreover, the sine function [42] is used for position encoding. The image patch size is set to 4×4 in all experiments, with 256 image tokens used for the input resolution of 256×256 and 576 for 384×384. The token dimension d is set to 192. The number of motion tokens is set to 10, as in [35,37]. The Adam optimizer [17] is adopted, where the initial learning rate is set as 2×10^{-4} and dropped by a factor of 10 at the end of $60th$ and $90th$ epoch. We train the entire networks on eight NVIDIA V100 GPU cards for 100 epochs.

4.1   Comparison with State-of-the-Art

Model Capacity: Under the general image animation framework, we analyze the difference between motion estimators from the model capacity view. As listed in Table 1, the proposed motion estimator has slightly less parameters than FOMM and MRAA, while the FLOPs of it are much heavier, which is caused by the high-resolution (4 times lower than the input image resolution) computation in the CNN encoder and in the global attention process in the vision transformer layers. It should be noted here, compared to the image generator that is always the same between our method and existing methods, the motion estimator tends to take a small computation cost in the whole image animation process. While being considerably lighter than the image generator, the motion estimator is proved to be efficient and effective for improving image animation performance, as in our method and recent works [35,37].

Quantitative Comparison: The video reconstruction results are presented in Table 2. As can be seen from the table, our method generally performs the best across all evaluation metrics, as well as across all benchmark datasets with object types including human body, human face and animal etc., reflecting the superiority of the motion transformer to perform general image animation. More specifically, a lower L1 distance straightforwardly indicates better video reconstruction quality achieved by our method. It is also worth noting that our method achieves considerable improvements in terms of AKD on the three datasests, which strongly suggests that our method achieves better transferred motion. This can be further validated in the qualitative results of Fig. 2. Moreover, our method also achieves the best performance on the AED metric, indicating that the identity information can be better preserved using our method for conducting image animation. The superiority on the AED metric is even more obvious on the VoxCeleb dataset, we draw reason that the identity information is especially important for a human face, while our method generally learns the global motion pattern for the human face, which enables it to better capture the global face structure.

Table 1. Parameters comparison of the proposed motion estimator with that of FOMM and MRAA. For clearness, we also list the parameters of the image generator (the encoder-decoder generator together with the mask predictor). The model parameters and FLOPs are computed with input image resolution 256×256.

	Parameters	FLOPs
ImageGenerator	45.57M	53.64G
MotionEstimator-FOMM	14.21M	1.28G
MotionEstimator-MRAA	14.20M	1.26G
MotionEstimator-Ours	12.23M	7.54G

Table 2. Quantitative comparisons with FOMM [35] and MRAA [37] on the video reconstruction task. We present results on four benchmarks, our method generally achieves the best performance on all datasets across all metrics.

	TaiChiHD			TEDTalks			VoxCeleb			MGIF
	L1	(AKD, MKR)	AED	L1	(AKD, MKR)	AED	L1	AKD	AED	L1
FOMM	0.057	(6.649, 0.036)	0.172	0.029	(4.382, 0.008)	0.127	0.041	1.29	0.133	0.0224
MRAA	0.048	(5.246, 0.024)	0.150	0.027	(3.955, **0.007**)	0.118	0.040	1.28	0.133	0.0274
Ours	**0.045**	(**4.670, 0.021**)	**0.148**	**0.026**	(**3.456, 0.007**)	**0.113**	**0.038**	**1.18**	**0.116**	**0.0200**

User Preference: To evaluate the cross-identity image animation, we conduct a user study with fifty participants. In more detail, we first prepare fifty comparison videos, each of which is a concatenation of a source image, a driving video featuring a different-identity, and videos generated by the three methods. Note that the spatial locations of the generated videos are randomly placed. Participants are required to evaluate these three videos according to the transferred motion and identity preservation. The results in Table 3 show that our method is clearly awarded more user preferences than other existing methods.

Qualitative Comparison: In Fig. 2, we present representative animation examples on the TaichiHD, Voxceleb1 and TEDTalks dataset. As the figure shows, our method is generally better at handling both global and local motions. In more detail, for the human face, despite the fact that both FOMM and MRAA can capture the head rotation, our method can synthesize the most realistic and detailed expression information.

Our analysis suggests that it is often the case that a rigid human face turns from left to right, and occlusion occurs in this kind of rigid or global motion. Our method can effectively learn global motion patterns for a human face; accordingly, this makes it easier to detect the occlusion caused by the head rotation and then guide the image generator to inpaint this occluded face structure. In FOMM and MRAA, the motion is learned in a relatively local manner, which makes it more difficult to capture the global face structure. For the human body, it can also be observed that our method synthesizes the most motion-stable results, while FOMM and MRAA often fail to capture the driving motions. We believe this occurs because, lacking awareness of the global motion information, FOMM

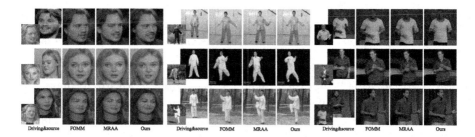

Fig. 2. Qualitative comparisons on the cross-identity image animation task. We show results on three datasets (from left to right: VoxCeleb, TaichiHD and TEDTalks), each with three paired examples.

Table 3. User preferences of our method against FOMM and MRAA on the TaichiHD, TEDTalks, and Voxceleb dataset.

	TaiChiHD	TEDTalks	VoxCeleb
FOMM	96.5%	66.4%	60.8%
MRAA	68.5%	57.1%	69.8%

Table 4. Performance comparison on the TaichiHD dataset with and without position encoding, denoted as w PE and w/o PE.

	L1	(AKD, MKR)	AED
w/o PE	0.047	(5.482, 0.028)	0.158
w PE	**0.045**	**(4.670, 0.021)**	**0.148**

and MRAA are easier to be affected by large motions and intervention of background features. By contrast, our motion transformer learn the part motions in a global-assisted fashion, enabling it to learn the more stable part motions.

4.2 Ablation Study and Parameter Analysis

In this section we study the influence of different components of our motion transformer. More specifically, we conduct video reconstruction experiments on the TaichiHD dataset for the purpose of quantitative analysis.

Position Encoding: As can be seen Table 4, it is crucial to add position encoding to the image tokens in our experiments. We can explain the importance of position encoding from two angles. On one hand, the motion (*a.k.a.*, keypoint and its corresponding affine matrix) estimation is a highly position-sensitive task, in which image tokens equipped with position encoding ease the learning process. On the other hand, in order to learn geometry-consistent keypoint and affine matrix representations in an unsupervised manner, the equivariance loss (*i.e.*, Eq. (10)) is considered to be more effective with position encoding, since the order of the image patches are shuffled by a geometric transformation; however, the position encoding is invariant to the shuffling process, thus the consistency loss can enforce the network to better capture useful image patch features.

CNN Encoder: To explore the influence of the image feature representation, we implement the motion transformer with different CNN backbones. As can be seen from Table 5, compared to our basic setting (*i.e.*, HR-w32), a light-weight CNN (*i.e.*, Stem net [39], a widely used CNN for quickly down-sampling the

Table 5. Performance comparison on the TaichiHD dataset with different CNN backbones of the motion transformer.

CNN	Param.	L1	(AKD, MKR)	AED
Stem	5.56M	0.048	(6.056, 0.030)	0.161
HR-W32	12.23M	**0.045**	**(4.670, 0.021)**	**0.148**
HR-W48	21.30M	**0.045**	(4.829, **0.020**)	0.149

Table 6. Performance comparison on the TaichiHD dataset with respect to different numbers of transformer layers.

Layers	L1	(AKD, MKR)	AED
4	0.046	(5.320, 0.027)	0.155
8	0.046	(5.226, 0.025)	0.154
12	**0.045**	**(4.670, 0.021)**	**0.148**

image by a scale factor of 4) yields worse performance, while a heavy CNN (*i.e.*, HR-w48 [39]) brings no significant improvement. We accordingly conclude that in the absence of a good image feature representation, the motion transformer can't work well; at the same time, the promotion of the CNN backbone to the final performance is limited, which we attribute to the lack of supervision in the unsupervised image animation.

Vision Transformer Layers: We further conduct experiments to explore the influence of different numbers of vision transformer layers used in our motion transformer. As can be seen from Table 6, with smaller numbers of layers, the performance declines considerably (especially on AKD and MKR, which evaluate the motion quality). This reflects the fact that the motion transformer with relatively deeper transformer layers facilitates to learn better motion embeddings for the regression of the motion information.

4.3   Visualization

In this section, we visualize intermediate results to analyze how the motion transformer learns global motions with different object types and what motion patterns have been learned. Samples are randomly chosen; while our observations suggest that the model behaves similarly on the entire dataset.

Visual Attention: We visualize the attention maps between motion tokens and image tokens, to reveal how the motion transformer learns the global and local motion. The results on the VoxCeleb and TEDTalks dataset are presented in Fig. 3. For human faces, it can be seen that the whole face region tends to be attended by all different motion tokens; this implies that each motion part is learned with awareness of the global motion, which is in line with our motivation. For human bodies, we first observe that the global motion is effectively captured; as can be seen in the third row, the motion token learned with a global pattern attends almost the whole regions of the object in the image. Moreover, we find that in the initial transformer layers, the local motions of the woman's hands are well captured. As the depth increases, motion tokens can also find some other meaningful relationship. For example, as illustrated in the attention results of the TEDTalks sample in Fig. 3, the motion token representing the woman's right hand (*i.e.*, the first row) actually shows that it is related to her upper body and head, as in the attention map of the sixth transformer layer. This is reasonable because, when a woman presents something in the talk, her body may move together with the hand gestures to communicate with the whole body language.

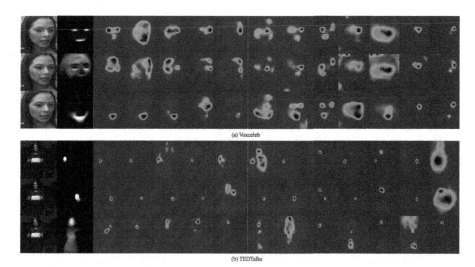

(a) Voxceleb

(b) TEDTalks

Fig. 3. Visualizations of visual attention maps between motion tokens and image tokens on the VoxCeleb and TEDTalks datasets. From left to right of in each row, the presented content respectively represents the driving image, the corresponding motion mask and the visual attention maps of each transformer layer. We present visual attention maps of three representative motion tokens for each dataset. Note that we reshape and resize the sequence attention values to the original image sizes.

5 Conclusion

We propose a new method, called the motion transformer, under the general image animation framework for unsupervised image animation. The motion transformer introduces both image tokens and learnable motion tokens. To encourage the interactions between image and motion tokens, our motion transformer network employs multiple transformer layers, which take those tokens as input in order to learn the underlying motion relationship and obtain better motion embeddings. We further conduct extensive experiments on four benchmark datasets. Our experimental results validate the effectiveness of capturing the global motion information in our motion transformer.

Acknowledgement. This work is supported by the Major Project for New Generation of AI under Grant No. 2018AAA0100400, the National Natural Science Foundation of China (Grant No. 62176047), Sichuan Science and Technology Program (No. 2021YFS0374, 2022YFS0600), Beijing Natural Science Foundation (Z190023), and Alibaba Group through Alibaba Innovation Research Program. This work is also partially supported by the Science and Technology on Electronic Information Control Laboratory.

References

1. Balakrishnan, G., Zhao, A., Dalca, A.V., Durand, F., Guttag, J.: Synthesizing images of humans in unseen poses. In: Proceedings of the IEEE Conference on Computer Vision and Pattern Recognition, pp. 8340–8348 (2018)
2. Bhatnagar, B.L., Tiwari, G., Theobalt, C., Pons-Moll, G.: Multi-garment net: learning to dress 3D people from images. In: Proceedings of the IEEE/CVF International Conference on Computer Vision, pp. 5420–5430 (2019)
3. Burkov, E., Pasechnik, I., Grigorev, A., Lempitsky, V.: Neural head reenactment with latent pose descriptors. In: Proceedings of the IEEE/CVF Conference on Computer Vision and Pattern Recognition, pp. 13786–13795 (2020)
4. Carion, N., Massa, F., Synnaeve, G., Usunier, N., Kirillov, A., Zagoruyko, S.: End-to-end object detection with transformers. In: Vedaldi, A., Bischof, H., Brox, T., Frahm, J.-M. (eds.) ECCV 2020. LNCS, vol. 12346, pp. 213–229. Springer, Cham (2020). https://doi.org/10.1007/978-3-030-58452-8_13
5. Chan, C., Ginosar, S., Zhou, T., Efros, A.A.: Everybody dance now. In: Proceedings of the IEEE/CVF International Conference on Computer Vision, pp. 5933–5942 (2019)
6. Chen, X., Song, J., Hilliges, O.: Unpaired pose guided human image generation. In: Conference on Computer Vision and Pattern Recognition (CVPR 2019). Computer Vision Foundation (CVF) (2019)
7. Chopra, A., Jain, R., Hemani, M., Krishnamurthy, B.: Zflow: gated appearance flow-based virtual try-on with 3d priors. In: Proceedings of the IEEE/CVF International Conference on Computer Vision, pp. 5433–5442 (2021)
8. Deng, J., Dong, W., Socher, R., Li, L.J., Li, K., Fei-Fei, L.: Imagenet: a large-scale hierarchical image database. In: 2009 IEEE Conference on Computer Vision and Pattern Recognition, pp. 248–255. IEEE (2009)
9. Dosovitskiy, A., et al.: An image is worth 16x16 words: transformers for image recognition at scale. In: International Conference on Learning Representations (2021)
10. Gao, Q., Wang, B., Liu, L., Chen, B.: Unsupervised co-part segmentation through assembly. In: International Conference on Machine Learning (2021)
11. Geng, Z., Cao, C., Tulyakov, S.: 3D guided fine-grained face manipulation. In: Proceedings of the IEEE/CVF Conference on Computer Vision and Pattern Recognition, pp. 9821–9830 (2019)
12. Ha, S., Kersner, M., Kim, B., Seo, S., Kim, D.: Marionette: few-shot face reenactment preserving identity of unseen targets. In: Proceedings of the AAAI Conference on Artificial Intelligence, vol. 34, pp. 10893–10900 (2020)
13. Huang, Z., Han, X., Xu, J., Zhang, T.: Few-shot human motion transfer by personalized geometry and texture modeling. In: Proceedings of the IEEE/CVF Conference on Computer Vision and Pattern Recognition, pp. 2297–2306 (2021)
14. Jiang, Y., Yang, S., Qiu, H., Wu, W., Loy, C.C., Liu, Z.: Text2human: text-driven controllable human image generation. ACM Trans. Graphics (TOG) **41**(4), 1–11 (2022). https://doi.org/10.1145/3528223.3530104
15. Johnson, J., Alahi, A., Fei-Fei, L.: Perceptual losses for real-time style transfer and super-resolution. In: Leibe, B., Matas, J., Sebe, N., Welling, M. (eds.) ECCV 2016. LNCS, vol. 9906, pp. 694–711. Springer, Cham (2016). https://doi.org/10.1007/978-3-319-46475-6_43

16. Kim, H., et al.: Deep video portraits. ACM Trans. Graphics (TOG) **37**(4), 163 (2018)
17. Kingma, D.P., Ba, J.: Adam: a method for stochastic optimization. In: Bengio, Y., LeCun, Y. (eds.) 3rd International Conference on Learning Representations, ICLR 2015, San Diego, CA, USA, 7–9 May 2015, Conference Track Proceedings (2015)
18. Li, Y., et al.: Tokenpose: learning keypoint tokens for human pose estimation. In: IEEE/CVF International Conference on Computer Vision (ICCV) (2021)
19. Li, Y., Huang, C., Loy, C.C.: Dense intrinsic appearance flow for human pose transfer. In: Proceedings of the IEEE/CVF Conference on Computer Vision and Pattern Recognition, pp. 3693–3702 (2019)
20. Li, Y., He, J., Zhang, T., Liu, X., Zhang, Y., Wu, F.: Diverse part discovery: occluded person re-identification with part-aware transformer. In: Proceedings of the IEEE/CVF Conference on Computer Vision and Pattern Recognition, pp. 2898–2907 (2021)
21. Liu, W., Piao, Z., Min, J., Luo, W., Ma, L., Gao, S.: Liquid warping GAN: a unified framework for human motion imitation, appearance transfer and novel view synthesis. In: Proceedings of the IEEE/CVF International Conference on Computer Vision, pp. 5904–5913 (2019)
22. Lorenz, D., Bereska, L., Milbich, T., Ommer, B.: Unsupervised part-based disentangling of object shape and appearance. In: Proceedings of the IEEE/CVF Conference on Computer Vision and Pattern Recognition, pp. 10955–10964 (2019)
23. Ma, L., Jia, X., Sun, Q., Schiele, B., Tuytelaars, T., Van Gool, L.: Pose guided person image generation. In: Advances in Neural Information Processing Systems, vol. 30, pp. 406–416 (2017)
24. Ma, L., Sun, Q., Georgoulis, S., Van Gool, L., Schiele, B., Fritz, M.: Disentangled person image generation. In: Proceedings of the IEEE Conference on Computer Vision and Pattern Recognition, pp. 99–108 (2018)
25. Nagrani, A., Chung, J.S., Zisserman, A.: Voxceleb: a large-scale speaker identification dataset. arXiv preprint arXiv:1706.08612 (2017)
26. Neverova, N., Alp Güler, R., Kokkinos, I.: Dense pose transfer. In: Ferrari, V., Hebert, M., Sminchisescu, C., Weiss, Y. (eds.) ECCV 2018. LNCS, vol. 11207, pp. 128–143. Springer, Cham (2018). https://doi.org/10.1007/978-3-030-01219-9_8
27. Nirkin, Y., Keller, Y., Hassner, T.: FSGAN: subject agnostic face swapping and reenactment. In: Proceedings of the IEEE/CVF International Conference on Computer Vision, pp. 7184–7193 (2019)
28. Pumarola, A., Agudo, A., Martinez, A.M., Sanfeliu, A., Moreno-Noguer, F.: GANimation: anatomically-aware facial animation from a single image. In: Ferrari, V., Hebert, M., Sminchisescu, C., Weiss, Y. (eds.) ECCV 2018. LNCS, vol. 11214, pp. 835–851. Springer, Cham (2018). https://doi.org/10.1007/978-3-030-01249-6_50
29. Ren, J., Chai, M., Tulyakov, S., Fang, C., Shen, X., Yang, J.: Human motion transfer from poses in the wild. In: Bartoli, A., Fusiello, A. (eds.) ECCV 2020. LNCS, vol. 12537, pp. 262–279. Springer, Cham (2020). https://doi.org/10.1007/978-3-030-67070-2_16
30. Ren, J., Chai, M., Woodford, O.J., Olszewski, K., Tulyakov, S.: Flow guided transformable bottleneck networks for motion retargeting. In: Proceedings of the IEEE/CVF Conference on Computer Vision and Pattern Recognition, pp. 10795–10805 (2021)
31. Ren, Y., Yu, X., Chen, J., Li, T.H., Li, G.: Deep image spatial transformation for person image generation. In: Proceedings of the IEEE/CVF Conference on Computer Vision and Pattern Recognition, pp. 7690–7699 (2020)

32. Ronneberger, O., Fischer, P., Brox, T.: U-Net: convolutional networks for biomedical image segmentation. In: Navab, N., Hornegger, J., Wells, W.M., Frangi, A.F. (eds.) MICCAI 2015. LNCS, vol. 9351, pp. 234–241. Springer, Cham (2015). https://doi.org/10.1007/978-3-319-24574-4_28

33. Sarkar, K., Mehta, D., Xu, W., Golyanik, V., Theobalt, C.: Neural re-rendering of humans from a single image. In: Vedaldi, A., Bischof, H., Brox, T., Frahm, J.-M. (eds.) ECCV 2020. LNCS, vol. 12356, pp. 596–613. Springer, Cham (2020). https://doi.org/10.1007/978-3-030-58621-8_35

34. Siarohin, A., Lathuilière, S., Tulyakov, S., Ricci, E., Sebe, N.: Animating arbitrary objects via deep motion transfer. In: Proceedings of the IEEE/CVF Conference on Computer Vision and Pattern Recognition, pp. 2377–2386 (2019)

35. Siarohin, A., Lathuilière, S., Tulyakov, S., Ricci, E., Sebe, N.: First order motion model for image animation. In: Advances in Neural Information Processing Systems (2019)

36. Siarohin, A., Sangineto, E., Lathuiliere, S., Sebe, N.: Deformable GANs for pose-based human image generation. In: Proceedings of the IEEE Conference on Computer Vision and Pattern Recognition, pp. 3408–3416 (2018)

37. Siarohin, A., Woodford, O., Ren, J., Chai, M., Tulyakov, S.: Motion representations for articulated animation. In: CVPR (2021)

38. Simonyan, K., Zisserman, A.: Very deep convolutional networks for large-scale image recognition. In: Bengio, Y., LeCun, Y. (eds.) 3rd International Conference on Learning Representations, ICLR 2015, San Diego, CA, USA, 7–9 May 2015, Conference Track Proceedings (2015)

39. Sun, K., Xiao, B., Liu, D., Wang, J.: Deep high-resolution representation learning for human pose estimation. In: Proceedings of the IEEE/CVF Conference on Computer Vision and Pattern Recognition, pp. 5693–5703 (2019)

40. Tao, J., et al.: Structure-aware motion transfer with deformable anchor model. In: Proceedings of the IEEE/CVF Conference on Computer Vision and Pattern Recognition, pp. 3637–3646 (2022)

41. Tripathy, S., Kannala, J., Rahtu, E.: Facegan: Facial attribute controllable reenactment gan. In: Proceedings of the IEEE/CVF Winter Conference on Applications of Computer Vision. pp. 1329–1338 (2021)

42. Vaswani, A., et al.: Attention is all you need. In: Advances in Neural Information Processing Systems, pp. 5998–6008 (2017)

43. Wan, Z., Zhang, J., Chen, D., Liao, J.: High-fidelity pluralistic image completion with transformers. In: Proceedings of the IEEE/CVF International Conference on Computer Vision (ICCV), pp. 4692–4701, October 2021

44. Wang, T.C., Mallya, A., Liu, M.Y.: One-shot free-view neural talking-head synthesis for video conferencing. In: Proceedings of the IEEE/CVF Conference on Computer Vision and Pattern Recognition, pp. 10039–10049 (2021)

45. Watchareeruetai, U., et al.: Lotr: face landmark localization using localization transformer. arXiv preprint arXiv:2109.10057 (2021)

46. Wei, D., Xu, X., Shen, H., Huang, K.: C2f-FWN: coarse-to-fine flow warping network for spatial-temporal consistent motion transfer. In: Proceedings of the AAAI Conference on Artificial Intelligence, vol. 35, no. 4, 2852–2860, May 2021. https://ojs.aaai.org/index.php/AAAI/article/view/16391

47. Wei, Y., Liu, M., Wang, H., Zhu, R., Hu, G., Zuo, W.: Learning flow-based feature warping for face frontalization with illumination inconsistent supervision. In: Vedaldi, A., Bischof, H., Brox, T., Frahm, J.-M. (eds.) ECCV 2020. LNCS, vol. 12357, pp. 558–574. Springer, Cham (2020). https://doi.org/10.1007/978-3-030-58610-2_33

48. Wiles, O., Koepke, A.S., Zisserman, A.: X2Face: a network for controlling face generation using images, audio, and pose codes. In: Ferrari, V., Hebert, M., Sminchisescu, C., Weiss, Y. (eds.) ECCV 2018. LNCS, vol. 11217, pp. 690–706. Springer, Cham (2018). https://doi.org/10.1007/978-3-030-01261-8_41

49. Xu, B., et al.: Move as you like: image animation in e-commerce scenario. In: Proceedings of the 29th ACM International Conference on Multimedia, pp. 2759–2761 (2021)

50. Yang, S., Quan, Z., Nie, M., Yang, W.: Transpose: keypoint localization via transformer. In: IEEE/CVF International Conference on Computer Vision (ICCV) (2021)

51. Yao, G., et al.: One-shot face reenactment using appearance adaptive normalization. In: Proceedings of the AAAI Conference on Artificial Intelligence, vol. 35, pp. 3172–3180 (2021)

52. Yoon, J.S., Liu, L., Golyanik, V., Sarkar, K., Park, H.S., Theobalt, C.: Pose-guided human animation from a single image in the wild. In: Proceedings of the IEEE/CVF Conference on Computer Vision and Pattern Recognition (CVPR), pp. 15039–15048, June 2021

53. Yu, X., Rao, Y., Wang, Z., Liu, Z., Lu, J., Zhou, J.: Pointr: diverse point cloud completion with geometry-aware transformers. In: Proceedings of the IEEE/CVF International Conference on Computer Vision, pp. 12498–12507 (2021)

54. Zablotskaia, P., Siarohin, A., Zhao, B., Sigal, L.: DWNET: dense warp-based network for pose-guided human video generation. In: BMVC, p. 51 (2019)

55. Zakharov, E., Shysheya, A., Burkov, E., Lempitsky, V.: Few-shot adversarial learning of realistic neural talking head models. In: Proceedings of the IEEE/CVF International Conference on Computer Vision, pp. 9459–9468 (2019)

56. Zhang, J., Li, K., Lai, Y.K., Yang, J.: PISE: person image synthesis and editing with decoupled GAN. In: Proceedings of the IEEE/CVF Conference on Computer Vision and Pattern Recognition, pp. 7982–7990 (2021)

57. Zhao, L., Peng, X., Tian, Yu., Kapadia, M., Metaxas, D.: Learning to forecast and refine residual motion for image-to-video generation. In: Ferrari, V., Hebert, M., Sminchisescu, C., Weiss, Y. (eds.) ECCV 2018. LNCS, vol. 11219, pp. 403–419. Springer, Cham (2018). https://doi.org/10.1007/978-3-030-01267-0_24

58. Zheng, S., et al.: Rethinking semantic segmentation from a sequence-to-sequence perspective with transformers. In: Proceedings of the IEEE/CVF Conference on Computer Vision and Pattern Recognition, pp. 6881–6890 (2021)

59. Zhu, X., Su, W., Lu, L., Li, B., Wang, X., Dai, J.: Deformable detr: Deformable transformers for end-to-end object detection. In: International Conference on Learning Representations (2021)

60. Zhu, Z., Huang, T., Shi, B., Yu, M., Wang, B., Bai, X.: Progressive pose attention transfer for person image generation. In: Proceedings of the IEEE/CVF Conference on Computer Vision and Pattern Recognition, pp. 2347–2356 (2019)

NÜWA: Visual Synthesis Pre-training for Neural visUal World creAtion

Chenfei Wu[1]([✉]), Jian Liang[2], Lei Ji[1], Fan Yang[1], Yuejian Fang[2], Daxin Jiang[1], and Nan Duan[1]

[1] Microsoft Research Asia, Beijing, China
chewu@microsoft.com
[2] Peking University, Beijing, China

Abstract. This paper presents a unified multimodal pre-trained model called NÜWA that can generate new or manipulate existing visual data (i.e., images and videos) for various visual synthesis tasks. To cover language, image, and video at the same time for different scenarios, a 3D transformer encoder-decoder framework is designed, which can not only deal with videos as 3D data but also adapt to texts and images as 1D and 2D data, respectively. A 3D Nearby Attention (3DNA) mechanism is also proposed to consider the nature of the visual data and reduce the computational complexity. We evaluate NÜWA on 8 downstream tasks. Compared to several strong baselines, NÜWA achieves state-of-the-art results on text-to-image generation, text-to-video generation, video prediction, etc. Furthermore, it also shows surprisingly good zero-shot capabilities on text-guided image and video manipulation tasks.

1 Introduction

Nowadays, the Web is becoming more visual than ever before, as images and videos have become the new information carriers and have been used in many practical applications. With this background, visual synthesis is becoming a popular research topic, which aims to generate new or manipulate existing visual data (i.e., images and videos) for various visual scenarios.

Auto-regressive models [21,29,35,39] play an important role in visual synthesis tasks, due to their explicit density modeling and stable training compared with GANs [3,26,33,41]. Earlier visual auto-regressive models like PixelCNN [21], PixelRNN [35], Image Transformer [24], iGPT [4], and Video Transformer [38], performed visual synthesis in a "pixel-by-pixel" manner. However, due to their high computational cost on high-dimensional visual data, such methods can be applied to low-resolution images or videos only and are hard to scale up.

Recently, with the arise of VQ-VAE [22] as a discrete visual tokenization approach, efficient and large-scale pre-training can be applied to visual synthesis

---

C. Wu and J. Liang—Both authors contributed equally to this research.

Supplementary Information The online version contains supplementary material available at https://doi.org/10.1007/978-3-031-19787-1_41.

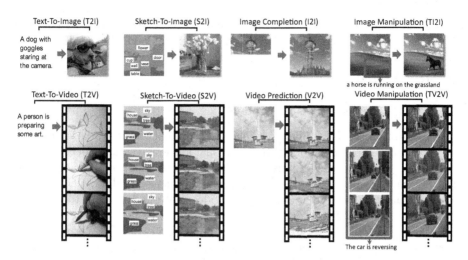

Fig. 1. Examples of 8 typical visual generation and manipulation tasks supported by the NÜWA model.

tasks for images (e.g., DALL-E [29] and CogView [7]) and videos (e.g., GODIVA [39]). To model the locality of images and videos, sparse attentions are commonly used to reduce computation and improve the performance. Although achieving great success, such solutions still have the following two limitations:

On the one hand, from the pre-training perspective, current works treat images and videos separately and focus on generating either of them. This limits the models to benefit from both image and video data.

On the other hand, from the model perspective, current works use block-sparse attention or axial-sparse attention as the pre-training backbone, both considering only part of visual locality. Block-sparse attention limits attention in a fixed 3D block and axial-sparse attentions limits attention in axes, both failed to fully model the locality of images and videos.

To handle the above issues, we propose NÜWA, with a 3D decoder to share information from both images and videos and a 3D Nearby-sparse Attention (3DNA) to model the full spatial and temporal locality. We verify NÜWA on 8 downstream visual synthesis, as shown in Fig. 1. The main contributions of this work are three-fold:

- We propose NÜWA, a general 3D transformer encoder-decoder framework, which covers language, image, and video at the same time for different visual synthesis tasks. It consists of an adaptive encoder that takes either text or visual sketch as input, and a decoder shared by 8 visual synthesis tasks.
- We propose a 3D Nearby Attention (3DNA) mechanism in the framework to consider the locality characteristic for both spatial and temporal axes. 3DNA not only reduces computational complexity but also improves the visual quality of the generated results.
- Compared to several strong baselines, NÜWA achieves state-of-the-art results on text-to-image generation, text-to-video generation, video prediction, etc.

Furthermore, NÜWA shows surprisingly good zero-shot capabilities not only on text-guided image manipulation, but also text-guided video manipulation.

2   Related Works

2.1   Visual Auto-regressive Models

The method proposed in this paper follows the line of visual synthesis research based on auto-regressive models. Earlier visual auto-regressive models [4, 21, 24, 35, 38] performed visual synthesis in a "pixel-by-pixel" manner. However, due to the high computational cost when modeling high-dimensional data, such methods can be applied to low-resolution images or videos only, and are hard to scale up.

Recently, VQ-VAE-based [22] visual auto-regressive models were proposed for visual synthesis tasks. By converting images into discrete visual tokens, such methods can conduct efficient and large-scale pre-training for text-to-image generation (e.g., DALL-E [29] and CogView [7]), text-to-video generation (e.g., GODIVA [39]), and video prediction (e.g., LVT [27] and VideoGPT [42]), with higher resolution of generated images or videos. However, none of these models was trained by images and videos together. But it is intuitive that these tasks can benefit from both types of visual data.

Compared to these works, NÜWA is a unified auto-regressive visual synthesis model that is pre-trained by the visual data covering both images and videos and can support various downstream tasks. We also verify the effectiveness of different pretraining tasks in Sect. 4.3. Besides, VQ-GAN [9] instead of VQ-VAE is used in NÜWA for visual tokenization, which, based on our experiment, can lead to better generation quality.

2.2   Visual Sparse Self-Attention

How to deal with the quadratic complexity issue brought by self-attention is another challenge, especially for tasks like high-resolution image synthesis or video synthesis.

Similar to NLP, sparse attention mechanisms have been explored to alleviate this issue for visual synthesis. [27, 38] split the visual data into different parts (or blocks) and then performed block-wise sparse attention for the synthesis tasks. However, such methods dealt with different blocks separately and did not model their relationships. [11, 29, 39] proposed to use axial-wise sparse attention in visual synthesis tasks, which conducts sparse attention along the axes of visual data representations. This mechanism makes training very efficient and is friendly to large-scale pre-trained models like DALL-E [29], CogView [7], and GODIVA [39]. However, the quality of generated visual contents could be harmed due to the limited contexts used in self-attention. [5, 24, 28] proposed to use local-wise sparse attention in visual synthesis tasks, which allows the models to see more contexts. But these works were for images only.

Compared to these works, NÜWA proposes a 3D nearby attention that extends the local-wise sparse attention to cover both images to videos. We also

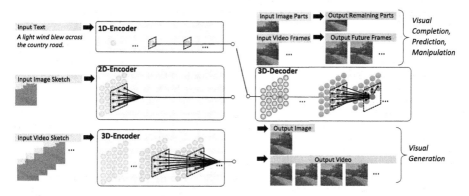

Fig. 2. Overview structure of NÜWA. It contains an adaptive encoder supporting different conditions and a pre-trained decoder benefiting from both image and video data. For image completion, video prediction, image manipulation, and video manipulation tasks, the input partial images or videos are fed to the decoder directly.

verify that local-wise sparse attention is superior to axial-wise sparse attention for visual generation in Sect. 4.3 (Fig. 2).

3   Method

3.1   3D Data Representation

To cover all texts, images, and videos or their sketches, we view all of them as tokens and define a unified 3D notation $X \in \mathbb{R}^{h \times w \times s \times d}$, where h and w denote the number of tokens in the spatial axis (height and width respectively), s denotes the number of tokens in the temporal axis, and d is the dimension of each token. In the following, we introduce how we get this unified representation for different modalities.

Texts are naturally discrete, and following Transformer [36], we use a lower-cased byte pair encoding (BPE) to tokenize and embed them into $\mathbb{R}^{1 \times 1 \times s \times d}$. We use placeholder 1 because the text has no spatial dimension.

Images are naturally continuous pixels. Input a raw image $I \in \mathbb{R}^{H \times W \times C}$ with height H, width W and channel C, VQ-VAE [22] trains a learnable codebook to build a bridge between raw continuous pixels and discrete tokens, as denoted in Eq. (1)~(2):

$$z_i = \arg\min_j ||E(I)_i - B_j||^2, \tag{1}$$

$$\hat{I} = G(B[z]), \tag{2}$$

where E is an encoder that encodes I into $h \times w$ grid features $E(I) \in \mathbb{R}^{h \times w \times d_B}$, $B \in \mathbb{R}^{N \times d_B}$ is a learnable codebook with N visual tokens, where each grid of $E(I)$ is searched to find the nearest token. The searched result $z \in \{0, 1, ..., N-1\}^{h \times w}$ are embedded by B and reconstructed back to $\hat{I}$ by a decoder G. The training loss of VQ-VAE can be written as Eq. (3):

$$L^V = ||I - \hat{I}||_2^2 + ||sg[E(I)] - B[z]||_2^2 + ||E(I) - sg[B[z]]||_2^2, \qquad (3)$$

where $||I - \hat{I}||_2^2$ strictly constraints the exact pixel match between I and $\hat{I}$, which limits the generalization ability of the model. Recently, VQ-GAN [9] enhanced VQ-VAE training by adding a perceptual loss and a GAN loss to ease the exact constraints between I and $\hat{I}$ and focus on high-level semantic matching, as denoted in Eq. (4)~(5):

$$L^P = ||CNN(I) - CNN(\hat{I})||_2^2, \qquad (4)$$

$$L^G = logD(I) + log(1 - D(\hat{I})). \qquad (5)$$

After the training of VQ-GAN, $B[z] \in \mathbb{R}^{h \times w \times 1 \times d}$ is finally used as the representation of images. We use placeholder 1 since images have no temporal dimensions.

Videos can be viewed as a temporal extension of images, and recent works like VideoGPT [42] and VideoGen [45] extend convolutions in the VQ-VAE encoder from 2D to 3D and train a video-specific representation. However, this fails to share a common codebook for both images and videos. In this paper, we show that simply using 2D VQ-GAN to encode each frame of a video can also generate temporal consistency videos and at the same time benefit from both image and video data. The resulting representation is denoted as $\mathbb{R}^{h \times w \times s \times d}$, where s denotes the number of frames.

For image sketches, we consider them as images with special channels. An image segmentation matrix $\mathbb{R}^{H \times W}$ with each value representing the class of a pixel can be viewed in a one-hot manner $\mathbb{R}^{H \times W \times C}$ where C is the number of segmentation classes. By training an additional VQ-GAN for image sketch, we finally get the embedded image representation $\mathbb{R}^{h \times w \times 1 \times d}$. Similarly, for video sketches, the representation is $R^{h \times w \times s \times d}$.

3.2   3D Nearby Self-Attention

In this section, we define a unified 3D Nearby Self-Attention (3DNA) module based on the previous 3D data representations, supporting both self-attention and cross-attention. We first give the definition of 3DNA in Eq. (6), and introduce detailed implementation in Eq. (7)–(11):

$$Y = 3DNA(X, C; W), \qquad (6)$$

where both $X \in \mathbb{R}^{h \times w \times s \times d^{in}}$ and $C \in \mathbb{R}^{h' \times w' \times s' \times d^{in}}$ are 3D representations introduced in Sec. 3.1. If $C = X$, 3DNA denotes the self-attention on target X and if $C \neq X$, 3DNA is cross-attention on target X conditioned on C. W denotes learnable weights.

We start to introduce 3DNA from a coordinate (i, j, k) under X. By a linear projection, the corresponding coordinate (i', j', k') under C is $\left(\lfloor i\frac{h'}{h} \rfloor, \lfloor j\frac{w'}{w} \rfloor, \lfloor k\frac{s'}{s} \rfloor \right)$. Then, the local neighborhood around (i', j', k') with a width, height and temporal extent $e^w, e^h, e^s \in \mathbb{R}^+$ is defined in Eq. (7),

$$N^{(i,j,k)} = \left\{ C_{abc} \Big| |a - i'| \leq e^h, |b - j'| \leq e^w, |c - k'| \leq e^s \right\}, \qquad (7)$$

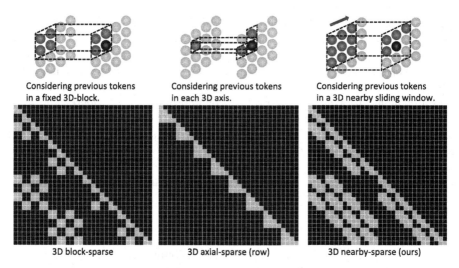

Fig. 3. Comparisons between different 3D sparse attentions. All samples assume that the size of the input 3D data is $4 \times 4 \times 2 = 32$. The illustrations in the upper part show which tokens (blue) need to be attended to generate the target token (orange). The matrices of the size 32×32 in the lower part show the attention masks in sparse attention (black denotes masked tokens). (Color figure online)

where $N^{(i,j,k)} \in \mathbb{R}^{e^h \times e^w \times e^s \times d^{in}}$ is a sub-tensor of condition C and consists of the corresponding nearby information that (i, j, k) needs to attend. With three learnable weights $W_Q, W_K, W_V \in \mathbb{R}^{d^{in} \times d^{out}}$, the output tensor for the position (i, j, k) is denoted in Eq. (8)~(11):

$$Q^{(i,j,k)} = XW^Q \tag{8}$$

$$K^{(i,j,k)} = N^{(i,j,k)}W^K \tag{9}$$

$$V^{(i,j,k)} = N^{(i,j,k)}W^V \tag{10}$$

$$y_{ijk} = softmax\left(\frac{(Q^{(i,j,k)})^\mathsf{T} K^{(i,j,k)}}{\sqrt{d^{in}}}\right) V^{(i,j,k)} \tag{11}$$

where the (i, j, k) position queries and collects corresponding nearby information in C. This also handles $C = X$, then (i, j, k) queries the nearby position of itself.

Figure 3 shows comparisons between different 3D sparse attentions. Assume we have 3D data with the size of $4 \times 4 \times 2$, the idea of 3D block-sparse attention is to split the 3D data into several fixed blocks and handle these blocks separately. There are many ways to split blocks, such as splitting in time, space, or both. The 3D block-sparse example in Fig. 3 considers the split of both time and space. The 3D data is divided into 4 parts, each has the size of $2 \times 2 \times 2$. To generate the orange token, 3D block-sparse attention considers previous tokens inside the fixed 3D block. Although 3D block-sparse attention considers both spatial and temporal axes, this spatial and temporal information is limited and fixed in the

3D block especially for the tokens along the edge of the 3D block. Only part of nearby information is considered since some nearby information outside the 3D block is invisible for tokens inside it. The idea of 3D axial-sparse attention is to consider previous tokens along the axis. Although 3D axis-sparse attention considers both spatial and temporal axes, this spatial and temporal information is limited along the axes. Only part of nearby information is considered and some nearby information that does not in the axis will not be considered in the 3D axis attention. In this paper, we propose a 3D nearby-sparse, which considers the full nearby information and dynamically generates the 3D nearby attention block for each token. The attention matrix also shows the evidence as the attended part (blue) for 3D nearby-sparse is more smooth than 3D block-sparse and 3D axial-sparse.

3.3   3D Encoder-Decoder

In this section, we introduce 3D encode-decoder built based on 3DNA. To generate a target $Y \in \mathbb{R}^{h \times w \times s \times d^{out}}$ under the condition of $C \in \mathbb{R}^{h' \times w' \times s' \times d^{in}}$, the positional encoding for both Y and C are updated by three different learnable vocabularies considering height, width, and temporal axis, respectively in Eq. (12)–(13):

$$Y_{ijk} := Y_{ijk} + P_i^h + P_j^w + P_k^s \tag{12}$$

$$C_{ijk} := C_{ijk} + P_i^{h'} + P_j^{w'} + P_k^{s'} \tag{13}$$

Then, the condition C is fed into an encoder with a stack of L 3DNA layers to model the self-attention interactions, with the lth layer denoted in Eq. (14):

$$C^{(l)} = 3DNA(C^{(l-1)}, C^{(l-1)}), \tag{14}$$

Similarly, the decoder is also a stack of L 3DNA layers. The decoder calculates both self-attention of generated results and cross-attention between generated results and conditions. The lth layer is denoted in Eq. (15).

$$Y_{ijk}^{(l)} = 3DNA(Y_{<i,<j,<k}^{(l-1)}, Y_{<i,<j,<k}^{(l-1)}) + 3DNA(Y_{<i,<j,<k}^{(l-1)}, C^{(L)}), \tag{15}$$

where $< i, < j, < k$ denote the generated tokens for now. The initial token $V_{0,0,0}^{(1)}$ is a special $< bos >$ token learned during the training phase.

3.4   Training Objective

We train our model on three tasks, Text-to-Image (T2I), Video Prediction (V2V) and Text-to-Video (T2V). The training objective for the three tasks are cross-entropys denoted as three parts in Eq. (16), respectively:

$$\mathcal{L} = -\sum_{t=1}^{h \times w} log\, p_\theta \left(y_t|y_{<t}, C^{text}; \theta\right) - \sum_{t=1}^{h \times w \times s} log\, p_\theta \left(y_t|y_{<t}, c; \theta\right) - \sum_{t=1}^{h \times w \times s} log\, p_\theta \left(y_t|y_{<t}, C^{text}; \theta\right) \tag{16}$$

For T2I and T2V tasks, C^{text} denotes text conditions. For the V2V task, since there is no text input, we instead get a constant 3D representation c of the special word "None". θ denotes the model parameters.

4 Experiments

Based on Sect. 3.4 we first pre-train NÜWA on three datasets: Conceptual Captions [16] for text-to-image (T2I) generation, which includes 2.9M text-image pairs, Moments in Time [20] for video prediction (V2V), which includes 727K videos, and VATEX dataset [37] for text-to-video (T2V) generation, which includes 241K text-video pairs. In the following, we first introduce implementation details in Sect. 4.1 and then compare NÜWA with state-of-the-art models in Sect. 4.2, and finally conduct ablation studies in Sect. 4.3 to study the impacts of different parts.

4.1 Implementation Details

In Sect. 3.1, we set the sizes of 3D representations for text, image, and video as follows. For text, the size of 3D representation is $1 \times 1 \times 77 \times 1280$. For image, the size of 3D representation is $21 \times 21 \times 1 \times 1280$. For video, the size of 3D representation is $21 \times 21 \times 10 \times 1280$, where we sample 10 frames from a video with 2.5 fps. Although the default visual resolution is 336×336, we pre-train different resolutions for a fair comparison with existing models. For the VQ-GAN model used for both images and videos, the size of grid feature $E(I)$ in Eq. (1) is 441×256, and the size of the codebook B is $12,288$.

Different sparse extents are used for different modalities in Sec. 3.2. For text, we set $(e^w, e^h, e^s) = (1, 1, \infty)$, where ∞ denotes that the full text is always used in attention. For image and image sketches, $(e^w, e^h, e^s) = (3, 3, 1)$. For video and video sketches, $(e^w, e^h, e^s) = (3, 3, 3)$.

We pre-train on 64 A100 GPUs for two weeks with the layer L in Eq. (14) set to 24, an Adam [13] optimizer with a learning rate of 1e−3, a batch size of 128, and warm-up 5% of a total of 50M steps. The final pre-trained model has a total number of 870M parameters.

4.2 Comparison with State-of-the-art

Text-to-Image (T2I) Fine-Tuning: We compare NÜWA on the MSCOCO [16] dataset quantitatively in Table 1 and qualitatively in Fig. 4. Following DALL-E [29], we use k blurred FID score (FID-k) and Inception Score (IS) [31] to evaluate the quality and variety respectively, and following GODIVA [39], we use CLIPSIM metric, which incorporates a CLIP [25] model to calculate the semantic similarity between input text and the generated image. For a fair comparison, all the models use the resolution of 256×256. We generate 60 images for each text and select the best one by CLIP [25]. In Table 1, NÜWA significantly outperforms CogView [7] with FID-0 of 12.9 and CLIPSIM of 0.3429. Although XMC-GAN [44] reports a significant FID score of 9.3, we find NÜWA generates more realistic images compared with the exact same samples in XMC-GAN's paper (see Fig. 4). Especially in the last example, the boy's face is clear and the balloons are correctly generated.

Fig. 4. Qualitative comparison with state-of-the-art models for Text-to-Image (T2I) task on MSCOCO dataset.

Table 1. Qualitative comparison with the state-of-the-art models for Text-to-Image (T2I) task on the MSCOCO (256 × 256) dataset.

Model	FID-0↓	FID-1	FID-2	FID-4	FID-8	IS↑	CLIPSIM↑
AttnGAN [41]	35.2	44.0	72.0	108.0	100.0	23.3	0.2772
DM-GAN [46]	26.0	39.0	73.0	119.0	112.3	**32.2**	0.2838
DF-GAN [32]	26.0	33.8	55.9	91.0	97.0	18.7	0.2928
DALL-E [29]	27.5	28.0	45.5	83.5	85.0	17.9	–
CogView [7]	27.1	19.4	**13.9**	19.4	23.6	18.2	0.3325
XMC-GAN [44]	**9.3**	–	–	–	–	30.5	–
NÜWA(scratch)	*Full Attention*						
	17.1	16.5	16.3	18.5	20.9	22.7	0.3257
NÜWA(scratch)	*Axial Attention*						
	18.7	18.4	19.2	20.3	21.3	22.8	0.3253
NÜWA(scratch)	*3D Nearby Attention (ours)*						
	16.9	15.6	16.5	18.9	**20.2**	23.1	0.3276
NÜWA(finetune)	*Pretrain on CC Dataset*						
	14.2	15.2	16.9	20.5	24.7	25.8	0.3424
NÜWA(finetune)	*Pretrain on CC, Moments and Vatex Dataset*						
	12.9	**13.8**	15.7	19.3	24	27.2	**0.3429**
NÜWA(zeroshot)	*Pretrain on CC, Moments and Vatex Dataset*						
	22.6	18.6	17.2	**17.4**	24.8	24.5	0.3331

To validate the effectiveness of our proposed 3DNA, we train NÜWA from scratch and Table 1 shows 3DNA outperforms axial attentions and full attentions. To validate the effectiveness of joint pretraining both images and videos, we pre-train NÜWA on pure image dataset (CC) and mixed image and video dataset

(CC, Moments, Vatex). Table 1 shows Text-to-Video task also helps Text-to-Image task. This is interesting as videos provides external motion knowledge to help the model better build the connection between text and image.

Fig. 5. Quantitative comparison with state-of-the-art models for Text-to-Video (T2V) task on Kinetics dataset.

Table 2. Quantitative comparison with state-of-the-art models for Text-to-Video (T2V) task on Kinetics dataset.

Model	Acc↑	FID-img↓	FID-vid↓	CLIPSIM↑
T2V (64 × 64) [15]	42.6	82.13	14.65	0.2853
SC (128 × 128) [1]	74.7	33.51	7.34	0.2915
TFGAN (128 × 128) [1]	76.2	31.76	7.19	0.2961
NÜWA(scratch)	77.4	29.32	7.08	0.3007
NÜWA(finetune)	**77.9**	**28.46**	**7.05**	**0.3012**

Table 3. Quantitative comparison with state-of-the-art models for Video Prediction (V2V) task on BAIR (64 × 64) dataset.

Model	Cond	FVD↓
DVD-GAN-FP [6]	1	110
Video Transformer (S) [38]	1	106 ± 3
TriVD-GAN-FP [17]	1	103
CCVS [19]	1	99 ± 2
Video Transformer (L) [38]	1	94 ± 2
NÜWA(scratch)	1	87.6
NÜWA(finetune)	1	**86.9**

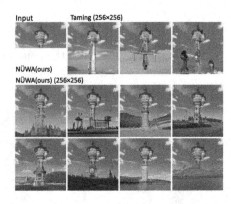

Fig. 6. Quantitative comparison with state-of-the-art models for Sketch-to-Image (S2I) task on MSCOCO-Stuff.

Fig. 7. Qualitative comparison with the state-of-the-art model for image completion (I2I) task in a zero-shot manner.

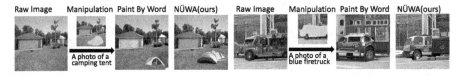

Fig. 8. Quantitative comparison with state-of-the-art models for text-guided image manipulation (TI2I) in a zero-shot manner.

Text-to-Video (T2V) Fine-Tuning: We compare NÜWA on the Kinetics [12] dataset quantitatively in Table 2 and qualitatively in Fig. 5. Following TFGAN [1], we evaluate the visual quality on FID-img and FID-vid metrics and semantic consistency on the accuracy of the label of generated video. To ensure NÜWA is trained with the same size information of Kinetics as other methods, images are first bilinear interpolated into 128×128 before resized into 336×336. As shown in Table 2, NÜWA achieves the best performance. In Fig. 5, we also show the strong zero-shot ability for generating unseen text, such as "playing golf at swimming pool" or "running on the sea".

Video Prediction (V2V) Fine-Tuning: We compare NÜWA on BAIR Robot Pushing [8] dataset quantitatively in Table 3. Cond. denotes the number of frames given to predict future frames. For a fair comparison, all the models use 64×64 resolutions. Although given only one frame as condition (Cond.), NÜWA still significantly pushes the state-of-the-art FVD [34] score from 94 ± 2 to 86.9.

Sketch-to-Image (S2I) Fine-Tuning: We compare NÜWA on MSCOCO stuff [16] qualitatively in Fig. 6. NÜWA generates realistic buses of great varieties compared with Taming-Transformers [9] and SPADE [23]. Even the reflection of the bus window is clearly visible.

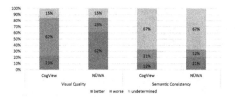
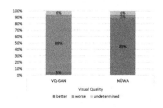

Fig. 9. Human evaluation on MSCOCO dataset for text-to-image (T2I) task.

Fig. 10. Human evaluation on MSCOCO dataset for image completion (I2I) task.

Image Completion (I2I) Zero-Shot Evaluation: We compare NÜWA in a zero-shot manner qualitatively in Fig. 7. Given the top half of the tower, compared with Taming Transformers [9], NÜWA shows richer imagination of what could be for the lower half of the tower, including buildings, lakes, flowers, grass, trees, mountains, etc.

Text-Guided Image Manipulation (TI2I) Zero-Shot Evaluation: We compare NÜWA in a zero-shot manner qualitatively in Fig. 8. Compared with Paint By Word [2], NÜWA shows strong manipulation ability, generating high-quality text-consistent results while not changing other parts of the image. For example, in the third row, the blue firetruck generated by NÜWA is more realistic, while the behind buildings show no change. This is benefited from real-world visual patterns learned by multi-task pre-training on various visual tasks. Another advantage is the inference speed of NÜWA, practically 50 s to generate an image, while Paint By Words requires additional training during inference, and takes about 300 s to converge.

Sketch-to-Video (S2V) fine-tuning and **Text-Guided Video Manipulation (TV2V) Zero-Shot Evaluation:** Since there are no current benchmarks for these two tasks, we thus arrange them in Ablation Study in Sect. 4.3.

Human Evaluation. Figure 9 presents human comparison results between CogView [7] and our NÜWA on the MSCOCO dataset for Text-to-Image (T2I) task. We randomly selected 2000 texts and ask annotators to compare the generated results between two models including both visual quality and semantic consistency. The annotators are asked to choose among three options: better, worse, or undetermined. NÜWA achieves 62% votes for visual quality and 21% votes for semantic consistency. Figure 10 shows another human comparison between VQ-GAN [9] and our NÜWA model on the MSCOCO dataset for the Image Completion (I2I) task.

4.3 Ablation Study

The above part of Table 4 shows the effectiveness of different VQ-VAE (VQ-GAN) settings. We experiment on ImageNet [30] and OpenImages [14]. R

Table 4. Effectiveness of different VQ-VAE (VQ-GAN) settings.

Model	$R \to D$	Rate	SSIM	FID
Trained on ImageNet Dataset				
VQ-VAE	$256^2 \to 16^2$	F16	0.7026	13.3
VQ-GAN	$256^2 \to 16^2$	F16	0.7105	6.04
VQ-GAN	$256^2 \to 32^2$	F8	0.8285	2.03
VQ-GAN	$336^2 \to 21^2$	F16	0.7213	4.79
Trained on OpenImages Dataset				
VQ-GAN	$336^2 \to 21^2$	F16	0.7527	4.31

Model	$R \to D$	Rate	PA	FWIoU
Trained on COCO-Stuff Dataset				
V.G-Seg	$336^2 \to 21^2$	F16	96.82	93.91
Trained on VSPW Dataset				
V.G-Seg	$336^2 \to 21^2$	F16	95.36	91.82

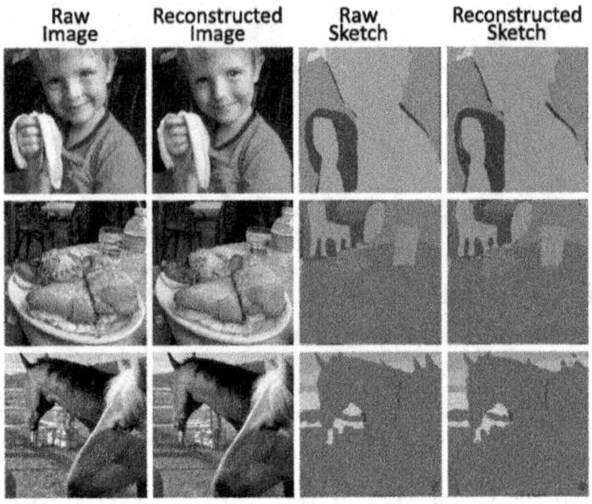

Fig. 11. Reconstruction samples of VQ-GAN and VQ-GAN-Seg.

Table 5. Effectiveness of multi-task pre-training for Text-to-Video (T2V) generation task on MSRVTT dataset.

Model	Pre-trained tasks	FID-vid↓	CLIPSIM↑
NÜWA-TV	T2V	52.98	0.2314
NÜWA-TV-TI	T2V+T2I	53.92	0.2379
NÜWA-TV-VV	T2V+V2V	51.81	0.2335
NÜWA	T2V+T2I+V2V	**47.68**	**0.2439**

Table 6. Effectiveness of 3D nearby attention for Sketch-to-Video (S2V) task on VSPW dataset.

Model	Encoder	Decoder	FID-vid↓	DetectedPA↑
NÜWA-FF	Full	Full	35.21	0.5220
NÜWA-NF	Nearby	Full	33.63	0.5357
NÜWA-FN	Full	Nearby	32.06	0.5438
NÜWA-AA	Axis	Axis	29.18	0.5957
NÜWA	Nearby	Nearby	**27.79**	**0.6085**

denotes raw resolution, D denotes the number of discrete tokens. The compression rate is denoted as Fx, where x is the quotient of $\sqrt{R}$ divided by $\sqrt{D}$. Comparing the first two rows in Table 4, VQ-GAN shows significantly better Fréchet Inception Distance (FID) [10] and Structural Similarity Matrix (SSIM) scores than VQ-VAE. Comparing Row 2–3, we find that the number of discrete tokens is the key factor leading to higher visual quality instead of compress rate. Although Row 2 and Row 4 have the same compression rate F16, they have different FID scores of 6.04 and 4.79. So what matters is not only how much we compress the original image, but also how many discrete tokens are used for representing an image. This is in line with cognitive logic, it's too ambiguous to represent human faces with just one token. And practically, we find that 16^2 discrete tokens usually lead to poor performance, especially for human faces, and 32^2 tokens show the best performance. However, more discrete tokens mean more computing, especially for videos. We finally use a trade-off version for our pre-training: 21^2 tokens. By training on the Open Images dataset, we further improve the FID score of the 21^2 version from 4.79 to 4.31.

The below part of Table 4 shows the performance of VQ-GAN for sketches. VQ-GAN-Seg on MSCOCO [16] is trained for Sketch-to-Image (S2I) task and VQ-GAN-Seg on VSPW [18] is trained for Sketch-to-Video (S2V) task. All the

Fig. 12. Samples of different manipulations on the same video.

above backbone shows good performance in Pixel Accuracy (PA) and Frequency Weighted Intersection over Union (FWIoU), which shows a good quality of 3D sketch representation used in our model. Figure 11 also shows some reconstructed samples of 336 × 336 images and sketches.

Table 5 shows the effectiveness of multi-task pre-training for the Text-to-Video (T2V) generation task. We study on a challenging dataset, MSR-VTT [40], with natural descriptions and real-world videos. Compared with training only on a single T2V task (Row 1), training on both T2V and T2I (Row 2) improves the CLIPSIM from 0.2314 to 0.2379. This is because T2I helps to build a connection between text and image, and thus helpful for the semantic consistency of the T2V task. In contrast, training on both T2V and V2V (Row 3) improves the FVD score from 52.98 to 51.81. This is because V2V helps to learn a common unconditional video pattern, and is thus helpful for the visual quality of the T2V task. The default setting, training on three tasks, achieves the best performance.

Table 6 shows the effectiveness of 3D nearby attention for the Sketch-to-Video (S2V) task on the VSPW [18] dataset. We study on the S2V task because both the encoder and decoder of this task are fed with 3D video data. To evaluate the semantic consistency for S2V, we propose a new metric called Detected PA, which uses a semantic segmentation model [43] to segment each frame of the generated video and then calculate the pixel accuracy between the generated segments and input video sketch. The default NÜWA setting with both nearby encoder and nearby decoder, achieves the best FID-vid and Detected PA. The performance drops if either encoder or decoder is replaced by full attention, showing that focusing on nearby conditions and nearby generated results is better than simply considering all the information.

We compare nearby-sparse and axial-sparse in two-folds. Firstly, the computational complexity of nearby-sparse is $O\left((hws)\left(e^h e^w e^s\right)\right)$ and axis-sparse attention is $O\left((hws)(h+w+s)\right)$. For generating long videos (larger s), nearby-sparse will be more computational efficient. Secondly, nearby-sparse has better performance because it attends to "nearby" locations containing interactions between both spatial and temporal axes, while axis-sparse handles different axis separately and only consider interactions on the same axis.

Figure 12 shows a new task "Text-Guided Video Manipulation (TV2V)" proposed in this paper. TV2V aims to change the future of a video starting from a selected frame guided by text. All samples start to change the future of the video from the second frame. The first row shows the original video frames, where a diver is swimming in the water. After feeding "The diver is swimming to the surface" into NÜWA's encoder and providing the first video frame, NÜWA successfully generates a video with the diver swimming to the surface in the second row. The third row shows another successful sample that lets the diver swim to the bottom. What if we want the diver flying to the sky? The fourth row shows that NÜWA can make it as well, where the diver is flying upward, like a rocket.

5 Conclusion

In this paper, we present NÜWA as a unified pre-trained model that can generate new or manipulate existing images and videos for 8 visual synthesis tasks. Several contributions are made here, including (1) a general 3D encoder-decoder framework covering texts, images, and videos; (2) a nearby-sparse attention mechanism that considers the nearby characteristic of both spatial and temporal axes; (3) comprehensive experiments on 8 synthesis tasks. This is our first step towards building an AI platform to enable visual world and help creators.

Acknowledgements. The paper is supported by the National Key Research and Development Project (Grant No. 2020AAA0106600).

References

1. Balaji, Y., Min, M.R., Bai, B., Chellappa, R., Graf, H.P.: Conditional GAN with discriminative filter generation for text-to-video synthesis. In: IJCAI, pp. 1995–2001 (2019)
2. Bau, D., et al.: Paint by word. arXiv preprint arXiv:2103.10951 (2021)
3. Brock, A., Donahue, J., Simonyan, K.: Large scale GAN training for high fidelity natural image synthesis. arXiv:1809.11096 [cs, stat], February 2019
4. Chen, M., et al.: Generative pretraining from pixels. In: International Conference on Machine Learning, pp. 1691–1703. PMLR (2020)
5. Child, R., Gray, S., Radford, A., Sutskever, I.: Generating long sequences with sparse transformers. arXiv preprint arXiv:1904.10509 (2019)
6. Clark, K., Khandelwal, U., Levy, O., Manning, C.D.: What does bert look at? an analysis of bert's attention. arXiv preprint arXiv:1906.04341 (2019)
7. Ding, M., et al.: CogView: mastering text-to-image generation via transformers. arXiv:2105.13290 [cs], May 2021
8. Ebert, F., Finn, C., Lee, A.X., Levine, S.: Self-supervised visual planning with temporal skip connections. In: CoRL, pp. 344–356 (2017)
9. Esser, P., Rombach, R., Ommer, B.: Taming transformers for high-resolution image synthesis. arXiv:2012.09841 [cs], June 2021
10. Heusel, M., Ramsauer, H., Unterthiner, T., Nessler, B., Hochreiter, S.: Gans trained by a two time-scale update rule converge to a local nash equilibrium. Adv. Neural Inf. Process. Syst. **30** (2017)

11. Ho, J., Kalchbrenner, N., Weissenborn, D., Salimans, T.: Axial attention in multi-dimensional transformers. arXiv preprint arXiv:1912.12180 (2019)
12. Kay, W., et al.: The kinetics human action video dataset. arXiv preprint arXiv:1705.06950 (2017)
13. Kingma, D., Ba, J.: Adam: a method for stochastic optimization. arXiv preprint arXiv:1412.6980 (2014)
14. Kuznetsova, A., et al.: The open images dataset v4. Int. J. Comput. Vis. **128**(7), 1956–1981 (2020)
15. Li, Y., Min, M., Shen, D., Carlson, D., Carin, L.: Video generation from text. In: Proceedings of the AAAI Conference on Artificial Intelligence, vol. 32 (2018)
16. Lin, T.-Y., et al.: Microsoft COCO: common objects in context. In: Fleet, D., Pajdla, T., Schiele, B., Tuytelaars, T. (eds.) ECCV 2014. LNCS, vol. 8693, pp. 740–755. Springer, Cham (2014). https://doi.org/10.1007/978-3-319-10602-1_48
17. Luc, P., et al.: Transformation-based adversarial video prediction on large-scale data. arXiv preprint arXiv:2003.04035 (2020)
18. Miao, J., Wei, Y., Wu, Y., Liang, C., Li, G., Yang, Y.: VSPW: a large-scale dataset for video scene parsing in the wild. In: Proceedings of the IEEE/CVF Conference on Computer Vision and Pattern Recognition, pp. 4133–4143 (2021)
19. Moing, G.L., Ponce, J., Schmid, C.: CCVS: context-aware controllable video synthesis. arXiv preprint arXiv:2107.08037 (2021)
20. Monfort, M., et al.: Moments in time dataset: one million videos for event understanding. IEEE Trans. Pattern Anal. Mach. Intell. **42**(2), 502–508 (2019)
21. van den Oord, A., Kalchbrenner, N., Vinyals, O., Espeholt, L., Graves, A., Kavukcuoglu, K.: Conditional image generation with pixelcnn decoders. arXiv preprint arXiv:1606.05328 (2016)
22. van den Oord, A., Vinyals, O., Kavukcuoglu, K.: Neural discrete representation learning. arXiv preprint arXiv:1711.00937 (2017)
23. Park, T., Liu, M.Y., Wang, T.C., Zhu, J.Y.: Semantic image synthesis with spatially-adaptive normalization. In: Proceedings of the IEEE/CVF Conference on Computer Vision and Pattern Recognition, pp. 2337–2346 (2019)
24. Parmar, N., et al.: Image transformer. arXiv preprint arXiv:1802.05751 (2018)
25. Radford, A., et al..: Learning transferable visual models from natural language supervision. arXiv:2103.00020 [cs], February 2021
26. Radford, A., Metz, L., Chintala, S.: Unsupervised representation learning with deep convolutional generative adversarial networks. arXiv preprint arXiv:1511.06434 (2015)
27. Rakhimov, R., Volkhonskiy, D., Artemov, A., Zorin, D., Burnaev, E.: Latent video transformer. arXiv preprint arXiv:2006.10704 (2020)
28. Ramachandran, P., Parmar, N., Vaswani, A., Bello, I., Levskaya, A., Shlens, J.: Stand-alone self-attention in vision models. arXiv preprint arXiv:1906.05909 (2019)
29. Ramesh, A., et al.: Zero-shot text-to-image generation. arXiv:2102.12092 [cs], February 2021
30. Russakovsky, O., et al.: ImageNet large scale visual recognition challenge. arXiv:1409.0575 [cs], January 2015
31. Salimans, T., Goodfellow, I., Zaremba, W., Cheung, V., Radford, A., Chen, X.: Improved techniques for training GANs. Adv. Neural Inf. Process. Syst. **29**, 2234–2242 (2016)
32. Tao, M., et al.: Df-gan: deep fusion generative adversarial networks for text-to-image synthesis. arXiv preprint arXiv:2008.05865 (2020)

33. Tulyakov, S., Liu, M.Y., Yang, X., Kautz, J.: Mocogan: decomposing motion and content for video generation. In: Proceedings of the IEEE Conference on Computer Vision and Pattern Recognition, pp. 1526–1535 (2018)
34. Unterthiner, T., van Steenkiste, S., Kurach, K., Marinier, R., Michalski, M., Gelly, S.: Towards accurate generative models of video: a new metric & challenges. arXiv preprint arXiv:1812.01717 (2018)
35. Van Oord, A., Kalchbrenner, N., Kavukcuoglu, K.: Pixel recurrent neural networks. In: International Conference on Machine Learning, pp. 1747–1756. PMLR (2016)
36. Vaswani, A., et al.: Attention is all you need. In: Advances in Neural Information Processing Systems, pp. 5998–6008 (2017)
37. Wang, X., Wu, J., Chen, J., Li, L., Wang, Y.F., Wang, W.Y.: Vatex: a large-scale, high-quality multilingual dataset for video-and-language research. In: Proceedings of the IEEE/CVF International Conference on Computer Vision, pp. 4581–4591 (2019)
38. Weissenborn, D., Täckström, O., Uszkoreit, J.: Scaling autoregressive video models. In: ICLR (2020)
39. Wu, C., et al.: GODIVA: generating open-DomaIn videos from nAtural descriptions. arXiv:2104.14806 [cs], April 2021
40. Xu, J., Mei, T., Yao, T., Rui, Y.: Msr-vtt: a large video description dataset for bridging video and language. In: Proceedings of the IEEE Conference on Computer Vision and Pattern Recognition, pp. 5288–5296 (2016)
41. Xu, T., Zhang, P., Huang, Q., Zhang, H., Gan, Z., Huang, X., He, X.: Attngan: fine-grained text to image generation with attentional generative adversarial networks. In: Proceedings of the IEEE Conference on Computer Vision and Pattern Recognition, pp. 1316–1324 (2018)
42. Yan, W., Zhang, Y., Abbeel, P., Srinivas, A.: VideoGPT: video generation using VQ-VAE and transformers. arXiv preprint arXiv:2104.10157 (2021)
43. Yuan, Y., Chen, X., Wang, J.: Object-contextual representations for semantic segmentation. In: Vedaldi, A., Bischof, H., Brox, T., Frahm, J.-M. (eds.) ECCV 2020. LNCS, vol. 12351, pp. 173–190. Springer, Cham (2020). https://doi.org/10.1007/978-3-030-58539-6_11
44. Zhang, H., Koh, J.Y., Baldridge, J., Lee, H., Yang, Y.: Cross-modal contrastive learning for text-to-image generation. In: Proceedings of the IEEE/CVF Conference on Computer Vision and Pattern Recognition, pp. 833–842 (2021)
45. Zhang, Y., Yan, W., Abbeel, P., Srinivas, A.: VideoGen: generative modeling of videos using VQ-VAE and transformers, September 2020
46. Zhu, M., Pan, P., Chen, W., Yang, Y.: Dm-gan: dynamic memory generative adversarial networks for text-to-image synthesis. In: Proceedings of the IEEE/CVF Conference on Computer Vision and Pattern Recognition, pp. 5802–5810 (2019)

EleGANt: Exquisite and Locally Editable GAN for Makeup Transfer

Chenyu Yang[1] , Wanrong He[1] , Yingqing Xu[1(✉)] , and Yang Gao[1,2(✉)]

[1] Tsinghua University, Beijing, China
{yangcy19,hwr19}@mails.tsinghua.edu.cn,
{yqxu,gaoyangiiis}@tsinghua.edu.cn
[2] Shanghai Qi Zhi Institute, Shanghai, China

Abstract. Most existing methods view makeup transfer as transferring color distributions of different facial regions and ignore details such as eye shadows and blushes. Besides, they only achieve controllable transfer within predefined fixed regions. This paper emphasizes the transfer of makeup details and steps towards more flexible controls. To this end, we propose Exquisite and locally editable GAN for makeup transfer (EleGANt). It encodes facial attributes into pyramidal feature maps to preserves high-frequency information. It uses attention to extract makeup features from the reference and adapt them to the source face, and we introduce a novel Sow-Attention Module that applies attention within shifted overlapped windows to reduce the computational cost. Moreover, EleGANt is the first to achieve customized local editing within arbitrary areas by corresponding editing on the feature maps. Extensive experiments demonstrate that EleGANt generates realistic makeup faces with exquisite details and achieves state-of-the-art performance. The code is available at https://github.com/Chenyu-Yang-2000/EleGANt.

Keywords: Makeup transfer · Sow-attention · GAN controls

1 Introduction

Makeup transfer aims to transfer the makeup from an specific reference image to a source image. It has tremendous value in practical scenarios, for instance, cosmetics try-on and marketing. The goal of makeup transfer is two-fold: (1) Precisely transferring the makeup attributes from the reference to the source. The attributes include low-frequency color features and high-frequency details such as the brushes of eye shadow and blushes on the cheek. (2) Preserving the identity of the source, involving shapes, illuminations, and even subtle wrinkles. Controllable transfer further meets the requirement in practice: it allows users to design customized makeup according to their preferences.

Supplementary Information The online version contains supplementary material available at https://doi.org/10.1007/978-3-031-19787-1_42.

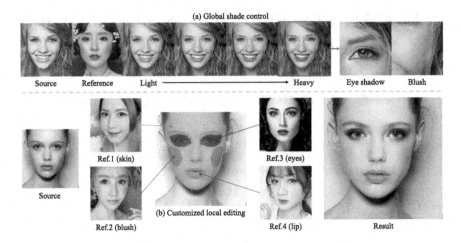

Fig. 1. Our proposed EleGANt generates makeup faces with exquisite details. It supports flexible control such as (a) global shade control and (b) customized local editing.

Deep learning approaches, especially GAN-based models [3,8,12,18,22], have been widely employed in this task. They mainly adopt the CycleGAN [44] framework that is trained on unpaired non-makeup and with-makeup images, with extra supervision, e.g., a makeup loss term [3,22,32], to guide the reconstruction of a specific makeup on the source face. Notwithstanding the demonstrated success, existing approaches mainly view makeups as color distributions and largely ignore the spatial, high-frequency information about details. Some methods [12,22] cannot tackle the pose misalignment between the two faces, while others represent makeups by matrices of limited size [18] or 1D-vectors [8], resulting in high-frequency attributes being smoothed. Meanwhile, the commonly used objective, histogram matching [22], only imposes constraints on color distributions without incorporating any spatial information. Besides, existing controllable models [8,18] only achieve editing the makeup within some fixed regions, for instance, skin, lip, and eyes, instead of arbitrary customized regions.

To address these issues, we propose Exquisite and locally editable GAN for makeup transfer (EleGANt). On one hand, we focus on high-frequency information to synthesize makeup faces with rich and delicate details. EleGANt encodes facial attributes of different frequencies into feature maps of a pyramid structure. To tackle misaligned head poses, it uses QKV-attention to extract makeup features from the reference and adapt them to the source face by pixel-wise correspondence. We employ a high-resolution feature map to preserve high-frequency attributes and further propose a novel Sow-Attention Module to reduce the computational cost. It computes attention efficiently within local windows and uses shifted overlapped windowing schemes to ensure the continuity of the output. The network is trained with a newly designed pseudo ground truth which comprises both color and spatial information. On the other hand, the high-res makeup feature maps support precise editing to control the makeup style and shade within arbitrary customized areas. To our best knowledge, EleGANt is the first makeup transfer network to achieve this free-style local editing.

Figure 1 exhibits the great capability and controllability of EleGANt. It generates realistic makeup images with high-fidelity colors and high-quality details. Global shade control (Fig. 1(a)) and customized local editing (Fig. 1(b)) can be realized by manipulating the makeup feature maps. Extensive experiments demonstrate the superiority of EleGANt compared with the existing methods.

The contributions of this paper can be summarized as follows:

- We propose EleGANt, a fully automatic makeup transfer network with the most flexible control among existing methods. To our best knowledge, it is the first to achieve customized local editing of makeup style and shade.
- EleGANt uses a pyramid structure with a high-resolution feature map to preserve high-frequency makeup features beyond color distributions. It achieves state-of-the-art performance, especially in processing makeup details.
- A novel Sow-Attention Module that computes attention within shifted overlapped windows is introduced, which guarantees the continuity of the output and reduces the computational cost for high-resolution inputs.

2   Related Work

2.1   Makeup Transfer

Makeup transfer has been studied in computer vision for a decade. Traditional methods [13,21,24,30,40] utilized image processing techniques. Later, Cycle-GAN [44] and its variants [2,6] were widely used for image-to-image translation tasks such as facial attribute transfer. However, these methods focus on domain-level rather than instance-level transfer, and they do not well maintain some facial attributes, e.g., shape and pose, that keep unchanged during makeup.

Inspired by the successes in GANs, makeup transfer was formulated as an asymmetric domain translation problem in PairedCycleGAN [3]. They also employed an additional discriminator to guide makeup transfer with pseudo transferred images. BeautyGAN [22] introduced a dual input/output GAN for simultaneous makeup transfer and removal and a color histogram matching loss for instance-level makeup transfer. BeautyGlow [5] utilized the Glow framework to disentangle the latent features into makeup and non-makeup components. LADN [12] leveraged multiple and overlapping local discriminators to ensure local details consistency. PSGAN [18] and FAT [32] proposed to use attention mechanism to handle misaligned facial poses and expressions. Lately, SCGAN [8] attempted to eliminate the spatial misalignment problem by encoding makeup styles into component-wise style-codes.

However, the existing methods have limitations in processing makeup details: some methods [12,22] cannot tackle the pose misalignment; [8] encodes the makeups into 1D-vectors, thus discarding a large proportion of spatial information; the high cost of pixel-wise attention in [18,32] limited the size of feature maps then harmed the preservation of details. EleGANt surpasses these approaches by using a high-res feature map and an efficient Sow-Attention Module. Besides, unlike previous models [8,18] that can only adjust the makeup in a fixed set of regions, our EleGANt supports customized local editing in arbitrary regions.

2.2 Style Transfer

Style transfer can be regarded as a general form of makeup transfer, and it has been investigated extensively since the rise of deep convolutional neural networks [10,19,23,26]. However, these methods either require a time-consuming optimization process or can only transfer a fixed set of styles. Then [16] proposed adaptive instance normalization (AdaIN) that matched the mean and variance of the content features with those of the style features and achieved arbitrary style transfer. Since style transfer methods do not consider the face-specific semantic correspondence and lack local manipulation and controllability, even the state-of-the-art algorithms [1,20,33] cannot fit makeup transfer applications.

2.3 Attention Mechanism

Attention mechanism has been widely used in the computer vision area. Early works [35,39] simply apply it to images where each pixel attends to every other pixel, but they do not scale to large input sizes due to a quadratic cost in the number of pixels. Many variants have been tried so far to make attention more efficient and applicable to high-resolution images, e.g., tokenization on patches [9,41], attention in local regions [7,25,27,43], and pyramid architecture [4,14,34]. Since these modules are plugged into discriminative networks for image representations, they are not exactly suitable for generative tasks. Meanwhile, sparse attention such as block-wise [28,36], axial-wise [15,37], and nearby attention [38] has been introduced into visual synthesis models, but neither global nor regional attributes of the makeup can be completely encoded due to the sparsity.

For makeup transfer, naive pixel-wise attention employed by [18,32] suffers from significant computational overhead and is hard to scale to inputs of larger size. To address this problem, EleGANt uses a novel Sow-Attention Module that performs attention within shifted overlapped windows.

3 Methodology

3.1 Formulation

Let X and Y be the non-makeup image domain and the makeup image domain, where $\{x^n\}_{n=1,\ldots,N}, x^n \in X$ and $\{y^m\}_{m=1,\ldots,M}, y^m \in Y$ denote the examples of two domains respectively. We assume no paired data is available, i.e., the non-makeup and makeup images have different identities. Our proposed EleGANt aims to learn a transfer function $\mathcal{G}$: given a source image x and a reference image y, $\hat{x} = \mathcal{G}(x, y)$, where $\hat{x}$ is the transferred image with the makeup style of y and the face identity of x.

3.2   Network Architecture

Overall. The architecture of EleGANt is shown in Fig. 2, which consists of three components: (1) *Facial Attribute Encoder.* FAEnc encodes the facial attributes into feature maps of pyramid structure. High-res feature maps X_H, Y_H mainly contain high-frequency information about sharp edges and details, while low-res ones X_L, Y_L contain more low-frequency information related to colors and shadows. (2) *Makeup Transfer Module.* In MTM, an Attention Module is applied to the low-res feature maps and yields a low-res makeup feature map Γ_L, while a Sow-Attention Module is for the high-res one Γ_H. These two modules extract makeup features of different frequencies and utilize attention to make them spatially aligned with the source face to tackle the misalignment between the two faces. (3) *Makeup Apply Decoder.* MADec applies the two makeup feature maps Γ_L and Γ_H to corresponding feature maps of the source respectively by element-wise multiplication and generates the final result.

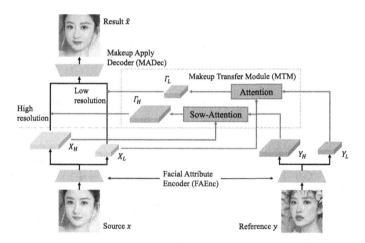

Fig. 2. Overall structure of our proposed EleGANt. Facial Attribute Encoder (FAEnc) constructs pyramidal feature maps. Makeup Transfer Module (MTM) yields low-res and high-res makeup feature maps, Γ_L and Γ_H, by Attention and Sow-Attention modules respectively. Makeup Apply Decoder (MADec) applies the makeup feature maps to the source to generate the final result.

Attention Module. Since the two faces may have discrepancies in expressions and poses, the makeup attributes of the reference face need to be adapted to the source face. We employ a QKV-cross-attention similar to Transformer [31] to model the pixel-wise correspondence between the two faces. Formally, given a pair of feature maps extracted from the source and reference, $X, Y \in \mathbb{R}^{HW \times C}$, where C, H and W are the number of channels, height and width of the feature map, we compute the attentive matrix $A \in \mathbb{R}^{HW \times HW}$ to specify how a pixel on X corresponds to its counterparts on Y:

$$A = softmax \left(\frac{\widetilde{X}Q(\widetilde{Y}K)^T}{\sqrt{C}} \right) \qquad (1)$$

where $K, Q \in \mathbb{R}^{C \times C}$ are learnable parameters, and $\widetilde{X}, \widetilde{Y}$ are the feature maps combined with positional embedding to introduce spatial information. Here, we adopt Landmark Embedding [32] that concatenates a vector representing the relative positions to the facial landmarks with the visual features for each pixel.

Makeup features are extracted by a 1×1-Conv with weights $V \in \mathbb{R}^{C \times C}$ from Y. They consequently maintain the spatial correspondence with Y. Then the attentive matrix A is applied to align them with the spatial distribution of X. The aligned features are represented by a makeup feature map $\Gamma \in \mathbb{R}^{HW \times C}$:

$$\Gamma = A(YV) \qquad (2)$$

After that, the makeup feature map Γ becomes the input of MADec and then is applied to the source feature map X by element-wise multiplication:

$$\widehat{X} = \Gamma \odot X \qquad (3)$$

Before being fed into MADec, Γ can be globally or locally manipulated to achieve controllable transfer, which will be discussed in detail in Sect. 4.4.

Sow-Attention Module. To avoid high-frequency information being smoothed and support precise local editing, we utilize feature maps of high resolution. However, the above pixel-wise attention is not practicable here due to the high quadratic cost. We propose a novel shifted overlapped windowing attention (Sow-Attention) to reduce the complexity. As illustrated in Fig. 3, the Sow-Attention Module obtains the makeup feature map Γ by three steps:

(1) *Coarse alignment.* We employ Thin Plate Splines (TPS) to warp Y into Y' to be coarsely aligned with the source face. Specifically, TPS is determined by N control points whose coordinates in the original space and the target space are denoted as $C = \{c_i\}_{i=1}^N$ and $C' = \{c_i'\}_{i=1}^N$, respectively. Here, C is set to be the coordinates of N landmark points [42] of the reference face and C' to be those of the source face, and then TPS warps Y into Y' to fit C into C'. We use the parameterized grid sampling [17] to perform 2D TPS that is differentiable w.r.t input Y and the formulation from [29] to obtain required parameters.

(2) *Attention.* Since X and Y' are coarsely aligned, local attention is enough for a point on X to capture makeup information from the neighbor region on Y'. To avoid the boundary issue of non-overlapped windows that leads to artificial edges in the output image (see Sect. 4.5), we perform attention in shifted overlapped windows. As depicted in Fig. 3, w_1, w_2, w_3 and w_4

represent 4 partitioning schemes that split the feature maps the into overlapped windows of size S with an $S/2$ shift. A QKV-attention is shared
by all windows, and the result computed within window w_j is denoted as
$\Gamma^{w_j} \in \mathbb{R}^{S^2 \times C}$, where C is the number of channels. An alternative view is
that a pixel cross-attends to the four windows it belongs to, for instance, the
pixel x_i marked in Fig. 3 attends to the windows $\{w_j\}_{j=1}^{4}$ on Y' and obtains
four vectors $\Gamma^{w_j}(x_i) \in \mathbb{R}^C, j = 1, 2, 3, 4$.

(3) *Aggregation.* For each pixel x_i, the four vectors derived from previous attention, $\{\Gamma^{w_j}(x_i), j : x_i \in w_j\}$, are aggregated into one vector as the final
output. We conduct this by a weighted sum:

$$\Gamma(x_i) = \sum_{j: x_i \in w_j} \Gamma^{w_j}(x_i) \cdot W(x_i, w_j) \tag{4}$$

where the weight $W(x_i, w_j)$ is determined by the relative position of x_i to w_j.
$W(x_i, w_j)$ should guarantee that the output is spatially continuous both inter
and intra windows. Besides, if x_i is closer to the center of w_j, $\Gamma^{w_j}(x_i)$ will
contain more information about its neighbor region, and $W(x_i, w_j)$ is expected
to be larger. We choose a "bilinear" form that works well in practice:

$$W(x_i, w_j) = \frac{\left| \left(S - 2 \left(\mathrm{x}(x_i) - \mathrm{x}(c_{w_j}) \right) \right) \left(S - 2 \left(\mathrm{y}(x_i) - \mathrm{y}(c_{w_j}) \right) \right) \right|}{S^2} \tag{5}$$

where c_{w_j} denotes the center of window w_j, $\mathrm{x}(\cdot)$ and $\mathrm{y}(\cdot)$ is the x-coordinate and
y-coordinate of a point respectively. Equation 4 and Eq. 5 can also be interpreted
as a kind of "bilinear interpolation": the attention results from different windows
are "interpolated" regarding their centers as anchor points.

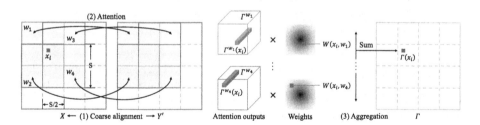

Fig. 3. Illustration of the Sow-Attention Module. Attention is computed within shifted
overlapped windows across coarsely aligned feature maps, and the outputs are aggregated by weighted sum. A darker color indicates a larger weight.

Sow-Attention reduces the cost of pixel-wise attention from $O\left((HW)^2\right)$ to
$O\left(HWS^2\right)$. Generally, $S = H/8$ would be enough, and then the complexity is
reduced by a factor of 16 since the attention is performed with four partitioning
schemes and the cost of each is $(1/8)^2$ of the original attention.

3.3   Makeup Loss with Pseudo Ground Truth

Due to the lack of paired makeup and non-makeup images, we adopt the Cycle-GAN [44] framework to train the network in an unsupervised way. Nevertheless, normal GAN training only drives the generator to produce realistic images with a general makeup. Therefore, to guide the reconstruction of specific makeup attributes of the reference on the source face, extra supervision, i.e., a makeup loss term for the generator, is introduced into the total objective L_{total}:

$$L_{total} = \lambda_{adv}(L_{\mathcal{G}}^{adv} + L_{\mathcal{D}}^{adv}) + \lambda_{cyc}L_{\mathcal{G}}^{cyc} + \lambda_{per}L_{\mathcal{G}}^{per} + \lambda_{make}L_{\mathcal{G}}^{make} \qquad (6)$$

where $L^{adv}, L^{cyc}, L^{per}$ are adversarial loss [11], cycle consistency loss [44], and perceptual loss [19] (formulations are summarized in Appendix C). The makeup loss $L_{\mathcal{G}}^{make}$ is defined with pseudo ground truth (PGT), namely, images that are synthesized independently of the generator and serve as the training objective:

$$L_{\mathcal{G}}^{make} = \|\mathcal{G}(x,y) - PGT(x,y)\|_1 + \|\mathcal{G}(y,x) - PGT(y,x)\|_1. \qquad (7)$$

where $PGT(x,y)$ has the makeup of y and the face identity of x. As shown in Fig. 4, typical strategies for PGT generation include histogram matching [18,22] and TPS warping [32]. Histogram matching equalizes the color distribution of the source face with that of the reference, but it suffers from extreme color differences and discards all spatial information. TPS warps the reference face into the shape of the source face by aligning detected landmarks, but it may result in artifacts of stitching and distortion and also mix in unwanted shadows. The imprecision of these PGTs will consequently cause a sub-optimal transfer.

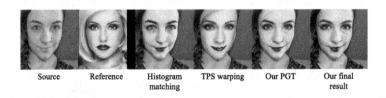

| Source | Reference | Histogram matching | TPS warping | Our PGT | Our final result |

Fig. 4. Different strategies for pseudo ground truth (PGT) generation. Our PGT has more accurate details (e.g., eye shadows) than histogram matching and fewer artifacts and distortions than TPS warping. The final generated image has better quality than PGTs, e.g., there are no jags on the edge of the lip and artifacts on the forehead.

To address these issues, we propose a novel strategy that incorporates both color and spatial information and avoids misleading signals. Although our PGT is not comparable with generated images in quality, it provides sufficient guidance complementary to the GAN training. As illustrated in Fig. 5, it consists of two stages: color matching and detail matching.

(1) *Color Matching.* We adopt histogram matching [22] to replicate the makeup color of the reference y to the source x. The color distributions in the skin, lip, and eye shadow regions are separately equalized between x and y.

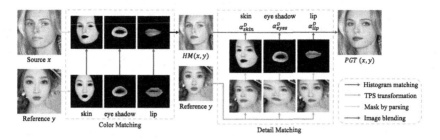

Fig. 5. The pipeline of our PGT generation. It utilizes histogram matching for color matching, TPS warping for detail matching, and annealing factors for image blending.

(2) *Detail Matching.* We employ TPS transformation to incorporate spatial information into the PGT. Specifically, skin, lip, and eye shadow regions of the reference y are separately warped to fit the source x using corresponding facial landmarks, and blended with their counterparts on the color matching result. The blending factors α_{skin}^{D}, α_{eyes}^{D}, and α_{lip}^{D} anneal during the training process to emphasize colors or details in different stages of training for better results. See Appendix D.3 for implementation details.

4   Experiments

4.1   Experiment Settings

We use the MT (Makeup Transfer) dataset [22] which contains 1115 non-makeup images and 2719 makeup images to train our model. We follow the strategy of [22] to split the train/test set. All images are resized to 256×256 before training. More implementation details and results are given in supplementary materials.

We conduct comparisons of EleGANt with general style transfer methods: DIA [23], CycleGAN [44] as well as makeup transfer methods: BeautyGAN [22], BeautyGlow [5], LADN [12], PSGAN [18], Spatial FAT [32], and SCGAN [8]. Since the implementation of BeautyGlow and Spatial FAT is not released, we follow [22] and take the results from corresponding papers. Table 1 summarizes the functions of open-source makeup transfer models. Our EleGANt demonstrates the greatest capability and flexibility among all methods. It can precisely transfer makeup details and is the first to achieve customized local editing.

4.2   Qualitative Comparison

Figure 6 shows the results on images with frontal faces in neutral expressions and light makeups. The results of DIA have unnatural colors on the hair and shadows on the face. CycleGAN synthesizes realistic images, but it simply performs domain transfer without recovering any makeup of the reference. BeautyGlow fails to transfer the correct color of skin and lip. Severe artifacts and blurs exist

Table 1. Analyses of EleGANt with existing open-source methods. "Misalign.": robust transfer with large spatial misalignment between the two faces. "Detail": precise transfer with high-quality details. "Shade": shade-controllable transfer. "Part": partial transfer for lip, eye, and skin regions. "Local.": local editing within arbitrary areas.

Method	Capability		Controllability		
	Misalign.	Detail	Shade	Part	Local.
BeautyGAN [22]					
LADN [12]			✓		
PSGAN [18]	✓		✓	✓	
SCGAN [8]	✓		✓	✓	
EleGANt (ours)	✓	✓	✓	✓	✓

in the results of LADN. Recent works such as PSGAN, SCGAN, and Spatial FAT can generate more visually acceptable results. However, PSGAN and SCGAN suffer from the color bleeding problem, e.g., blurs at the edge of the lips. Spatial FAT generates results with richer details, but there are still apparent artifacts. Compared to existing methods, our proposed EleGANt generates the most realistic images with natural light and shadow while synthesizing high-fidelity makeup colors and high-quality details.

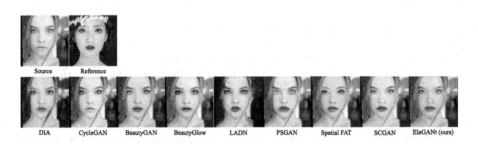

Fig. 6. Qualitative comparisons with existing methods. EleGANt generates the most precise transferred result with the desired makeup and high-quality details.

To test the effectiveness on complex makeups and robustness against spatial misalignment, we compare our method with makeup transfer models that have released code. The results are shown in Fig. 7. None of the existing methods can precisely transfer makeup details such as the shapes and colors of eye shadows (the 1st, 2nd, and 4th rows) due to the loss of high-frequency information. Our EleGANt synthesizes these details with the highest quality with the help of high-resolution feature maps. Existing methods fall short in cases that the two faces have large discrepancies in pose or illumination, resulting in unnatural colors and shadows (the 3rd and 4th rows) in the transferred images. The results of EleGANt have natural colors and shadows consistent with the faces. Besides,

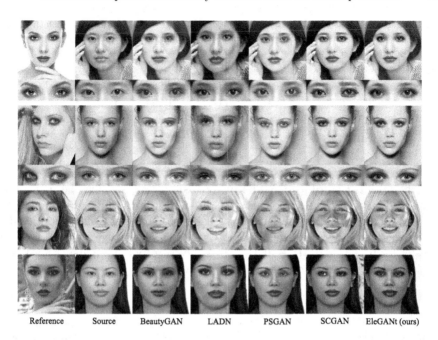

Reference   Source   BeautyGAN   LADN   PSGAN   SCGAN   EleGANt (ours)

Fig. 7. Comparison on images with misaligned poses and complex makeups. Our proposed EleGANt generates the most exquisite details (e.g., the shapes and colors of the eye shadows), the most natural colors and shadows, and the fewest artifacts.

images generated by EleGANt have the fewest artifacts (4th row) among all of the methods. More samples of comparison are provided in Appendix E.2.

4.3 Quantitative Comparison

We conduct a user study to quantitatively evaluate the generation quality and the transfer precision of different models. For a fair comparison, we compare our EleGANt with the methods whose code and pre-train model are available: BeautyGAN, LADN, PSGAN, and SCGAN. We randomly selected 20 generated images from the test split of the MT dataset. Totally 40 participants were asked to evaluate these samples in three aspects: "visual quality", "detail processing" (the quality and precision of transferred details), and "overall performance" (considering the visual quality, the fidelity of transferred makeup, etc.). They then selected the best one in each aspect. Table 2 demonstrates the results of the user study. Our EleGANt outperforms other methods in all aspects, especially in synthesizing makeup details.

4.4 Controllable Makeup Transfer

Since the makeup feature maps Γ_H and Γ_L spatially correspond to the source face and are combined with the source feature maps by element-wise multiplication, controllability can be achieved by interpolating those makeup feature maps.

Table 2. User study results (ratio (%) selected as the best). "Quality", "Detail" and "Overall" denote the three aspects for evaluation: visual quality, detail processing, and overall performance.

	BeautyGAN	LADN	PSGAN	SCGAN	EleGANt (ours)
Quality	6.75	1.88	11.13	21.25	**59.00**
Detail	5.38	2.75	11.25	14.38	**66.25**
Overall	4.75	2.75	9.88	20.38	**62.25**

Partial and Interpolated Makeup Transfer. Transferring the makeup in predefined parts of the face, for instance, lip and skin, can be realized automatically by masking makeup feature maps using face parsing results. Specifically, let x denote the source image, y_i denote the reference image we would like to transfer for part i, and $M_i^x \in [0,1]^{H \times W}$ is the corresponding parsing mask, the partial transferred feature maps are calculated as:

$$\Gamma_* = M_i^x \odot \Gamma_*^{y_i} + (1 - M_i^x) \odot \Gamma_*^x \tag{8}$$

where $* \in \{H, L\}$, Γ_*^y is the makeup feature map with x being the source and y being the reference, the mask is expanded along the channel dimension, and $\odot$ denotes the Hadamard product. Interpolating the makeup feature maps can control the shade of makeup or fuse makeup from multiple references. Given a source image x, two reference images y_1, y_2, and a coefficient $\alpha^S \in [0,1]$, we first get the makeup feature maps of y_1 and y_2 w.r.t. x, and interpolate them by

$$\Gamma_* = \alpha^S \Gamma_*^{y_1} + (1 - \alpha^S)\Gamma_*^{y_2} , \ * \in \{H, L\} \tag{9}$$

If we want to adjust the shade of makeup using a single reference style, just simply set $y_2 = x$, then α^S indicates the intensity. We also perform partial and interpolated makeup transfer simultaneously by leveraging both area masks and shade intensity, and the results are shown in Fig. 8.

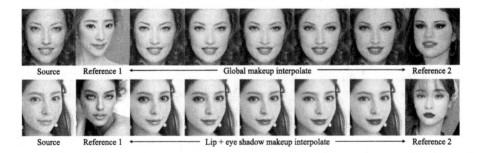

Fig. 8. Partial and interpolated makeup transfer. The first row applies a global transfer. The second row only transfers the lipsticks and eye shadows of the two references.

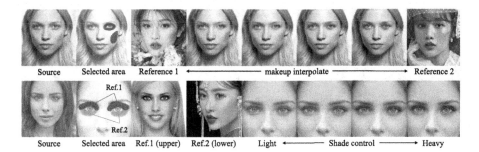

Fig. 9. Customized local editing. In the first row, we select the areas around the eyes and cheek to adjust the eye shadow and blush. In the second row, we select the areas of upper and lower eye shadow and assign them with different references.

Customized Local Editing. The general controls can be formulated as: given a set of k reference images $\{y_i\}_{i=1}^k$ with corresponding masks $\{M_i^x\}_{i=1}^k$ for the area to apply makeup and coefficients $\{\alpha_i^S\}_{i=1}^k$ to specify the shade, the fused makeup feature maps are computed by

$$\Gamma_* = \sum_{i=1}^k \alpha_i^S M_i^x \odot \Gamma_*^{y_i} + \left(1 - \sum_{i=1}^k \alpha_i^S M_i^x\right) \odot \Gamma_*^x \ , \ * \in \{H, L\} \qquad (10)$$

Unlike SCGAN [8] that restrict the areas for partial transfer to skin, lip and eye shadow, our EleGANt is more interactive and flexible. In Eq. 10, M_i^x can be the mask of any arbitrary area, and it has the same size as the makeup feature map. Therefore, we can specify customized regions to edit their makeup styles and shades, and the paint-board for region selection is of the same resolution as the high-res makeup feature map Γ_H. Though the masks need to be down-sampled when applied to the low-res Γ_L, the makeup details are dominantly determined by Γ_H, whose high resolution guarantees the precision of the controls. Figure 9 illustrates customized local editing of makeup style and shade.

4.5 Ablation Study

Architecture Design. Our EleGANt utilizes high-resolution feature maps to preserve high-frequency information as well as attention in shifted overlapped windows which reduces the complexity and avoids the boundary issue. Here we conduct an ablation study to evaluate their effectiveness. As shown in Fig. 10, without the high-res feature maps, detailed facial attributes, e.g., eye shadows of the reference (the 1st and 2nd rows) and freckles of the source (2nd row), are lost or smoothed during the transfer. When replacing Sow-attention with attention in non-overlapped windows, the outputs are discontinuous on the boundaries of the windows. Figure 10 shows that there are color blocks nearby the eyebrow (1st row) and on the forehead (2nd row).

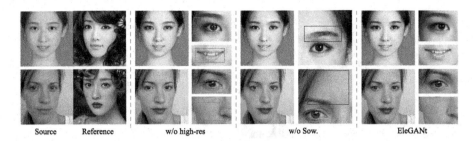

Fig. 10. Ablation study of architecture design. "w/o Sow." denotes performing attention in non-overlapped windows instead of shifted overlapped windows. Unnatural colors are marked by red boxes. Detail missing and blurs are marked by green boxes. (Color figure online)

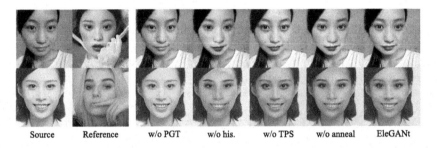

Fig. 11. Ablation study of PGT generation. We train the network with different PGT settings and compare the final results generated by the network. "w/o his." denotes generating PGT without histogram matching. "w/o anneal" indicates using fixed blending factors rather than annealing during the training process.

PGT Generation. To confirm our proposed strategy for pseudo ground truth generation, we conduct ablation studies on all techniques we employ. Qualitative comparisons are demonstrated in Fig. 11. The model will only yield images with general instead of particular makeups if we do not use PGT for supervision. Spatial information cannot be injected into PGT without TPS transformation, so the model fails to transfer makeup details (e.g., the eye shadows are missing in the first row and inaccurate in the second row.) due to the lack of corresponding supervision. Without histogram matching or annealing, artifacts introduced by TPS warping and image blending remain in the PGT, which are then learned by the model (e.g., stitching traces nearby the lip of the first row and on the forehead of the second row). Besides, the model learns to directly copy the shadows of the reference to the source face when only guided by TPS-warped images.

5   Conclusion

In this paper, we emphasize that makeup transfer is beyond transferring color distributions. We propose Exquisite and locally editable GAN for makeup transfer (EleGANt) to improve the synthesis of details and step towards more flexible controls. It utilizes high-res feature maps to preserve high-frequency attributes, and a novel Sow-Attention Module performs attention within shifted overlapped to reduce the computational cost. The model is trained with a newly designed objective that leverages both color and spatial information. Besides partial and interpolated makeup transfer, it is the first to achieve customized local makeup editing within arbitrary regions. Extensive experiments demonstrate the superiority of EleGANt compared with existing approaches. It can generate realistic images with exquisite details, despite various facial poses and complex makeup styles. Besides, we believe that the attention scheme with shifted overlapped windowing (Sow-Attention) would be helpful for other tasks and networks.

Although our model succeeds in daily makeup transfer, it fails in some cases such as extreme makeup. The MT dataset [22] we currently use has limited resolution and diversity of makeup styles and skin tones. This may be addressed if more data is available and we leave this for future work.

Acknowledgements. This work is supported by the Ministry of Science and Technology of the Peoples Republic of China, the 2030 Innovation Megaprojects "Program on New Generation Artificial Intelligence" (Grant No. 2021AAA0150000). This work is also supported by a grant from the Guoqiang Institute, Tsinghua University. Thanks to Steve Lin for his pre-reading and constructive suggestions.

References

1. An, J., Xiong, H., Huan, J., Luo, J.: Ultrafast photorealistic style transfer via neural architecture search. In: Proceedings of the AAAI Conference on Artificial Intelligence, vol. 34, pp. 10443–10450 (2020)
2. Cai, M., Zhang, H., Huang, H., Geng, Q., Li, Y., Huang, G.: Frequency domain image translation: more photo-realistic, better identity-preserving. In: Proceedings of the IEEE/CVF International Conference on Computer Vision (ICCV), pp. 13930–13940 (2021)
3. Chang, H., Lu, J., Yu, F., Finkelstein, A.: PairedCycleGAN: asymmetric style transfer for applying and removing makeup. In: Proceedings of the IEEE/CVF Conference on Computer Vision and Pattern Recognition (CVPR), pp. 40–48 (2018)
4. Chen, C.F., Panda, R., Fan, Q.: RegionViT: regional-to-local attention for vision transformers. In: Proceedings of the International Conference on Learning Representations (ICLR) (2022)
5. Chen, H.J., Hui, K.M., Wang, S.Y., Tsao, L.W., Shuai, H.H., Cheng, W.H.: BeautyGlow: on-demand makeup transfer framework with reversible generative network. In: Proceedings of the IEEE/CVF Conference on Computer Vision and Pattern Recognition (CVPR), pp. 10042–10050 (2019)
6. Choi, Y., Uh, Y., Yoo, J., Ha, J.W.: StarGAN v2: diverse image synthesis for multiple domains. In: Proceedings of the IEEE/CVF Conference on Computer Vision and Pattern Recognition (CVPR), pp. 8188–8197 (2020)
7. Chu, X., et al.: Twins: revisiting the design of spatial attention in vision transformers. In: Proceedings of the International Conference on Neural Information Processing Systems (NIPS), pp. 9355–9366 (2021)

8. Deng, H., Han, C., Cai, H., Han, G., He, S.: Spatially-invariant style-codes controlled makeup transfer. In: Proceedings of the IEEE/CVF Conference on Computer Vision and Pattern Recognition (CVPR), pp. 6549–6557 (2021)

9. Dosovitskiy, A., et al.: An image is worth 16x16 words: transformers for image recognition at scale. In: Proceedings of the International Conference on Learning Representations (ICLR) (2021)

10. Gatys, L.A., Ecker, A.S., Bethge, M.: Image style transfer using convolutional neural networks. In: Proceedings of the IEEE Conference on Computer Vision and Pattern Recognition (CVPR), pp. 2414–2423 (2016)

11. Goodfellow, I., et al.: Generative adversarial nets. In: Proceedings of the International Conference on Neural Information Processing Systems (NIPS) (2014)

12. Gu, Q., Wang, G., Chiu, M.T., Tai, Y.W., Tang, C.K.: LADN: local adversarial disentangling network for facial makeup and de-makeup. In: Proceedings of the IEEE/CVF International Conference on Computer Vision (ICCV), pp. 10481–10490 (2019)

13. Guo, D., Sim, T.: Digital face makeup by example. In: Proceedings of the IEEE Conference on Computer Vision and Pattern Recognition (CVPR), pp. 73–79 (2009)

14. Heo, B., Yun, S., Han, D., Chun, S., Choe, J., Oh, S.J.: Rethinking spatial dimensions of vision transformers. In: Proceedings of the IEEE/CVF International Conference on Computer Vision (ICCV), pp. 11936–11945 (2021)

15. Ho, J., Kalchbrenner, N., Weissenborn, D., Salimans, T.: Axial attention in multidimensional transformers. arXiv preprint arXiv:1912.12180 (2019)

16. Huang, X., Belongie, S.: Arbitrary style transfer in real-time with adaptive instance normalization. In: Proceedings of the IEEE International Conference on Computer Vision (ICCV), pp. 1501–1510 (2017)

17. Jaderberg, M., Simonyan, K., Zisserman, A., Kavukcuoglu, K.: Spatial transformer networks. In: Proceedings of the International Conference on Neural Information Processing Systems (NIPS), pp. 2017–2025 (2015)

18. Jiang, W., et al.: PSGAN: pose and expression robust spatial-aware GAN for customizable makeup transfer. In: Proceedings of the IEEE/CVF Conference on Computer Vision and Pattern Recognition (CVPR), pp. 5194–5202 (2020)

19. Johnson, J., Alahi, A., Fei-Fei, L.: Perceptual losses for real-time style transfer and super-resolution. In: Proceedings of the European Conference on Computer Vision (ECCV), pp. 694–711 (2016)

20. Kim, S.S., Kolkin, N., Salavon, J., Shakhnarovich, G.: Deformable style transfer. In: Proceedings of the European Conference on Computer Vision (ECCV), pp. 246–261 (2020)

21. Li, C., Zhou, K., Lin, S.: Simulating makeup through physics-based manipulation of intrinsic image layers. In: Proceedings of the IEEE Conference on Computer Vision and Pattern Recognition (CVPR), pp. 4621–4629 (2015)

22. Li, T., et al.: BeautyGAN: instance-level facial makeup transfer with deep generative adversarial network. In: Proceedings of the 26th ACM International Conference on Multimedia, pp. 645–653 (2018)

23. Liao, J., Yao, Y., Yuan, L., Hua, G., Kang, S.B.: Visual attribute transfer through deep image analogy. ACM Trans. Graph. 36(4), 1–15 (2017)

24. Liu, L., Xing, J., Liu, S., Xu, H., Zhou, X., Yan, S.: Wow! You are so beautiful today! ACM Trans. Multim. Comput. Commun. Appl. (TOMM) 11(1s), 1–22 (2014)

25. Liu, Z., et al.: Swin transformer: Hierarchical vision transformer using shifted windows. In: Proceedings of the IEEE/CVF International Conference on Computer Vision (ICCV), pp. 10012–10022 (2021)
26. Luan, F., Paris, S., Shechtman, E., Bala, K.: Deep photo style transfer. In: Proceedings of the IEEE Conference on Computer Vision and Pattern Recognition (CVPR). pp. 4990–4998 (2017)
27. Parmar, N., et al.: Image transformer. In: Proceedings of the International Conference on Machine Learning (ICML), pp. 4055–4064 (2018)
28. Rakhimov, R., Volkhonskiy, D., Artemov, A., Zorin, D., Burnaev, E.: Latent video transformer. In: Proceedings of the 16th International Joint Conference on Computer Vision, Imaging and Computer Graphics Theory and Applications (VISIGRAPP), pp. 101–112 (2021)
29. Shi, B., Yang, M., Wang, X., Lyu, P., Yao, C., Bai, X.: Aster: an attentional scene text recognizer with flexible rectification. IEEE Trans. Pattern Anal. Mach. Intell. **41**(9), 2035–2048 (2018)
30. Tong, W.S., Tang, C.K., Brown, M.S., Xu, Y.Q.: Example-based cosmetic transfer. In: Proceedings of the 15th Pacific Conference on Computer Graphics and Applications (PG), pp. 211–218 (2007)
31. Vaswani, A., et al.: Attention is all you need. In: Proceedings of the International Conference on Neural Information Processing Systems (NIPS), pp. 6000–6010 (2017)
32. Wan, Z., Chen, H., An, J., Jiang, W., Yao, C., Luo, J.: Facial attribute transformers for precise and robust makeup transfer. In: Proceedings of the IEEE/CVF Winter Conference on Applications of Computer Vision (WACV), pp. 1717–1726 (2022)
33. Wang, H., Li, Y., Wang, Y., Hu, H., Yang, M.H.: Collaborative distillation for ultra-resolution universal style transfer. In: Proceedings of the IEEE/CVF Conference on Computer Vision and Pattern Recognition (CVPR), pp. 1860–1869 (2020)
34. Wang, W., et al.: Pyramid vision transformer: A versatile backbone for dense prediction without convolutions. In: Proceedings of the IEEE/CVF International Conference on Computer Vision (ICCV), pp. 568–578 (2021)
35. Wang, X., Girshick, R., Gupta, A., He, K.: Non-local neural networks. In: Proceedings of the IEEE Conference on Computer Vision and Pattern Recognition (CVPR), pp. 7794–7803 (2018)
36. Weissenborn, D., Täckström, O., Uszkoreit, J.: Scaling autoregressive video models. In: Proceedings of the International Conference on Learning Representations (ICLR) (2020)
37. Wu, C., et al.: Godiva: generating open-domain videos from natural descriptions. arXiv preprint arXiv:2104.14806 (2021)
38. Wu, C., et al.: N\" UWA: visual synthesis pre-training for neural visual world creation. arXiv preprint arXiv:2111.12417 (2021)
39. Xu, K., et al.: Show, attend and tell: Neural image caption generation with visual attention. In: Proceedings of the International Conference on Machine Learning (ICML), pp. 2048–2057 (2015)
40. Xu, L., Du, Y., Zhang, Y.: An automatic framework for example-based virtual makeup. In: Proceedings of the IEEE International Conference on Image Processing (ICIP), pp. 3206–3210 (2013)
41. Yuan, L., et al.: Tokens-to-token VIT: Training vision transformers from scratch on imageNet. In: Proceedings of the IEEE/CVF International Conference on Computer Vision (ICCV), pp. 558–567 (2021)

42. Zhang, K., Zhang, Z., Li, Z., Qiao, Y.: Joint face detection and alignment using multitask cascaded convolutional networks. IEEE Signal Process. Lett. **23**(10), 1499–1503 (2016)
43. Zhang, P., et al.: Multi-scale vision longformer: a new vision transformer for high-resolution image encoding. In: Proceedings of the IEEE/CVF International Conference on Computer Vision (ICCV), pp. 2998–3008 (2021)
44. Zhu, J.Y., Park, T., Isola, P., Efros, A.A.: Unpaired image-to-image translation using cycle-consistent adversarial networks. In: Proceedings of the IEEE International Conference on Computer Vision (ICCV), pp. 2223–2232 (2017)

Author Index

Printed in the United States
by Baker & Taylor Publisher Services